SPECIAL NOTATION

Properties

Uppercase	Extensive	$K : V, G, U, H, S, \ldots$
Lowercase	Intensive (molar)	$k = \frac{K}{n} = v, g, u, h, s, \ldots$
Circumflex, lowercase	Intensive (specific)	$\hat{k} = \frac{K}{m} = \hat{v}, \hat{g}, \hat{u}, \hat{h}, \hat{s}, \ldots$

Mixtures

Subscript i	Pure species property	$K_i : V_i, G_i, U_i, H_i, S_i, \ldots$
		$k_i : v_i, g_i, u_i, h_i, s_i, \ldots$
Bar, subscript i	Partial molar property	$\overline{K}_i : \overline{V}_i, \overline{G}_i, \overline{U}_i, \overline{H}_i, \overline{S}_i, \ldots$
As is	Total solution property	$K : V, G, U, H, S, \ldots$
		$k : v, g, u, h, s, \ldots$
Delta, subscript *mix*	Property change of mixing:	$\Delta K_{mix} : \Delta V_{mix}, \Delta H_{mix}, \Delta S_{mix}, \ldots$
		$\Delta k_{mix} : \Delta v_{mix}, \Delta h_{mix}, \Delta s_{mix}, \ldots$

Other

Dot	Rate of change	$\dot{Q}, \dot{W}, \dot{n}, \dot{V}, \ldots$
Overbar	Average	$\overline{V^2}, \overline{c}_P, \ldots$

A complete set of notation used in this text can be found on page (ix)

Engineering and Chemical Thermodynamics

Engineering and Chemical Thermodynamics

Milo D. Koretsky
Department of Chemical Engineering
Oregon State University

WILEY

JOHN WILEY & SONS, INC.

Executive Editor   *Bill Zobrist*
Project Editor   *Jenny Welter*
Marketing Manager   *Ilsa Wolfe*
Senior Production Editor   *Norine M. Pigliucci*
Design Director   *Maddy Lesure*
Editorial Assistant   *Mary Moran*
Production Management Services   *Hermitage Publishing Services*

The cover depicts a turbine, a common process that is analyzed using thermodynamics. A cutaway of the physical apparatus reveals the conceptual tools that will be developed in this text. This includes a hypothetical thermodynamic pathway marked by dashed arrows. You will learn how to construct such pathways to solve a variety of problems. The figure also contains a "molecular dipole," which is drawn in the *PT* plane associated with the real fluid. This conceptually illustrates the use of molecular concepts to reinforce thermodynamic principles as they are developed, another important thematic link to the text.

The book was set in New Caledonia by Hermitage Publishing Services and printed and bound by Hamilton Printing. The cover was printed by Phoenix Color Corp.

This book is printed on acid free paper. ∞

To order books or for customer service please, call 1(800)-CALL-WILEY (225-5945).

Library of Congress Cataloging in Publication Data:
Koretsky, Milo D.
    Engineering and chemical thermodynamics / Milo D. Koretsky.
        p. cm.
    Includes bibliographical references.
    ISBN 978-0-471-38586-8 (cloth: acid-free paper)
    1. Thermodynamics. 1. Title

    TP155.2.T45K67 2004
    6601.2969–dc21

                                                                2003053848

ISBN 0-471-38586-7
WIE ISBN: 0-471-45237-8

Printed in the United States of America

15   14   13   12

For Nicole and Moses

Preface

You see, I have made contributions to biochemistry. There were no courses in molecular biology. I had no courses in biology at all, but I am one of the founders of molecular biology. I had no courses in nutrition or vitaminology. Why? Why am I able to do these things? You see, I got such a good basic education in the fields where it is difficult for most people to learn by themselves.

Linus Pauling
On his ChE education

▶ AUDIENCE

Engineering and Chemical Thermodynamics is intended for use in the undergraduate thermodynamics course(s) taught in the sophomore or junior year in most Chemical Engineering (ChE) Departments. For the majority of ChE undergraduate students, chemical engineering thermodynamics, concentrating on the subjects of phase equilibria and chemical reaction equilibria, is one of the most abstract and difficult core courses in the curriculum. In fact, it has been noted by more than one thermodynamics guru (e.g., Denbigh, Sommerfeld) that this subject cannot be mastered in a single encounter. Understanding comes at greater and greater depths with every skirmish with this subject. Why another textbook in this area? This textbook is targeted specifically at the sophomore or junior undergraduate who must, for the first time, grapple with the treatment of equilibrium thermodynamics in sufficient detail to solve the wide variety of problems that chemical engineers must tackle. It is a *conceptually* based text, meant to provide students with a solid foundation in this subject in a single iteration. Its intent is to be both *accessible* and *rigorous*. Its accessibility allows students to retain as much as possible through their first pass while its rigor provides them the foundation to understand more advanced treatises and forms the basis of commercial computer simulations such as ASPEN®, HYSIS®, and CHEMCAD®.

▶ GOALS AND METHODOLOGY

The text was developed from course notes that have been used in the undergraduate chemical engineering classes at Oregon State University since 1994. It uses a logically consistent development whereby each new concept is introduced in the context of a framework laid down previously. This textbook has been specifically designed to accommodate students with different learning styles. Its conceptual development, worked-out examples, and numerous end-of-chapter problems are intended to promote *deep learning* and provide students the ability to apply thermodynamics to real-world engineering problems. Two major threads weave throughout the text: (1) a common methodology for approaching topics, be it enthalpy or fugacity, and (2) the reinforcement of classical thermodynamics with molecular principles. Whenever possible, intuitive and qualitative arguments complement mathematical derivations.

The basic premise on which the text is organized is that *student learning is enhanced by connecting new information to prior knowledge and experiences*. The approach is to introduce new concepts in the context of material that students *already know*. For example, the second law of thermodynamics is formulated analogously to the first law, as a generality to many observations of nature (as opposed to the more common approach of using specific statements about obtaining work from heat through thermodynamic cycles). Thus, the experience students have had in learning about the thermodynamic property energy, which they have already encountered in several classes, is applied to introduce a new thermodynamic property, entropy. Moreover, the underpinnings of the second law—reversibility, irreversibility, and the Carnot cycle—are introduced with the first law, a context with which students have more experience; thus they are not new when the second law is introduced.

▶ LEARNING STYLES

There has been recent attention in engineering education to crafting instruction that targets the many ways in which students learn. For example, in their landmark paper "Learnings and Teaching Styles in Engineering Education,"[1] Richard Felder and Linda Silverman define specific dimensions of learning styles and corresponding teaching styles. In refining these ideas, the authors have focused on four specific dimensions of learning: sequential vs. global learners; active vs. reflective learners; visual vs. verbal learners; and sensing vs. intuitive learners. This textbook has been specifically designed to accommodate students with different learning styles by providing avenues for students with each style and, thereby, reducing the mismatches between its presentation of content and a student's learning style. The objective is to create an effective text that enables students to access new concepts. For example, each chapter contains learning objectives at the beginning and a summary at the end. These sections do not parrot the order of coverage in the text, but rather are presented in a *hierarchical* order from the most significant concepts down. Such a presentation creates an effective environment for global learners (who should read the summary before embarking on the details in a chapter). On the other hand, to aid the sequential learner, the chapter is developed in a logical manner, with concepts constructed step by step based on previous material. Identified key concepts are presented schematically to aid visual learners. Questions about key points that have been discussed previously are inserted periodically in the text to aid both active and reflective learners. Examples are balanced between those that emphasize concrete, numerical problem solving for sensing learners and those that extend conceptual understanding for intuitive learners.

In the cognitive dimension, we can form a taxonomy of the hierarchy of knowledge that a student may be asked to master. For example, a modified Bloom's taxonomy includes: remember, understand, apply, analyze, evaluate, and create. The tasks are listed from lowest to highest level. To accomplish the lower-level tasks, surface learning is sufficient, but the ability to perform at the higher levels requires *deep learning*. In deep learning, students look for patterns and underlying principles, check evidence and relate it to conclusions, examine logic and argument cautiously and critically, and through this process become actively interested in course content. In contrast, students practicing surface learning tend to memorize facts, carry out procedures algorithmically, find it difficult to make sense of new ideas, and end up seeing little value in a thermodynamics course. While it is reinforced throughout the text, promotion of deep learning is most significantly influenced by what a student is expected to do. End-of-chapter problems have been constructed to cultivate a deep understanding of the material. Instead of merely finding the right equation to "plug and chug," the student is asked to search for connections and patterns in the material, understand the physical meaning of the equations, and creatively apply the fundamental principles that have been covered to entirely new problems. The belief is that only through this deep learning is a student able to synthesize information from the university classroom and creatively apply it to new problems in the field.

▶ SOLUTION MANUAL

The Solutions Manual is available for instructors who have adopted this book for their course. Please visit the Instructor Companion site located at www.wiley.com/college/koretsky to register for a password.

▶ MOLECULAR CONCEPTS

While outside the realm of classical thermodynamics, the incorporation of molecular concepts is useful on many levels. In general, by the time undergraduate thermodynamics is taught, the chemical engineering student has had many chemistry courses, so why not take advantage of this experience! Thermodynamics is inherently abstract. Molecular concepts reinforce the text's

---

[1] Felder, Richard M., and Linda K. Silverman, *Engr. Education*, **78**, 674 (1988).

intuitive approach, providing more access to the typical undergraduate student than could a mathematical derivation alone. A molecular approach is also becoming important on a technological level, with the increased development of molecular-based simulations. Finally, molecular understanding allows the student to form a link between the understanding of equilibrium thermodynamics and other fundamental engineering sciences, such as transport processes.

▶ THERMOSOLVER SOFTWARE

The accompanying ThermoSolver software has been specifically designed to complement the text. This integrated, menu-driven program is easy to use and learning-based. ThermoSolver readily allows students to perform more complex calculations, giving them opportunity to explore a wide range of problem solving in thermodynamics. Equations used to perform the calculations can be viewed within the program and use nomenclature consistent with the text. Since the equations from the text are integrated into the software, students are better able to connect the concepts to the software output, reinforcing learning. The ThermoSolver software may be downloaded for free from the student companion site located at www.wiley.com/college/koretsky.

▶ ACKNOWLEDGMENTS

I would like to acknowledge and offer thanks to those individuals who have provided thoughtful input: Wayne Anderson, Connelly Barnes, Hugo Caran, Chih-hung (Alex) Chang, John Falconer, Dennis Hess, P. K. Lim, Erik Muehlenkamp, Jeff Reimer, Skip Rochefort, Wyatt Tenhaeff, Darrah Thomas, and David Wetzel. Last, but not least, I am tremendously grateful to the students with whom, over the years, I have shared the thermodynamics classroom.

▶ NOTATION

The study of thermodynamics inherently contains detailed notation. Below is a summary of the notation used in this text. The list includes: special notation, symbols, Greek symbols, subscripts, superscripts, operators and empirical parameters. Due to the large number of symbols as well as overlapping by convention, the same symbol sometimes represents different quantities. In these cases, you will need to deduce the proper designation based on the context in which a particular symbol is used.

Special Notation

Properties

Uppercase	Extensive	$K : V, G, U, H, S, \ldots$
Lowercase	Intensive (molar)	$k = \dfrac{K}{n} = v, g, u, h, s, \ldots$
Circumflex, lowercase	Intensive (specific)	$\hat{k} = \dfrac{K}{m} = \hat{v}, \hat{g}, \hat{u}, \hat{h}, \hat{s}, \ldots$

Mixtures

Subscript i	Pure species property	$K_i : V_i, G_i, U_i, H_i, S_i, \ldots$
		$k_i : v_i, g_i, u_i, h_i, s_i, \ldots$
Bar, subscript i	Partial molar property	$\overline{K}_i : \overline{V}_i, \overline{G}_i, \overline{U}_i, \overline{H}_i, \overline{S}_i, \ldots$
As is	Total solution property	$K : V, G, U, H, S, \ldots$
		$k : v, g, u, h, s, \ldots$
Delta, subscript *mix*	Property change of mixing:	$\Delta K_{mix} : \Delta V_{mix}, \Delta H_{mix}, \Delta S_{mix}, \ldots$
		$\Delta k_{mix} : \Delta v_{mix}, \Delta h_{mix}, \Delta s_{mix}, \ldots$

Other

Dot	Rate of change	$\dot{Q}, \dot{W}, \dot{n}, \dot{V}, \ldots$
Overbar	Average	$\overline{V^2}, \bar{c}_P, \ldots$

Symbols

$a, b \ldots, i, \ldots$	Generic species in a mixture	K_i	K-value
a, A	Helmholtz energy	L	Flow rate of liquid
A, B	Labels for processes to be compared	m	Number of chemical species
		m	Mass
A	Area	MW	Molecular weight
a_i	Activity of species i	n	Number of moles
A_i	Species i in a chemical reaction	n	Concentration of electrons in a semiconductor
b_j	Element vector	n_i	Intrinsic carrier concentration
c_P	Heat capacity at constant pressure	N	Number of molecules in the system or in a given state
c_v	Heat capacity at constant volume	N_A	Avagadro's number
		OF	Objective function
c_i	Molal concentration of species i	p	Concentration of holes in a semiconductor
C_i	Mass concentration of species i	P	Pressure
$[i]$	Molar concentration of species i	p_i	Partial pressure of species i in an ideal gas mixture
COP	Coefficient of performance	P_i^{sat}	Saturation pressure of species i
D_{i-j}	Bond $i-j$ dissociation energy	q, Q	Heat
e, E	Energy	Q	Electric charge
e_k, E_K	Kinetic energy	r	Distance between two molecules
e_p, E_P	Potential energy	R	Gas constant
$\vec{E}$	Electric field	R	Number of independent chemical reactions
F	Force		
F	Flow rate of feed	s	Stoichiometric constraints
F	Faraday's constant	s, S	Entropy
$\mathfrak{I}$	Degrees of freedom	t	Time
f_i	Fugacity of pure species i	T	Temperature
$\hat{f}_i$	Fugacity of species i in a mixture	T_b	Temperature at the boiling point
f	Total solution fugacity	T_m	Temperature at the melting point
g, G	Gibbs energy		
g	Gravitational acceleration	T_u	Upper consulate temperature
h, H	Enthalpy		
$\Delta \tilde{h}_s$	Enthalpy of solution	u, U	Internal energy
$\mathcal{H}_i$	Henry's law constant of solute i	v, V	Volume
		V	Flow rate of vapor
i	Interstitial	V	Vacancy
I	Ionization energy	$\vec{V}$	Velocity
I	Ionic strength	w, W	Work
k, K	Generic representation of any thermodynamic property except P or T	w_{flow}, W_{flow}	Flow work
		w_s, W_S	Shaft work
		w^*, W^*	Non-Pv work
k	Boltzmann's constant	w_i	Weight fraction of species i
k	Heat capacity ratio (c_P/c_v)		
k	Spring constant	x	Quality (fraction vapor)
K	Equilibrium constant	x	Position along x-axis
k_{ij}	Binary interaction parameter between species i and j	x_i	Mole fraction of liquid species i

X_i	Mole fraction of solid species i	z	Valence of an ion in solution
y_i	Mole fraction of vapor species i	$1, 2 \ldots$	Labels of specific states
z	Compressibility factor		of a system
z	Position along z-axis	$1, 2 \ldots$	Generic species in a mixture

Greek Symbols

α_i	Polarizability of species i	η	Efficiency factor
β	Thermal expansion coefficient	λ_i	Lagrangian multiplier
β_{ij}	Formula coefficient matrix	Γ	Molecular potential
E	Electrochemical potential		energy
φ_i	Fugacity coefficient of pure species i	Γ_i	Activity coefficient of solid species i
$\hat{\varphi}_i$	Fugacity coefficient of species i in a mixture	Γ_{ij}	Molecular potential energy between species i and j
φ	Total solution fugacity coefficient	κ	Isothermal compressibility
γ_i	Activity coefficient of species i	μ_i	Dipole moment of species i
$\gamma_i^{Henry's}$	Activity coefficient using a Henry's law reference state	μ_i	Chemical potential of species i
		μ_{JT}	Joule-Thomson coefficient
		π	Phases
γ_i^m	Molality based activity coefficient	Π	Osmotic pressure
		ρ	Density
$\gamma_{\pm}$	Mean activity coefficient of anions and cations in solution	ν_i	Stiochiometric coefficient
		ω	Pitzer acentric factor
		ξ	Extent of reaction

Subscripts

$a, b, \ldots, i, \ldots$	Generic species in a mixture	net	Net heat or work transferred
atm	Atmosphere	out	Flow stream out of the system
c	Critical point		
C	Cold thermal reservoir	products	Products of a chemical reaction
calc	Calculated		
cycle	Property change over a thermodynamic cycle	pc	Pseudocritical
		r	Reduced property
exp	Experimental	reactants	Reactants in a chemical reaction
f	Property value of formation (with Δ)		
		real gas	Real gas
fus	Fusion	rev	Reversible process
E	External	rxn	Reaction
H	Hot thermal reservoir	sub	Sublimation
high	High value (e.g. in interpolation)	surr	Surroundings
		sys	System
ideal gas	Ideal gas	univ	Universe
in	Flow stream into the system	v	Vapor
inerts	Inerts in a chemical reaction	vap	Vaporization
irrev	Irreversible process	z	In the z direction
l	Liquid	$1, 2 \ldots$	Labels of specific states of a system
low	Low value (e.g. in interpolation)		
		$1, 2 \ldots$	Generic species in a mixture
mix	Equation of state parameter of a mixture		

Superscripts

dep	Departure function (with Δ)	s	Solid
E	Excess property	sat	At saturation
ideal	Ideal solution	v	Vapor
ideal gas	Ideal gas	α, β	Generic phases
molecular	Molecular		(in equilibrium)
l	Liquid	γ	Volume exponential of
o	Value at the reference state		a polytropic process
real	Real fluid with	∞	At infinite dilution
	intermolecular	(0)	Simple fluid term
	interactions	(1)	Correction term

Operators

d	Total differential	δ	Inexact (path dependent)
∂	Partial differential		differential
Δ	Difference between the final	ln	Natural (base e) logarithm
	and initial value of	log	Base 10 logarithm
	a state property	$\prod$	Cumulative product
∇	Gradient operator		operator
$\int$	Integral	$\sum$	Cumulative sum operator

Empirical parameters

a, b	van der Waals or Redlich-Kwong attraction and size parameter, respectively
$a, b, \alpha, \kappa \ldots$	Empirical parameters in various cubic equations of state
A	Two-suffix Margules activity coefficient model parameter
A_{ij}	Three-suffix Margules activity coefficient model parameters (one form)
A, B	Three-suffix Margules or van Laar activity coefficient model parameters
A, B	Debye-Huckel parameters
A, B, C	Empirical constants for the Antoine equation
A, B, C, D, E	Empirical constants for the heat capacity equation
B, C, D	Second, third and fourth virial coefficients
B', C', D'	Second, third and fourth virial coefficient in the pressure expansion
C_6	Constant of van der Waals or Lennard-Jones attraction
C_n	Constant of intermolecular repulsion potential of power r^{-n}
ε	Lennard-Jones energy parameter
Λ_{ij}	Wilson activity coefficient model parameters
σ	Distance parameter in hard sphere, Lennard-Jones and other potential functions

Contents

►**CHAPTER 1**

Measured Thermodynamic Properties And Other Basic Concepts   1

Learning Objectives   **1**
1.1   Thermodynamics   **2**
1.2   Preliminary Concepts—The Language
       of Thermo   **3**
1.3   Measured Thermodynamic Properties   **7**
       Volume (Extensive or Intensive)   **7**
       Temperature (intensive)   **8**
       Pressure (intensive)   **11**
1.4   Equilibrium   **12**
1.5   Independent and Dependent Thermodynamic
       Properties   **15**
1.6   The PvT Surface and Its Projections for Pure
       Substances   **17**
1.7   Thermodynamic Property Tables   **23**
1.8   The Ideal Gas   **26**
1.9   Summary   **28**
1.10 Problems   **29**

►**CHAPTER 2**

The First Law Of Thermodynamics   31

Learning Objectives   **31**
2.1   The First Law of Thermodynamics   **32**
       Forms of Energy   **32**
       Work and Heat: Transfer of Energy
         Between the System and the
         Surroundings   **37**
2.2   Construction of Hypothetical Paths   **41**
2.3   Reversible and Irreversible Processes   **42**
2.4   The First Law of Thermodynamics for
       Closed Systems   **49**
       Integral Balances   **49**
       Differential Balances   **51**
2.5   The First Law of Thermodynamics for Open
       Systems   **52**
2.6   Thermochemical Data for U and H **58**
       Heat Capacity: c_v and c_P **58**
       Latent Heats   **67**
       Enthalpy of Reactions   **70**
2.7   Reversible Processes in Closed Systems   **76**
       Reversible, Isothermal Expansion
         (Compression)   **76**
       Adiabatic Expansion (Compression) with
         Constant Heat Capacity   **78**
       Summary   **79**

2.8   Open-System Energy Balances on Process
       Equipment   **80**
2.9   Thermodynamic Cycles and the Carnot Cycle   **86**
2.10 Summary   **92**
2.11 Problems   **93**

►**CHAPTER 3**

Entropy And The Second Law Of Thermodynamic   103

Learning Objectives   **103**
3.1   Directionality of Processes/Spontaneity   **104**
3.2   Reversible and Irreversible Processes (Revisited) and
       Their Relationship to Directionality   **105**
3.3   Entropy, the Thermodynamic Property   **107**
3.4   The Second Law of Thermodynamics   **115**
3.5   Other Common Statements of the Second Law of
       Thermodynamics   **117**
3.6   The Second Law of Thermodynamics for Closed and
       Open Systems   **118**
       Calculation of Δs for Closed Systems   **119**
       Calculation of Δs for Open Systems   **123**
3.7   Calculation of Δs for an Ideal Gas   **126**
3.8   The Mechanical Energy Balance and the Bernoulli
       Equation   **135**
3.9   Vapor-Compression Power and Refrigeration
       Cycles   **138**
       The Rankine Cycle   **138**
       The Vapor-Compression Refrigeration
         Cycle   **143**
3.10 Molecular View of Entropy   **146**
       Maximizing Molecular Configurations over
         Space   **148**
       Maximizing Molecular Configurations over
         Energy   **149**
3.11 Summary   **152**
3.12 Problems   **154**

►**CHAPTER 4**

Equations Of State And Intermolecular Forces   164

Learning Objectives   **164**
4.1   Introduction   **165**
       Motivation   **165**
       The Ideal Gas   **166**
4.2   Intermolecular Forces   **166**
       Internal (Molecular) Energy   **166**
       Attractive Forces   **168**
       Intermolecular Potential Functions and
         Repulsive Forces   **177**

Principle of Corresponding States   180
Chemical Forces   182
4.3   Equations of State   186
The Van Der Waals Equation of State   186
Cubic Equations of State   192
The Virial Equation of State   194
Equations of State for Liquids and Solids   196
4.4   Generalized Compressibility Charts   197
4.5   Determination of Parameters for Mixtures   200
4.6   Summary   204
4.7   Problems   205

▶CHAPTER 5
The Thermodynamic Web   211

Learning Objectives   211
5.1   Types of Thermodynamic Properties   211
5.2   Thermodynamic Property Relationships   212
Dependent and Independent Properties   212
Fundamental Property Relations   214
Maxwell Relations   216
Other Useful Mathematical Relations   217
Using the Thermodynamic Web to Access
    Reported Data   218
5.3   Calculation of Δs, Δu, and Δh Using Equations
    of State   220
Relation of ds in Terms of Independent
    Variables T and v and Independent Variables
    T and P 220
Relation of du in Terms of Independent Variables
    T and v 221
Relation of dh in Terms of Independent Variables
    T and P 224
5.4   Departure Functions   230
5.5   Joule–Thomson Expansion and Liquefication   237
5.6   Summary   243
5.7   Problems   243

▶CHAPTER 6
Phase Equilibria I: Problem Formulation   250

Learning Objectives   250
6.1   Introduction   250
6.2   Pure Species Phase Equilibrium   253
Gibbs Energy as a Criterion for Chemical
    Equilibrium   253
Roles of Energy and Entropy in Phase
    Equilibria   255
The Relationship Between Saturation Pressure
    and Temperature: The Clapeyron
    Equation   258
Pure Component Vapor–Liquid Equilibrium:
    The Clausius–Clapeyron Equation   260
6.3   Thermodynamics of Mixtures   263
Introduction   263

Partial Molar Properties   264
The Gibbs–Duhem Equation   269
Summary of the Different Types of
    Thermodynamic Properties   270
Property Changes of Mixing   271
Determination of Partial Molar Properties   280
Relations Among Partial Molar Quantities   288
6.4   Multicomponent Phase Equilibria   289
The Chemical Potential—The Criteria for
    Chemical Equilibrium   289
Temperature and Pressure Dependence
    of μ_i 292
6.5   Summary   293
6.6   Problems   294

▶CHAPTER 7
Phase Equilibria II: Fugacity   302

Learning Objectives   302
7.1   Introduction   302
7.2   The Fugacity   303
Definition of Fugacity   303
Other Forms of Fugacity   305
Criteria for Chemical Equilibria in Terms
    of Fugacity   306
7.3   Fugacity in the Vapor Phase   307
Fugacity and Fugacity Coefficient of Pure
    Gases   307
Fugacity and Fugacity Coefficient of Species
    i in a Gas Mixture   313
The Lewis Fugacity Rule   319
Property Changes of Mixing for Ideal Gases   320
7.4   Fugacity in the Liquid Phase   322
Reference States for the Liquid Phase   322
Thermodynamic Relations Between γ_i 330
Models for γ_i Using g^E 336
Equation of State Approach to the Liquid
    Phase   353
7.5   Fugacity in the Solid Phase   353
Pure Solids   353
Solid Solutions   353
Interstitials and Vacancies in Crystals   353
7.6   Summary   354
7.7   Problems   356

▶CHAPTER 8
Phase Equilibria III: Phase Diagrams   364

Learning Objectives   364
8.1   Vapor–Liquid Equilibrium (VLE)   365
Raoult's Law (Ideal Gas and Ideal Solution)   365
Nonideal Liquids   372
Azeotropes   381
Fitting Activity Coefficient Models
    with VLE Data   386

Solubility of Gases in Liquids **391**
8.2 Liquid(α)–Liquid(β) Equilibrium: LLE **397**
8.3 Vapor–Liquid(α)–Liquid(β) Equilibrium: VLLE **403**
8.4 Solid–Liquid and Solid–Solid Equilibrium:
    SLE and SSE **407**
    Pure Solids **407**
    Solid Solutions **410**
8.5 Colligative Properties **412**
8.6 Summary **419**
8.7 Problems **420**

►**CHAPTER 9**
Chemical Reaction Equilibria **433**

Learning Objectives **433**
9.1 Introduction **434**
9.2 Chemical Reaction and Gibbs Energy **435**
9.3 Equilibrium for a Single Reaction **438**
9.4 Calculation of K from Thermochemical Data **442**
    Calculation of K from Gibbs Energy of
    Formation **443**
    The Temperature Dependence of K **444**
9.5 Relationship Between the Equilibrium Constant
    and the Concentrations of Reacting Species **448**
    The Equilibrium Constant for a Gas-Phase
    Reaction **449**
    The Equilibrium Constant for a Liquid-Phase
    (or Solid-Phase) Reaction **456**
    The Equilibrium Constant for a Heterogeneous
    Reaction **457**
9.6 Equilibrium in Electrochemical Systems **459**
9.7 Multiple Reactions **467**
    Extent of Reaction and Equilibrium
    Constant for R Reactions **467**
    Gibbs Phase Rule for Chemically Reacting
    Systems **469**
    Solution of Multiple Reaction Equilibria
    by Minimization of Gibbs Energy **475**
9.8 Reaction Equilibria of Point Defects in
    Crystalline Solids **478**
    Atomic Defects **478**
    Electronic Defects **481**
    Effect of Gas Partial Pressure on
    Defect Concentrations **485**
9.9 Summary **489**
9.10 Problems **491**

►**APPENDIX A**
Physical Property Data **499**

A.1 Critical Constants, Acentric Factors, and
    Antoine Coefficients **499**

A.2 Heat Capacity Data **501**
A.3 Enthalpy and Gibbs Energy of Formation
    at 298 K and 1 bar **503**

►**APPENDIX B**
Steam Tables **507**

B.1 Saturated Water: Temperature Table **508**
B.2 Saturated Water: Pressure Table **510**
B.3 Saturated Water: Solid–Vapor **512**
B.4 Superheated Water Vapor **513**
B.5 Subcooled Liquid Water **519**

►**APPENDIX C**
Lee–Kesler Generalized Correlation Tables **520**

C.1 Values for $z^{(0)}$ **520**
C.2 Values for $z^{(1)}$ **522**
C.3 Values for $\left[\dfrac{h_{T_r,P_r} - h_{T_r,P_r}^{\text{ideal gas}}}{RT_c} \right]^{(0)}$ **524**
C.4 Values for $\left[\dfrac{h_{T_r,P_r} - h_{T_r,P_r}^{\text{ideal gas}}}{RT_c} \right]^{(1)}$ **526**
C.5 Values for $\left[\dfrac{s_{T_r,P_r} - s_{T_r,P_r}^{\text{ideal gas}}}{R} \right]^{(0)}$ **528**
C.6 Values for $\left[\dfrac{s_{T_r,P_r} - s_{T_r,P_r}^{\text{ideal gas}}}{R} \right]^{(1)}$ **530**
C.7 Values for $\log [\phi^{(0)}]$ **532**
C.8 Values for $\log [\phi^{(1)}]$ **534**

►**APPENDIX D**
Unit Systems **536**

D.1 Common Dimensions Used in Thermodynamics
    and Their Associated Units **536**
D.2 Conversion Between Gaussian and SI Units **539**

►**APPENDIX E**
ThermoSolver Software **540**

E.1 Software Description **540**
E.2 Corresponding States using the Lee–Kesler
    Equation of State **543**

►**APPENDIX F**
References **545**

F.1 Sources of Thermodynamic Data **545**
F.2 Textbooks and Monographs **546**

Index **549**

Engineering and Chemical Thermodynamics

CHAPTER 1

Measured Thermodynamic Properties and Other Basic Concepts

The Buddha, the Godhead, resides quite as comfortably in the circuits of a digital computer or the gears of a cycle transmission as he does on the top of a mountain or the petal of a flower. To think otherwise would be to demean the Buddha — which is to demean oneself. This is what I want to talk about in this Chautauqua.

–*Zen and the Art of Motorcycle Maintenance*, by Robert M. Pirsig

Learning Objectives

To demonstrate mastery of the material in Chapter 1, you should be able to:

▶ Define the following terms in your own words:

- Universe, system, surroundings, and boundary
- Open system, closed system, and isolated system
- Thermodynamic property, extensive and intensive properties
- Thermodynamic state, state and path functions
- Thermodynamic process; adiabatic, isothermal, isobaric, and isochoric processes
- Phase and phase equilibrium
- Macroscopic, microscopic, and molecular-length scales
- Equilibrium and steady-state

Ultimately, you need to be able to apply these concepts to formulate and solve engineering problems.

▶ Relate the measured thermodynamic properties of temperature and pressure to molecular behavior. Describe phase and chemical reaction equilibrium in terms of dynamic molecular processes.

▶ Apply the state postulate and the phase rule to determine the appropriate independent properties to constrain the state of a system that contains a pure species.

▶ Given two properties, identify the phases present on a PT or a Pv phase diagram, including solid, subcooled liquid, saturated liquid, saturated vapor, and superheated vapor and two-phase regions. Identify the critical point and

> triple point. Describe the difference between saturation pressure and vapor pressure.
>
> ► Use the steam tables to identify the phase of a substance and find the value of desired thermodynamic properties with two independent properties specified, using linear interpolation if necessary.
>
> ► Use the ideal gas model to solve for an unknown measured property given measured property values.

► 1.1 THERMODYNAMICS

Science changes our perception of the world and contributes to an understanding of our place in it. Engineering can be thought of as a profession that creatively applies science to the development of processes and products to benefit humankind. Thermodynamics, perhaps more than any other subject, interweaves both these elements, and thus its pursuit is rich with practical as well as aesthetic rewards. It embodies engineering science in its purest form. As its name suggests, thermodynamics originally treated the conversion of heat to motion. It was first developed in the nineteenth century to increase the efficiency of engines—specifically, where the heat generated from the combustion of coal was converted to useful work. Toward this end, the two primary laws of thermodynamics were postulated. However, in extending these laws through logic and mathematics, thermodynamics has evolved into an engineering science that comprises much greater breadth. In addition to the calculation of heat effects and power requirements, thermodynamics can be used in many other ways. For example, we will learn that thermodynamics forms the framework whereby a relatively limited set of collected data can be efficiently used in a wide range of calculations. We will learn that you can determine certain useful properties of matter from measuring other properties and that you can predict the physical (phase) changes and chemical reactions that species undergo. A tribute to the wide applicability of this subject lies in the many fields that consider thermodynamics part of their core knowledge base. Such disciplines include biology, chemistry, physics, geology, oceanography, materials science, and, of course, engineering.

Thermodynamics is a self-contained, logically consistent theory, resting on a few fundamental postulates that we call **laws**. A law, in essence, compresses an enormous amount of experience and knowledge into one general statement. We test our knowledge through experiment and use laws to extend our knowledge and make predictions. The laws of thermodynamics are based on observations of nature and taken to be true on the basis of our everyday experience. From these laws, we can derive the whole of thermodynamics using the rigor of mathematics. Thermodynamics is self-contained in the sense that we do not need to venture outside the subject itself to develop its fundamental structure. On one hand, by virtue of their generality, the principles of thermodynamics constitute a powerful framework for solving a myriad of real-life engineering problems. However, it is also important to realize the limitations of this subject. Equilibrium thermodynamics tells us *nothing* about the mechanisms or rates of physical or chemical processes. Thus, while the final design of a chemical process requires the study of the kinetics of chemical reactions and rates of transport, thermodynamics defines the driving force for the process and provides us with a key tool in engineering analysis and design.

We will pursue the study of thermodynamics from both conceptual and applied viewpoints. The conceptual perspective enables us to construct a broad intuitive foundation that provides us the ability to address the plethora of topics that thermodynamics spans. The applied approach shows us how to actually use these concepts to solve problems of practical interest and, thereby, also enhances our conceptual understanding. Synergistically, these two tacks are intended to impart a *deep understanding* of thermodynamics.[1] In demonstrating a deep understanding, you will need to do more than regurgitate isolated facts and find the right equation to "plug and chug." Instead, you will need to search for connections and patterns in the material, understand the physical meaning of the equations you use, and creatively apply the fundamental principles that have been covered to entirely new problems. In fact, it is through this depth of learning that you will be able to take the synthesized information you are learning in the classroom and usefully and creatively apply it to new problems in the field or in the lab as a professional chemical engineer.

▶ 1.2 PRELIMINARY CONCEPTS—THE LANGUAGE OF THERMO

In engineering and science, we try to be precise with the language that we use. This exactness allows us to translate the concepts we develop into quantitative, mathematical form.[2] We are then able to use the rules of mathematics to further develop relationships and solve problems. This section introduces some fundamental concepts and definitions that we will use as a foundation for constructing the laws of thermodynamics and quantifying them with mathematics.

In thermodynamics, the **universe** represents all measured space. It is not very convenient, however, to consider the entire universe every time we need to do a calculation. Therefore, we break down the universe into the region in which we are interested, the **system**, and the rest of the universe, the **surroundings**. The system is usually chosen so that it contains the substance of interest, but not the physical apparatus itself. It may be of fixed volume, or its volume may change with time. Similarly, it may be of fixed composition, or the composition may change due to mass flow or chemical reaction. The system is separated from the surroundings by its **boundary**. The boundary may be real and physical, or it may be an imaginary construct. There are times when a judicious choice of the system and its boundary saves a great deal of computational effort.

In an **open** system both mass and energy can flow across the boundary. In a **closed** system no mass flows across the boundary. We call the system **isolated** if neither mass nor energy crosses its boundaries. You will find that some refer to an open system as a **control volume** and its boundary as a **control surface**.

For example, say we wish to study the piston–cylinder assembly in Figure 1.1. The usual choice of system, surroundings, and boundary are labeled. The boundary is depicted by the dashed line just inside the walls of the cylinder and below the piston. The system contains the gas within the piston–cylinder assembly but not the physical housing. The surroundings are on the other side of the boundary and comprise the rest of the universe. Likewise the system, surroundings, and boundary of an open system are labeled in Figure 1.2. In this case, the inlet and outlet flow streams, labeled "in" and "out," respectively, allow mass to flow into and out of the system, across the system boundary.

The substance contained within a system can be characterized by its **properties**. These include measured properties of volume, pressure, and temperature. The properties of the gas in Figure 1.1 are labeled as T_1, the temperature at which it exists; P_1,

---

[1] For more discussion on deep learning vs. shallow learning in engineering education, see Philip C. Wancat, "Engineering Education: Not Enough Education and Not Enough Engineering," 2nd International Conference on Teaching Science for Technology at the Tertiary Level, Stockholm, Sweden, June 14, 1997.

[2] It can be argued that the ultimate language of science and engineering is mathematics.

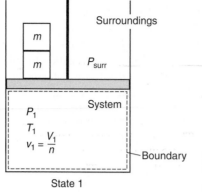

State 1

Figure 1.1 Schematic of a piston–cylinder assembly. The system, surroundings, and boundary are delineated.

its pressure; and v_1, its molar volume. The properties of the open system depicted in Figure 1.2 are also labeled, T_{sys} and P_{sys}. In this case, we can characterize the properties of the fluid in the inlet and outlet streams as well, as shown in the figure. Here $\dot{n}$ represents the molar flow rate into and out of the system. As we develop and apply the laws of thermodynamics, we will learn about other properties; for example, internal energy, enthalpy, entropy, and Gibbs energy are all useful thermodynamic properties.

Thermodynamic properties can be either **extensive** or **intensive**. Extensive properties depend on the size of the system while intensive properties do not. In other words, extensive properties are additive; intensive properties are not additive. An easy way to test whether a property is intensive or extensive is to ask yourself, "Would the value for this property change if I divided the system in half?" If the answer is "no," the property is intensive. If the answer is "yes," the property is extensive. For example, if we divide the system depicted in Figure 1.1 in half, the temperature on either side remains the same. Thus, the value of temperature does not change, and we conclude that temperature is intensive. Many properties can be expressed in both extensive and intensive forms. We must be careful with our nomenclature to distinguish between the different forms of these properties. We will use a capital letter for the extensive form of such a thermodynamic property. For example, extensive volume would be V of $[m^3]$. The intensive form will be lowercase. We denote molar volume with a lowercase v $[m^3/mol]$ and specific volume by $\hat{v}$ $[m^3/kg]$. On the other hand, pressure and temperature are always intensive and are written P and T, by convention.

The thermodynamic **state** of a system is the condition in which we find the system at any given time. The state fixes the values of a substance's intensive properties. Thus,

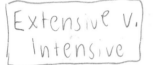

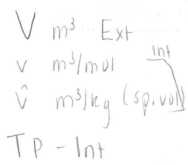

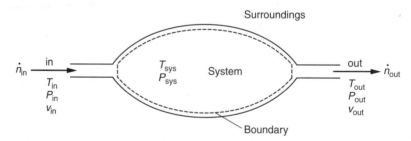

Figure 1.2 Schematic of an open system into and out of which mass flows. The system, surroundings, and boundary are delineated.

two systems comprised of the same substance whose intensive properties have identical values are in the same state. The system in Figure 1.1 is in state 1. Hence, we label the properties with a subscript "1." A system is said to undergo a **process** when it goes from one thermodynamic state to another. Figure 1.3 illustrates a process instigated by removing a block of mass m from the piston of Figure 1.1. The resulting force imbalance will cause the gas to expand and the piston to rise. As the gas expands, its pressure will drop. The expansion process will continue until the forces once again balance. Once the piston comes to rest, the system is in a new state, state 2. State 2 is defined by the properties T_2, P_2 and v_2. The expansion process takes the system from state 1 to state 2. As the dashed line in Figure 1.3 illustrates, we have chosen our system boundary so that it expands with the piston during the process. Thus, no mass flows across the boundary and we have a closed system. Alternatively, we could have chosen a boundary that makes the volume of the system constant. In that case, mass would flow across the system boundary as the piston expands, making it an open system. In general, the former choice is more convenient for solving problems.

Similarly, a process is depicted for the open system in Figure 1.2. However, we view this process slightly differently. In this case, the fluid enters the system in the inlet stream at a given state "in," with properties T_{in}, P_{in}, and v_{in}. It undergoes the process in the system and changes state. Thus, it exits in a different state, with properties T_{out}, P_{out}, and v_{out}. During a process, at least some of the properties of the substances contained in the system change. In an **adiabatic** process, no heat transfer occurs across the system boundary. In an **isothermal** process, the temperature of the system remains constant. Similarly, **isobaric** and **isochoric** processes occur at constant pressure and volume, respectively.

The values of thermodynamic properties do not depend on the process (i.e., path) through which the system arrived at its state; they depend *only* on the state itself. Thus, the change in a given property between states 1 and 2 will be the same for *any* process that starts at state 1 and ends at state 2. This aspect of thermodynamic properties is very useful in solving problems; we will exploit it often. We will devise *hypothetical* paths between thermodynamic states so that we can use data that are readily available to more easily perform computation. Thus, we may choose the following hypothetical path to calculate the change in any property for the process illustrated in Figure 1.3: We first

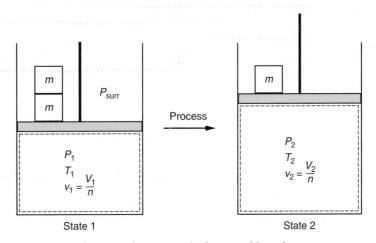

Figure 1.3 Schematic of a piston–cylinder assembly undergoing an expansion process from state 1 to state 2. This process is initiated by removal of a block of mass m.

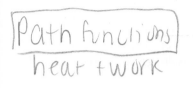

consider an isothermal expansion from P_1, T_1 to P_2, T_1. We then execute an isobaric cooling from P_2, T_1 to P_2, T_2. The hypothetical path takes us to the same state as the real process—so all the properties must be identical. Since properties depend only on the state itself, they are often termed **state functions**. On the other hand, there are quantities that we will be interested in, such as heat and work, that depend on path. These are referred to as **path functions**. When calculating values for these quantities, we must use the real path the system takes during the process.

A given **phase** of matter is characterized by both uniform physical structure and uniform chemical composition. It can be solid, liquid, or gas. The bonds between the atoms in a solid hold them in a specific position relative to other atoms in the solid. However, they are free to vibrate about this fixed position. A solid is called crystalline if it has a long-term, periodic order. The spatial arrangement in which the atoms are bonded is termed the lattice structure. A given substance can exist in several different crystalline lattice structures. Each different crystal structure represents a different phase, since the physical structure is different. For example, solid carbon can exist in the diamond phase or the graphite phase. A solid with no long-range order is called amorphous. Like a solid, molecules within the liquid phase are in close proximity to one another due to intermolecular attractive forces. However, the molecules in a liquid do not have specific, directional bonds; rather, they are in motion, free to move relative to one another. Multicomponent liquid mixtures can form different phases if the composition of the species differs in different regions. For example, while oil and water can coexist as liquids, they are considered separate liquid phases, since their compositions differ. Similarly, solids of different composition can coexist in different phases. Gas molecules show relatively weak intermolecular interactions. They move about to fill the entire volume of the container in which they are housed. This movement occurs in a random manner as the molecules continually change direction as they collide with one another and bounce off the container surfaces.

More than one phase can coexist within the system at equilibrium. When this phenomenon occurs, a **phase boundary** separates the phases from each other. One of the major topics in chemical thermodynamics, **phase equilibrium**, is used to determine the chemical compositions of the different phases that coexist in a given mixture at a specified temperature and pressure.

In this text, we will refer to three volume scales: the macroscopic, microscopic, and molecular. The **macroscopic** scale is the largest; it represents the bulk systems we observe in everyday life. We will often consider the entire macroscopic system to be in a uniform thermodynamic state. In this case, its properties (e.g., T, P, v) are uniform throughout the system. By **microscopic**, we refer to **differential** volume elements that are too small to see with the naked eye; however, each volume element contains enough molecules to be considered as having a continuous distribution of matter, which we call a continuum. Thus, a microscopic volume element must be large enough for temperature, pressure, and molar volumes to have meaningful values. Microscopic balances are performed over differential elements, which can then be integrated to describe behavior in the macroscopic world. We often use microscopic balances when the properties change over the volume of the system or with time. The **molecular**[3] scale is that of individual atoms and molecules. At this level the continuum breaks down and matter can be viewed as discrete elements. We cannot describe individual molecules in terms of temperature, pressure, or molar volumes. Strictly speaking, the word *molecule* is

---

[3] Some fields of science such as statistical mechanics use the term *microscopic* for what we call *molecular*.

outside the realm of classical thermodynamics. In fact, all of the concepts developed in this text can be developed based entirely on observations of macroscopic phenomena. This development does not require any knowledge of the molecular nature of the world in which we live. However, we are chemical engineers and can take advantage of our chemical intuition. Molecular concepts do account qualitatively for trends in data as well as magnitudes. Thus, they provide a means of understanding many of the phenomena encountered in classical thermodynamics. Consequently, we will often refer to molecular chemistry to explain thermodynamic phenomena.[4] The objective is to provide an intuitive framework for the concepts about which we are learning.

By this time, you are probably experienced in working with units. Most science and engineering texts have a section in the first chapter on this topic. In this text, we will mainly use the Système International, or **SI units**. The SI unit system uses the *primary* dimensions m, s, kg, mol, and K. Details of different unit systems can be found in Appendix D. One of the easiest ways to tell that an equation is wrong is that the units on one side do not match the units on the other side. Probably the most common errors in solving problems result from dimensional inconsistencies. The upshot is: Pay close attention to units! Try not to write a number down without the associated units. You should be able to convert between unit systems. It is often easiest to put all variables into the same unit system before solving a problem.

How many different units can you think of for length? For pressure? For energy?

▶ 1.3 MEASURED THERMODYNAMIC PROPERTIES

We have seen that if we specify the property values of the substance(s) in a system, we define its thermodynamic state. It is typically the **measured thermodynamic properties** that form our gateway into characterizing the particular state of a system. Measured thermodynamic properties are those that we obtain through direct measurement in the lab. These include volume, temperature, and pressure.

Volume (Extensive or Intensive)

Volume is related to the size of the system. For a rectangular geometry, volume can be obtained by multiplying the measured length, width, and height. This procedure gives us the extensive form of volume, V, in units of [m³] or [gal]. We purchase milk and gasoline in volume with this form of units. Volume can also be described as an intensive property, either as molar volume, v [m³/mol], or specific volume, $\hat{v}$ [m³/kg]. The specific volume is the reciprocal of density, ρ [kg/m³]. If a substance is distributed continuously and uniformly throughout the system, the intensive forms of volume can be determined by dividing the extensive volume by the total number of moles or the total mass, respectively. Thus,

$$v = \frac{V}{n} \qquad (1.1)$$

and

$$\hat{v} = \frac{V}{m} = \frac{1}{\rho} \qquad (1.2)$$

---

[4] While this objective can often be achieved formally and quantitatively through statistical mechanics and quantum mechanics, we will opt for a more qualitative and descriptive approach reminiscent of the chemistry classes you have taken.

If the amount of substance varies throughout the system, we can still refer to the molar or specific volume of a microscopic control volume. However, its value will change with position. In this case, the molar volume of any microscopic element can be defined:

$$v = \lim_{V \to V'} \left(\frac{V}{n} \right) \qquad (1.3)$$

where V' is the smallest volume over which the continuum approach is still valid and n is the number of moles.

Temperature (Intensive)

Temperature, T, is loosely classified as the *degree of hotness* of a particular system. No doubt, you have a good intuitive feel for what temperature is. When the temperature is 90°F in the summer, it is *hotter* than when it is 40°F in the winter. Likewise, if you bake potatoes in an oven at 400°F, they will cook faster than at 300°F, apparently since the oven is *hotter*. In general, to say that object A is *hotter* than object B is to say $T_A > T_B$. In this case, A will spontaneously transfer energy via heat[5] to B. Likewise if B is *hotter* than A, $T_A < T_B$, and energy will transfer spontaneously from B to A. When there is no tendency to transfer energy via heat in either direction, A and B must have equal hotness and $T_A = T_B$.[6] A logical extension of this concept says that if two bodies are at equal *hotness* to a third body, they must be at the same temperature themselves. This principle forms the basis for thermometry, where a judicious choice of the third body allows us to measure temperature. Any substance with a measurable property that changes as its temperature changes can then serve as a thermometer. For example, in the commonly used mercury in glass thermometer, the change in the volume of mercury is correlated to temperature. For more accurate measurements, the pressure exerted by a gas or the electric potential of junction between two different metals can be used.

On the molecular level, temperature is proportional to the average *kinetic* energy of the individual atoms (or molecules) in the system. All matter contains atoms that are in motion.[7] Species in the gas phase, for example, move chaotically through space with finite velocities. (What would happen to the air in a room if its molecules weren't moving?) They can also vibrate and rotate. Figure 1.4 illustrates individual molecular velocities. The piston–cylinder assembly depicted to the left schematically displays the velocities of a set of individual molecules. Each arrow represents the velocity vector with the size of the arrow proportional to a given molecule's speed. The velocities vary widely in magnitude and direction. Furthermore, the molecules constantly redistribute their velocities among themselves when they elastically collide with one another. In an elastic collision, the total kinetic energy of the colliding atoms is conserved. On the other hand, a particular molecule will change its velocity; as one molecule speeds up via collision, however, its collision partner slows down.

Since the molecules in a gas move at great speeds, they collide with one another billions of times per second at room temperature and pressure. An individual molecule frequently speeds up and slows down as it undergoes these elastic collisions. However, within a short period of time the *distribution* of speeds of *all* the molecules in

---

[5] In Chapter 2, we will more carefully define heat.

[6] This relation for temperature is often referred to as the "zeroth law of thermodynamics." However, in the spirit of Rudolph Clausius, we will view thermodynamics in terms of two fundamental laws of nature that are represented by the first and second laws of thermodynamics.

[7] Except in the ideal case of a perfect solid at a temperature of absolute zero.

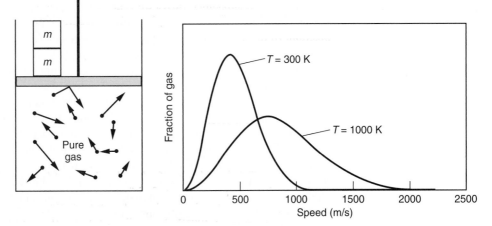

Figure 1.4 A schematic representation of the different speeds molecules have in the gas phase. The left-hand side shows molecules flying around in the system. The right-hand side illustrates the Maxwell–Boltzmann distributions of O_2 molecules at 300 K and 1000 K.

a given system becomes constant and well defined. It is termed the Maxwell–Boltzmann distribution and can be derived using the kinetic theory of gases. The right-hand side of Figure 1.4 shows the Maxwell–Boltzmann distributions for O_2 at 300 K and 1000 K. The y-axis plots the fraction of O_2 molecules at the speed given on the x-axis. At a given temperature, the fraction of molecules at any given speed does not change.[8] In fact, the temperature of a gas is only strictly defined for an aggregate of gas molecules that have obtained this characteristic distribution. Similarly, for a microscopic volume element to be considered a continuum, it must have enough molecules for the gas to approximate this distribution. The distribution at the higher temperature has shifted to higher speeds and flattened out. Kinetic theory shows the temperature is proportional to the average translational molecular kinetic energy, $e_K^{molecular}$, which is related to the mean-square molecular velocity:

$$T \approx \frac{1}{2}m\overline{\vec{V}^2} = \overline{e_K^{molecular}} \qquad (1.4)$$

where m is the mass of an individual molecule and $\overline{\vec{V}^2}$ is the mean-square velocity. $\overline{e_K^{molecular}}$ represents the average kinetic energy of the "center-of-mass" motion of the molecules. Diatomic and polyatomic molecules can have vibrational and rotational energy as well. The higher the temperature, the faster the atoms move and the higher the average kinetic energy. Temperature is independent of the nature of the particular substance in the system. Thus, when we have two different gases at the same temperature, the average kinetic energies of the molecules in each gas are the same.

This principle can be extended and applied to the liquid and solid phases as well. The temperature in the condensed phases is also a measure of the average kinetic energy of the molecules. For molecules to remain in the liquid or solid phase, however, the

[8] The macroscopic and the molecular scales present an interesting juxtaposition. At a well-defined temperature, there is one distinct distribution of molecular speeds. Thus, we say we have only one macrostate possible. However, if we keep track of all the individual molecules, we see there are many ways to arrange them within this one macrostate; that is, any given molecule can have many possible speeds. In Chapter 3, we will see that entropy is a measure of how many different molecular configurations a given macrostate can have.

potential energy of attraction between the molecules must be greater than their kinetic energy. Thus, species condense and freeze at lower temperature when the kinetic energy of the molecules is lower and the potential energy of attraction dominates.

As you know, if we allow two solid objects with different initial temperatures to contact each other, and we wait long enough, their temperatures will become equal. How can we understand this phenomenon in the context of average atomic kinetic energy? In the case of solids, the main mode of molecular kinetic energy is in the form of vibrations. The atoms of the hot object are vibrating with more kinetic energy and, therefore, moving faster than the atoms of the cold object. At the interface, the faster atoms vibrating in the hot object transfer more energy to the cold object than the slower-moving atoms in the cold object transfer to the hot object. Thus, with time, the cold object gains atomic kinetic energy (vibrates more vigorously) and the hot object loses atomic kinetic energy. This transfer of energy occurs until their average atomic kinetic energies become equal. At this point, their temperatures are equal and they transfer equal amounts of energy to each other, so their temperature does not change any further. This case illustrates that temperature and molecular kinetic energy are intimately linked. We will learn more about these molecular forms of energy when we discuss the conservation of energy in Chapter 2.

To assign quantitative values to temperature, we need an agreed upon **temperature scale**. Each unit of the scale is then called a **degree**(°). Since the temperature is linearly proportional to the average kinetic energy of the atoms and molecules in the system, we just need to specify the constant of proportionality to define a temperature scale. By convention, we choose $(3/2)\,k$ where k is Boltzmann's constant. Thus we can define T in a particular unit system by writing.[9]

$$e_K^{\text{molecular}} \equiv (3/2)kT \qquad (1.5)$$

Since temperature is defined as the average kinetic energy *per molecule*, it does not depend on the size of the system. Hence temperature is always intensive. The scale resulting from Equation (1.5) defines the **absolute temperature scale** in which the temperature is zero when there is no molecular kinetic energy. In SI units, degrees Kelvin [K] is used as the temperature scale and $k = 1.38 \times 10^{-23}$ [J/(molecule K)]. The temperature scale in English units is degrees Rankine [°R]. Conversion between the SI and English systems can be achieved by realizing that the scale in English units is 9/5 times greater than that in SI. Hence

$$T[°R] = (9/5)T\,[K] \qquad (1.6)$$

No substance can have a temperature below zero on an absolute temperature scale, since that is the point where there is no molecular motion. However, *absolute zero*, as it is called, corresponds to a temperature that is very, very cold. It is often more convenient to define a temperature scale around those temperatures more commonly found in the natural world. The Celsius temperature scale [°C] uses the same scale per degree as the Kelvin scale; however, the freezing point of pure water is 0°C and the boiling point of pure water as 100°C. It shifts the Kelvin scale by 273.15, that is,

$$T[K] = T[°C] + 273.15 \qquad (1.7)$$

[9] In fact, in the limit of very low temperature, quantum effects can become measurable for some gases, and Equation (1.5) breaks down. However, these effects can be reasonably neglected for our purposes.

In this case, the temperature of no molecular motion (absolute zero) occurs at $-273.15°C$. Similarly, the Fahrenheit scale [°F] uses the same scale per degree as the Rankine scale, but the freezing point of pure water is 32°F and the boiling point of pure water is 212°F. Thus,

$$T[°R] = T[°F] + 459.67 \tag{1.8}$$

Absolute zero then occurs at $-459.67°F$. It is straightforward to show that conversion between Celsius and Fahrenheit scales can be accomplished by

$$T[°F] = (9/5)T[°C] + 32 \tag{1.9}$$

Finally, we note the measurement of temperature is actually indirect, but we have such a good feel for T, that we classify it as a measured parameter.

Pressure (Intensive)

Pressure is the normal force per unit area exerted by a substance on its boundary. The boundary can be the physical boundary that defines the system. If the pressure varies spatially, we can also consider a hypothetical boundary that is placed within the system. Let us consider again the piston–cylinder assembly. As illustrated in Figure 1.5, the pressure of the gas on the piston within the piston–cylinder assembly can be conceptualized in terms of the force exerted by the molecules as they bounce off the piston. We consider the molecular collisions with the piston to be elastic. According to Newton's second law, the time rate of change of momentum equals the force. Each molecule's velocity in the z direction, $\vec{V}_z$ changes sign as a result of collision with the piston, as illustrated in Figure 1.5. Thus, the change in momentum for a molecule of mass m that hits the piston is given by

$$\left\{ \frac{\text{change in momentum}}{\text{molecule that hits the piston}} \right\} = m\vec{V}_z - (-m\vec{V}_z) = 2m\vec{V}_z \tag{1.10}$$

This momentum must then be absorbed by the piston. The total pressure the gas exerts on the face of the piston results from summing the change in z momentum of all N of the individual atoms (or molecules) impinging on the piston and dividing by the area, A:

$$P = \frac{1}{A} \sum_{i=1}^{N} \left[\frac{d(m\vec{V}_z)}{dt} \right]_i \tag{1.11}$$

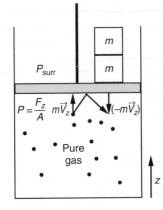

Figure 1.5 Schematic of how the gas molecules in a stationary piston–cylinder assembly exert pressure on the piston through transfer of z momentum.

In other words, the pressure is equal to momentum change per second per area that the impinging molecules deliver to the piston. The large number of molecules in a macroscopic system effectively makes this exerted force constant across the piston. We can examine how certain processes affect the pressure of a gas through examination of Equation (1.11). If we increase the number of molecules, N, in the system, more molecules will impinge on the piston surface per time and the pressure will increase. Likewise, if we increase the velocity of the molecules, $\vec{V}_z$, through an increase in temperature, the pressure will go up. If the piston is stationary, a force balance shows that the pressure of the gas must be identical to the force per area exerted on the other side by the surroundings; that is, $P = P_{surr}$ where the subscript "surr" refers to the surroundings. On the other hand, if the pressure of the surroundings is greater than that of the system, the forces will not balance and the piston will move down in a compression process. The pressure in the piston will increase until the forces again balance and compression stops. There is another effect we must also consider during compression. The atoms will pick up additional speed from momentum transfer with the moving piston. In a sense, the piston will "hit" the molecules much like a bat hits a baseball. The additional momentum transferred to the molecules will cause the z component of velocity leaving the piston to be greater than the incoming velocity. Therefore, the compression process causes the average molecular kinetic energy in the system to rise, and, consequently, T will increase. A similar argument can be made for why temperature decreases during expansion. We will discuss these effects in compression and expansion processes in more detail when we look at work in Chapter 2.

We can think of force as the extensive version of pressure. If we double the area of the piston in Figure 1.5, we double the force. On the other hand, the pressure is intensive and stays the same. In this text, we will usually refer to pressure and seldom to force, since pressure can be extended to the context within a system and to microscopic volume elements. The SI unit of pressure is the Pa. It has the following equivalent dimensional forms:

$$1\,[\text{Pa}] = 1\big[\text{kg/ms}^2\big] = 1\big[\text{N/m}^2\big] = 1\big[\text{J/m}^3\big] \tag{1.12}$$

Since pressure represents the force per area, the unit of $[\text{N/m}^2]$ is most straightforward; however, in the context of the energy balances we will be addressing in this text, the unit of $[\text{J/m}^3]$ is often more useful.

▶ 1.4 EQUILIBRIUM

A large part of thermodynamics deals with predicting the state that systems will reach at equilibrium. **Equilibrium** refers to a condition in which the state neither changes with time nor has a tendency to spontaneously change. At equilibrium, there is no net *driving force* for change. In other words, all opposing *driving forces* are counterbalanced. We use *driving force* as a generic term that represents some type of influence for a system to change. If the equilibrium state is **stable**, the system will return to that state when a small disturbance is imposed upon it. A system that has mass being supplied or removed cannot be at equilibrium, since a net driving force must exist to move the species about. Hence, equilibrium can only occur in a *closed* system. In general, any system subject to net fluxes cannot be in equilibrium.

We can distinguish between a system in an equilibrium state and a process at steady-state. If the state of an *open* system does not change as it undergoes a process, it is said to be at **steady-state**; however, it is not at equilibrium since there must be a net driving force to get the mass into and out of the system. For example, consider the open

The triple point is labeled on the *PT* diagram in Figure 1.6. In this state, a pure substance can have vapor, liquid, and solid phases all coexisting together. The phase rule tells us that each phase has zero degrees of freedom. Consequently, both the system temperature and pressure are fixed as a point on the *PT* diagram. The *Pv* projection shows the three-phase region as a line, the **triple line,** since the molar volume changes as the proportion of each phase changes.

The projections of *PvT* surfaces are useful for identifying the thermodynamic state of a system. To illustrate this point, we show five different states, all at identical pressures, in Figure 1.7. On the left of the figure, each state is identified in the context of a piston–cylinder assembly. If the system represented by state 1 undergoes a set of isobaric processes whereby energy is input to the system, it will go from state 1 to 2 to 3 to 4 to 5. These states are also identified by number on the *Pv* diagram and the *PT* diagram on the right of the figure. Note that the lower half of the *Pv* diagram is omitted for clarity. State 1 represents **subcooled liquid,** where pressure and temperature are independent properties. *As energy is put into* the system, the temperature will rise until the liquid becomes saturated, as illustrated on the *PT* diagram. The volume also increases; however, the magnitude of the change is small, since the volume of a liquid is relatively insensitive to temperature. The substance is known to be in a **saturated** condition when it is in the two-phase region at vapor–liquid equilibrium. A **saturated liquid** is "ready" to boil; that is, any more energy input will lead to a bubble of vapor. It is labeled as state 2 on the left

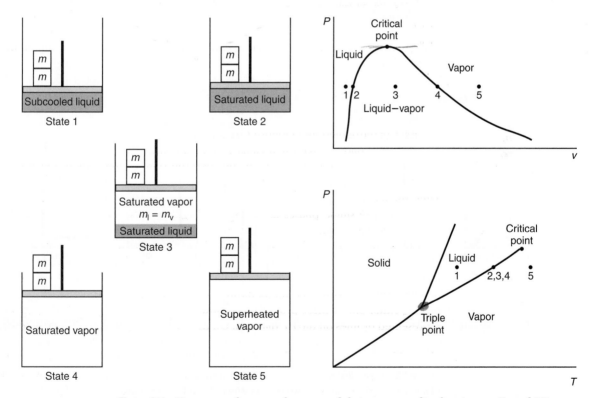

Figure 1.7 Five states of a pure substance and their corresponding locations on *Pv* and *PT* projections. All states are at the same pressure. State 1 is a subcooled liquid; state 2, a saturated liquid; state 3, a saturated liquid–vapor mixture; state 4, a saturated vapor; and state 5, a superheated vapor. The volume of liquid is exaggerated for clarity.

$T_{bp} = T_{sat}$

of the liquid–vapor dome in the Pv diagram. Since we are now in a two-phase region, the temperature is no longer independent. At a given pressure, the temperature at which a pure substance boils is known as the **saturation temperature**. The saturation temperature at any pressure is given by the line in the PT diagram on which state 2 is labeled. The temperature at which a species boils at 1 atm is termed the **normal boiling point**. State 3 represents the state where half the mass in the system has vaporized. Therefore, it is identified as halfway across the liquid–vapor dome in the Pv projection. The molar volume represented by state 3 is not realized by either phase of the system; rather, it is an average of the liquid and vapor that we use to characterize the molar volume of the system. In fact, the fraction of mass in the liquid phase in state 3 has the same molar volume as that in state 2. Similarly, state 4 is **saturated vapor**, the point at which any energy that is removed would cause a drop of liquid to condense. Note that states 2, 3, and 4 are represented by an identical point on the PT diagram, since P and T are not independent in the two-phase region—illustrating the fact that we cannot use P and T to constrain the state of the system. Finally, state 5 is **superheated vapor**, which exists at a higher temperature than the saturated vapor. The increase in volume with temperature in the vapor phase is much more pronounced than it was for the liquid. How would you draw this process on a Tv diagram?

Sat P vs
Vap p

The **saturation pressure** is the pressure at which a *pure* substance boils at a given temperature. A related quantity, the **vapor pressure** of a substance, is its contribution to the total pressure in a *mixture* at a given temperature. As we will see, this contribution is equal to the partial pressure of the substance in an ideal gas mixture.[17] Figure 1.8 provides a schematic representation of each of these quantities. The two piston–cylinder assemblies depicted on the left represent cases for which the saturation pressure is defined. In these systems, pure species a is in vapor–liquid equilibrium at temperatures T_1

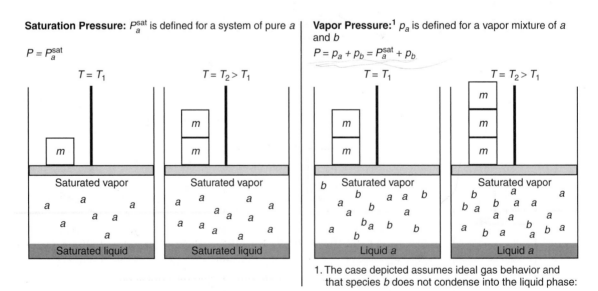

Saturation Pressure: P_a^{sat} is defined for a system of pure a

$P = P_a^{sat}$

$T = T_1$ $T = T_2 > T_1$

Saturated vapor   Saturated vapor

Saturated liquid   Saturated liquid

Vapor Pressure:[1] p_a is defined for a vapor mixture of a and b

$P = p_a + p_b = P_a^{sat} + p_b$

$T = T_1$ $T = T_2 > T_1$

Saturated vapor   Saturated vapor

Liquid a Liquid a

1. The case depicted assumes ideal gas behavior and that species b does not condense into the liquid phase:

Figure 1.8 Graphical representation of the saturation pressure of pure a and the vapor pressure of a in a mixture of a and b. Two temperatures, T_1 and T_2 are shown.

---

[17] The ideal gas model will be covered in Section 1.8.

and T_2, respectively, where T_2 is greater than T_1. In each case, there is a unique pressure at which the two phases can be in equilibrium—defined as the saturation pressure, P_a^{sat}. For example, pure water at 293 K (20°C) has a saturation pressure of 2.34 kPa. Said another way, for *pure* water to boil at 293 K, the pressure of the system must be 2.34 kPa. If the pressure is higher, water will exist only as a liquid. Conversely, if the pressure is lower, it will be a single-phase vapor. At a higher temperature, the saturation pressure will be higher, as depicted for T_2 in Figure 1.8. For example, at 303 K (30°C), water has a saturation pressure of 4.25 kPa. This incremental increase in temperature nearly doubles the saturation pressure.

A schematic illustrating when we use vapor pressure is shown for the two systems depicted on the right in Figure 1.8, where the vapor phase contains a mixture of species a and b. The vapor pressure of species a represents its contribution to the total pressure of the mixture. The two temperatures shown, T_1 and T_2, are identical to those for pure species a on the left of the figure. For convenience, we assume species b is does not noticeably condense in the liquid and that the vapor behaves as an ideal gas.[18] Then the vapor pressure of species a is identical to the corresponding saturation pressure at the same temperature. For example, now consider an open container of water sitting in a room at 293 K and 1 bar. Some of the water will **evaporate** and go into the air. The partial pressure of water at equilibrium with the air will be equal to the saturation pressure of pure a, 2.34 kPa. Since water is but one of many components in the mixture, we say water has a vapor pressure of 2.34 kPa. In contrast, the total pressure of the system is around 1 atm. The vapor pressure presented in Figure 1.8 depends only on the temperature of the water, not on the total pressure of the system. In other words, the vapor pressure of a is independent of how much b is present. While we can use saturation pressures to determine the vapor pressure in a given mixture, the term *saturation pressure* refers to the *pure* species. You should learn the difference between saturation pressure and vapor pressure because they are often confused.

A magnified view of the upper part of the *Pv* phase diagram is shown in Figure 1.9. Four isotherms are shown. Along all four isotherms, the volume increases as the pressure decreases. At the lowest two temperatures, the isotherms start in the liquid phase. In the liquid phase, the volume change is relatively small as the pressure drops. Along a given isotherm, the pressure decreases until it reaches the saturation pressure. This point is marked by the intersection with the left side of the liquid–vapor dome. At this point, any increase in volume leads to a two-phase liquid–vapor mixture, where the value of liquid volume is given by the intersection of the isotherm with the left side of the dome and the vapor volume is given by the intersection of the isotherm with the right side of the dome. The pressure remains constant in the two-phase region, since P and T are no longer independent. After complete vaporization, the pressure again decreases. The corresponding increase in volume of the vapor is noticeably larger than that of the liquid. As the temperatures of the isotherms increase, the saturated liquid volumes get larger and the saturated vapor volumes get smaller. Finally, at the **critical point**, located at the top of the liquid–vapor dome, the values of v_l and v_v become identical. The critical point represents a unique state and is identified with the subscript "c." Thus, it is constrained by the critical temperature, T_c and the critical pressure, P_c. Values for these critical properties of many pure substances are reported in Appendix A. The critical point represents the point at which liquid and vapor regions are no longer distinguishable. The critical point is also labeled in the depictions in Figure 1.6. The critical isotherm goes

---

[18] We will learn how to treat the more general case in Chapters 7 and 8.

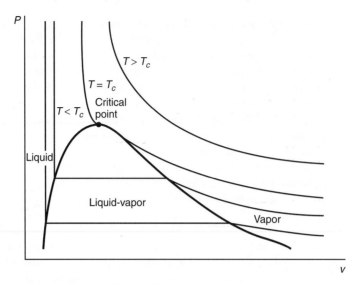

Figure 1.9 Magnified view of the Pv diagram. Four isotherms are shown—two below the critical temperature (subcritical), one at the critical temperature, and one above the critical temperature (supercritical).

through an inflection point at the critical point. Mathematically, this condition can be written as

$$\left(\frac{\partial P}{\partial v}\right)_{T_c} = 0 \tag{1.23}$$

and

$$\left(\frac{\partial^2 P}{\partial v^2}\right)_{T_c} = 0 \tag{1.24}$$

The partial derivatives in Equations (1.23) and (1.24) specify that we need to keep the temperature constant at its value at the critical point.

The isotherm above the critical point is representative of a **supercritical** fluid. This isotherm continuously decreases in pressure as the volume increases. A supercritical fluid has partly liquidlike characteristics (e.g., high density) and partly vaporlike characteristics (compressibility, high-diffusivity). Not surprisingly, there are many interesting engineering applications for substances in this state. There can be confusion between the terms *gas* and *vapor*. We refer to a gas as any form of matter that fills the container; it can be either subcritical or supercritical. When we speak of vapor, it is gas that if isothermally compressed will condense into a liquid and is, therefore, always subcritical.

▶ **EXAMPLE 1.1**
Determination of
Location of a Two-Phase
System on a Phase
Diagram

Consider a two-phase system at a specified T that contains 20% vapor, by mass, and 80% liquid. Identify the state on a Tv phase diagram. Explain why graphical determination of the state is termed the *lever* rule.

SOLUTION The quality of the system, defined as the fraction of matter in the vapor, is 0.2. The molar volume can be written in terms of the quality according to Equation (1.22):

$$v = v_l + x(v_v - v_l) = v_l + 0.2(v_v - v_l) \tag{E1.1A}$$

The fraction of *vapor* is found by solving Equation (E1.1A):

$$0.2 = \frac{v - v_l}{v_v - v_l} \tag{E1.1B}$$

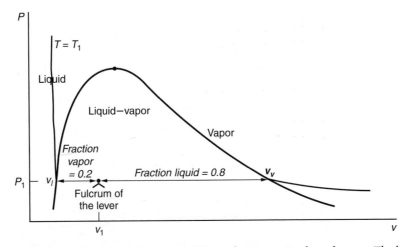

Figure E1.1 The state of the system of Example 1.1 on a Pv phase diagram. The lever rule is indicated.

Equation (E1.1B) can be interpreted as follows: The fraction of vapor is the ratio of the difference between the system volume and the volume of the *other* phase to the difference between vapor and liquid volumes. Similarly, the fraction of *liquid* is given by

$$0.8 = \frac{v_v - v}{v_v - v_l}$$

The fraction of liquid is the ratio of the difference between the volume of the *other* phase and the system volume to the difference between vapor and liquid volumes. This result is graphically presented in Figure E1.1. The overall composition of the system in the two-phase region is shown on top of the fulcrum. The intersection of the horizontal line with the liquid vapor dome gives the molar volumes of the liquid and vapor phases. The horizontal line is referred to as the tie line. The fraction of each phase present is obtained by taking the length of the line segment to the *other* phase and dividing by the total length of the line between one phase and the other. The line segment representing the liquid is four times greater than that representing the vapor.

▶ 1.7 THERMODYNAMIC PROPERTY TABLES

As we have seen, if we specify two independent intensive properties of a pure substance, the state of the system is constrained. Thus, any other thermodynamic property is restricted to only one possible value. Not surprisingly, for commonly used substances, such as water, tables of thermodynamic properties have been constructed that tabulate a set of useful properties. Appendix B reproduces a portion of the "steam tables," where six intensive properties of water are tabulated.[19] Recall that we use the term *water* to indicate the chemical species H_2O in any phase: solid, liquid, or gas. The tabulated thermodynamic properties include the measured properties T, P, and $\hat{v}$, as well as three other properties we will learn about in Chapters 2 and 3—the specific internal energy, $\hat{u}$, the specific enthalpy, $\hat{h}$, and the specific entropy, $\hat{s}$. The values reported for the latter three properties are not reported as absolute values but rather as the change in that property relative to a well-defined reference state. The reference state used for the steam tables is as a liquid at the triple point of water. At this state, both internal energy and entropy are

[19] J. H. Keenan, F. G. Keys, P. G. Hill, J. G. Moore, *Steam Tables* (New York: Wiley, 1969).

defined as zero.[20] Similar property tables are available in the literature or the National Institute of Standards and Technology (NIST) website[21] for many other common species as well, including Ar, N_2, O_2, CH_4, C_2H_4, C_2H_6, C_3H_8, C_4H_{10}, and several refrigerants (NH_3, R-12, R-13, R-14, R-21, R-22, R-23, R-113, R-114, R-123, R-134a ...). The reference state must be consistent when doing a thermodynamic calculation. Care should be used when taking values for a substance from different sources, since they sometimes use different reference states.

The steam tables are organized according to the phase in which water exists in the state of interest. Figure 1.10 shows a PT diagram for water with the corresponding appendices where the thermodynamic property data are located. Appendices B.1 and B.2 report data for the saturated vapor–liquid region. Since the pressure and temperature are no longer independent in this two-phase region, if we specify either property, we fix the other. Appendix B.1 presents data for saturated vapor and liquid water at even intervals of temperature. We use this appendix when the value of temperature is known. Appendix B.2 also presents data for saturated water, but in terms of round numbers of pressure. For each of the other properties in the tables ($\hat{v}, \hat{u}, \hat{h}$, and $\hat{s}$), values are presented for the liquid, l; the vapor, v; and the difference between the vapor and the liquid; $\Delta = v - l$. By analogy to Equation (1.22), any property value of the system is scaled by the quality. For example the specific internal energy can be found according to

$$\hat{u} = (1 - x)\hat{u}_l + x\hat{u}_v = \hat{u}_l + x\hat{u}_\Delta \qquad (1.25)$$

Property values for superheated water vapor and subcooled liquid water are presented in Appendices B.4 and B.5, respectively. In these regions, T and P are independent, so if we specify both, we constrain the state of the system. The tables are organized first

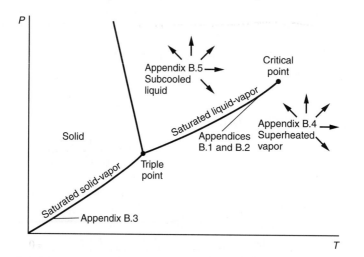

Figure 1.10 Illustration of the steam tables available for different phases of H_2O.

[20] In fact, the "third law" of thermodynamics specifies that the entropy of a perfect crystal is zero at a temperature of absolute zero. This principle allows absolute entropies to be defined and calculated. However, the entropies presented in the steam tables do not make use of the third law; thus, the values presented are relative.

[21] P. J. Linstrom and W. G. Mallard, Eds., **NIST Chemistry WebBook, NIST Standard Reference Database Number 69**, March 2003, National Institute of Standards and Technology, Gaithersburg MD, 20899 (http://webbook.nist.gov/chemistry/fluid).

► **EXAMPLE 1.3**
Determination of
Ideal Gas Law
from the Definition
of *P* and *T*

Show that the ideal gas model can be derived from the molecular definition of pressure, Equation (1.11). The molecular relationship between temperature and kinetic energy, Equation (1.5), is also useful.

SOLUTION   The definition of *pressure* is given by Equation (1.11):

$$P = \frac{1}{A} \sum_{i=1}^{N} \left[\frac{d(m\vec{V}_z)}{dt} \right]_i \qquad (1.11)$$

The rate of change in z momentum of any particular molecule can be separated into two parts, as follows:

$$\frac{d(m\vec{V}_z)}{dt} = \left\{ \frac{\text{change in momentum}}{\text{collision of molecule with the piston}} \right\} \left\{ \frac{\text{collisions}}{\text{time}} \right\} \qquad (E1.3A)$$

The first term on the right-hand side of Equation (E1.3A) is given by Equation (1.10):

$$\left\{ \frac{\text{change in momentum}}{\text{collision of molecule with the piston}} \right\} = m\vec{V}_z - (-m\vec{V}_z) = 2m\vec{V}_z \qquad (1.10)$$

The second term can be obtained if we realize a molecule must travel a length, l, to collide with the piston. Hence the rate of collisions can be approximated by

$$\left\{ \frac{\text{collisions}}{\text{time}} \right\} = \frac{\vec{V}_z}{l} = \frac{A\vec{V}_z}{V} \qquad (E1.3B)$$

Substituting Equations (E1.3B) and 1.10 into Equation (E1.3A), and then using Equation (E1.3A) in Equation (1.11) gives

$$P = \frac{1}{A} \sum_{i} \left[\frac{d(m\vec{V}_z)}{dt} \right]_i = \frac{1}{A} \sum_{i=1}^{N/2} \left[\frac{2mA\vec{V}_z^2}{V} \right]_i \qquad (E1.3C)$$

where we have divided the total number of molecules in the system by 2 since the molecules heading away from the piston with a velocity $(-\vec{V}_z)$ will not hit it. Thus, we do not count them in the calculation of the pressure. We can rewrite Equation (E1.3C) by using the average mean speed instead of summing over all the individual velocities. The average mean speed is given by the following relation:

$$\sum_{i=1}^{N/2} (\vec{V}_z^2)_i = \frac{N}{2} \left(\overline{\vec{V}_z^2} \right)$$

Inserting this expression into Equation (E1.3C), we get

$$P = \frac{mN}{V} \left(\overline{\vec{V}_z^2} \right) \qquad (E1.3D)$$

Since the molecules are equally likely to move in any of three directions, we can replace the speed in the z direction with the total speed $\vec{V}$, as follows:

$$\overline{\vec{V}_z^2} = (1/3)\overline{\vec{V}^2} \qquad (E1.3E)$$

The factor of 3 arises since there are three possible directions of motion. Plugging in Equation (E1.3E) into (E1.3D) gives

$$P = \frac{2N}{3V} \left(\frac{1}{2} m\overline{\vec{V}^2} \right) = \frac{2N}{3V} \left(\overline{e_K^{\text{molecular}}} \right)$$

where Equation (1.4) was used. Finally substituting Equation (1.5) gives the ideal gas relation:

$$P = \frac{NkT}{V} = \frac{nRT}{V}$$

where $R = kN_A$ and N_A is Avogadro's number.

▶ **EXAMPLE 1.4**
Improvement in
Interpolation of
Example 1.2

Based on the ideal gas law, reestimate the specific volume of water at $P = 1.4$ MPa and $T = 333°C$ using data from the steam tables.

SOLUTION   If we apply the ideal gas model, we have

$$v = \frac{RT}{P}$$

The molar volume, v, is proportional to T. The first linear interpolation of Example 1.2 is consistent with this result, so again we have $\hat{v}_{T=333} = 0.18086\,[\text{m}^3/\text{kg}]$. However, the ideal gas law shows molar volume is inversely proportional to pressure, so it is better to interpolate in $(1/P)$. Thus, the second interpolation becomes:

$$\hat{v}_{P=1.4} = \hat{v}_{P=1} + (\hat{v}_{P=1.5} - \hat{v}_{P-1})\left(\frac{\frac{1}{P} - \frac{1}{P_{\text{low}}}}{\frac{1}{P_{\text{high}}} - \frac{1}{P_{\text{low}}}}\right) = 0.19418\,[\text{m}^3/\text{kg}]$$

The difference in the value found in Example 1.2 and Example 1.4 is:

$$\%_{\text{diff}} = \frac{0.19951 - 0.19418}{0.19418} \times 100 = 2.7\%$$

This example illustrates the effectiveness of letting physical principles guide our mathematical procedures so that we can make better engineering estimates.

▶ **1.9 SUMMARY**

The material in Chapter 1 forms the conceptual foundation on which we will construct our understanding of thermodynamics. We will formulate thermodynamics by identifying the **state** that a **system** is in and by looking at **processes** by which a system goes from one state to another. We are interested in both **closed systems**, which can attain thermodynamic **equilibrium**, and **open systems**. The **state postulate** and the **phase rule** allow us to identify which independent, **intensive** thermodynamic **properties** we can choose to constrain the state of the system. If we also know the amount of matter present, we can determine the **extensive** properties in the system. Thermodynamic properties are also called **state functions**. Since they do not depend on path, we may devise a convenient hypothetical path to calculate the change in their values between two states. Conversely, other quantities, such as heat or work, are **path functions**.

The measured properties $\boldsymbol{T}$, $\boldsymbol{P}$, and $\boldsymbol{v}$ are especially useful in determining the thermodynamic state since we have access to them in the lab. Projections of the PvT surface, in the form of PT, Pv, and Tv phase diagrams, allow us to identify whether the system is in a single **phase** or is in **phase equilibrium** between two or three different phases. The pressures and temperatures of each of the phases in equilibrium are identical. Moreover, when a pure species contains two phases, T and P are not independent; therefore, the **saturation pressure** takes a unique value for any given temperature. The saturation pressure of a pure species can be related to its **vapor pressure** in a mixture. The **ideal gas** model allows us to relate P, v, and T for gases at low pressure or high temperature.

On a molecular level, temperature is proportional to the average kinetic energy of the individual atoms (or molecules) in the system. Pressure can be viewed as the normal force per unit area exerted by the molecules as they elastically collide with a system boundary. Phase equilibrium can be described as a dynamic process on the molecular level, where the number of molecules leaving the surface of one phase is exactly balanced by the number arriving. Likewise, chemical reaction equilibrium can be described as the dynamic balance of forward and reverse reactions.

▶ 1.10 PROBLEMS

1.1 Go to the teaching or research labs at your university and determine three ways temperature is measured and three ways pressure is measured.

1.2 Estimate the speed at which the average oxygen molecule is moving in the room that you are in.

1.3 Consider a binary mixture of a light gas a with mass m_a and a heavy gas b with mass m_b at temperature T. How does the mean-square velocity of the two species compare? Which species, on average, moves faster?

1.4 The Reamur temperature scale uses the normal freezing and boiling points of water to define $0°$ and $80°$, respectively. What is the value of room temperature ($22°C$) on the Reamur scale?

1.5 Consider the system sketched below:

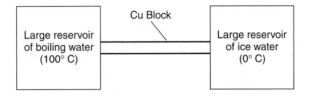

(a) After a short time, is this system in equilibrium?
(b) After a long time?
(c) After a *very* long time?

1.6 At what temperature does water boil on the top of Mount Everest, elevation $z = 8848$ m? Recall that the dependence of pressure with altitude is given by

$$P = P_{atm}\exp\left(-\frac{MWgz}{RT}\right)$$

where, P_{atm} is atmospheric pressure, g is the gravitational acceleration, and MW is the molecular weight of the gas.

1.7 Consider a system containing water in the following states. What phases are present?

(a) $P = 10$ [bar]; $T = 170$ [°C]
(b) $\hat{v} = 3$ [m^3/kg]; $T = 70$ [°C]
(c) $P = 60$ [bar]; $\hat{v} = 0.05$ [m^3/kg]
(d) $P = 5$ [bar]; $s = 7.0592$ [kJ/(kg K)]

1.8 Water is cooled in a rigid closed container from the critical point to 10 bar. Determine the quality of the final state.

1.9 Using linear interpolation, estimate the specific volume of water under the following conditions using data from the steam tables:

(a) $P = 1.9$ [MPa]; $T = 250$[°C]
(b) $P = 1.9$ [MPa]; $T = 300$[°C]
(c) $P = 1.9$ [MPa]; $T = 270$[°C]

Look up the specific volumes of water that correspond to cases (a), (b), and (c) on the website http://webbook.nist.gov/ chemistry/fluid/ and report their values. Comment on the agreement between the two sources.

1.10 Determine the mass of 1 L of saturated liquid water at 25°C. How do you think this value compares to the mass of 1 L of subcooled liquid water at 25°C and atmospheric pressure?

1.11 Determine the temperature, quality, and internal energy of 5 kg of water in a rigid container of volume 1 m^3 at a pressure of 2 bar.

1.12 A rigid container of volume 1 m^3 contains saturated water at 1 MPa. If the quality is 0.10, what is the volume occupied by the vapor?

1.13 Use the data in the steam tables to plot the vapor–liquid dome on a Pv diagram. It is useful to plot v on a log scale.

1.14 Calculate the volume of water using the ideal gas model under the following conditions. Then report the percent error when compared to the values reported in the steam tables.

(a) $P = 1.01$ [bar]; $T = 100$ [°C]

(b) $P = 1$ [bar]; $T = 500$ [°C]

(c) $P = 100$ [bar]; $T = 500$ [°C]

(d) $P = 100$ [bar]; $T = 1000$ [°C]

1.15 How many moles of air are in the room in which you are sitting? What is its mass?

1.16 Consider a gas at 20°C and 1 bar. The molecules may be considered to be hard spheres with a diameter of 3 Å. Estimate the percentage of the available volume they occupy.

1.17 You want to keep your house dry enough so that water does not condense on your walls at night. If the temperature gets down to 40°F at night, what is the maximum allowable density of water in the room during the day when the room is at 70°F?

1.18 I thoroughly inflated a bag of soccer balls last summer. However, when I brought them out to play this winter, they all were underinflated. Discuss the possible reasons.

1.19 Consider a rigid, thick-walled tube that is filled with H_2O liquid and vapor at 0.1 MPa. After it is sealed, it is heated so that it passes through its critical point. What fraction of the mass in the tube is liquid?

1.20 Relative humidity is defined as the ratio of the mass of water in air divided by the mass of water at saturation. Compare the water content in the air on a day on which the temperature is 10°C with 90% relative humidity to a day at 30°C and 50% relative humidity. Which day has higher water content?

1.21 When a system contains regions that differ in physical structure or chemical composition, an overall value can be assigned to its properties. Consider the system, system 1, shown below. It contains n_a molecules in state a and n_b molecules in state b.

(a) Develop an expression for the extensive volume V_1 in terms of n_a, n_b, and the volumes of each homogeneous region V_a and V_b.

(b) Develop an expression for the intensive molar volume v_1 in terms of n_a, n_b and the molar volumes of each homogeneous region v_a and v_b.

(c) Generalize the result of part (a) to come up with an expression for any extensive property K_1 in terms of n_a, n_b, and the extensive properties K_a and K_b.

(d) Generalize the result of part (b) to come up with an expression for the intensive form of the property in part (c), k_1^i, in terms of n_a, n_b, and the intensive properties k_a and k_b.

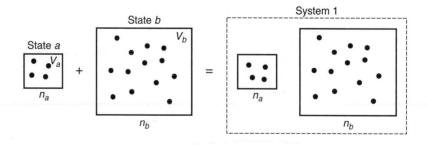

CHAPTER **2**

The First Law of Thermodynamics

Learning Objectives

To demonstrate mastery of the material in Chapter 2, you should be able to:

▶ State and illustrate by example the first law of thermodynamics—that is, the conservation of mass and energy—and its basic concepts, including conversion of energy from one form to another and the transfer of energy from the surroundings to the system by heat, work, and flow of mass.

▶ Write the integral and differential forms of the first law for (1) closed systems and (2) open systems under steady-state and transient (uniform-state) conditions. Convert these equations between intensive and extensive forms and between mass-based and molar forms. Given a physical problem, determine which terms in the equation are important and which terms are negligible or zero and determine whether the ideal gas model or property tables are needed to solve the problem.

▶ Apply the first law of thermodynamics to identify, formulate, and solve engineering problems for the following systems: rigid tank, adiabatic or isothermal expansion/compression in a piston–cylinder assembly, nozzle, diffuser, turbine, pump, heat exchanger, throttling device, filling or emptying of a tank, and Carnot power and refrigeration cycles.

▶ Describe the molecular basis for internal energy, heat transfer, work, and heat capacity.

▶ Describe the difference between a reversible process and an irreversible process, and, given a process, identify whether it is reversible or irreversible.

▶ Explain why it is convenient to use the thermodynamic property enthalpy for (1) streams flowing into and out of open systems and (2) closed systems at constant pressure. Describe the role of flow work and shaft work in open-system energy balances.

▶ Describe the energy changes associated with sensible heat, latent heat, and chemical reaction on both a macroscopic and a molecular level. Calculate their enthalpy changes using available data such as heat capacity, enthalpies of vaporization, fusion and sublimation, and enthalpies of formation.

▶ 2.1 THE FIRST LAW OF THERMODYNAMICS

The first law of thermodynamics states that while energy can be changed from one form to another, *the total quantity of energy, E, in the universe is constant.*[1] This statement can be quantitatively expressed as follows:

$$\Delta E_{univ} = 0 \tag{2.1}$$

However, it is very inconvenient to consider the entire universe every time we need to do a calculation. As we have seen, we can break down the universe into the region in which we are interested (the system) and the rest of the universe (the surroundings). The system is separated from the surroundings by its boundary. We can now restate the first law by saying that *the energy change of the system must be equal to the energy transferred across its boundaries from the surroundings.* Energy can be transferred by heat, Q, by work, W, and, in the case of open systems, by the energy associated with the mass that flows into and out of the system. In essence, the first law then lets us be accountants of the energy in the system, by tracking the "deposits" to and "withdrawals" from the surroundings in much the same way as you would account for the balance of money in your bank account. We will consider explicit forms of the first law for closed and open systems shortly.

Handwritten margin note: $\Delta E_{syst} = - \Delta E_{surr}$

Handwritten margin note: Energy = a system's capacity to do work

Forms of Energy

The energy within a system can be transformed from one form to another.

▶ **EXERCISE**

Name the three common forms that energy is divided into. See if you can define each form:

Energy is classified according to three specific forms: (1) The macroscopic **kinetic** energy, E_K is the energy associated with the bulk (macroscopic) motion of the system as a whole. For example, an object of mass m moving at velocity $\vec{V}$ has a kinetic energy given by

$$E_K = \frac{1}{2}m\vec{V}^2 \tag{2.2}$$

(2) The macroscopic **potential** energy, E_P, is the energy associated with the bulk (macroscopic) position of the system in a potential field. For example, an object in the Earth's gravitational field has a potential energy given by

$$E_P = mgz \tag{2.3}$$

where z is the height above the surface of the Earth and g is the gravitational constant.[2] (3) The **internal** energy, U, is the energy associated with the motion, position, and chemical-bonding configuration of the individual molecules of the substances within the system.

Energy is not an absolute quantity but rather is only defined relative to a reference state, so we must be careful to identify the particular reference state that we are using. As you read this text, what is your kinetic energy (assuming you are not riding the bus)?

---

[1] Nuclear reaction presents an interesting case where energy and mass are coupled. However, we will not address this case in this text.

[2] Potentials due to surface tension or electric or magnetic fields can also be included.

Work and Heat: Transfer of Energy Between the System and the Surroundings

You may use the terms *work* and *heat* in a variety of ways in casual day-to-day conversations such as, "I can't go to the movies tonight because I have to *work* on my thermodynamics problem set" or "If you can't stand the *heat*, get out of the kitchen." However, in thermodynamics these terms have very precise meanings. Both terms refer to the **transfer of energy** between the system and surroundings.[5] In a closed system, the transfer of energy between the system and the surroundings can only be accomplished by heat or by work. Heat is the dissipation of energy by a temperature gradient, whereas all other forms of energy transfer in a closed system occur via work. We generally associate work with something useful being done by (or to) the system. We will examine these terms in more detail below.

There are many forms of work, for example, mechanical (expansion/compression, rotating shaft), electrical, and magnetic. The most common case of work in engineering thermodynamics is when a force causes a displacement in the boundary of a system. In the case of expansion, for example, the system needs to push the surroundings out of the way to increase the boundary; in this process, it expends energy. Thus, the system exchanges energy with the surroundings in the form of work. The work, W, can be described mathematically by the line integral of the external force, F_E, with respect to the direction of displacement, dx:

$$W = \int F_E \cdot dx \qquad (2.6)$$

Since work refers to the transfer of energy between the system and the surroundings, it has the same units as energy, such as joules, ergs, BTU, and so on. To complete the definition, we need to choose a sign convention for work. In this text, we will say work is positive when energy is transferred from the surroundings to the system and work is negative when energy in transferred from the system to the surroundings. The definition given by Equation (2.6) is consistent with this sign convention. You should be aware that this sign convention is completely arbitrary. We choose this convention to be consistent with today's convention. However, when the first and second laws of thermodynamics were originally formulated, in the context of powering the steam engine, the opposite sign convention was used: Work from the system to the surroundings was defined as positive (since the engineering objective was to get work out of the system to power a train!). When you go to other sources, be careful to note which sign convention is chosen for work or you may get tripped up.

A plot of F_E vs. x for a general process is shown in Figure 2.1a. The work associated with the process in Figure 2.1a can be obtained from the area under the curve [which is equivalent to graphically integrating the expression in Equation (2.6)]. If the boundary of the system does not move, no work has been done, no matter how large the force is. In contrast to thermodynamic properties, the work on a system depends not only on the initial state, 1, and the final state, 2, of the system, but also on the specific **path** that it takes. Whenever we calculate the work, we must account for the real path that the system takes. If the external force is acting on a surface of cross-sectional area A, we can rewrite Equation (2.6) as follows:

$$W = \int \frac{F_E}{A} \cdot d(Ax) = \int P_E dV \cos \theta = -\int P_E dV \qquad (2.7)$$

---

[5] In the sciences we need to be very careful about how we use language and define terms such as *work*, *heat*, and *energy*

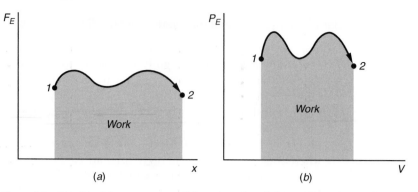

Figure 2.1   Graphical determination of the value of work for a system that undergoes a process between states 1 and 2 by integrating: (a) F_E vs. x; (b) P_E vs. V.

where P_E is the external pressure to the surface. The negative sign in Equation (2.7) results, since the external force and displacement vectors are in opposite directions. Again, the work can be obtained from the area under the appropriate curve, as shown in Figure 2.1b. If Equation (2.7) is written on a molar basis (J/mol), we get

$$ w = -\int P_E dv \tag{2.8}$$

Equation (2.8) is often encountered in thermodynamics; the work described by this equation will be referred to as "Pv work." On a molecular scale, the energy transfer by Pv work can be understood in terms of momentum transfer of the molecules in the system when they bounce off the moving boundary, as discussed in Section 1.3. A piston–cylinder assembly is a common system that is used to obtain work (e.g., in your automobile). Example 2.3 illustrates how work is calculated for such a system.

▶ **EXAMPLE 2.3**
Calculation of *Pv* Work in a Piston–Cylinder Assembly

Consider the constant pressure expansion that is illustrated in Figure E2.3. Initially the system contains 1 mole of gas A at 2 bar within a volume of 10 L. The expansion process is initiated by releasing the latch. The gas in the cylinder expands until the pressure of the gas matches the pressure of the surroundings. The final volume is 15.2 L. Calculate work done by the system during this process.

SOLUTION   The amount of work done can be calculated by applying Equation (2.7):

$$ W = -\int_{V_1}^{V_2} P_E dV \tag{2.7}$$

Since the *external* pressure is constant, it can be pulled out of the integral:

$$ W = -P_E \int_{V_1}^{V_2} dV = -P_E(V_2 - V_1) = -1\,\text{bar}\left[\frac{10^5\,\text{Pa}}{\text{bar}}\right](15.2-10)\text{L}\left[\frac{10^{-3}\,\text{m}^3}{\text{L}}\right] = -520\,\text{J}$$

In this case, the value for work is negative since the system loses energy to the surroundings as a result of this process. The units of pressure and volume have been converted to their SI equivalents in this calculation [1 Pa m^3 = 1 J].   ◀

Shaft work, W_s, is another important type of work encountered in engineering practice. Often a rotating shaft is used to deliver energy between the system and the

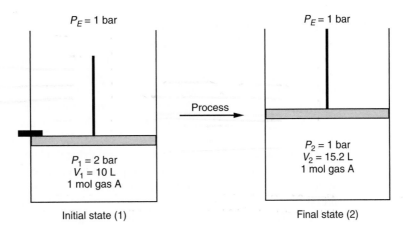

Figure E2.3 Example of a process in which energy is transferred from the system to the surroundings by Pv work: expansion of a gas in a piston–cylinder assembly. The surroundings are maintained at 1 bar.

surroundings. For example, consider the turbine shown in Figure 2.2. It is designed to convert the internal energy of the working fluid into useful work by means of a shaft. In this case, as the fluid passes through the turbine, it expands and cools, leading to the rotation of the shaft at the end of the turbine. A magnet placed on the end of the turbine also rotates. The changing magnetic field induces an electrical potential, which is used to generate a current that can charge a battery and store energy. Note that depending on how we draw the boundary to our system, the transfer of energy from the system to the surroundings (the shaft work) can be mechanical, magnetic, or electric. However, in all cases we are converting the internal energy of the fluid into useful work. In fact, when we generically use the term *shaft work*, it may indeed be any of these forms of work.

Heat, Q, refers to the transfer of energy between the system and the surroundings where the driving force is provided by a temperature gradient. Energy will transfer spontaneously from the high-temperature region to the low-temperature region. Sometimes this form of energy transfer is part of an engineering design (such as "heating" up your house on a cold day). Often, however, heat provides a path for the unwanted dissipation of energy (such as when your coffee gets cold or your soda gets warm). In the latter case, it is useful to try to insulate the system as well as possible to eliminate unwanted

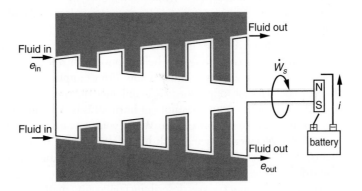

Figure 2.2 Schematic of a turbine converting the energy from a flowing fluid into shaft work. The rotating shaft is coupled to a battery by means of a magnet affixed to its end.

transfer of energy. In the ideal case, the transfer of energy by heat would be reduced to zero. We term such a process **adiabatic** ($Q = 0$). Unlike for work, the sign convention for heat is unambiguous. A positive value indicates that energy is transferred from the surroundings to the system or, colloquially, the system is "heating up." Alternatively, you may think of a positive value for Q to correspond with an increase in internal energy (ΔU) of the system (ignoring work). In Section 2.1 we referred to the effects of changes in U on the system. Those changes that manifested as changes in temperature were termed *sensible heat* while changes in phase were termed *latent heat*. This nomenclature is somewhat misleading, since we are referring to changes in the internal energy of the system and not "heat," which more precisely refers to transfer of energy across the system boundary in a specified way. However, this nomenclature is deeply rooted in the literature.

There are three modes through which energy can be transferred by a temperature gradient: conduction, convection, and radiation. You will learn how to quantify the rates of these processes in your heat-transfer (or transport processes) class; however, the underlying mechanisms of each process will be described briefly here. It is easiest to think of conduction in terms of a solid body. If you expose the front side of a chunk of quartz glass, for example, to temperature T_{high} and the back side to a temperature T_{low}, energy will transfer through the glass. On a molecular scale, the phonons on the hot side (remember, phonons refer to solid vibrations) will be vibrating at a higher velocity, that is, with greater energy. However, these atoms are connected to those next to them on the crystal lattice and the region of high-energy lattice vibration will "spread out." With time, the phonons on the front side will vibrate less vigorously (thus reducing their temperature) while the phonons on the back side increase in energy. The end result is a transfer of energy from T_{high} to T_{low}. The rate at which the energy transfers via conduction—that is, the rate at which the lattice vibration spreads out—is proportional to a property of the material called the thermal conductivity, κ. Glass is not very conductive; it has a thermal conductivity of around 42 W m^{-1} °C^{-1}. On the other hand, metals tend to conduct well. In the case of metals, there is an additional mechanism for conduction of energy—drift of free electrons in the valence band. A thermally conducting material like copper may conduct energy an order of magnitude faster than glass, having a thermal conductivity of 385 W m^{-1} °C^{-1}. Wood, on the other hand, is an effective thermal insulator, having a thermal conductivity around 0.1 W m^{-1} °C^{-1}. Liquids and gases may also transfer energy through conduction. The thermal conductivity of liquids tends to be lower than that of solids, and gases have even lower thermal conductivities than liquids. Typical values range from 0.06–0.6 W m^{-1} °C^{-1} for most liquids and 0.01–0.07 W m^{-1} °C^{-1} for most gases. Can you provide a molecular explanation as to why gases have much lower thermal conductivity than solids?

Convection is another mechanism by which energy can be transferred between the system and the surroundings via heat. Convection refers to the case of enhanced heat transfer through coupling with fluid flow. For example, consider the example of wanting to cool a bowl of hot soup. One way to enhance the transfer of energy (so that you can eat quickly and your tongue does not get burned) is by blowing on the soup in your spoon. This is an example of convection. When you blow on the soup, the flow of gas carries away hot molecules (moving with high velocity) and replaces them with colder fluid. Thus, the temperature difference between the soup and the neighboring gas—that is, the driving force for energy transfer—is greater, and cooling occurs more quickly than by conduction alone. Clearly describing convection mathematically is more difficult than describing conduction. Convection depends not only on the conductive properties of the soup and the air but also on the type of flow patterns that are set up.

position, leading to a molar volume less than 0.08 [m³/mol], where it is then again turned around and expands, and so on and so on. This process will inevitably contain a frictional dissipative mechanism that causes the piston to come to rest at state 2. Since the external pressure is the same during these oscillations, the contribution of the oscillating expansions and contractions to the work will exactly cancel, leading to the same value as calculated by Equation (2.16).[6,7] In this text, we will ignore the oscillatory behavior of these types of processes and approximate them in the simpler context where the system does not overshoot its final equilibrium state. While it is only an approximation of the real behavior, this simplification proves useful in allowing us to compare irreversible processes with reversible processes.

We next want to calculate the work needed to compress the gas from state 2 back to state 1. This process is illustrated in Figure 2.5 and is labeled process B. In process B,

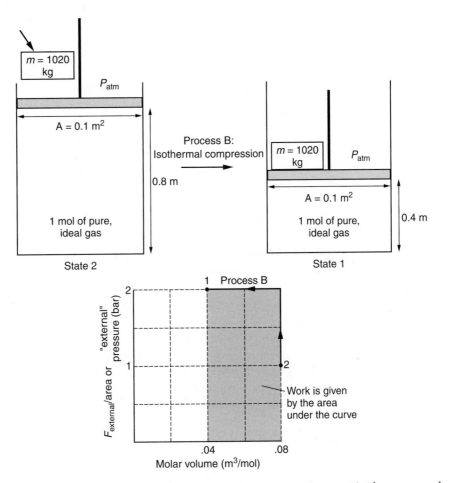

Figure 2.5 Schematic of an isothermal compression process (process B). The corresponding plot of the process on a $P_E v$ diagram is shown at the bottom.

---

[6] We assume all the energy dissipation occurs within the system.

[7] On the other hand, if there were no dissipative mechanism, the piston would oscillate forever. Its kinetic energy could be accounted for by the difference in force between the system pressure and the external pressure.

we drop the 1020-kg mass back on the piston, originally in state 2. The external pressure now exceeds the pressure of the gas initiating the compression process. The piston goes down until the pressures equilibrate, at state 1. The external pressure vs. molar volume is plotted in Figure 2.5. In this case, the external pressure consists of contributions from both the block and the atmosphere. Again, we are not representing the system pressure in this graph, but rather the force per area that acts on the piston. The work is found by Equation (2.8):

$$w = -\int_{v_2}^{v_1} P_E dv = -\left(P_{atm} + \frac{mg}{A}\right)\int_{v_2}^{v_1} dv = 8000 \left[\frac{J}{mol}\right] \qquad (2.17)$$

This value can also be found from the area under the curve. Comparing process A and process B, we see it costs us more work to compress the piston back to state 1 than we got from expanding it to state 2. The net difference in work ($8000-4000 = 4000$ [J]) in going from state 1 to state 2 and back to state 1 results in a "net effect on the surrounding." Examining our definition of a reversible process, we see that processes A and B are irreversible.

Next we again consider expansion from state 1 to 2 (process C) and compression from state 2 to 1 (process D), but now we use two 510-kg masses instead of one larger 1020-kg mass. The expansion is carried out as follows: The system is originally in state 1 when the first 510-kg mass is removed. This causes the gas to expand to an intermediate state at a pressure of 1.5 bar and a molar volume of 0.053 m^3/mol. The second 510-kg mass is then removed, completing the expansion to state 2. These three states are shown in the Pv diagram in Figure 2.6. The expansion process is labeled process C and follows arrows from state 1 to the intermediate state to state 2. The work the system delivers to the surroundings is given by

$$w = -\int_{v_1}^{v_2} P_E dv = -\left[\left(P_{atm} + \frac{\frac{m}{2}g}{A}\right)\int_{v_1}^{v_{int}} dv + P_{atm}\int_{v_{int}}^{v_2} dv\right] = -4667 \left[\frac{J}{mol}\right] \qquad (2.18)$$

Again, the work can be found graphically from the area under the curve. Process C is "better" than process A in that it allows us to extract more work from the system. We want to get the most work *out* of a system as possible.

The compression process is the opposite of the expansion. With the system in state 2, a 510-kg block is placed on the piston until it compresses to the intermediate state, followed by placement of the second block. The work is found by

$$w = -\int_{v_2}^{v_1} P_E dv = -\left[\left(P_{atm} + \frac{\frac{m}{2}g}{A}\right)\int_{v_2}^{v_{int}} dv + \left(P_{atm} + \frac{mg}{A}\right)\int_{v_{int}}^{v_1} dv\right] = 6667 \left[\frac{J}{mol}\right]$$

$$(2.19)$$

In analogy to the expansion process, process D is "better" than process B in that it costs us less work to compress the system back to state 1. When we have to put work *into* a system, we want it to be as small as possible. However, it still costs us more work to compress from state 2 to state 1 than we get out of the expansion, so these processes are still irreversible.

We did "better" in both expansion and compression processes when we divided the 1020-kg mass into two parts. Presumably we would do better by dividing it into four

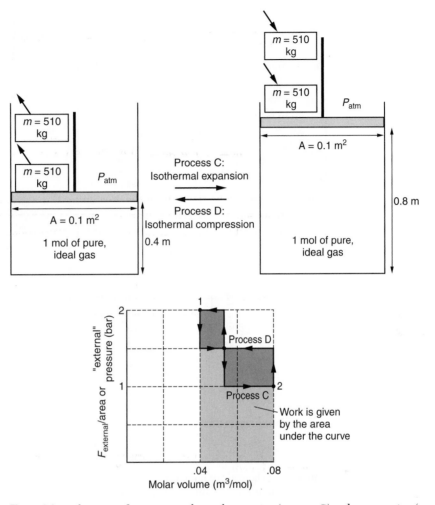

Figure 2.6 Schematic of two-step isothermal expansion (process C) and compression (process D) processes. The corresponding plots of the processes on a $P_E v$ diagram is shown at the bottom.

parts, and even better by dividing it into eight parts, and so on. If we want to do the best possible, we can divide the 1020-kg mass into many "differential" units and take them off one at a time for expansion or place them on one at a time for compression. These processes are labeled process E and process F, respectively, and are illustrated in Figure 2.7. At each differential step, the system pressure is no more than $\partial mg/A$ different than the external pressure. Thus, to a close approximation

$$\text{✕} \quad P = P_E \qquad \underline{When\ reversible} \tag{2.20}$$

The process paths are illustrated in the Pv diagram in Figure 2.7. To find the work, we integrate over the external pressure. However, since the external pressure is equal to the system pressure, we get

$$w = -\int_{v_1}^{v_2} P_E dv = -\int_{v_1}^{v_2} P dv = -\int_{v_1}^{v_2} \frac{P_1 v_1}{v} dv = -P_1 v_1 \ln\left(\frac{v_2}{v_1}\right) = -5545 \left[\frac{J}{mol}\right] \tag{2.21}$$

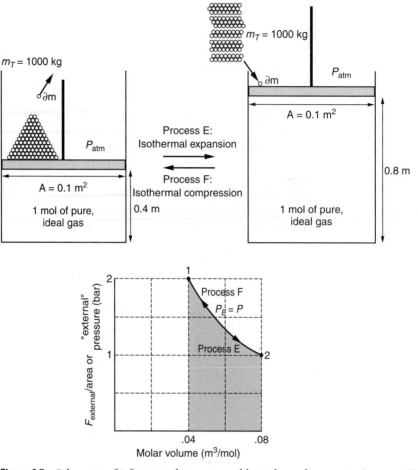

Figure 2.7 Schematic of infinitesimal-step, reversible isothermal expansion (process E) and compression (process F) processes. The corresponding plots of the processes on a $P_E v$ diagram is shown at the bottom.

Similarly the work of compression is

$$w = -\int_{v_1}^{v_2} P_E dv = -\int_{v_2}^{v_1} P dv = -\int_{v_2}^{v_1} \frac{P_1 v_1}{v} dv = P_1 v_1 \ln\left(\frac{v_1}{v_2}\right) = 5545 \left[\frac{J}{mol}\right] \quad (2.22)$$

In processes E and F, we can return to state 1 by supplying the same amount of work that we got from the system in the expansion process. Hence, we can go from state 1 to 2 and back to state 1 without a net effect on the surroundings. From our definition, we see that these processes are **reversible**. In a reversible process, we are never more than slightly out of equilibrium. At any point during the expansion, we could turn the process around the other way and compress the piston by adding differential masses instead of removing them. Moreover, we get more work out of the reversible expansion than the irreversible expansions. Similarly, the reversible compression costs us less work than the irreversible processes. The reversible case represents the limit of what is possible in the real world—it gives us the most work we can get out or the least work we have to put in! Moreover, *only* in a reversible process can we substitute the system pressure for the external pressure.

Why do we get less work out of the irreversible expansion (process A) than the reversible expansion (process E)? Work is the transfer of energy between the system and the surroundings. In this case, as the gas molecules bounce off the piston, their change in net z momentum between before and after a collision is determined by the movement of the piston. This change of momentum with time represents the net energy transferred between the system and surroundings; that is, it is the work. The irreversible expansion, process A, never has a mass on the piston; thus the molecules of the gas are running into something "smaller" and will not transfer as much energy. On the other hand, in the irreversible compression process, the mass on the piston is larger than the corresponding reversible process. It therefore imparts more energy to those molecules and costs more work. Those of you who are baseball fans may consider an analogy to the size of a hitter's bat. The heavier the bat, the more energy can be transferred to the baseball. The greater the force the piston exerts on a given molecule as it rebounds off the piston, the more the molecule's speed will increase and the higher its kinetic energy. If we sum up all the molecules, we see that the net energy transfer (work) is greater.

We can compare the amount of work required in an irreversible process to that of a reversible process by defining the efficiency factor, η. For an expansion process, we compare how much work we actually get out to the idealized, reversible process. Thus, the efficiency of expansion, η_{exp}, is

$$\eta_{\text{exp}} = \frac{w_{\text{irrev}}}{w_{\text{rev}}} \tag{2.23}$$

For example, the efficiency of Process A would be

$$\eta_{\text{exp}} = \frac{w_A}{w_E} = \frac{-4000}{-5545} = 0.72 \tag{2.24}$$

where w_i represents the work of process i. Thus, we say process A is 72% efficient. On the other hand, to determine the efficiency of a compression process, η_{comp} we compare the reversible work to the work we must actually put in:

$$\eta_{\text{comp}} = \frac{w_{\text{rev}}}{w_{\text{irrev}}} \tag{2.25}$$

For example, the efficiency of process B would be

$$\eta_{\text{comp}} = \frac{w_F}{w_B} = \frac{5545}{8000} = 0.69 \tag{2.26}$$

or 69% efficient. In both cases, efficiencies are defined so that if we can operate a process reversibly, we would have 100% efficiency, while the real processes are less efficient. Our strategy for actual, irreversible processes is often solving a problem for the idealized, reversible process and then correcting for the irreversibilities using an assigned efficiency factor.

▶ 2.4 THE FIRST LAW OF THERMODYNAMICS FOR CLOSED SYSTEMS

Integral Balances

In this section, we consider energy balances for closed systems. In the next section, open systems will be treated. Figure 2.8 shows a schematic of a closed system that undergoes

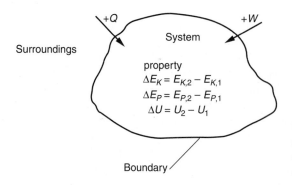

Figure 2.8 Illustration of closed system and sign conventions for heat and work. All three forms of energy are considered.

a process from initial state 1 to final state 2. In this figure, the system, surroundings, and boundary are delineated. In a closed system, mass cannot transfer across the system boundary. There are two ways in which to catalog the amount of material in the system—by mass or by moles. Each way can be convenient. For a pure species or a mixture of constant composition, the two forms are equivalent and can be interconverted using the molecular weight. When we address systems undergoing chemical reaction, care must be taken. While the total mass must be conserved, the number of moles or the mass of a particular component may change. In the absence of chemical reaction, the number of moles remains constant:

$$n_1 = n_2 \tag{2.27}$$

Since mass cannot enter or leave a closed system, the changes in the energy within the system (Δ = final − initial) are equal to the energy transferred from the surroundings by either heat or work. Figure 2.8 also illustrates the sign convention that we have defined for heat and work, namely, positive for energy transfer from the surroundings to the system. Writing down the first law in quantitative terms, we get:[8]

$$\left\{ \begin{array}{l} \text{change in} \\ \text{energy in} \\ \text{system} \end{array} \right\} = \left\{ \begin{array}{l} \text{energy transferred} \\ \text{from surroundings} \\ \text{to system} \end{array} \right\}$$

$$\underbrace{\Delta U + \Delta E_K + \Delta E_P}_{\substack{property: \\ \text{depends only} \\ \text{on state 1 and 2}}} = \underbrace{Q + W}_{\substack{\text{depends on} \\ path}} \tag{2.28}$$

The properties on the left-hand side of Equation (2.28) depend only on the initial and final states. They can be calculated using the real path or any hypothetical path we create. The terms on the right-hand side are process dependent and the real path of the system must be used.

Since the composition of a closed system remains constant (barring chemical reaction), we can rewrite Equation (2.28) using intensive variables by dividing through by

---

[8] Heat and work already refer to the amount of energy transferred; hence, it would be would be wrong to write them ΔQ or ΔW. We reserve the Δ for state function that depends just on the initial and final state of the system.

the total number of moles:[9]

$$\Delta u + \Delta e_K + \Delta e_P = q + w \qquad (2.29)$$

We often *neglect macroscopic kinetic and potential energy*. For this case, the extensive and intensive forms of the closed system energy balances become

$$\Delta U = Q + W \qquad (2.30)$$

and

$$\Delta u = q + w \qquad (2.31)$$

respectively.

Differential Balances

Isolated system $\Delta U = 0$

If the transfer of energy from (or to) the surroundings changes as the process proceeds, the first law must be written for each differential step in time during the process. Numerical solutions are obtained by integration of the resulting differential energy balance. Common forms of energy balance over a differential element can be written by analogy to the equations just presented.

$$dU + dE_K + dE_P = \delta Q + \delta W \qquad \text{extensive } \left[\text{J}\right] \qquad (2.32)$$

$$du + de_K + de_P = \delta q + \delta w \qquad \text{intensive } \left[\text{J/mol}\right]$$

or, neglecting kinetic and potential energy,

$$dU = \delta Q + \delta W \qquad \text{extensive } \left[\text{J}\right] \qquad (2.33)$$

$$du = \delta q + \delta w \qquad \text{intensive } \left[\text{J/mol}\right]$$

We use the exact differential d with the energy terms to indicate that they depend only on the final and initial states; in contrast, we use the inexact differential δ with heat and work to remind us that we must keep track of the path when we integrate to get these quantities.

The energy balances above are often differentiated with respect to time, yielding

$$\frac{dU}{dt} + \frac{dE_K}{dt} + \frac{dE_P}{dt} = \dot{Q} + \dot{W} \qquad \text{extensive } \left[\text{W}\right] \qquad (2.34)$$

$$\frac{du}{dt} + \frac{de_K}{dt} + \frac{de_P}{dt} = \dot{q} + \dot{w} \qquad \text{intensive } \left[\text{W/mol}\right]$$

where the rate of heat transfer and the rate of work [J/s or W] are denoted with a dot over the corresponding variable. Again, we often neglect kinetic and potential energy to give:

$$\frac{dU}{dt} = \dot{Q} + \dot{W} \qquad \text{extensive } \left[\text{W}\right] \qquad (2.35)$$

$$\frac{du}{dt} = \dot{q} + \dot{w} \qquad \text{intensive } \left[\text{W/mol}\right]$$

---

[9] We will write balance equations on a molar basis, utilizing the appropriate intensive thermodynamic properties. For example, internal energy will be u [J/mol]. You should be able to convert any equation to a mass basis that uses the corresponding specific property, for example, $\hat{u}$ [J/kg].

▶ **EXERCISE** Consider the six processes depicted Figures 2.4 through 2.7. What is the heat transferred in each case. Can you explain the difference in relation to the efficiency factor?

▶ 2.5 THE FIRST LAW OF THERMODYNAMICS FOR OPEN SYSTEMS

In open systems, mass can flow into and out of the system. A generic open system with two streams in and two streams out is shown in Figure 2.9. In setting up the balance equations, it is convenient to discuss rates: molar flow rates [mol/s], mass flow rates [kg/s], and rates of energy transfer [J/s or W]. We must keep track of the mass in the system since it can change with time. In the general case of many streams *in* and many streams *out*, we must sum over all the *in* and *out* streams. We first consider conservation of mass, on a molar basis, for a nonreacting system. The accumulation of moles in the system is equal to the difference of the total rate of moles in minus the total rate of moles out. Thus the mole balance can be written:

$$\left(\frac{dn}{dt}\right)_{sys} = \sum_{in} \dot{n}_{in} - \sum_{out} \dot{n}_{out} \tag{2.36}$$

where $\dot{n}$ the molar flow rate in [mol/s]. For a stream flowing through a cross-sectional area A, at velocity $\vec{V}$, the molar flow rate can be written as:

$$\dot{n} = \frac{A\vec{V}}{v} \tag{2.37}$$

The energy balance for an open system contains all the terms associated with an energy balance for a closed system, but we must also account for the energy change in the system associated with the inlet and outlet streams. To accomplish this task, we consider the case of the generic open system illustrated in Figure 2.9. This open system happens to have two streams in and two streams out; however, the balances developed here will be true for any number of inlet or outlet streams. Lets look at the contribution to the energy balance from the inlet stream labeled stream 1. Compared to the closed-system analysis we performed in Section 2.4, there are two additional ways in which energy can be transferred from the surroundings. First, the molecules flowing into the

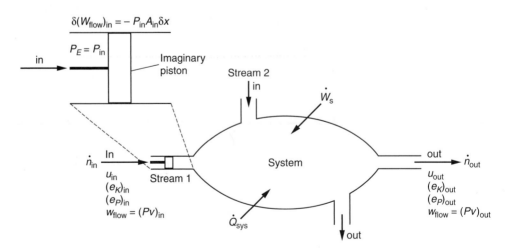

Figure 2.9 Schematic of an open system with two streams in and two streams out. The piston shown in the plot is hypothetical; it illustrates the point that flow work is always associated with fluid flowing into or out of the system.

Steady state
moves in =
moves out

Flow work:

$(\dot{w}_{flow})_{in} = n_{in}$
$\qquad (P_{in} v_{in})$

system carry their own energy; typically the most important form of energy is internal u_{in}, but the flowing streams can also have (macroscopic) kinetic energy, $(e_K)_{in}$, and potential energy, $(e_P)_{in}$, associated with them. Second, the inlet stream adds energy into the system by supplying work, the so-called **flow work**. Flow work is the work the inlet fluid must do on the system to displace fluid within the system so that it can enter. It may be visualized by placing an imaginary piston in front of the material that is about to enter the system, as depicted in Figure 2.9. The imaginary piston acts like the real piston shown in Figure 2.5. The rate of work exerted by the fluid to enter the system is, therefore, given by

$$\times \quad (\dot{W}_{flow})_{in} = -P_{in} A_{in} \frac{dx}{dt} = P_{in} A_{in} \vec{V}_{in} = P_{in} \dot{n}_{in} v_{in} \qquad (2.38)$$

where Equation (2.37) was used. We have set $P_E = P_{in}$ and eliminated the negative sign, since the direction of velocity is the negative of the normal vector from the piston. If we divide by molar flow rate, we can write the flow work in intensive form:

$$(w_{flow})_{in} = \frac{(\dot{W}_{flow})_{in}}{\dot{n}_{in}} = P_{in} v_{in} \qquad (2.39)$$

We conclude that the flow work of any inlet stream is given by the term $(Pv)_{in}$.

By a similar argument, we can show the flow work of any outlet stream is given by

$$(\dot{W}_{flow})_{out} = -\dot{n}_{out} P_{out} v_{out} \qquad (2.40)$$

We can write the total energy transfer due to work in the system in terms of shaft work, W_s, and flow work, as follows:

$$\dot{W} = \dot{W}_{shaft} + \dot{W}_{flow} = \left[\dot{W}_s + \sum_{in} \dot{n}_{in}(Pv)_{in} + \sum_{out} \dot{n}_{out}(-Pv)_{out} \right] \qquad (2.41)$$

The shaft work is representative of the *useful* work that is obtained from the system. While shafts are commonly used to get work out of an open system, as discussed earlier, $\dot{W}_s$ generically represents any possible way to achieve useful work; therefore, it does not include flow work! The flow work does not provide a source of power; it is merely the "cost" of pushing fluid into or out of the open system.

We can now include the new ways in which energy can exchange between the system and surroundings in our energy balance for open systems. In the general case of many streams *in* and many streams *out*, we must sum over all the *in* and *out* streams. First, we consider a system at steady-state; that is, there is *no* accumulation of energy or mass in the system with time. The energy balance is written as [on a rate basis, by analogy to Equation (2.35)]:

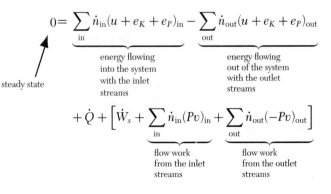

$$0 = \underbrace{\sum_{in} \dot{n}_{in}(u + e_K + e_P)_{in}}_{\substack{\text{energy flowing} \\ \text{into the system} \\ \text{with the inlet} \\ \text{streams}}} - \underbrace{\sum_{out} \dot{n}_{out}(u + e_K + e_P)_{out}}_{\substack{\text{energy flowing} \\ \text{out of the system} \\ \text{with the outlet} \\ \text{streams}}}$$

steady state

$$+ \dot{Q} + \left[\dot{W}_s + \underbrace{\sum_{in} \dot{n}_{in}(Pv)_{in}}_{\substack{\text{flow work} \\ \text{from the inlet} \\ \text{streams}}} + \underbrace{\sum_{out} \dot{n}_{out}(-Pv)_{out}}_{\substack{\text{flow work} \\ \text{from the outlet} \\ \text{streams}}} \right]$$

Rearranging, we get

$$0 = \sum_{\text{in}} \dot{n}_{\text{in}}[(u + Pv) + e_K + e_P]_{\text{in}} - \sum_{\text{out}} \dot{n}_{\text{out}}[(u + Pv) + e_K + e_P]_{\text{out}} + \dot{Q} + \dot{W}_s \quad (2.42)$$

The inlet streams that flow into open systems always have terms associated with *both* internal energy and flow work; therefore, it is convenient to group these terms together (so that we don't forget one). We give it the name **enthalpy**, h [J/mol]:

enthalpy considers flow work

enthalpy

$$h \equiv u + Pv \quad (2.43)$$

Since u, P, and v are all properties, this new group, enthalpy, is also a property. Thus, the additional energy associated with the flowing inlet stream is given by h_{in} [as well as $(e_K)_{\text{in}}$ and $(e_P)_{\text{in}}$]. The term h_{in} includes both the internal energy of the stream and the flow work it adds to enter the system. Similarly, the combined internal energy and flow work leaving the system as a result of the exiting streams are given by h_{out}. In summary, the **steady-state** energy balances can be written:

$$\cancel{0} = \sum_{\text{in}} \dot{n}_{\text{in}}(h + e_K + e_P)_{\text{in}} - \sum_{\text{out}} \dot{n}_{\text{out}}(h + e_K + e_P)_{\text{out}} + \dot{Q} + \dot{W}_s \quad (2.44)$$

In cases where we can **neglect (macroscopic) kinetic and potential energy**, the steady-state, integral energy balance becomes

$$0 = \sum_{\text{in}} \dot{n}_{\text{in}} h_{\text{in}} - \sum_{\text{out}} \dot{n}_{\text{out}} h_{\text{out}} + \dot{Q} + \dot{W}_s \quad (2.45)$$

Another useful form of the energy balance for open systems is for **unsteady-state** conditions. Unsteady-state is important, for example, in start-up as the equipment "warms up." In the case when the inlet and outlet streams stay constant with time, the unsteady-state energy balance becomes

$$\left(\frac{dU}{dt} + \frac{dE_K}{dt} + \frac{dE_P}{dt}\right)_{\text{sys}} = \sum_{\text{in}} \dot{n}_{\text{in}}(h + e_K + e_P)_{\text{in}} - \sum_{\text{out}} \dot{n}_{\text{out}}(h + e_K + e_P)_{\text{out}} + \dot{Q} + \dot{W}_s$$

$$(2.46)$$

In cases where we can **neglect (macroscopic) kinetic and potential energy**, the unsteady-state, integral energy balance becomes:

$$\left(\frac{dU}{dt}\right)_{\text{sys}} = \sum_{\text{in}} \dot{n}_{\text{in}} h_{\text{in}} - \sum_{\text{out}} \dot{n}_{\text{out}} h_{\text{out}} + \dot{Q} + \dot{W}_s \quad (2.47)$$

In Equations (2.46) and (2.47), the left-hand side represents accumulation of energy within the system. There is no flow work associated with this term; hence, the appropriate property is U. On the other hand, the first two terms on the right-hand side account for energy flowing in and out of the system, respectively. These terms must account for both the internal energy and flow work of the flowing streams. In this case, h is appropriate. It is worthwhile for you to take a moment and reconcile the use of U and h above. It will save you many mistakes down the road!

Since the internal energy of an ideal gas depends only on T, application of the definition of enthalpy gives

$$h_{\text{ideal gas}} = u + Pv = u + RT = f(T \text{ only}) \quad (2.48)$$

since $Pv = RT$ for an ideal gas. Thus, like the internal energy, the enthalpy of an ideal gas depends only on T.

▶ EXERCISE

(a) Simplify Equations (2.44) and (2.46) for the case of one stream in and one stream out.

(b) What are ways to find h of a system?

(c) For the energy balance depicted in Equation (2.46), which terms correspond to accumulation? To energy in? To energy out?

▶ **EXAMPLE 2.4**
Calculation of Work
from the First Law

Steam enters a turbine with a mass flow rate of 10 kg/s. The inlet pressure is 100 bar and the inlet temperature is 500°C. The outlet contains saturated steam at 1 bar. At steady-state, calculate the power (in kW) generated by the turbine.

SOLUTION   The steady-state energy balance, in terms of specific (mass-based) properties, is written by analogy to Equation (2.45):

$$0 = \sum_{in} \dot{m}_{in} \hat{h}_{in} - \sum_{out} \dot{m}_{out} \hat{h}_{out} + \dot{Q} + \dot{W}_s \qquad (E2.4A)$$

For a turbine, there is one stream in and one stream out. Moreover, heat dissipation is negligible, since $\dot{Q} \ll \dot{W}_s$. If we label the inlet stream "1" and the outlet stream "2," Equation (E2.4A) becomes:

$$0 = \dot{m}_1 \hat{h}_1 - \dot{m}_2 \hat{h}_2 + \dot{W}_s \qquad (E2.4B)$$

At steady-state, the mass balance can be written as:

$$\dot{m}_1 = \dot{m}_2 = \dot{m} \qquad (E2.4C)$$

Plugging in Equation (E2.4C) into (E2.4B) gives:

$$\dot{W}_s = \dot{m} \left(\hat{h}_2 - \hat{h}_1 \right) \qquad (E2.4D)$$

We can look up values for specific enthalpy from the steam tables. For state 2, we use saturated steam at 1 bar ($= 100$ kPa):

$$\hat{h}_2 = 2675.5 \ [kJ/kg]$$

while state 1 is superheated steam at 500°C and 100 bar ($= 10$ MPa):

$$\hat{h}_1 = 3373.6 \ [kJ/kg]$$

Plugging these numerical values into Equation (E2.4D) gives

$$\dot{W}_s = 10 \ [kg/s](2675.5 - 3373.6) \left[kJ/kg \right] = -6981 \ [kW]$$

Thus, this turbine generates approximately 7 MW of power. Note the negative sign, which indicates that we are getting useful work from the system. This is the equivalent power to that delivered by approximately 70 automobiles running simultaneously.

▶ **EXAMPLE 2.5**
Calculation of Final
Temperature for
a Transient Process

Steam at 10 MPa, 450°C is flowing in a pipe, as shown in Figure E2.5A. Connected to this pipe through a valve is an evacuated tank. The valve is opened and the tank fills with steam until the pressure is 10 MPa, and then the valve is closed. The process takes place adiabatically.

(a) Determine the final temperature of the steam in the tank.

(b) Explain why the final temperature in the tank is not the same as that of the steam flowing in the pipe.

SOLUTION **(a)** This problem can be solved several ways. We first solve it with the fixed boundary illustrated in Figure (E2.5A). An alternative solution with a moving system boundary is then presented.

Unsteady-State Analysis
For the choice of system shown in Figure E2.5A, we must use an *unsteady-state* energy balance since the mass and energy inside the tank (system) increase with time. Let us define the initial state of the tank (empty), state 1, and the final state (filled to 10 MPa), state 2. The inlet flow stream will be designated by "in."

The conservation of mass can be written in analogy to Equation (2.36). Since we have one stream in and no streams out, we get:

$$\left(\frac{dm}{dt}\right)_{\text{sys}} = \dot{m}_{\text{in}}$$

Separating variables and integrating gives

$$\int_{m_1=0}^{m_2} dm = \int_0^t \dot{m}_{\text{in}} dt$$

Since state 1 identically has a value for m of 0, we get

$$m_2 = \int_0^t \dot{m}_{\text{in}} dt \tag{E2.5A}$$

The unsteady energy balance, in mass (specific) units, is written in analogy to Equation (2.47):

$$\left(\frac{dU}{dt}\right)_{\text{sys}} = \dot{m}_{\text{in}}\hat{h}_{\text{in}} - \overset{0}{\cancel{\dot{m}_{\text{out}}\hat{h}_{\text{out}}}} + \overset{0}{\cancel{\dot{Q}}} + \overset{0}{\cancel{\dot{W}_s}}$$

where the terms associated with flow out, heat, and work are zero. Separating variables and integrating both sides with respect to time from the initial empty state to the final state when the tank is at a pressure of 10 MPa gives

$$\int_{U_1=0}^{U_2} dU = \int_0^t \dot{m}_{\text{in}}\hat{h}_{\text{in}} dt = \hat{h}_{\text{in}} \int_0^t \dot{m}_{\text{in}} dt \tag{E2.5B}$$

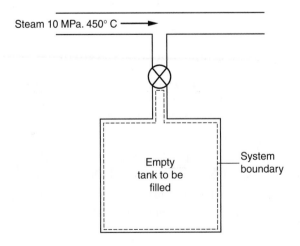

Figure E2.5A Flow of superheated steam from a supply line to fill an empty tank. The system boundary is indicated with dashed lines.

We have moved the specific enthalpy out of the integral, since the properties of the inlet stream remain constant throughout the process. Applying Equation (E2.5A) and integrating, Equation (E2.5B) becomes

$$m_2 \hat{u}_2 = m_2 \hat{h}_{in} \qquad \text{(E2.5C)}$$

since state 1 identically has a value for u of 0. Equation (E2.5C) simplifies to:

$$\hat{u}_2 = \hat{h}_{in} \qquad \text{(E2.5D)}$$

From the steam tables, steam at $P = 10$ MPa and $T = 450°$C has a specific enthalpy

$$\hat{h}_{in} = 3241 \left[\text{kJ/kg}\right]$$

According to Equation (E2.5D), the inlet specific enthalpy must be equal to the final specific internal energy of the system. Hence, the final state has two independent intensive properties specified: $P = 10$ MPa and $\hat{u} = 3241$ [kJ/kg]. From the steam tables, for steam at $P = 10$ MPa and $\hat{u} = 3241$ [kJ/kg], the final temperature of the system is:

$$T_2 = 600°\text{C}$$

Closed-System Analysis:

Alternatively, we can use a different choice of system to solve this problem. We consider the mass that starts in the pipe and eventually ends up in the tank as part of the system. The system boundary in this case is illustrated on the left side of Figure E2.5B. As mass flows into the tank, the boundary contracts. With this choice of system, no mass flow crosses the boundary; thus, we have a closed system. An equivalent closed system, in terms of our familiar piston–cylinder assembly, is shown on the right of Figure E2.5B. The adiabatic chamber is separated by a diaphragm that plays the same role as the valve in the original system. Above the diaphragm, the cylinder contains steam at 10 MPa and 450°C, identical to the inlet steam. Below the diaphragm is a vacuum. The process is initiated by removing the diaphragm, and the system is taken from state "in" to state 2. An external pressure of 10 MPa, representative of the steam in the pipe *outside* the system boundary, acts against the piston until the pressures equilibrate. During the compression process, the surroundings transfer energy to the system via work. Can you see that the two processes depicted in Figure E2.5B are equivalent?

A mass balance on either closed system depicted in Figure (E2.5B) gives

$$m_{in} = m_2 \qquad \text{(E2.5E)}$$

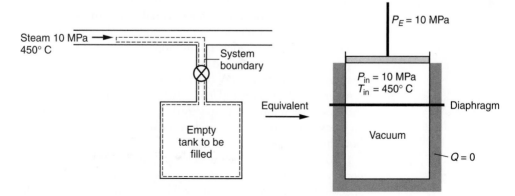

Figure E2.5B A closed system approach to solving the problem in Example 2.5.

Likewise, the energy balance is

$$\Delta U = m_2 \hat{u}_2 - m_{in} \hat{u}_{in} = Q + W \tag{E2.5F}$$

Since the process is adiabatic, $Q = 0$. Work is given by

$$W = -\int_{V + m_{in}\hat{v}_{in}}^{V} P_E dv = m_{in} P_{in} \hat{v}_{in} \tag{E2.5G}$$

Substituting Equations (E2.5E) and (E2.5G) into (E2.5F) and rearranging gives:

$$m_2 \hat{u}_2 = m_{in}(\hat{u}_{in} + P_{in}\hat{v}_{in}) = m_{in}\hat{h}_{in} = m_2\hat{h}_{in}$$

which is identical to the result we obtained for the unsteady-state open system of Figure E2.5.A., given by Equation (E2.5D).

(b) The fluid in the system receives **flow work** from the fluid behind it to get into the system. This work component adds energy from the incoming stream to the system, increasing T from 450°C to 600°C. The flow work of steam flowing into the system shown on the left of Figure E2.5B is equivalent to work done by an equivalent constant external pressure acting on the piston shown on the right. Thus, the closed-system analysis helps us see the importance of accounting for flow work in energy balances on open systems.

▶ 2.6 THERMOCHEMICAL DATA FOR U AND H

Heat Capacity: c_v and c_P

In order to perform energy balances on both closed and open systems, it is necessary to be able to determine how the energy (or enthalpy) of the species in the system changes during a process. As we learned in Section 1.3, the internal energy, u, and the enthalpy, h, for a pure species are constrained by specifying two independent intensive properties. Moreover, since u is a thermodynamic property, we can specify a hypothetical path to calculate the change in internal energy, Δu. It does not have to be the path of the actual process. Likewise for Δh. While any thermodynamic properties can be used, it is often convenient to choose measured properties (T, P, or v) as the independent variables. Temperature is almost always chosen as one of the independent properties, since it can be measured in the lab (or field), and there is a direct relationship between T and u; that is, temperature is a measure of the molecular kinetic energy, which is one component of u (see Section 2.1). In fact, for an ideal gas, it is only this component that contributes to u. The other variable is typically also a measurable property. Either P or v can then be chosen as the other independent variable according to convenience.

Figure 2.3 illustrates a common hypothetical path used to calculate Δu. In this case, T and v are chosen as independent properties. In step 1, we must know the temperature dependence of u to calculate Δu, as we go from T_1 to T_2 at constant volume. This information is often obtained in the form of heat capacity (or specific heat). Similarly, the temperature dependence of h used to find Δh can be found through reported heat capacity values. Therefore, heat capacity data are crucial in this problem-solving methodology. In this section, we will explore how heat capacities are experimentally determined and how they are reported.

To measure the heat capacity at constant volume, c_v, an experimental setup as conceptually shown in Figure 2.10a can be used. This closed system consists of pure species A within a rigid container. The container is connected to a heat source (in this case, a resistance heater) and is otherwise well insulated. The experiment is conducted

[Handwritten margin notes:
ep and e_k left out
closed ΔU = Q + W
open
$\frac{d}{dt}(U + E_K + E_P)_{sys} =$
$\Sigma \dot{n}_{in} h_{in} - \Sigma \dot{n}_{out} h_{out} +$
$\dot{Q} + \dot{W}_s$
ep and e_k assumed smaller than h
State point all
u, h → can use hypothetical path. State functions!]

as follows: As a known amount of heat, q, is provided through the resistive heater, the temperature, T, of the system is measured. As heat is supplied, the molecules of A move faster, and the temperature increases. A typical data set for pure species A is also shown in Figure 2.10a. Since we can "sense" the result of the heat input with the thermocouple, this type of energy change is labeled *sensible heat*. This setup is known as a *constant-volume calorimeter*. Since there is no work done in this process, we can apply the first law [Equation (2.31)] to this system and get

$$\Delta u = q \qquad \text{closed system, const V} \tag{2.49}$$

Note that in Figure 2.10a, the amount of heat supplied is plotted on the y-axis and the temperature measured is on the x-axis. However, Equation (2.49) shows heat input is identical to Δu. We define the heat capacity at constant volume, c_v, as

$$c_v \equiv \left(\frac{\partial u}{\partial T}\right)_v \tag{2.50}$$

Hence, the slope of the curve gives us the heat capacity at any temperature. In these data, the heat capacity at T_1 is less than that at T_2. Typically, heat capacity changes with T.

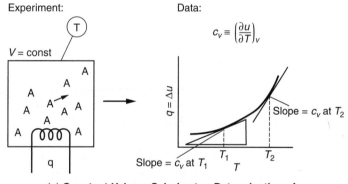

(a) **Constant-Volume Calorimeter: Determination of c_v**

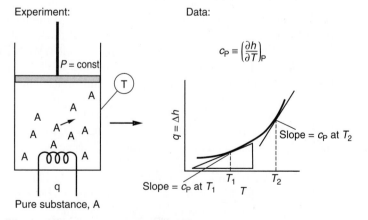

(b) **Constant-Pressure Calorimeter: Determination of c_P**

Figure 2.10 Schematics of the experimental determination of heat capacity. (a) constant-volume calorimeter to obtain c_v; (b) constant-pressure calorimeter to determine c_P.

By taking the slope of this curve as a function of temperature, we can get

$$c_v = c_v(T) \tag{2.51}$$

We can then fit the data to a polynomial expression of the form:

$$c_v = a + BT + CT^2 + DT^{-2} + ET^3 \tag{2.52}$$

Parameters a, B, C, D, and E are then tabulated and can be used any time we want to know how the internal energy of species A changes with temperature at constant volume. We can then find Δu by integration:

$$\Delta u = \int_{T_1}^{T_2} c_v dT = \int_{T_1}^{T_2} \left[a + BT + CT^2 + DT^{-2} \right] dT \tag{2.53}$$

Consider an ideal gas. Equation (2.50) shows us that if we increase T, we increase u. The amount the internal energy increases with temperature is quantified by the heat capacity according to Equation (2.53). The more energy it gains, the larger c_v. On a molecular level, we may want to know how this increase in energy manifests itself. We can associate the increase in molecular kinetic energy and, therefore, in u with temperature to three possible modes in which the molecules can obtain kinetic energy. The first mode is related to the center-of-mass motion of the molecules through space. In Chapter 1, we saw that the Maxwell–Boltzmann distribution characterizes the velocities of the molecules at a given temperature. This translational energy contributes $kT/2$ per molecule (or $RT/2$ per mole) to the kinetic energy in each direction that the molecule moves. Since molecules translate through space in three directions, the translational motion contributes $3RT/2$ per mole to the internal energy of the molecules. The contribution of translational motion to c_v, given by the derivative of the internal energy with respect to temperature, is $3R/2$. In fact, monatomic gases have heat capacities given by this value.[10]

Diatomic and polyatomic molecules can manifest kinetic energy in rotational and vibrational modes as well. Except at extremely low temperature, the additional kinetic energy due to rotational motion for linear and nonlinear molecules is RT and $3RT/2$ per mole, respectively. The kinetic energy due to vibration is much more interesting. It is related to the specific quantized energy levels of the molecule. The distribution of these levels depends on temperature. To account for the vibrational modes of kinetic energy, we would need to resort to quantum mechanics. We will not formally address that problem here; however, it is useful to realize that the temperature dependence of the heat capacity indicated by Equation (2.52) manifests itself in the vibrational mode. At low temperature, the vibrational contribution goes to zero and the heat capacity is given by the translational and rotational modes only. At high temperature, where the vibrational motion is fully active, the contribution is R per mole. In summary, c_v can be attributed to molecular structure and the ways in which each species exhibits translational, rotational, and vibrational kinetic energy.

Heat capacities for gases, liquids, and solids can be obtained in this manner; however, heat capacity should only be used for temperature changes between the same phase. When a phase change occurs, the latent heat must also be considered, as discussed shortly.

[10] Except at very high temperature when electrons occupy excited states.

[handwritten: constant, assume external = syst pressure]

[handwritten left margin: $W = \int -P_E \, dV$]

[handwritten left margin: $W = -P\Delta V$]

The heat capacity at constant pressure, c_P, is measured in a similar manner, only gas A is no longer held within a rigid container, but the system can expand as it is heated to keep the pressure constant. A conceptual representation of the experimental setup to measure c_P is shown in Figure 2.10b. While the actual apparatus may look different, this depiction is in terms of the piston–cylinder assembly that we have previously examined. Since the system is now doing Pv work as it expands, the energy balance contains a term for work:

$$\Delta u = q + w = q - P\Delta v \tag{2.54}$$

So Equation (2.54) can be rewritten as

[handwritten: 1.30 ✱]

[handwritten: $\Delta(Pv) = P\Delta V + V\Delta P \; {}^{\to 0}$]

$$\Delta u + \Delta(Pv) = q \tag{2.55}$$

since at constant pressure, $\Delta P = 0$, and

$$\Delta(Pv) = P\Delta v + v\Delta P = P\Delta v \tag{2.56}$$

Applying the definition for enthalpy [Equation (2.43)], we get:

[handwritten: intensive, see extensive on next page]

$$\Delta h = q \qquad \text{closed system, const } P \tag{2.57}$$

Hence, in this case, an energy balance tells us the heat supplied at constant pressure is just equal to the change in the thermodynamic property, enthalpy. Therefore, we define the heat capacity at constant pressure as

$$c_P \equiv \left(\frac{\partial h}{\partial T}\right)_P \tag{2.58}$$

Again, typical data for species A are presented in Figure 2.10b and can be fit to the polynomial form:

$$c_P = A + BT + CT^2 + DT^{-2} + ET^3 \tag{2.59}$$

The parameters A, B, C, D, and E are reported for some ideal gases in Appendix A.2. Heat capacity parameters at constant pressure of some liquids and solids are reported in this appendix.

[handwritten left margin: Δh: considers internal energy and work]

Recall that we "constructed" the property enthalpy to account for both the internal energy and flow work for streams flowing into and out of open systems. However, inspection of Equation (2.57) suggests a second common use of enthalpy. This equation holds, in general, for closed systems at constant P. In this case, it accounts for both the change in internal energy and the Pv work as the system boundary changes in order to keep pressure constant. In both cases, the property h couples internal energy and work. *Therefore, experiments that are conveniently done in closed systems at constant P are reported using the thermodynamic property enthalpy.* For example, the energetics of a chemical reaction, the so-called *enthalpy of reaction*, is reported in terms of the property Δh_{rxn}. In this way, the experimentally measured heat can be related directly to a thermodynamic property.

[handwritten left margin: $\Delta h \to c_p$]

By comparing Figure 2.10a to Figure 2.10b, we can estimate the difference in c_v and c_P for the different phases of matter. If species A is in the liquid or solid phase, its volume expansion upon heating should be relatively small; that is, the molar volumes of liquids and solids do not change much with temperature. Hence, the piston depicted in Figure 2.10b will not move significantly and the value of work in Equation (2.54) will be small compared to q. Thus Equation (2.58) and Equation (2.50) are approximately equivalent and, consequently:[11]

$$c_P \approx c_v \qquad \text{liquids and solids} \tag{2.60}$$

On the other hand, the volume expansion of a gas will be significant. For the case of an ideal gas, we can figure out the relationship between c_P and c_v by applying the ideal gas law to the definition for c_P as follows:

$$c_P \equiv \left(\frac{\partial h}{\partial T}\right)_P = \left[\frac{\partial(u + Pv)}{\partial T}\right]_P = \left(\frac{\partial u}{\partial T}\right)_P + \left(\frac{\partial RT}{\partial T}\right)_P = \left(\frac{\partial u}{\partial T}\right)_P + R \tag{2.61}$$

since $Pv = RT$ for an ideal gas. However, for an ideal gas, the internal energy depends on temperature *only*; that is, the only change in molecular energy is in molecular kinetic energy. Therefore

$$\left(\frac{\partial u}{\partial T}\right)_P = \frac{du}{dT} = \left(\frac{\partial u}{\partial T}\right)_v \tag{2.62}$$

Plugging Equation (2.62) into (2.61) gives:

$$c_P = c_v + R \qquad \text{for ideal gases} \tag{2.63}$$

Values of heat capacity for gases are almost always reported for the ideal gas state. Thus, when doing calculations using these data, you must choose a hypothetical path where the change in temperature occurs when the gas behaves ideally.

For many gases, heat capacity data are often reported in terms of the mean heat capacity, $\bar{c}_P$. Use of $\bar{c}_P$ eliminates the need for integration and can make the mechanics of problem solving easier. As its name suggests, the mean heat capacity is the average of c_P between two temperatures. It is usually reported between 298 K and a given temperature, T. Hence, the enthalpy change becomes

$$\Delta h = \bar{c}_P(T - 298) \tag{2.64}$$

Solving for $\bar{c}_P$ gives

$$\bar{c}_P = \frac{\int_{298}^{T} c_P \, dT}{T - 298} \tag{2.65}$$

Note that Equation (2.65) is also, by definition, the mathematical average of the continuous function c_P over temperature between 298 K and T.

▶ **EXAMPLE 2.6**
Heat Input Calculations Using Different Data Sources

Consider heating 2 moles of steam from 200°C and 1 MPa to 500°C and 1 MPa. Calculate the heat input required using the following sources for data:

(a) Heat capacity

(b) Steam tables

SOLUTION (a) Since this process occurs at constant pressure, the system will expand as T increases. In accordance with the discussion above, enthalpy is the appropriate property to calculate the heat input. The extensive version of Equation (2.57) can be written as

$$Q = n\Delta h$$

[11] We will revisit the relation between c_P and c_v of liquids in Chapter 5, Problems 5.9 and 5.10.

Substituting Equations (E2.8B) and (E2.8C) into (E2.8A) and using the values from Table E2.7, we get

$$\dot{Q} = \dot{n}\Delta h = 10[30.71(900 - 298) - 29.97(600 - 298)] = 94,363 \,[\text{J/min}]$$

▶ **EXAMPLE 2.9**
Spring-Assembled
Piston–Cylinder

Air is contained within a piston–cylinder assembly, as shown in Figure E2.9A. The cross-sectional area of the piston is 0.1 m². Initially the piston is at 1 bar and 25°C, 10 cm above the base of the cylinder. In this state, the spring exerts no force on the piston. The system is then reversibly heated to 100°C. As the spring is compressed, it exerts a force on the piston according to

$$F = -kx$$

where $k = 50,000$ [N/m] and x is the displacement length from its uncompressed position.

(a) Determine the work done.

(b) Determine the heat transferred.

SOLUTION (a) Since the process is reversible, the system pressure is always balanced by the external pressure and the work done is given by

$$W = -\int_{V_1}^{V_2} P\,dV \tag{E2.9A}$$

We can draw a free-body diagram to determine how all the forces acting on the piston balance, as shown in Figure E2.9A.

The displacement of the spring, x, can be written in terms of the change in volume:

$$x = \frac{V - V_1}{A} = \frac{\Delta V}{A}$$

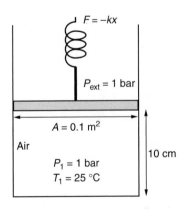

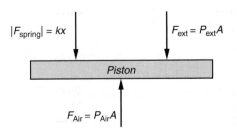

Figure E2.9A Piston–cylinder assembly with a spring attached to the piston. The initial state of the system for Example 2.9 is shown.

Figure E2.9B Schematic of the forces acting on the piston in Figure E2.9A as the gas expands.

where $\Delta V = V - V_1$. We can then equate the force per area acting on each side of the piston to get

$$P = P_{ext} + \frac{kx}{A} = P_{ext} + \frac{k\Delta V}{A^2} \qquad \text{(E2.9B)}$$

Plugging Equation (E2.9B) into (E2.9A) and integrating gives

$$W = - \int_{V_1}^{V_2} P_{ext}dV - \int_{0}^{\Delta V = (V_2 - V_1)} \frac{k\Delta V}{A^2}d(\Delta V) = -P_{ext}(V_2 - V_1) - \frac{k(V_2 - V_1)^2}{2A^2} \qquad \text{(E2.9C)}$$

To solve Equation (E2.9C), we must find V_2. Applying the ideal gas law gives

$$\frac{P_1 V_1}{T_1} = \frac{P_2 V_2}{T_2} = \frac{V_2}{T_2}\left(P_{ext} + \frac{k(V_2 - V_1)}{A^2}\right)$$

Solving this quadratic equation for V_2 gives

$$V_2 = 0.00116 \, [\text{m}^3]$$

which can be plugged back into Equation (E2.9C) to get

$$W = -166 \, [\text{J}]$$

(b) To find the heat transferred during this process, we can apply the first law to this closed system:

$$\Delta U = Q + W \qquad \text{(E2.9D)}$$

The change in internal energy is given by

$$\Delta u = \int_{T_1}^{T_2} c_v dT = \int_{T_1}^{T_2} (c_P - R)dT = R\int_{T_1}^{T_2} \left[(A-1) + BT + DT^{-2}\right]dT$$

$$= R\left[(A-1)T + \frac{B}{2}T^2 - \frac{D}{T}\right]_{T_1}^{T_2}$$

The heat capacity parameters for air can be found in Appendix A.2:

$$A = 3.355, \qquad B = 0.575 \times 10^{-3}, \qquad \text{and } D = -0.016 \times 10^5$$

Thus,

$$\Delta u = 1,580 \, [\text{J/mol}]$$

and

$$\Delta U = n\Delta u = \left(\frac{P_1 V_1}{RT_1}\right)\Delta u = 638 \, [\text{J}]$$

We can now solve for the heat transfer from Equation (E2.9D):

$$Q = \Delta U - W = 803 \, [\text{J}]$$

Latent Heats

[margin note: purpose of latent heat]

When a substance undergoes a phase change, there is a substantial change in internal energy associated with it (see Section 2.1). We need to quantify this energy if we want to apply the first law to a process involving a phase change. Like heat capacities, the energetics characteristic of a given phase change are reported based on accessible measured data.

For example, consider the vaporization of a liquid. Liquids are held together by attractive forces between the molecules. To vaporize a liquid, we must supply enough energy to overcome the forces of attraction. A typical experimental setup is schematically shown in Figure 2.11. A given amount of liquid substance A is placed in a well-insulated closed system at constant pressure. A measured amount of heat is supplied until A becomes all vapor. We chose a system at constant pressure, as depicted in Figure 2.11, since the temperature stays constant during the phase change. Examination of Equation (2.57) shows that if we measure the *heat* absorbed as A changes phase, it is equal to the enthalpy of the vapor minus the enthalpy of the liquid. Correspondingly, the term to describe the change of enthalpy during a phase transition at constant pressure is *latent heat*. It is called "latent" because we cannot "sense" the heat input by detecting the temperature change, as is the case with "sensible" heat, described previously. A schematic for the data acquired in this experiment is presented on the right of Figure 2.11. The temperature of the subcooled liquid and the superheated vapor increases as energy is supplied via heat. It is only the heat input at constant temperature during the phase transition that is reported as the enthalpy of vaporization.

[margin note: $q_p = \Delta h$]

[margin note: sign counts]

Latent heats reported for transitions between a liquid and a vapor are reported as enthalpies of vaporization, Δh_{vap}. If we need to calculate the energetics of a vapor condensing to a liquid, we simply use the negative of this value. Similarly, changes between the liquid and solid phases are enthalpies of fusion, Δh_{fus}, and changes between the solid and vapor phases are enthalpies of sublimation, Δh_{sub}. How would you find the internal energy of vaporization, Δu_{vap}, given Δh_{vap}?

[margin note: latent occurs only @ a specific T and P]

Latent heats change with temperature. We typically know, for example, the enthalpy of vaporization at 1 bar, the so-called **normal** boiling point, T_b. If we need Δh_{vap} at another pressure, we must construct an appropriate thermodynamic pathway, based on the measured value to which we have access. Figure 2.12 illustrates a path for the calculation of $\Delta h_{vap, T}$ at any T. It consists of three steps. In step 1, we calculate the

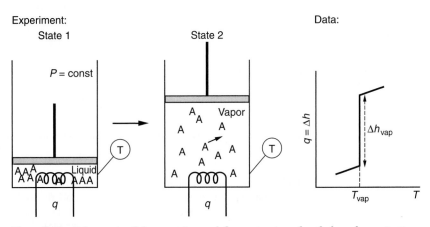

Experiment:

State 1   State 2   Data:

P = const

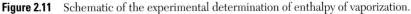

Figure 2.11 Schematic of the experimental determination of enthalpy of vaporization.

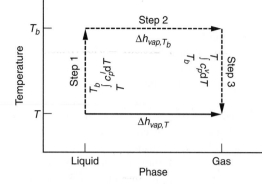

Figure 2.12 Hypothetical path to calculate Δh_{vap} at temperature T from data available at T_b and heat capacity data.

change in enthalpy of the liquid from T to T_b, using heat capacity data. In step 2, we vaporize the liquid at the normal boiling point, since this is the value we have for Δh_{vap}. In step 3, we calculate the change in enthalpy of the vapor from the normal boiling point to T. Adding together the three steps, we get

$$\Delta h_{vap,T} = \int_{T}^{T_b} c_P^l\, dT + \Delta h_{vap,T_b} + \int_{T_b}^{T} c_P^v\, dT = \Delta h_{vap,T_b} + \int_{T_b}^{T} \Delta c_P^{vl}\, dT \qquad (2.66)$$

where we used the following definition:

$$\Delta c_P^{vl} = c_P^v - c_P^l \qquad (2.67)$$

This procedure can likewise be applied to determine the values of Δh_{fus} and Δh_{sub} at different temperatures than that at which they are reported.

▶ **EXAMPLE 2.10**
Determination of Heat Required to Evaporate Hexane

10 mol/sec of liquid hexane flows into a steady-state boiler at 25°C. It exits as a vapor at 100°C. What is the required heat input to the heater? Take the enthalpy of vaporization at 68.8°C to be

$$\Delta h_{vap,68.8°C} = 28.88\,[\text{kJ/mol}]$$

SOLUTION This process occurs in an open system with one stream in and one stream out. Can you draw a schematic? We label the inlet state as "state 1" and the outlet as "state 2." In this case, Equation (2.45) becomes

$$0 = \dot{n}_1(h + \overset{0}{e_K} + \overset{0}{e_P})_1 - \dot{n}_2(h + \overset{0}{e_K} + \overset{0}{e_P})_2 + \dot{Q} + \overset{0}{\dot{W}_s} \qquad (\text{E2.10A})$$

where we have set the macroscopic kinetic and potential energy and the shaft work equal to zero. A mole balance gives:

$$\dot{n}_1 = \dot{n}_2 = \dot{n} \qquad (\text{E2.10B})$$

Plugging (E2.10B) into (E2.10A) and solving for the heat-transfer rate, we get

$$\dot{Q} = \dot{n}(h_2 - h_1) \qquad (\text{E2.10C})$$

The enthalpy change can be divided into three parts: (1) the sensible heat required to raise liquid hexane to its boiling point; (2) the enthalpy of vaporization, that is, latent heat; and (3) the sensible heat required to raise hexane in the vapor state to 100°C. Thus, the enthalpy difference becomes

$$h_2 - h_1 = \underset{l,25°C \rightarrow 68.8°C}{\Delta h} + \Delta h_{vap,68.8°C} + \underset{v,68.8°C \rightarrow 100°C}{\Delta h} \qquad (\text{E2.10D})$$

Values for the first and third terms can be found from the appropriate heat capacity data:

$$\Delta h_{l,25°C \rightarrow 68.8°C} = \int_{298.2}^{342} 216.3 dT = 9485 \text{ [J/mol]} = 9.49 \text{ [kJ/mol]}$$

and

$$\Delta h_{v, 68.8°C \rightarrow 100°C} = \int_{342}^{373.2} R(A + BT + CT^2) dT$$

$$= 8.314 \left[3.025(373.2 - 342) + \frac{53.722 \times 10^{-3}}{2}(373.2^2 - 342^2) \right.$$

$$\left. - \frac{16.791 \times 10^{-6}}{3}(373.2^3 - 342^3) \right]$$

$$= 5.20 \text{ [kJ/mol]}$$

Summing together values for enthalpy in Equation (E2.10D) and plugging the result into Equation (E2.10C) gives the rate at which heat must be supplied:

$$\dot{Q} = \left(10 \text{ [mol/s]}\right)\left(9.49 + 28.85 + 5.20\right) \text{ [kJ/mol]}\right) = 435 \text{ [kJ/s]}$$

✗very helpful → not an easy problem

▶ **EXAMPLE 2.11**
Determination of Heat Required to Evaporate H_2O

A rigid vessel contains 50.0 kg of saturated liquid water and 4.3 kg of saturated vapor. The system pressure is at 10 kPa. What is the minimum amount of heat needed to evaporate all the liquid?

SOLUTION A schematic of the process is shown in Figure E2.11. We label the initial state as "state 1" and the final state as "state 2." The left-hand side of the figure shows the physical process while the right-hand side represents it on a Pv diagram.

This heating process occurs in a closed system at constant volume. As water boils, the pressure in the vapor phase will increase. The increase in pressure will require an increase in the temperature of the system for boiling to proceed. (Remember, a given pressure constrains the temperature of the two-phase region.) Thus, energy is required for both the evaporation of water (latent heat) and the increase in temperature (sensible heat). Since bulk kinetic and potential energy are negligible, we can write the first law according to Equation (2.30):

$$\Delta U = Q + W \tag{E2.11}$$

Since our system is at constant volume, there is no work done and the heat needed equals the change in internal energy. This result contrasts with Figure 2.11, where *enthalpy* is used, since, in that case, there is Pv work. A mass balance gives:

assumed? rigid?

$$m_2^v = m_1^l + m_1^v$$

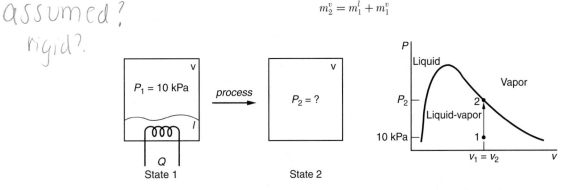

Figure E2.11 Schematic of evaporation of saturated liquid H_2O in a rigid, closed system.

The internal energy in state 1 must account for that of both the saturated liquid and the saturated vapor. Hence, we can write Equation (E2.11) as

$$U_2 - U_1 = (m_2^v \hat{u}_2^v) - (m_1^l \hat{u}_1^l + m_1^v \hat{u}_1^v) = Q$$

State 1 is completely constrained. Looking up values for internal energy in the steam tables for saturated water: pressure (Appendix B.2) gives

$$\hat{u}_1^l = 191.8 \,[\text{kJ/kg}]$$

$$\hat{u}_1^v = 2437.9 \,[\text{kJ/kg}]$$

We now need to constrain state 2. We know we have saturated vapor, but we need to find a value for a thermodynamic property. Since the container is rigid, this process occurs at constant volume, as illustrated on the Pv diagram in Figure E2.11. We can find the volume of the vessel as follows:

$$V_1 = V_2 = m_1^l \hat{v}_1^l + m_1^v \hat{v}_1^v = (50)(0.001) + (4.3)(14.67) = 63.1 \text{ m}^3$$

We can now solve for the specific volume of state 2:

$$\hat{v}_2^v = \frac{V_2}{m_2^v} = \frac{63.1 \text{ m}^3}{54.3 \text{ kg}} = 1.16 \left[\frac{\text{m}^3}{\text{kg}}\right]$$

Looking up the value, we find the specific volume of saturated vapor matches at

$$P_2 = 0.15 \text{ MPa}$$

If this value did not match a table entry, we would have to interpolate. Looking up the internal energy of state 2 gives

$$\hat{u}_2^v = 2519.6 \,[\text{kJ/kg}]$$

Solving Equation (E2.10) for heat gives

$$Q = (m_2^v \hat{u}_2^v) - (m_1^l \hat{u}_1^l + m_1^v \hat{u}_1^v) = (54.3)(2519.6) - [(50)(191.8) - (4.3)(2437.9)]$$

$$= 117 \times 10^6 \text{ J}$$

Enthalpy of Reactions

A large amount of energy is "stored" in the chemical bonds within molecules. When the atoms in molecules rearrange by undergoing a chemical reaction, the energy stored within the bonds of the products is typically different from that of the reactants. Thus, significant amounts of energy can be absorbed or liberated during chemical reactions. The energy change upon reaction is an important component in applying the first law to reacting systems. It can be characterized by a change in internal energy, Δu_{rxn}; however, it is more common to report the change in enthalpy of reaction, Δh_{rxn}, since these experiments are more conveniently executed at constant pressure.

For example, consider the reaction of two molecules of hydrogen gas with one molecule of oxygen to form gaseous water at 298 K and 1 bar. The reaction stoichiometry can be expressed as follows:

$$2\text{H}_2(g) + \text{O}_2(g) \longrightarrow 2\text{H}_2\text{O}(g) \tag{2.68}$$

The reactants contain three bonds per molecule of oxygen reacting: single bonds between hydrogen atoms in each of the two H_2 molecules and a double bond between the oxygen

TABLE 2.1   Bond Dissociation Energies of H,O Bonds

Bond	Energy [eV/molecule]
H–H	4.50
O=O	5.13
O–H	4.41

Source: Derived from average bond enthalpies reported in G. C. Pimentel and R. D. Spratley, *Understanding Chemical Thermodynamics* (San Francisco: Holden-Day, 1969).

atoms in O_2. The two product H_2O molecules have four oxygen–hydrogen single bonds. These bonds are covalent in nature and vary in energy based on how the interacting valence electrons overlap. Bond energies for the three different types of bonds in this system are reported in Table 2.1.

Let's consider the energetics of this reaction by the following path: We first pull apart the reactant molecules into their constituent atoms and then have the atoms recombine to form the product molecules. This path, in essence, defines a reference state as the atomic form of each element in the system. A schematic of the energetics in this reaction path is shown in Figure 2.13. The energy differences between states depicted by the arrows are based on the bond energies. It takes 14.1 eV of energy ($1\,\text{eV} = 1.6 \times 10^{-19}$ J) to dissociate two molecules of H_2 and one molecule of O_2 into four H atoms and two O atoms, respectively. However, when these atoms re-form into two water molecules, they release 17.1 eV of energy. The net energy change represents the internal energy of reaction and is found to be -3.5 eV for every two molecules of water produced by this reaction. The negative sign indicates that the products are more stable than the reactants, and, consequently, energy is released. Reactions that release energy are said to be *exothermic*, while reactions that absorb energy are termed *endothermic*.

To generalize to any reaction, we introduce the stoichiometric coefficient, ν_i. The stoichiometric coefficient equals the proportion of a given species consumed or produced in a reaction relative to the other species. It can be found as the number before the corresponding species in a balanced chemical reaction. By convention, it is unitless and positive for products, $\nu_{\text{products}} > 0$, negative for reactants, $\nu_{\text{reactants}} < 0$, and zero for inerts, $\nu_{\text{inerts}} = 0$. For example, in Reaction(2.68):

$$\nu_{H_2} = -2, \quad \nu_{O_2} = -1, \quad \text{and} \quad \nu_{H_2O} = 2$$

In the previous discussion, we used atoms for our reference state, since this choice made it straightforward to see how the energy of molecules changes with atomic rearrangement. However, this reference state is inconvenient in practice, since, in nature,

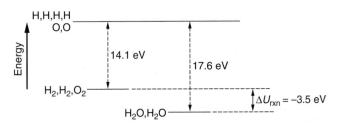

Figure 2.13   Schematic representation of the energy needed to dissociate the reactants and products of Reaction (2.68) into their atoms. The resulting energy difference characterizes the internal energy change of reaction, ΔU_{rxn}.

species seldom exist as atoms at 298 K and 1 bar. We are free to pick any reference state that we desire as long as we stick to it. Consequently, the reference state we usually use is the pure elemental constituents, as found in nature, of the species of interest. The enthalpy difference between a given molecule and the reference state is defined as the enthalpy of formation, Δh_f. The enthalpy of formation can be represented as:

$$\text{elements} \xleftrightarrow{\Delta h_f} \text{species } i \tag{2.69}$$

The enthalpy of formation of a species containing only one element, as it is found in nature, is identically zero. Enthalpies in the form of Δh_f are the most common thermochemical data available to calculate the enthalpy of reaction; Appendix A.3 shows some representative values for 25°C and 1 atm. For example, the enthalpy of formation of liquid water is defined by the reaction

$$H_2(g) + \frac{1}{2}O_2(g) \Leftrightarrow H_2O(l) \tag{2.70}$$

since the elements found in water, hydrogen and oxygen are found as diatomic gases at 25°C and 1 atm. The value for the enthalpy of formation for this reaction is found in Appendix A.3 to be $\Delta h^\circ_{f,298} = -285.83\,[\text{kJ/mol}]$. Appendix A.3 also reports the enthalpy of reaction for gaseous water at 298 K and 1 atm. Although water cannot physically exist in this state, the enthalpy of formation is representative of a hypothetical (but important!) change of state that is often useful. For example, we may be interested in a system in which water is reacting at higher temperatures, where it is a vapor. The first step in obtaining the enthalpy of reaction at the system T would be finding it at 298 K. Example 2.14 illustrates such a calculation.

With the enthalpies of formation available, it is straightforward to calculate the enthalpy of reaction. Such a calculation path for the enthalpy of reaction at 298 K is illustrated in Figure 2.14. In the dashed (calculation) path, the reactants are first decomposed into their constituent elements, as found in nature. This part of the path is labeled Δh_1. The constituent elements are then allowed to react to form products, as given by Δh_2. Note that the stoichiometric coefficients of the reactants are negative, making the signs for Δh_1 consistent with the definition of enthalpy of formation above. Equating the two paths yields

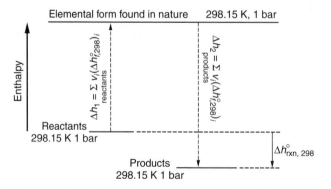

Figure 2.14 Calculation path of $\Delta h^\circ_{\text{rxn}}$ from the standard enthalpies of formation, $(\Delta h^\circ_f)_i$.

$$\Delta h^{\circ}_{\text{rxn},298} = \Delta h_1 + \Delta h_2 = \sum_{\text{reactants}} v_i \left(\Delta h^{\circ}_{f,298}\right)_i + \sum_{\text{products}} v_i \left(\Delta h^{\circ}_{f,298}\right)_i$$

$$= \sum_i v_i \left(\Delta h^{\circ}_{f,298}\right)_i \tag{2.71}$$

Thus, if enthalpies of formation are available for all the species in the chemical reaction of interest, the enthalpy of reaction can be determined by scaling each species' Δh_f by its stoichiometric coefficient. In summary,

$$\Delta h^0_{\text{rxn}} = \sum v_i h^{\circ}_i = \sum v_i \left(\Delta h^{\circ}_f\right)_i \tag{2.72}$$

Often a reaction does not go to completion; that is, there remain some reactants in the outlet stream. For incomplete reactions, we must account only for the enthalpy of reaction for the species that *did* react in our energy balance. Indeed, in Chapter 9 we will learn how to quantify the extent to which a chemical reaction proceeds at equilibrium.

▶ **EXAMPLE 2.12**
Determination of
Enthalpy of Reaction

Calculate the enthalpy of reaction at 298 K for the following reaction:

$$H_2O(g) + CH_3OH(g) \Longleftrightarrow CO_2(g) + 3H_2(g)$$

SOLUTION   The enthalpy of reaction can be found from the enthalpy of formation data presented in Appendix A.3. In this case, Equation (2.72) can be written as follows:

$$\Delta h^0_{\text{rxn}} = \sum v_i \left(\Delta h^{\circ}_{f,298}\right)_i = \left(\Delta h^{\circ}_{f,298}\right)_{CO_2} + 3\left(\Delta h^{\circ}_{f,298}\right)_{H_2} - \left(\Delta h^{\circ}_{f,298}\right)_{H_2O} - \left(\Delta h^{\circ}_{f,298}\right)_{CH_3OH}$$

The enthalpy of formation of H_2 is zero by definition, since that is the form hydrogen takes at 298 K and 1 bar. Taking values for the others from Appendix A.3:

$$\Delta h^{\circ}_{298} = (-393.51) + 3(0) - (-241.82) - (-200.66) = 49.0 \, [\text{kJ/mol}]$$

The sign of the enthalpy of reaction is positive, indicating that this reaction is endothermic.

▶ **EXAMPLE 2.13**
Adiabatic Flame
Temperature
Calculation

max temp
a reactor can
reach for combustion
of a given fuel
@ const p

Propane enters an adiabatic, constant-pressure combustion chamber at 25°C. It is mixed with a stoichiometric amount of air. Assume complete combustion and that the carbon distribution in the product stream contains 90% CO_2 and 10% CO. What is the exit temperature?

SOLUTION   A balanced equation of the chemical reaction can be written as follows:

$$10C_3H_8 + \frac{97}{2}O_2 \longrightarrow 27CO_2 + 3CO + 40H_2O \tag{E2.13A}$$

A schematic of this process, assuming Reaction (E2.13A) goes to completion, is shown in Figure E2.13.A. If we chose a basis of 10 moles of propane, we have (97/2) moles of oxygen and

$$n_{N_2} = \left(\frac{0.79}{0.21}\right)\left(\frac{97}{2}\right) = 182.5 \, [\text{mol}]$$

initially. Inspection of reaction (E2.13A) shows that the final state contains

$$n_{CO_2} = 27 \, [\text{mol}], \; n_{CO} = 3 \, [\text{mol}], \; \text{and} \; n_{H_2O} = 40 \, [\text{mol}]$$

while the number of moles of N_2 remains unchanged. An energy balance on the closed system at constant pressure gives [see Equation (2.57)]:

$$\Delta H = \sum (n_i h_i)_2 - \sum (n_i h_i)_1 = Q = 0 \tag{E2.13B}$$

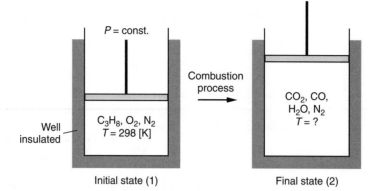

Figure E2.13A Schematic of complete combustion of propane in a stoichiometric mixture of air at constant pressure. The closed system is adiabatic.

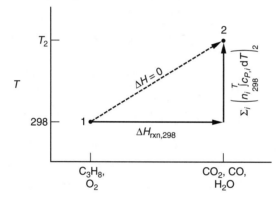

Figure E2.13B. Hypothetical path for calculation of enthalpy in complete combustion of propane.

We now need a path to calculate ΔH. A convenient choice is illustrated by the solid lines in Figure E2.13B. That figure also shows the overall energy balance constraint of Equation (E2.13B) as a dashed line. Since enthalpy-of-reaction data are available at 298 K (Appendix A.3), we choose to first completely combust propane at 298 K, then heat the products to a temperature T_2, which makes the enthalpy change between states 1 and 2 zero. Using this hypothetical path, Equation (E2.13B) becomes

$$\Delta H_{\text{rxn},298} + \int_{298}^{T_2} \sum (n_i)_2 (c_P)_i \, dT = 0$$

The extensive enthalpy of reaction can be found using the number of moles determined based on the reaction stoichiometry above:

$$\Delta H_{\text{rxn},298} = \sum n_i \left(\Delta h_f^{\circ} \right)_i = n_{\text{CO}_2} \left(\Delta h_f^{\circ} \right)_{\text{CO}_2} + n_{\text{CO}} \left(\Delta h_f^{\circ} \right)_{\text{CO}} + n_{\text{H}_2\text{O}} \left(\Delta h_f^{\circ} \right)_{\text{H}_2\text{O}}$$
$$- n_{\text{C}_3\text{H}_8} \left(\Delta h_f^{\circ} \right)_{\text{C}_3\text{H}_8} - n_{\text{O}_2} \left(\Delta h_f^{\circ} \right)_{\text{O}_2}$$

Using values from Appendix A.3, we get

$$\Delta H_{\text{rxn},298} = (27)(-393.51) + 3(-110.53) + 40(-241.82) - 10(-103.85) - 0 = -19{,}591 \text{ [kJ]}$$

The enthalpy of reaction must be counteracted by an equal but opposite increase in the sensible heat. This effect will allow us to calculate T_2. The sensible heat can be found through a summation of heat capacities, as follows:

$$\int_{298}^{T_2} \sum (n_i)_2 \, (c_P)_i \, dT = \int_{298}^{T_2} n_{CO_2} \, (c_P)_{CO_2} \, dT + \int_{298}^{T_2} n_{CO} \, (c_P)_{CO} \, dT$$

$$+ \int_{298}^{T_2} n_{H_2O} \, (c_P)_{H_2O} \, dT + \int_{298}^{T_2} n_{N_2} \, (c_P)_{N_2} \, dT \qquad \text{(E2.13C)}$$

Values for heat capacity parameters can be found in Appendix A.3. They are summarized in Table E2.13.

Integrating each term on the right of Equation (E2.13C) gives:

$$\int_{298}^{T_2} n_{CO_2} \, (c_P)_{CO_2} \, dT = (27) \, R \left[A_{CO_2}(T_2 - 298) + \frac{B_{CO_2}}{2} \left(T_2^2 - (298)^2 \right) - D_{CO_2} \left(\frac{1}{T_2} - \frac{1}{298} \right) \right]$$

$$\int_{298}^{T_2} n_{CO} \, (c_P)_{CO} \, dT = (3) \, R \left[A_{CO}(T_2 - 298) + \frac{B_{CO}}{2} \left(T_2^2 - (298)^2 \right) - D_{CO} \left(\frac{1}{T_2} - \frac{1}{298} \right) \right]$$

$$\int_{298}^{T_2} n_{H_2O} \, (c_P)_{H_2O} \, dT = (40) \, R \left[A_{H_2O}(T_2 - 298) + \frac{B_{H_2O}}{2} \left(T_2^2 - (298)^2 \right) - D_{H_2O} \left(\frac{1}{T_2} - \frac{1}{298} \right) \right]$$

$$\int_{298}^{T_2} n_{N_2} \, (c_P)_{N_2} \, dT = (182.5) \, R \left[A_{N_2}(T_2 - 298) + \frac{B_{N_2}}{2} \left(T_2^2 - (298)^2 \right) - D_{N_2} \left(\frac{1}{T_2} - \frac{1}{298} \right) \right]$$

The only unknown, T_2, can be solved for implicitly by the value at which the sum of the four equations above equals $-\Delta H_{rxn,298}$, 19,591 [kJ]. T_2 is found to be

$$T_2 = 2610 \, [\text{K}]$$

This value, known as the adiabatic flame temperature, is quite large. *The adiabatic flame temperature indicates the maximum temperature a reactor can reach for a given fuel.* If there is heat transfer out of the system, the temperature will be lower. If you used excess air, do you think the adiabatic flame temperature would be higher, lower, or the same? How about if pure O_2 is used instead of air?

▶ **EXAMPLE 2.14**
Enthalpy of Reaction at Different T

How would you calculate of the enthalpy of reaction, Δh_{rxn}, at any temperature T, given data for enthalpy of formation at 298 K and heat capacity parameters, available in Appendix A?

TABLE E2.13 Heat Capacity Parameters for Species in State 2

Species	A	B	D
CO_2	5.457	1.05×10^{-3}	-1.16×10^5
CO	3.376	5.57×10^{-4}	-3.10×10^3
H_2O	3.470	1.45×10^{-3}	1.21×10^4
N_2	3.280	5.93×10^{-4}	4.00×10^3

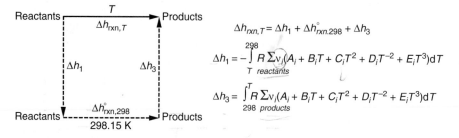

$$\Delta h_{rxn,T} = \Delta h_1 + \Delta h^{\circ}_{rxn.298} + \Delta h_3$$

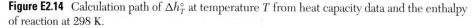

$$\Delta h_1 = -\int_{T}^{298} R\sum_i v_i (A_i + B_iT + C_iT^2 + D_iT^{-2} + E_iT^3)dT$$
reactants

$$\Delta h_3 = \int_{298}^{T} R\sum_i v_i (A_i + B_iT + C_iT^2 + D_iT^{-2} + E_iT^3)dT$$
products

Figure E2.14 Calculation path of Δh°_T at temperature T from heat capacity data and the enthalpy of reaction at 298 K.

SOLUTION Since enthalpy is a thermodynamic property, we can construct a hypothetical path that utilizes the available data. The enthalpy of reaction at any temperature T can then be found from the path illustrated in Figure E2.14. The reactants are first brought to 298 K. They are then allowed to react under standard conditions to make the desired products. The products are then brought back up to the system temperature, T. Adding these three steps gives the following integral:

$$\Delta h_{rxn,T} = \Delta h_{rxn,298} + \int_{298}^{T} \left(\sum_i v_i c_{p,i} \right) dT$$

Substituting in Equations (2.72) and (2.59) gives

$$\Delta h_{rxn,T} = \sum_i v_i \left(\Delta h^{\circ}_{f,298} \right)_i + \int_{298}^{T} \left(R\sum_i v_i \left(A_i + B_iT + C_iT^2 + \frac{D_i}{T^2} + E_iT^3 \right) \right) dT \qquad (E2.14)$$

When standard enthalpies of formation and heat capacity parameters are available, Equation (E2.14) can be solved explicitly for $\Delta h_{rxn,T}$ at any given T. Note the similarity between the paths in Figure 2.12 and Figure E2.14. In thermodynamics we can often apply concepts developed to solve one type of problem to many other cases.

▶ 2.7 REVERSIBLE PROCESSES IN CLOSED SYSTEMS

One useful application of thermodynamics is in the calculation of work and heat effects for many different processes by applying the first law. This information allows engineers to use energy more efficiently, saving costs and resources. Since heat and work are path dependent, the specific process must be defined in order to perform the necessary calculations. In this section, we go through two such examples of these types of calculations using an **ideal gas** undergoing **reversible** processes. We will look at nonideal gases in Chapter 5. The intent is to gain some experience with applying the first law to get values for work and heat as well as to develop expressions that are useful in understanding the Carnot cycle (Section 2.9).

Reversible, Isothermal Expansion (Compression)

Consider a **reversible, isothermal** expansion of an **ideal gas**. A schematic of a piston–cylinder assembly undergoing such a process is shown in Figure 2.15. The gas is kept at constant temperature by keeping it in contact with a **thermal reservoir**. A thermal reservoir contains enough mass so that its temperature does not noticeably change during the process. Can you predict the signs of $\Delta U, Q,$ and W?

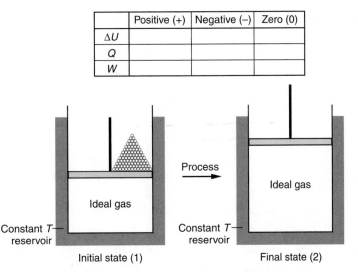

	Positive (+)	Negative (−)	Zero (0)
ΔU			
Q			
W			

Figure 2.15 An ideal gas in a piston–cylinder assembly undergoing a reversible, isothermal expansion. See if you can predict the signs of $\Delta U, Q$, and W for this process in the table.

Since the internal energy of an ideal gas is only a function of temperature,

$$Q = -W \qquad \Delta U = 0 \rightarrow \text{isothermal} \qquad (2.73)$$

For a reversible process, we can integrate over the system pressure (see Section 2.3):

$$W = -\int P dV \qquad (2.74)$$

Applying the ideal gas relationship

$$V = \frac{nRT}{P} \qquad (2.75)$$

the differential in volume can be transformed into a differential in pressure (remembering, for this case, that T is constant):

$$dV = -\frac{nRT}{P^2} dP \qquad (2.76)$$

Substituting Equation (2.76) into Equation (2.74) and integrating gives

$$W = \int_1^2 \frac{nRT}{P} dP = nRT \ln\frac{P_2}{P_1} \qquad (2.77)$$

Now applying the first law, we get

$$Q = \Delta U - W = -nRT \ln\frac{P_2}{P_1} \qquad (2.78)$$

Since $P_2 < P_1$, the sign for W is negative and for Q is positive. Did you get the sign right in the table in Figure 2.15? How do Equations (2.77) and (2.78) change if the gas undergoes a compression process instead of an expansion?

Handwritten annotations:

$P_2 > P_1$

$\ln \frac{P_2}{P_1} \oplus \quad W \oplus$

↑

compression

$W \oplus \quad Q \ominus$

expansion

$W \ominus \quad Q \oplus$

$W = \int -P_E dV$
always true.
when reversible
$W = \int -P dV$

$PV = const$
(less
steep
slope)

$W = nRT \ln\left(\frac{P_2}{P_1}\right)$

Assumes, PV-work, closed syst, reversible, ideal gas

Adiabatic Expansion (Compression) with Constant Heat Capacity

Consider when the same **ideal** gas undergoes an **adiabatic, reversible** expansion (as opposed to isothermal). We will assume that the heat capacity of this gas does not change with temperature, that is, **constant heat capacity**. This process is illustrated in Figure 2.16. Again, can you predict the signs of ΔU, Q, and W?

Neglecting macroscopic kinetic and potential energy, the first law for a closed system in differential form is obtained from Equation (2.33):

$$dU = \overset{0}{\delta Q} + \delta W \tag{2.79}$$

where the heat transfer was set to zero, since this process is adiabatic. From Equation (2.50) we get

$$dU = nc_v dT \tag{2.80}$$

and for a reversible process

$$\delta W = -PdV \tag{2.81}$$

Substituting Equations (2.80) and (2.81) into Equation (2.79) yields

$$nc_v dT = -PdV \tag{2.82}$$

We can use the ideal gas law to relate the measured properties T, V, and P:

$$d(nRT) = d(PV) = PdV + VdP \tag{2.83}$$

where we applied the product rule. Solving Equation (2.83) for dT and then plugging back into Equation (2.82) and rearranging gives

$$c_v VdP = -(c_v + R)PdV = -c_P PdV \tag{2.84}$$

(Handwritten notes, left margin:)

$\Delta U = nC_v \Delta T$

$\int nC_v \Delta T\, dT = -\int PdV$

$nC_v \int dT = -\int PdV$

$PV = nRT$

$dT = \dfrac{PdV + VdP}{nR}$ 1.30*

$\dfrac{C_v}{R} \int PdV + \int VdP = -\int PdV$

$\dfrac{C_v}{R} + 1 \int PdV = -\dfrac{C_v}{R}\int VdP$

$\dfrac{C_p + R}{R} \int \dfrac{dV}{V} = \dfrac{C_v}{R}\int \dfrac{dP}{P}$

$C_p \ln \dfrac{V_2}{V_1} = C_v \ln \dfrac{P_1}{P_2}$

	Positive (+)	Negative (−)	Zero (0)
ΔU			
Q			
W			

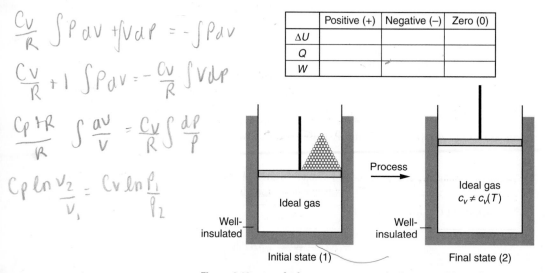

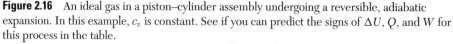

Figure 2.16 An ideal gas in a piston–cylinder assembly undergoing a reversible, adiabatic expansion. In this example, c_v is constant. See if you can predict the signs of ΔU, Q, and W for this process in the table.

Separating variables in Equation (2.84):

$$-\frac{c_P}{c_v}\frac{dV}{V} = \frac{dP}{P} \tag{2.85}$$

Now we integrate Equation (2.85) from the initial state 1 to the final state 2,

$$-k\ln\left(\frac{V_2}{V_1}\right) = \ln\left(\frac{P_2}{P_1}\right) \tag{2.86}$$

where $k = c_P/c_v$. Applying mathematical relationships of the natural logarithm, we can rewrite the left-hand side of Equation (2.86) as

$$-k\ln\left(\frac{V_2}{V_1}\right) = \ln\left(\frac{V_2}{V_1}\right)^{-k} = \ln\left(\frac{V_1}{V_2}\right)^{k} \tag{2.87}$$

so

$$\ln(P_1V_1^k) = \ln(P_2V_2^k) \tag{2.88}$$

or

$$PV^k = \text{const} \tag{2.89}$$

Now integrating for work:

$$W = -\int P\,dV = -\int \text{const}\,V^{-k}\,dV = \frac{\text{const}}{k-1}\left[\frac{1}{V_2^{k-1}} - \frac{1}{V_1^{k-1}}\right]$$

$$= \frac{1}{k-1}[P_2V_2 - P_1V_1] = \frac{nR}{k-1}[T_2 - T_1] \tag{2.90}$$

From the first law:

$$\Delta U = W = \frac{1}{k-1}[P_2V_2 - P_1V_1] = \frac{nR}{k-1}[T_2 - T_1] \tag{2.91}$$

Summary

A summary of the two cases presented in this section is shown in Table 2.2. In both cases, expansion of a piston provides useful energy to the surroundings in the form of work. However, each case represents a limit. In the isothermal process, all the energy delivered as work is provided by the surroundings in the form of heat. On the other hand, for the adiabatic case, the energy for work is provided by the internal energy of the gas in the

TABLE 2.2 Summary of Expressions for Change in Internal Energy, Heat, and Work for an Ideal Gas Undergoing a Reversible Process

	Isothermal	Adiabatic, $c_v \neq c_v(T)$
ΔU	0	$\frac{nR}{k-1}[T_2 - T_1]$
Q	$-nRT\ln\frac{P_2}{P_1}$	0
W	$nRT\ln\frac{P_2}{P_1}$	$\frac{nR}{k-1}[T_2 - T_1]$

system. An intermediate case where there is some heat adsorbed from the surroundings as well as some "cooling" of the gas in the system is also possible.

A process is defined as **polytropic** if it follows the relation:

$$A \, did \quad PV^\gamma = \text{const} \qquad (2.92)$$
$$\gamma = K$$

Isotherm $PV = const$

Both the processes in this section can be considered polytropic. The isothermal expansion of an ideal gas follows Equation (2.92) with $\gamma = 1$ while the reversible, adiabatic expansion of an ideal gas with constant heat capacity has $\gamma = k = c_P/c_v$. Can you think of another example of a polytropic process?

▶ 2.8 OPEN-SYSTEM ENERGY BALANCES ON PROCESS EQUIPMENT

In this section, we will examine examples of how to apply the first law to common types of process equipment. These systems will be analyzed at steady-state, when the properties at any place in the system do not change with time. Most cases will consist of one stream in and one stream out, which will be labeled streams 1 and 2, respectively. For these cases, the mole balance becomes

$$\dot{n}_1 = \dot{n}_2 \qquad (2.93)$$

and the energy balance, Equation (2.44), becomes

$$0 = \dot{n}_1(h + e_K + e_P)_1 - \dot{n}_2(h + e_K + e_P)_2 + \dot{Q} + \dot{W}_s \qquad (2.94)$$

It is important to remember that the examples in this section are restricted to cases when steady-state can be applied. If we are interested in start-up or shutdown of these processes, or the case where there are fluctuations in feed or operating conditions, we must use the unsteady form of the energy balance.

Nozzles and Diffusers

These process devices convert between internal energy and kinetic energy by changing the cross-sectional area through which a fluid flows. In a nozzle the flow is constricted, increasing e_K. A diffuser increases the cross-sectional area to decrease the bulk flow velocity. An example of a process calculation through a diffuser follows.

▶ **EXAMPLE 2.15**
Diffuser Final
Temperature
Calculation

The intake to the engine of a jet airliner consists of a *diffuser* that must reduce the air velocity to zero so that it can enter the compressor. Consider a jet flying at a cruising speed of 350 m/s at an altitude of 10,000 m where the temperature is 10°C. What is the temperature of the air upon exiting the diffuser and entering the compressor?

SOLUTION A schematic diagram of the system, including the information that we know, is shown in Figure E2.15.

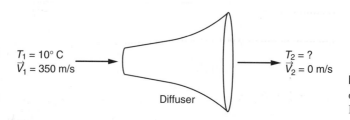

$T_1 = 10°$ C
$\vec{V}_1 = 350$ m/s

$T_2 = ?$
$\vec{V}_2 = 0$ m/s

Diffuser

Figure E2.15 Schematic of the diffuser in Example 2.15.

This steady-state process occurs in an open system with one stream in and one stream. In this case, we can write the first law using Equation (2.94):

$$0 = \dot{n}_1(h + e_K + \overset{0}{e_P})_1 - \dot{n}_2(h + \overset{0}{e_K} + \overset{0}{e_P})_2 + \overset{0}{Q} + \overset{0}{W_s} \tag{E2.15A}$$

where the negligible terms have been set to zero. Note that the reference state for potential energy is set at 10,000 m. A mole balance gives

$$\dot{n}_1 = \dot{n}_2$$

so that Equation (E2.15A) can be simplified to

$$e_K = \frac{1}{2}(MW)\vec{V}_1^2 = (h_2 - h_1) = \int_{T_1}^{T_2} c_{P,\text{air}}\,dT \tag{E2.15B}$$

Looking up the value for heat capacity for air in Appendix A.3, we get

$(h + e_k)_{in} = (h + e_k)_{out}$

$$A = 3.355, \qquad B = 0.575 \times 10^{-3}, \qquad \text{and } D = -0.016 \times 10^5$$

Using the definition of heat capacity, we get the following integral expression:

$$\int_{T_1}^{T_2} c_P\,dT = R\int_{283}^{T_2}[A + BT + DT^{-2}]dT = R\left[AT + \frac{B}{2}T^2 - \frac{D}{T}\right]_{283}^{T_2} \tag{E2.15C}$$

Using Equation (E2.15C), Equation (E2.15B) becomes

E_K calc
$$\frac{1}{2}(MW)\vec{V}_1^2 = R\left[A(T_2 - 283) + \frac{B}{2}(T_2^2 - 283^2) - D\left(\frac{1}{T_2} - \frac{1}{283}\right)\right]$$

We now have one equation with one unknown, T_2, which can be solved implicitly to give

$$T_2 = 344 \text{ [K]}$$

The temperature of the air increases because the kinetic energy of the inlet stream is being converted to internal energy.

Turbines and Pumps (or Compressors)

These processes involve the transfer of energy via shaft work. A turbine serves to generate power as a result of a fluid passing through a set of rotating blades. They are commonly found in power plants and used to produce energy locally as part of chemical plants. This process was described in Section 2.1 and illustrated in Example 2.4. Pumps and compressors use shaft work to achieve a desired outcome. The term *compressor* is reserved for gases, since they are compressible. Typically they are used to raise the pressure of a fluid. However, they can also be used to increase its potential energy, as illustrated in Example 2.16.

▶ **EXAMPLE 2.16**
Pump Power
Calculation

You wish to pump 0.001 m³/s of water from a well to your house on a mountain, 250 m above. Calculate the minimum power needed by the pump, neglecting the friction between the flowing water and the pipe.

SOLUTION Can you draw a schematic of this process? We need to write the energy balance. This system is at steady-state, with one stream in and one stream out. When working with macroscopic potential energy, it is often convenient to write the balance on a mass (rather than mole) basis. We will neglect the bulk kinetic energy of the water at the inlet and outlet and the heat loss through the pipe. Since there are no frictional losses, the exit temperature is the same as the inlet; therefore, their enthalpy is equal. Thus, the first law simplifies to

$$0 = \dot{m}_1(\cancel{h} + \cancel{\hat{e}_k} + \cancel{\hat{e}_P})_1 - \dot{m}_2(\cancel{h} + \cancel{\hat{e}_k} + \hat{e}_P)_2 + \cancel{\dot{Q}} + \dot{W}_s$$

or, rearranging,

$$\dot{W}_s = \dot{m}_2(\hat{e}_P)_2 = \frac{\dot{V}_2}{\hat{v}_2} g z_2$$

where $\dot{V}$ is the volumetric flow rate. Solving for shaft work gives

$$\dot{W}_s = \left[\frac{(0.001[\mathrm{m}^3/\mathrm{s}])}{(0.001[\mathrm{m}^3/\mathrm{kg}])} \right] (9.8[\mathrm{m/s}^2])(250\,[\mathrm{m}]) = 2.5\,[\mathrm{kW}]$$

Note that the sign for work is positive. Why? The actual work needed would be greater due to frictional losses.

Heat Exchangers

These processes are designed to "heat up" or "cool down" fluids through thermal contact with another fluid at a different temperature. The radiator in your automobile is an example of a heat exchanger. In this application, energy is removed from the engine block to keep it from overheating during combustion. The most common design is when the two streams are separated from each other by a wall through which energy, but not mass, can pass. A calculation on a system employing this design is given in Example 2.17. An alternative design allows the fluids to be mixed directly. An example of such an open feedwater heater is given in Example 2.18.

▶ **EXAMPLE 2.17**
Heat Exchanger
Flow Rate Calculation

You plan to use a heat exchanger to bring a stream of saturated liquid CO_2 at 0°C to a superheated vapor state at 10°C. The flow rate of CO_2 is 10 mol/min. The hot stream available to the heat exchanger is air at 50°C. The air must leave no cooler than 20°C. The enthalpy of vaporization for CO_2 at 0°C is given by

$$\Delta\hat{h}_{\mathrm{vap,\ CO_2}} = 236\,[\mathrm{kJ/kg}]\ \text{at}\ 0°C$$

What is the required flow rate of air?

SOLUTION First, let's draw a diagram of the system including the information that we know, shown in Figure E2.17A.

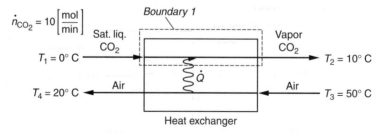

Figure E2.17A Schematic of heat exchanger with boundary 1 depicted.

There are several possible choices for our system boundary. We will choose a boundary around the CO_2 stream, labeled "boundary 1" in Figure E2.17A. In this case, the heat transferred from the air stream to evaporate and warm the CO_2 stream is labeled $\dot{Q}$. We next need to perform a first-law balance around boundary 1. The appropriate energy balance is for an open system at steady-state with one stream in and one stream out is

$$0 = \dot{n}_1(h + \cancelto{0}{e_K} + \cancelto{0}{e_P})_1 - \dot{n}_2(h + \cancelto{0}{e_K} + \cancelto{0}{e_P})_2 + \dot{Q} + \cancelto{0}{\dot{W}_s}$$

where we have set the bulk kinetic and potential energies and shaft work to zero. A mole balance yields

$$\dot{n}_1 = \dot{n}_2 = \dot{n}_{CO_2}$$

so the first-law balance on boundary 1 simplifies to

$$\dot{Q} = \dot{n}_{CO_2}(h_2 - h_1) \qquad (E2.17A)$$

To determine the change in enthalpy, we must account for the latent heat (vaporization) and the sensible heat of the CO_2 stream, that is:

$$(h_2 - h_1) = \Delta h_{vap,CO_2} + \int_{T_1}^{T_2} c_{P,CO_2}\,dT \qquad (E2.17B)$$

These can be found, in [J/mol], as follows. The latent heat is given by:

$$\Delta h_{vap,CO_2} = \left(236\left[\frac{kJ}{kg}\right]\right)\left(44\left[\frac{kg}{kmol}\right]\right) = 10{,}400\left[\frac{J}{mol}\right]$$

and the sensible heat is given by

$$\int_{T_1}^{T_2} c_{P,CO_2}\,dT = R\left[A(T_2 - T_1) + \frac{B}{2}(T_2^2 - T_1^2) - D\left(\frac{1}{T_2} - \frac{1}{T_1}\right)\right] = 353\left[\frac{J}{mol}\right]$$

where the numerical values for the heat capacity parameters, A, B, and D, are given in Appendix A.3. Thus, the energy transferred via heat to boundary 1 is

$$\dot{Q} = 100{,}753\,[J/min]$$

Now that we know the rate at which energy must be supplied to the CO_2 stream, we can find the flow required for the air. We do this by choosing a different system boundary in the heat exchanger, which is labeled boundary 2 in Figure E2.17B.

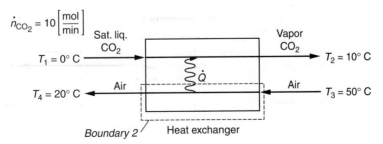

Figure E2.17B Schematic of heat exchanger with boundary 2 depicted.

A balance similar to that above yields

$$-\dot{Q} = \dot{n}_{air}(h_4 - h_3) \tag{E2.17C}$$

Note that we must be careful about signs! We have included a negative sign on $\dot{Q}$ since the heat that *enters* boundary 1 must *leave* boundary 2. Rearranging Equation (E2.17C) gives

$$\dot{n}_{air} = -\frac{\dot{Q}}{(h_4 - h_3)} = -\frac{\dot{Q}}{\displaystyle\int_{T_3}^{T_4} c_{P,air}dT} = \frac{\dot{Q}}{R\left[A(T_4 - T_3) + \frac{B}{2}(T_4^2 - T_3^2) - D\left(\frac{1}{T_4} - \frac{1}{T_3}\right)\right]}$$

Looking up values for the heat capacity parameters in Appendix A.3, we get

$$\dot{n}_{air} = \frac{(100,753\,[\text{J/min}])}{877\,[\text{J/mol}]} = 123\,[\text{mol/min}]$$

Alternatively, this problem could have been solved with a system boundary around the entire heat exchanger. In that case, a first-law balance would give:

$$0 = \dot{n}_{CO_2}(h_2 - h_1) + \dot{n}_{air}(h_4 - h_3)$$

which could then be solved for $\dot{n}_{air}$.

▶ **EXAMPLE 2.18**
Open Feedwater
Heater Calculation

Superheated water vapor at a pressure of 200 bar, a temperature of 500°C, and a flow rate of 10 kg/s is to be brought to a saturated vapor state at 100 bar in an open feedwater heater. This process is accomplished by mixing this stream with a stream of liquid water at 20°C and 100 bar. What flow rate is needed for the liquid stream?

SOLUTION   The first step is to draw a diagram of the system with the known information, as shown in Figure 2.18.

This example has two inlet streams in, so Equation (2.94) does not apply. If we assume that the rate of heat transfer and the bulk kinetic energy of the streams are negligible and the bulk potential energy and shaft work are set to zero, an energy balance reduces to:

$$0 = \dot{m}_1\hat{h}_1 + \dot{m}_2\hat{h}_2 - \dot{m}_3\hat{h}_3 \tag{E2.18A}$$

Similarly, a mass balance at steady-state gives

$$0 = \dot{m}_1 + \dot{m}_2 - \dot{m}_3 \tag{E2.18B}$$

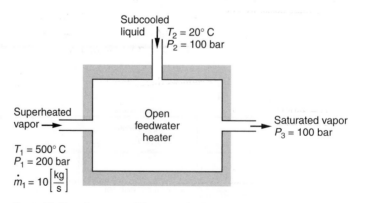

Figure E2.18   Schematic of the open feedwater heater in Example 2.15.

Rearranging Equation (E2.18B) and substituting into (E2.18A) gives

$$0 = \dot{m}_1 \hat{h}_1 + \dot{m}_2 \hat{h}_2 - (\dot{m}_1 + \dot{m}_2)\hat{h}_3 \qquad \text{(E2.18C)}$$

We can look up values for the enthalpies from the steam tables (Appendix B). For state 1, the superheated steam is at 500°C and 200 bar ($= 20$ MPa), so

$$\hat{h}_1 = 3238.2 \, [\text{kJ/kg}]$$

For state 2, we use subcooled liquid at 20°C and 100 bar:

$$\hat{h}_2 = 93.3 \, [\text{kJ/kg}]$$

and the saturated vapor at 100 bar (10 MPa) for state 3 is

$$\hat{h}_3 = 2724.7 \, [\text{kJ/kg}]$$

Finally, rearranging Equation (E2.18C) and plugging in values gives

$$\dot{m}_2 = \frac{\dot{m}_1(\hat{h}_1 - \hat{h}_3)}{(\hat{h}_3 - \hat{h}_2)} = 1.95 \left[\frac{\text{kg}}{\text{s}} \right]$$

Throttling Devices

These components are used to reduce the pressure of flowing streams. The pressure reduction can be accomplished by simply placing a restriction in the flow line, such as a partially opened valve or a porous plug. Since these devices occupy a relatively small volume, the residence time of the passing fluid is small. Hence, there is little energy loss by the transfer of heat. Consequently, we can neglect heat transfer. Since there is also no shaft work, the energy balance reduces to a very simple equation, as the next example illustrates.

▶ **EXAMPLE 2.19**
Throttling Device
Calculation

Water at 350°C flows into a porous plug from a 10-MPa line. It exits at 1 bar. What is the exit temperature?

SOLUTION First, let's draw a diagram of the system, as shown in see Figure E2.19. A steady-state energy balance with one stream in and one stream out is appropriate for this system. We will assume that the bulk kinetic energy of the stream is negligible and that the porous plug is sufficiently small as not to allow a significant rate of heat transfer. Rewriting Equation (2.94) on a mass basis, we get

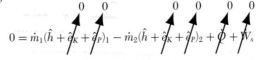

Hence, the energy balance reduces this system to an *isenthalpic* process:[12]

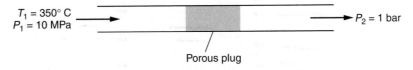

Figure E2.19 Schematic of the throttling device in Example 2.19.

[12] In general, the energy balance of throttling processes reduces to this simple form.

$$\hat{h}_1 = \hat{h}_2$$

Looking up the value for the inlet stream from Appendix B.4, we find

$$\hat{h}_1 = 2923.4 \ [\text{kJ/kg}]$$

Since the enthalpy of stream 2 equals that of stream 1, we have two intensive properties to constrain the exit state: $\hat{h}_2$ and P_2. To find the temperature of stream 2, we must use linear interpolation. Inspection of the steam tables shows that T_2 is somewhere between 200 and 250°C. The following are taken from the superheated steam table at 100 kPa.

$P = 100 \ \text{kPa}$

$T[°C]$	$\hat{h}[\text{kJ/kg}]$
200	2875.3
250	2974.3

Interpolation gives

$$T_2 = 200 + [\Delta T]\left[\frac{\hat{h}_2 - \hat{h}_{\text{at } 200}}{\hat{h}_{\text{at } 250} - \hat{h}_{\text{at } 200}}\right] = 200 + [50]\left[\frac{2923.4 - 2875.3}{2974.3 - 2875.3}\right] = 224[°C]$$

▶ 2.9 THERMODYNAMIC CYCLES AND THE CARNOT CYCLE

A thermodynamic cycle describes a set of processes through which a system returns to the same state that it was in initially. Since the system returns to its initial state after the cycle has been completed, all the properties have the same values they had originally. Typically cycles are used to produce power or provide refrigeration. The advantage of executing a thermodynamic cycle is that by having the system return to its initial state, we can repeat the cycle continuously. There are many different examples of thermodynamic cycles; in this section, we examine one such cycle—the Carnot cycle.[13] In Chapter 3, we will learn that a Carnot cycle represents the most efficient type of cycle we can possibly have.

Figure 2.17 shows an ideal gas in a piston–cylinder assembly undergoing a Carnot cycle. In this cycle, the gas goes through four *reversible* processes through which it returns to its initial state. Two processes occur isothermally, alternating with two adiabatic processes. These processes were analyzed, individually, in Section 2.7. Consider a gas that is initially in state 1 at a pressure P_1 and a temperature T_1 as shown at the top of Figure 2.17. The first step of a Carnot cycle is a reversible isothermal expansion, in which the gas is exposed to a hot reservoir at temperature, T_H; it gains energy via heat, Q_H, as indicated on the diagram. During this process, which takes the system from state 1 to state 2 (at P_2 and T_2), the pressure decreases while the temperature stays the same. The piston–cylinder assembly is then transferred into an adiabatic (well-insulated) environment and expanded further to state 3. In this step, both T and P decrease. In both expansion processes, work is done by the system on the surroundings; that is, we get useful work out. The system then undergoes two reversible compression processes. First, it is isothermally compressed by being placed in contact with a cold thermal reservoir at temperature T_C. The gas loses an amount of energy via heat, Q_C to the cold reservoir. This process takes the system to state 4 (P_4, T_4). The system returns to its initial state (state 1) through an adiabatic compression.

---

[13] This cycle was conceived in 1824 by Sadi Carnot, a French engineer, to explore the maximum possible efficiency that could be obtained by the steam engines of his time.

$W_{isotherm}$

$W = nRT \ln\left(\frac{P_2}{P_1}\right)$

$\Delta U = 0$

$Q = -W$

Adid

$nC_V \Delta T = \frac{nR}{k-1}\Delta T$

$Q_H > Q_C$

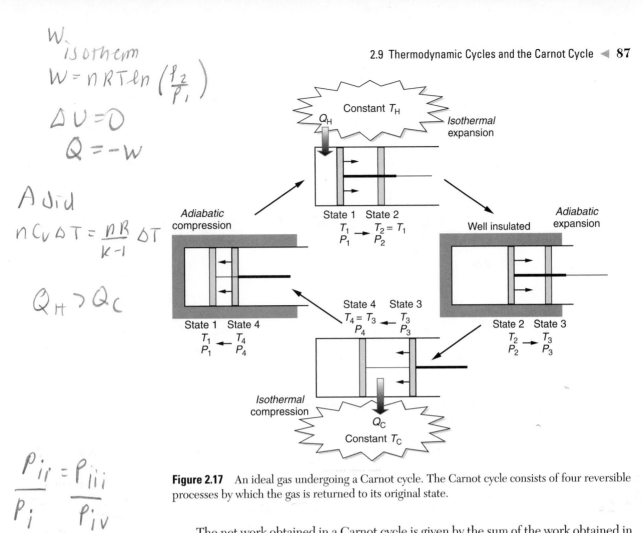

Figure 2.17 An ideal gas undergoing a Carnot cycle. The Carnot cycle consists of four reversible processes by which the gas is returned to its original state.

$\frac{P_{ii}}{P_i} = \frac{P_{iii}}{P_{iv}}$

$\frac{P_i}{P_{iv}} = \frac{P_{ii}}{P_{iii}}$

The net work obtained in a Carnot cycle is given by the sum of the work obtained in all four processes:

$$-W_{net} = |W_{12}| + |W_{23}| - |W_{34}| - |W_{41}| \qquad (2.95)$$

Since the overall effect of the power cycle is to deliver work from the system to the surroundings, the sign of W_{net} is negative. The subscript "ij" on the terms for work in Equation (2.95) refers to the work obtained in going from state i to state j. Absolute values are used to explicitly distinguish the steps where we get work out from those where we must put work in.

The net work obtained from a Carnot cycle can also be calculated by applying the first law to the entire cycle. Since the cycle returns the system to its original state, its internal energy must have the same value as at the start of the cycle. Thus

$$\Delta U_{cycle} = 0 = W_{net} + Q_{net} \qquad (2.96)$$

Comparing Equations (2.95) and (2.96), we see that

$W_{net} = -Q_H$

$$-W_{net} = Q_{net} = \overset{Q_H}{\cancel{Q_{12}}} + \overset{0}{\cancel{Q_{23}}} + \overset{Q_C}{\cancel{Q_{34}}} + \overset{0}{\cancel{Q_{41}}} = |Q_H| - |Q_C| \qquad (2.97)$$

We see that the net work obtained is the difference in heat absorbed from the hot reservoir, Q_H, and expelled to the cold reservoir, Q_C. An alternative way of schematically

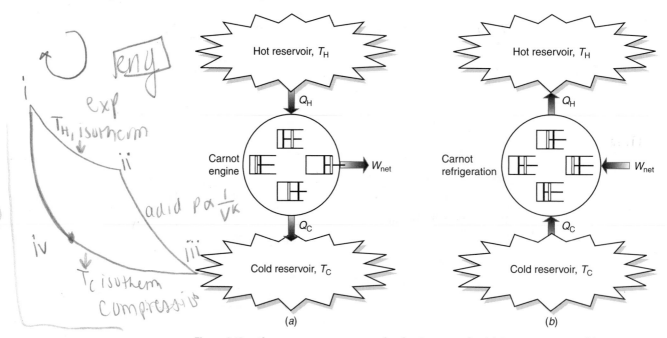

Figure 2.18 Alternative representation for the Carnot cycle. (*a*) Carnot "engine"; (*b*) Carnot refrigerator.

representing a Carnot cycle is shown in Figure 2.18*a*. This schematic gives an overview of the energy transferred between the Carnot engine and the surroundings. Inside the circle labeled "Carnot engine" are the four processes depicted in Figure 2.17. The efficiency, η, of the cycle is defined as the net work obtained divided by the heat absorbed from the hot reservoir:

$$\eta \equiv \frac{\text{net work}}{\text{heat absorbed from the hot reservoir}} = \frac{W_{net}}{Q_H} \tag{2.98}$$

For a given amount of energy available from the hot reservoir via Q_H, the greater the efficiency, the more work we obtain. For example, say the high temperature in the hot reservoir is obtained from the combustion of coal. A high efficiency means we can reduce the amount of coal we need to combust to produce a given amount of work.[14]

A **refrigeration cycle** allows us to cool a system so that we can store some ice cream and so on. In this case we want to expel heat from a cold reservoir. It takes work from the surroundings to accomplish this task. Thus, your freezer at home needs electricity to keep the ice cream cold. Figure 2.18*b* shows a schematic way of representing the energy transferred in a refrigeration cycle. We supply work to the cycle in order to absorb energy via heat Q_C *from* the cold reservoir. We then expel the energy via heat Q_H *to* the hot reservoir. Thus, the direction of heat transfer is opposite that of the power cycle depicted in Figure 2.18*a*. The effectiveness of a refrigeration cycle is measured by its coefficient of performance, COP, which is defined as follows:

$$\text{COP} = \frac{Q_C}{W_{net}} \tag{2.99}$$

[14] The steam engine was patented by James Watt in 1765. These first steam engines had efficiencies of only about 1%. Indeed, we can see there was much engineering that remained to be done!

We can see from Equation (2.99) that the higher the COP, the less work it takes to produce a desired level of cooling.

Can you draw the analogous cycle to Figure 2.17 that goes in the circle labeled "Carnot refrigerator"?

▷ **EXAMPLE 2.20**
Carnot Cycle
Efficiency

Consider 1 mole of an ideal gas in a piston–cylinder assembly. This gas undergoes a Carnot cycle, which is described below. The heat capacity is constant, $c_v = (3/2)R$.

   (i)   A reversible, isothermal expansion from 10 bar to 0.1 bar.
   (ii) A reversible, adiabatic expansion from 0.1 bar and 1000 K to 300 K.
   (iii) A reversible, isothermal compression at 300 K.
   (iv) A reversible, adiabatic compression from 300 K to 1000 K and 10 bar.

Perform the following analysis:

   (a) Calculate Q, W, and ΔU for each of the steps in the Carnot cycle.
   (b) Draw the cycle on a Pv diagram.
   (c) Calculate the efficiency of the cycle.
   (d) Compare η to $1 - (T_c/T_H)$.
   (e) If what is found in part (d) is true, in general, suggest two ways to make the above process more efficient.

SOLUTION

   (a) We will analyze each of the steps separately, with a little help from the results of Section 2.7. We label each state in a manner consistent with Figure 2.17.

   (i) The first process is a reversible, isothermal expansion at 1000 K from state 1 at 10 bar to state 2 at 0.1 bar. By definition, the internal energy change for an ideal gas at constant temperature is

$$\Delta U = 0$$

We can calculate the work using a result from Section 2.7:

$$W = \int \frac{nRT}{P}\,dP = nRT \ln \frac{P_2}{P_1} = -38,287 \ [\text{J}]$$

The negative sign indicates that the system is performing work on the surroundings (we are getting useful work out). To find the heat, we apply the first law:

$$Q_H = \Delta U - W = 38,287 \ [\text{J}]$$

   (ii) The second process is a reversible, adiabatic expansion from 0.1 bar and 1000 K to state 3 at 300 K. The pressure decreases during this process. By the definition of an *adiabatic* process:

$$Q = 0$$

At constant heat capacity, the change in internal energy becomes

$$\Delta U = nc_v(T_3 - T_2) = -8730 \ [\text{J}]$$

Applying the first law gives

$$W = \Delta U = -8730 \ [\text{J}]$$

$\overset{+}{W}_{net} = |Q_H| - |Q_c|$

$-(-) = \oplus$

(iii) The third process is a reversible, isothermal compression at 300 K. Again

$$\Delta U = 0$$

and

$$W = -\int P \mathrm{d}V = \int \frac{nRT}{P} \mathrm{d}P = nRT \ln \frac{P_4}{P_3} \qquad (E2.20A)$$

However, we now need to find P_3 and P_4. From Section 2.7, we know $PV^k = \text{const}$ for the polytropic, adiabatic processes (ii) and (iv). We first find k:

$$k = \frac{c_P}{c_v} = \frac{c_v + R}{c_v} = 1.67$$

Setting PV^k equal for states 2 and 3 gives:

$$PV^k = P_2 V_2^{1.67} = \frac{(nRT_2)^{1.67}}{P_2^{0.67}} = 7347 = \frac{(nRT_3)^{1.67}}{P_3^{0.67}}$$

Solving for P_3, we get:

$$P_3 = \left[\frac{(nRT_3)^{1.67}}{7,347} \right]^{1.5} = 0.0052 \text{ bar}$$

Similarly for P_4:

$$PV^k = P_1 V_1^{1.67} = \frac{(nRT_1)^{1.67}}{P_1^{0.67}} = 335 = \frac{(nRT_4)^{1.67}}{P_4^{0.67}}$$

and

$$P_4 = \left[\frac{(nRT_4)^{1.67}}{335} \right]^{1.5} = 0.52 \text{ bar}$$

Thus, the work given by Equation (E2.20A) is:

$$W = nRT \ln \frac{0.52 \text{ bar}}{0.0052 \text{ bar}} = 11,486 \, [\text{J}]$$

The work is positive for this compression process. From the first law,

$$Q_C = \Delta U - W = -11,486 \, [\text{J}]$$

(iv) The fourth process is a reversible, adiabatic compression from state 4 at 300 K and 0.52 bar back to state 1 at 1000 K and 10 bar (process 4 $\longrightarrow$ 1). After this process, the gas can repeat steps (i), (ii) ... Again, for this adiabatic compression:

$$Q = 0$$

At constant heat capacity, the change in internal energy becomes

$$\Delta U = nc_v(T_1 - T_4) = 8730 \, [\text{J}]$$

Applying the first law gives

$$W = \Delta U = 8730 \, [\text{J}]$$

TABLE E2.20A **Results of Calculations for Carnot Cycle in Example 2.20**

Process	ΔU [J]	W [J]	Q [J]
(i) State 1 to 2	0	−38,287	38,287
(ii) State 2 to 3	−8,730	−8,730	0
(iii) State 3 to 4	0	11,486	−11,486
(iv) State 4 to 1	8,730	8,730	0
Total	0	−26,800	26,800

TABLE E2.20B **T, P, and v for Carnot Cycle in Example 2.20**

State	T [K]	P [bar]	v [m³/mol]
1	1000	10	0.0083
2	1000	0.1	0.8314
3	300	0.0052	4.80
4	300	0.52	0.048

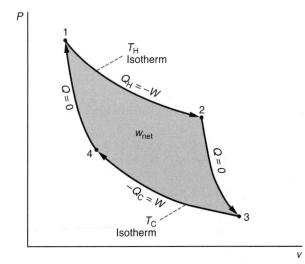

Figure E2.20 Pv diagram of a Carnot cycle. The shaded area represents the net work obtained from one cycle.

A summary of ΔU, W, and Q for the four processes and the totals for the cycle are presented in Table E2.20A. We get a net work of 26.8 kJ after one cycle.

(b) To sketch this process on a Pv diagram, we first calculate the molar volume at each state using the ideal gas law. The results are presented in Table E2.20B. A sketch (not to scale) of the Pv diagram is presented in Figure E2.20. The work for a reversible process is given by the area under the Pv curve; hence, the net work is given by the shaded area in the box in Figure E2.20. Isotherms, T_H and T_C, for processes (i) and (iii) are also labeled.

(c) The efficiency is given by Equation 2.98:

$$\eta \equiv \frac{\text{net work}}{\text{heat absorbed from the hot reservoir}} = \frac{26,800}{38,287} = 0.70 \qquad \text{(E2.20B)}$$

In practice, electrical power plants have efficiencies around 40%.

(d) Applying the relation in the problem statement, we get:

$$1 - \frac{T_C}{T_H} = 1 - \frac{300}{1000} = 0.7 \qquad \text{(E2.20C)}$$

Comparing the values of Equations (E2.20B) and (E2.20C), we get:

$$\eta = 1 - \frac{T_C}{T_H} \qquad \text{(E2.20D)}$$

(e) If Equation (E2.20D) holds, the process can be made more efficient by raising T_H or lowering T_C. Note that these options will push the isotherms depicted in Figure E2.20 up and down, respectively. Thus either raising T_H or lowering T_C will serve to make the shaded box, which represents net work, larger. We will learn in Chapter 3 that, indeed, Equation (E2.20D) is true in general. However, we can also reach this conclusion by realizing that the isotherms in Figure E2.20 are fixed on the Pv plane.

▶ 2.10 SUMMARY

The first law of thermodynamics states that the total energy in the universe is a constant. Energy balances have been developed for closed systems and for open systems. For example, the integral equation of the first law for a **closed system**, written in extensive form, is

$$\Delta U + \Delta E_K + \Delta E_P = Q + W \qquad \text{(2.28)}$$

For **open systems**, it is convenient to write the first law on a rate bases. The integral equation, in extensive form, is

$$\left(\frac{dU}{dt} + \frac{dE_K}{dt} + \frac{dE_P}{dt} \right)_{\text{sys}} = \sum_{\text{in}} \dot{n}_{\text{in}}(h + e_K + e_P)_{\text{in}} - \sum_{\text{out}} \dot{n}_{\text{out}}(h + e_K + e_P)_{\text{out}} + \dot{Q} + \dot{W}_s \qquad \text{(2.46)}$$

We have also developed these equations in intensive forms, on a mass and a molar basis, and for differential increments. Given a physical problem, we must determine which terms in these equations are important and which terms are negligible or zero. We must also identify whether the ideal gas model or property tables are needed to solve the problem.

We applied the first law to many engineering systems. Examples of closed systems included rigid tanks and adiabatic or isothermal expansion/compression in a piston–cylinder assembly. Steady-state open systems were illustrated by nozzles, diffusers, turbines, pumps, heat exchangers, and throttling devices. Transient open-system problems included filling or emptying of a tank, while the Carnot power and refrigeration cycles provided examples of thermodynamic cycles. However, you should understand the concepts well enough so that you are not restricted to the systems discussed above but rather are able to apply the first law to any system of interest.

A process is **reversible** if, after the process occurs, the system can be returned to its original state without any net effect on the surroundings. This result occurs only when the driving force is infinitesimally small. The reversible case represents the limit of what is possible in the real world— it gives us the most work we can get out or the least work we have to put in. Moreover, *only* in a reversible process can we substitute the system pressure for the external pressure in calculating Pv work. Real processes are **irreversible**. They have friction and are carried out with finite *driving forces*. In an irreversible process, if the system is returned to its original state, the surroundings must be altered. We can compare the amount of work required in an irreversible process to that of a reversible process by defining the **efficiency factor**, η. Our strategy for actual, irreversible processes is often solving a problem for the idealized, reversible process and then correcting for the irreversibilities using an assigned efficiency factor.

The thermodynamic property **enthalpy**, h, is defined as:

$$h \equiv u + Pv \qquad \text{(2.43)}$$

We first saw the utility of h in describing streams flowing into and out of open systems. In such cases, we must account for both the internal energy of the stream and the **flow work** associated with it entering or leaving the system. Enthalpy accounts for both these effects. Enthalpy also describes the combined effects of internal energy and Pv work for closed systems at **constant P**. Therefore, experiments conveniently done in closed systems at constant pressure are reported using enthalpy. For example, experimental data for the energetics associated with phase changes and chemical reaction are typically reported using enthalpy.

On a **molecular** level, internal energy encompasses the kinetic and potential energies of the molecules. Changes in internal energy present themselves in several **macroscopic** manifestations, including changes in temperature, changes in phase, and changes in chemical structure, that is, chemical reaction. A change in internal energy that is manifested as a change in temperature is often termed **sensible heat**. We can associate the increase in molecular kinetic energy and, therefore, in u with temperature to three possible modes in which the molecules can obtain kinetic energy: translational, vibrational, and rotational. Data for the change in energy due to sensible heat can be obtained using **heat capacity** or using property tables. The heat capacity for a species is often fit to a polynomial in temperature of the form

$$c_P = A + BT + CT^2 + DT^{-2} + ET^3 \qquad (2.59)$$

The parameters A, B, C, D, and E are reported for some ideal gases in Appendix A.2. Heat capacity parameters at constant pressure of some liquids and solids are also reported in this appendix. An ideal gas presents a special case: All its internal energy is realized as molecular kinetic energy—it is a function of T only. We refer to a change in energy that results in phase transformations as **latent heat**. This energy is associated with the different degrees of attraction between the molecules in the different phases. Data for latent heat are reported as enthalpies of vaporization, fusion, and sublimation. These values are typically reported at 1 bar; however, using these values, we can construct hypothetical paths to find the latent heats at any pressure. The internal energy changes associated with chemical reactions can be attributed to the energetic differences between the chemical bonds of the reactants and the products. The enthalpy of a given chemical reaction can be determined from reported enthalpies of formation.

▶ 2.11 PROBLEMS

2.1 Consider that towel you used to dry yourself with after your last shower. After being washed it must be dried so it can be used again. Estimate how much energy it takes to dry the towel after it has been washed. State all assumptions that you make. Now consider the dryer in your dwelling (or at the laundromat). If the towel is the *only* item placed in the dryer, estimate how much energy is used to actually dry the towel. What is the efficiency of the process? Can you suggest any ways to make it more efficient?

2.2 Consider a cup of cold water. Come up with and sketch as many ways as you can think of to raise the temperature of the water.

2.3 Consider the compression of a spring by placing a large mass on it. The degree to which the spring compresses is related to its spring constant, k. The force exerted by the spring on the mass is given by:

$$F = -kx$$

where x represents the distance the spring compresses from its relaxed position. Does the compression of a spring represent potential or internal energy?

2.4 Take a thick rubber band and expand it by stretching it. If you hold it to your lips, you will sense that it is hotter. However, we have seen that the temperature of a gas in a piston–cylinder assembly cools upon expansion. Explain these opposite results in the context of an energy balance.

2.5 If you sprinkle water on a very hot skillet, it will evaporate. However, you get the paradoxical result that at higher temperatures the water drops take longer to evaporate than at lower temperatures. Explain this result.

2.6 On a hot summer day, your roommate suggests that you open the refrigerator to cool off your apartment. Choosing the entire apartment as the system, perform a first-law analysis to decide whether this idea has merit.

2.7 You are making plans to stay warm in the winter. Due to your busy schedule, you are typically away from your house all day. You know it costs a lot to operate the electric heaters to keep your house warm. However, you have been told that it is more efficient to leave your house warm all day rather than turn off the heat during the day and reheat the house when you get home at night. You think that thermodynamics may be able to resolve this issue. Draw a schematic of the system, the surroundings, and the boundary. Illustrate the alternative processes. What is your choice to save power? Justify your answer.

2.8 Consider boiling water to make a pot of tea. Say it takes roughly 10 min to bring 1 L of H_2O taken from the tap at 25°C to boil. What is the total heat input, Q? What is the rate of heat input, $\dot{Q}$?

2.9 Consider a process that takes steam from an initial state where $P = 1$ bar and $T = 400$°C to a state where $P = 0.5$ bar and $T = 200$°C. Calculate the change in internal energy for this process using the following sources for data: (a) the steam tables; (b) ideal gas heat capacity.

2.10 Consider a piston–cylinder assembly that contains 2.5 L of an *ideal gas* at 30°C and 8 bar. The gas reversibly expands to 5 bar.

(a) Write an energy balance for this process (you may neglect changes in potential and kinetic energy).

(b) Suppose that the process is done isothermally. What is the change in internal energy, ΔU, for the process? What is the work done, W, during the process? What is the heat transferred, Q?

(c) If the process is done adiabatically (instead of isothermally), will the final temperature be greater than, equal to, or less than 30°C? Explain.

2.11 Five moles of nitrogen are expanded from an initial state of 3 bar and 88°C to a final state of 1 bar and 88°C. You may consider N_2 to behave as an ideal gas. Answer the following questions for each of the following *reversible* processes:

(a) The first process is isothermal expansion. (i) Draw the path on a Pv diagram and label it path A. (ii) Calculate the following: w, q, Δu, Δh.

(b) The second process is heating at constant pressure followed by cooling at constant volume. (i) Draw the path on the same Pv diagram label it path B. (ii) Calculate the following: w, q, Δu, Δh.

2.12 A 5-kg aluminum block sits in your lab, which is at 21°C. You wish to increase the temperature of the block to 50°C. How much heat in [J] must be supplied?

2.13 A piston–cylinder assembly contains 3 kg of steam at a pressure of 100 bar and a temperature of 400°C. It undergoes a process whereby it expands against a constant pressure of 20 bar, until the forces balance. During the process, the piston generates 748,740 [J] of work. Water is *not* an ideal gas under these conditions. Determine the final temperature in K and the heat transferred (in [J]) during the process.

2.14 Consider a piston–cylinder assembly that contains 1 mole of ideal gas, A. The system is well insulated. Its initial volume is 10 L and initial pressure, 2 bar. The gas is allowed to expand against a constant external pressure of 1 bar until it reaches mechanical equilibrium. Is this a reversible process? What is the final temperature of the system? How much work was obtained? For gas A: $c_V = (5/2) R$.

2.15 For the well-insulated piston–cylinder assembly containing 1 mole of ideal gas described in Problem 2.14, describe the process by which you can obtain the maximum work from the system. Calculate the value for the work. What is the final temperature? Why is T lower than that calculated in Problem 2.14?

2.16 The insulated vessel shown on the next page has two compartments separated by a membrane. On one side is 1 kg of steam at 400°C and 200 bar. The other side is evacuated. The membrane ruptures, filling the entire volume. The final pressure is 100 bar. Determine the final temperature of the steam and the volume of the vessel.

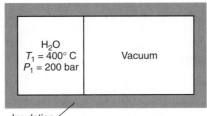

Insulation

2.17 A membrane divides a rigid, well-insulated 2-m³ tank into two equal parts. The left side contains an ideal gas [$c_P = 30$ J/(mol K)] at 10 bar and 300 K. The right side contains nothing; it is a vacuum. A small hole forms in the membrane, gas slowly leaks out from the left side, and eventually the temperature in the tank equalizes. What is the final temperature? What is the final pressure?

2.18 Consider a piston–cylinder assembly containing 10 kg of water. Initially the gas has a pressure of 20 bar and occupies a volume of 1.0 m³. Under these conditions, water does *not* behave as an ideal gas.

(a) The system now undergoes a **reversible** process in which it is compressed to 100 bar. The pressure–volume relationship is given by:

$$Pv^{1.5} = \text{constant}$$

What is the final temperature and internal energy of the system? Sketch this process on a Pv diagram. Draw the area that represents the work for this process. Calculate the work done during this process. How much heat was exchanged?

(b) Consider a different process by which the system gets to the *same final state as in part (a)*. In this case, a large block is placed on the piston, forcing it to compress. Sketch this process on a Pv diagram. Draw the area that represents the work for this process. Calculate the work done during this process. How much heat was exchanged?

(c) Can you think of a process by which the system could go from the initial state to the final state with no *net* heat exchange with the surroundings? Describe such a process, putting in numerical values wherever possible, and sketch it on a Pv diagram.

2.19 Consider the piston–cylinder assembly containing a pure gas shown below. The initial volume of the gas is 0.05 m³, the initial pressure is 1 bar, and the cross-sectional area of the piston is 0.1 m². Initially the spring exerts no force on the very *thin* piston.

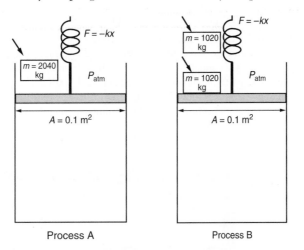

Process A   Process B

(a) A block of mass $m = 2040$ kg is then placed on the piston. The final volume is 0.03 m³, and the final pressure is 2×10^5 Pa. You may assume that the force exerted by the spring on the piston varies linearly with x and that the spring is very "tight," so that the volume of the gas is never less than the final volume. Precisely draw the process on a PV diagram, labeling it "process A." Draw the area that represents the work for this process. What is the value of the work? [*Hint:* First find the value for the spring constant, k.]

(b) Consider instead a process in which a block of mass 1020 kg is placed on the piston in the *original initial state*, and after the gas inside has been compressed another block of mass 1020 kg is placed on the piston. Draw the process on the same PV diagram, labeling it "process B." Draw the area that represents the work for this process. What is the value of the work?

(c) Describe the process in which it will take the least amount of work to compress the piston. Draw the process on the same PV diagram, labeling it "process C." Draw the area that represents the work for this process. What is the value of the work?

2.20 A rigid tank of volume 0.5 m³ is connected to a piston–cylinder assembly by a valve, as shown below. Both vessels contain pure water. They are immersed in a constant-temperature bath at 200°C and 600 kPa. Consider the tank and the piston–cylinder assembly as the system and the constant-temperature bath as the surroundings. *Initially* the valve is closed and both units are in equilibrium with the surroundings (the bath). The rigid tank contains saturated water with a quality of 95% (i.e., 95% of the mass of water is vapor). The piston–cylinder assembly initially has a volume of 0.1 m³.

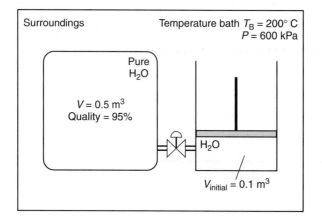

The valve is then opened. The water flows into the piston–cylinder until equilibrium is obtained. For this process, determine the work done, W, by the piston; the change in internal energy, ΔU, of the system; and the heat transferred, Q.

2.21 Consider the well-insulated, rigid container shown below. Two compartments, A and B, contain H₂O and are separated by a thin metallic piston. Side A is 10 cm long. Side B is 50 cm long. The cross-sectional area is 0.1 m². The left compartment is initially at 20 bar and 250°C; the right compartment is initially at 10 bar and 700°C. The piston is initially held in place by a latch. The latch is removed, and the piston moves until the pressure and temperature in the two compartments become equal. Determine the final temperature, the final pressure, and the distance that the piston moved.

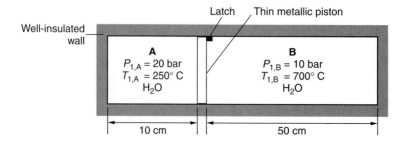

2.22 When you open a can of soda (or beer), compressed CO_2 expands irreversibly against the atmosphere as it bubbles up through the drink. Assume that the process is adiabatic and that the CO_2 has an initial pressure of 3 bar. Take CO_2 to be an ideal gas, with a constant heat capacity of $c_P = 37$ [J/(mol K)]. What is the final temperature of the CO_2 after it has reached atmospheric pressure?

2.23 Find the Pv work required to blow up a balloon to a diameter of 1 ft. Does the value you calculate account for all the work that is required? Explain.

2.24 You have a rigid container of volume 0.01 m³ that you wish to contain water at its critical point. To accomplish this task, you start with pure saturated water at 1 bar and heat it.
(a) How much water do you need?
(b) What is the quality of water that you need to begin this process?
(c) How much heat in [J] will it take?

2.25 In an attempt to save money to compensate for a budget shortfall, it has been determined that the steam in ChE Hall will be shutdown at 6:00 P.M. and turned back on at 6:00 A.M., much to the chagrin of a busy class of chemical engineers who have an outrageously long computer project due the following day. The circulation fans will stay on, however, leaving the entire building at approximately the same temperature. Well, things aren't going as quickly as you might have hoped and it is getting cold in the computer lab. You look at your watch; it is already 10:00 P.M. and the temperature has already fallen halfway from the comfortable 22°C it was maintained at during the day to the 2°C of the outside temperature (i.e., the temperature is 12°C at 10:00 P.M.). Seeing as you will probably be there all night and you need a diversion, you decide to estimate what the temperature will be at 6:00 A.M. You may assume the outside temperature stays constant at 2°C from 6:00 P.M. to 6:00 A.M. You may take the heat transfer to be given by the following expression:

$$q = h(T - T_{\text{surr}})$$

where h is a constant.
(a) Plot what you think the temperature will be as a function of time. Explain.
(b) Calculate the temperature at 6:00 A.M.

2.26 Explain why ice often forms on the valve of a tank of compressed gas (high pressure) when it is opened to the atmosphere and the gas escapes. Where does the H_2O come from?

2.27 Steam at 6 MPa, 400°C is flowing in a pipe. Connected to this pipe through a valve is a tank of volume 0.4 m³. This tank initially contains saturated water vapor at 1 MPa. The valve is opened and the tank fills with steam until the pressure is 6 MPa, and then the valve is closed. The process takes place adiabatically. Determine the temperature in the tank right as the valve is closed.

2.28 Consider filling a cylinder of compressed argon from a high-pressure supply line as shown below. Before filling, the cylinder contains 10 bar of argon at room temperature. The valve is then opened, exposing the tank to a 50 bar line at room temperature until the pressure of the cylinder

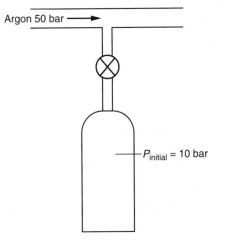

Argon 50 bar

$P_{\text{initial}} = 10$ bar

reaches 50 bar. The valve is then closed. For argon take $c_P = (5/2) R$ and the molecular weight to be 40 kg/kmol. You may use the ideal gas model.

(a) What is the temperature right after the valve is closed?

(b) If the cylinder sits in storage for a long time, how much heat is transferred (in kJ/kg)?

(c) What is the pressure of the cylinder when it is shipped (after it was stored for a long time)?

2.29 A well-insulated piston–cylinder assembly is connected to a CO_2 supply line by a valve, as shown below. Initially there is no CO_2 in the piston–cylinder assembly. The valve is then opened, and CO_2 flows in. What is the temperature of the CO_2 when the volume inside the piston–cylinder assembly reaches 0.1 m^3? How much CO_2 has entered the tank? Take CO_2 to be an ideal gas, with a constant heat capacity of $c_P = 37$ [J/(mol K)].

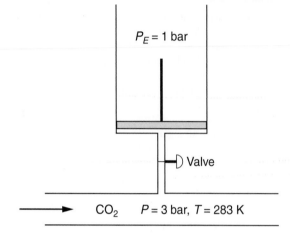

2.30 A rigid tank has a volume of 0.01 m^3. It initially contains saturated water at a temperature of 200°C and a quality of 0.4. The top of the tank contains a pressure-regulating valve that maintains the vapor at constant pressure. This system undergoes a process whereby it is heated until all the liquid vaporizes. How much heat (in [kJ]) is required? You may assume that there is no pressure drop in the exit line.

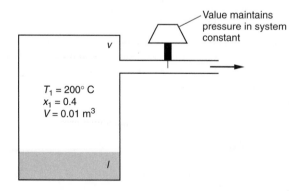

2.31 You wish to measure the temperature and pressure of steam flowing in a pipe. To do this task, you connect a well-insulated tank of volume 0.4 m^3 to this pipe through a valve. This tank initially is at vacuum. The valve is opened, and the tank fills with steam until the pressure is 9 MPa. At this point the pressure of the pipe and tank are equal, and no more steam flows through the valve. The valve is then closed. The temperature right after the valve is closed is measured to be 800°C. The process takes place adiabatically. Determine the temperature (in [K]) of the steam flowing in the pipe. You may assume the steam in the pipe stays at the same temperature and pressure throughout this process.

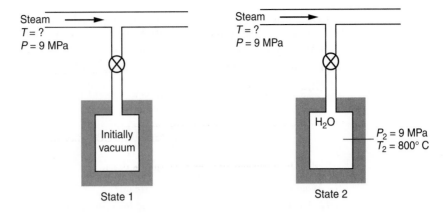

State 1   State 2

2.32 An electric generator coupled to a waterfall produces an average electric power output of 5 kW. The power is used to charge a storage battery. Heat transfer from the battery occurs at a constant rate of 1 kW.

(a) Determine the total amount of energy stored in the battery, (in [kJ]) in 10 hours of operation.

(b) If the water flow rate is 200 kg/s and conversion of kinetic energy to electric energy has an efficiency of 50%, what is the average velocity (in m/s) of the water? Where does the rest of the energy go?

2.33 Air enters a well-insulated turbine operating at steady, state with negligible velocity at 4 MPa, 300°C. The air expands to an exit pressure of 100 kPa. The exit velocity and temperature are 90 m/s and 100°C, respectively. The diameter of the exit is 0.6 m. Determine the power developed by the turbine (in kW). You may assume air behaves like an ideal gas throughout the process.

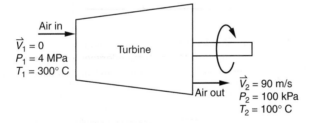

2.34 Ethylene (C_2H_4) at 100°C passes through a heater and emerges at 200°C. Compute the heat supplied into the unit per mole of ethylene that passes through. You may assume ideal gas behavior.

2.35 Propane at 350°C and 600 cm³/mol is expanded in a turbine. The exhaust temperature is 308°C, and the exhaust pressure is atmospheric. What is the work obtained? You may assume ideal gas behavior.

2.36 Consider 20 mol/s of CO flowing through a heat exchanger at 100°C and 0.5 bar.

(a) At what rate must heat be added (in kW) to raise the stream to 500°C?

(b) Consider the same molar flow rate of n-hexane and the same rate of heat input calculated in part (a). Without doing any calculations, explain whether you expect the final temperature of n-hexane to be greater or less than 500°C.

2.37 Steam enters a well-insulated nozzle at 10 bar and 200°C. It exits as saturated vapor at 100 kPa. The mass flow rate is 1 kg/s. What is the steady-state exit velocity? What is the outlet cross-sectional area?

2.38 Propane enters a nozzle at 5 bar and 200°C. It exits at a velocity of 500 m/s. At steady-state, what is the exit temperature? Assume ideal gas conditions.

2.39 Consider a diffuser operating at steady-state with an outlet twice the area of the inlet. Air flows in with a velocity of 300 m/s, a pressure of 1 bar, and a temperature of 70°C. The outlet is at 1.5 bar. What is the exit temperature? What is the exit velocity?

2.40 A stream of air is compressed in an adiabatic, steady-state flow process at 50 mol/s. The inlet is at 300 K and 1 bar. The outlet is at 10 bar. Estimate the minimum power that the compressor uses. You may assume air behaves as an ideal gas.

2.41 Sulfur dioxide (SO_2) with a volumetric flow rate of 5000 cm^3/s at 1 bar and 100°C is mixed with a second SO_2 stream flowing at 2500 cm^3/s at 2 bar and 20°C. The process occurs at steady-state. You may assume ideal gas behavior. For SO_2, take the heat capacity at constant pressure to be

$$\frac{c_P}{R} = 3.267 + 5.324 \times 10^{-3}T$$

(a) What is the molar flow rate of the exit stream?

(b) What is the temperature of the exit stream?

2.42 A mass flow controller (MFC) is used to accurately control the molar flow rate of gases into a system. A schematic of an MFC is shown below. It consists of a main tube and a sensor tube, to which a constant fraction of the flowing gas is diverted. In the sensor tube, a constant amount of heat is provided to the heating coil. The temperature difference is measured by upstream and downstream temperature sensors, as shown. A control valve can then be opened or closed to ensure the desired flow rate.

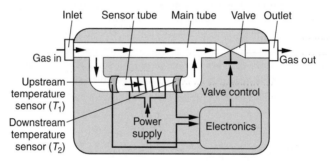

(a) Flow rates are typically reported as *standard cubic centimeters per minute* (SCCM), which represents the volume the gas would have at a "standard" pressure of 1.0135 bar and a "standard" temperature of 0°C. What molar flow rate (in [mol/s]), does 1 SCCM correspond to?

(b) Consider controlling the flow of N_2. Develop an equation for the molar flow rate of N_2 in SCCM in terms of the measured temperature difference, the heat input to the heating coil, and the fraction of gas diverted to the sensor tube. State the assumptions that you make.

(c) Instead of recalibrating the MFC for any gas that is used, conversion factors allow you to correct the MFC readout for different gases. Consider controlling SiH_4 instead of N_2. What conversion factor must be applied?

2.43 (a) What requires more heat input: to raise the temperature of a gas in a constant-pressure cylinder or in a constant-volume bomb? Explain.

(b) Explain why you feel less comfortable on a hot summer day when the (relative) humidity is higher even though the temperature is the same.

2.44 Using data from the steam tables, come up with an expression for the ideal gas heat capacity of H_2O in the form

$$c_v = A + BT$$

Compare your answer to the values in Appendix A.2.

2.45 Steam at 8 MPa and 500°C flows through a throttling device, where it exits at 100 kPa. What is the exit temperature?

2.46 A monatomic, ideal gas expands reversibly from 500 kPa and 300 K to 100 kPa in a piston–cylinder assembly. Calculate the work done if this process is (a) isothermal; (b) adiabatic.

2.47 Compare the change in internal energy for the following two processes:

(a) Water is heated from its freezing point to its boiling point at 1 atm.

(b) Saturated liquid water is vaporized at 1 atm.

Repeat for the change in enthalpy.

2.48 Calculate the values of the heat capacity of Ar, O_2, and NH_3 at 300 K. Account for their relative magnitudes in terms of the three modes (translational, rotational, and vibrational) in which molecules can exhibit kinetic energy.

2.49 Two kilograms of water, initially saturated liquid at 10 kPa, are heated to saturated vapor while the pressure is maintained constant. Determine the work and the heat transfer for the process, each in kJ.

2.50 A rigid vessel contains 5 kg of saturated liquid water and 0.5 kg of saturated vapor at 10 kPa. What amount of heat must be transferred to the system to evaporate all the liquid?

2.51 Consider the cooling of a glass of tap water by the addition of ice. The glass contains 400 ml of tap water at room temperature, to which 100 gm of ice is added. Assume the glass is adiabatic and that thermal equilibrium is obtained. The ice is originally at $-10°C$ when removed from the freezer and put in the glass. For ice, $\Delta h_{fus} = -6.0 \, [kJ/mol]$.

(a) What is the final temperature?

(b) What % of the cooling is achieved by latent heat (the melting of ice)?

2.52 One mole of saturated liquid propane and 1 mole of saturated vapor are contained in a rigid container at $0°C$ and 4.68 bar. How much heat must be supplied to evaporate all of the propane. At $0°C$,

$$\Delta h_{vap} = 16.66 \, [kJ/mol]$$

You may treat propane as an ideal gas.

2.53 Calculate the enthalpy of reaction at 298 K for the following reactions:

(a) $CH_4(g) + 2O_2(g) \Leftrightarrow CO_2(g) + 2H_2O(g)$

(b) $CH_4(g) + 2O_2(g) \Leftrightarrow CO_2(g) + 2H_2O(l)$

(c) $CH_4(g) + H_2O(g) \Leftrightarrow CO(g) + 3H_2(g)$

(d) $CO(g) + H_2O(g) \Leftrightarrow CO_2(g) + H_2(g)$

(e) $4NH_3(g) + 5O_2(g) \Leftrightarrow 4NO(g) + 6H_2O(g)$

2.54 Calculate the adiabatic flame temperature of acetylene gas at a pressure of 1 bar under the following conditions. The reactants are initially at 298 K. Assume that the acetylene reacts completely to form CO_2 and H_2O:

(a) It is combusted in a stoichiometric mixture of O_2.

(b) It is combusted in a stoichiometric mixture of air.

(c) It is combusted with twice the amount of air as needed stoichiometrically.

2.55 Calculate the adiabatic flame temperature of the following species in a stoichiometric mixture of air at a pressure of 1 bar The reactants are initially at 298 K. Assume that they react completely to form CO_2 and H_2O. Compare the answers and comment.

(a) propane

(b) butane

(c) pentane

2.56 In an experiment, methane is burned with the theoretically required amount of oxygen for complete combustion. Because of faulty operation of the adiabatic burner, the reaction does not proceed to completion. You may assume all of the methane that does react forms H_2O and CO_2. If the reactants are fed into the reactor at $25°C$ and 1 bar and the exiting gases leave at $1000°C$ and 1 bar, determine the percentage of methane that passes through the reactor unburned.

2.57 The following data are from a system that undergoes a thermodynamic cycle among three states. Fill in the missing values for ΔU, Q, and W. Is this a power cycle or a refrigeration cycle?

Process	ΔU [kJ]	W [kJ]	Q [kJ]
State 1 to 2			350
State 2 to 3	800	800	
State 3 to 1	−750		−500

2.58 One mole of air undergoes a Carnot cycle. The hot reservoir is at 800°C and the cold reservoir is at 25°C. The pressure ranges between 0.2 bar and 60 bar. Determine the net work produced and the efficiency of the cycle.

2.59 One mole of air undergoes a Carnot refrigeration cycle. The hot reservoir is at 25°C and the cold reservoir is at −15°C. The pressure ranges between 0.2 bar and 1 bar. Determine the coefficient of performance.

2.60 A Rankine cycle is shown below. This cycle is used to generate power with water as the working fluid. It consists of four unit processes in a thermodynamic cycle: a turbine, a condenser, a compressor, and a boiler. The state of each stream is labeled on the plot and defined in the table below. The mass flow rate of water is 100 kg/s. Kinetic and potential energy effects are negligible. Answer the following questions:

State	1	2	3	4
T [°C]	520			80
P [bar]	100	0.075	0.075	100
Quality		90% sat. vapor	sat. liquid	

Rankine cycle

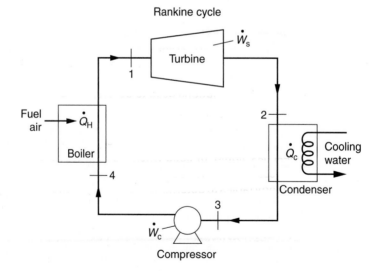

(a) Sketch all four processes on a Pv diagram. Include the vapor–liquid dome.

(b) From the plot above, explain why the net power is negative.

(c) Determine the heat-transfer rates in the boiler, $\dot{Q}_H$, and the condenser, $\dot{Q}_C$.

(d) Determine the *net* power developed in the cycle.

(e) What is the thermal efficiency, η of the cycle?

$$\eta \equiv \frac{\text{net power delivered by the plant}}{\text{rate of heat transfer in the boiler}}$$

CHAPTER 3

Entropy and the Second Law of Thermodynamics

Learning Objectives

To demonstrate mastery of the material in Chapter 3, you should be able to:

▶ State and illustrate by example the second law of thermodynamics, that is, entropy analysis and its basic concepts, including directionality, reversibility and irreversibility, and efficiency.

▶ Write the integral and differential forms of the second law for (1) closed systems and (2) open systems, under steady-state and transient (uniform-state) conditions. Convert these equations between intensive and extensive forms and between mass-based and molar forms.

▶ Apply the second law of thermodynamics to identify, formulate, and solve engineering problems, including for the following systems: rigid tank, adiabatic or isothermal expansion/compression in a piston–cylinder assembly, nozzle, diffuser, turbine, pump, heat exchanger, throttling device, filling or emptying of a tank, and vapor compression power and refrigeration cycles.

▶ State the assumptions used to develop the Bernoulli equation, and apply this equation when appropriate to solve engineering problems.

▶ Develop a hypothetical reversible path to calculate the entropy change between any two states. Write the expression for the entropy change of an ideal gas, liquid, or solid when the heat capacity is known. Use these expressions to calculate the entropy difference between two states. Use property tables to determine the entropy change between two states. Calculate entropy changes for species undergoing phase transformation or chemical reaction.

▶ Describe how a vapor-compression power cycle and a refrigeration cycle work. Identify the key issues in selecting the working fluid. Solve for the net power obtained and the efficiency of a reversible power cycle and the coefficient of performance of a reversible refrigeration cycle. Correct those values for real units by using isentropic efficiencies.

▶ Describe the molecular basis for entropy, including its relation to spatial and energetic configurations. Relate macroscopic directional processes to molecular mixing.

▶3.1 DIRECTIONALITY OF PROCESSES/SPONTANEITY

Thermodynamics rests largely on the consolidation of many observations of nature into two fundamental postulates or laws. Chapter 2 addressed the first law—the energy of the universe is conserved. We cannot *prove* this statement, but based on over a hundred years of observation, we believe it to be true. In order to use this law quantitatively—that is, to make numerical predictions about a system—we cast it in terms of a thermodynamic property: internal energy, u. Likewise, the second law summarizes another set of observations about nature. We will see that to quantify the second law, we need to use a different thermodynamic property: entropy, s. Like internal energy, entropy is a conceptual property that allows us to quantify a "law" of nature and solve engineering problems. This chapter examines the observations on which the second law is based; explores how the property s quantifies these observations; illustrates ways we can use the second law to make numerical predictions about closed systems, open systems, and thermodynamic cycles; and discusses the molecular basis of entropy.

We first examine several examples of the type of observation on which the second law is based—the directionality of processes. These examples are taken from scenarios with which you are probably familiar. First, consider the tank of compressed gas shown in Figure 3.1a as the system. It is initially in state 1. The surroundings are at atmospheric temperature and pressure. When the valve is opened, gas will spontaneously flow from the system to the surroundings until the pressure in the tank reaches 1 atm. With time, the system reaches state 2, where the pressure inside the cylinder is equal to the pressure outside. During this process, energy is conserved. Hence, if we consider the first law, the energy of the universe is identical in each of the two states. However, there is clearly a **direction** in which this process occurs spontaneously. It would be absurd to declare that the gas will spontaneously flow from the atmosphere (state 2) into the cylinder (state 1). Since the driving force that pushes the system from state 1 to state 2 is pressure, we label this as an example of mechanical directionality.

Similarly, there is a clear directionality if we place a high-temperature block in a room at 25°C, as illustrated in Figure 3.1b. With time, the block cools to room temperature (state 2). Again, the system will spontaneously go from state 1 to state 2, but it will not go spontaneously from state 2 to state 1. Again, energy is conserved. Since a temperature gradient provides the driving force for this process, we label this example "thermal directionality". Figure 3.1c illustrates a system in which two different gases, gas A and gas B, are initially separated by a diaphragm in state 1. Upon removing the diaphragm, the gases will eventually mix completely, obtaining state 2. Again, we do not observe a mixed gas to spontaneously separate into pure species. This provides an example of chemical directionality.

In all three examples, the first law says nothing about which direction the system spontaneously will go. However, there is a clear directionality associated with each process. Assigning a direction to these processes is easy, since we have experience with them. In other cases, the direction in which a process will go may not be so obvious. For example, consider the following scenario: An excess amount of zinc has been found in the groundwater near the former site of a metal-plating plant. You have been tasked with developing a process to clean up the groundwater. It has been suggested to form a zinc precipitate through reaction with lime, $Ca(OH)_2$. Is this a reasonable approach? How much Zn can you expect to remove? Another way to phrase these questions is, "To what extent can zinc react with lime so that it does not violate my experience about the directionality of nature?" As we will soon see, the second law of thermodynamics provides a quantitative statement about the directionality of nature and allows us to predict

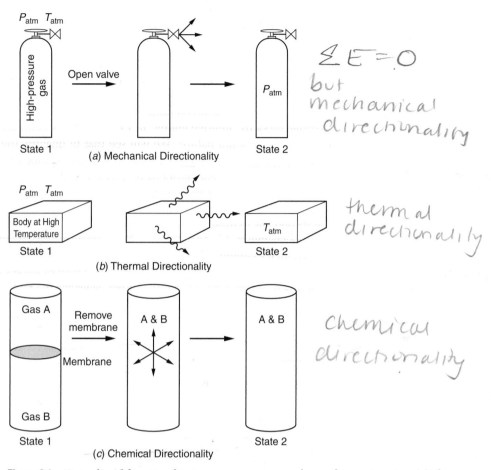

(handwritten annotations: $\Sigma E = 0$ but mechanical directionality; thermal directionality; chemical directionality)

Figure 3.1 Examples of directionality in common processes observed in engineering. *(a)* The expansion of a compressed gas illustrates mechanical directionality. *(b)* The cooling of a hot block exhibits thermal directionality. *(c)* The mixing of gas A and gas B represents chemical directionality.

which way a process, such as the one described above, will spontaneously go. Thus, with knowledge of the second law of thermodynamics, you can evaluate whether the proposed solution to clean up groundwater is possible.

▶**EXERCISE**

In 10 minutes, come up with as many examples as you can that illustrate the directionality of nature.

▶3.2 REVERSIBLE AND IRREVERSIBLE PROCESSES (REVISITED) AND THEIR RELATIONSHIP TO DIRECTIONALITY

From one perspective, the second law of thermodynamics addresses directionality. From another, it is about the reversibility and irreversibility of processes. In this section, we review examples of when mechanically and thermally driven processes are reversible and when they are irreversible.

Process I: Mechanical Process

Let's review the mechanical process from which we learned about reversibility and irreversibility. In Section 2.3, we considered a piston–cylinder assembly that underwent an isothermal expansion/compression, as shown in Figure 3.2a. When we remove the 1020-kg block, as depicted on the left, the piston expands irreversibly. Likewise, when we replace the block on the piston, it compresses irreversibly. In this case, the driving force for change is a pressure difference. The processes are illustrated on the Pv curve at the bottom. Notice that the irreversible processes have a definite directionality. The arrows that describe the expansion process do not overlap with those that describe the compression process. As we saw, to compress the assembly requires more work than we got out of expanding it and is represented by a very different directional process (with different arrows and different shaded area on the Pv curve). The irreversible expansion and compression processes are distinct and different.

The reversible process is executed by changing the force that acts on the piston by differential amounts, as shown to the right. In this case, the expansion and the compression curves on the Pv diagram meet. The reversible process can be reversed at any point in the process and, therefore, does not have a directionality like those real processes illustrated in Section 3.1. Remember, a reversible process is an idealization and represents the limiting case where a process is perfectly executed. In terms of work, a reversible

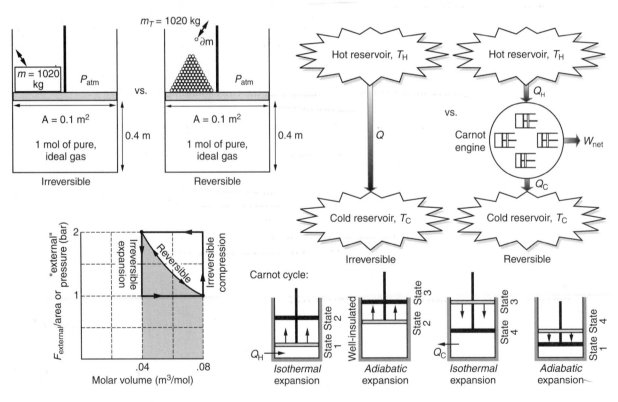

(a) Process I: Isothermal Expansion/Compression   (b) Process II: Thermal Heat Engine

Figure 3.2 Illustration of irreversible and reversible processes. (a) Mechanical process of isothermal expansion/compression. (b) Thermal process in which work can be obtained from a Carnot engine.

best we can do

Rev process
exp: upper lim of
work we
get out
comp: lower limit
we must put in

Carnot rev →
temp diff
drives cycle

Carnot →
max heat
achieved

How to exp +
comp follow
same paths?

process represents the upper limit of the v...
and the lower limit for the work we must p...
represents the best we can do and serves as...
real, irreversible processes. Additionally, f...
expansion is exactly the same as that requ...
expansion and compression processes foll...

expansion and

Process II: Thermal Heat Engine

In Section 2.9, we learned how to execu...
analogy to the mechanical process desc...
for change is a temperature difference. ...
Figure 3.2*b*. We know that energy will spo...
cold body in the form of heat. In the irreversible process illustrated on the left-hand...
an amount of heat, Q, flows spontaneously from the hot reservoir to the cold reservoir,
and we get no work out. Again, this process is directional in that energy will not flow
spontaneously from the cold reservoir to the hot reservoir. If instead we insert a Carnot
engine between these two reservoirs, we can use a reversible transport of energy via
heat to obtain work. Again, the reversible process (the Carnot heat engine) represents
the maximum work we can get out of the system. The process can also be reversed. By
reversing the work and putting it into the Carnot cycle, we transfer the heat from the
cold body to the hot body in the form of a Carnot refrigerator.

In an intermediate case to those discussed above, we place an irreversible (real) heat
engine between these two reservoirs. In this case, the work obtained would be less than
that provided by the reversible Carnot cycle. Similarly, a real refrigeration cycle would
not represent the reverse of the real heat engine but rather would require more work to
get a desired level of refrigeration than the corresponding Carnot refrigerator.

Let's summarize this discussion:

- Irreversible processes are distinct and show directionality.
- Reversible processes do not show directionality and represent the best that we
  can do (e.g., the maximum work we get out or minimum work we put in).

In each example in this section, we can see the *driving force* for an irreversible process
and the *direction* that each process wants to go. In each example, we see how to make
the process reversible so it produces the maximum work (or consumes the minimum).
In more complex systems, these effects may not be as obvious. In such cases, we look for
answers using the second law of thermodynamics and the related property, entropy.

▶ 3.3 ENTROPY, THE THERMODYNAMIC PROPERTY

We would like to generalize our experience with the directionality of nature (and the
limits of reversibility) into a quantitative statement that allows us to do calculations and
draw conclusions about what is possible, what is not possible, and whether we are close
to or far away from the idealization represented by a reversible process. Indeed, it
would be nice if we had a thermodynamic property (i.e., a state function) which would
help us to quantify directionality, just as internal energy, u, was central in quantifying the
conservation of energy (the first law of thermodynamics). It turns out the thermodynamic
property entropy, s, allows us to accomplish this goal.

Three historical milestones have established three corresponding distinct contextual
paradigms for entropy. First, the property entropy was conceived by Rudolph Clausius

U → quantifies
conservation of
energy

in 1865, based largely on Sadi Carnot's work on maximizing the efficiency of cyclic processes. He coined the term *entropy* from the Greek word meaning "transformation," deliberately choosing a word that sounded like energy to emphasize its equal importance. Clausius related entropy to reversible heat transfer and temperature. This definition is the basis for entropy in classical thermodynamics and will be presented below.

In 1877, Ludwig Boltzmann conceptualized entropy in terms of the behavior of molecules. This formulation is the basis of statistical mechanics. In this context, entropy is related, in the most general sense, to molecular probability and statistics.[1] States that can exhibit a larger number of different molecular configurations are more probable and have greater entropy. Since macroscopic systems contain such a large number of atoms, knowledge of the probable behavior of the molecules in a system leads to knowledge about how the system will behave as a whole. Based on this view, entropy is often interpreted as the degree of disorder in a system, or, as J. Willard Gibbs prefers the "mixed-up-ness."[2] We will learn more about the molecular origins of entropy in Section 3.10.

According to the molecular concept of entropy, there are more molecular configurations accessible when a system has high entropy and is disordered than when it has low entropy and is more ordered. We can view this axiom in terms of information; since there are more possibilities from which to choose in the disordered state, we have less of a chance of guessing the precise molecular configuration of the disordered state as compared to its more ordered counterpart. In analogy, Claude Shannon in 1948 conceived of "entropy" associated with missing information and thus gave birth to the field of information theory. In information theory, "entropy" is seen as a measure of the uncertainty of the true content of a message. In fact, Shannon mathematically defined information "entropy" for bits of information using the identical formula that Boltzmann applied to molecular configurations. Similarly, "entropy"-based arguments have expanded into such diverse fields as economics, theology, sociology, art, and philosophy.

We can gain some insight about how entropy is defined for the macroscopic systems in classical thermodynamics by borrowing from Boltzmann's molecular views, where entropy can be seen as the degree of disorder. Consider a closed system where energy transfer can occur by work or by heat. Energy transfer by work occurs in a very directed way. For example, in the rotation of a shaft or the movement of a piston, the interaction of the system with the surroundings occurs via a boundary that moves in a specific and well-defined direction; that is, all the molecules in the shaft or the piston have the same (angular) speed and are moving in the same direction. Similarly, electrical work is achieved by directed flow of electrons in a wire. On the other hand, energy transfer via heat is driven by temperature, which can be related to the random motion of molecules and therefore can be related to a "disordered" form of energy transfer. In Boltzmann's terms, the effect of energy transfer by work is directed and ordered and should not affect the entropy. Conversely, we should be able to relate entropy to the disordered energy transfer by heat.

Correspondingly, the thermodynamic property entropy, *s*, is *defined* in terms of heat absorbed during a *reversible* process. In differential form, the change in entropy of a substance undergoing a reversible process is equal to the incremental heat it absorbs

---

[1] In fact, the equation for his statistical based entropy, $S = k \log \mathbf{\Omega}$, serves as Boltzmann's epitaph and is engraved on his tombstone. In the equation above, Ω is the number of distinct,different molecular configurations to which a macroscopic state has access.

[2] We will use the common term *disordered* to represent what may be more objectively viewed as greater spread or dispersion.

divided by the temperature:

$$ds \equiv \frac{\delta q_{rev}}{T} \tag{3.1}$$

We can integrate Equation (3.1) between the initial and final states to give

$$\Delta s = \int_{initial}^{final} \frac{\delta q_{rev}}{T} \tag{3.2}$$

For entropy to be considered a thermodynamic property, the entropy change from the initial state to the final state must have the same value no matter what path is taken. Since the definition in Equation (3.2) is written in terms of the path-dependent property q_{rev}, the path independence of s is not obvious. However, we can logically show that entropy is indeed a thermodynamic property, independent of path, and that it is defined by Equation (3.1). Thus, any process that goes between the initial state and the final state has the entropy defined by Equation (3.2), be it reversible or irreversible. The proof can be demonstrated either by using arbitrarily small Carnot cycles[3] or, more formally, through the general examination of reversible and adiabatic surfaces using the principle of Carathéodory.[4] Those interested in the specific details of these arguments are referred to the sources cited.

The property s quantitatively tells us about the directionality of nature, reversibility vs. irreversibility, and the maximum work that we can get out of a process (or the minimum work we need to put in). We cannot deduce this supposition directly from the definition in Equation (3.1), but rather, as we shall see, it just works out that way! Specifically, we will need to determine the **entropy change of the universe** for a particular process of interest. Recall that the universe is comprised of the system together with the surroundings.

In this section, we will illustrate how the entropy change of the universe relates to directionality and reversibility through two cases: adiabatic expansion/compression (case I), and thermodynamic cycles (case II). For each case, we will look both at the reversible processes that represent the best we can do and at the corresponding irreversible processes. These cases provide an interesting juxtaposition. In case I, the entropy change to the surroundings is always zero; therefore, we can determine the entropy change of the universe solely by examining the entropy change of the system. Conversely, in case II, the entropy change to the system is always zero, allowing us to examine the entropy change of the universe through the entropy change of the surroundings. The conclusions from these two cases will be generalized in Section 3.4 to form the second law of thermodynamics. In Example 3.1, we verify the second law for a set of isothermal expansion/compression processes where both the entropy change of the system and that of the surroundings need to be taken into account.

Case I: Adiabatic Expansion/Compression

To illustrate how entropy tells us about the directionality of nature, we first pick a set of four mechanical processes similar to those described in Figure 3.2a: (1) reversible

[3] See, e.g., Kenneth Denbigh, *The Principles of Chemical Equilibrium*, 3rd ed. (New York: Cambridge University Press, 1971).

[4] See, e.g., Adrian Bejan, *Advanced Engineering Thermodynamics*, 2nd ed. (New York: Wiley, 1997).

expansion, (2) irreversible expansion, (3) reversible compression, and (4) irreversible compression. In this case, however, we choose *adiabatic* rather than isothermal processes. The adiabatic process represents the limit of no heat transfer between the system and the surroundings. If you look at Equation (3.1), you might induce why an adiabatic process was chosen for this first illustrative case. In Example 3.1, we show that identical conclusions are formed for isothermal expansion/compression processes, which represent the limit of rapid heat transfer. As in our discussion of Figure 3.2, the reversible expansion gives the *most* work we can get out of the system while the irreversible expansion has a definite directionality. Similarly for compression, the reversible process defines the *least* amount of work that we must put into the system while the irreversible compression process has a definite directionality.

For each process, we will calculate three forms of entropy: Δs_{sys}, the entropy change of the system (note that we will often omit the subscript "sys" and write it as Δs); Δs_{surr}, the entropy change of the surroundings; and Δs_{univ}, the entropy change of the universe. These three forms are related by

$$\Delta s_{univ} = \Delta s_{sys} + \Delta s_{surr} \tag{3.3}$$

First consider a reversible, adiabatic expansion of a piston–cylinder assembly in which the system expands from state 1 to state 2. This process is marked "reversible" on the *PT* diagram shown in Figure 3.3. The entropy change for the system is given by

rev adia exp

$$\Delta s = s_2 - s_1 = \int_{initial}^{final} \frac{\delta q_{rev}}{T} = 0 \tag{3.4}$$

isentropic definition

since there is no heat transfer. This process is termed **isentropic**. *For a reversible, adiabatic process, the entropy of the system remains constant.* The entropy change for the surroundings is zero, since there is no heat transferred to the surroundings. (For closed systems, all adiabatic processes, reversible or irreversible, result in $\Delta s_{surr} = 0$, since q ___ = 0). Inspection of Equation (3.3) shows that the entropy change of the zero. The changes in entropy for the reversible, adiabatic expansion are Table 3.1.

rev → max work?
irrev → max Δs!
on next pag

balance gives

$$\Delta u = u_2 - u_1 = w_{rev} \tag{3.5}$$

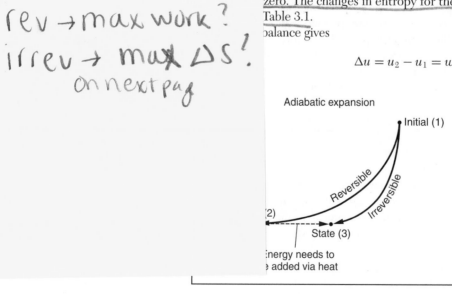

Adiabatic expansion

Initial (1)

Reversible

Irreversible

(2)

State (3)

Energy needs to be added via heat

T

Figure 3.3 *PT* diagrams of adiabatic reversible and irreversible expansions. Note that after an irreversible expansion, heat must be expelled to bring the system to the same state as after the reversible expansion.

TABLE 3.1   Summary of Entropy Change for Reversible and Irreversible Expansion (or Compression) of the Adiabatic Piston–Cylinder Assembly in Case I

Reversible Process	Irreversible Process
$s_2 = s_1$	$s_3 > s_1$
$q = 0$	$q = 0$
$\Delta s = 0$	$\Delta s > 0$
$\Delta s_{surr} = 0$	$\Delta s_{surr} = 0$
$\Delta s_{univ} = 0$	$\Delta s_{univ} > 0$

Since energy from the system is required in the form of Pv work to execute the expansion, w_{rev} is negative and state 2 has a lower internal energy and, consequently, a lower temperature.

How about an *irreversible* process? The irreversible process brings the system to a new state that we will label "3" on the PT diagram. We wish to ask, "Does the entropy of the system go up, go down, or remain the same?" Well, a first-law energy balance on the irreversible expansion gives

$$\Delta u = u_3 - u_1 = w_{irrev} \tag{3.6}$$

We know that $|w_{irrev}| < |w_{rev}|$ because a reversible process gives us the maximum possible work out from an expansion. If we now compare Equations (3.5) and (3.6), we can see that

$$u_3 > u_2 \tag{3.7}$$

So, for an ideal gas

$$T_3 > T_2 \tag{3.8}$$

as shown in Figure 3.3. How about the entropy change? Our definition of entropy requires that we have a reversible process. So to calculate entropy, we must construct a *reversible* process to go from state 1 to state 3.[5] In this case, we can first go reversibly and adiabatically from state 1 to state 2. Second, we can reversibly and isobarically transfer heat into the system to bring the gas from state 2 to state 3. From state 2 to state 3, q_{rev} is positive, so

$$\Delta s = s_3 - s_2 = \int_{initial}^{final} \frac{\delta q_{rev}}{T} = \int_{T_2}^{T_3} \frac{c_P}{T} dT > 0 \tag{3.9}$$

From Equation (3.9), we see that $s_3 > s_2 = s_1$. Therefore, for the irreversible expansion,

$$\Delta s = s_3 - s_1 > 0 \tag{3.10}$$

Thus, for an irreversible, adiabatic expansion, the entropy of the system increased. Again, the entropy change to the surroundings is zero, since $q_{surr} = 0$. Summing together the entropy changes of the system and surroundings, we find the entropy change of the

---

[5] We will frequently calculate entropy changes of irreversible processes by constructing an alternative reversible process between the initial state and the final state.

universe increases for the irreversible process. The entropy changes for this case are summarized in Table 3.1.

We will now apply the same analysis to an adiabatic compression. The PT diagrams for this process are shown in Figure 3.4. The reversible compression brings the system from state 1 to state 2. The entropy change for the system is given by

$$\Delta s = s_2 - s_1 = \int_{\text{initial}}^{\text{final}} \frac{\delta q_{\text{rev}}}{T} = 0 \tag{3.11}$$

Again, for a reversible, adiabatic process the entropy of the system remains constant. Likewise, the entropy changes of the surroundings and universe are zero. We see that the results for reversible, adiabatic compression are identical to the results presented in Table 3.1 for reversible, adiabatic expansion. An energy balance gives

$$\Delta u = u_2 - u_1 = w_{\text{rev}} \tag{3.12}$$

In this case, state 2 is at a higher temperature, since we are adding energy to the system via work to compress the gas.

How about an *irreversible* process? A first-law energy balance on an irreversible compression gives

$$\Delta u = u_3 - u_1 = w_{\text{irrev}} > w_{\text{rev}} \tag{3.13}$$

where the inequality results because we know a reversible process takes the least amount of work for a compression. If we now compare Equations (3.12) and (3.13), we can see that

$$u_3 > u_2 \tag{3.14}$$

So

$$T_3 > T_2 \tag{3.15}$$

Thus state 3 is shown at a higher temperature in Figure 3.4. To find the entropy change of the system, again let's go from state 1 to state 3 by a *reversible* process. As with the expansion, we can first go adiabatically from state 1 to state 2. Second, we must isobarically transfer heat into the system to bring the gas from state 2 to state 3. From state 2 to state 3, q_{rev} is positive and

$$\Delta s = s_3 - s_2 = \int_{\text{initial}}^{\text{final}} \frac{\delta q_{\text{rev}}}{T} > 0 \tag{3.16}$$

Figure 3.4 *PT* diagrams of adiabatic reversible and irreversible compression. Note that after an irreversible compression, heat must be expelled to bring the system to the same state as after the reversible compression.

The results from Equations (3.11) and (3.16) show that for the irreversible compression,

$$s_3 > s_2 \tag{3.17}$$

Thus, for an irreversible, adiabatic compression the entropy of the system also increased!
The entropy change to the surroundings is again zero while the entropy change to the universe has increased. In summary, the results for the irreversible, adiabatic compression are also identical to the irreversible expansion presented in Table 3.1.

If we generalize, we see that:

> *For an adiabatic* reversible *process (whether compression or expansion), the entropy of the* system *remains unchanged while for an* irreversible *process (whether compression or expansion), the entropy of the system* increases. *In both cases, the entropy changes for the surroundings are zero. Therefore, the entropy change of the system is equal to the entropy change of the universe.*

The mechanical expansion/compression process described above is convenient since it is adiabatic, so the entropy change of the *surroundings* is, by definition, zero. We have seen from Equation (3.3) that the entropy change of the *universe* is the sum of the change of the *system* and the *surroundings*, so our next question is, "What happens if the entropy change of the *surroundings* is not zero?" In our next case, we will consider a set of cyclic processes in which there are only entropy changes in the surroundings and see that we come to a similar generalization as above.

Case II: Carnot Cycle

We wish to calculate the entropy change for the Carnot cycle, the reversible process by which we converted heat into work as illustrated in Figure 2.17. We then analyze an irreversible cyclic process that has definite directionality. Since this is a cyclic process in which the system returns to its initial state, *all properties must return to their initial value*, that is, for the *system* $\Delta u_{\text{cycle}} = 0$, and similarly $\Delta s_{\text{cycle}} = 0$. The entropy change to the *surroundings* for the entire cycle equals the entropy change of each of the four steps:

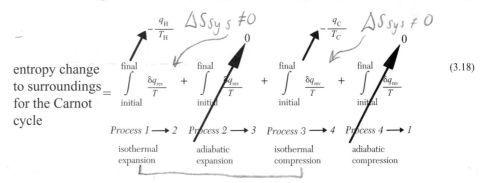

$$\text{entropy change to surroundings for the Carnot cycle} = \int_{\text{initial}}^{\text{final}} \frac{\delta q_{\text{rev}}}{T} + \int_{\text{initial}}^{\text{final}} \frac{\delta q_{\text{rev}}}{T} + \int_{\text{initial}}^{\text{final}} \frac{\delta q_{\text{rev}}}{T} + \int_{\text{initial}}^{\text{final}} \frac{\delta q_{\text{rev}}}{T} \tag{3.18}$$

| Process 1 → 2 | Process 2 → 3 | Process 3 → 4 | Process 4 → 1 |
| isothermal expansion | adiabatic expansion | isothermal compression | adiabatic compression |

where negative signs are used for q because the heat transfer to the surroundings is the negative of the heat transfer to the system; that is, if energy *enters* the system, it must *leave* the surroundings. Equation (3.18) reduces to:

$$\Delta s_{\text{surr}} = \Delta s_{\text{surr,H}} + \Delta s_{\text{surr,C}} = -\frac{q_{\text{H}}}{T_{\text{H}}} - \frac{q_{\text{C}}}{T_{\text{C}}} \tag{3.19}$$

We will now come up with alternative expressions for the right-hand side of Equation (3.19) based on our first-law analysis of the reversible processes presented in Section 2.7.

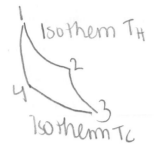

Isothem T_H

Isothem T_C

We can rewrite Equation (2.78) as follows:

Isothemm equns

$$q_H = -RT_H \ln \frac{P_2}{P_1} \tag{3.20}$$

and

$$q_C = -RT_C \ln \frac{P_4}{P_3} \tag{3.21}$$

For the two adiabatic processes ($2 \rightarrow 3$ and $4 \rightarrow 1$), Equation (2.89) can be applied:

$$PV^k = \text{const} \tag{3.22}$$

Hence, for the adiabatic process $2 \rightarrow 3$:

$$\frac{P_2}{P_3} = \left(\frac{V_3}{V_2}\right)^k = \left(\frac{nRT_C}{nRT_H}\right)^k \left(\frac{P_2}{P_3}\right)^k \tag{3.23}$$

Rearranging Equation (3.23):

$$\frac{P_2}{P_3} = \left(\frac{T_C}{T_H}\right)^{k/1-k} \tag{3.24}$$

Similarly, we find the pressures of states 1 and 4 can be related by

$$\frac{P_1}{P_4} = \left(\frac{T_C}{T_H}\right)^{k/1-k} \tag{3.25}$$

Equating Equations (3.24) and (3.25), we get

$$\frac{P_2}{P_3} = \frac{P_1}{P_4} \tag{3.26}$$

or rearranging:

$$\frac{P_2}{P_1} = \frac{P_3}{P_4} \tag{3.27}$$

Dividing Equation (3.20) by (3.21) and using Equation (3.27) gives

$$\frac{q_H}{q_C} = -\frac{T_H}{T_C} \left[\frac{\ln \frac{P_2}{P_1}}{\ln \frac{P_3}{P_4}}\right] = -\frac{T_H}{T_C} \tag{3.28}$$

or

$$-\frac{q_H}{T_H} - \frac{q_C}{T_C} = 0 \tag{3.29}$$

For both Equations (3.29) and (3.19) to be true, the total entropy change to the *surroundings* for the four *reversible* processes in the Carnot cycle must be zero:

$$\Delta s_{H,surr} + \Delta s_{C,surr} = 0 \tag{3.30}$$

As an aside, if we look at the efficiency of the Carnot cycle (see Section 2.9), we get

$$\eta \equiv \frac{\text{net work}}{\text{heat absorbed from the hot reservoir}} = \frac{|W_{net}|}{|q_H|} = \frac{|q_H| - |q_C|}{|q_H|} = 1 - \frac{|q_C|}{|q_H|} \tag{3.31}$$

Carnot pressure ratios

Refrigerator

$COP = \dfrac{q_C}{w_{net}}$

$= \dfrac{q_C}{|q_H| - |q_C|}$

$= \dfrac{1}{\dfrac{|q_H|}{q_C} - 1} = \dfrac{1}{\dfrac{T_H}{T_C} - 1} = \dfrac{T_C}{T_H - T_C}$

TABLE 3.2 Summary of Entropy Change for the Reversible Carnot Cycle and the Irreversible Cycle in Case II

Reversible Cycle	Irreversible Cycle
$\Delta s_{cycle} = 0$	$\Delta s_{cycle} = 0$
$\Delta s_{surr} = 0$	$\Delta s_{surr} > 0$
$\Delta s_{univ} = 0$	$\Delta s_{univ} > 0$

We can relate the heat transferred to the temperature by Equation (3.29), giving:

$$\eta = 1 - \frac{T_C}{T_H} \tag{3.32}$$

Equation (3.32) represents the highest efficiency that we can possibly have in operating between a hot reservoir at T_H and a cold reservoir at T_C. To improve the efficiency further requires a hotter energy source or a colder energy sink. We first encountered this relationship in Example 2.20.

Now let's consider an irreversible cycle. We know that an irreversible process produces less work than a reversible one. If the heat absorbed from the hot bath is identical, less net work means more heat must be discarded to the cold bath (since $w_{net} = q_H + q_C$):

$$(q_C)_{irrev,surr} > (q_C)_{rev,surr} \tag{3.33}$$

So for an *irreversible* cycle, $(\Delta s_C)_{surr}$ is greater than for the Carnot cycle; thus,

$$\Delta s_{surr} = (\Delta s_H)_{surr} + (\Delta s_C)_{surr} > 0 \tag{3.34}$$

So we find that the entropy change to the *surroundings* for an *irreversible cycle* is greater than zero. The changes in entropy for the reversible Carnot cycle and the irreversible power cycle are given in Table 3.2.

We can summarize this analysis as follows:

For the set of reversible processes described in the Carnot cycle, the entropy of the surroundings remains unchanged, while for an equivalent irreversible cycle the entropy of the surroundings increases. In both cases, the entropy changes for the system are zero. Therefore, the entropy change of the surroundings is equal to the entropy change of the universe.

Comparison of Tables 3.1 and 3.2 shows a common theme. The entropy change for the universe is zero for the reversible process and greater than zero for the irreversible process.

3.4 THE SECOND LAW OF THERMODYNAMICS

In case I of Section 3.3, we looked at the adiabatic expansion and compression of a piston–cylinder assembly. In these processes, the entropy change to the surroundings is zero. We saw that the entropy change of the system is zero for a reversible process and that the entropy change for the system is greater than zero for an irreversible process. In case II, we looked at the entropy change associated with a thermodynamic cycle, where the entropy change to the system is, by definition, zero. The entropy change of the surroundings is zero for a reversible process, and the entropy change for the surroundings is greater than zero for an irreversible process. It turns out that we can generalize our observations above for *all* processes by considering the entropy change of the universe. Equation (3.3) indicates that the entropy change of the universe is the sum of the entropy change to the system and the entropy change to the surroundings. We call this generalization the second law of thermodynamics:

For any reversible process, the entropy of the universe remains unchanged, while for any irreversible process, the entropy of the universe increases.

or, in other words,

Entropy is time's arrow—the larger the entropy of the universe, the more recent the event.

Just as with the first law, we believe these statements because all the observations we have made and used to test this statement are consistent with it (just as in the two specific cases above). If we quantify the statement above (so that we can solve problems such as determining whether we can remove zinc with lime), we have, for the second law of thermodynamics,

$$\Delta s_{univ} \geq 0 \tag{3.35}$$

For a *reversible* process,

$$\Delta s_{univ} = 0 \tag{3.36}$$

and for an irreversible process,

$$\Delta s_{univ} > 0 \tag{3.37}$$

Equation (3.36) sets the limit of reversibility. This case represents the best we can possibly do for a given design. Equation (3.37) tells us about directionality. If you can determine the entropy of the *universe* for two states, the one with the higher entropy associated with it happened more recently. If the entropy of the universe is the same, the process is reversible and can occur in any direction.

▶ **EXAMPLE 3.1**
Calculation of Δs_{univ} for Processes With Contributions From Both Δs_{sys} and Δs_{surr}

In Section 2.3, we learned about reversible and irreversible processes in the context of a piston–cylinder assembly undergoing isothermal expansion and compression processes. Four of these processes are summarized in Figure 3.2:

(i) Irreversible expansion, process A
(ii) Irreversible compression, process B
(iii) Reversible expansion, processes E
(iv) Reversible compression, processes F

Calculate Δs_{univ} for each of these isothermal processes and show that the result is consistent with the second law of thermodynamics.

SOLUTION In this example, we must consider both the entropy change of the system and the entropy change of the surroundings. We outline the solution methodology with process A, the isothermal, irreversible expansion from state 1 at 2 bar and 0.04 m³/mol to state 2 at 1 bar and 0.08 m³/mol. Since the process is isothermal, we assume the temperature of the surroundings and that of the system are identical. Their values can be found using the ideal gas law:

$$T = T_{surr} = \frac{Pv}{R} = 962 \, [K]$$

To calculate the entropy change of the system, we must devise a reversible process between the same states, 1 and 2. Such a process is depicted in Figure 2.7 by process E. Since the property entropy is independent of path, the entropy change of the *system* for process E and process A must be identical. Applying the definition provided by Equation (3.2) to the isothermal process, we get

$$\Delta s_{sys} = \int \frac{\delta q_{rev}}{T} = \frac{q_{rev}}{T} \tag{E3.1A}$$

We can find the heat transfer by applying the first law to the reversible process:

$$\Delta u = q_{rev} + w_{rev}$$

The internal energy change for an ideal gas undergoing an isothermal process is zero; thus,

$$q_{rev} = -w_{rev} = 5545 \left[J/mol \right] \tag{E3.1B}$$

where we have used the value reported by Equation (2.21). Substituting Equation (E3.1B) into (E3.1A) gives:

$$\Delta s_{sys} = \frac{5545}{962} = 5.76 \left[J/mol\ K \right] \tag{E3.1C}$$

This value is reported for Δs_{sys} for both process A and process E in Table E3.1.

We next calculate the entropy change for the surroundings. To compute this value, a conceptual argument is useful. *The change in entropy of the surroundings is identical for reversible heat transfer and for irreversible heat transfer, as long as the magnitude of q is the same.* Macroscopically, we can envision the same effect on the surroundings for a given amount of heat transfer. Heat represents the transfer of energy by a temperature gradient. From a molecular perspective, temperature is representative of the random motion of molecules. Thus, energy transfer via heat represents the greatest possible increase in disorder, that is, increase in entropy—whether it occurs reversibly or irreversibly. In contrast, when energy is transferred via work, molecules are specifically directed. Therefore, the entropy does not increase. Hence, for this calculation we use the actual magnitude of q found for process A. Since the heat transfer to the surroundings is the negative of the heat transfer to the system—that is, heat that leaves the surroundings enters the system—we get

$$\Delta s_{surr} = \frac{-q_A}{T_{surr}} = \frac{w_A}{T_{surr}} = -\frac{4000}{962} = -4.14 \left[J/mol\ K \right] \tag{E3.1D}$$

where the value of work for process A is given by Equation (2.16). The entropy change of the surroundings is recorded in Table E3.1. Finally, the entropy change of the universe is obtained by adding the values in Equations (E3.1C) and (E3.1D):

$$\Delta s_{univ} = \Delta s_{sys} + \Delta s_{surr} = 1.61 \left[J/(mol\ K) \right]$$

We see that the entropy change for the universe is greater than zero for this irreversible process, as the second law requires. Values for Δs_{sys}, Δs_{surr}, and Δs_{univ} for processes B, E, and F can be found in a similar manner. The values that are obtained are reported in Table E3.1. *For both irreversible processes, the entropy change for the universe is greater than zero; on the other hand, the entropy change of the universe equals zero for the reversible processes.* In the compression processes, the entropy change of the system is *negative*. This result is possible as long as the entropy change of the surroundings is suitably positive. ◀

▶ 3.5 OTHER COMMON STATEMENTS OF THE SECOND LAW OF THERMODYNAMICS

Historically, the second law was construed in the context of producing power (work) from cyclic processes in which heat was absorbed from a hot reservoir and rejected to a

Table E3.1 Summary of Entropy Change for the Isothermal Expansion/Compression Processes Described in Section 2.3

	Irreversible Processes		**Reversible Processes**	
	Expansion (Process A)	Compression (Process B)	Expansion (Process E)	Compression (Process F)
$\Delta s_{sys}\left[\text{J/(mol K)}\right]$	5.76	−5.76	5.76	−5.76
$\Delta s_{surr}\left[\text{J/(mol K)}\right]$	−4.14	8.32	−5.76	5.76
$\Delta s_{univ}\left[\text{J/(mol K)}\right]$	**1.61**	**2.56**	**0**	**0**

cold reservoir. Thus, the second law is often stated in terms of work produced and heat rejected. For example, the second law is often defined as follows:

Heat cannot be caused to flow from a cooler body to a hotter body without producing some other effect.

Clausius

We can see that the Clausius statement is consistent with the general definition presented in Section 3.4. If a positive quantity of heat, q, flows reversibly from a cool body at T_C to a hot body at T_H, with no other effect, the associated entropy chance is equal to

$$\Delta s = -\frac{q}{T_C} + \frac{q}{T_H} < 0 \tag{3.38}$$

The quantity on the right-hand side of Equation (3.38) is less than zero, since $T_C < T_H$. Since Δs is negative, it violates the second law of thermodynamics, unless there is "some other effect" that has a positive Δs large enough to offset this transfer of energy via heat.

Another common form of the second law is given by:

It is impossible to build an engine which operating in a cycle can convert all the heat it absorbs into work.

Kelvin and Planck

Again, we can see that this statement is consistent with the general definition presented in Section 3.4. Let's consider the most efficient engine possible, the Carnot engine. If $q_C = 0$. Equation (3.19) gives

$$\Delta s_{univ} = \Delta s_{surr} = -\frac{q_H}{T_H} < 0 \tag{3.39}$$

which also violates the second law of thermodynamics. There are many other specific forms of the second law. It is clear that the statements above can be viewed in terms of the generalizations presented in Section 3.4, but the reverse is not so clear.

▶ 3.6 THE SECOND LAW OF THERMODYNAMICS FOR CLOSED AND OPEN SYSTEMS

In general, there are two ways to apply the second law of thermodynamics:

$$\Delta S_{univ} \geq 0 \tag{3.40}$$

1. We may determine whether a process is possible (and estimate how efficient it is). This applies to real, irreversible processes for which the entropy change of the universe is greater than 0:

$$\Delta S_{univ} > 0 \qquad (3.41)$$

2. The second law can also be used to provide an additional constraint to solve a problem; that is, it provides us another equation to use. To apply the second law in this way, we must assume the process is reversible. In this case, we apply

$$\Delta S_{univ} = 0 \qquad (3.42)$$

Calculation of ∆s for Closed Systems

In a *closed* system, mass cannot transfer across the system boundary. We can write Equation (3.40) as

$$\Delta S_{sys} + \Delta S_{surr} \geq 0 \qquad (3.43)$$

The change in entropy of the surroundings is identical for reversible heat transfer and for irreversible heat transfer, as long as the magnitude of Q is the same (see the discussion in Example 3.1). If the surroundings are at **constant temperature**, T_{surr}, we can write the entropy change to the surroundings as

$$\Delta S_{surr} = \int_{initial}^{final} \frac{\delta Q_{surr}}{T_{surr}} = \frac{1}{T_{surr}} \int_{initial}^{final} \delta Q_{surr} = \frac{Q_{surr}}{T_{surr}} \qquad (3.44)$$

We must be careful about sign conventions. If heat flows into the system, it must flow out of the surroundings, that is:

$$Q_{surr} = -Q \qquad (3.45)$$

Substitution of Equation (3.44) and (3.45) into Equation 3.43 gives

$$\Delta S_{sys} + \frac{Q_{surr}}{T_{surr}} = n(s_{final} - s_{initial}) + \frac{Q_{surr}}{T_{surr}} \geq 0 \qquad (3.46)$$

$$= n(s_{final} - s_{initial}) - \frac{Q}{T_{surr}} \geq 0$$

As with the first law, the second law can be written in *differential* form:

$$nds + \frac{\delta Q_{surr}}{T_{surr}} \geq 0 \qquad (3.47)$$

Where do we obtain values of s?

▶ **EXAMPLE 3.2**
Calculation of ΔS_{univ} by Selecting a Hypothetical Reversible Process

An insulated tank ($V = 1.6628$ L) is divided into two equal parts by a thin partition. On the left is an ideal gas at 100 kPa and 500 K; on the right is a vacuum. The partition ruptures with a loud bang.

(a) What is the final temperature in the tank?
(b) What is Δs_{univ} for the process?

SOLUTION A schematic of the system is shown on the left-hand side of Figure E3.2.

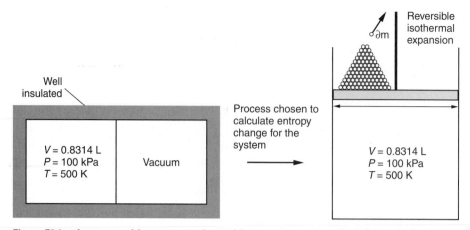

Figure E3.2 The irreversible process in the problem statement is on the left-hand side. The hypothetical reversible path between the same initial and final states shown on the right-hand side is used to calculate the entropy change of the system.

(a) We can apply the first law of thermodynamics to find the final temperature. If we choose the entire tank as the system, there is no work or heat transfer across its boundaries during this process. Hence,

$$\Delta U = Q + W = 0 \tag{E3.2}$$

The internal energy in the final state is the same as it was initially. For an ideal gas, $u = u$ (T only); thus, to follow Equation (E3.2), the temperature does not change:

$$T_2 = 500 \text{ K}$$

(b) We must apply the second law of thermodynamics. Since s is a property of the system, it depends only on the final state and the initial state, *not* on the process (or path). To calculate ΔS_{sys}, we choose a reversible path by which the system goes from the same initial state to the same final state as the real system. We can then use the definition for change in entropy, since it involves a reversible process, $\left(\int_{\text{initial}}^{\text{final}} \dfrac{\delta q_{rev}}{T} \equiv \Delta s \right)$. Such a hypothetical path is illustrated on the right-hand side of Figure E3.2. For this hypothetical *reversible* process, we have:

$$\Delta U = Q_{rev} + W_{rev} = 0 \qquad \text{← no PV work}$$

or

$$Q_{rev} = -W_{rev} = \int_{0.8314 \text{ L}}^{1.6628 \text{ L}} P dV = \int \frac{nRT}{V} dV$$

Solving with numerical values, we get

$$q_{rev} = \frac{Q_{rev}}{n} = RT \ln\left(\frac{V_2}{V_1}\right) = \left(8.314 \left[\frac{\text{J}}{\text{mol K}}\right]\right) (500 \,[\text{K}]) \ln 2 = 2881 \left[\frac{\text{J}}{\text{mol}}\right]$$

Plugging into the definition for Δs:

$$\Delta s_{sys} = \int \frac{\delta q_{rev}}{T} = \frac{q_{rev}}{T} = \frac{2881 \,[\text{J/mol}]}{500 \,[\text{K}]} = 5.76 \left[\frac{\text{J}}{\text{mol K}}\right]$$

where we pulled T out of the integral since the hypothetical process is isothermal. Since there is no heat transferred to the surroundings in the real irreversible process,

$$\Delta s_{\text{surr}} = 0$$

Finally, the entropy change of the universe can be found by adding that of the system to the surroundings according to Equation (3.3):

$$\Delta s_{\text{univ}} = \Delta s_{\text{sys}} + \Delta s_{\text{surr}} = 5.76 \text{ J/(mol k)}$$

Note that this value is grater than zero, indicating an irreversible process, as we might expect for this very spontaneous process:

$$\Delta s_{\text{univ}} > 0 \qquad \qquad \blacktriangleleft$$

▶ **EXAMPLE 3.3**
Entropy Change in
Obtaining Thermal
Equilibrium

Consider an isolated system containing two blocks of copper with equal mass. One block is initially at 0°C while the other is at 100°C. They are brought into contact with each other and allowed to thermally equilibrate. What is the entropy change for the system during this process? The heat capacity for copper is found to be:

$$c_P = 24.5 \left[\text{J/(mol K)} \right]$$

*Thermal
equilibrium →
think about
enthalpy
change
to find
entropy
change*

SOLUTION We label the cooler block "block I" and the hotter block "block II" as illustrated in Figure E3.3. The blocks go from state 1, in which they are at different temperatures, to state 2, in which their temperatures are equal. The pressure remains constant during the process.

We must first find the final temperature, T_2. From the first law for an isolated system at constant pressure, we get

$$\Delta H = n_{\text{I}}\Delta h_{\text{I}} + n_{\text{II}}\Delta h_{\text{II}} = 0 \qquad \text{(E3.3A)}$$

where the subscripts "I" and "II" refer to the appropriate block. For constant heat capacity, Equation E3.1 can be written

$$n_{\text{I}}c_P \left(T_2 - T_{1,\text{I}}\right) + n_{\text{II}}c_P \left(T_2 - T_{1,\text{II}}\right) = 0 \qquad \text{(E3.3B)}$$

Since the blocks have equal mass,

$$n_{\text{I}} = n_{\text{II}} = n \qquad \text{(E3.3C)}$$

Solving Equation (E3.3B) subject to (E3.3C) gives

$$T_2 = \frac{T_{1,\text{I}} + T_{1,\text{II}}}{2} = 323 \text{ [K]} \qquad \textit{Finding } T_2 \textit{ w/}$$

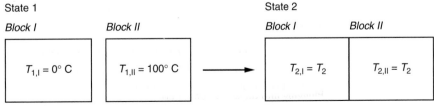

State 1		State 2	
Block I	*Block II*	*Block I*	*Block II*
$T_{1,\text{I}} = 0°$ C	$T_{1,\text{II}} = 100°$ C	$T_{2,\text{I}} = T_2$	$T_{2,\text{II}} = T_2$

Figure E3.3 Initial (state 1) and final (state 2) states of two copper blocks that obtain thermal equilibrium.

To get the total entropy change for the system, we can add the changes of each block individually:

$$\Delta S = n_I \Delta s_I + n_{II} \Delta s_{II} \tag{E3.3D}$$

We can apply Equation (3.1) to solve for Δs_I and Δs_{II}. We first need to find δq_{rev}. We do that by constructing a reversible path at constant pressure from T_1 to T_2. Such a process can be conceived by placing each block in thermal contact with a series of reservoirs that are each just differentially hotter (or cooler) than the block. Then q_{rev} is given by:

@ constant $q = \Delta h$ $q_{rev} = \Delta h = \int_{T_1}^{T_2} c_P dT$
pressure

or
$$\delta q_{rev} = c_P dT$$

Thus, the entropy change is described by

$$\Delta s = \int \frac{\delta q_{rev}}{T} = \int_{T_1}^{T_2} \frac{c_P d\,T}{T} \tag{E3.3E}$$

Applying Equation (E3.3E) to the system of two blocks with constant heat capacity, we get

$$\Delta S = n_I c_P \ln\left(\frac{T_2}{T_{1,I}}\right) + n_{II} c_P \ln\left(\frac{T_2}{T_{1,II}}\right)$$

Rearranging and plugging in numerical values yields

$$\Delta s = \frac{\Delta S}{2n} = \frac{c_P}{2}\left[\ln\left(\frac{T_2}{T_{1,I}}\right) + \ln\left(\frac{T_2}{T_{1,II}}\right)\right] = \frac{c_P}{2} \ln\left(\frac{T_2^2}{T_{1,I}, T_{1,II}}\right)$$

$$= \left(\frac{24.5}{2}\left[\frac{J}{mol\,K}\right]\right) \ln\left(\frac{323^2}{(273)(373)}\right) = 0.30 \left[\frac{J}{mol\,K}\right]$$

Since this is an isolated system, the entropy change to the surroundings is zero. The positive value we obtained confirms that energy spontaneously flows from a hot block to a cold block.   ◀

▶ **EXAMPLE 3.4**
Entropy Calculation for a Phase Change

Calculate the change in entropy when 1 mole of saturated ethanol vapor condenses at its normal boiling point.

SOLUTION   For ethanol, we can find that the enthalpy of vaporization and normal boiling point are

$$\Delta h_{vap,C_2H_6O} = 38.56 \left[kJ/mol\right] \text{ and } T_b = 78.2 \,[^\circ C]$$

The boiling point is, by definition, the temperature at which the phase change occurs reversibly. Hence, we can apply it directly to the definition of entropy, Equation (3.2). From an energy balance at constant P, we get

$$\Delta h = h_l - h_v = q_{rev} = -\Delta h_{vap,C_2H_6O} \tag{E3.4}$$

We then substitute Equation (E3.4) into the definition for entropy. Realizing that this process occurs at constant temperature T_b, we get

$$\Delta s = \int_{vapor}^{liquid} \frac{\delta q_{rev}}{T} = \frac{-\Delta h_{vap,C_2H_6O}}{T_b}$$

Plugging in the values above gives

$$\Delta s = \frac{-38.56\,[\text{kJ/mol}]}{351.4\ \text{K}} = -0.1098 \left[\frac{\text{kJ}}{\text{mol K}}\right]$$

The entropy change for the system has a negative value since the liquid is in a more ordered state than the vapor. However, this process does not violate the second law, since it is the entropy change of the *universe* that must be greater than zero. Therefore, we can also say that the entropy change of the surroundings must be at least $-0.1098\,[\text{kJ/(mol K)}]$ for this process to occur. ◀

Calculation of Δs for Open Systems

In **open** systems, mass can flow into and out of the system. Thus, the entropy from flowing streams transfers from the surroundings to the system. An example of an open system with two streams in and two streams out is shown in Figure 3.5. As was the case with the first law, it is often convenient to discuss rates of flow [mol/sec] and energy transfer [J/s] or [W]. Dividing Equation (3.43) by Δt and taking the limit as the time step becomes zero, the second law becomes

$$\left(\frac{dS}{dt}\right)_{\text{univ}} = \left(\frac{dS}{dt}\right)_{\text{sys}} + \left(\frac{dS}{dt}\right)_{\text{surr}} \geq 0 \tag{3.48}$$

At **steady-state**, the entropy change of the system is zero:

$$\left(\frac{dS}{dt}\right)_{\text{sys}} = 0 \tag{3.49}$$

In addition to the heat exchanged with the surroundings, the entropy change in the surroundings is affected by the mass flow out of or into the system. Each mole of an outlet stream contains a quantity of entropy, s_{out}, that it adds to the surroundings, while each inlet stream takes a quantity of entropy, s_{in}, away from the surroundings. Therefore, at *constant temperature*, T_{surr}, the rate of entropy change with the surroundings can be written

$$\left(\frac{dS}{dt}\right)_{\text{surr}} = \sum_{\text{out}} \dot{n}_{\text{out}} s_{\text{out}} - \sum_{\text{in}} \dot{n}_{\text{in}} s_{\text{in}} + \frac{\dot{Q}_{\text{surr}}}{T_{\text{surr}}}$$

$$= \sum_{\text{out}} \dot{n}_{\text{out}} s_{\text{out}} - \sum_{\text{in}} \dot{n}_{\text{in}} s_{\text{in}} - \frac{\dot{Q}}{T_{\text{surr}}} \tag{3.50}$$

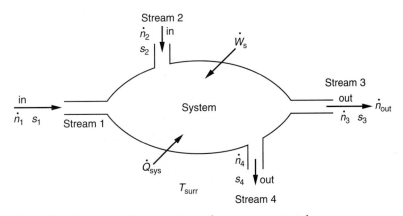

Figure 3.5 Schematic of open system with two streams in and two streams out.

where Equation (3.45) was used. Applying Equations (3.49) and (3.50), at steady-state Equation (3.48) becomes

$$\sum_{out} \dot{n}_{out} s_{out} - \sum_{in} \dot{n}_{in} s_{in} - \frac{\dot{Q}}{T_{surr}} \geq 0 \tag{3.51}$$

In the case of one stream in and one stream out, Equation (3.51) can be written

$$\dot{n}(s_{out} - s_{in}) - \frac{\dot{Q}}{T_{surr}} \geq 0 \tag{3.52}$$

where the outlet and inlet molar flow rates are identically $\dot{n}$. In *differential* form, the steady-state entropy balance becomes:

$$\dot{n}ds - \frac{\delta Q}{T_{surr}} \geq 0 \tag{3.53}$$

For unsteady-state problems, the entropy change of the system, $(dS/dt)_{sys}$, must be included. Adding this term to Equation (3.51) gives:

$$\frac{dS}{dt} + \sum_{out} \dot{n}_{out} s_{out} - \sum_{in} \dot{n}_{in} s_{in} - \frac{\dot{Q}}{T_{surr}} \geq 0 \tag{3.54}$$

▶ **EXAMPLE 3.5**
Using Entropy to Help Calculate the Exit Velocity From a Nozzle

Steam enters a nozzle at 300 kPa and 700°C with a velocity of 20 m/s. The nozzle exit pressure is 200 kPa. Assuming this process is reversible and adiabatic determine **(a)** the exit temperature and **(b)** the exit velocity.

SOLUTION   First we draw a schematic of the process including all the information we know. We label the inlet state "1" and the outlet state "2," as shown in Figure 3.5.
    A first-law balance for an open system at steady-state gives

$$0 = \dot{m}_1(\hat{h} + \hat{e}_K + \hat{e}_P)_1 - \dot{m}_2(\hat{h} + \hat{e}_K + \hat{e}_P)_2 + \dot{Q} + \dot{W}_s$$

We have written this equation on a mass basis in anticipation of using the steam tables for thermodynamic property data. Using a mass balance, $\dot{m}_1 = \dot{m}_2$, and the definition for kinetic energy, we get

$$\hat{h}_1 + \frac{\vec{V}_1^2}{2} = \hat{h}_2 + \frac{\vec{V}_2^2}{2} \tag{E3.5A}$$

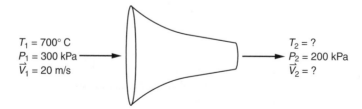

$T_1 = 700°$ C
$P_1 = 300$ kPa
$\vec{V}_1 = 20$ m/s

$T_2 = ?$
$P_2 = 200$ kPa
$\vec{V}_2 = ?$

Figure E3.5 Nozzle of Example 3.5 with known and unknown properties of steam delineated.

Equation (E3.5A) has two unknowns, $\hat{h}_2$ and $\vec{V}_2$; therefore, we need another equation. For a reversible adiabatic process, the second law gives $\Delta s = 0$, or

$$\hat{s}_1 = \hat{s}_2 \qquad \text{(E3.5B)}$$

It is worthwhile, at this point, to reflect on the overall solution methodology. State 1 is thermo-dynamically determined, since we know T_1 and P_2. Thus, we can find any other property, including $\hat{s}_1$. Once the specific entropy of state 1 is determined, we know $\hat{s}_2$ from Equation (E3.5B). This value, along with P_2, constrains state 2, so we can, in principle, find any other property, including temperature and enthalpy.

From the steam tables, for state 1, we get

$$\hat{h}_1 = 3927.1 \left[\text{kJ/kg} \right]$$

and

$$\hat{s}_1 = 8.8319 \left[\text{kJ/(kg K)} \right] = \hat{s}_2$$

We now go to the table for $P = 200$ kPa and determine (by interpolation) that steam has $\hat{s}_2 = 8.8319 \left[\text{kJ/(kg K)} \right]$ when

$$T_2 = 623[°C]$$

and

$$\hat{h}_2 = 3754.7 \left[\text{kJ/kg} \right]$$

Now we can solve Equation (E3.5A) for the outlet velocity:

$$\vec{V}_2 = \sqrt{2(\hat{h}_1 - \hat{h}_2) + \vec{V}_1^2} = \sqrt{2\left(172.4 \times 10^3\right)\left[\text{J/kg}\right] + 400\left[\text{m}^2/\text{s}^2\right]} = 587.5\left[\text{m/s}\right]$$

The steam lost enough internal energy in cooling from 700°C to 623°C to manifest itself in a very high exit velocity (575 m/s). Again, this example illustrates the large amounts of energy manifest in u.

▶ **EXAMPLE 3.6**
Entropy Generated by an Open Feedwater Heater

In Example 2.18, we analyzed an open feedwater heater. Superheated water vapor at a pressure of 200 bar, a temperature of 500°C, and a flow rate of 10 kg/s was brought to a saturated vapor state at 100 bar by mixing this stream with a stream of liquid water at 20°C and 100 bar. The flow rate for the liquid stream was found to be 1.95 kg/s. What is the entropy generated during this process?

SOLUTION First let's draw a diagram of the system, as shown in Figure E3.6. If we assume the rate of heat transfer is negligible, the entropy change reduces to the sum of the outlet stream minus the inlet streams:

$$\left(\frac{dS}{dt}\right)_{\text{univ}} = \dot{m}_3\hat{s}_3 - (\dot{m}_1\hat{s}_1 + \dot{m}_2\hat{s}_2) \qquad \text{(E3.6A)}$$

A mass balance at steady-state gives

$$\dot{m}_3 = \dot{m}_1 + \dot{m}_2 = 11.95\left[\text{kg/s}\right] \qquad \text{(E3.6B)}$$

We can look up values for the enthalpies from the steam tables (Appendix B). For state 1 [superheated steam is at 500°C and 200 bar ($=20$ MPa)],

$$\hat{s}_1 = 6.1400 \left[\frac{\text{kJ}}{\text{kg K}}\right]$$

For state 2, we use subcooled liquid at 20°C and 100 bar:

$$\hat{s}_2 = 0.2945 \left[\frac{\text{kJ}}{\text{kg K}}\right]$$

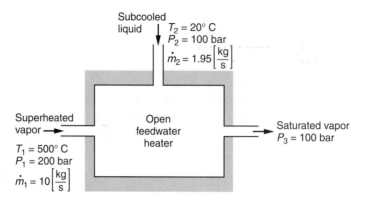

Figure E3.6   Schematic of the open feedwater heater.

and saturated vapor at 100 bar (10 MPa) for state 3:

final state

$$\hat{s}_3 = 5.6140 \left[\frac{kJ}{kg\ K} \right]$$

Finally, rearranging Equation (E3.6A) and plugging in values gives

$$\left(\frac{dS}{dt} \right)_{univ} = \left\{ \left(11.95 \left[\frac{kg}{s} \right] \right) \left(5.6140 \left[\frac{kJ}{kg\ K} \right] \right) \right\}$$

$$- \left\{ \left(10.0 \left[\frac{kg}{s} \right] \right) \left(6.1400 \left[\frac{kJ}{kg\ K} \right] \right) \right.$$

$$+ \left(1.95 \left[\frac{kg}{s} \right] \right) \left(0.2945 \left[\frac{kJ}{kg\ K} \right] \right) \right\} = 5.11 \left[\frac{kW}{K} \right]$$

The entropy increases, as we would expect for this spontaneous process. ◀

▶ 3.7 CALCULATION OF Δs FOR AN IDEAL GAS

This section illustrates how to calculate the change in entropy of an ideal gas between two states if P and T for each state are known. We will define the initial state as state 1, at P_1 and T_1, and the final state as state 2, at P_2 and T_2. Since entropy is a state function, we can construct any path that is convenient between state 1 and state 2 to calculate Δs. Figure 3.6 illustrates such a hypothetical path. We choose a reversible process for our hypothetical path so that we can apply the definition of entropy. The first step consists of isothermal expansion, while the second step is isobaric heating. To find Δs, we will calculate the entropy change for each step and add them together. Details of the analysis for each step follow.

Step 1:   Reversible, isothermal expansion

The change in internal energy for an ideal gas is, by definition, zero; thus, the differential energy balance is

$$du = \delta q_{rev} + \delta w_{rev} = 0 \tag{3.55}$$

Solving Equation (3.55) for the differential heat transfer, we get

$$\delta q_{rev} = -\delta w_{rev} = P dv \tag{3.56}$$

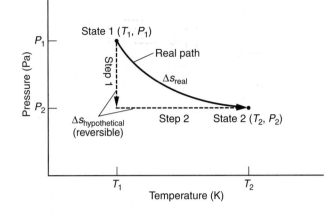

Figure 3.6 Plot of a process in which a system goes from state 1 to state 2 in TP space. The change in entropy is calculated along a reversible hypothetical path.

Applying Equation (3.56) to the definition of entropy, Equation (3.2), gives

$$\Delta s_{step1} = \int \frac{\delta q_{rev}}{T} = \int \left(\frac{P}{T}\right) dv \qquad (3.57)$$

Using the ideal gas law ($P/T = R/v$) and recognizing that the volume at the end of step 1 is RT_1/P_2 gives

$$\Delta s_{step1} = \int_{v_1}^{RT_1/P_2} \left(\frac{R}{v}\right) dv = R \ln\left[\frac{RT_1/P_2}{v_1}\right] = R \ln\left(\frac{P_1}{P_2}\right) = -R \ln\left(\frac{P_2}{P_1}\right) \qquad (3.58)$$

Step 2: Reversible, isobaric heating

Again, we want to solve for the reversible heat transfer, q_{rev}, and use it in the definition of entropy. Applying the first law gives

$$du = \delta q_{rev} + \delta w_{rev} = \delta q_{rev} - Pdv \qquad (3.59)$$

Solving Equation (3.59) for δq_{rev}:

$$\delta q_{rev} = du + Pdv + vdP = d(u + Pv) = dh \qquad (3.60)$$

where the term vdP was added to the right-hand side since $dP = 0$ for a constant-pressure process. Thus the entropy change is equal to

$$\Delta s_{step2} = \int \frac{\delta q_{rev}}{T} = \int \frac{dh}{T} = \int_{T_1}^{T_2} \frac{c_P}{T} dT \qquad (3.61)$$

where the definition of heat capacity, Equation (2.58), is used.

Adding together the two steps of our hypothetical reversible path gives

$$\Delta s = \Delta s_{step2} + \Delta s_{step1} = \int_{T_1}^{T_2} \frac{c_P}{T} dT - R \ln\left(\frac{P_2}{P_1}\right) \qquad (3.62)$$

Equation (3.62) is true, in general, for the entropy change associated with an ideal gas between state 1 and state 2. *Note*: in this expression, the **property**,

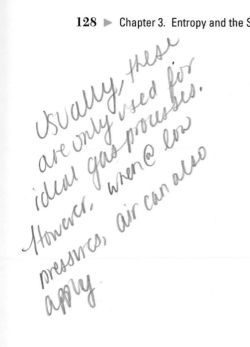

Δs just depends on other properties—that is, c_P, T, P—so it is independent of path. Therefore, Equation (3.62) can be applied to *any process*, be it reversible or irreversible. It is not limited only to the reversible processes for which it was developed.

If c_P is constant, Equation (3.62) becomes

$$\Delta s = c_P \ln \left(\frac{T_2}{T_1} \right) - R \ln \left(\frac{P_2}{P_1} \right) \qquad (c_P \text{ is constant}) \qquad (3.63)$$

In a similar manner, it can be shown that for an ideal gas, the entropy change between (T_1, v_1) and (T_2, v_2) is given by (see Problem 3.1):

$$\Delta s = \int_{T_1}^{T_2} \frac{c_v}{T} dT + R \ln \left(\frac{v_2}{v_1} \right) \qquad (3.64)$$

With c_v constant, Equation (3.64) becomes

$$\Delta s = c_v \ln \left(\frac{T_2}{T_1} \right) + R \ln \left(\frac{v_2}{v_1} \right) \qquad (c_v \text{ is constant}) \qquad (3.65)$$

▶ **EXAMPLE 3.7**
Calculation of Entropy Change for an Irreversible, Isothermal Compression

A piston–cylinder device initially contains 0.50 m³ of an ideal gas at 150 kPa and 20°C. The gas is subjected to a constant external pressure of 400 kPa and compressed in an isothermal process. Assume the surroundings are at 20°C. Take $c_P = 2.5R$ and assume the ideal gas model holds.
(a) Determine the heat transfer (in kJ) during the process.
(b) What is the entropy change of the system, surroundings, and universe?
(c) Is the process reversible, irreversible, or impossible?

SOLUTION (a) To determine the heat transfer during the process, we can apply the first law. The change in internal energy of an ideal gas undergoing an isothermal process is zero:

$$\Delta U = Q + W = 0$$

Solving for Q gives

$$Q = -W = P_{ext}(V_2 - V_1) = P_2 \left(\frac{P_1 V_1}{P_2} - V_1 \right) = V_1(P_1 - P_2)$$

Finally, plugging in numerical values:

$$Q = \left(0.50 \, [m^3] \right) \left(-250 \, [kPa] \right) = -125 \, [kJ]$$

(b) To find the entropy change of the system, we can apply Equation (3.63):

$$\Delta s_{sys} = c_P \ln \frac{T_2}{T_1} - R \ln \frac{P_2}{P_1}$$

The first term is zero, since the process is at constant temperature. Thus,

$$\Delta s_{sys} = n \Delta s_{sys} = \left(\frac{PV}{T} \right)_1 \left(-\ln \frac{P_2}{P_1} \right)$$

Plugging in numerical values gives

$$\Delta S_{sys} = \left(\frac{150 \, [kPa] 0.5 \, [m^3]}{293 \, [K]} \right) \left(-\ln \frac{400}{150} \right) = -0.25 \left[\frac{kJ}{K} \right]$$

The entropy change of the surroundings is given by

$$\Delta S_{\text{surr}} = \frac{Q_{\text{surr}}}{T} = \frac{-Q}{T} = \frac{125\,[\text{kJ}]}{293\,[\text{K}]} = 0.43 \left[\frac{\text{kJ}}{\text{K}}\right]$$

The entropy change of the universe is equal to that of the system plus the surroundings:

$$\Delta S_{\text{univ}} = \Delta S_{\text{sys}} + \Delta S_{\text{surr}} = 0.18\,[\text{kJ/K}]$$

(c) Since entropy of the universe increases, this process is irreversible. The irreversibility arises from the finite pressure difference.   ◄

► **EXAMPLE 3.8**
Calculation of
Entropy Change
Between Two States

Calculate the entropy change when 1 mole of air is heated and expanded from 25°C and 1 bar to 100°C and 0.5 bar.

SOLUTION   At these low pressures, we can assume that air is an ideal gas. The entropy change between the initial state (1) and final state (2) is given by Equation (3.62):

$$\Delta s = \int_{T_1}^{T_2} \frac{c_P}{T}\, dT - R \ln\left(\frac{P_2}{P_1}\right) \qquad \text{IG eqⁿ} \qquad \text{(E3.8A)}$$

We can integrate the first term on the right-hand side as follows:

$$\int_{T_1}^{T_2} \frac{c_P}{T}\, dT = R \int_{298}^{373} \left[AT^{-1} + B + DT^{-3}\right] dT = R\left[A\ln T + BT - \frac{D}{2T^2}\right]_{298}^{373} = 0.793R \qquad \text{(E3.8B)}$$

where the following parameters were used for air:

$$A = 3.355, \qquad B = 0.575 \times 10^{-3}, \qquad \text{and } D = -0.016 \times 10^5$$

Substituting Equation (E3.8B) into (E3.8A) gives

$$\Delta s = R(0.793 - \ln 2) = 0.83\left[\text{J/(mol K)}\right]$$

An alternative approach is to use the mean heat capacity values in Example 2.7

$$\Delta s = \bar{c}_P \ln\left(\frac{T_2}{T_1}\right) - R\ln\left(\frac{P_2}{P_1}\right) = 29.37 \ln\left(\frac{373}{298}\right) - 8.314\ln\left(\frac{1}{0.5}\right) = 0.83\left[\frac{\text{J}}{\text{mol K}}\right]$$

We have interpolated between 300 K and 400 K to get $\bar{c}_P = 29.37\,[\text{J/(mol K)}]$. In this case, these two approaches give identical values. Do you think that is always true?   ◄

► **EXAMPLE 3.9**
Calculation of
Entropy Change
of Mixing

One mole of pure N_2 and 1 mole of pure O_2 are contained in separate compartments of a rigid container at 1 bar and 298 K. The gases are then allowed to mix. Calculate the entropy change of the mixing process.

SOLUTION   A schematic of the mixing process represented by ΔS_{mix} is shown in Figure E3.9A. We are free to choose any path to calculate the change in entropy of the mixing process. One possible solution path is shown in Figure 3.9B. In the first step (step I), each of the pure species, N_2 and O_2, is reversibly and isothermally expanded to the size of the container of the mixture. During this process, the pressure drops to p_{O_2} and p_{N_2} respectively.

The entropy change for each step is given by Equation (3.62):

$$\Delta S_{O_2}^I = n_{O_2}\left(s_{O_2}(\text{at }p_{O_2}, T) - s_{O_2}(\text{at }P, T)\right) = -n_{O_2}R\left(\ln\frac{p_{O_2}}{P}\right)$$

and

$$\Delta S_{N_2}^I = n_{N_2}\left(s_{N_2}(\text{at }p_{N_2}, T) - s_{N_2}(\text{at }P, T)\right) = -n_{N_2}R\left(\ln\frac{p_{N_2}}{P}\right)$$

where the partial pressures p_{O_2} and p_{N_2} equal 0.5 bar. The next step (step II) is to superimpose both these expanded systems. Since O_2 behaves as an ideal gas, it does not know N_2 is there and vice versa; thus, the properties of each individual species do not change. Therefore,

$$\Delta S^{II} = n_{O_2}\Delta s_{O_2}^{II} + n_{N_2}\Delta s_{N_2}^{II} = 0$$

so

$$\Delta S = \Delta S_{O_2}^I + \Delta S_{N_2}^I + \Delta S^{II} = -n_{O_2}R\left(\ln\frac{p_{O_2}}{P}\right) - n_{N_2}R\left(\ln\frac{p_{N_2}}{P}\right) = 11.53\left[\frac{J}{K}\right]$$

The path shown in Figure E3.9B has interesting implications in terms of the molecular viewpoint of the entropy increase for each species. Increased entropy dose not directly result because N_2 mixes in O_2, but rather because O_2 has more room to move around in, so that there is more uncertainty where it is; therefore, information is lost, and the entropy is higher. ◀

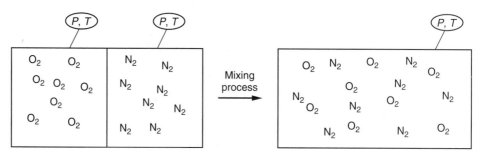

Figure E3.9A Mixing of N_2 and O_2 in a rigid container at constant P and T.

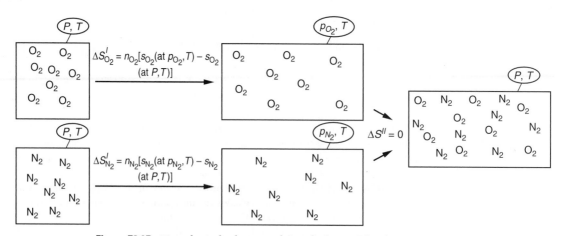

Figure E3.9B Hypothetical solution path to calculate ΔS for the mixing process.

▶ **EXAMPLE 3.10**
Reformulation of
Polytropic Process
in Terms of Entropy

In Section 2.7.2, we came up with the following expression for a reversible, adiabatic process on an ideal gas with constant heat capacity, based on first-law analysis:

$$PV^k = \text{const}$$

Come up with the same equation based on second-law analysis, starting from Equation (3.63).

SOLUTION Equation (3.63) can be set to zero for a reversible, adiabatic process:

$$\Delta s = c_P \ln\left(\frac{T_2}{T_1}\right) - R \ln\left(\frac{P_2}{P_1}\right) = 0$$

Rearranging, we get

$$\ln\left(\frac{T_2}{T_1}\right)^{c_P} = \ln\left(\frac{P_2}{P_1}\right)^{R} \tag{E3.10}$$

Applying the ideal gas law and the relation $R = c_P - c_v$, Equation (E3.10) becomes

$$\left(\frac{T_2}{T_1}\right)^{c_P} = \left(\frac{\left(\frac{PV}{nR}\right)_2}{\left(\frac{PV}{nR}\right)_1}\right)^{c_P} = \left(\frac{P_2}{P_1}\right)^{c_P - c_v}$$

However, since $n_1 = n_2$, we get

$$\left(\frac{P_2}{P_1}\right)^{c_v} = \left(\frac{V_1}{V_2}\right)^{c_P}$$

or $$PV^k = \text{const}$$

with $$k = \frac{c_p}{c_v}$$

The analysis is much easier using what we have learned about entropy. ◀

▶ **EXAMPLE 3.11**
Entropy Change for
the Expansion of Argon
From a Compressed
Cylinder

Consider a cylinder containing 4 moles of compressed argon at 10 bar and 298 K. The cylinder is housed in a big lab maintained at 1 bar and 298 K. The valve develops a leak and Ar escapes to the atmosphere until the pressure and temperature equilibrate. After a sufficient amount of time, the argon in the room reaches its background mole fraction of 0.01. Estimate, as closely as possible, the entropy change for the universe. You may assume that argon behaves as an ideal gas. Do you think this expansion process as a whole can be treated as reversible?

SOLUTION A schematic of the process is shown in Figure E3.11A. We label the initial state "state 1" and the final state "state 4." We are tempted to solve this problem as an open system where our boundary is contained by the walls of the cylinder; however, instead we will choose a system containing all the gas in the cylinder. Our problem is then reduced to that of a fixed amount of gas expanding in a closed system, with some of the gas then mixing with the environment. Our solution path is shown in Figure E3.11B. In step 1, we isothermally expand the gas to 1 bar. Our calculation for this part will be similar to the approach in Example 3.7.[6] In step 2, we divide the gas into two parts: that which remains in the cylinder and that which is in the air-filled room. The entropy change for this step is zero. Step 3 involves calculating the entropy change of the mixing of the Ar with the other gases in the atmosphere (such as we calculated in Example 3.9). We will assume that the Ar mixes until it reaches its composition in air (about 1%) and that no air diffuses back into the cylinder; that is, the cylinder remains pure Ar.

[6] Alternatively, the open-system approach to step 1 will be presented at the end of this example.

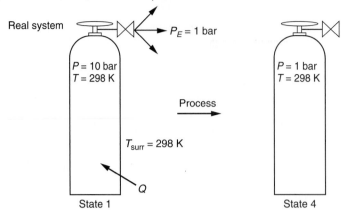

Figure E3.11A Real system of a gas undergoing an expansion from a cylinder.

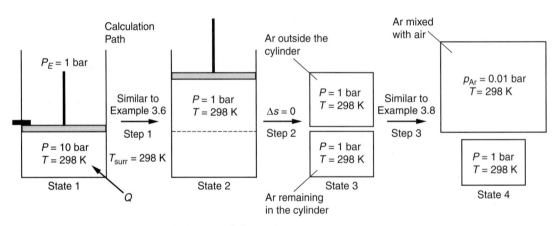

Figure E3.11B Calculation path for Δs from state 1 to state 4.

Step I: State 1 to state 2

To solve for the entropy change of the universe in step I, we must determine the entropy change of the system as well as the entropy change of the surroundings. Since the initial and final states are constrained, the entropy change of the system can be calculated from Equation (3.63):

$$\Delta s_{\text{sys}} = c_P \ln\left(\frac{T_2}{T_1}\right)^{\!0} - R\ln\left(\frac{P_2}{P_1}\right)$$

Thus, the value for the entropy change of the system is given by

$$\Delta s_{\text{sys}} = -R\ln\left(\frac{P_2}{P_1}\right) = -\left(8.314\left[\frac{\text{J}}{\text{mol K}}\right]\right)\ln\left(\frac{1}{10}\right) = 17.1\left[\frac{\text{J}}{\text{mol K}}\right]$$

As the argon does work on the surroundings during the expansion process, it will lose energy. An equal amount of energy must be transferred in via heat to keep the temperature constant at 298 K. To find the entropy change to the surroundings, we need to determine the amount of heat transferred. A first-law balance gives

$$\Delta u = q - P_{\text{ext}}(v_2 - v_1) = q - P_2(v_2 - v_1) = 0$$

where we have set the change in internal energy equal to zero, since we are treating Ar as an ideal gas at constant temperature. Solving for q, we get the entropy change of the surroundings to be

$$\Delta s_{\text{surr}} = \frac{q_{\text{surr}}}{T_{\text{surr}}} = -\frac{q}{T_2} = \frac{-P_2(v_2 - v_1)}{T_2} = -R\left[1 - \frac{P_2}{P_1}\right]$$

$$= -(8.314)(0.9) = -7.483\left[\frac{\text{J}}{\text{mol K}}\right]$$

Thus, the entropy change of the universe is then given by

$$\Delta s_{\text{univ}}^{\text{I}} = \Delta s_{\text{sys}} + \Delta s_{\text{surr}} = 17.1 - 7.5 = 9.6\left[\text{J/(mol K)}\right]$$

or, for the extensive value,

$$\Delta S_{\text{univ}}^{\text{I}} = n(\Delta s_{\text{univ}}) = 38.4\left[\text{J/K}\right] \tag{E3.11A}$$

Step II: State 2 to state 3

The entropy change for step II is zero:

$$\Delta S_{\text{univ}}^{\text{II}} = 0$$

The number of moles remaining in the cylinder can be found using the ideal gas law (with T and V constant)

$$(n_{\text{cyl}})_3 = (n_{\text{cyl}})_1 \frac{P_3}{P_1} = 0.4\,[\text{mol}]$$

Step III: State 3 to state 4

The entropy change for the system from state 3 to state 4 is similar to that of the gases in Example 3.9. The entropy of mixing of argon outside the cylinder is given by

$$\Delta S_{\text{sys}}^{\text{III}} = n_{\text{outside}}(s(\text{at } p_{\text{Ar}}, T) - s(\text{at } P, T)) = -n_{\text{outside}}R\left(\ln\frac{p_{\text{Ar}}}{P}\right)$$

Plugging in numbers, we get

$$\Delta S_{\text{sys}}^{\text{III}} = -(3.6)(8.314)(\ln(0.01)) = 137.8\left[\text{J/K}\right]$$

Assuming the composition of the air in the room does not noticeably change by the dilute addition of argon, the entropy change to the surroundings is zero and[7]

$$\Delta S_{\text{univ}}^{\text{III}} = 137.8[\text{J/K}]$$

Adding together the three steps in the solution path, we get:

$$\Delta S_{\text{univ}} = \Delta S_{\text{univ}}^{\text{I}} + \Delta S_{\text{univ}}^{\text{II}} + \Delta S_{\text{univ}}^{\text{III}} = 176.2\left[\text{J/K}\right]$$

The value is greater than zero, as we would expect for this very irreversible process. It should be noted that solutions to this kind of problem are often presented where reversibility is assumed if the leak is "slow enough." However, this approach is wrong. No matter how slow the leak, the driving force for the expansion is finite (as opposed to differential), so this process is not reversible. It would be just as absurd as to invoke

---

[7] Alternatively, this problem can be solved by mixing the argon with an infinite amount of air, with the same result.

reversibility in Example 3.9 by saying that the mixing of the gases in the two compartments occurred through a vary small hole and, therefore, very slowly.

[Step 1] alternative: Open-system analysis

We begin with the unsteady-state open-system entropy equation, Equation (3.54), with no inlet and one exit stream:

$$\left(\frac{dS}{dt}\right)_{univ} = \left(\frac{dS}{dt}\right)_{sys} + \dot{n}_e s_e + \frac{\dot{Q}_{surr}}{T_{surr}} = \left(\frac{dS}{dt}\right)_{sys} + \dot{n}_e s_e - \frac{\dot{Q}}{T_{surr}} \tag{E3.11B}$$

If the exit state is assumed to be uniform, we can integrate Equation (E3.11A) to give

$$\Delta S_{univ} = n_2 s_2 - n_1 s_1 + s_e \int_0^t \dot{n}_e dt - \frac{Q}{T_{surr}} \tag{E3.11C}$$

However, since the exit state and state 2 are identical,

$$s_e = s_2 \tag{E3.11D}$$

and by a mole balance:

$$\int_0^t \dot{n}_e dt = n_1 - n_2 \tag{E3.11E}$$

Plugging Equations (E3.11D) and (E3.11E) into Equation (E3.11C) gives

$$\Delta S_{univ} = n_1(s_2 - s_1) - \frac{Q}{T_{surr}} \tag{E3.11F}$$

We can find the entropy difference between states 1 and 2 by using Equation (3.63):

$$s_2 - s_1 = \int_{T_1}^{T_2} \frac{c_P}{T} dT - R \ln\left(\frac{P_2}{P_1}\right) = 0 - \left(8.314 \left[\frac{J}{mol\ K}\right]\right) \ln\left(\frac{1}{10}\right)$$

$$= 17.1 \left[\frac{J}{mol\ K}\right] \tag{E3.11G}$$

We must now solve for the heat transfer Q, by an energy balance. Integrating Equation (2.47), we get

$$n_2 u_2 - n_1 u_1 = -\left[\int_0^t \dot{n}_e dt\right] h_e + Q = -\left[\int_0^t \dot{n}_e dt\right] (u_e + P_e v_e) + Q \tag{E3.11H}$$

Applying Equation (E3.11E) and solving Equation (E3.11H) for Q gives

$$Q = \left[\int_0^t \dot{n}_e dt\right] P v_e = (n_1 - n_2) P_e v_e = n_1 \left(1 - \frac{n_2}{n_1}\right) P_e v_e = n_1 \left(1 - \frac{P_2}{P_1}\right) R T_2$$

Therefore, we get

$$\frac{Q}{T_{surr}} = n_1 R \left[1 - \frac{P_2}{P_1}\right] = 4(8.314)(0.9) = 29.93 \left[\frac{J}{K}\right] \tag{E3.11I}$$

Plugging the results from Equations (E3.11G) and (E3.11I) into Equation (E3.11F):

$$\Delta S_{univ} = 4(17.1) - 29.9 = 38.5\,[J/K] \tag{E3.11J}$$

The results for the closed- and open-system analysis, Equations (E3.11A) and (E3.11J), are identical to within round-off error. ◀

► 3.8 THE MECHANICAL ENERGY BALANCE AND THE BERNOULLI EQUATION

We next consider a special case—flow processes that are at **steady-state** and **reversible**, with one stream in and one stream out. We wish to come up with an expression to evaluate the work in such a process. The first-law balance, in differential form, is given by

[handwritten margin notes: Open Steady stream reversible 1 stream in and 1 stream out — 1st law]

$$0 = -\dot{n}[d(h + e_K + e_P)] + \delta \dot{Q}_{sys} + \delta \dot{W}_s \qquad (3.66)$$

The second law is

$$\dot{n}ds + \frac{\delta \dot{Q}_{surr}}{T} = 0 \qquad (3.67)$$

However, since

$$\delta \dot{Q}_{surr} = -\delta \dot{Q}_{sys} \qquad (3.68)$$

Equation (3.67) can be substituted into (3.66) to give

$$0 = -\dot{n}[d(h + e_K + e_P)] + \dot{n}Tds + \delta \dot{W}_s \qquad (3.69)$$

Rearranging Equation (3.69) to solving for work:

$$\frac{\delta \dot{W}_s}{\dot{n}} = [dh - Tds + de_K + de_P] \qquad (3.70)$$

We can simplify Equation (3.70) even further. For a reversible process, we can write

$$du = \delta q_{rev} + \delta w_{rev} = Tds - Pdv \qquad (3.71)$$

If we add $d(Pv)$ to both sides, we get

$$dh = d(u + Pv) = Tds + vdP \qquad (3.72)$$

Solving Equation (3.72) for $(dh - Tds)$ and plugging into Equation (3.70) gives:

$$\frac{\delta \dot{W}_s}{\dot{n}} = [vdp + de_K + de_P] \qquad (3.73)$$

Equation (3.73) is termed the differential mechanical energy balance. It is a useful form, since the work is written in terms of the measured variables P and v as well as bulk potential and kinetic energy. It is applicable to reversible, steady-state processes with one stream in and one stream out. If we integrate Equation (3.73), we get

$$\dot{W}_s/\dot{n} = \int_1^2 vdP + (e_{K_2} - e_{K_1}) + (e_{P_2} - e_{P_1})$$

$$= \int_1^2 vdP + MW\left[(\vec{V}_2^2 - \vec{V}_1^2)/2\right] + MWg(z_2 - z_1) \qquad (3.74)$$

where MW is the molecular weight. There are two frequent cases where Equation (3.74) is applied

Case 1: No Work (Nozzles, Diffusers)

$$0 = \int_1^2 v dP + MW \left(\frac{\vec{V}_2^2 - \vec{V}_1^2}{2} \right) + MW g(z_2 - z_1) \qquad (3.75)$$

Equation (3.75) is the celebrated Bernoulli equation.

Case 2: No e_K, e_P (Turbines, Pumps)

$$\frac{\dot{W}_s}{\dot{n}} = \int_1^2 v dP \qquad (3.76)$$

▶ **EXAMPLE 3.12**
Power Generated by a Turbine

An ideal gas enters a turbine with a flow rate of 250 mol/s at a pressure of 125 bar and a specific volume of 500 cm³/mol. The gas exits at 8 bar. The process operates at steady-state. Assume the process is reversible and polytropic with

$$Pv^{1.5} = \text{const}$$

Find the power generated by the turbine.

SOLUTION Since this process is at steady-state and is reversible, we can use Equation (3.76):

$$\frac{\dot{W}_s}{\dot{n}} = \int_{P_1}^{P_2} v dP \qquad (E3.12A)$$

Since both v and P vary during the process, we must write v in terms of P to perform the integral in Equation (E3.12A). This polytropic process is described by

$$Pv^{1.5} = \text{const} = P_1 v_1^{1.5}$$

where we have written the constant in terms of state 1, since we know both v and P. Solving for v gives

$$v = \left(\frac{P_1 v_1^{1.5}}{P} \right)^{2/3} \qquad (E3.12B)$$

Substituting Equation (E3.12B) into (E3.12A) and integrating leaves

$$\frac{\dot{W}_s}{\dot{n}} = (P_1)^{0.6667} v_1 \int_{P_1}^{P_2} P^{-(2/3)} dP = 3(P_1)^{0.6667} v_1 \left[P^{0.3333} \right]_{P_1}^{P_2}$$

Plugging in numerical values, we get

$$\frac{\dot{W}s}{\dot{n}} = 3 \left(1.25 \times 10^7 [\text{Pa}] \right)^{0.6667} \left(5 \times 10^{-4} [\text{m}^3/\text{mol}] \right) \left(\sqrt[3]{8 \times 10^5 [\text{Pa}]} - \sqrt[3]{1.25 \times 10^7 [\text{Pa}]} \right)$$

$$= -11,250 \, [\text{J/mol}]$$

Finally, solving for power gives

$$\dot{W}_s = (250 \, [\text{mol/s}]) \left(-11,250 \, [\text{J/mol}] \right) = -2.8 \, [\text{MW}]$$

The sign is negative, since we get useful work out of the turbine. Note, for comparison, that a coal-fired power plant generates on the order of 1000 MW. ◀

▶ **EXAMPLE 3.13**
Correction for
Efficiency in a
Real Turbine

In an actual expansion through the turbine of Example 3.12, −2.1 [MW] of power is obtained. What is the isentropic efficiency, η_{turbine}, for the process? The isentropic efficiency is given by

$$\eta_{\text{turbine}} = \frac{(\dot{W}_s)_{\text{real}}}{(\dot{W}_s)_{\text{rev}}}$$

where $(\dot{W}_s)_{\text{rev}}$ represents the power obtained in the reversible process from the same inlet state and outlet pressure and $(\dot{W}_s)_{\text{real}}$ is the power of the actual process.

SOLUTION   We run into many different definitions of efficiency, depending on the context. **Isentropic efficiencies** compare the actual performance of a process operation with the performance it would obtain if it operated reversibly; that is, we compare the given process to the best it could do. For a turbine, the same inlet state and same exit pressure are used in the calculation. Figure E3.13 shows both the actual process (solid line) and the ideal, reversible process (dashed line) on a Ts diagram. Both processes start at the same state and end at the same pressure. The final temperature of the actual process is higher than the reversible process, since less energy is removed via work.

The isentropic efficiency is calculated to be

$$\eta_{\text{turbine}} = \frac{-2.1[\text{MW}]}{-2.8[\text{MW}]} = 0.75$$

We obtain an isentropic efficiency of 75%.

We can use a similar approach to determine isentropic efficiencies of other unit processes, such as pumps or nozzles. The isentropic efficiency of a pump compares the minimum work needed from the same inlet state and an outlet at the same pressure to the actual work:

$$\eta_{\text{pump}} = \frac{(\dot{W}_s)_{\text{rev}}}{(\dot{W}_s)_{\text{real}}}$$

Can you draw a figure analogous to Figure E3.13 for a pump? The isentropic efficiency of a nozzle compares the actual exit kinetic energy to that the fluid would obtain in a reversible process. Values between 70% and 90% are typical for turbines and pumps, while nozzles typically obtain 95% or better. ◀

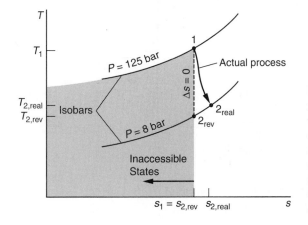

Figure E3.13   A Ts diagram illustrating the states between which the isentropic efficiency is calculated. The actual process is shown by the solid line, between states 1 and 2_{real}, while the ideal, reversible process is shown by the dashed line between states 1 and 2_{rev}.

► 3.9 VAPOR-COMPRESSION POWER AND REFRIGERATION CYCLES

In this section, we will examine the basic elements of common industrial power and refrigeration systems. These systems employ a thermodynamic cycle in which a working fluid is alternatively vaporized and condensed as it flows through a set of four processes. Recall from Section 2.9 that, after completing a cycle, the working fluid returns to the same state it was in initially so that the cycle can be repeated. We will use the principles of energy conservation and entropy to analyze the performance of power and refrigeration cycles. We will first examine the Rankine cycle, which is used to convert a fuel source to electrical power. We will then look at a vapor-compression cycle operated in "reverse" to expel heat from a cold reservoir and produce refrigeration.

The Rankine Cycle

[handwritten note: Assumptions
- Turbine and compressor operate adiabatically and reversibly
- boiler + compressor operate @ constant pressure]

Say we wish to convert a fossil-fuel, nuclear, or solar energy source into net electrical power. To accomplish this task, we can use a Rankine cycle. The Rankine cycle is an idealized vapor power system that contains the major components found in more detailed, practical steam power plants. While hydroelectric and wind are possible alternative sources, the steam power plant is presently the dominant producer of electrical power.

A schematic of the Rankine cycle is shown in Figure 3.7. The left-hand side shows the four unit processes in order: a turbine, a condenser, a compressor, and a boiler. States 1, 2, 3, and 4 are labeled. The right-hand side identifies each of these states on a *Ts* diagram. Each of the four individual processes operates as an open system at steady-state, such as those modeled in Section 2.8. Moreover, these processes are assumed to be reversible; hence, the efficiency we calculate will be the best possible for a given design scenario. The working fluid that flows through these processes is usually water. We will formulate our analysis on a mass basis, in anticipation of using the steam tables for thermodynamic data. Electrical power is generated by the turbine, while the energy from combustion of the fuel is input via heat transfer in a boiler. Energy transfer between the surroundings and the system is further needed to return the system to its initial state and complete the cycle. This energy transfer occurs via heat expulsion in the condenser and via work

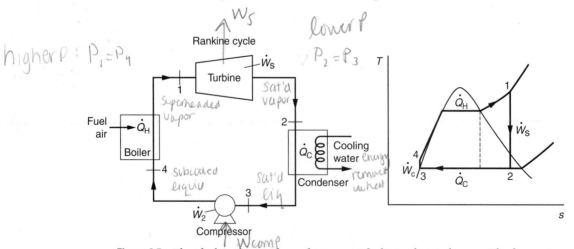

Figure 3.7 The ideal Rankine cycle used to convert fuel into electrical power. The four unit processes are sketched on the left, while the path on a *Ts* diagram is shown on the right. The working fluid is typically water.

input to the compressor. A more detailed analysis of the four processes in the Rankine cycle follows.

We start from state 1 in the diagram in Figure 3.7, where the working fluid enters the turbine as a superheated vapor. As it goes through the turbine, it expands and cools while producing work. It exits the turbine at state 2. The rate of work produced can be determined by applying the first law [Equation (2.94)]. Assuming that bulk kinetic and potential energy and heat transfer are negligible, the power produced by the turbine becomes

$$\dot{W}_s = \dot{m}(\hat{h}_2 - \hat{h}_1) \tag{3.77}$$

where $\dot{m}$ is the flow rate of the working fluid. Since the process is reversible with negligible heat transfer, the entropy remains constant, as depicted by the vertical line in the Ts diagram:

$$\hat{s}_1 = \hat{s}_2 \tag{3.78}$$

Because the steam enters as a superheated vapor, it does not condense significantly in the turbine. If the steam is saturated as it enters the turbine, a significant fraction of liquid is formed as the temperature drops isentropically. The dashed line on the Ts diagram illustrates this case. This option is impractical, since too much liquid causes erosion and wear of the turbine blades.

The steam next enters a condenser; it exits in state 3 as saturated liquid water. The change of phase occurs at constant pressure and requires that energy be removed from the flowing stream via heat. Thus, a low-temperature reservoir is needed. A first-law balance around the condenser gives

$$\dot{Q}_C = \dot{m}(\hat{h}_3 - \hat{h}_2) \tag{3.79}$$

Next it is desired to raise the pressure of the liquid, which is accomplished using a compressor. High-pressure water exits the compressor in state 4. The work required to compress the liquid is given by

$$\dot{W}_c = \dot{m}(\hat{h}_4 - \hat{h}_3) = \dot{m}\hat{v}_l(P_4 - P_3) \tag{3.80}$$

where Equation (3.76) was integrated assuming $\hat{v}_l$ is constant. Since the molar volume of the liquid is significantly less than that of the vapor, the work required by the compressor is much less than that produced by the turbine. Typically, a small fraction of the power produced by the turbine is used to compress the liquid, and the remaining power is the net power obtained by the cycle. The liquid that enters the compressor is saturated, by design, since most compressors cannot handle a two-phase mixture. Finally, the high pressure liquid is brought back to a superheated vapor state in the boiler. It is in this step that energy released from the combustion of fuel is transferred to the working fluid. The fuel provides the high-temperature reservoir for the boiler. The boiler isobarically heats the liquid to saturation, vaporizes it, and then superheats the vapor. The rate of heat transfer in the boiler is given by

$$\dot{Q}_H = \dot{m}(\hat{h}_1 - \hat{h}_4) \tag{3.81}$$

The vapor exits the boiler in state 1 and the cycle is repeated.

We define the efficiency of the cycle as the ratio of the net work obtained divided by the heat absorbed from the boiler:

$$\eta_{\text{Rankine}} = \frac{\left|\dot{W}_s\right| - \dot{W}_c}{\dot{Q}_H} = \frac{\left|(\hat{h}_2 - \hat{h}_1)\right| - (\hat{h}_4 - \hat{h}_3)}{(\hat{h}_1 - \hat{h}_4)} \tag{3.82}$$

In this definition, it is assumed that the heat absorbed is proportional to the amount of fuel consumed. The Ts diagram provides a useful graphical aid in interpreting the Rankine cycle. From the definition of entropy,

$$q_{\text{rev}} = \int T\,ds \qquad (3.83)$$

Since we are assuming reversibility for the cycle, the heat absorbed by the water in the boiler, q_{H}, and the heat expelled in the condenser, q_{C}, are equal to the respective area under the Ts curve. These graphical depictions are illustrated in the first two diagrams of Figure 3.8. The net work produced by the cycle is given by the difference of these two quantities:

$$q_{\text{H}} - |q_{\text{C}}| = |w_s| - w_c = w_{\text{net}} \qquad (3.84)$$

Thus, the net work is equal to the area of the box in the third diagram. If we can make this box bigger relative to q_{H}, we increase the efficiency. Can you think of ways to accomplish this?

▶ **EXAMPLE 3.14**
Calculation of the
Power and Efficiency
of a Rankine Power Cycle

Steam enters the turbine in a power plant at 600°C and 10 MPa and is condensed at a pressure of 100 kPa. Assume the plant can be treated as an ideal Rankine cycle. Determine the power produced per kg of steam and the efficiency of the cycle. How does the efficiency compare to a Carnot cycle between these two temperatures?

SOLUTION We can refer to Figure 3.7 to identify the states of water as it goes through the cycle. It is useful to refer to this figure as we are solving the problem. Examining Equation (3.82), we see that we need to determine the enthalpies in the four states to solve for the efficiency. Steam enters the turbine at 600°C and 10 MPa. Looking up values from the steam tables (Appendix B), we get

$$\hat{h}_1 = 3625.3\,[\text{kJ/kg}]$$

and

$$\hat{s}_1 = 6.9028\,[\text{kJ/(kg K)}] = \hat{s}_2$$

Equation (3.78) was used to relate the entropy at state 1 to that at state 2. We also know the pressure at state 2, $P_2 = 100$ kPa. Thus, state 2 is completely constrained, and we can determine

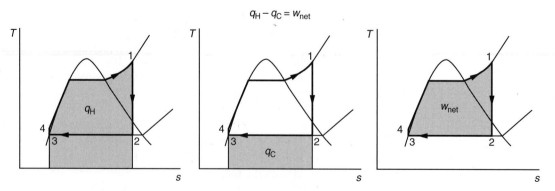

$q_{\text{H}} - q_{\text{C}} = w_{\text{net}}$

Figure 3.8 Graphical representation of the heat absorbed in the boiler, q_{H}, the heat expelled in the condenser, q_{C}, and the net work, w_{net}, in an ideal Rankine cycle.

its enthalpy from the steam tables. Since it exists as a liquid–vapor mixture, we must determine the quality, x, of the steam as follows:

$$\hat{s}_2 = 6.9028[\text{kJ}/(\text{kg K})] = (1 - x)\hat{s}_l + x\hat{s}_v = (1 - x)(1.3025\,[\text{kJ/kg K}]) + x(7.3593\,[\text{kJ/kg K}])$$

Solving for x gives

$$x = 0.925$$

Therefore, the enthalpy in state 2 is given by

$$\hat{h}_2 = (1 - x)\hat{h}_l + x\hat{h}_v = (0.075)(417.44\,[\text{kJ/kg}]) + 0.925(2675.5\,[\text{kJ/kg}]) = 2505.3\,[\text{kJ/kg}]$$

The power generated by the turbine is found by the enthalpy difference between state 2 and state 1:

$$\hat{w}_s = \hat{h}_2 - \hat{h}_1 = 2505.3 - 3625.3 = -1120.0\,[\text{kJ/kg}]$$

The enthalpy of state 3 is a saturated liquid at 100 kPa:

$$\hat{h}_3 = \hat{h}_l = 417.44\,[\text{kJ/kg}]$$

The enthalpy at state 4 can be determined from Equation (3.80):

$$\hat{h}_4 = \hat{h}_3 + \hat{v}_l(P_4 - P_3) = 427.34\,[\text{kJ/kg}]$$

Note that since we are increasing the pressure of water in the liquid state, the work required is only 9.9 kJ/kg. This value is significantly less than the work produced by expansion of vapor through the turbine (1120 kJ/kg). The net work is given by

$$\boxed{\hat{w}_{\text{net}} = \hat{w}_s + \hat{w}_c = -1120.0 + 9.9 = -1110.1\,[\text{kJ/kg}]}$$

Solving for the efficiency from Equation (3.82), we get

$$\boxed{\eta_{\text{Rankine}} = \left(|\dot{W}_s| - \dot{W}_c\right)/\dot{Q}_{\text{H}} = \left[\left|\left(\hat{h}_2 - \hat{h}_1\right)\right| - \left(\hat{h}_4 - \hat{h}_3\right)\right]/(\hat{h}_1 - \hat{h}_4) = 0.347}$$

The 34.7% efficiency is the best-case scenario for this cycle, since we assumed reversible processes. In reality, we would not even achieve this value!

The Carnot efficiency is given by Equation (3.32):

$$\eta_{\text{Carnot}} = 1 - \frac{T_{\text{C}}}{T_{\text{H}}} = 1 - \frac{373}{873} = 0.573$$

The Rankine efficiency is lower than the Carnot efficiency. We can see the basis if we compare the net work graphically, as we did in Figure 3.8. The box representing net work for each cycle is shown in Figure E3.14. Recall that the steam that enters the turbine of the Rankine cycle is superheated to eliminate wear and corrosion on the turbine blades. In modifying the cycle in this way, we "crop off" a significant portion of the rectangle that represents the Carnot cycle. ◄

► **EXAMPLE 3.15**
Modification of
Rankine Analysis for
Nonisentropic Steps

Redo the analysis of the Rankine cycle of Example 3.14 but include isentropic efficiencies of 85% in the pump and turbine. Determine the net power and the overall efficiency of the power cycle.

SOLUTION Recall the discussion of isentropic efficiencies from Example 3.13. The isentropic efficiency of the turbine is given by:

$$\eta_{\text{turbine}} = \frac{(\dot{W}_s)_{\text{actual}}}{(\dot{W}_s)_{\text{rev}}} = \frac{(\hat{w}_s)_{\text{actual}}}{(\hat{w}_s)_{\text{rev}}}$$

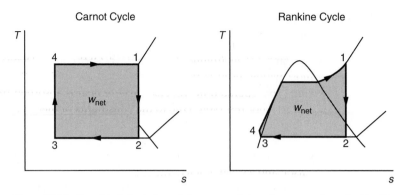

Figure E3.14 Graphical depiction of the net work in a Carnot cycle and a Rankine cycle.

The reversible work was found in Example 3.14. Solving for actual work gives

$$(\hat{w}_s)_{\text{actual}} = \eta_{\text{turbine}}(\hat{w}_s)_{\text{rev}} = 0.85(-1120.0) = -952.0 \, [\text{kJ/kg}]$$

As we suspect, the work we get out of the turbines in the real, irreversible process is less than that for the reversible process. Solving the energy balance around the turbine gives the enthalpy in state 2 as:

$$(\hat{h}_2)_{\text{actual}} = (\hat{w}_s)_{\text{actual}} + \hat{h}_1 = -952.0 + 3625.3 = 2673.3 \, [\text{kJ/kg}]$$

This value is higher than that found in Example 3.14, indicating that the temperature entering the condenser for the actual process is higher than it would be in the reversible case. State 3 remains the same:

$$\hat{h}_3 = \hat{h}_1 = 417.44 \, [\text{kJ/kg}]$$

Since the reversible work represents the best we can possibly do, the actual work needed in the compressor must be greater than the reversible value. Hence, the isentropic efficiency in the compressor is given by

$$(\hat{w}_c)_{\text{actual}} = \frac{(\hat{w}_c)_{\text{rev}}}{\eta_{\text{compressor}}} = \frac{9.9}{0.85} = 11.6 \, [\text{kJ/kg}]$$

Solving for the enthalpy at the exit of the compressor gives

$$(\hat{h}_4)_{\text{actual}} = (\hat{w}_c)_{\text{actual}} + \hat{h}_3 = 11.6 + 417.44 = 429.1 \, [\text{kJ/kg}]$$

The net work can be found by adding the actual work generated in the turbine to that consumed in the compressor:

$$\boxed{\hat{w}_{\text{net}} = \hat{w}_s + \hat{w}_c = -952.0 + 11.6 = -940.4 [\text{kJ/kg}]}$$

Similarly, the actual efficiency must be calculated using these values:

$$\boxed{\eta_{\text{Rankine}} = (|\dot{W}_s| - \dot{W}_c)/\dot{Q}_{\text{H}} = \left[\left| (\hat{h}_2 - \hat{h}_1) \right| - (\hat{h}_4 - \hat{h}_3) \right] \Big/ (\hat{h}_1 - \hat{h}_4) = 0.294}$$

Introducing irreversibilities in turbine and pump reduced efficiency from 34.7% to 29.4%. ◀

The Vapor-Compression Refrigeration Cycle

Refrigeration systems are important in industrial and home use when temperatures less than the ambient environment are required. Of the several types of refrigeration systems, the most widely used is the vapor-compression refrigeration cycle. It is essentially a Rankine cycle operated in "reverse," where heat is absorbed from a cold reservoir and rejected to a hot reservoir. Due to the constraints of the second law, this process can be accomplished only with a concomitant consumption of power.

A schematic of the ideal vapor-compression cycle is shown in Figure 3.9. The left-hand side shows the four unit processes in order: an evaporator, a compressor, a condenser, and a valve. Each of the four individual processes operates as an open system at steady-state. States 1, 2, 3, and 4 are labeled. The right-hand side identifies each of these states on a Ts diagram. Unlike in the Rankine cycle, the work required for refrigeration is not represented by the area enclosed on the Ts diagram because the expansion through the valve is irreversible.

The working fluid is termed the *refrigerant*. In choosing a refrigerant, we must realize that both the evaporation and condensation processes contain phase transformations. Thus, in each of these processes, T and P are not independent. Specifying the temperature at which these processes occur, restricts the pressure for a given choice of refrigerant. For example, the evaporator temperature is determined by the temperature, T_C, required from our refrigeration system. For a given working fluid, constraining T_C also constrains the evaporator P. We typically want a species that boils at lower temperatures than water. Ideally, we choose a species that provides the desired refrigeration temperature at a pressure slightly above atmospheric. In that way there is a positive pressure against the environment. Common choices are CCl_2F_2 (refrigerant 12), CCl_3F (refrigerant 11), CH_2FCF_3 (refrigerant 134a), and NH_3. The first two species, the chlorofluorocarbons, are very stable if released to the environment. They have mostly been phased out of use because they lead to depletion of the ozone layer and also contribute to the greenhouse effect, which leads to global warming.

An analysis of the four processes in the vapor-compression refrigeration cycle follows. We start from state 1 in the diagram in Figure 3.9, where the working fluid enters the evaporator. In the evaporator, heat is transferred from the refrigerated unit to the working fluid. This occurs at temperature T_C. The working fluid absorbs $\dot{Q}_C$ as it changes phase.

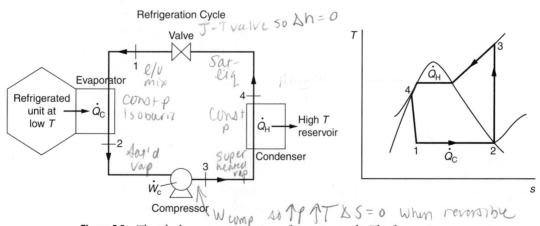

Figure 3.9 The ideal vapor-compression refrigeration cycle. The four unit processes are sketched on the left, while the path on a Ts diagram is shown on the right.

It emerges in state 2, where it is a vapor. The heat transferred is given by

$$\dot{Q}_C = \dot{n}(h_2 - h_1) \tag{3.85}$$

where $\dot{n}$ is the molar flow rate of the refrigerant.

We must then compress the refrigerant to a high enough pressure that it condenses at the temperature of the hot reservoir available to us, T_H. The choice of refrigerant determines the outlet pressure required of the compressor. Since we are performing the compression in the vapor phase, where molar volumes are large, a significant amount of work is needed. The higher the pressure, the more work is required for a given refrigeration effect. The power of compression is given by

$$\dot{W}_c = \dot{n}(h_3 - h_2) \tag{3.86}$$

If the compression is assumed to be reversible,

$$s_3 = s_2 \tag{3.87}$$

The high-pressure vapor is then condensed at T_H, expelling heat $\dot{Q}_H$ to the hot reservoir.

$$\dot{Q}_H = \dot{n}(h_4 - h_3) \tag{3.88}$$

The high-pressure liquid is then expanded in a valve back to state 1 so the cycle can be repeated. A valve is used instead of the turbine that was used in the Rankine cycle. The amount of work that would be produced by a turbine is small, so we replace it with a valve to reduce the complexity. This step is represented by a throttling process, where

$$h_4 = h_1 \tag{3.89}$$

Since the pressure decreases as the refrigerant passes through the valve, its entropy increases, as shown in Figure 3.9. Can you locate the evaporator and condenser on the refrigerator you have at home?

The coefficient of performance, COP, of a refrigeration cycle measures its performance. It is defined as the ratio of the heat absorbed from the cold reservoir (the refrigeration effect) to the work required:

$$COP = \frac{\dot{Q}_C}{\dot{W}_c} = \frac{h_2 - h_1}{h_3 - h_2} \tag{3.90}$$

In real refrigeration systems, a finite temperature difference is needed to get practical heat-transfer rates in the evaporator and the condenser. Thus, the evaporator must operate at a *lower* temperature than the desired refrigeration temperature, while the condenser must operate at a higher temperature than the ambient heat reservoir. Thus, more work is required to obtain a given refrigeration effect. Moreover, irreversibilities in the compressor must be considered, also adding to required work load and further decreasing the COP. COPs of well-designed real refrigeration systems typically fall between 2 and 5.

▶ **EXAMPLE 3.16**
Estimation of the *COP* of a Vapor-Compression Refrigeration Cycle

It is desired to produce 10 kW of refrigeration from a vapor-compression refrigeration cycle. The working fluid is refrigerant 134a. The cycle operates between 120 kPa and 900 kPa. Assuming an ideal cycle, determine the *COP* and the mass flow rate of refrigerant needed. Properties of refrigerant 134a can be found at http://webbook.nist.gov/chemistry/fluid/. Data can be viewed in an HTML table.

SOLUTION   A sketch of the process on a *Ts* diagram is shown in Figure E3.16.

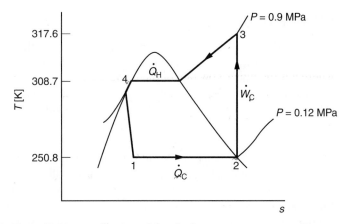

Figure E3.16 A *Ts* diagram of the ideal vapor-compression refrigeration cycle of Example 3.15.

The following saturated data for refrigerant 134a are obtained from the NIST site (see citation Pg. 24):

| P | T | h_l | h_v | s_l | s_v |
[MPa]	[K]	[kJ/mol]	[kJ/mol]	[J/(mol K)]	[J/(mol K)]
0.12	250.84	17.412	$39.295 = h_2$	90.649	$177.89 = s_2$
0.90	308.68	$24.486 = h_4$	42.591	$119.32 = s_4$	174.74

To get the amount of refrigeration, we can combine Equations (3.85) and (3.89):

$$q_C = (h_2 - h_1) = (h_2 - h_4) = 39.295 - 25.486 = 13.809 \,[\text{kJ/mol}]$$

To find the state at the exit of the compressor, we use Equation (3.87):

$$s_3 = s_2$$

At a pressure of 0.90 MPa, from the NIST website, we get the entropies to match state 3 at:

| P | T | h_v | s_v |
[MPa]	[K]	[kJ/mol]	[J/(mol K)]
0.90	317.62	43.578	177.89

The work required by the compressor is

$$w_c = (h_3 - h_2) = 43.578 - 39.295 = 4.283 \,[\text{kJ/mol}]$$

Solving for the coefficient of performance gives

$$\text{COP} = \frac{\dot{Q}_C}{\dot{W}_c} = \frac{h_2 - h_1}{h_3 - h_2} = 3.22$$

To get the desired 10 kW of refrigeration, we need

$$\dot{n} = \frac{\dot{Q}_C}{q_C} = \frac{10 \,[\text{kW}]}{13.809 \,[\text{kJ/mol}]} = 0.73 \,[\text{mol/s}]$$

The temperature of the evaporator is around 250 K, below the freezing point of water. On the other hand, the condenser operates between 308 K and 317 K. This temperature is warm enough to expel heat to the ambient environment. ◀

▶ 3.10 MOLECULAR VIEW OF ENTROPY

When we discussed internal energy, u, in the context of the first law, we gained insight through understanding it on a molecular level. To that end, we discussed the molecular components of kinetic and potential energy that atoms and molecules possess. In analogy, we may ask, what is the molecular basis of entropy, s? The molecular view of entropy relates, in the most general sense, to molecular probability. The more different *molecular configurations* a state exhibits, the more likely that state will exist and the greater its entropy.

To examine the idea of molecular probability and configurations, consider a system with two different ideal gaseous species, which we will call atom A and atom B. Entropy is a measure of how many different ways we can configure these species. To illustrate this concept, let's start with a system whose center is divided by a porous membrane. To simplify the problem, but still get our central idea across, we will assign roughly equal volume elements to each atom in the system. We can relax this constraint later if we wish. We will begin modestly, by considering there to be four atoms (two A atoms and two B atoms) in the system. We must realize, however, that we need to extend this concept to roughly 10^{23} atoms to understand the macroscopic systems that we encounter in our physical world. Initially, we will say that the system contains pure gas A on the left of the membrane and pure gas B on the right of the membrane, as shown in Figure 3.10a. In this case, we know exactly where all the A atoms are located (and where all the B atoms are located). Next we allow the atoms to randomly distribute. Figure 3.10b illustrates that six possible distinguishable *molecular configurations* result. It seems logical to assume that all these configurations are equally likely;[8] hence, there is a 1/6 chance the A atoms will be found in their initial state with two atoms on the left-hand side of the membrane, a 4/6 chance there will be 1 A to the left of the membrane, and a 1/6 chance there will be no A. The most probable state of the system is the one in which there is 1 A and 1 B on each side, or, in other words, when A and B are *mixed*! Moreover, in this mixed state, we are not certain exactly where the A atom is; we have lost some information about the system.

This effect of mixing becomes even more dramatic as we increase the size of the system. We now wish to extend the size of the system toward macroscopic amounts. Figure 3.11 shows a stick diagram of the probabilities of the number of A atoms on the left-hand side of a porous membrane for system consisting of ideal gases A and B, as above. The upper-left-hand corner of this figure shows the case depicted in Figure 3.10, that is, a system consisting of 2 A atoms and 2 B atoms. The case with 1 A on the left-hand side is represented by a stick four times as great as the case of 0 A atoms or 2 A atoms, since its probability is four times higher. Let's double the size of the system so that we have 4 A atoms and 4 B atoms. There are now five possibilities for A on the left-hand side: 0, 1, 2, 3, or 4. The probability for this system is presented in the next diagram to the right in Figure 3.11. If we consider each configuration equally likely, the number of *configurations* with 2 A atoms to the left-hand side represents 36 out of 70 possible configurations. Again, we find the case of complete mixing most probable. Again, in its most probable state, we become less certain about the exact positions of the A atoms. On the other hand, only one *configuration* in 70 leads to pure A on the left-hand side, where we know, with certainty, where all the A atoms are. Similar stick diagrams are shown for systems consisting of 8 A atoms and 8 B atoms, 16 A atoms and 16 B atoms,

---

[8] This assumption forms the basis for the *ergodic hypothesis* upon which statistical thermodynamics is constructed.

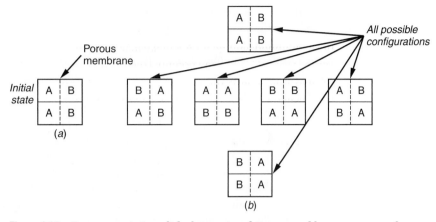

Figure 3.10 System consisting of ideal gases A and B separated by a porous membrane. (*a*) Initial state of the system with all gas A on the left and all gas B on the right. (*b*) Possible configurations after the atoms are allowed to randomly redistribute.

32 A atoms and 32 B atoms, and 128 A atoms and 128 B atoms. As the number of species increases, the mixing becomes more pronounced; that is, the likelihood of having roughly identical numbers of A atoms and B atoms on each side of the membrane is great and the probability of finding all the A atoms on a given side becomes negligible. For example, in the case of 128 A atoms and 128 B atoms, only 1 in 5.8×10^{75} configurations leads to pure A on the left-hand side. Thus, if all *molecular configurations* are equally likely, finding pure A by chance is so improbable that it essentially will not happen. On the other hand, the configurations between 52 and 76 A atoms on the left-hand side of the membrane occurs 99.8% of the time.

What does this argument about molecular probability say about macroscopic systems in which we have on the order of 10^{23} A atoms and 10^{23} B atoms? Regardless of the initial

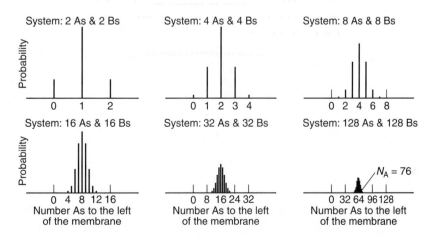

Figure 3.11 Conditional probabilities of the number of A atoms on the left-hand side of a system consisting of ideal gases A and B separated by a porous membrane. Cases are shown for a system consisting of 2 A atoms and 2 B atoms (as depicted in Figure 3.10), 4 A atoms and 4 B atoms, 8 A atoms and 8 B atoms, 16 A atoms and 16 B atoms, 32 A atoms and 32 B atoms, and 128 A atoms and 128 B atoms. As the number of species increases, the mixing becomes more pronounced.

state, if all the *molecular configurations* in this system are allowed to occur with equal probability, the system will evolve to the point where, for all practical purposes, roughly the same number of A atoms and B atoms are on each side of the membrane. We can relate our understanding of molecular probability to our macroscopic observation of the second law if we say that the number of *molecular configurations* a system can take is proportional to its entropy.[9] Thus, if we start with a macroscopic system with pure A on the left-hand side of a porous membrane and pure B on the right-hand side, it has only one possible molecular configuration, so its entropy is low. On the other hand, if it completely mixes, it has a large number of *molecular configurations* and its entropy is high. Thus, the spontaneous mixing of pure gases can be related on a molecular level to molecular probability and on a macroscopic level to entropy. In short, we can relate the irreversibility of macroscopic processes to *mixing* on a molecular level, where we become less certain of the exact molecular state of the system, because it can exist in many equivalent molecular configurations.

We can relate the above discussion to the commonly expressed viewpoint that entropy relates to "disorder." As entropy increases, we become less certain about the exact molecular state of a system; that is, there are more equivalent molecular configurations in which it can exist. Therefore, in a loose sense, we would view it as more disordered. Thus, if we relate the degree of disorder to the number of configurations a system can have, higher entropy means more disorder.

Several examples of directional processes were presented in Section 3.1. We know that the directionality of an irreversible process relates to an increase in entropy. Let's consider a few examples of how it also corresponds to an increase in molecular configurations, that is, mixing.

Maximizing Molecular Configurations over Space

Example 1: Chemical Directionality
The example discussed above directly corresponds to chemical directionality with the mixing of molecules A and B. As Figure 3.11 illustrates, the more completely the two species mix, the greater the number of equivalent spatial configurations and, consequently, the greater the probability the system will be found in that state. This mixing occurs over space, as is illustrated in Figure 3.12.

Example 2: Mechanical Directionality
Another example of an increase in spatial configurations is given by a gas expanding from a compressed cylinder. In this case, the molecules increase in entropy because there are more configurations available in the larger volume after the gas has expanded; that is, we are less certain exactly where a particular molecule is located. This type of "mixing" is illustrated in Figure 3.13. In this figure, the box inset is an expanded molecular view partly within the cylinder and partly outside of it. We can see we have greater knowledge of where the species are in state 1 than in state 2; hence, state 1 has a lower value of entropy. In fact, the spatial mixing depicted in Figure 3.13 is very similar to the species mixing of Figure 3.12. If we examine the hypothetical solution path we used in Example 3.9 (Figure E3.9B), we see that the increase in entropy of a given species as ideal gases mix arises from the fact that it has more space in which to move, whether another species is there or not.

---

[9] It is actually mathematically proportional to e^s.

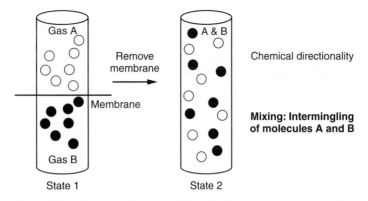

Figure 3.12 Example of chemical directionality as the mixing of molecules A and B over space.

Maximizing Molecular Configurations over Energy

In determining the number of possible configurations of a set of molecules, we must consider not only where they are in space but also how their energy is distributed. This factor results from the quantized nature of energy on the molecular scale. The energy within a given system must be assigned to quantized energy levels in much the same way the mass in the system must be assigned to a specific space. The filling of energy levels is constrained by the total energy of the system. We can understand the role of how the molecules distribute over energy in analogy to the way we related the spatial configurations and probability in the discussion with Figures 3.12 and 3.13. The greater the number of equivalent configurations with which a set of molecules can distribute their energy, the greater the state's entropy. Thus, an increase in entropy can be characterized by either "mixing" over space or "mixing" over energy. Mixing over energy decreases our knowledge of where the energy in the system is.

Example 3: Thermal Directionality
In addition to mixing spatially, irreversible processes are sometimes driven by a mixing over energy levels. The thermal directionality of the cooling of a hot block can be thought

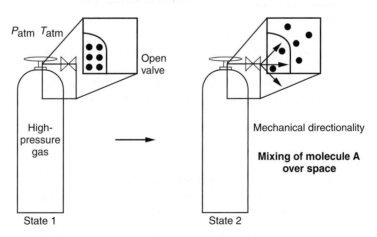

Figure 3.13 Example of mechanical directionality as the mixing of molecule A over space as the gas expands. The molecular projection in this illustration consists of a region within the cylinder and a region outside the cylinder.

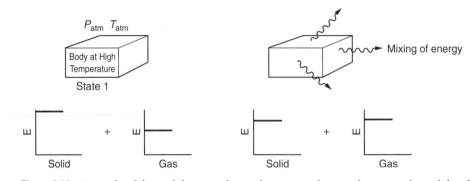

Figure 3.14 Example of thermal directionality as the mixing of energy between a hot solid and the gas surroundings.

of as energetic mixing, as shown in Figure 3.14. There are more configurations of energy as the temperatures of the solid and the gas equilibrate; hence, the probability the system will be in that state is greater.

Example 4: Kinetic Energy to Friction

Other entropic processes can also be viewed of in terms of mixing of energy. Consider a block that is sliding across a surface. It slows down as macroscopic kinetic energy (E_K) is converted into internal energy by friction. This irreversible process can be thought of as mixing of energy from a directed form to dissipation in all directions. Figure 3.15 illustrates this concept. There is only one configuration the block's kinetic energy can be in—that of directed, forward motion. After the block slows down, the increase in temperature leads to greater lattice vibrations in which the energy can be distributed in many configurations. Hence, the conversion of kinetic energy to internal energy by friction represents an irreversible process in which entropy increases.

Example 5: Chemical Reaction

Let's look at the reaction we studied when we learned about the energetics of reaction in Section 2.6. Two molecules of hydrogen gas will spontaneously react with one molecule of oxygen to give two molecules of water. The rearrangement of chemical bonds liberates 3.5 eV of energy. Suppose this reaction occurs adiabatically, so that the entropy of the surroundings does not change. The products will manifest the bond energy liberated by an increase in temperature. Since this process is spontaneous and irreversible, the second law tells us the products must have greater entropy than the reactants. We can understand this in terms of energy configurations.

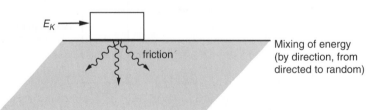
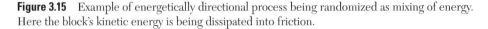

Figure 3.15 Example of energetically directional process being randomized as mixing of energy. Here the block's kinetic energy is being dissipated into friction.

Recall that there are two types of molecular energy: potential energy, such as that stored within a chemical bond, and kinetic energy, which includes translational, vibrational, and rotational motion. In this discussion we will not consider vibrational and rotational motion; however, even with their inclusion, the essence of the argument does not change. Before reaction, the 3.5 eV of energy stored in the reactants can only exist in one configuration—that of bond energy between atoms in the H_2 and O_2 molecules. In other words, each bond has a specified and known strength. Once they react, the two molecules of water must share this exact amount of energy; however, they do not need to share it equally. In fact, there are many combinations of increased velocity these two molecules can have to sum up to the total 3.5 eV of energy. If we pick one of the water molecules, we are not sure of its exact speed. We have lost information. There are more energetic configurations and higher entropy. If we consider four molecules of H_2 and two molecules of O_2, the situation becomes even more drastic. Again, we can specify completely the energy configuration of the reactants. However, now the products are free to distribute 7.0 eV between four molecules. There are many configurations of molecular kinetic energy that can accomplish the feat. If we extend this concept to macroscopic sizes, we have on the order of 10^5 J of energy to distribute among 1 mole of water produced. The 10^{23} molecules will have many configurations of velocity (molecular kinetic energy) to distribute the energy of reaction. This state is much more probable than that of the reactants which distribute this energy in a specific way, that is, in the form of bond energy—so the entropy of reaction is positive and the reaction proceeds spontaneously.

The above discussion focuses on the energy component of chemical reaction. However, there is also a spatial component, which counteracts the energy component. If the system completely reacts, it contains pure water. On the other hand, if some reactant remains, we have a mixture of three species. Those species have many more spatial configurations than pure water. Hence, as the reaction approaches completion, there is a trade-off between the entropy it gains by an increase in the energy configurations with an increase in temperature and the spatial configurations it loses due to "unmixing" in going to pure water.[10] In general, the trade-off between these two components will determine how far a reaction will proceed. We will learn to quantify the extent to which species chemically react in Chapter 9.

In summary, as a system evolves to states with more possible configurations (spatially or energetically) of its molecular states (more probable), its entropy increases.

▶ **EXAMPLE 3.17**
Refrigeration by Adiabatic Demagnetization

Magnetic refrigeration cycles can be used to achieve supercold temperatures. They typically operate between a "hot" reservoir at liquid helium temperature (4.4 K) and a cold reservoir at very low temperature (as low as 0.0065 K). One configuration consists of a paramagnetic working material, such as gadolinium gallium garnet (GGG) or ferric ammonium alum (FAU), in the form of a rim of a wheel. The wheel is rotated between the high-temperature (4.4 K!) reservoir and the low-temperature reservoir. In the high-temperature reservoir, the working material expels heat as it is subjected to a large magnetic field (7 tesla). As the working material is rotated into the low-temperature reservoir, the field slowly becomes smaller and eventually zero. This *demagnetization* process may be assumed to be adiabatic and reversible. As the working material demagnetizes, it cools off. When the working material is then exposed to a low-temperature reservoir, it absorbs heat. From a molecular basis, explain how adiabatic demagnetization cools the working material.

SOLUTION A paramagnetic material consists of many unpaired electrons, each of which has a definite spin state (spin-up or spin-down). In the absence of a magnetic field, the energies of each

[10] For every three molecules that react, only two molecules form; this effect also contributes to the decrease in the spatial component of entropy.

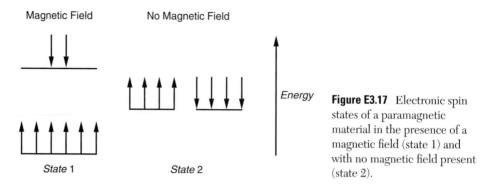

Figure E3.17 Electronic spin states of a paramagnetic material in the presence of a magnetic field (state 1) and with no magnetic field present (state 2).

of the electron states are identical. Thus, an equal number of electrons will occupy each state. When a magnetic field is applied to the material, the energies of the spin-up and spin-down states are no longer identical. Consequently, more electrons occupy the lower energy state then the upper energy state. A schematic of the spin states with a magnetic field (state 1) and without a magnetic field (state 2) is illustrated in Figure E3.17.

The entropy of this system can be thought of as having two components, that due to magnetization, $s_{\text{magnetization}}$, and that due to temperature, $s_{\text{temperature}}$. We first consider $s_{\text{magnetization}}$. Inspection of Figure E3.17 shows that there are fewer configurations possible (less disorder) in state 1, with the magnetic field applied, than in state 2. Thus, the entropy due to magnetization increases during the demagnetization process as the working material goes from state 1 to 2:

$$\Delta s_{\text{magnetization}} > 0$$

However, since the overall process is reversible and adiabatic, the entropy change is zero:

$$\Delta s = 0 = \Delta s_{\text{magnetization}} + \Delta s_{\text{temperature}}$$

Thus,
$$\Delta s_{\text{temperature}} < 0$$

The entropy of the temperature component is lowered as the temperature decreases, that is,

$$T_2 < T_1$$

It is interesting to compare the two states in an adiabatic demagnetization cycle (Figure E3.17) with those in the conventional vapor-compression refrigeration cycle described in Section 3.9. Both cycles rely on the working substance absorbing heat from a cold reservoir as the working substance goes from a more ordered state to a less ordered state. In this example, the liquid-to-vapor transition of conventional refrigeration is replaced with the demagnetization of the paramagnetic material. Similarly, in both cycles, heat is rejected to a hot reservoir as the working substance transits to a more ordered state (liquid or magnetic). This example illustrates one approach to creatively developing engineering processes—applying the same fundamental approach in a new way that is better suited for the given application. Clearly, the super-low temperatures obtained by adiabatic demagnetization refrigeration systems could not be obtained by a conventional liquid–vapor transition, since vapors do not exist at these temperatures. However, in developing this process, the same fundamental type of process (i.e., ordered-to-disordered transition) is exploited. It is just accomplished in a way that is appropriate for the application, where the ordered-to-disordered transition occurs within a solid. Many clever engineering processes have been created by applying analogous fundamental mechanisms in this manner. ◀

▶ 3.11 SUMMARY

The second law of thermodynamics states that the total **entropy of the universe** increases or, at best, remains the same. It never decreases. The entropy of the universe remains unchanged for a **reversible** process, while it increases for an **irreversible** process. Entropy balances have been developed for closed systems and for open systems. In each case, we account for the entropy change of the universe by adding together the entropy change for the system and the entropy change for the surroundings. For example, the integral equation of the second law for a **closed system**, written in extensive form, is

$$\Delta S_{univ} = n(s_{final} - s_{initial}) - \frac{Q}{T_{surr}} \geq 0 \qquad (3.46)$$

In Equation (3.45), the surroundings are at constant temperature, T_{surr}. For **open systems**, it is convenient to write the second law on a rate bases. The integral equation, in extensive form, is

$$\frac{dS}{dt} + \sum_{out} \dot{n}_{out}s_{out} - \sum_{in} \dot{n}_{in}s_{in} - \frac{\dot{Q}}{T_{surr}} \geq 0 \qquad (3.54)$$

The first term in Equation (3.54) describes the rate of entropy change of the system, while the next three terms describe the entropy change of the surroundings due to mass flow from the system, mass flow to the system, and heat transfer. The first and second laws can be combined for flow processes that are at **steady-state** and **reversible**, with one stream in and one stream out, to give the Bernoulli equation:

$$\frac{\dot{W}_s}{\dot{n}} = \int_1^2 vdP + MW\left(\frac{\vec{V}_2^2 - \vec{V}_1^2}{2}\right) + MWg(z_2 - z_1) \qquad (3.74)$$

We have also used these equations in intensive forms, on a mass and a molar basis, and for differential increments.

We applied the second law to many engineering systems. Given a physical problem, we must determine which terms in these equations are important and which terms are negligible or zero. Examples of closed systems included the rigid tank and adiabatic or isothermal expansion/compression in a piston–cylinder assembly. Steady-state open systems include nozzles, diffusers, turbines, pumps, heat exchangers, and throttling devices. Transient open-system problems can entail filling or emptying of a tank. Finally, the vapor-compression power and refrigeration cycles provided useful examples of thermodynamic cycles. In this case, the performance of the cycles is quantified by the **efficiency** and the **coefficient of performance**, respectively. You should understand the concepts well enough that you are not restricted to the systems discussed above but rather can apply the second law to any system that interests you.

When we perform calculations on a reversible process, the second law provides an additional constraint that allows us to determine an unknown state or to calculate quantities such as heat or work. We can use these values to estimate the corresponding values of real processes. The **isentropic efficiency** compares the actual performance of a process operation with the performance it would obtain if it operated reversibly. When applied to a general process, a positive entropy change of the universe shows that the process is possible. If this quantity is zero, the process is reversible, while a negative entropy change of the universe shows the process is not possible.

The change in entropy between state 1 and state 2 is defined as

$$\Delta s = \int_1^2 \frac{\delta q_{rev}}{T} \qquad (3.2)$$

Since entropy is *defined* in terms of heat absorbed during a *reversible* process, we can calculate the entropy between any two states by constructing a path that follows a reversible process from state 1 to state 2. In this way, we found that the entropy change of an **ideal gas** is given by

$$\Delta s = \int_{T_1}^{T_2} \frac{c_P}{T}dT - R\ln\left(\frac{P_2}{P_1}\right) \qquad (3.62)$$

Equation (3.60) is true, in general, for the entropy change associated with an ideal gas between state 1 and state 2; it is not limited only to the reversible process that we chose to develop it. For an ideal gas, the entropy change between (T_1, v_1) and (T_2, v_2) is analogously given by Equation (3.62). Alternatively, the entropy change between two states can be directly obtained from property tables, if they are available.

The molecular view of entropy relates, in the most general sense, to molecular probability. The more different **molecular configurations** a state exhibits, the more likely that state will exist and the greater its entropy. In determining the number of possible configurations of a set of molecules, we must consider not only where they are in space but also how their energy is distributed. This factor results from the quantized nature of energy on the molecular scale. The greater the number of equivalent configurations with which a set of molecules can distribute their energy, the greater the state's entropy. Thus, an increase in entropy can be characterized by either "mixing" over space or "mixing" over energy. These molecular-based concepts have spread into many other fields. Information theory mathematically defines information "entropy" for bits of information using the identical formula that Boltzmann applied to molecular configurations. Similarly, "entropy"-based arguments have expanded into such diverse fields as economics, theology, sociology, art, and philosophy.

▶ 3.12 PROBLEMS

3.1 Develop a general expression for Δs_{sys} for an ideal gas that goes from (v_1, T_1) to (v_2, T_2) based on the path below.

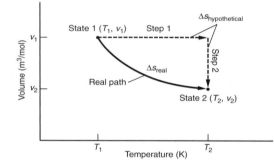

3.2 Develop a general expression for Δs_{sys} for an ideal gas that goes from (P_1, T_1) to (P_2, T_2) where heat capacity is given by:

$$c_P = A + BT + CT^2$$

3.3 A rigid vessel contains 10 kg of steam. The steam is initially at 10 bar and 300°C. After a period of time, the pressure in the vessel is reduced to 1 bar due to heat transfer with the surroundings. The surroundings are at a constant temperature of 20°C. Determine the change in entropy of the system, the surroundings, and the universe during this process.

3.4 A 10-kg block of copper is initially at 100°C. It is thrown in a very large lake that is at 280 K. What is the entropy change of the copper? What is the entropy change of the universe?

3.5 Calculate the change in entropy for the system for each of the following cases. Explain the sign that you obtain by a physical argument.

(a) A gas undergoes a reversible, adiabatic expansion from an initial state at 500 K, 1 MPa, and 8.314 L to a final volume of 16.628 L.

(b) One mole of methane vapor is condensed at its boiling point, 111 K; $\Delta h_v = 8.2\,[\text{kJ/mol}]$.

(c) One mole of liquid water is cooled from 100 °C to 0°C. Take the average heat capacity of water to be 4.2 JK^{-1} g^{-1}.

(d) Two blocks of the same metal with equal mass are at different temperatures, 200°C and 100°C. These blocks are brought together and allowed to come to the same temperature. Assume that these blocks are isolated from their surroundings. The average heat capacity of the metal is $24 \, \mathrm{J \, K^{-1} \, mol^{-1}}$.

3.6 Calculate the change in entropy of the universe for the process described in Problem 2.14. Repeat for Problem 2.15.

3.7 Determine the change in entropy of an ideal gas with constant heat capacity, $c_P = (7/2)R$, between the following states:

(a) $P_1 = 1 \, \mathrm{bar}, T_1 = 300 \, \mathrm{K}; P_2 = 0.5 \, \mathrm{bar}, T_2 = 500 \, \mathrm{K}$

(b) $v_1 = 0.05 \, \mathrm{m^3/mol}, T_1 = 300 \, \mathrm{K}; v_2 = 0.025 \, \mathrm{m^3/mol}, T_2 = 500 \, \mathrm{K}$

(c) $P_1 = 1 \, \mathrm{bar}, T_1 = 300 \, \mathrm{K}; v_2 = 0.025 \, \mathrm{m^3/mol}, T_2 = 500 \, \mathrm{K}$

3.8 Compare the change in entropy (a) when water is heated from its freezing point to its boiling point at 1 atm and (b) when saturated liquid water is vaporized at 1 atm.

3.9 You have just cooled a glass of tap water at 20°C by adding ice, at −10°C. The glass originally contains 400 mL of tap water, to which 100 g of ice is added. Assume that the glass is adiabatic. Calculate the change in entropy of the universe after thermal equilibrium has been obtained. For ice, take $\Delta h_{fus} = -6.0 \, \mathrm{[kJ/mol]}$.

3.10 Consider a piston–cylinder assembly that initially contains 0.5 kg of steam at 400°C and 100 bar. For the isothermal expansion of the steam in this system to a final pressure of 1 bar, determine the following:

(a) What is the maximum possible work (in [kJ]) that can be obtained during this process and the entropy change of the surroundings (in [kJ/K])?

(b) Repeat part (a) using the ideal gas model for steam. Compare your answers.

3.11 Consider the piston–cylinder assembly shown below. It is well insulated and initially contains two 5000-kg blocks at rest on the 0.05-m² piston. The initial temperature is 500 K. The ambient pressure is 5 bar. One mol of an ideal gas is contained in the cylinder. This gas is compressed in a process in which another 5000-kg block is added. The heat capacity of the gas at constant volume can be taken to have a constant value of $(5/2) \, R$, where R is the gas constant.

(a) What are the initial and final pressures of the gas in the system?

(b) Do you expect the temperature to rise or fall? Explain.

(c) What is the final temperature? (This is *not* necessarily a polytropic process!)

(d) Calculate Δs_{sys} and Δs_{surr}. [*Hint:* You may want to refer to the Carnot cycle to get an idea of a possible set of reversible processes to pick.]

(e) Does this process violate the second law of thermodynamics? Explain.

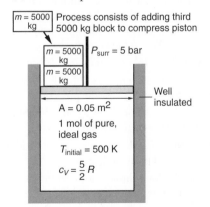

3.12 Problem 3.11 consists of an irreversible process in which an ideal gas with constant heat capacity was compressed in a piston–cylinder assembly. As part of this problem, you were asked to calculate Δs_{sys} for this process. Entropy change is defined for a *reversible* process as

$$\Delta s = \int_{initial}^{final} \frac{\delta q_{rev}}{T}$$

Since entropy is a property, the change in entropy depends only on the final and initial states of the system, not on the path the system went through for a particular process. Therefore, *we can pick any reversible path we want, as long as it takes us from the initial state to the final state*. Calculate Δs_{sys} for the process depicted in Problem 3.11, using each of the following paths:

(a) a reversible, adiabatic compression, followed by a reversible, isothermal expansion (two of the four steps in the Carnot cycle)

(b) a reversible, isobaric heating followed by a reversible, isothermal compression

(c) a reversible, isochoric (constant-volume) heating followed by a reversible, isothermal compression

3.13 Consider a *well-insulated* piston–cylinder assembly. O_2, initially at 250 K and 1 bar, undergoes a *reversible* compression to 12.06 bar. You may assume oxygen is an ideal gas. Answer the following questions:

(a) What is the entropy change for this process?

(b) What is the final temperature of the oxygen?

(c) What is the value of work for this process?

(d) If the oxygen in this system had undergone an *irreversible* compression to 12.06 bar, would the final temperature be higher than or lower than that calculated in part (b)? Explain.

3.14 The insulated vessel shown below has two compartments separated by a membrane. On one side is 1 kg of steam at 400°C and 200 bar. The other side is evacuated. The membrane ruptures, filling the entire volume. The final pressure is 100 bar. Determine the entropy change for this process.

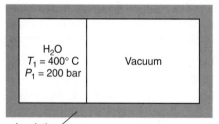

3.15 A partition divides a rigid, well-insulated 1-m³ tank into two equal parts. The left side contains an ideal gas $[c_P = (5/2)R]$ at 10 bar and 300 K. The right side contains nothing; it is a vacuum. A small hole forms in the partition, gas slowly leaks out from the left side, and eventually the temperature in the tank equalizes. What is the entropy change?

3.16 An insulated tank is divided by a thin partition.

(a) On the left is 0.79 mole of N_2 at 1 bar and 298 K; on the right is 0.21 mole of O_2 at 1 bar and 298 K. The partition ruptures. What is ΔS_{univ} for the process?

(b) On the left is 0.79 mole of N_2 at 2 bar and 298 K; on the right is 0.21 mole of O_2 at 1 bar and 298 K. The partition ruptures. What is ΔS_{univ} for the process?

[*Hint*: Consider the entropy change of each gas separately and add them together. To do this, you will need to use the concept of partial pressure.]

3.17 Steam at 8 MPa and 500°C flows through a throttling device, where it exits at 100 kPa. Determine the entropy change for this process.

3.18 A fast-talking salesperson comes to your doorstep and says she is down on her luck and is willing to sell you the patent rights to her most glorious invention. She brings out a mysterious black box and says it can take an inlet stream of ideal gas at 2 kg/s and 4 bar and cool part of it (0.5 kg/s) from 50°C to −10°C with no external parts, as shown below.

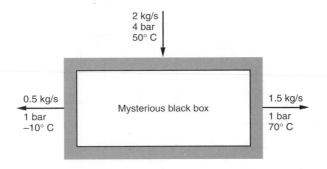

You are feeling somewhat adventurous and are tempted by this offer but must ask the fundamental question: "Can it work?" Can it? Explain.

3.19 Steam enters a nozzle at 4 MPa and 640°C with a velocity of 20 m/s. This process may be considered **reversible** and **adiabatic**. The nozzle exit pressure is 0.1 MPa.

(a) Draw a sketch of this process. Include all known information.

(b) What is the entropy change of the steam?

(c) What is the exit temperature?

(d) What is the exit velocity?

3.20 Propane at 350°C and 600 cm³/mol is expanded in a turbine. The exhaust pressure is atmospheric. What is the lowest possible exhaust temperature? How much work is obtained? You may assume ideal gas behavior and that heat transfer to the surroundings is negligible.

3.21 What is the minimum amount of work required to separate an inlet stream of air flowing at 20°C and 1 bar into exit streams of pure O_2 and pure N_2 at 20°C and 1 bar?

3.22 Consider the well-insulated container shown below. Two gases, gas A and gas B, are separated by a metallic piston. The piston is initially held in place by a latch 10 cm from the left of the container.

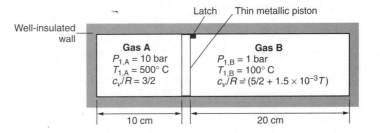

Gas A, which is located in the left compartment, is initially at 10 bar and 500°C. The heat capacity of gas A is constant: $(c_{v,A}/R) = 3/2$. Gas B is located in the right compartment and is initially at 1 bar and 100°C. The heat capacity of gas B is given by $(c_{v,B}/R) = 5/2 - 1.5 \times 10^{-3}\,T$ where T is in Kelvin. You may use the *ideal gas model* for both gases.

(a) The latch is removed and the piston moves until the pressure and temperature in the two compartments become equal. What are the final pressure and temperature? State any assumptions that you make.

(b) Calculate the entropy change of the universe. Is this process possible?

3.23 A rigid tank of volume 0.5m³ is connected to a piston–cylinder assembly by a valve as shown below. Both vessels contain pure water. They are immersed in a constant-temperature bath at 200°C and 600 kPa. Consider the tank and the piston–cylinder assembly as the system and the constant-temperature bath as the surroundings. *Initially* the valve is closed, and both units are in equilibrium with the surroundings (the bath). The rigid tank contains saturated water with a quality of 95% (i.e., 95% of the mass of water is vapor). The piston–cylinder assembly initially has a volume of 0.1 m³. The valve is then opened. The water flows into the piston–cylinder assembly until equilibrium is obtained. For this process, calculate the change in entropy for the system, the surroundings, and the universe.

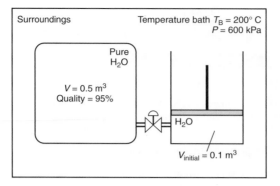

3.24 A rigid, well-insulated container is initially divided into three compartments. The top compartment contains a vacuum. It is separated from the middle compartment A by a frictionless mass of 1000 kg and area 0.098 m². Compartment A contains 2 moles of ideal gas at 300 K and is separated from compartment B, on the bottom of the container, by a rigid partition. Compartment B initially contains 2 moles of the same ideal gas at 300 K and occupies a volume of 0.1 m³. A process is initiated by removing the partition. The mass then reequilibrates in the container. What is the change in entropy? Take $c_P = (5/2)R$.

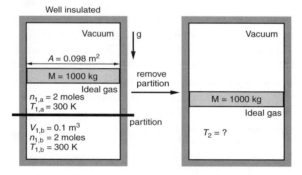

3.25 Consider the system shown below. Tank A has a volume of 0.3 m³ and initially contains an ideal diatomic gas at 700 kPa, 40°C. Cylinder B has a piston resting on the bottom, at which point the spring exerts no force on the piston. The piston–cylinder has a cross-sectional area of 0.065 m², the piston has a mass of 40 kg, and the spring constant is 3500 N/m. Atmospheric pressure is 100 kPa. Tanks A and B are well insulated and do not transfer heat between each other. The valve is opened and gas flows into the cylinder until pressures in A and B become equal and the valve is closed. You may assume constant heat capacity. Determine the final pressure in the system. Assuming the gas in A has undergone a **reversible, adiabatic** expansion, find the final temperature in cylinder A. The temperatures in tank A and B are not necessarily equal.

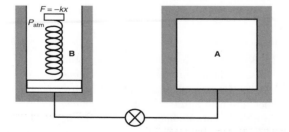

3.26 A *steam* turbine in a small electric power plant is designed to accept 4500 kg/hr of steam at 60 bar and 500°C and exhaust the steam at 10 bar. Heat transfers to the surroundings ($T_{\text{surr}} = 300\ K$) at a rate of 69.86 kW. Answer the following questions:

(a) Calculate the *maximum* power ($\dot{W}_{max}$) that the turbine can generate.

(b) In this case, what is the exit temperature of the steam?

(c) You know that the isentropic efficiency of the turbine is actually 66.5%. What is the actual power produced?

(d) Do you expect the exit temperature to be higher or lower than that calculated in part (a). Explain. (Assume the heat transfer does not change.)

(e) What is the actual exit temperature?

3.27 Air flowing at 1 m³/s enters an adiabatic compressor at 20°C and 1 bar. It exits at 200°C. The isentropic efficiency of the compressor is 80%. Calculate the exit pressure and the power required.

3.28 Steam enters a turbine at 10 MPa and 500°C and leaves at 100 kPa. The isentropic efficiency of the turbine is 85%. Calculate the exit temperature and the work generated per kg of steam flowing through.

3.29 Nitrogen gas at 27°C flows into a well-insulated device operating at steady-state. There is no shaft work. The device has two exit streams. Two-thirds of the nitrogen, by mass, exits at 127°C and 1 bar. The remainder exits at an unknown temperature and 1 bar. Find the exit temperature of the third stream. What is the minimum possible pressure of the inlet stream? Assume ideal gas behavior.

3.30 Consider a *well-insulated* piston–cylinder assembly containing 5 kg of water vapor, initially at 540°C and 60 bar, that undergoes a *reversible* expansion to 20 bar. The surroundings are at 1 bar and 25°C. Answer the following questions:

(a) What is the entropy change (Δs_{univ}, Δs_{surr}, and Δs_{sys}) for this process?

(b) What is the final temperature of the water?

(c) What is the value of work for this process?

(d) What is the final volume of the system?

3.31 Consider a well-insulated, rigid tank containing 5 kg of water vapor in the same initial state as in Problem 3.30 (540°C, 60 bar). Again, the surroundings are at 1 bar and 25°C, as shown below. A tiny leak develops, and water slowly escapes until the pressure reaches 20 bar.

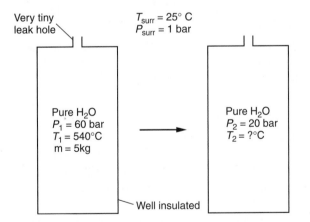

Do you expect the final temperature to be higher than, lower than, or the same as that calculated in part (b) of Problem 3.30? Explain your answer.

3.32 Consider filling a "type A" gas cylinder from with water from a high-pressure supply line as shown on next page. Before filling, the cylinder is empty (vacuum). The valve is then opened, exposing the tank to a 3-MPa line at 773 K until the pressure of the cylinder reaches 3 MPa. The valve is then closed. The volume of a "type A" cylinder is 50 L.

(a) What is the change in entropy of the universe immediately after the valve is closed?

(b) If the cylinder then sits in storage at 293 K for a long time, what is the entropy change of the universe?

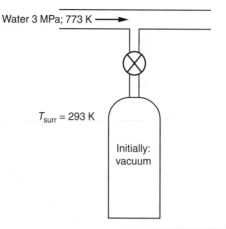

Water 3 MPa; 773 K ⟶

T_{surr} = 293 K

Initially:
vacuum

3.33 A rigid tank has a volume of 0.01 m³. It initially contains saturated water at a temperature of 200°C and a quality of 0.4. The top of the tank contains a pressure-regulating valve that maintains the vapor at constant pressure. This system undergoes a process whereby it is heated until all the liquid vaporizes. You may assume there is no pressure drop or heat transfer in the exit line. The surroundings are at 200°C. What is the entropy change of the universe?

3.34 Consider the system sketched below in which a turbine is placed between two rigid tanks. Tank 1 initially contains an ideal gas at 10 bar and 1000 K. Its volume is 1 m³. Tank 2 is 9 m³ and is initially at vacuum. The heat transfer with the surroundings is negligible. Determine the maximum work (in [J]) obtainable by the turbine. You may take the heat capacity of the gas to be $c_P = (5/2)R$. You may neglect the volume in the turbine and assume the final temperature in the two tanks is equal.

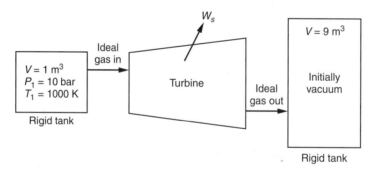

W_s

$V = 9$ m³

Ideal
gas in

$V = 1$ m³
$P_1 = 10$ bar
$T_1 = 1000$ K

Turbine

Ideal
gas out

Initially
vacuum

Rigid tank

Rigid tank

3.35 A hot reservoir is available at 500°C and a cold reservoir at 25°C. Calculate the maximum possible efficiency of a power cycle that operates between these two reservoirs.

3.36 An ideal Rankine cycle operates with the following design: 100 kg/s of steam enters the turbine at 30 bar and 500°C and is condensed at 0.1 bar. Determine the power produced and the efficiency of the cycle.

3.37 Come up with four ways in which you can make the power cycle of Problem 3.36 more efficient. Illustrate how your ideas achieve increased efficiency using sketches like that in Figure 3.8.

3.38 An ideal Rankine cycle produces 100 MW of power. If steam enters the turbine at 100 bar and 500°C and is condensed at 1 bar, determine the mass flow rate of steam. Recalculate the mass flow rate assuming that the isentropic efficiency of the turbine and the compressor are 80%.

3.39 You are considering building a solar power plant which uses CCl_2F_2 as its working fluid. It enters the turbine as a saturated vapor at 1.7 MPa and leaves at 0.7 MPa. Based on the ideal Rankine cycle, determine the efficiency. Property data for CCl_2F_2 may be found at http://webbook.nist.gov/chemistry/fluid/.

3.40 Consider a refrigeration system based on an ideal vapor-compression cycle using R-134a as the refrigerant. It operates between 0.7 MPa and 0.12 MPa with a flow rate of 0.5 mol/s. Calculate the following:

(a) the reate of heat removal from the refrigerated unit

(b) the power input needed to the compressor

(c) the COP

The properties of R-134a can be found at http://webbook.nist.gov/chemistry/fluid/.

3.41 If the throttling valve in Problem 3.40 is replaced by an isentropic turbine, what is the COP? Is this modification practical? Explain.

3.42 A two-stage cascade refrigeration system is shown below. The refrigerant is $R134a$. It consists of two ideal vapor-compression cycles with heat exchange between the condenser of the lower-temperature cycle and the evaporator of the higher-temperature cycle. The hotter cycle operates between 0.7 MPa and 0.35 MPa, while the cooler cycle operates between 0.35 MPa and 0.12 MPa. If the flow rate in the hotter cycle is 0.5 mol/s, determine the following:

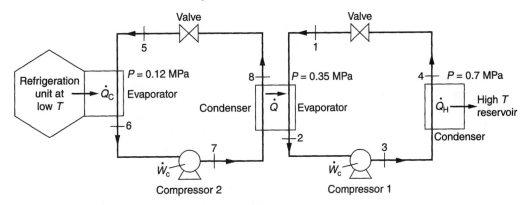

(a) What is the flow rate in the cooler cycle?

(b) What is the rate of heat removal from the refrigerated unit?

(c) What is the power input needed to the compressors?

(d) What is the COP?

(e) Compare the performance with the cycle in Problem 3.40.

3.43 Design a vapor-compression refrigeration system to cool a system to $-5°C$ with the capability for up to 20 kW of cooling. You have a reservoir at 20°C to reject heat to. Refrigerants and their properties can be found at http://webbook.nist.gov/chemistry/fluid/.

3.44 Modify the vapor-compression refrigeration system presented in Section 3.9 to apply to a refrigerator for home use. This system needs to provide cooling to two units: the freezer at $-15°C$ and the main compartment at 5°C. Take the refrigeration capacity, Q_C, of each compartment to be equal. You are limited to one compressor and one condenser. Draw a schematic of the process and the associated Ts diagram. Select an appropriate refrigerant and define the states of the system. Refrigerants and their properties can be found at http://webbook.nist.gov/chemistry/fluid/.

3.45 Consider an ideal, reversible magnetic refrigeration cycle shown in the figure on next page. A paramagnetic working material in the form of the rim of a wheel is rotated between a high-temperature reservoir and a low-temperature reservoir. In the high-temperature reservoir, the working material expels heat into the reservoir as it is subjected to a high magnetic field. As the working material moves into the low-temperature reservoir, the field becomes smaller and eventually zero. The demagnetization process causes the working fluid to absorb heat from the low-temperature reservoir. The working material is gadolinium sulphate octahydrate. On the upper-left-hand side of the figure, helium enters the porous wheel at temperature 1.1 K and a magnetic field of 0.9 tesla [T] and is forced to flow in heat exchange with the moving wheel. The wheel absorbs heat from the helium as it is demagnetized. The temperature of the helium drops 0.2 K during this process. The helium then absorbs heat Q_C from the load and reenters the heat exchanger area at 1.1 K. Similarly, in the lower-right-hand side of the figure, helium enters the porous wheel at 8 K and a magnetic field of 1.6 T. The wheel deposits heat into helium as the material is magnetized. The temperature rises to 9.5 K at 6.4 T.

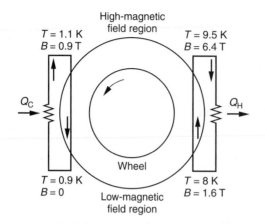

Approximately how much heat is being expelled by the cold reservoir? How much heat is being absorbed by the hot reservoir? Estimate the coefficient of performance for this process. How does the numerical value compare with a conventional refrigeration cycle? How is work being supplied to perform the refrigeration process? A Ts diagram for gadolinum sulphate octa-hydrate is provided.

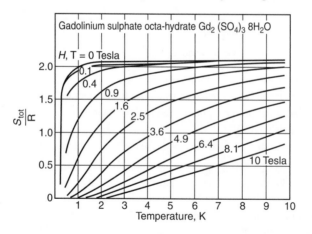

3.46 In Problem 2.4, you explained why a thick rubber band heats when stretched. Offer an alternative explanation in the context of entropy. You may consider this process isentropic.

3.47 Consider the oxidation of cuprous oxide to form cupric oxide by the following reaction:

$$2Cu_2O(s) + O_2(g) \leftrightarrow 4CuO(s)$$

Calculate Δs_{rxn}. This task can be done in the same type of path described in Section 2.6 for Δh_{rxn}. You can calculate the values of entropies of formation from the data in Appendix A.3 by applying the following relationship:

$$\Delta s_f^o = \frac{\Delta h_f^o - \Delta g_f^o}{T}$$

Physically explain the sign of Δs_{rxn}. Does the formation of CuO violate the second law of thermodynamics? Explain.

3.48 CdTe forms a II–VI compound semiconductor. This solid forms a well-ordered single crystal where Cd atoms and Te atoms sit in distinct sites adjacent to each other in a crystal lattice. Estimate the entropy of mixing from an initial state of 1 mole of pure solid Cd and 1 mole of pure solid Te to a final state of 1 mole of CdTe at constant temperature and pressure.

3.49 In his book *The Trouble Waters*, Henry Morris proclaims:

> The Law of Increasing Entropy *is an impenetrable barrier which no evolutionary mechanism yet suggested has ever been able to overcome. Evolution and entropy are opposing and mutually exclusive concepts. If the entropy principle is really a universal law, then evolution must be impossible.*

Is this argument scientifically sound? Explain.

3.50 "Four of a kind" is one of the best hands you can have in poker. Can you relate this statement to the concept of entropy? Would you say this hand has a high value of s?

3.51 The concept of entropy was developed in the nineteenth century, in order to study the efficiency of the steam engine, largely through the work of Sadi Carnot, Rudolph Clausius, and Lord Kelvin. However, it has had major implications well beyond the realm of engineering, including impacting the thought, philosophy, and theology of nineteenth-century Europe. Go to the library or the Web and find a nonengineering topic in which entropy plays a major role. Describe it in about a hundred words and cite your source(s).

CHAPTER 4

Equations of State and Intermolecular Forces

Learning Objectives

To demonstrate mastery of the material in Chapter 4, you should be able to:

▶ Given a chemical species, identify which intermolecular interactions are significant. Given different species, qualitatively compare the magnitude of their dipole moments, polarizabilities, intermolecular interactions, Lennard-Jones parameters ε and σ, and van der Waals parameters a and b.

▶ Given two of the measured properties P, v, and T, calculate the value of the third using: cubic equations of state (e.g., van der Waals, Redlich–Kwong, Peng–Robinson), the virial equation, generalized compressibility charts, and ThermoSolver software. Apply the Rackett equation, the thermal expansion coefficient, and the isothermal compressibility to find molar volumes of liquids and solids.

▶ State the molecular components that contribute to internal energy. Describe and illustrate by example the following intermolecular interactions: point charges, dipoles, induced dipoles, dispersion (London) interactions, repulsive forces, and chemical effects. Define a van der Waals force, and relate it to the dipole moment and polarizability of a molecule. Ultimately, we want to relate macroscopic thermodynamic behaviors to their molecular origins as much as possible.

▶ Define a potential function. Write equations for the ideal gas, hard sphere, Sutherland, and Lennard–Jones potentials and relate the terms to intermolecular interactions.

▶ State the molecular assumptions of an ideal gas. Describe how the terms in the van der Waals equation relax these assumptions. Identify how the general form of cubic equations of state accounts for attractive and repulsive interactions in a similar manner.

▶ State the principle of corresponding states on molecular and macroscopic levels. Apply this principle to develop expressions to solve for parameters in equations of state from the critical property data of a given species. Describe why the acentric factor was introduced and its role in constructing the generalized compressibility charts.

▶ Write the van der Waals mixing rules for coefficients a and b. Explain their functionality in terms of molecular interactions. Write the mixing rules for the virial coefficients and for pseudocritical properties using Kay's rules.

Apply these mixing rules to solve for *P*, *v*, or *T* of a mixture using equations of state or generalized compressibility charts.

▶ 4.1 INTRODUCTION

Motivation

The intensive thermodynamic properties that can be experimentally measured are pressure, temperature, molar volume, and composition. For any *pure* species, only two intensive properties are independent; thus, we can graphically map a "surface" from experimental data using P, v, and T as coordinates and plot it as we did in Section 1.6. Alternatively, we could tabulate the data as was done for water in the steam tables (Section 1.7). However, in solving problems, it is often inconvenient to have to resort to graphs or tables for numerical values (as well as slopes for derivatives and areas for integrals). We therefore seek to come up with an equation that relates these measured variables by fitting experimental data. In the language of math, we want an equation of the form

$$f(P, v, T) = 0 \tag{4.1}$$

Such an equation is fit to experimental data and is known as an **equation of state** since it allows us to calculate the unknown measured property from the two that constrain the state. Equations of state can be explicit in pressure, that is,

$$P = f(T, v) \tag{4.2}$$

in molar volume,

$$v = f(T, P) \tag{4.3}$$

or in terms of the dimensionless compressibility factor, z,

$$z = \frac{Pv}{RT} = f(T, v) \tag{4.4a}$$

or

$$z = \frac{Pv}{RT} = f(T, P) \tag{4.4b}$$

In developing an equation of state, the goal is to come up with an equation that fits experimental data as accurately as possible; there may or may not be a physical basis for its form. An equation of this type is called *constitutive* (as opposed to a *fundamental* equation such as the first law). What are other constitutive equations that you have encountered as a chemical engineer?

In practice, there are hundreds of analytical equations of this form in the literature from which to choose! It would be quite burdensome (and very impractical) to examine every one to decide when to use a particular form. Consequently, we will take a different tack. We will start with the "friendliest" equation of state, the ideal gas model. After examining its limitations, we will explore how we can describe deviations from ideal gas behavior. We will investigate the generalities of these different forms as well as try to develop an intuition about what equations to use and when to use them. To this end, we will examine the molecular origins of macroscopic thermodynamic property behavior. Perilous straits indeed, but a journey with rich rewards!

The Ideal Gas

As you know well, the most common equation of state is the ideal gas model. It can be written explicitly for pressure in terms of the intensive properties v and T as follows:

$$P = \frac{RT}{v} \tag{4.5}$$

The ideal gas model is of the form presented in Equation (4.1) in that it relates the measured variables P, T, and v. The ideal gas equation can be derived directly from the kinetic theory of gases for a gas consisting of molecules that are *infinitesimally small, hard round spheres that occupy negligible volume and exert forces upon each only other through collisions*. Stated more concisely, the assumptions of the ideal gas model are that molecules:

1. Occupy *no* volume
2. Exert *no* intermolecular forces

As we shall see in Section 4.2, the absence of intermolecular forces leads to the internal energy being independent of pressure. It depends only on temperature, that is, the molecular kinetic energy of the molecules. Hence,

$$u_{\text{ideal gas}} = f(T \text{ only}) \tag{4.6}$$

As the pressure goes to zero, all gases approach ideal gas behavior.

▶ 4.2 INTERMOLECULAR FORCES

Internal (Molecular) Energy

"Molecular" energy, or internal energy, u, can be divided into two parts: *molecular kinetic energy* and *molecular potential energy*. Kinetic energy results from the translational, rotational, and vibrational motion of the molecules; hence, kinetic energy manifests itself by the molecules' **velocities** and is directly related to the measured variable temperature via Maxwell–Boltzmann statistics. The potential energy results from the **position** of one atom or molecule relative to others in the system. As we saw in Section 2.6, a significant component of molecular potential energy is related to the covalent bonds between atoms of the *same* molecule. We refer to this type of potential energy as *intramolecular* potential energy. When these bonds rearrange in a chemical reaction, there can be large changes in the molecular potential energy. In this section, we will examine another component of molecular potential energy—that related to interactions between *different* molecules (or atoms that are not covalently bonded), the *intermolecular* potential energy. Since the intermolecular potential energy depends on position, we can relate it directly to the measured variable pressure. At constant temperature, as the pressure increases, the average distance between molecules decreases and their positions relative to one another get closer. In an ideal gas, there are no intermolecular forces, so the position of the molecules relative to one another does not matter. Consequently, the internal energy of an ideal gas is independent of pressure and depends *only* on temperature.

We wish to relax the ideal gas assumption so that we can develop more general equations of state. To accomplish this task, we need to establish the relationship for the internal energy of a system as a function of distance between the molecules and their orientations. We are specifically interested in the intermolecular potential energy component of internal energy. The intermolecular interactions between species arise from the electronic and quantum nature of atoms. An atom can be viewed as containing a fixed, positively charged nucleus surrounded by a relatively mobile, negatively charged electron cloud. When a molecule is in close enough proximity to another molecule, the electrically charged structure of its atoms can lead to attractive and repulsive forces. The attractive forces include electrostatic forces between point charges or permanent dipoles, induction forces, and dispersion forces.

Since intermolecular interactions result from the electric nature of atoms, it is often useful to apply the concept of an electric field when discussing the effect of a given species on the system. Recall the electric field intensity, $\vec{E}$, is defined as the force per unit charge exerted on a positive test charge, Q, in the field. It is related to the negative gradient of the molecular potential energy, Γ, by:

$$\vec{E} = \frac{F}{Q} = \frac{-\nabla \Gamma}{Q} \tag{4.7}$$

The electric field intensity from a given molecule is the same regardless of the species with which it interacts. The principle of superposition says the total electric field in the system is given by the vector sum of the individual electric fields of all the species in a system. Therefore, if we understand the behavior of a single molecule, we can quantify its contribution to the energy of the macroscopic system as a whole. Equation (4.7) relates the electric field, $\vec{E}$, to the intermolecular forces, F, and to the intermolecular potential energy, Γ. In the discussion that follows, we will refer to all of these quantities, as appropriate, to characterize the intermolecular interactions that cause the system to deviate from ideal gas behavior.[1]

It is important to realize that there is presently no direct quantitative relationship between molecular physics and classical thermodynamics except in very simple systems. So why are we studying molecular physics? This topic warrants study for four reasons:

1. To strengthen our intuition about nonideal behavior and our judgment about what equations of state to use.

2. To understand why some mathematical relationships for $f(P, v, T)$ work better than others.

3. While there is no quantitative connection between molecular physics and classical thermodynamics, with the increased computational power of modern computers, and the development of such techniques as density functional theory and Monte Carlo simulations, this quantitative connection may soon be realized.

4. To develop "mixing rules" for how thermodynamic properties of a mixture depend on composition. This ability allows us to extend data of pure systems to mixtures. Note that while thermodynamic data for pure species are readily available, the availability of data for the particular mixture you may be interested in is much less likely since there is an infinite number of permutations of mixtures.

---

[1] Unit systems related to electrical quantities can be confusing. Depending on the unit system, these equations can take different forms. The equations below are written for CGS (Gaussian) units as opposed to SI.

Equations that are denoted with "CGS units" are valid *only* in those units. See Appendix D for further discussion.

Attractive Forces

Electrostatic Forces

Electrostatic interactions between molecules can result from the net charges of ions and also from permanent charge separation in neutral species. In this section, we will examine the nature of these interactions.

Point charges The simplest electrostatic interaction results from the force of attraction between two species with nonzero charge. There are two types of electric charge, positive and negative. If two species have the same type of charge, they will repel each other, while species of unlike charge attract. Consider species i and j which have positive charges, Q_i and Q_j, respectively, and are separated by a distance r. The resulting repulsive force is illustrated in Figure 4.1. If the length r is much greater than the radius of i or j, these charges can be treated as *point charges*. Recall from basic physics that the force between point charges is inversely proportional to the square of the distance between the charges, that is:

$$F_{ij} = \frac{Q_i Q_j}{r^2} \quad \text{CGS units} \tag{4.8}$$

Equation (4.8) is written for CGS units and is known as Coulomb's law. See Appendix D for the SI units equivalent of Coulomb's law. The potential energy between species i and j is found by rearranging Equation (4.7), and using the expression for force given in Equation (4.8) yields

$$\Gamma_{ij} = -\int F_{ij} dr = \frac{Q_i Q_j}{r} \quad \text{CGS units} \tag{4.9}$$

Examination of Equation (4.9) shows the potential energy is negative for unlike charges. A lower (more negative) energy indicates a more stable system, that is, an attractive force. For like charges, the potential is positive, indicating a repulsive interaction. At large separations ($r \to \infty$), the potential goes to zero, indicating that the point charges do not interact.

The effect of a given point charge on its neighbors can also be visualized in terms of electric fields. Electric field lines for positive and negative point charges are shown in Figure 4.2. The field lines represent "lines of force" and point in the direction a positive *test* charge would move if it were put in that field. The density of field lines is proportional to their strength; the closer the lines are spaced, the stronger the field at that point. Examination of Figure 4.2 shows that the positive *test* charge would move away from the positive point charge on the left and toward the negative point charge on the right and that the force exerted by the field falls off as the distance away from the point charge increases. These qualitative observations are consistent with the analytical relation given by Equation (4.9). If the "lines of force" given by the electric field sufficiently affect the bulk behavior of the system, intermolecular forces are significant and we can no longer treat the system using the ideal gas model.

Figure 4.1   The coulombic repulsion between two like point charges separated by a distance r.

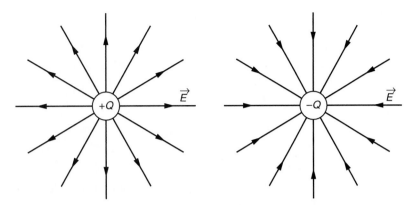

Figure 4.2   Electric field lines from positive and negative point charges.

Point charges exert strong forces and fall off relatively slowly with distance. It is uncommon for an isolated net charge to exist in nature since it will typically find an oppositely charged species and combine. However, some molecular examples exist, including the following:

1. *Ionic solids (e.g.* NaCl *crystals)*   Ionic solids are made up of positively charged and negatively charged ions within a crystal lattice. The bond energy of the solid results from attractive forces of the oppositely charged ions, as given by Equation (4.9). For example, in table salt, NaCl, a positively charged sodium ion, Na^+, is electrostatically attracted to the negatively charged chloride ions, Cl^-, that surround it. In Problem 4.12, you will calculate the bond strength of this ionic solid.

2. *Electrolytes (e.g.,* 18 M H_2SO_4)   Net charges exist in the liquid phase in electrolyte solutions and molten salts. In an H_2SO_4 acid bath, for example, H^+ and SO_4^{2-} exist in the liquid and exert Coulombic forces. The polar structure of water allows charged species to be stable. The water forms an electrostatic cloud that shields the ions from one another. The electrostatic interactions of the charged species within the electrolyte solution form a key component in the behavior of electrochemical systems.[2]

3. *Ionized gases or plasmas*   Point charges exist in the gas phase in the form of plasmas. Plasmas exist in many forms, from the Earth's ionosphere to the glow discharge plasmas used in etching to the deposition of thin films in integrated circuit manufacturing.

The thermodynamic properties of point charges requires particular attention to these strong electrical forces and will not be covered further in this text.

Electric Dipoles   While they have a net neutral charge, some molecules are configured so that there is an overall separation of charge. These molecules may be treated as an electric dipole (two poles). In a dipole, there is a region of positive charge ($+Q$) next to a region of negative charge ($-Q$). The magnitudes of positive and negative charges are

---

[2] For a treatment of the behavior of ions in electrochemical solutions, see J. O. M. Bockris and A. K. N. Reddy, *Modern Electrochemistry* (New York: Plenum Press, 1970).

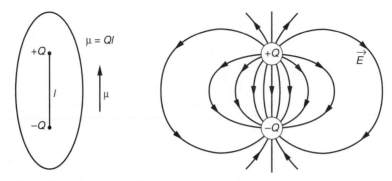

Figure 4.3   Schematic of charge separation leading to an electric dipole. Electric field lines are depicted on the sketch to the right.

equal. A dipole is illustrated on the left of Figure 4.3. The electric field lines associated with the dipole are shown on the right. We can see from these field lines that dipoles can exert forces on other species in their vicinity. The strength of the dipole is characterized by the dipole moment, μ, a vector that points from the negative charge to the positive charge. The magnitude of the dipole moment is equal to the product of the magnitude of charge, Q, and the distance by which the positive charge and the negative charge are separated, l. The common unit of the dipole moment is the debye [D]:

$$\left[1\mathrm{D} = 10^{-18}\left(\mathrm{erg}\ \mathrm{cm}^3\right)^{1/2} \right]$$

Dipole moments are commonly found in nature on the molecular level. Molecules are formed by covalent bonds between the valence electrons of their atoms. When the sharing of electrons in a covalent bond is unequal, an atom can gain electron density at the expense of the atom to which it is bonded. For example, consider an HCl molecule. The Lewis dot structure is depicted on the left of Figure 4.4. The valence electrons of the chlorine atom are indicated by dots, while the valence electron of hydrogen is an "x." The bond in this molecule is formed by the sharing of one electron from chlorine and one from hydrogen. However, the Cl is highly electronegative since it needs an electron to complete its outer shell. It pulls the electron from H much more strongly than hydrogen pulls the electron from Cl. Thus, the electrons are not shared equally;

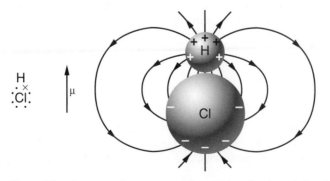

Figure 4.4   The Lewis dot structure of an HCl molecule and the electric field lines emanating from its dipole.

rather, the lone hydrogen electron spends more time close to the Cl atom than the shared Cl electron does next to hydrogen. The result is a net charge separation with the Cl obtaining a net negative charge and the H a net positive charge, as depicted on the right of Figure 4.4. This separation of charge leads to a permanent dipole moment. In the case of HCl, the magnitude of the dipole moment is around 1.1 D. Diatomic molecules with electronegative (F, Cl) and electropositive (Li, Na) atoms have large dipole moments.

Diatomic molecules with dissimilar atoms, such as HCl, will always exhibit some degree of charge separation; however, it may be small. Thus, these molecules will have permanent dipoles. On the other hand, for polyatomic molecules with more than two atoms, we must look at the molecular structure to see whether a dipole exists. Dipole moments in these molecules are caused by nonsymmetric distributions of the electron cloud in the molecule. Symmetric molecules have no dipole moment. The greater the molecular asymmetry, the greater the dipole moment. For example, the electron cloud in CH_3Cl is pulled strongly to the electronegative Cl and exhibits a dipole of 1.87 D. On the other hand, CH_4 is symmetric and does not have a permanent dipole moment. There are many other molecular examples of dipoles: H_2O, HF, and so on. Can you think of some? Will CO_2 exhibit a dipole? Dipole moments of several representative molecules are presented in Table 4.1.

The interaction of one dipole with a neighboring dipole is termed a dipole–dipole interaction. Such an interaction can lead to deviations from ideal gas behavior. The dipole–dipole interaction depends on the relative orientation of the molecules. However, to relate this type of electrostatic force to macroscopic behavior, we must average over all orientations. For example, consider the case of two HCl molecules in proximity to each other. As the electric field from the dipole depicted in Figure 4.4 shows, each HCl can exert an electrical force on its neighbor. The bottom of Figure 4.5 illustrates dipoles in the two extreme limits of orientation. In the lowest energy configuration, the negative side of one dipole is aligned next to the positive side of the other. This leads to electrostatic attraction, as depicted in Figure 4.5. Conversely, in the highest energy configuration, like-charged sides of the dipole align, leading to electrostatic repulsion. Dipoles may take any orientation in between these two limits. If the orientation of the two dipoles were completely random, the average force would be zero, since attractive and repulsive orientations would occur equally. As the top of the Figure 4.5 shows, however, molecular dipoles in gases and liquids are free to rotate. This movement allows the energetically favored lower-energy attractive interactions to occur more frequently, as the dipoles tend to rotate to align. On the other hand, thermal energy leads to a randomization of orientation. The trade-off between these two effects can be quantified according to the Boltzmann factor, $e^{-\Gamma/(kT)}$. Thus, the ratio of the number of dipoles in any two states, N_1 and N_2 is related to their potential energy difference and the temperature according to:

$$\frac{N_1}{N_2} = e^{-\left[(\Gamma_1 - \Gamma_2)/(kT)\right]} \tag{4.10}$$

We can see that Equation (4.10) implies that more dipoles will align in lower-energy orientations. Averaging the potential energy of the dipole–dipole interaction between species i and j over all possible orientations gives the following expression:

$$\overline{\Gamma}_{ij} = -\frac{2}{3}\frac{\mu_i^2 \mu_j^2}{r^6 \, kT} \quad \text{CGS units} \tag{4.11}$$

TABLE 4.1 Dipole Moments, Polarizabilities, and Ionization Energies

Molecule	μ [D]	α [cm$^3 \times 10^{25}$]	I [eV]
H_2	0	8.19	15.42
He	0	2.06	24.59
N_2	0	17.7	15.58
O_2	0	16	12.07
Ne	0	3.97	21.56
Cl_2	0	46.1	11.5
Ar	0	16.6	15.76
Kr	0	25.3	14
Xe	0	41.1	12.13
HF	1.91	5.1	16.03
HCl	1.08	26.3	12.74
HBr	0.8	36.1	11.68
HI	0.42	54.5	10.39
H_2O	1.85	14.8	12.62
H_2S	0.9	37.8	10.46
CH_3OH	1.7	32.3	10.84
NH_3	1.47	22.2	10.07
NO	0.2	17.4	9.26
N_2O	0.2	30	12.89
SF_6	0	44.7	15.32
SO_2	1.63	38.9	12.35
CH_4	0	26	12.61
CH_3F	1.85	26.1	12.5
CH_3Cl	1.87	45.3	11.26
CH_3Br	1.81	55.5	10.54
CH_2F_2	1.97	27.3	12.71
CH_2Cl_2	1.8	64.8	11.33
CHF_3	1.65	28	13.86
$CHCl_3$	1.1	85	11.37
CF_4	0	28.5	16
$CFCl_3$	0.45	82.4	11.68
CCl_4	0	105	11.47
CO	0.12	19.8	14.01
CO_2	0	26.3	13.78
CS_2	0	87.4	10.07
C_2H_6	0	44.7	11.52
C_2H_4	0	42.2	10.51
C_2H_2	0	34.9	11.4
C_3H_8	0	62.9	10.94
HCN	3	25.9	13.6
$(CN)_2$	0.2	50.1	13.37
CH_3OCH_3	1.7	51.6	10.02

TABLE 4.1   Continued

Molecule	μ [D]	α [cm$^3 \times 10^{25}$]	I [eV]
$(CH_2)_3$	0.4	56.4	9.86
$CH_3(CO)CH_3$	2.9	63.3	9.7
C_6H_6	0	104	9.24
C_6H_5Cl	1.69	122.5	9.07
$C_6H_5NO_2$	4	129.2	9.94
o-$C_6H_4Cl_2$	2.5	141.7	9.06
m-$C_6H_4Cl_2$	1.72	142.3	9.1

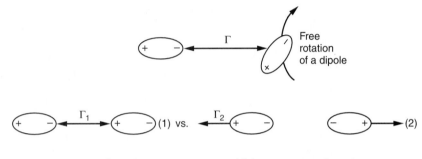

Lowest energy configuration   Highest energy configuration

Figure 4.5   Different possible orientations of a freely rotating electric dipole–dipole pair.

where $\overline{\Gamma}$ is the average potential energy of the dipole–dipole interaction. The potential energy between dipoles has a $1/r^6$ dependence on position; it falls off much more quickly than the Coulomb interaction. It is also proportional to the square of the dipole moment of each species. The (kT) term in the denominator results from the averaging. At higher temperatures, the orientations are more randomly distributed and the attractive force decreases.

In addition to permanent dipoles, species can exhibit higher-order terms in the multipolar expansion, such as quadrapoles (four poles), octapoles (eight poles), or higher-order multipoles. However, these higher-order terms do not measurably influence macroscopic property behavior.

Induction Forces

Induction results when the electric field from a dipole affects the electric structure of its neighbor. Since the negatively charged electrons are free to move about the atom, the electrons in molecule i can be displaced due to a neighboring dipole, molecule j. This displacement "induces" the separation of charge in molecule i, causing a dipole to form. Consequently i and j are attracted to each other. This phenomenon is termed **induction**. Dipoles can be induced in both polar and nonpolar species. As an example, let's consider the effect of the electric field from the dipole HCl on an Ar atom in close proximity to it. Without any external influences, Ar has no net charge separation. However, as shown in Figure 4.6, the electrons in Ar will respond to the dipole field of HCl. Recall that the electric field lines show the direction in which a positive test charge will go. Hence, the negatively charged electrons will be attracted to the top of the Ar atom in Figure 4.6, leading to an induced dipole. The induced dipole in Ar will then be attracted to the permanent HCl dipole.

Since the nature of induction is the same as a dipole–dipole attraction, we also expect a $1/r^6$ dependence. Again, the magnitude of induction depends on the orientation of dipole j. The average over all orientations of the potential energy between an induced dipole i and the permanent dipole j is given by:

$$\overline{\Gamma}_{ij} = -\frac{\alpha_i \mu_j^2}{r^6} \quad \text{CGS units} \tag{4.12}$$

where α is the polarizability of molecule i.

Polarizability is a parameter that characterizes the ease with which a molecule's electron cloud can be displaced by the presence of an electric field. A larger displacement induces a stronger dipole. Electrons position themselves around the positively charged nucleus of an atom. Since the valence electrons of larger atoms are farther away from the nucleus, they are less rigidly held and the atom is more polarizable. Thus, in general, the larger the atom, the larger the value for α. Polarizability is also roughly additive; that is, the value of polarizability scales with the number of atoms. For example, ozone, O_3, has roughly 1.5 times the polarizability of oxygen, O_2, since the ratio of oxygen atoms is 3:2. A rough estimation of the polarizability of a molecule can be obtained by adding together the polarizability of all the atoms in the molecule.[3] Hence molecules with more atoms have greater polarizabilities. Polarizabilities of several representative molecules are reported in Table 4.1.

Dispersion (London) Forces

Nonpolar molecules, such as N_2 and O_2, show forces of attraction that lead them to condense and freeze at low enough temperatures. Yet they do not have a dipole moment, and the pure species are not subject to dipole–dipole interactions and induction. The interactions between nonpolar molecules result from a third type of interaction: dispersion (or London) forces. Dispersion is inherently a quantum-mechanical phenomenon; we would need to understand quantum electrodynamics to develop a rigorous model of dispersion. However, it can be viewed "classically" as follows: Nonpolar molecules are really only nonpolar when the electron cloud is averaged over time. In a given

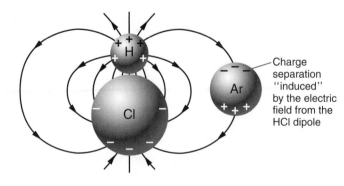

Charge separation "induced" by the electric field from the HCl dipole

Figure 4.6 The induction of a dipole in argon from the electric field due to a HCl dipole.

---

[3] A more accurate approach is to proportion polarizability to the type and number of covalent bonds in a molecule. Values for the contribution of several types of bonds (e.g., C–C, C=C, C–H, N–H, C–Cl, etc.) and molecular groups (e.g., C–O–H, C=O, C–O–C, etc) can be found in J. O. Hirshfelder, C. F. Curtiss, and R. B. Bird, *Molecular Theory of Gases and Liquids*, (New York: Wiley, 1954).

"snapshot" of time, the molecule has a temporary dipole moment. Dispersion forces result from the instantaneous nonsymmetry of the electron cloud surrounding a nucleus. The instantaneous dipole moment induces a dipole in a neighboring molecule, leading to an attractive force. Using quantum mechanics and perturbation theory, London developed the following expression for the energy of attraction of symmetric molecules i and j:

$$\Gamma_{ij} \approx -\frac{3}{2} \frac{\alpha_i \alpha_j}{r^6} \left(\frac{I_i I_j}{I_i + I_j} \right) \quad \text{CGS units} \tag{4.13}$$

where I is the first ionization potential, that is, the energy required for the following reaction: $M \rightarrow M^+ + e$. Ionization potentials of common species are presented in Table 4.1. These "temporary dipole" interactions also have a $1/r^6$ dependence on position. They also depend on the polarizability of each species involved, since the extent of the instantaneous dipole is related to the looseness of the nucleus's control of the valence electrons; similarly, the induction in the neighboring molecule depends on its polarizability. While Equation (4.13) was developed for nonpolar species, polar molecules are subject to dispersion interactions as well.

The classical explanation of dispersion above would leave us to believe that dispersion is a relatively small force. However, as is often the case, our classical intuition is defied by quantum mechanics. It turns out that *dispersion forces are surprisingly large in magnitude*, as we will see in Examples 4.1 and 4.2.

Dipole–dipole, induction, and dispersion forces are collectively referred to as van der Waals forces. The intermolecular potential energy for each of these interactions falls off as the sixth power of position. Thus, all three van der Waals interactions are of the form:

$$\Gamma_{ij} = -\frac{C_6}{r^6} \tag{4.14}$$

The magnitude of the constant C_6 is proportional to the strength of the attractive force. The subscript "6" indicates that the potential falls off as the sixth power of distance.

► **EXAMPLE 4.1**
Comparison of van der Waals Forces for Pure Species

As best as you can, compare the strength of the dipole–dipole, induction, and dispersion interactions for each of the following species at 298 K: H_2O, NH_3, CH_4, CH_3Cl, CCl_4. Discuss the results.

SOLUTION The dipole moment, μ, the polarizability, α, and the ionization energy, I, can be obtained from Table 4.1. They are summarized in Table E4.1. With these molecular parameters, we can determine the approximate dipole–dipole, induction, and dispersion potentials. Putting Equations (4.11), (4.12), and (4.13) in the form of Equation (4.14), for any pure species i, we get:

$$\text{dipole–dipole:} \quad (C_6)_{\text{dipole–dipole}} = \frac{2}{3} \frac{\mu_i^4}{kT} \quad \text{CGS units} \tag{E4.1A}$$

$$\text{induction:} \quad (C_6)_{\text{induction}} = 2\alpha_i \mu_i^2 \quad \text{CGS units} \tag{E4.1B}$$

$$\text{dispersion:} \quad (C_6)_{\text{dispersion}} \approx \frac{3}{4} \alpha_i^2 I_i \quad \text{CGS units} \tag{E4.1C}$$

Note Equation (E4.1B) is multiplied by 2, since each species in a two-body interaction can induce a dipole in its neighbor. Equation (E4.1C) is rigorously valid only for the symmetric species CH_4

Table E4.1 Relative Van Der Waals Interactions for Species of Example 4.1

Molecule	μ [D]	α [cm$^3 \times 10^{25}$]	I [eV]	$C_6 \times 10^{60}$ [erg cm^6]	$(C_6)_{\text{dipole–dipole}}$	$(C_6)_{\text{induction}}$	$(C_6)_{\text{dispersion}}$
H_2O	1.85	14.8	12.62	233	190	10	33
NH_3	1.47	22.2	10.07	145	76	10	60
CH_4	0	26	12.61	102	0	0	102
CH_3Cl	1.87	45.3	11.26	507	198	32	277
CCl_4	0	105	11.47	1517	0	0	1517

and CCl_4. However, we assume is provides a valid estimate for the other three species. The total van der Waals interaction is given by

$$C_6 = (C_6)_{\text{dipole–dipole}} + (C_6)_{\text{induction}} + (C_6)_{\text{dispersion}} \qquad \text{(E4.1D)}$$

The values for C_6 calculated from Equations (E4.1D), (E4.1A), (E4.1B), and (E4.1C) for each of the five species are presented in Table E4.1.

Even though it is non-polar, CCl_4 exhibits the largest intermolecular forces, approximately five times greater than the strongly polar CH_3Cl. The magnitude results from the large polarizability associated with the four Cl atoms in CCl_4. *It is curious to note how large the dispersion forces can be*! Although it is roughly the same size as ammonia or methane, water is more apt to be nonideal due to its more polar structure and concomitant dipole–dipole interactions. However, CH_3Cl, which has a similar dipole moment to water, demonstrates much larger van der Waals interactions (over twice the value for C_6) since this larger molecule is much more easily polarized and, consequently, has a larger dispersion interaction. ◀

▶ **EXAMPLE 4.2**
Magnitude of van der Waals Potentials in a Mixture

Consider a mixture of Ar and HCl. Predict the relative importance of the van der Waals interactions of the different "two-body" interactions in the mixture. Compare the unlike interactions to that which is obtained from prediction using the geometric mean of the like interactions.

SOLUTION In a binary mixture, there are three possible interactions: Ar–Ar, HCl–HCl, and Ar–HCl. The like species interactions, Ar–Ar and HCl–HCl can be found using the same approach as in Example 4.1. Thus, Equations (E4.1A) through (E4.1D) are used. For the Ar–HCl interaction, Equations (4.11), (4.12), and (4.13) can be used to get

$$(C_6)_{\text{dipole–dipole}} = \frac{2}{3}\frac{\mu_{\text{Ar}}^2 \mu_{\text{HCl}}^2}{kT} \quad \text{CGS units}$$

$$(C_6)_{\text{induction}} = \alpha_{\text{Ar}}\mu_{\text{HCl}}^2 + \alpha_{\text{HCl}}\mu_{\text{Ar}}^2 \quad \text{CGS units}$$

(with a "0" arrow pointing to the $\alpha_{\text{HCl}}\mu_{\text{Ar}}^2$ term)

and

$$(C_6)_{\text{dispersion}} = \frac{3}{2}\alpha_{\text{Ar}}\alpha_{\text{HCl}}\left(\frac{I_{\text{Ar}}I_{\text{HCl}}}{I_{\text{Ar}} + I_{\text{HCl}}}\right) \quad \text{CGS units}$$

Again, the total van der Waals interaction is given by

$$C_6 = (C_6)_{\text{dipole–dipole}} + (C_6)_{\text{induction}} + (C_6)_{\text{dispersion}}$$

The values of C_6 obtained from these equations are shown in Table E4.2. The magnitude of the unlike interaction, Ar–HCl, falls in between each of the like-species interactions. Often we estimate

TABLE E4.2   Relative Magnitudes of Attractive Forces Between Different Molecules

Molecule–Molecule	$C_6 \times 10^{60}$ [erg cm^6]	$(C_6)_{\text{dipole–dipole}}$	$(C_6)_{\text{induction}}$	$(C_6)_{\text{dispersion}}$
Ar–Ar	52	0	0	52
HCl–HCl	134	22	6	106
Ar–HCl	76	0	2	74

the magnitude of the unlike interaction as the geometric mean of the like interactions. In this case we would get

$$(C_6)_{\text{Ar–HCl}} = \sqrt{(C_6)_{\text{Ar–Ar}}(C_6)_{\text{HCl–HCl}}} = 83 \times 10^{-60} [\text{erg/cm}^6]$$

This value is different from that reported in Table E4.2 by 9%. In every case, the dispersion interaction is the largest. ◀

▶ **EXAMPLE 4.3**
Prediction of the Relative Size of van der Waals Forces Based on Molecular Structure

Consider the following molecules: $CCl_4, CF_4, SiCl_4$. List these species in order of their total van der Waals forces of attraction, C_6, from the largest value to the smallest. Explain your choice based on molecular arguments.

SOLUTION   The molecular parameters for $SiCl_4$ are not reported in Table 4.1. However, we can use qualitative molecular arguments to solve this problem. In general, attractive interactions include dispersion, dipole–dipole, and induction forces. The three species listed—$CCl_4, CF_4, SiCl_4$—are all nonpolar and, therefore, exhibit only dispersion forces. The magnitude of these forces is related to the polarizability, α, of these species. Each atom in CF_4 is from the second row of the periodic table. In $SiCl_4$ each atom is in the third row. Thus, the valence electrons in CF_4 are held in toward the nuclei the most tightly (least "sloshy"), so this molecule has the smallest dispersion forces. Conversely, the electrons in $SiCl_4$ are the farthest away and most easily polarized. So we would expect that for these species:

$$(C_6)_{\text{SiCl}_4} > (C_6)_{\text{CCl}_4} > (C_6)_{\text{CF}_4}$$ ◀

Intermolecular Potential Functions and Repulsive Forces

To account for nonideal gas behavior, we want to describe how the intermolecular potential energy depends on the position between molecules. A function or plot of potential energy vs. molecular separation is called a *potential function*. The potential function is ultimately what determines how the internal energy of a gas depends on pressure. There are many models of potential functions that are used to approximate the relation between intermolecular energy and position. These models include both attractive and repulsive interactions. For net neutral species, the attractive forces can be described by the van der Waals forces discussed previously. We next examine two ways in which the repulsive forces are approximated and the potential functions which result.

The Hard Sphere Model and the Sutherland Potential
Repulsive forces among molecules result from their finite sizes. In the simplest model, the hard sphere model, we consider the molecules to be finite hard spheres of diameter σ. Thus, molecules act like billiard balls; when they physically get close together, they run into each other and repel. Therefore, the potential between a pair of molecules is zero until the two molecules' diameters touch, where the potential increases to infinity. Mathematically, the potential function is described by

$$\Gamma = \begin{cases} 0 & \text{for } r > \sigma \\ \infty & \text{for } r \leq \sigma \end{cases} \tag{4.15}$$

A plot of the hard sphere potential function is presented in Figure 4.7a.

The Sutherland model adds the van der Waals attractive term proportional to r^{-6} to the hard sphere model. Therefore, the potential function is mathematically described by

$$\Gamma = \begin{cases} \dfrac{-(C_6)}{r^6} & \text{for } r > \sigma \\ \infty & \text{for } r \leq \sigma \end{cases} \tag{4.16}$$

The Sutherland model is illustrated in Figure 4.7b. The Sutherland potential accounts for both attractive and repulsive forces; however, there is discontinuity right at $r = \sigma$.

What would the ideal gas model look like on the plots in Figure 4.7?

The Lennard-Jones Potential

In reality, molecules are not rigid but rather are bounded by diffuse electron clouds. Repulsive interactions occur when the molecules get so close that their electron clouds overlap, leading to coulombic repulsion as well as a possible violation of the Pauli exclusion principle. This effect leads to a violent repulsion of the two molecules. Quantum mechanics says the repulsion should have an exponential dependence on position, since atomic wavefunctions fall off exponentially at large distances. However, it is more convenient to represent the repulsive potential empirically in terms of an inverse power law expression, as follows:

$$\Gamma_{ij} = \frac{(C_n)}{r^n} \qquad \text{where } 8 < n < 16 \tag{4.17}$$

where C_n is a constant proportional to the magnitude of the repulsive force that falls off as the inverse of separation to the power n. If we consider both van der Waals attractive forces and quantum (repulsive) effects, we come up with an expression for the molecular potential energy of the form:

$$\Gamma_{ij} = \frac{(C_n)}{r^n} - \frac{(C_6)}{r^6} \tag{4.18}$$

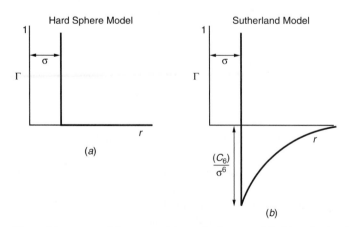

Figure 4.7 Potential functions. (a) Hard sphere model; (b) Sutherland model.

The attractive part is negative, since it lowers the energy, while the repulsive part is positive, raising the energy.

Lennard-Jones recognized a mathematically convenient form of this equation came about if $n = 12$, resulting in the Lennard-Jones potential function:

$$\Gamma = 4\varepsilon \left[\left(\frac{\sigma}{r} \right)^{12} - \left(\frac{\sigma}{r} \right)^{6} \right] \tag{4.19}$$

where $\qquad\qquad C_{12} = 4\varepsilon\sigma^{12} \qquad$ and $\qquad C_6 = 4\varepsilon\sigma^6$

A plot of the Lennard-Jones potential function is given in Figure 4.8. The parameters ϵ and σ can be physically interpreted as an energy parameter and a distance parameter, respectively. As illustrated in Figure 4.8, the energy parameter, ϵ, is given by the depth of the potential well, while the distance parameter, σ, is given by the distance at which attractive and repulsive potentials are equal and is characteristic of the molecular size. Some typical values of Lennard-Jones parameters are given in Table 4.2.[4] The size parameter, σ, increases with molecular size, and the energy parameter, ϵ, scales with the magnitude of the van der Waals interaction.

Figure 4.9*a* plots Lennard-Jones potential functions for O_2, Cl_2, and C_6H_6. These species are all nonpolar; the only van der Waals forces of attraction are from dispersion. Thus, their potential interactions depend only on the distance of separation between two molecules, not on their relative orientation. The relative contribution of attractive and repulsive interactions in these three species approximately scales with size; that is, as these species size get larger, their attractive forces get proportionately larger. We term these species *simple* molecules. For comparison, the Lennard-Jones potential functions for CH_3OH and SO_2 are included with the three simple species in Figure 4.9*b*. CH_3OH has a much larger force of attraction in proportion to its size—presumably due to its polar structure. Moreover, its polar structure leads to a dependence on orientation of one CH_3OH relative to another. Thus, the interactions with this species are inherently more "complex." A more accurate potential function would include orientation as a variable. The potential function depicted in Figure 4.9*b* is an approximation of the average over all the orientations one methanol molecule can have relative to another. We say that CH_3OH belongs to different *class*. SO_2 has large repulsive interactions but relatively weak attractive forces.

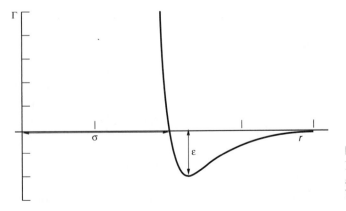

Figure 4.8 Plot of the Lennard-Jones potential as a function of distance between molecules.

[4] From Hirshfelder et al., *Molecular Theory of Gases and Liquids.*

TABLE 4.2   Lennard-Jones Parameters for Several Species

Gas	ε/k (K)	σ (Å)
He	10.2	2.58
H_2	35.7	2.94
C_2H_4	205	4.23
C_6H_6	440	5.27
F_2	112	3.65
Cl_2	307	4.62
O_2	101	3.5
N_2	86	3.7
CCl_4	327	5.88
CH_4	137	3.8
Ne	31.6	2.8
Ar	120	3.4
Xe	229	4.1
CH_3OH	507	3.6
SO_2	252	4.3

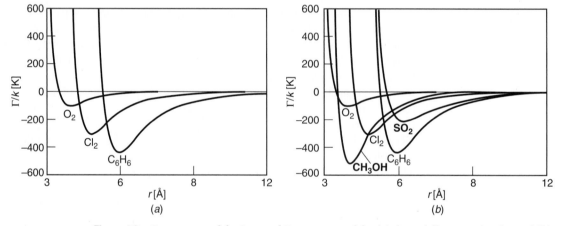

Figure 4.9   Comparison of the Lennard-Jones potential for (*a*) three different molecules and (*b*) five different molecules.

Principle of Corresponding States

We now wish to generalize our treatment of the intermolecular interactions that lead to nonideal gas behavior. We first consider nonpolar molecules. Inspection of Figure 4.9*a* suggests that if we scale the intermolecular potential appropriately, we can come up with a universal expression that applies to all nonpolar molecules. The ability to scale intermolecular interactions in this way leads to the *principle of corresponding states*:

> *The dimensionless potential energy is the same for all species.*

In quantitative form, it says there exists a universal function that applies to all species if we scale the potential energy to the energy parameter and the distance between molecules to the size parameter. Therefore, we can write

$$\mathbf{F}\left(\frac{\Gamma_{ii}}{\epsilon_i}, \frac{r}{\sigma_i}\right) = 0 \tag{4.20}$$

Equation (4.20) is not restricted to the Lennard-Jones potential function but rather it says the dimensionless potential energy is some universal function of the dimensionless distance.

We can extend the principle of corresponding states to macroscopic thermodynamic properties. In this form, we can write a general equation of state that applies to all species if we scale the measured properties P, v, and T appropriately. Van der Waals recognized that, for a given species, it was particularly suitable to scale the values to those at its critical point. The critical point represents a unique state, and that state is determined by the intermolecular interactions characteristic of a given species. Thus, we can construct a "reduced" coordinate system with the following three dimensionless groupings:[5]

$$T_r = \frac{T}{T_c}, \qquad P_r = \frac{P}{P_c}, \qquad \text{and} \qquad v_r = \frac{v}{v_c} \tag{4.21}$$

The principle of corresponding states says that there is some universal function that is the same (i.e., the same form and the same constants) for *all* substances:

$$\mathbf{F}\left(\frac{T}{T_c}, \frac{P}{P_c}, \frac{v}{v_c}\right) = 0 \tag{4.22}$$

Alternatively, one of the dimensionless groupings might be the compressibility factor, z. Thus

$$z = \frac{Pv}{RT} = \mathbf{F}\left(\frac{T}{T_c}, \frac{P}{P_c}\right) \tag{4.23}$$

Equation (4.23) illustrates the macroscopic version of the principle of corresponding states:

> *All fluids at the same reduced temperature and reduced pressure have the same compressibility factor.*

We based the discussion above on nonpolar species. Figure 4.9*b* illustrates that there are different classes of molecules based on the particular nature of the intermolecular interactions involved. For example, CH_3OH, with its strong dipole moment, behaves differently from the nonpolar species depicted in Figure 4.9*a*. We can improve the principle of corresponding states if we group molecules according to class and assert that within any one class intermolecular interactions scale similarly. To accomplish this objective, we introduce a third parameter characteristic of classes of molecules. There are many ways to introduce a parameter for classes of molecules; we will explore only one—the Pitzer acentric factor, ω. It characterizes how "nonspherical" a molecule is, thereby assigning it to a class. The definition of ω is somewhat arbitrary:

$$\omega \equiv -1 - \log_{10}[P^{\text{sat}}(T_r = 0.7)/P_c], \tag{4.24}$$

---

[5] P_c, T_c, and v_c are not all independent, since we need only two properties to constrain the state of the critical point. Thus both molecular and macroscopic versions of the principle of corresponding states have two independent scaling parameters.

where $P^{sat}(T_r = 0.7)$ is the saturation pressure at a reduced temperature of 0.7. This definition for a third parameter is convenient, since it gives a value of zero for the simple fluids Ar, Kr, and Xe. Moreover, other fluids have positive values less than 1. Since tabulated data for ω are usually available, we seldom need Equation (4.24)—we just know where to look ω up. Appendix A.1 presents acentric factors for a number of common species.

With the introduction of the acentric factor to categorize classes of molecules, the general macroscopic equation is of the form

$$\mathbf{F}\left(\frac{T}{T_c}, \frac{P}{P_c}, \frac{v}{v_c}, \omega\right) = 0 \tag{4.25}$$

Equation (4.25) is often written in the form

$$z = \mathbf{F}^0(T_r, P_r) + \omega \mathbf{F}^1(T_r, P_r) \tag{4.26}$$

where $\mathbf{F}^0$ and $\mathbf{F}^1$ depend only on the reduced pressure and temperature and the acentric factor is used to modulate the effect of the $\mathbf{F}^1$ term. Thus a perfectly "spherical" molecule (such as Ar) depends only on $\mathbf{F}^0$.

Chemical Forces

The physical forces described above aptly account for most molecular interactions in the gas phase. We now direct our discussion toward the condensed phases. Solids and liquids form when the net attractive intermolecular forces are stronger than the thermal energy in the system and, consequently, hold the molecules together. While the force of attraction can sometimes be attributed to the electrostatic and van der Waals interactions described above, **chemical** forces also frequently play a role in condensed phases.[6] Chemical forces are based on the nature of covalent electrons, the concept of the chemical bond, and the formation of new chemical species. The main difference between chemical and physical forces is that chemical forces **saturate** whereas physical forces do not, since chemical interactions are specific to the electronic wavefunctions of the chemical species involved. Indeed, a complete quantitative description of chemical interactions involves solution of the Schrödinger equation to describe the overlap of the molecular orbitals involved. We will consider chemical interactions only qualitatively. The goal of this discussion is to realize that there may be other important forces that govern the behavior of solids and liquids and to get a flavor of what these forces might be.

The most prevalent chemical effects are due to hydrogen bonds and acid–base complexes. In both cases, there exists a sharing of valence electrons between different molecules. Hydrogen bonding is the "chemical bond" that results between an electronegative atom (usually F, O, or N) and a hydrogen atom bonded to another electronegative atom in a second molecule. Figure 4.10 illustrates the hydrogen bond that forms between the electronegative oxygen in water molecule 1 and the adjacent hydrogen in water molecule 2. Since the electronegative oxygen atom in water 2 pulls the hydrogen atom's electron away from its nucleus, a partial charge separation results. The resulting positive charge on the hydrogen atom can be attracted to a partial negative charge of an electronegative atom in the adjacent oxygen atom in water 1, leading to an attractive force. So far, this sounds like the van der Waals interactions we just described, and we might be tempted to say that hydrogen bonds are caused by the dipole–dipole interactions. However, the mechanism of attraction is *not purely electrostatic* but rather has a significant

---

[6] In some unique gas systems, chemical forces also manifest themselves.

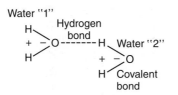

Figure 4.10 Hydrogen bonding in water.

amount of sharing of lone-pair electrons characteristic of a covalent bond. This leads to fundamental differences between hydrogen bonds and dipole–dipole forces. Hydrogen bonds form a relatively strong, highly directional interaction with characteristic saturation. Typically, this force is two orders of magnitude stronger than the van der Waals forces $(1/r^6)$ described above but one order of magnitude weaker than a covalent bond. In other words, if you plugged numbers into Equation (4.11) to try to account for the strength of a hydrogen bond, you would get a number whose magnitude is far below the strength experimentally observed. Moreover, the distance between the hydrogen-bonded atoms is considerably less than that predicted based on the hard sphere model. Thus, the hydrogen is actually penetrating into the electron cloud of the electronegative atom in the other molecule. The hydrogen bond leads to interesting thermodynamic phenomena. For example, the extensive network of hydrogen bonds leads to the open structure of ice. Thus, unlike most species, water expands when it freezes.[7]

To illustrate how chemical interactions can affect thermodynamic properties, consider two types of chemical behavior: solvation and association. Solvation is the tendency of *unlike* molecules to form chemical complexes. It is generically represented by the following reaction:

$$A + B \Longleftrightarrow AB \tag{4.27}$$

Association is the tendency of *like* molecules to form complexes (polymerize) and can be represented as follows:

$$A + A \Longleftrightarrow A_2 \tag{4.28}$$

Naturally, the ability of a molecule to solvate or associate is intimately linked to its electronic structure. Hydrogen bonding can lead to either of these behaviors. An example of solvation is given by a mixture of chloroform and acetone. Hydrogen bonding causes the unlike molecules to form a complex:

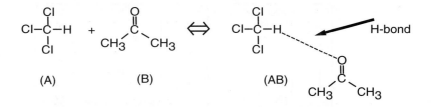

Acid–base pairs also solvate. The dimerization of acetic acid illustrates association:

---

[7] Imagine what our world would look like if ice were denser than water. All the ice in the ocean would sink to the bottom and, consequently, be insulated from energy input from the sun. Thus, most of the ocean would be permanently frozen.

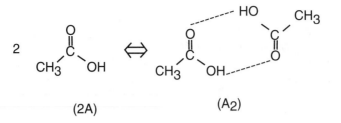

(2A)   (A$_2$)

Can you think of other possible solvation and association reactions?

To see how chemical effects can change the equilibrium behavior of a system, let's examine Figure 4.11, which depicts species A and B in liquid–vapor equilibrium. We will assume ideal gas behavior in the vapor. We wish to compare the composition of the vapor phase for three scenarios occurring in the liquid:

(i)   A and B are the only species present in the system and Raoult's law applies.

(ii)   Species A and B solvate in the liquid phase, but we are unaware of this chemistry.

(iii)   Species A associates in the liquid phase, but we are unaware of this chemistry.

We wish to examine how solvation [scenario (ii)] or association [scenario (iii)] changes our perception of this system. In scenario (ii), species A and B solvate in the liquid phase. This reaction depletes the liquid of species A and B, since the complex AB is formed. Since AB is a different chemical species from A or B, some A and B in the vapor phase will then condense to compensate. This leads to lower total system pressure than in the ideal case [scenario (i)]; thus solvation effects lead to a "negative" deviation from Raoult's law.

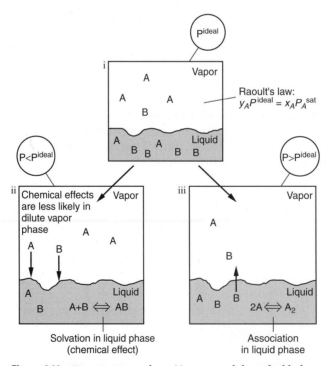

Figure 4.11   Binary system where (i) system exhibits ideal behavior; (ii) species A and species B solvate in the liquid phase; (iii) species A associates in the liquid phase.

Next we wish to examine what happens in scenario (iii), where species A associates in the liquid phase. The dimer is a different chemical species and assumed involatile. Let's consider the point of view of molecule B. The association causes a higher mole fraction of B to be in the liquid than if A did not associate. Thus B will evaporate to compensate. This will lead to a higher system pressure than the ideal case. Association leads to "positive" deviations from Raoult's law.

The above arguments have provided a qualitative and intuitive feel for what we will spend a large percentage of the text developing: ways to predict equilibrium behavior and deviations from ideality in chemical systems.

▷ **EXAMPLE 4.4**
Comparison of P^{sat} for H_2O and CH_3OH

Determine the saturation pressure, P^{sat}, of water and methanol at 100°C, 50°C, and 25°C. Report values in [Pa]. Based on intermolecular forces explain (i) why the vapor pressure of methanol at a given temperature is greater than water and (ii) why vapor pressure increases with temperature.

SOLUTION We can obtain the saturation pressure of water from the steam tables (Appendix B.1). The saturation pressure of methane is given by the Antoine equation,

$$\ln(P^{sat}\,[\text{bar}]) = A - \frac{B}{T[\text{K}] + C} = 11.9673 - \frac{3626.55}{T - 34.29}$$

where the constants were found consulting Appendix A.1. The resulting values of P^{sat} are reported in Table E4.4:

(i) Vapor pressure is a function of how easily molecules can "escape" from the liquid phase. This is dictated by the strength of the intermolecular interactions of the species involved given a certain thermal energy. The weaker the intermolecular forces, the larger P^{sat}.

Both methanol and water form hydrogen bonds in the liquid phase. Hydrogen bonds are much stronger than van der Waals interactions, so they will control the behavior of the vapor pressure for these two species. Water can form two hydrogen bonds/ molecule. as shown in Figure E4.1, while CH_3OH can form only one. Thus at a given temperature, water has stronger attractive forces in the liquid, and a lower vapor pressure.

(ii) Explanation #1: Temperature is a measure of the molecular kinetic energy (part of the internal energy, u). While it is representative of the average molecular kinetic energy, species at thermal

TABLE E4.4. Values of P^{sat} for Water and Methanol

$T[°C]$	P^{sat} [Pa]	
	Water	Methanol
25	3.169×10^3	1.69×10^4
50	1.235×10^4	5.56×10^4
100	1.014×10^5	3.54×10^5

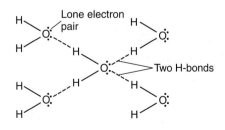

Figure E4.1 Schematic of two H-bonds formed per H_2O molecule.

equilibrium have a distribution of energies. This distribution is given by the Maxwell–Boltzmann equation. A certain fraction of species (water or methanol in this case) will have enough kinetic energy to overcome the attractive forces (H-bonds) keeping them in the liquid phase. As this fraction increases, more molecules enter the vapor phase and P^{sat} increases. Since the Maxwell-Boltzmann distribution depends exponentially on temperature, P^{sat} also increases exponentially with temperature.

Explanation #2: You may be tempted to use following explanation for the temperature dependence of P^{sat}. As T increases, more molecules would hit the walls of the container. This can be seen, for example, in the ideal gas equation $P = RT/v$. Thus P^{sat} increases with T. This explanation is not wrong but it is incomplete! This would predict a linear relation between P^{sat} and T not the exponential relation we observe experimentally.

Viewed in another way, we may ask, do the number of molecules in the vapor phase increase as T increases? Explanation #1 asserts that they do. In explanation #2, however, you could get a higher P^{sat} without adding any more species to the vapor. Viewing it the latter way is wrong. ◀

▶ 4.3 EQUATIONS OF STATE

The van der Waals Equation of State

We will now use our knowledge of intermolecular interactions to modify the ideal gas model for situations when potential interactions between the species are important. In this section, we will use the Sutherland potential function to describe the intermolecular interactions, that is, use the hard sphere model to account for repulsive forces and van der Waals interactions to describe attractive forces. This development leads to the van der Waals equation of state. This equation is particularly well suited for illustrating how the molecular concepts we learned about in Section 4.2 can be related to macroscopic property data. However, it should be emphasized that more accurate equations of state have been developed and will be covered next.

First, let's consider the "size" of the molecules based on the hard sphere model. The entire volume of the system will no longer be available to the molecules. We can account for this effect by replacing the volume term in the ideal gas model with one for available volume. Recall that in the hard sphere model, the molecules have a diameter σ. Thus, the center of one molecule cannot approach another molecule closer than a distance σ. The excluded volume of the two molecules is then $(4/3)\pi\sigma^3$. Dividing by 2 and multiplying by Avogadro's number, N_A, we get one mole of molecules occupying a volume $b = (2/3)\pi\sigma^3 N_A$. To correct for size, we modify the ideal gas model to include only the unoccupied molar volume, $(v - b)$. Hence, we get:

$$P = \frac{RT}{v - b} \tag{4.29}$$

since one molecule cannot occupy the space in which another molecule already sits.

We still need to take into account attractive intermolecular forces. In the absence of net electric charge, the attractive forces in the gas phase can include dispersion, dipole–dipole, and induction, all of which have an r^{-6} dependence. However, we do not have "distance" as a parameter in our equations, but rather volume, which is proportional to the cube of the distance ($v \approx r^3$). We can say, therefore, that all of these terms are proportional to v^{-2}. But how do we incorporate this into our equation of state? As we saw in Section 4.2, the variable most related to potential energy is pressure. So we correct the pressure by including a term that accounts for attractive forces. Attractive forces should *decrease* the pressure, since the molecules will not bang into the container as readily; hence, we subtract a correction term as follows:

$$P = \frac{RT}{v - b} - \frac{a}{v^2} \tag{4.30}$$

This equation was first proposed by the Dutch physicist van der Waals in 1873. Since it assumes a $1/r^6$ dependence for all attractive forces, any force with this functionality (be it dispersion, dipole–dipole, or induction) has been termed a "van der Waals force." The parameter a in Equation (4.30) can be related to molecular constants by integrating the Sutherland potential function. This calculation gives $a = (2\pi N_A^2 C_6)/(3\sigma^3)$. In practice, a and b are treated as empirical constants that account for the magnitude of the attractive and repulsive forces. Can you think of how we might find values for the constants a and b?

We can rewrite Equation (4.30) as follows:

$$Pv^3 - (RT + Pb)v^2 + av - ab = 0 \tag{4.31}$$

Equation (4.31) is termed a *cubic* equation of state since there are three roots for volume for fixed values of T and P given the values of the parameters a and b. We will consider other cubic equations of state shortly. Figure 4.12 illustrates the general characteristics of this equation for various isotherms. A cubic equation has three roots in volume for a given pressure and temperature. However, these roots have different characteristics above the critical point than below it. Above the critical point there is one positive, real root and two roots containing negative or imaginary numbers. Only the positive, real root represents a physical value—the volume of the supercritical fluid. The other roots are mathematical artifacts with no physical basis. Below the critical point, however, there may be three real, positive roots. We can ascribe the lowest root to the molar volume of the liquid state and the highest root to the vapor state. We throw out the middle root where $dP/dv > 0$ on physical grounds. Can you think of a physical reason why this relationship is not possible at constant temperature? In reality, isotherms are horizontal in the two-phase envelope where vapor and liquid coexist. This discontinuity eludes description by cubic equations of state.

One feature of the two-phase region can be determined by cubic equations. Maxwell's "equal-area rule" (which will be verified in Chapter 6) provides a graphical means to determine P^{sat} for a given T. It states that the saturation pressure is the pressure at which a horizontal line equally divides the area between the real isobar and the

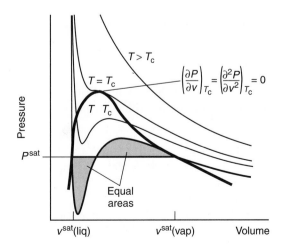

Figure 4.12 Pv behavior of the van der Waals equation. This behavior is representative of other cubic equations of state.

solution given by the cubic equation. Such a construct is illustrated in Figure 4.12, where the equal areas above and below the isobar fix the value for P^{sat}. This procedure can be achieved by trial and error. If a higher saturation pressure were predicted, the upper area would be too small. Conversely, too low a value for P^{sat} would make the upper area too large.

To use the van der Waals equation, the parameters a and b must be determined for a species of interest. The most accurate values are obtained by fitting experimental PvT data. However, when these data are not available, we can use the principle of corresponding states (see Section 4.2). Recall that the principle of corresponding state scales property data to that at the critical point. We can relate the van der Waals parameters to the temperature and pressure at the critical point by noting that there is an inflection point on the critical isotherm, as shown in Figure 4.12. Mathematically, we can say:

$$\left(\frac{\partial P}{\partial v}\right)_{T_c} = \left(\frac{\partial^2 P}{\partial v^2}\right)_{T_c} = 0 \qquad (4.32)$$

Thus, at the critical point we have

$$P_c = \frac{RT_c}{v_c - b} - \frac{a}{v_c^2} \qquad (4.33)$$

$$\left(\frac{\partial P}{\partial v}\right)_{T_c} = 0 = \frac{-RT_c}{(v_c - b)^2} + \frac{2a}{v_c^3} \qquad (4.34)$$

and

$$\left(\frac{\partial^2 P}{\partial v^2}\right)_{T_c} = 0 = \frac{2RT_c}{(v_c - b)^3} - \frac{6a}{v_c^4} \qquad (4.35)$$

If we multiply Equation (4.34) by 2 and Equation (4.35) by $(v - b)$ and add them together, we get:

$$0 = \frac{4a}{v_c^3} - \frac{6a(v_c - b)}{v_c^4} \qquad (4.36)$$

Equation (4.36) can be solved to give:

$$v_c = 3b \qquad (4.37)$$

If we plug Equation (4.37) back into Equation (4.34) and solve for a, we get:

$$a = \frac{9}{8} v_c RT_c \qquad (4.38)$$

Finally, if we plug Equation (4.38) back into Equation (4.33), we can solve for the van der Waals constants in terms of the critical temperature and the critical pressure:

$$\boxed{a = \frac{27}{64} \frac{(RT_c)^2}{P_c}} \qquad (4.39)$$

and

$$\boxed{b = \frac{(RT_c)}{8P_c}} \qquad (4.40)$$

Example 4.7 shows another approach to obtaining van der Waals parameters from properties at the critical point. We now have an equation of state for which we only need critical property data to solve for the parameters. This approach will not be as accurate as fitting PvT data to get the constants, a and b, but then again you do not have to go through the expense of laboratory measurements, since values of critical properties are usually known and readily available.

We can use the results above to write the van der Waals equation in terms of the reduced variables T_r, P_r, and v_r. We start with Equation (4.30):

$$P = \frac{RT}{v - b} - \frac{a}{v^2} \tag{4.30}$$

We can substitute for a and b and rearrange using Equations (4.38), (4.37), and (4.40) to give

$$\left(\frac{P}{P_c}\right) = \frac{8\left(\dfrac{T}{T_c}\right)}{3\left(\dfrac{v}{v_c}\right) - 1} - \frac{3}{\left(\dfrac{v}{v_c}\right)^2} \tag{4.41}$$

Equation (4.41) can be written in reduced form as

$$P_r = \frac{8T_r}{3v_r - 1} - \frac{3}{v_r^2} \tag{4.42}$$

If we compare Equation (4.42) to (4.22), we see we have defined a universal function for the reduced pressure in terms of the reduced temperature and the reduced volume. Thus, we have a specific expression to delineate the corresponding states between species.

We can calculate the compressibility factor at the critical point from Equations (4.37) and (4.40):

$$z_c = \frac{P_c v_c}{RT_c} = \frac{3}{8} \tag{4.43}$$

Thus, applying the principle of corresponding states to the van der Waals equation leads to a value of 0.375 for the compressibility factor at the critical point for all species. Experimental values for the compressibility factor at the critical point are around 0.29 for simple species and usually less for complex species. Thus, the value predicted by the van der Waals equation is considerably high—indicating its limitations in predicting PvT behavior.

▶ **EXAMPLE 4.5**
Prediction of the Relative Size of van der Waals Forces Based on Molecular Structure

Consider the following molecules: CCl_4, CF_4, $SiCl_4$, $SiCl_3H$.
(a) List these species in order of their van der Waals constant a, from the largest value of a to the smallest. Explain your choice based on molecular arguments.
(b) Repeat for the van der Waals constant b. Explain your choice based on molecular arguments.

SOLUTION **(a)** The van der Waals a constant is representative of attractive interactions due to dispersion, dipole–dipole, and induction forces. The first three species listed—CCl_4, CF_4, $SiCl_4$—are all nonpolar and exhibit only dispersion forces. The magnitude of these forces is related to the polarizability, α, of these species. In CF_4, the valence electrons are held in toward the nuclei the most tightly (least "sloshy"), so this has the smallest dispersion forces. Conversely, the electrons in $SiCl_4$ are the farthest away and most easily polarized. So we would expect that for these species:

$$a_{SiCl_4} > a_{CCl_4} > a_{CF_4}$$

The fourth species, $SiCl_3H$, has two forces, dispersion and dipole–dipole, which add together. So the question becomes where we stick this species in the hierarchy above. This is a tough call. We expect a fairly strong dipole (> 1 D), as shown below:

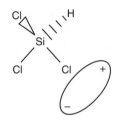

However, polarizabilities are additive among atoms in a molecule, and this species replaces a very polarizable atom (Cl) with an almost nonpolarizable atom (H). We may say that the dipole wins out and:

$$a_{SiCl_3H} > a_{SiCl_4} > a_{CCl_4} > a_{CF_4}$$

however,

$$a_{SiCl_4} > a_{SiCl_3H} > a_{CCl_4} > a_{CF_4} \text{ and even } a_{SiCl_4} > a_{CCl_4} > a_{SiCl_3H} > a_{CF_4}$$

are possible.

(b) The van der Waals constant b is representative of the volume a molecule occupies. $SiCl_4$ is certainly the largest and CF_4 the smallest, but how about $SiCl_3H$ vs. CCl_4? Si is bigger than C, but Cl is bigger than H. If you imagine how the atoms stack—starting with a triangle of Cl and then either an Si in the middle with an H on top or a C in the middle with a Cl on top—you can see that CCl_4 is larger. So for b:

$$b_{SiCl_4} > b_{CCl_4} > b_{SiCl_3H} > b_{CF_4}$$

◀

▶ **EXAMPLE 4.6**
Calculation of van der Waals Constants from Critical Properties

Calculate the van der Waals parameters from critical point data for the following gases: benzene, toluene, cyclohexane. Explain the relative magnitudes of a and b from a physical basis.

SOLUTION The van der Waals parameters are calculated from the critical pressure and temperature as follows:

	P_c [bar]	T_c [K]	$a = \dfrac{27}{64}\dfrac{(RT_c)^2}{P_c}\left[\dfrac{\text{Jm}^3}{\text{mol}^2}\right]$	$b = \dfrac{(RT_c)}{8P_c}\left[\dfrac{\text{m}^3}{\text{mol}}\right]$
Benzene	49.1	562	1.88	1.19×10^{-4}
Toluene	42.0	594	2.45	1.47×10^{-4}
Cyclohexane	40.4	553	2.21	1.42×10^{-4}

The attractive interactions of all three compounds are dominated by dispersion interactions (parameter a), while size affects parameter b. Toluene has the highest values for a and b. Toluene has seven carbon atoms, whereas the other molecules have only six. This results in the largest polarizability as well as the largest size. Cyclohexane's electrons are freer to move than the tight resonance structure exhibited by benzene. This leads to a greater polarizability than benzene. In fact the magnitude of dispersion forces is closer to toluene than benzene. Finally, cyclohexane has a three-dimensional structure, while the other two are planar and flat. Hence, the constant b, representative of size, is almost as large for cyclohexane as it is for benzene.

◀

► **EXAMPLE 4.7**
Alternative Determination
of van der Waals
Constants from
Corresponding States

At the critical point, the three roots in volume to a cubic equation must converge. Thus,

$$(v - v_c)^3 = 0 \tag{E4.7A}$$

Use Equation (E4.7A) to write the van der Waals parameters a and b in terms of the critical pressure and temperature.

SOLUTION   We can rewrite Equation (4.31) as follows:

$$v^3 - \left[\frac{RT}{P} + b\right] v^2 + \frac{a}{P} v - \frac{ab}{P} = 0$$

or at the critical point

$$v^3 - \left[\frac{RT_c}{P_c} + b\right] v^2 + \frac{a}{P_c} v - \frac{ab}{P_c} = 0 \tag{E4.7B}$$

Expanding Equation (E4.7A) gives:

$$v^3 - 3v^2 v_c + 3v v_c^2 - v_c^3 \tag{E4.7C}$$

We can now set each of the terms in volume from Equation (E4.7B) equal to those from Equation (E4.7C). For the root of v^0, we have

$$\frac{ab}{P_c} = v_c^3 \tag{E4.7D}$$

For the root of v^1, we have

$$3v_c^2 = \frac{a}{P_c} \tag{E4.7E}$$

and for the root of v^2, we have

$$3v_c = \left[\frac{RT_c}{P_c} + b\right] \tag{E4.7F}$$

We can solve Equation (E4.7E) for the parameter a:

$$a = 3v_c^2 P_c \tag{E4.7G}$$

and then Equation (E4.7D) for b:

$$b = \frac{P_c v_c^3}{a} = \frac{v_c}{3} \tag{E4.7H}$$

Finally, solving Equation (E4.7F) for b and substituting the result of Equation (E4.7H) gives

$$v_c = \frac{3RT_c}{8P_c} \tag{E4.7I}$$

We can solve for the parameters a and b by substituting Equation (E4.7I) into Equations (E4.7G) and (E4.7H), respectively:

$$a = \frac{27(RT_c)^2}{64P_c}$$

and
$$b = \frac{RT_c}{8P_c}$$

Note the expressions we obtained for the parameters a and b in this example match those given by Equations (4.39) and (4.40). ◄

Cubic Equations of State

The van der Waals equation is an example of a cubic equation of state, since it goes as volume to the third power; however, it is not as accurate as more recent cubic equations of state. The van der Waals equation was presented because of the clear way in which it incorporates the potential interactions we have discussed. Keep in mind that if you need an accurate answer, there are better equations to use. In fact, we will see that many modern cubic equations have the same basic form as the van der Waals equation. There are hundreds of different cubic equations of state. All these equations are approximate. They merely fit experimental data. Yet, in general, they can provide reasonable values for both the vapor and liquid regions of hydrocarbons and the vapor region for many other fluids. We will not attempt to go through a critical review of all the available equations; rather, we will illustrate the scientific concepts and engineering application through a few simpler, commonly used cubic equations. The majority of equations that have been proposed are variations of the forms we will study. The most rigorous test for an equation of state is near the critical point. When greater accuracy is needed, a higher-order equation of state with more fitting parameters may be used. Equations of state have been proposed with more than 50 parameters! However, these equations can become mathematically cumbersome.

The general form of a cubic equation is

$$v^3 + f_1(T,P)v^2 + f_2(T,P)v + f_3(T,P) = 0 \tag{4.44}$$

where $f_i(T,P)$ represent a function that can contain fitting parameters as well as the properties T and P. The three characteristic roots in volume follow the same trends as those we discussed with the van der Waals equation in relation to Figure 4.12. Table 4.3 illustrates some examples of the form

$$P = \frac{RT}{v-b} - \text{Attr} \tag{4.45}$$

All these equations use the same "repulsive" term as the van der Waals equation. The term indicated by "Attr" quantifies attractive interactions. In general, these terms are empirically established to best fit experimental data.

The Redlich–Kwong equation, the Soave–Redlich–Kwong[8] equation, and the Peng–Robinson equation are all commonly used. **The Redlich–Kwong equation of state is** given by

TABLE 4.3   Parameters for Some Popular Cubic Equations of State of the Form $P = RT/(v-b) - \text{Attr}$

Equation	Year	Attr
van der Waals	1873	$\dfrac{a}{v^2}$
Redlich–Kwong	1949	$\dfrac{a/\sqrt{T}}{v(v+b)}$
Soave–Redlich–Kwong	1972	$\dfrac{a\alpha(T)}{v(v+b)}$
Peng–Robinson	1976	$\dfrac{a\alpha(T)}{v(v+b)+b(v-b)}$

[8] This equation is also referred to as the Redlich–Kwong–Soave equation of state.

$$P = \frac{RT}{v - b} - \frac{a}{T^{1/2}v(v + b)} \tag{4.46}$$

The relationships for parameters a and b can be written in terms of critical temperature and critical pressure using the same methodology that we applied to the van der Waals equation. In this case, the alternative method illustrated in Example 4.7 is convenient to implement (Problem 4.28). After working through the math, we get:

$$a = \left(\frac{1}{9(\sqrt[3]{2} - 1)}\right) \frac{R^2 T_c^{2.5}}{P_c} = \frac{0.42748 R^2 T_c^{2.5}}{P_c} \tag{4.47}$$

and

$$b = \left(\frac{\sqrt[3]{2} - 1}{3}\right) \frac{RT_c}{P_c} = \frac{0.08664\, RT_c}{P_c} \tag{4.48}$$

Note the parameters a and b in the Redlich–Kwong equation are different from those for the van der Waals parameters and *cannot* be interchanged. The Redlich–Kwong equation works well over a wide range of conditions but departs significantly from measured values near the critical point. In reduced form, the Redlich–Kwong Equation gives:

$$P_r = \frac{3T_r}{v_r - 0.2599} - \frac{1}{0.2599\sqrt{T_r}v_r(v_r + 0.2599)} \tag{4.49}$$

and the compressibility factor at the critical point is found to be:

$$z_c = \frac{1}{3} \tag{4.50}$$

While this value is closer to experimental values than the van der Waals equation, it is still too high.

The **Peng–Robinson equation of state** is given by:

$$P = \frac{RT}{v - b} - \frac{a\alpha(T)}{v(v + b) + b(v - b)} \tag{4.51}$$

with

$$a = 0.45724\frac{R^2 T_c^2}{P_c} \tag{4.52}$$

$$b = 0.07780\frac{RT_c}{P_c} \tag{4.53}$$

$$\alpha(T) = \left[1 + \kappa(1 - \sqrt{T_r})\right]^2 \tag{4.54}$$

how spherical

$$\kappa = 0.37464 + 1.54226\omega - 0.26992\omega^2 \tag{4.55}$$

The compressibility factor at the critical point is found to be: $z_c = 0.307$. The Peng–Robinson equation is an option in the equation of state menu of the ThermoSolver software that comes with the text.

The Redlich–Kwong equation with critical constant estimation of parameters uses a "two-parameter" corresponding states expression represented, in general, by Equation (4.22). On the other hand, the Peng–Robinson equation utilizes the third parameter, ω; thus, we expect it to be better suited for different classes of molecules. The Soave–Redlich–Kwong equation of state is a three-parameter equation similar in form to the Peng–Robinson equation and is also commonly used.

The Virial Equation of State

The virial equation of state has a sound theoretical foundation; it can be derived from first principles using statistical mechanics. This equation is given by a power series expansion for the compressibility factor in density (or the reciprocal of volume) about $1/v = 0$:

$$z = \frac{Pv}{RT} = 1 + \frac{B}{v} + \frac{C}{v^2} + \frac{D}{v^3} + \cdots \tag{4.56}$$

Here B, C, ... are called the second, third, ... virial coefficients; these parameters depend only on temperature (and composition for mixtures). An alternative expression for the virial equation is a power series expansion in pressure:

$$z = \frac{Pv}{RT} = 1 + B'P + C'P^2 + D'P^3 + \cdots \tag{4.57}$$

By solving Equation (4.56) for P and substituting into Equation (4.57), it is straightforward to show the two sets of coefficients are related by:

$$B' = \frac{B}{RT}$$

$$C' = \frac{C - B^2}{(RT)^2} \quad \text{and so on} \tag{4.58}$$

A common question is: What power series expansion do I use? Well, first of all, one is explicit in pressure and the other is explicit in volume, so if you need an expression that is explicit in one of these variables (so you can take a derivative, for example), use the appropriate form. The next issue is a question of accuracy. It turns out that at moderate pressures (up to about 15 bar) when you keep only the second virial coefficient, the power series expansion in pressure is better:

$$z = \frac{Pv}{RT} = 1 + B'P = 1 + \frac{BP}{RT} \tag{4.59}$$

From 15 to 50 bar, the virial equation should contain three terms, and the expansion in density is more accurate:

$$z = 1 + \frac{B}{v} + \frac{C}{v^2} \tag{4.60}$$

Using statistical mechanics, we can relate the virial coefficients to intermolecular potentials. We will leave the derivation to a physical chemistry course and merely present the results. The second virial coefficient, B, results from all the "two-body" interactions in the system, that is, all the interactions between two molecules; the third virial coefficient, C, results from all the "three-body" interactions in the system; and so on. From this point of view, can you see why you need to include more and more terms as the pressure increases? Additionally, if the pressure is so low that not even two-body interactions affect the system properties, we have an ideal gas. As an example, consider the case of spherically symmetric molecules. According to statistical mechanics, the second virial coefficient is given by the following expression:

$$B = 2\pi N_A \int_0^\infty \left(1 - e^{-\Gamma(r)/(kT)}\right) r^2 dr \tag{4.61}$$

The principle of corresponding states is often applied to the truncated virial equation given by Equation (4.59). It can be put in the form given by Equation (4.26):

$$B_r = B^{(0)} + \omega B^{(1)} \tag{4.62}$$

where

$$B_r = \frac{BP_c}{RT_c} \tag{4.63}$$

Several correlations of parameters $B^{(0)}$ and $B^{(1)}$ to reduced temperature have been proposed.[9] For example, Abbott found that they can be calculated by:

$$B^{(0)} = 0.083 - \frac{0.422}{T_r^{1.6}} \tag{4.64}$$

and

$$B^{(1)} = 0.139 - \frac{0.172}{T_r^{4.2}} \tag{4.65}$$

respectively.

The **Beattie–Bridgeman equation of state** is a specific version of the virial equation:

$$z = \frac{Pv}{RT} = 1 + \frac{B}{v} + \frac{C}{v^2} + \frac{D}{v^3} \tag{4.66}$$

where

$$B = B_0 - \frac{A_0}{RT} - \frac{c}{T^3} \tag{4.67}$$

$$C = -B_0 b + \frac{A_0 a}{RT} - \frac{cB_0}{T^3} \tag{4.68}$$

$$D = \frac{bcB_0}{T^3} \tag{4.69}$$

where $A_0, B_0, a, b,$ and c are adjustable parameters.

The **Benedict–Webb–Rubin equation of state** modifies the virial equation by adding an exponential term. The form is given by

$$z = 1 + \left(B_0 - \frac{A_0}{RT} - \frac{C_0}{RT^3}\right)v^{-1} + \left(b - \frac{a}{RT}\right)v^{-2} + \frac{a\alpha}{RT}v^{-5} + \frac{\beta}{RT^3 v^2}\left(1 + \frac{\gamma}{v^2}\right)\exp\left(-\frac{\gamma}{v^2}\right) \tag{4.70}$$

It has been shown to model both liquid and vapor PvT behavior well even in the critical region. However, you must have values for all eight coefficients, and you must have the computational muscle to do the calculations. The extension of the Benedict–Webb–Rubin equation by Lee and Kessler is presented in Appendix E and forms the basis for the generalized compressibility charts discussed in Section 4.4.

[9] S. M. Walas, *Phase Equilibria in Chemical Engineering* (Boston: Butterworth, 1985).

Equations of State for Liquids and Solids

Liquid and solid molar volumes are straightforward to measure in the lab. For example, there is data available for the molar volume of many liquids at room temperature or at their normal boiling point. Table 4.4 reports the volume of a sample set of liquids and solids at 20°C and 1 bar. The volumes of condensed phases are also much less sensitive to temperature and pressure than gases. The measured values can be adjusted for temperature or pressure changes by using a Taylor series expansion on density. For liquids significantly below the critical temperature and for solids, we can neglect all terms but the first (linear) term of the Taylor expansion. This approach leads to quantification of the temperature and pressure dependencies of volume with the thermal expansion coefficient,[10] β, and the isothermal compressibility, κ, respectively. These quantities are defined as:

$$\beta \equiv \frac{1}{v}\left(\frac{\partial v}{\partial T}\right)_P \tag{4.71}$$

and

$$\kappa \equiv -\frac{1}{v}\left(\frac{\partial v}{\partial P}\right)_T \tag{4.72}$$

From inspection of Equations (4.71) and (4.72), it can be deduced that β has SI units of $[K^{-1}]$ and κ has Si units of $[Pa^{-1}]$. Representative values of β and κ are reported in Table 4.4. More extensive compilations are found in many engineering and materials handbooks.

Some of the equations of state discussed above are applicable to liquids as well as gases. For example, the Benedict–Webb–Rubin equation of state provides reasonable estimates for most hydrocarbons. The generalized compressibility charts that will be discussed in the next section are based on an extension of this equation of state and

TABLE 4.4 Molar Volume, Thermal Expansion Coefficient, and Isothermal Compressibility of Some Liquid and Solid Species at 20°C and 1 atm

	$v[\text{cm}^3/\text{mol}]$	$\beta[\text{K}^{-1}] \times 10^3$	$\kappa[\text{Pa}^{-1}] \times 10^{10}$
Liquid			
Acetone	73.33	1.49	12.7
Benzene	86.89	1.24	9.4
Methanol	39.56	1.12	12.1
Ethanol	58.24	1.12	11.1
n-Hexane	130.77		15.5
Mercury	14.75	0.181	0.40
Solid			
Aluminum	9.96	0.0672	0.145
Copper	7.11	0.0486	0.091
Iron	7.10	0.035	0.048
Diamond	3.42	0.0036	0.010

Source: R. H. Perry, D. W. Green, and J. O. Maloney (eds.), *Perry's Chemical Engineers' Handbook*, 7th ed. (New York: McGraw-Hill, 1997); D. R. Lide, *CRC Handbook of Chemistry and Physics*, 83rd ed. (Boca Raton, FL: CRC Press, 2002–2003).

[10] This quantity is also referred to as the volume expansivity.

can be used for both gas and liquid phases. Alternatively, correlations have been developed explicitly for the liquid phase. For example, the liquid volume at saturation is given by the Rackett equation:

$$v^{l,sat} = \frac{RT_c}{P_c}(0.29056 - 0.08775\omega)^{[1+(1-T_r)^{2/7}]} \tag{4.73}$$

▶ **EXAMPLE 4.8**
Temperature Correction for Molar Volume of Solid Cu

Determine the molar volume of copper at 500°C from the data in Table 4.4.

SOLUTION   We can rewrite Equation (4.71) as follows:

$$\left(\frac{\partial v}{\partial T}\right)_P = \beta v$$

Separation of variables leads to:

$$\frac{dv}{v} = \beta dT \tag{E4.8A}$$

Integration of Equation (E4.8A) from state 1 at 20°C to state 2 at 500°C gives

$$\ln\frac{v_2}{v_1} = \beta(T_2 - T_1)$$

Solving for the molar volume of solid in state 2, and plugging in values from Table 4.4, we get

$$v_2 = v_1\exp[\beta(T_2 - T_1)] = 7.28\,[cm^3/mol] \tag{E4.8B}$$

Since values for the thermal expansion coefficient are usually small, Equation (E4.8B) is often rewritten using a series expansion for the exponential:

$$v_2 \approx v_1[1 + \beta(T_2 - T_1)] \tag{E4.8C}$$

In this example, the use of the approximation given by Equation (E4.8C) results in an error of only 0.03% as compared to Equation (E4.8B).   ◀

▶ 4.4 GENERALIZED COMPRESSIBILITY CHARTS

The principle of corresponding states invokes a unique generalized relation between the compressibility factor and reduced temperature and pressure for a given class of molecules. It is sometimes convenient to have graphs or tables that quantify this relationship. In this section, we present charts and tabular data for the compressibility factor, z, in terms of P_r, T_r, and ω. To account for different classes of molecules, we use the form

$$z = z^{(0)} + \omega z^{(1)} \tag{4.74}$$

The first term on the right hand side of Equation 4.74, $z^{(0)}$, accounts for simple molecules, while the second term, $z^{(1)}$, is a correction factor for the "nonsphericity" of a species. Both $z^{(0)}$ and $z^{(1)}$ depend only on T_r and P_r.

Values for $z^{(0)}$ and $z^{(1)}$ vs. P_r at different values of T_r are shown in Figures 4.13 and 4.14, respectively. These charts are developed based on the Lee–Kesler equation of state.[11] The same data are reported in tabular form in Appendix C (Tables C.1 and C.2).[12]

---

[11] See Appendix E to see how these were calculated.
[12] Lee and Kesler's value for the critical compressibility factor ($P_r = 1$ and $T_r = 1$) is at the inflection point of the critical isotherm, while Tables C.1 and C.2 report the value obtained directly from the solution of their equation of state.

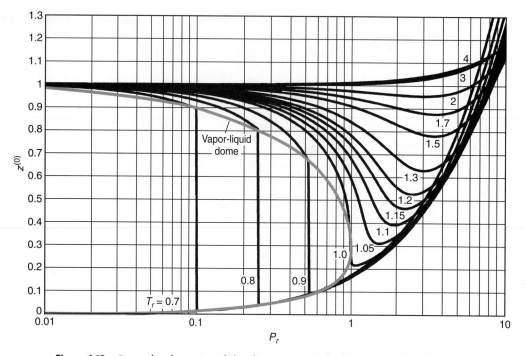

Figure 4.13 Generalized compressibility factor—simple fluid term. Based on the Lee–Kesler equation of state.

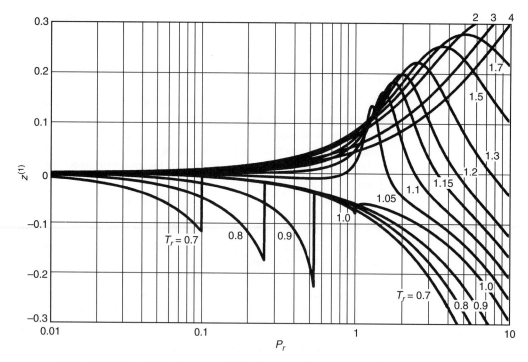

Figure 4.14 Generalized compressibility factor-correction term based on the Lee-Kesler equation of state.

If you want to find the volume at a specific temperature and pressure, it is straightforward to use these the graphs or tables. First, determine the reduced temperature (T/T_r) and reduced pressure (P/P_r) and look up the acentric factor. Then go to Figures 4.13 and 4.14 or Tables C.1 and C.2, and determine the compressibility factor by Equation (4.74). You can then calculate the volume from z. When P or T is unknown, a trial-and-error method must be used. The generalized compressibility factor using the Lee–Kesler equation is an option in the equation of state menu of the ThermoSolver software that comes with the text.

▶ **EXAMPLE 4.9**
Calculation of v by the Redlich–Kwong Equation of State and the Generalized Compressibility Charts

Calculate the volume occupied by 10 kg of butane at 50 bar and 60°C using the Redlich–Kwong equation and the generalized compressibility charts.

SOLUTION *Using the Redlich-Kwong equation of state*

We first find the Redlich–Kwong parameters a and b using critical properties:

$$a = \frac{0.42748R^2T_c^{2.5}}{P_c} = 29.08 \left[\frac{JK^{1/2}m^3}{mol^2} \right] \quad \text{and } b = \frac{0.08664RT_c}{P_c} = 8.09 \times 10^{-5} \left[\frac{m^3}{mol} \right]$$

We can use these with the Redlich–Kwong equation:

$$P = \frac{RT}{v - b} - \frac{a}{T^{1/2}v(v + b)}$$

Solving by trial and error, we get

$$v = 1.2 \times 10^{-4} \, [m^3/mol]$$

and

$$V = \frac{m}{MW} \times v = \frac{10}{.05812} \times 1.20 \times 10^{-4} = 0.021 \, [m^3]$$

Using the compressibility charts

We first find P_r, T_r, and ω:

$$P_r = \frac{P}{P_c} = \frac{50\,\text{bar}}{38.9\,\text{bar}} = 1.32, \qquad T_r = \frac{T}{T_c} = \frac{333.2\,K}{345.2\,K} = 0.78, \qquad \text{and } \omega = 0.199$$

From Tables C.1 and C.2, we get:

| | z^0 | | | z^1 | |
| | P_r | | | P_r | |
T_r	1.3	1.4	T_r	1.3	1.4
0.75	0.2142	0.2303	0.75	−0.0871	−0.0934
0.78	**0.2116**	**0.2274**		**−0.0843**	**−0.0903**
(interpolated)					
0.80	0.2099	0.2255	0.80	−0.0825	−0.0883

By double linear interpolation, where the first interpolation is performed in the table, we get

$$z^{(0)} = 0.2116 + \frac{0.02}{0.1}(0.2274 - 0.2116) = 0.2148$$

$$z^{(1)} = -0.0843 + \frac{0.02}{0.1}(-0.0903 - (-0.0843)) = -0.0855$$

Thus
$$z = z^{(0)} + \omega z^{(1)} = 0.198$$

The low value for the compressibility factor indicates that butane is a liquid. Now solving for volume:

$$v = \frac{zRT}{P} = \frac{0.198 \times 8.314 \times 333.15}{50 \times 10^5} = 1.1 \times 10^{-4} \left[\frac{m^3}{mol} \right]$$

and
$$V = \frac{m}{MW} \times v = \frac{10}{.05812} \times 1.1 \times 10^{-4} = 0.019 \, [m^3]$$

The compressibility charts and the Redlich–Kwong equation give similar values for liquid butane at 50 bar and 60°C. ◀

▶ 4.5 DETERMINATION OF PARAMETERS FOR MIXTURES

In chemical processes, we are usually concerned with mixtures. Since an infinite combination of mixtures is possible, the most practical approach to using equations of state is to form "mixing rules" whereby we develop equations for properties of the mixture based on pure component data. Only the virial equation provides a theoretical basis for mixing rules. Some mixing rules are ad hoc, generated as much by mathematical convenience as by any firm theory. Other mixing rules for equations of state, however, can be related to the physical origin of the terms involved.

Let's see how we develop mixing rules according to those originally proposed by van der Waals for his equation of state. As we have seen, the van der Waals a term is related to the attractive force between two molecules, while we can consider the van der Waals b term to be related to the volume that a species occupies. A schematic for a binary mixture of species 1 and 2 is shown in Figure 4.15. The van der Waals term a_1 represents the attractive interaction between two molecules of species 1, as shown in the figure. These interactions are based on a so-called two body interaction; that is one "1" molecule must find another "1." It will occur in proportion to the mole fraction of the first "1" times the mole fraction of the other "1," that is, y_1^2. Similarly the 2-2 interaction will occur in proportion to y_2^2. The unlike 1-2 interaction will occur in proportion to $y_1 y_2$ since a "1" molecule must find a "2" molecule, while the 2-1 interaction is in proportion to $y_2 y_1$. Summing these attractive interactions together to account for their relative proportions gives

$$a_{mix} = y_1^2 a_1 + y_1 y_2 a_{12} + y_2 y_1 a_{21} + y_2^2 a_2 \qquad (4.75)$$

However, the 1-2 and 2-1 interactions are equivalent, so

$$a_{12} = a_{21} \qquad (4.76)$$

so the mixing rule given by Equation 4.75 simplifies to

$$a_{mix} = y_1^2 a_1 + 2 y_1 y_2 a_{12} + y_2^2 a_2 \qquad (4.77)$$

The cross coefficient is often found from pure species data according to

$$a_{12} = \sqrt{a_1 a_2} \qquad (4.78)$$

or, if data are available for the binary pair in the form of the binary interaction parameter, k_{12}, the cross coefficient can be written

$$a_{12} = \sqrt{a_1 a_2}(1 - k_{12}) \qquad (4.79)$$

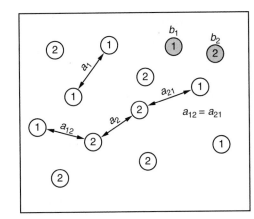

Figure 4.15   Van der Waals interactions in a binary mixture of species 1 and 2.

The van der Waals coefficient b represents excluded volume, so, on average, it is given by

$$b_{\text{mix}} = y_1 b_1 + y_2 b_2 \tag{4.80}$$

An extension of the above mixing rules to a multicomponent mixture gives

$$a_{\text{mix}} = \sum_i \sum_j y_i y_j a_{ij} \tag{4.81}$$

where $a_{ii} = a_i$, and

$$b_{\text{mix}} = \sum_i y_i b_i \tag{4.82}$$

The mixing rules defined by Equations (4.79), (4.81), and (4.82) can be applied to any van der Waals type cubic equation of state—such as the Redlich–Kwong equation, the Soave–Redlich–Kwong equation, or the Peng–Robinson equation. In the latter cases, Equation (4.81) is written as:

$$a_{\text{mix}} = \sum_i \sum_j y_i y_j [a\alpha(T)]_{ij} \tag{4.83}$$

Let's now consider application of mixing rules to the virial equation. Since there is a sound theoretical basis for the virial coefficients in terms of intermolecular interactions, we can relate the virial coefficients for mixtures in terms of intermolecular potentials via Equation (4.61) with *no* arbitrary assumptions; that is, these mixing rules are rigorous results from statistical mechanics. Consider first a binary mixture of 1 and 2. Again, there are three different types of "two-body" interactions characteristic of the second virial coefficient: 1-1 interactions characterized by Γ_{11} and therefore B_{11}, 2-2 interactions characterized by Γ_{22} and B_{22}, and 1-2 interactions characterized by Γ_{12} and B_{12}. The three second virial coefficients characteristic of these interactions depend *only* on the intermolecular potential; they are independent of density and composition. Thus, the second virial coefficient for a binary mixture is proportional to the number of different possible binary interactions weighted by the amount of species present. It is given by

$$B_{\text{mix}} = y_1^2 B_{11} + 2 y_1 y_2 B_{12} + y_2^2 B_{22} \tag{4.84}$$

where $B_{12} = B_{21}$. Note that B_{mix} refers to the parameter for the entire mixture and is different from B_{12}, which refers to a specific binary interaction. In general, for a mixture of n components, the second virial coefficient is given by

$$B_{mix} = \sum_{i=1}^{n}\sum_{j=1}^{n} y_i y_j B_{ij} \qquad (4.85)$$

Similarly, the third virial coefficient, which depends on three-body interactions, can be written as

$$C_{mix} = \sum_{i=1}^{n}\sum_{j=1}^{n}\sum_{k=1}^{n} y_i y_j y_k C_{ijk} \qquad (4.86)$$

So for our binary system, for example:

$$C_{mix} = y_1^3 C_{111} + 3y_1^2 y_2 C_{112} + 3y_1 y_2^2 C_{122} + y_2^3 C_{222} \qquad (4.87)$$

where $C_{112} = C_{121} = C_{211}$. Once again, it is the theoretical foundation of the virial equations from statistical mechanics that provides validity to this extension to mixtures and makes them such a powerful tool. In fact, the virial equation is the only equation of state for which rigorous mixing rules are available.

To apply corresponding states and generalized correlations to mixtures, we need the relationship of the pseudocritical properties, the critical properties of the mixture, to the pure component critical properties. Many mixing relationships have been proposed. The simplest and most commonly used approximation is known as Kay's rules. The pseudocritical temperature, T_{pc}, is given by averaging the critical temperature's of each species in proportion to the amount of that species present in the mixture:

$$T_{pc} = \sum y_i T_{c,i} \qquad (4.88)$$

Similarly, the pseudocritical pressure, P_{pc}, and acentric factor, ω_{pc}, become

$$P_{pc} = \sum y_i T_{c,i} \qquad (4.89)$$

and

$$\omega_{pc} = \sum y_i \omega_{c,i} \qquad (4.90)$$

respectively. There is absolutely *no* basis for these rules other than convenience. Alternatively, a geometric mean combining rule for critical temperature has been used:

$$T_{pc,ij} = \sqrt{T_{c,i} T_{c,j}} \qquad (4.91)$$

This has been extended to include an additional parameter the binary interaction parameter, k'_{ij}, to better fit experimental data:

$$T_{pc,ij} = \sqrt{T_{c,i} T_{c,j}}(1 - k'_{ij}) \qquad (4.92)$$

The understanding of the relationship between mixing rules and the thermodynamic properties of mixtures is still incomplete and warrants study.

▶ **EXAMPLE 4.10**
PvT Calculations for
Pure Species and
Mixtures

Calculate the following:

(a) The volume occupied by 20 kg of propane at 100°C and 70 bar

(b) The pressure needed to fill a 0.1 m^3-vessel at room temperature to store 50 mol of propane

(c) The pressure needed to fill a 0.1 m^3-vessel at room temperature to store a mixture of 20 mol of propane and 30 mol of ethane

As a general strategy, first check whether conditions represent ideal gas behavior. If they do, $Pv = RT$ can be used. If they do not, we must use an approach that incorporates nonidealities. If P and T are given, we can use the *compressibility charts* directly.

If T and v are given, an equation of state of the form $P = f(T, v)$ is easier, since we can calculate P directly. An accurate equation of state is the *Redlich–Kwong equation*:

$$P = \frac{RT}{v - b} - \frac{a}{T^{1/2}v(v + b)}$$

where the relationships for parameters a and b can be found using the principle of corresponding states:

$$a = \frac{0.42748R^2T_c^{2.5}}{P_c} \quad \text{and} \quad b = \frac{0.08664RT_c}{P_c}$$

Why did we choose this equation instead of the van der Waals equation?

(a) At 70 bar, propane is not an ideal gas. Since we are given T and P, we can use the compressibility charts directly. First, we need to find the reduced pressure and reduced temperature using the critical data available in Appendix A:

$$P_r = \frac{P}{P_c} = \frac{70\,\text{bar}}{42.4\,\text{bar}} = 1.65 \quad \text{and} \quad T_r = \frac{T}{T_c} = \frac{373\,\text{K}}{370\,\text{K}} = 1.01$$

We also have to look up the value of the acentric factor:

$$\omega = 0.153$$

Interpolating from Tables C.1 and C.2,

$$z = z^{(0)} + \omega z^{(1)} = 0.2822 + 0.153 \times (-0.0670) = 0.272$$

So

$$V = nv = \frac{m}{MW}\left(\frac{zRT}{P}\right) = \frac{20 \times 10^3}{44}\left(\frac{0.272 \times 8.314 \times 373}{70 \times 10^5}\right) = 0.0548\,\text{m}^3$$

(b) Here we are given T and v, so we can use the Redlich–Kwong equation. Plugging in constants ($P_c = 42.24$ bar and $T_c = 370$ K):

$$a = \frac{0.42748R^2T_c^{2.5}}{P_c} = 18.35\,\frac{\text{JK}^{1/2}\text{m}^3}{\text{mol}^2} \quad \text{and} \quad b = \frac{0.08664\,RT_c}{P_c} = 6.29 \times 10^{-5}\,\frac{\text{m}^3}{\text{mol}}$$

These parameters give (with room temperature = 295 K):

$$P = \frac{RT}{v - b} - \frac{a}{T^{1/2}v(v + b)} = 1.01\,\text{MPa}$$

(c) Now we have a mixture of propane (1) and ethane (2), so we must use mixing rules. We can use a_1 and b_1, for propane as above. For ethane ($P_c = 48.7$ bar and $T_c = 305.5$ K)

$$a_2 = \frac{0.42748R^2T_c^{2.5}}{P_c} = 9.90\,\frac{\text{JK}^{1/2}\text{m}^3}{\text{mol}^2} \quad \text{and} \quad b_2 = \frac{0.08664\,RT_c}{P_c} = 4.52 \times 10^{-5}\,\frac{\text{m}^3}{\text{mol}}$$

We will use the van der Waals mixing rules with $y_1 = 0.4$ and $y_2 = 0.6$:

$$a_\text{mix} = y_1^2 a_1 + 2y_1 y_2 \sqrt{a_1 a_2} + y_2^2 a_2 \qquad b_\text{mix} = y_1 b_1 + y_2 b_2$$

This gives

$$a_\text{mix} = 9.73\,\frac{\text{JK}^{1/2}\text{m}^3}{\text{mol}^2} \quad \text{and} \quad b_\text{mix} = 5.23 \times 10^{-5}\,\frac{\text{m}^3}{\text{mol}}$$

Plugging into the Redlich–Kwong equation:

$$P = \frac{RT}{v - b_\text{mix}} - \frac{a_\text{mix}}{T^{1/2}v(v + b_\text{mix})} = 1.12\,\text{MPa}$$

Note that the pressures reported in parts (b) and (c) are too large for the ideal gas law to be accurate.

▶ **4.6 SUMMARY**

In this chapter, we studied **equations of state**, which relate the measured properties P, v, and T. Examples include **cubic equations of state** (e.g., van der Waals, Redlich–Kwong, Peng–Robinson), the **virial equation** (with several specific forms), and the **generalized compressibility charts**. The Rackett equation allows us to estimate the molar volume of liquids at saturation, while the **thermal expansion coefficient** and the **isothermal compressibility** allow us to determine how to correct for the volumes of liquids and solids with temperature and pressure, respectively.

We developed an understanding of the underlying form of these equations by looking at the molecular behavior of chemical species. "Molecular" energy, or internal energy, u, can be divided into two parts: *molecular kinetic energy and molecular potential energy*. In Chapter 1, we saw that molecular kinetic energy is proportional to the macroscopic property temperature. In this chapter, we identified the basis for potential energy between molecules. Specifically, we related intermolecular interactions to: **point charges, dipoles, induced dipoles, dispersion (London) interactions, repulsive forces**, and **chemical effects**. Dipole–dipole, induced dipole, and dispersion interactions all demonstrate a r^{-6} dependence on the distance between the molecules and are collectively referred to as **van der Waals forces**. The molecular parameters, **dipole moment**, and **polarizability** determine the magnitude of these interactions.

The molecular assumptions of the ideal gas model were relaxed to develop the van der Waals equation of state, by including a r^{-6} attractive term and a hard sphere repulsive term. This equation heuristically illustrates how molecular concepts can be applied to developing an equation of state. In fact, it was shown that the accurate cubic equations that have been developed since van der Waals's time have the same general form. Alternatively, the virial equation results from a power series expansion of the compressibility factor, either in molar density or in pressure.

The parameters in a given equation of state must be determined before it can be applied. The best course is to fit these parameters with measured experimental data. When measured data are not available, we can resort to the **principle of corresponding states**. On a molecular scale, the principle of corresponding states asserts that the dimensionless potential energy is the same for all species. On a macroscopic scale, it translates to the statement that all fluids at the same reduced temperature and reduced pressure have the same compressibility factor. We applied the principle of corresponding states to relate the parameters of equations of state to the critical temperature and pressure by noting that there is an inflection point on the critical isotherm at the critical point. Relations were given for the following cubic equations: van der Waals [Equations (4.39) and (4.40)], Redlich–Kwong [Equations (4.47) and (4.48)], and Peng–Robinson [Equations (4.52) through (4.55)].

We can extend the principle of corresponding states to account for different classes of molecules, based on the particular nature of the intermolecular interactions involved. One way to accomplish this objective is by introducing a third parameter—the **Pitzer acentric factor**, ω. We then write the compressibility factor in terms of $z^{(0)}$, which accounts for simple molecules, and $z^{(1)}$, a correction factor for the "nonsphericity":

$$z = z^{(0)} + \omega z^{(1)} \tag{4.74}$$

Both $z^{(0)}$ and $z^{(1)}$ depend only on T_r and P_r. Values for $z^{(0)}$ and $z^{(1)}$ vs. P_r at different values of T_r are presented as **generalized compressibility charts** and shown in Figures 4.13 and 4.14, respectively. They are also reported in Tables C.1 and C.2 in Appendix C. These charts are based on the Lee–Kesler equation of state (see Appendix E).

We use **mixing rules** to extend equations of state to mixtures. The mixing rules allow us to extrapolate these equations to mixtures, from mostly pure component data. Mixing rules for van der Waals–type parameters a and b were developed based on a two-body attractive interaction and a hard sphere repulsion, respectively. The binary interaction parameter allows us to better describe the cross coefficient, a_{12}; however, data from the mixture are needed. Mixing rules for the viral coefficients arise from a theoretical basis. Mixing rules for the second virial coefficient, B, are based on two-body interactions; for the third virial coefficient, C, on three-body interactions; and so on. Finally, **Kay's rules** were presented, from which we can find the psuedocritical properties of the mixture from the pure component properties. These values allow us to apply the generalized compressibility charts to mixtures.

▶4.7 PROBLEMS

4.1 At very *high* temperatures, a gas can be ionized and remain in thermodynamic equilibrium. Consider the case of gas containing *only* ions, A^+. Your supervisor requests that you come up with a simple (one-parameter) equation of state for this gas. Your assistant leaves you a memo that she has fit the PvT data to an equation of state of the form:

$$\left(P_{A^+} + \frac{a}{v_{A^+}^n} \right) v_{A^+} = RT$$

She tells you the data fit this equation well but, unfortunately, leaves you no numbers. Your meeting with your supervisor is in 10 minutes, and your assistant is nowhere to be found! In order to be ready for the meeting, you need to answer the following questions:

(a) Is the form of this equation reasonable? Explain.

(b) What sign would you expect for the constant, a? Will this be a small or large number? Explain.

(c) What number will you use for n (it can be a fraction)? What are the units of a? Show your work.

4.2 Consider a mixture of O_2 (*a*) and C_3H_8 (*b*):

(a) Write expressions for the attractive interactions Γ_{aa}, Γ_{bb}, and Γ_{ab} as a function of distance between the molecules, r.

(b) How does Γ_{ab} compare to $\sqrt{\Gamma_{aa}\Gamma_{bb}}$?

(c) Write a general expression for the average attractive intermolecular interaction in a mixture as a function of mole fractions of O_2 and C_3H_8 represented by y_a and y_b, respectively.

4.3 Consider $BClH_2$. In each of the following cases, when do you expect the compressibility factor to be closer to one. Explain.

(a) At 300 K, 10 bar or at 300 K, 20 bar

(b) At 300 K, 20 bar or 1000 K, 20 bar

(c) Consider a mixture of $BClH_2$ and H_2 at 300 K, 10 bar. Qualitatively plot the compressibility factor vs. mole fraction $BClH_2$. Point out any important features.

4.4 While returning to your dorm late last night with a hot cup of coffee, the heat overcomes you and, much to your chagrin, you drop the paper cup, spilling its entire contents. As you had just spent your laundry money, this is somewhat upsetting, especially since you still have a good deal of thermo left to study and your last clean pair of pants are now covered with coffee.

You yearn for the old days of polystyrene (Styrofoam) cups, which never got hot. Being an ambitious student (and looking for a distraction), you decide to come up with a process to recycle polystyrene (Styrofoam) so that environmental concerns will no longer keep the coffee shop from using this very good insulating material.

After several hours, you have come up with what you think is a very reasonable process (you cannot wait to call the patent attorney!) and have just a few final issues to resolve. In the purification process, you believe you have reduced the polystyrene to its monomer, styrene, shown below:

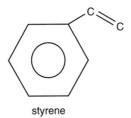

styrene

In this case, the reactor would consist of 100 moles of styrene in a volume of 30 L at a pressure of 10 bar. You are concerned that the temperature is beyond the limit for the decomposition of styrene, 289°C.

Since you are studying for the thermo exam, and have just gotten to the van der Waals equation, you want to decide whether this would be a good equation of state to use.

(a) What deviations from ideality would you expect at these reaction conditions? In order of importance, list the types of intermolecular forces you think contribute to nonideality. Is the van der Waals equation appropriate? Explain.

(b) Your search for experimental values for the van der Waals constants, a and b, is futile; you do, however, find values for the critical constants for styrene:

$$P_c = 39 \text{ bar}$$

$$T_c = 374°C$$

Calculate the temperature of the reactor, *using the van der Waals equation*. Will the styrene decompose?

(c) Your classmate, who's taking a polymers class, says the polystyrene probably has not reduced to a monomer but still exists as a reduced polymer chain, perhaps five monomers long:

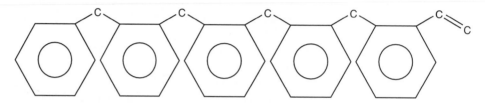

Using only the information above, what are reasonable values for the van der Waals constants, a and b, of this reduced polymer chain? Explain.

(d) Calculate the temperature in the reactor at the same reactor volume and pressure and initial Styrofoam mass as for part (b), except where you have a five-unit polymer instead of the monomer. Explain the difference in value to that calculated in part (b). Will decomposition occur (assume around 289°C)?

4.5 The London force is directly related to the polarizabilities of the corresponding molecules. Consider the following table of molecular polarizability, α:

Species	$\alpha \ (10^{25} \text{ cm}^3)$
CH_4	26
C_2H_6	44.7
C_3H_8	62.9
CH_3Cl	45.6
CH_2Cl_2	64.8
$CHCl_3$	82.3
CCl_4	105

From these data, come up with a model to account for the contribution of each atom to the polarizability of a molecule. Predict the polarizability of C_4H_{10} and C_2H_5Cl.

4.6 The Lennard-Jones potential function is often used to describe the molecular potential energy between two species. Rank each of the following sets of species, from largest to smallest, in terms of Lennard-Jones parameters σ and ϵ. If there is no noticeable difference, write that they are roughly the same. Explain your choice using molecular arguments.

(a) O_2, S_2, I_2

(b)

$$\underset{\text{n-butanol}}{H_3C-\overset{\overset{\displaystyle H}{|}}{\underset{\underset{\displaystyle H}{|}}{C}}-\overset{\overset{\displaystyle H}{|}}{\underset{\underset{\displaystyle H}{|}}{C}}-\overset{\overset{\displaystyle H}{|}}{\underset{\underset{\displaystyle H}{|}}{C}}-OH,} \quad \underset{\text{diethylether}}{H_3C-\overset{\overset{\displaystyle H}{|}}{\underset{\underset{\displaystyle H}{|}}{C}}-O-\overset{\overset{\displaystyle H}{|}}{\underset{\underset{\displaystyle H}{|}}{C}}-CH_3,} \quad \underset{\text{methyl ethyl ketone}}{H_3C-\overset{\overset{\displaystyle H}{|}}{\underset{\underset{\displaystyle H}{|}}{C}}-\overset{\overset{\displaystyle O}{||}}{C}-CH_3}$$

n-butanol   diethylether   methyl ethyl ketone

4.7 Using your knowledge of intermolecular forces, explain the following observation:

(a) At 300°C and 30 bar, the internal energy of water is less than at 300°C and 20 bar.

(b) At 300 K and 30 bar, the compressibility factor of isopropanol ($H_3CCOHCH_3$) is less than that of n-pentane (C_5H_{12}), but at 500 K and 30 bar, the compressibility factor of isopropanol ($H_3CCOHCH_3$) is greater than that of n-pentane (C_5H_{12}).

$$
\begin{array}{c}
H \\
| \\
O \\
| \\
H_3C-C-CH_3 \\
| \\
H
\end{array}
$$

isopropanol

4.8 Consider comparing 1 mole of NH_3 at 10 bar and 500 K behaving as a real gas (i.e., considering its intermolecular interactions) vs. 1 mole of NH_3 at 10 bar and 500 K behaving as an ideal gas (i.e., hypothetically "turning off" the intermolecular interactions.) Answer the following questions using *molecular* arguments. Explain your choice with diagrams and descriptions of the interactions involved.

(a) In which case is the compressibility factor, z, higher?

(b) In which case is the internal energy, u, higher?

(c) In which case is the entropy, s, higher? You need consider only the "spatial" contribution to entropy.

4.9 Consider comparing 1 mole of NH_3 at 10 bar and 500 K vs. 1 mole of Ne at 10 bar and 500 K. Answer the following questions using *molecular* arguments. Explain your choice with diagrams and descriptions of the interactions involved.

(a) In which case is the compressibility factor, z, higher?

(b) In which case is the entropy, s, higher? You need consider only the "spatial" contribution to entropy.

4.10 Consider 2 Ar atoms at 25 bar and 300 K:

(a) What is the average distance between them (in Å)?

(b) Calculate the potential energy due to *gravity* (between the two atoms).

(c) Calculate the potential energy due to London interactions.

(d) Compare the values obtained in parts (b) and (c).

4.11 As discussed in the text, the repulsive term in the Lennard-Jones potential should have an exponential dependence rather than r^{-12}. Graphically compare the features of the Lennard-Jones potential to one that has the same attractive term but whose repulsive term is given by:

$$
C_1 e^{C_2/r}
$$

C_1 and C_2 are constants that you need to choose so the term above fits as closely as possible to the Lennard-Jones potential (Figure 4.8). Comment on the differences between these potential functions.

4.12 Calculate the bond strength in [eV] of a sodium ion in a crystal of NaCl. For the salt lattice:

(a) Consider *only* the six nearest-neighbor Cl^- ions. The Cl^- ions are at a distance $r = 2.76$ Å from the Na^+ ion.

(b) In addition to the six nearest-neighbor Cl^- ions, include the twelve next-nearest-neighbor Na^+ ions at a distance $\sqrt{2}r$.

(c) Now include the eight next-next-nearest-neighbor Cl^- ions at a distance $\sqrt{3}r$.

(d) Finally, include the six next Na^+ ions at a distance of $2r$.

4.13 The normal boiling points of some halide silanes are reported below. Explain the order in terms of intermolecular forces.

Species	SiClF$_3$	SiBrF$_3$	SiCl$_3$F	SiBr$_3$F	SiICl$_3$
Boiling point [°C]	− 70.0	−41.7	12.2	83.8	114

4.14 Using data from Table 4.2, estimate the equilibrium bond length that would exist in a molecule of Xe$_2$.

4.15 Calculate the van der Waals parameters from critical point data for the following gases: He, CH$_4$, NH$_3$, and H$_2$O. Explain the relative magnitudes of a and b from a physical basis.

4.16 Calculate the van der Waals parameter b for CH$_4$, C$_6$H$_6$, and CH$_3$OH. Based on these values, estimate the molecular diameter of each species. Compare the values obtained with those in Table 4.2.

4.17 Calculate the van der Waals parameter a for CH$_4$, C$_6$H$_6$, and CH$_3$OH. Based on these values, estimate the value of C_6 for each species. Compare the values obtained with that calculated by Equation (4.13).

4.18 Table 4.3 compares the van der Waals (1873), Redlich–Kwong (1949), and Peng–Robinson (1976) equations of state in similar forms. Based on intermolecular interactions, qualitatively analyze how the progression of equations may have given more accurate results.

4.19 Consider a cylinder fitted with a piston that contains 2 mol of H$_2$O in a container at 1000 K. Calculate how much work is required to isothermally and reversibly compress this gas from 10 L to 1 L, in each of the following cases:

(a) Use the ideal gas model for water.

(b) Use the Redlich–Kwong equation to relate P, v, and T:

$$P = \frac{RT}{v - b} - \frac{a}{T^{1/2}v(v + b)}$$

where

$$a = 14.24 \left[(\text{JK}^{1/2}\text{m}^3)/\text{mol}^2 \right] \text{ and } b = 2.11 \times 10^{-5} \left[\text{m}^3/\text{mol} \right]$$

(c) Use the Steam tables.

Compare these three methods.

4.20 Write the van der Waals equation in the virial equation form (both in pressure and in molar volume).

4.21 Verify Equations (4.58) by rewriting the expansion of the virial equation in pressure [Equation (4.57)] in terms of the virial expansion in the reciprocal of molar volume [Equation (4.56)].

4.22 Consider the Berthelot equation of state given below. Show how to calculate the constants a and b using only critical point data.

$$P = \frac{RT}{v - b} - \frac{a}{Tv^2}$$

4.23 Find the reduced form of the Berthelot equation of state.

4.24 (a) Use the data in the steam tables to come up with an expression for the second virial coefficient for water vapor.

(b) Calculate the value of B_{H_2O} using the principle of corresponding states. Compare the value to that obtained in part (a). A helpful value is:

$$v_c = 56 \left[\text{cm}^3/\text{mol} \right]$$

4.25 Calculate the saturation pressure of n-pentane at 90°C by applying the "equal area" rule to (a) the Redlich–Kwong equation; (b) the Peng–Robinson equation. Compare these results to the measured value of 5.7 bar.

4.26 At −30°C, the saturation pressure of ethane is 10.6 bar. Calculate the densities of the liquid and vapor phases using the Peng–Robinson equation. Compare to the reported values for the liquid and vapor densities of 0.468 and 0.0193 g/cm^3.

4.27 Welcome to Beaver Gas Co.! Your first task is to calculate the annual gross sales of our superpure-grade nitrogen and oxygen gases.

(a) The total gross sales of N_2 is 30,000 units. Take the volume of the cylinder to be 43 L, the pressure to be 12,400 kPa, and the cost to be $6.1/kg. Compare your result to that you would obtain using the ideal gas model

(b) Repeat for 30,000 units of O_2 at 15,000 kPa and $9/kg.

4.28 Verify Equations (4.47) through (4.50) using the approach of Example 4.7.

4.29 Using the steam tables, estimate the values for the thermal expansion coefficient, β, and the isothermal compressibility, κ, of liquid water at 20°C and 100°C.

4.30 Use the Rackett equation to calculate the liquid-phase molar volume of each of the following species at the same temperature as the measured values reported. Which species had the greatest absolute percent error? The least? Can the trend be explained by molecular concepts?

(a) methane (CH_4), $\qquad v_{exp} = 37.7$ cm³/mol at 111 K

(b) ethane (C_2H_6), $\qquad v_{exp} = 54.8$ cm³/mol at 183 K

(c) n-octane (C_8H_{18}), $\qquad v_{exp} = 162.5$ cm³/mol at 293 K

(d) water (H_2O), $\qquad v_{exp} = 18.0$ cm³/mol at 293 K

(e) acetic acid ($C_2H_4O_2$), $\qquad v_{exp} = 57.2$ cm³/mol at 293 K

4.31 Calculate the following:

(a) the volume occupied by 20 kg of ethane at 70°C and 30 bar

(b) the pressure needed to fill a 0.1 m³-vessel at room temperature to store 40 kg of ethane

4.32 Calculate the volume occupied by 50 kg of propane at 35 bar and 50°C, using the following:

(a) the ideal gas model

(b) The Redlich–Kwong equation of state

(c) The Peng–Robinson equation of state

(d) The compressibility charts

(e) The textbook software, ThermoSolver

4.33 For a lecture-demonstration experiment, it is desired to construct a sealed glass vial containing a pure substance that can be made to pass through the critical point by heating the vial in a person's hand. Thus, at room temperature the vial should contain a liquid and its vapor.

(a) From the list of critical properties, select a suitable substance to be sealed within the vial.

(b) What magnitude of pressures must the vial withstand?

(c) For a vial of 100 cm³, how much of the substance should be enclosed in the vial?

(d) Describe the changes within the vial as it is heated if it contains an amount of substance that is less than that calculated in part (c).

4.34 Compare the compressibility factor of methane at $T_r = 1.1$ and $P_r = 1.2$ using the Peng–Robinson equation of state and the compressibility charts. Repeat the calculations for methanol.

4.35 Using the generalized compressibility charts, calculate the molar volume of ammonia at 92°C and 306.5 bar. What phase is ammonia in?

4.36 Use the Redlich–Kwong equation to calculate the size of vessel you would need to contain 30 kg of acetylene mixed with 50 kg of n-butane at 30 bar and 450 K. The binary interaction coefficient is given by $k_{12} = 0.092$.

4.37 You wish to use the Redlich–Kwong equation of state to describe a mixture of carbon dioxide (1) and toluene (2). To be as accurate as possible with the mixing rules, you want to include the binary interaction parameter, k_{12}. In the literature, you find reference to an experiment with the following conditions:

n_1	2.0 mol
n_2	3.0 mol
V	10.0 L
T	400.0 K
P	1.353 MPa

Using the data above and critical point property data, estimate k_{12}.

4.38 You are planning an experiment in which you have a mixture of 5 moles of hydrogen (H_2), 4 moles of water (H_2O), and 1 of mole ethane (C_2H_6). You want to calculate the pressure of this mixture to determine which material to use to construct the vessel to contain these gases. The vessel needs to be able to hold 12.5 L (0.0125 m³), and the maximum temperature in the laboratory is 27°C. You then go to the library and find the pure species parameters for the *van der Waals equation*, a and b. However, when you get back to the laboratory, and realize you forgot to label them.

(a) *Using only molecular arguments*, match each species to its appropriate set of parameters. *Explain your reasoning.*

a [Jm³/mol]	b(m³/mol)	Species
0.564	6.38×10^{-5}	
0.025	2.66×10^{-5}	
0.561	3.05×10^{-5}	

(b) Calculate the van der Waals parameters, a and b, for the mixture.

(c) Calculate the pressure of this mixture.

4.39 The following second virial coefficients have been reported for a mixture of n-butane (1) and carbon dioxide (2) at 313.2 K.

$$B_{11} = -625 \left[cm^3/mol \right]$$

$$B_{22} = -110 \left[cm^3/mol \right]$$

$$B_{12} = -153 \left[cm^3/mol \right]$$

From these data, do the following:

(a) Predict the molar volume of a mixture of 25 mol % butane in carbon dioxide at 313.2 K and 10 bar.

(b) Estimate the value of the binary interaction parameter, k_{12}, at 313.2 K.

4.40 If the diatomic gas of Problem 3.25 were nonideal at the pressures in the problem and attractive forces dominate, qualitatively describe how the final temperature in tank A would change from the answer you obtained in that problem.

4.41 Re-solve Example 4.9 using the text software, ThermoSolver. Compare your answer to the answer that is given in the example.

4.42 Solve the following using ThermoSolver:

(a) In *Species Database*, select **Ethane**. Report its critical temperature and pressure and $\Delta h^o_{f,298}$.

(b) In *Saturation Pressure Calculator*, find the saturation temperature of ethane at 40 bar.

(c) In *Equation of State Solver*, find the volume and compressibility factor of ethane in the following states using the Lee–Kesler equation (generalized correlations) and the Peng–Robinson equation. Report the value of each and the percent difference between the two methods: (i) $P = 40$ bar, $T = 290$ K; (ii) $P = 40$ bar; $T = 302$ K.

CHAPTER 5

The Thermodynamic Web

Learning Objectives

To demonstrate mastery of the material in Chapter 5, you should be able to:

▶ Use the *thermodynamic web* to relate measured, fundamental, and derived thermodynamic properties. In doing so, apply the fundamental property relations, Maxwell relations, the chain rule, derivative inversion, the cyclic relation, and Equations (5.22), (5.23), and (5.24). Use Figure 5.2 to rewrite partial derivatives with *T, P, s,* and *v* in more convenient forms.

▶ Develop hypothetical paths to calculate the change in a desired property between two states, using appropriate property data. Appropriate data may include heat capacity data, pressure or volume explicit equations of state, or thermal expansion coefficients and isothermal compressibilities.

▶ Write the exact differential for any intensive thermodynamic property in terms of partial derivatives of specified independent, intensive properties. For example, given $h = h(T,P)$, write dh. Define what is meant by independent variables and dependent variables.

▶ Write Δs, Δu, and Δh in terms of independent variables T and P or the independent variables T and v. Use these expressions to solve first- and second-law problems.

▶ Define a departure function. Use generalized enthalpy and entropy departure functions to solve first- and second-law problems for systems that exhibit nonideal behavior.

▶ Define Joule–Thomson expansion and the Joule–Thomson coefficient. Explain how Joule–Thomson expansion is used in liquefaction.

▶ 5.1 TYPES OF THERMODYNAMIC PROPERTIES

We have seen that the thermodynamic state of a system can be characterized by its properties. Our goal in this chapter is to develop mathematical expressions through which we can relate the properties of a system to one another and to forms in which data are typically reported. We begin by defining three distinct categories of thermodynamic properties: measured properties, fundamental properties, and derived properties.

Measured Properties
As we explored in Chapter 1, the measured properties are:

$$P, v, T, \quad composition$$

Measured properties are those properties that are directly accessible from measurements in the laboratory. Can you think of a couple of ways in which each of the properties above can be measured?

Fundamental Properties

Observations of nature led us to the two laws of thermodynamics presented in Chapters 2 and 3. These laws are quantified based on two previously unrecognized properties of matter:

$$u \qquad \text{(from conservation of energy)}$$

$$s \qquad \text{(from directionality of nature)}$$

Since internal energy and entropy come from the two *fundamental* postulates of thermodynamics—that energy is conserved (First law) and that entropy of the universe always increases (Second law)—we call them fundamental properties. These properties cannot be measured directly. In fact, it could be said that these are not real things (at least in the measurable sense) but rather constructs of our mind to generalize experimental observations.

Derived Thermodynamic properties

Finally, the most distant from direct experience are derived thermodynamic quantities. These cannot be measured in the lab, nor are they properties directly fundamental to the postulates that govern thermodynamics; they are merely some specific combination of the above two types of properties that are defined out of convenience. Consider, for example, enthalpy:

$$h = u + Pv \qquad (5.1)$$

In open systems, the mass that crosses the boundary between the surroundings and the system *always* contributes to two terms in the energy balance: internal energy and flow (Pv) work. Since these terms are always coupled, it is convenient to *define* a property that includes both terms. In this way we never need to explicitly account for flow work.

Two other convenient properties are the Helmholz energy,

$$a = u - Ts \qquad (5.2)$$

and the Gibbs energy,

$$g = h - Ts \qquad (5.3)$$

For the time being, we will not elucidate why a and g may be conveniently derived thermodynamic properties.[1] However, you should realize that because they are combinations of state functions, they, too, must be properties that are independent of path.

▶5.2 THERMODYNAMIC PROPERTY RELATIONSHIPS

Dependent and Independent Properties

In this section, we will develop a *web* of property relationships whereby we can relate the thermodynamic variables we need to solve problems to variables we can measure in the

---

[1] In Chapter 6, we will learn why Gibbs energy is so useful.

lab. We want relationships between fundamental and derived thermodynamic properties, such as u, s, h, a, and g, and things we can measure, such as measured properties P, v, T or quantities in which measured data are typically reported, for example, c_v, c_P, β, and κ. We will exploit the rigor of mathematics to allow us to form this intricate *web* of these relationships. As with searching for sites on the Internet, there is usually more than one way to obtain our final answer; some are quicker, while others are slower.

We restrict our present discussion to *constant composition* systems; we will learn about mixtures that can change in composition in Chapter 6. Recall that the state postulate says that for systems of constant composition, values of two independent, intensive properties completely constrain the state of the system. In mathematical terms, the change in any intensive thermodynamic property of interest, z, can be written in terms of partial derivatives of the two independent intensive properties, x and y, as follows:

$$dz = \left(\frac{\partial z}{\partial x}\right)_y dx + \left(\frac{\partial z}{\partial y}\right)_x dy \tag{5.4}$$

The total differential dz is exact; that is, the integral of dz is independent of path. On the other hand, the partial differential, ∂, indicates we are specifying a constraint to the path. For example, in taking the partial derivative, $(\partial z/\partial x)_y$, we must evaluate the change in the dependent variable z with respect to the independent variable x over a path where the independent variable y is constant. However, the value of the partial derivative may be different at different values of the constant y. Another way of viewing this relationship is in terms of a three-dimensional plot of x, y, and z. The partial derivative is the slope of z vs. x along a plane of constant y. Through the use of partial derivatives, we can isolate the effect of one independent property by holding the second independent property constant.

The form of Equation (5.4) is general and we can use it to express any of the properties we examined in Section 5.1 in terms of two independent properties. For example, say we want to calculate the change in internal energy for a first-law analysis of a closed system. We may choose to relate the differential change in internal energy, du, to the measured variables temperature, T, and molar volume, v. In the form of Equation (5.4), we write:

$$du = \left(\frac{\partial u}{\partial T}\right)_v dT + \left(\frac{\partial u}{\partial v}\right)_T dv \tag{5.5a}$$

In Equation (5.5a), the intensive properties T and v constrain the state of the system; we call these two properties the *independent* variables. For brevity, we will use the notation:

$$u = u(T, v) \tag{5.5b}$$

as an equivalent form of Equation (5.5a), that is, to indicate our choice of T and v as independent variables to constrain u. All the other properties in the system are *dependent* variables since they are all constrained by the two independent properties. In Equation (5.5a) the change in internal energy is written as an exact differential, du. The exact differential is used since once changes to *both* T and v are specified, the internal energy u can change in only one way, as constrained by the state postulate. We are free to specify any two independent properties to constrain the exact differential du. For example, we can use any of the following forms: $u = u(T, P)$, $u = u(T, s)$, $u = u(h, s)$, and so on. However, it turns out that the measured properties T and v are particularly convenient choices for the independent variables when looking at changes in u.

Now let's look at how the form of Equation (5.5a) helps us in solving our first-law problem. Knowing that u is a state function, we are free to choose any path we desire to calculate the change in this property for a given process. We can say the gas in the system undergoes a process from state 1 (T_1 and v_1) to state 2 (T_2 and v_2) where we have framed our problem in terms of independent variables T and v, that is, $u = u(T, v)$. Figure 5.1 presents a graphical representation of our problem on a Tv diagram. We often use sketches like this one that delineate the process in which we are interested in terms of our independent variables. Such diagrams are very useful in formulating solutions to problems. One possible path from state 1 to state 2 is to vary both T and v simultaneously, as occurs in the real process. This path is depicted by the solid line in Figure 5.1. However, another possibility is to develop a **hypothetical** path in which we change only one variable at a time (see Section 2.2). In the hypothetical path depicted in Figure 5.1, we first perform an isothermal expansion to a volume large enough for the system to behave as an ideal gas. In step 2, we perform a constant volume heating to the final temperature T_2. Finally, the third step consists of an isothermal compression to v_2. Since internal energy is a state function, it is independent of path; therefore, both paths will have the same value for Δu. In practice, the hypothetical process is easier for our calculations. In that case, we can use available data for ideal gas heat capacity to calculate Δu for step 2. It is not so easy to find heat capacity data for nonideal gases! Thus, the hypothetical path shown in Figure 5.1 is constructed to make use of data available in the literature. This theme will commonly recur as we solve problems in thermodynamics. We also need to calculate the changes in steps 1 and 3. This task will be easy once we have a *web* of thermodynamic relations, as we will see next. The example above illustrates an important strategy in solving thermodynamic problems. There are an infinite number of paths that connect one state to another. Solving a problem often reduces to picking the proper path. In considering which path to choose, two things should be considered. (1) What property data are available? (2) Within the constraints posed by consideration (1), what path yields the easiest calculation?

We wrote Equation (5.5a) in terms of $(\partial u/\partial T)_v$. Is there a difference between $(\partial u/\partial T)_v$ and $(\partial u/\partial T)_P$?

Fundamental Property Relations

Let's consider again the calculation of internal energy. This time we begin with the fundamental postulates of thermodynamics. For a closed system undergoing a reversible

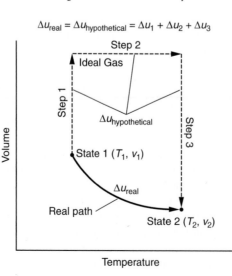

$$\Delta u_{real} = \Delta u_{hypothetical} = \Delta u_1 + \Delta u_2 + \Delta u_3$$

Figure 5.1 Computational paths for the change in internal energy from state 1 to state 2.

process with only Pv work, the relations developed in Sections 2.3 and 3.3 can be applied to the differential energy balance, Equation (2.33). Hence, the first law and second laws are combined to give

Formulas

for

fundamental

property relations

$$du = \delta q_{rev} + \delta w_{rev}$$

$$= Tds - Pdv \qquad (5.6)$$

We can apply the definition of the derived thermodynamic property h given by Equation (5.1) to get

$$dh = du + d(Pv)$$

$$= Tds + vdP \qquad (5.7)$$

Similarly, we can apply the definitions of the derived thermodynamic properties a and g given by Equations (5.2) and (5.3) to get

$$da = du - d(Ts)$$

$$= -sdT - Pdv \qquad (5.8)$$

and

$$dg = dh - d(Ts)$$

$$= -sdT + vdP \qquad (5.9)$$

respectively. Equations (5.6) through (5.9) are known as the **fundamental property relations**. Although Equation (5.6) was derived for a reversible process, the ultimate expressions that define the fundamental property relations are *only* between properties. Therefore, these equations can be applied to any process: reversible or irreversible. We can make this statement since properties are independent of path and, therefore, independent of process. So even though these equations were derived for the specific case of a reversible process, we can apply these equations even if the physical process that leads from one state to another is irreversible! However, in the case of irreversibility, Tds is no longer equal to the differential heat transferred across the boundary and $-Pdv$ no longer gives the value of δw.

If we apply the approach that we just developed, we can write the change in internal energy in terms of independent variables s and v, that is, $u = u(s, v)$:

$$du = \left(\frac{\partial u}{\partial s}\right)_v ds + \left(\frac{\partial u}{\partial v}\right)_s dv \qquad (5.10)$$

For both Equations (5.10) and (5.6) to be true, we must have:

$$\left(\frac{\partial u}{\partial s}\right)_v = T \quad \text{and} \quad \left(\frac{\partial u}{\partial v}\right)_s = -P \qquad (5.11)$$

The grouping represented by $u = u(s, v)$ results in the partial derivatives of Equation (5.10) corresponding to thermodynamic properties as defined in Equation (5.11). While any two properties can be used to constrain u, no other grouping of independent properties x and y, $u = u(x, y)$, allows us to write partial derivatives in terms of thermodynamic properties as we did in Equation (5.11). Therefore, we say that $\{u, s, v\}$ form a **fundamental grouping**. Additionally, from Equation (5.7), we can see that $\{h, s, P\}$ form a fundamental grouping. It follows that

$$\left(\frac{\partial h}{\partial s}\right)_P = T \qquad \text{and} \qquad \left(\frac{\partial h}{\partial P}\right)_s = v \qquad (5.12)$$

Likewise, the fundamental grouping $\{a, T, v\}$ results in

$$\left(\frac{\partial a}{\partial T}\right)_v = -s \qquad \text{and} \qquad \left(\frac{\partial a}{\partial v}\right)_T = -P \qquad (5.13)$$

while $\{g, T, P\}$ gives

$$\left(\frac{\partial g}{\partial T}\right)_P = -s \qquad \text{and} \qquad \left(\frac{\partial g}{\partial P}\right)_T = v \qquad (5.14)$$

The latter two fundamental groupings give us our first insight into the utility of g and a. These derived thermodynamic quantities are grouped exclusively with **measured properties**, T, P, and v. Thus, there is a direct link between the change of dependent variables g and a and what we can measure in the lab. Specifically, we may anticipate that the Gibbs energy will become useful when we approach phase equilibria. As we saw in Chapter 1, T and P form the criteria for thermal and mechanical equilibrium; thus, the fundamental grouping $\{g, T, P\}$ is of special interest. We will learn more about such things in Chapter 6.

Maxwell Relations

lead to interesting results w/ web

Additional relations between thermodynamic properties and their derivatives can be derived from the second derivatives of the fundamental property relationships. These relations are called *Maxwell relations* and can be obtained by noting that the order of partial differentiation of an exact differential does not matter. For example, we can equate the following two sets of partial derivatives of the exact differential du from the fundamental grouping $\{u, s, v\}$:

$$\left[\frac{\partial}{\partial v}\left(\frac{\partial u}{\partial s}\right)_v\right]_s = \left[\frac{\partial}{\partial s}\left(\frac{\partial u}{\partial v}\right)_s\right]_v \qquad (5.15)$$

Substitution of Equation (5.11) into Equation (5.15) gives

$$\left(\frac{\partial T}{\partial v}\right)_s = -\left(\frac{\partial P}{\partial s}\right)_v \qquad (5.16)$$

Similarly, from the other three fundamental property relationships we get

$$\left(\frac{\partial T}{\partial P}\right)_s = \left(\frac{\partial v}{\partial s}\right)_P \qquad (5.17)$$

$$\left(\frac{\partial s}{\partial v}\right)_T = \left(\frac{\partial P}{\partial T}\right)_v \qquad (5.18)$$

$$-\left(\frac{\partial s}{\partial P}\right)_T = \left(\frac{\partial v}{\partial T}\right)_P \qquad (5.19)$$

Can you verify these last three relations? In the Maxwell relations given by Equations (5.18) and (5.19), all the properties on the right-hand side are measured properties! These permit the calculation of entropy from the measured PvT data. The derivative relations of Equations (5.6), (5.7), and (5.9) then enable us to calculate u, h, and g.

Other Useful Mathematical Relations

In this section, we present three other mathematical relations that will be of use in helping us surf the thermodynamic web. The first relationship is the *chain rule*, which can be written in general as follows:

[handwritten margin note: Different calculus properties]

$$\left(\frac{\partial z}{\partial x}\right)_a = \left(\frac{\partial z}{\partial y}\right)_a \left(\frac{\partial y}{\partial x}\right)_a \tag{5.20}$$

Derivative inversion allows us to flip partial derivatives as follows:

$$\left(\frac{\partial x}{\partial z}\right)_y = \frac{1}{\left(\dfrac{\partial z}{\partial x}\right)_y} \tag{5.21}$$

We are not through yet. We can derive an additional relation based on the mathematical behavior of state functions. We begin with Equation (5.4) above:

$$dz = \left(\frac{\partial z}{\partial x}\right)_y dx + \left(\frac{\partial z}{\partial y}\right)_x dy \tag{5.4}$$

If we take the partial derivative of each term with respect to x at constant y, we get[2]

$$\left(\frac{\partial z}{\partial x}\right)_z = \left(\frac{\partial z}{\partial x}\right)_y \left(\frac{\partial x}{\partial x}\right)_z + \left(\frac{\partial z}{\partial y}\right)_x \left(\frac{\partial y}{\partial x}\right)_z \tag{5.22}$$

Since we cannot simultaneously change z and keep it constant,

$$\left(\frac{\partial z}{\partial x}\right)_z = 0 \tag{5.23}$$

Equation (5.23) is useful in its own right. It is easy to see that

$$\left(\frac{\partial x}{\partial x}\right)_z = 1 \tag{5.24}$$

Applying Equations (5.21), (5.23), and (5.24) to (5.22) and rearranging gives the *cyclic relation*:[3]

$$-1 = \left(\frac{\partial x}{\partial z}\right)_y \left(\frac{\partial y}{\partial x}\right)_z \left(\frac{\partial z}{\partial y}\right)_x \tag{5.25}$$

[2] The mathematical development of Equation (5.22) from Equation (5.4) is actually more complex than this heuristic explanation. However, viewing it in this simplified manner is convenient and, in general, works. We present it here since we will use this method in other places. See Example 5.1 for an alternative development of Equation (5.25).

[3] This relation is alternatively termed the triple product rule.

This expression is easy to remember; each variable appears in the numerator once, appears in the denominator once, and is held constant once. Hence, Equation (5.25) is termed the cyclic relation.

▶ **EXAMPLE 5.1**
Alternative Derivation for the Cyclic Rule

Develop an expression for the cyclic relation by equating $z = z(x, y)$ and $y = y(x, z)$

SOLUTION We can write the change in dependent variable z in terms of independent variables x and y according to Equation (5.4):

$$dz = \left(\frac{\partial z}{\partial x}\right)_y dx + \left(\frac{\partial z}{\partial y}\right)_x dy \tag{E5.1A}$$

Alternatively, we can choose y as the dependent variable and write it in terms of independent variables x and z:

$$dy = \left(\frac{\partial y}{\partial x}\right)_z dx + \left(\frac{\partial y}{\partial z}\right)_x dz \tag{E5.1B}$$

Solving Equation (E5.1A) for dy using derivative inversion gives

$$dy = -\left(\frac{\partial y}{\partial z}\right)_x \left(\frac{\partial z}{\partial x}\right)_y dx + \left(\frac{\partial y}{\partial z}\right)_x dz \tag{E5.1C}$$

The first term on the right-hand sides of Equations (E5.1B) and (E5.1C) must be equal; therefore,

$$\left(\frac{\partial y}{\partial x}\right)_z = -\left(\frac{\partial y}{\partial z}\right)_x \left(\frac{\partial z}{\partial x}\right)_y \tag{E5.1D}$$

Rearranging Equation (E5.1D) and applying derivative inversion, we get the cyclic rule

$$-1 = \left(\frac{\partial x}{\partial z}\right)_y \left(\frac{\partial y}{\partial x}\right)_z \left(\frac{\partial z}{\partial y}\right)_x \tag{5.25}$$

This derivation of the cyclic rule does not depend on the heuristic mathematical argument presented earlier; yet it gives the equivalent result. ◀

Using the Thermodynamic Web to Access Reported Data

We have seen that problem solving in thermodynamics frequently involves construction of hypothetical paths to find the change in a given property between two states. In applying this procedure, we often come up with a partial derivative of one property with respect to another, holding a third constant. In this section, we will use the thermodynamic web to translate partial derivatives to forms in which experimental data are routinely reported, such as c_v, c_P, β, κ, and derivatives of equations of state.

Figure 5.2 presents a way to navigate the thermodynamic web when partial derivatives with T, P, s, and v are encountered. It provides 12 permutations of partial derivatives between these properties. Derivative inversion can also be applied to form 12 additional relationships to make a complete set. Each of the terms in Figure 5.2 is related to other terms using either the cyclic rule or Maxwell relations and is delineated with the appropriate icon as defined in the upper right of the figure. In those cases where the cyclic rule is used, the term at the origin of the two arrows can be replaced by the product of the two terms at the other end of the arrows. For example, at the top of the figure, $-(\partial P/\partial v)_s$ can be replaced by the product $(\partial P/\partial s)_v (\partial s/\partial v)_P$. Any of the terms on one side of the double arrow of a Maxwell relation can be replaced by the term on the other side. For example,

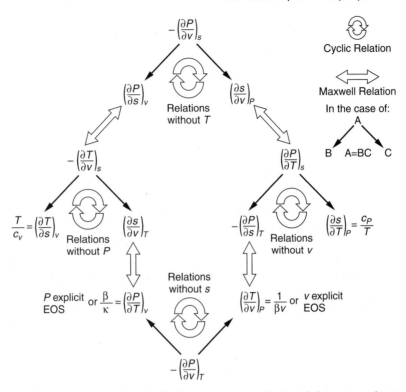

Figure 5.2 Roadmap through the thermodynamic web. Partial derivatives of $T, P, s,$ and v are related to each other and to reported properties $c_v, c_P, \beta,$ and κ.

we can use $-(\partial T/\partial v)_s$ for $(\partial P/\partial s)_v$. Thus, any derivative presented in Figure 5.2 can be rewritten by following a path in the diagram to ultimately lead to $c_v, c_P, \beta, \kappa,$ and derivatives of equations of state. For example, we can ultimately represent $-(\partial P/\partial v)_s$ by $-(c_P/v\kappa c_v)$.

If we have a pressure-explicit or volume-explicit equation of state, we can analytically assess the partial derivatives $(\partial P/\partial T)_v$ and $(\partial v/\partial T)_P$, respectively. Alternatively, we may use the thermal expansion coefficient, β, and the isothermal compressibility, κ, to give:

$$\left(\frac{\partial P}{\partial T}\right)_v = \frac{\beta}{\kappa} \tag{5.26}$$

and

$$\left(\frac{\partial v}{\partial T}\right)_P = \beta v \tag{5.27}$$

where the cyclic rule was used to obtain Equation (5.26) (see Problem 5.9). Recall $\beta \equiv (1/v)(\partial v/\partial T)_P$ and $\kappa \equiv -(1/v)(\partial v/\partial P)_T$ were defined in Section 4.3. Both alternatives are indicated in Figure 5.2.

Partial derivatives of s with T can be related to heat capacity by applying the fundamental property relations to the definition of heat capacity. Inserting Equation (5.6) into the definition of heat capacity at constant volume, we get

$$c_v = \left(\frac{\partial u}{\partial T}\right)_v = T\left(\frac{\partial s}{\partial T}\right)_v - P\left(\frac{\partial v}{\partial T}\right)_v = T\left(\frac{\partial s}{\partial T}\right)_v \tag{5.28}$$

Thus,

$$\left(\frac{\partial s}{\partial T}\right)_v = \frac{c_v}{T} \tag{5.29}$$

Similarly applying the fundamental property relation for h, Equation (5.7), to the definition of c_P gives

$$c_P = \left(\frac{\partial h}{\partial T}\right)_P = T\left(\frac{\partial s}{\partial T}\right)_P + v\left(\frac{\partial P}{\partial T}\right)_P = T\left(\frac{\partial s}{\partial T}\right)_P \tag{5.30}$$

or

$$\left(\frac{\partial s}{\partial T}\right)_P = \frac{c_P}{T} \tag{5.31}$$

Equations (5.29) and (5.31) are used in the appropriate places in Figure 5.2. See if you can justify the relations indicated in Figure 5.2. This exercise will pay dividends when you encounter these forms in solving problems with the thermodynamic web.

▶ 5.3 CALCULATION OF Δs, Δu, AND Δh USING EQUATIONS OF STATE

Relation of ds in Terms of Independent Variables T and v and Independent Variables T and P

The relations shown in Figure 5.2 allow us to express the dependent variable s in terms of measured properties. We can write s in terms of independent variables T and v by applying the form of Equation (5.4):

$$ds = \left(\frac{\partial s}{\partial T}\right)_v dT + \left(\frac{\partial s}{\partial v}\right)_T dv \tag{5.32}$$

Substitution of Equation (5.29) and the Maxwell relation (5.18) into Equation (5.32) gives

$$ds = \frac{c_v}{T}dT + \left(\frac{\partial P}{\partial T}\right)_v dv \tag{5.33}$$

Integration of Equation (5.33) gives

$$\Delta s = \int \frac{c_v}{T}dT + \int \left(\frac{\partial P}{\partial T}\right)_v dv \tag{5.34}$$

path needed

If we choose the measured properties T and P as our independent variables, we get

$$ds = \left(\frac{\partial s}{\partial T}\right)_P dT + \left(\frac{\partial s}{\partial P}\right)_T dP \tag{5.35}$$

Substituting in Equation (5.31) and the Maxwell relation (5.19) into Equation (5.35) gives

$$ds = \frac{c_P}{T}dT - \left(\frac{\partial v}{\partial T}\right)_P dP \tag{5.36}$$

Equation (5.36) is generally applicable to a system with constant composition. Integration of Equation (5.36) gives:

$$\Delta s = \int \frac{c_P}{T}dT - \int \left(\frac{\partial v}{\partial T}\right)_P dP \tag{5.37}$$

If we use the ideal gas model, Equation (5.37) simplifies to Equation (3.62). It is instructive to compare Equation (5.34) to Equation (5.37). Equation (5.34) was developed using

T and v as the independent variables, while the development of Equation (5.37) uses T and P. We see that in the former case we get a partial derivative in P. Therefore, this form is amenable to a pressure-explicit equation of state. Conversely, Equation (5.37) gives a partial derivative for v and is more amenable to a volume-explicit equation of state. This observation holds generally; that is, *T and v are convenient independent variables when we have a pressure-explicit equation of state, while T and P are convenient for a volume-explicit equation.* This rule of thumb is reinforced in Example 5.4, where the choices of independent variables (T,v) and (T,P) are compared for a calculation using the pressure-explicit Redlich–Kwong equation of state.

Relation of d*u* in Terms of Independent Variables *T* and *v*

We can now go back to the calculation for Δu illustrated by the hypothetical path in Figure 5.1. Recall that we can relate the differential change in internal energy to the independent variables T and v by Equation (5.5a):

$$du = \left(\frac{\partial u}{\partial T}\right)_v dT + \left(\frac{\partial u}{\partial v}\right)_T dv \qquad (5.5a)$$

The first term on the right-hand side of Equation (5.5a) can be written using the definition of heat capacity at constant volume:

$$\left(\frac{\partial u}{\partial T}\right)_v = c_v \qquad (5.38)$$

We usually want to evaluate this term under ideal gas conditions, since ideal gas heat capacity data are readily available for most gases. In Example 5.3, we will take another approach. The second term can be simplified using the fundamental property relation, Equation (5.6):

$$\left(\frac{\partial u}{\partial v}\right)_T = \left(\frac{Tds - Pdv}{dv}\right)_T = \left[T\left(\frac{\partial s}{\partial v}\right)_T - P\right] \qquad (5.39)$$

We can then use the Maxwell relation, Equation (5.13), to get

$$\left(\frac{\partial u}{\partial v}\right)_T = \left[T\left(\frac{\partial P}{\partial T}\right)_v - P\right] \qquad (5.40)$$

In general, we can add together the effect of changes in both independent variables to get

$$du(T,v) = c_v dT + \left[T\left(\frac{\partial P}{\partial T}\right)_v - P\right] dv \qquad (5.41)$$

Integrating Equation (5.41) gives

$$\Delta u(T,v) = \int c_v dT + \int \left[T\left(\frac{\partial P}{\partial T}\right)_v - P\right] dv \qquad (5.42)$$

We can solve the second integral on the right-hand side of Equation (5.42) with the appropriate PvT data or with data for the thermal expansion coefficient and the isothermal compressibility. For example, if an equation of state is available of the form $P = f(T,v)$, we can take the partial derivative with respect to T at constant v, multiply by T, subtract P, and integrate. If no equation is available, we could solve graphically.

The hypothetical path depicted in Figure 5.1 allows us to reduce Equation (5.42) to one term for each step. The first and third steps are isothermal, so the first term on the right-hand side goes to zero and we use only the second term. Conversely, step 2 is isochoric and we use only the first term. Furthermore, we constructed this path to carry out step 2 under ideal gas conditions, where heat capacity data are readily available. The ideal gas heat capacity depends *only* on the individual molecular structure of the species themselves, not on the interactions between them.[4] On the other hand, the second term relates to β and κ or an equation of state and its derivatives, and it depends on how the molecules interact with one another. In constructing these hypothetical thermodynamic paths—be it for u, s, h, or other properties—we often choose T for one independent variable and either P or v for the other independent variable. In doing so, we relate the T dependent term to c_P or c_v, which can be reduced to the individual molecular structure of the species in the system. The P or v term then accounts for the interactions of the species with one another.

► **EXAMPLE 5.2**
First-Law—Closed-System—Calculation Using the Thermodynamic Web

One mole of propane gas is to be expanded from 0.001 m³ to 0.040 m³ while in contact with a heating bath at 100°C. The expansion is not reversible. The heat extracted from the bath is 600 J. Using the van der Waals equation of state, determine the work for the expansion.

SOLUTION To find the work, we apply the first law:

$$\Delta u = q + w \tag{E5.2A}$$

Since we know $q = 600$ J/mol, we just need to find Δu. Writing u as a function of the independent variables T and v:

$$du = \left(\frac{\partial u}{\partial T}\right)_v dT + \left(\frac{\partial u}{\partial v}\right)_T dv = c_v dT + \left[T\left(\frac{\partial P}{\partial T}\right)_v - P\right] dv \tag{E5.2B}$$

where we have applied Equation (5.41). This process occurs at constant T, so integration of Equation (E5.2B) gives:

$$\Delta u = \int \left[T\left(\frac{\partial P}{\partial T}\right)_v - P\right] dv \tag{E5.2C}$$

From the van der Waals equation, we have

$$P = \frac{RT}{v - b} - \frac{a}{v^2} \tag{E5.2D}$$

and

$$\left(\frac{\partial P}{\partial T}\right)_v = \frac{R}{v - b} \tag{E5.2E}$$

Using Equations (E5.2D) and (E5.2E) in Equation (E5.2C) gives:

$$\Delta u = \int \left[\frac{a}{v^2}\right] dv$$

or

$$\Delta u = \left.\frac{-a}{v}\right|_{0.001 \text{ m}^3}^{0.040 \text{ m}^3} = -\frac{27(RT_c)^2}{64P_c}\left(\frac{1}{v_2} - \frac{1}{v_1}\right)$$

$$= -\left(0.96\left[\frac{\text{Jm}^3}{\text{mol}^2}\right]\right)\left(\frac{1}{0.04} - \frac{1}{0.001}\right)\left[\frac{\text{mol}}{\text{m}^3}\right] = 936\,\frac{\text{J}}{\text{mol}}$$

---

[4] Recall the discussion after Equation (2.52).

Using this value of Δu in Equation (E5.2A) gives:

$$w = \Delta u - q = 336\ [\text{J/mol}]$$

◀

▶ **EXAMPLE 5.3**
Alternative Calculation
Path for Δu

Develop a methodology for calculating Δu according to the path shown in Figure E5.3A, in which the change in T occurs when intermolecular interactions are important.

SOLUTION There are many ways to get from one thermodynamic state to another using the thermodynamic web. We consider the path shown in Figure E5.3A as an alternative calculation path for Δu to that presented in Figure 5.1. In this case, our hypothetical path consists of two steps: isochoric heating (step 1) followed by isothermal compression (step 2). However, for the temperature change in step 1, the gas no longer behaves as an ideal gas. Again, we can write the change in internal energy as a function of the independent variables T and v:

$$du = \left(\frac{\partial u}{\partial T}\right)_v dT + \left(\frac{\partial u}{\partial v}\right)_T dv = c_v^{real} dT + \left[T\left(\frac{\partial P}{\partial T}\right)_v - P\right] dv \qquad (\text{E5.3A})$$

We have used the results given by Equation (5.41). In calculating the internal energy change in step 1, we must now recognize that the heat capacity is no longer under ideal gas conditions but rather in the region where intermolecular interactions are significant; therefore, c_v now depends on both T and v, that is,

$$c_v^{real} = c_v(T, v_1)$$

In general, however, heat capacity data are available only at ideal gas conditions. Hence, we need to relate the real heat capacity at any given temperature along step 1 to the ideal gas heat capacity. Again, this task can be accomplished by utilizing the thermodynamic web. As illustrated in Figure E5.3B, we need to calculate $(\partial c_v / \partial v)_T$ and then integrate from the volume of an ideal gas to the volume of the real gas, v_1. We can develop this relationship by using the definition of heat capacity and then changing the order of differentiation (as we did in developing the Maxwell relations):[5]

$$\left(\frac{\partial c_v}{\partial v}\right)_T = \left[\frac{\partial}{\partial v}\left(\frac{\partial u}{\partial T}\right)_v\right]_T = \left[\frac{\partial}{\partial T}\left(\frac{\partial u}{\partial v}\right)_T\right]_v \qquad (\text{E5.3B})$$

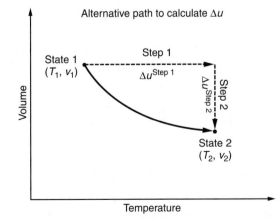

Alternative path to calculate Δu

Step 1

State 1
(T_1, v_1)

$\Delta u^{Step\ 1}$

$\Delta u^{Step\ 2}$ Step 2

State 2
(T_2, v_2)

Volume

Temperature

Figure E5.3A Alternative computational path to Figure 5.1 for the change in internal energy from state 1 to state 2.

[5] Again, we see a common theme of creative problem solving: being able to take what we learn in one place and apply it in a different context.

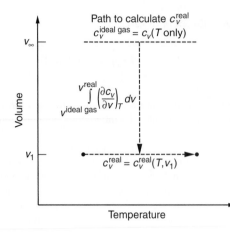

Figure E5.3B Computational paths to calculate c_v^{real} from $c_v^{\text{ideal gas}}$.

Using the value for $(\partial u / \partial v)_T$ obtained in Equation (5.40) and expanding, we get

$$\left(\frac{\partial c_v}{\partial v}\right)_T = \left[\frac{\partial}{\partial T}\left(\left[T\left(\frac{\partial P}{\partial T}\right)_v - P\right]\right)\right]_v = \left(\frac{\partial P}{\partial T}\right)_v + T\left(\frac{\partial^2 P}{\partial T^2}\right)_v - \left(\frac{\partial P}{\partial T}\right)_v = T\left(\frac{\partial^2 P}{\partial T^2}\right)_v$$

where we have used the product rule. The differential change in heat capacity at any given temperature is therefore given by:

$$dc_v = \left[T\left(\frac{\partial^2 P}{\partial T^2}\right)_v\right] dv \tag{E5.3C}$$

Integrating both sides of Equation (E5.3C) gives

$$\int_{\text{ideal gas}}^{\text{real}} dc_v = \int_{v_{\text{ideal gas}}}^{v_{\text{real}}} \left[T\left(\frac{\partial^2 P}{\partial T^2}\right)_v\right] dv$$

Solving for c_v^{real}, we get:

$$c_v^{\text{real}} = c_v(T, v_1) = c_v^{\text{ideal gas}} + \int_{v_{\text{ideal gas}}}^{v_1} \left[T\left(\frac{\partial^2 P}{\partial T^2}\right)_v\right] dv \tag{E5.3D}$$

Examining Equation (E5.3D), we see that if we have ideal gas heat capacity data and an appropriate equation of state, we can solve for the real gas heat capacity. We can then use this expression in Equation (E5.3A) to get:

$$du = \left\{c_v^{\text{ideal gas}} + \int_{v_{\text{ideal gas}}}^{v_1} \left[T\left(\frac{\partial^2 P}{\partial T^2}\right)_v\right] dv\right\} dT + \left[T\left(\frac{\partial P}{\partial T}\right)_v - P\right] dv$$

Relation of d*h* in Terms of Independent Variables *T* and *P*

We can obtain an expression for the change in enthalpy in terms of the independent variables T and P by applying the thermodynamic web in a manner similar to that used above. We first write the differential expression in the form of Equation (5.4):

$$dh = \left(\frac{\partial h}{\partial T}\right)_P dT + \left(\frac{\partial h}{\partial P}\right)_T dP \tag{5.43}$$

Applying the definition for heat capacity

$$\left(\frac{\partial h}{\partial T}\right)_P = c_P \tag{5.44}$$

and the fundamental property relation for h, Equation (5.7), and the Maxwell relation (5.14):

$$\left(\frac{\partial h}{\partial P}\right)_T = \left(\frac{T\partial s + v\partial P}{\partial P}\right)_T = T\left(\frac{\partial s}{\partial P}\right)_T + v = -T\left(\frac{\partial v}{\partial T}\right)_P + v \tag{5.45}$$

Substitution of Equations (5.44) and (5.45) into Equation (5.43) gives

$$dh = c_P dT + \left[-T\left(\frac{\partial v}{\partial T}\right)_P + v\right]dP \tag{5.46}$$

Integrating this expression gives

path needed

$$\Delta h = \int c_P dT + \int \left[-T\left(\frac{\partial v}{\partial T}\right)_P + v\right]dP \tag{5.47}$$

In analogy to our discussion for Δu, we usually construct a path whereby we take the system to a low enough pressure to apply the ideal gas heat capacity. Alternatively, we could calculate the real heat capacity at constant pressure using the same method we used to calculate c_v^{real} in Example 5.3. In Problem 5.14, you will verify that

$$c_P^{\text{real}} = c_P(T,P) = c_P^{\text{ideal gas}} - \int_{P_{\text{ideal gas}}}^{P_{\text{real}}} \left[T\left(\frac{\partial^2 v}{\partial T^2}\right)_P\right]dP \tag{5.48}$$

◄

► **EXAMPLE 5.4**
First-law—Open-System—Calculation Using the Thermodynamic Web

The first step in manufacturing isobutene from isomerization of n-butane is to compress the feed stream of n-butane. It goes into the compressor at 9.47 bar and 80°C and optimally leaves at 18.9 bar and 120°C, so that it can be fed into the isomerization reactor. The work supplied to the compressor is 2100 J/mol. Compute the heat that needs to be supplied into the unit per mole of n-butane that passes through.

SOLUTION First, the process is sketched in Figure E5.4A. To find the heat in, we will apply the first law (i.e., do an energy balance). Assuming steady-state, the open-system energy balance with one stream in and one stream out can be written:

$$0 = \dot{n}_1(h + \cancel{e_K} + \cancel{e_P})_1 - \dot{n}_2(h + \cancel{e_K} + \cancel{e_P})_2 + \dot{Q} + \dot{W}_s$$

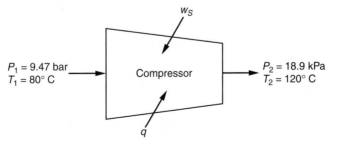

$P_1 = 9.47$ bar
$T_1 = 80°$ C

Compressor

w_S

q

$P_2 = 18.9$ kPa
$T_2 = 120°$ C

Figure E5.4A Schematic of the compressor for Example 5.3.

or
$$q = \frac{\dot{Q}}{\dot{n}} = h_2 - h_1 - \frac{\dot{W}_s}{\dot{n}}$$

We know the power used by the compressor. Thus this problem reduces to finding the change in the thermodynamic property enthalpy from the inlet to the outlet. We know two intensive properties at both the inlet and outlet, so the values for the other properties (like enthalpy!) are already constrained.

Since pressures on the order of 10 bar are being used, we do not expect the ideal gas law to hold. However, since enthalpy is a property and independent of path, we are free to choose whatever path is convenient.

First, it may be helpful to find some data for n-butane. From Appendix A.2, we have an expression for the *ideal gas* heat capacity:

$$\frac{c_P}{R} = 1.935 + 36.915 \times 10^{-3}\,T - 11.402 \times 10^{-6}\,T^2$$

with T in [K]. Since this expression is limited to ideal gases, any change in temperature must be under ideal conditions.

An equation of state may also be useful. Many equations are available; an accurate one is the Redlich–Kwong equation of state:

$$P = \frac{RT}{v - b} - \frac{a}{T^{1/2}v(v + b)}$$

We can find the parameters a and b using the principle of corresponding states. To do this, we first find critical parameters from Appendix A.1: ($T_c = 425.2$ K; $P_c = 37.9$ bar). Applying Equations (4.47) and (4.48):

$$a = \frac{0.42748\,R^2 T_c^{2.5}}{P_c} = 29.08\left[\frac{\mathrm{J\,K^{1/2}m^3}}{\mathrm{mol}^2}\right] \quad \text{and} \quad b = \frac{0.08664\,RT_c}{P_c} = 8.09 \times 10^{-5}\left[\frac{\mathrm{m^3}}{\mathrm{mol}}\right]$$

We are now free to choose a path to solve this problem. We are constrained in that any change in temperature should be carried out when C_4H_{10} behaves as an ideal gas. We first will solve this with T and v as independent variables, then show that it can be solved with T and P as independent variables.

Solution with T and v

The solution path is depicted in the Tv plane in Figure E5.4B.

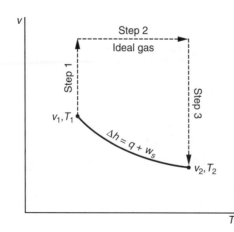

Figure E5.4B Solution path in the Tv plane for the compressor in Example 5.4.

To execute this path, we need to know the molar volumes of states 1 and 2. The Redlich–Kwong equation is implicit in v; solving for the three roots and taking the largest gives

$$v_1 = 2.59 \times 10^{-3} [\text{m}^3/\text{mol}]$$

and

$$v_2 = 1.27 \times 10^{-3} [\text{m}^3/\text{mol}]$$

Now we can write an expression for enthalpy in terms of our independent variables T and v:

$$dh = \left(\frac{\partial h}{\partial T}\right)_v dT + \left(\frac{\partial h}{\partial v}\right)_T dv$$

For steps 1 and 3, T is constant:

$$\Delta h_{1 \, or \, 3} = \int \left(\frac{\partial h}{\partial v}\right)_T dv \tag{E5.4A}$$

From the fundamental property relation, Equation (5.7), Equation (E5.4A) becomes

$$\Delta h = \int \left(\frac{T \partial s + v \partial P}{\partial v}\right)_T dv = \int \left[T \left(\frac{\partial s}{\partial v}\right)_T + v \left(\frac{\partial P}{\partial v}\right)_T\right] dv$$

If we use the Maxwell relation (5.18), we get

$$\Delta h = \int \left[T \left(\frac{\partial P}{\partial T}\right)_v + v \left(\frac{\partial P}{\partial v}\right)_T\right] dv \tag{E5.4B}$$

We can now use the Redlich–Kwong equation, which is explicit in pressure:

$$P = \frac{RT}{v - b} - \frac{a}{T^{1/2} v(v + b)} \tag{E5.4C}$$

Differentiating Equation (E5.4C) gives

$$\left(\frac{\partial P}{\partial T}\right)_v = \frac{R}{v - b} + \frac{a}{2T^{3/2} v(v + b)} \tag{E5.4D}$$

and

$$\left(\frac{\partial P}{\partial v}\right)_T = -\frac{RT}{(v - b)^2} + \frac{a(2v + b)}{T^{1/2}(v^2 + bv)^2} \tag{E5.4E}$$

Plugging Equations (E5.4D) and (E5.4E) back in to Equation (E5.4B) gives

$$\Delta h = \int \left[T \left[\frac{R}{v - b} + \frac{a}{2T^{3/2} v(v + b)}\right] + v \left[-\frac{RT}{(v - b)^2} + \frac{a(2v + b)}{T^{1/2}(v^2 + bv)^2}\right]\right] dv$$

Simplifying

$$\Delta h = \int \left[-\frac{bRT}{(v - b)^2} + \frac{a}{T^{1/2}} \left(\frac{1}{2v(v + b)} + \frac{v(2v + b)}{[v(v + b)]^2}\right)\right] dv \tag{E5.4F}$$

and integrating

$$\Delta h_{1 \, or \, 3} = \frac{bRT}{(v - b)} + \frac{a}{T^{1/2}} \left(\frac{3}{2b} \ln\left(\frac{v}{v + b}\right) - \frac{1}{(v + b)}\right) \Bigg|_{v_i}^{v_f}$$

We can now plug in numerical values. What should we use for v_{large}? For step 1, we use $T = 353.15$ K, getting

$$\Delta h_1 = 1368 \text{ [J/mol]}$$

For step 3, we use $T = 393.15$ K, getting

$$\Delta h_3 = -2542 \text{ [J/mol]}$$

For step 2, we have an ideal gas undergoing an isochoric process:

$$\Delta h_2 = \int \left(\frac{\partial h}{\partial T} \right)_v dT = \int \left(\frac{\partial (u + Pv)}{\partial T} \right)_v dT$$

or

$$\Delta h_2 = \int \left[\left(\frac{\partial u}{\partial T} \right)_v + \left(\frac{\partial Pv}{\partial T} \right)_v \right] dT = \int [c_v + R] dT = \int c_P dT$$

where we used the ideal gas law to simplify. Plugging in numbers:

$$\Delta h_2 = R \int_{353}^{393} [1.935 + 36.915 \times 10^{-3} T - 11.402 \times 10^{-6} T^2] dT$$

and integrating

$$\Delta h_2 = 8.314 \big[1.935(393 - 353) + 18.458 \times 10^{-3} (393^2 - 353^2) - 3.801 \times 10^{-6} (393^3 - 353^3) \big]$$

$$\Delta h_2 = 4696 \text{ [J/mol]}$$

Finally, summing together gives the heat input as

$$\Delta h = \Delta h_1 + \Delta h_2 + \Delta h_3 = q + w_s = 3522 \text{ [J/mol]}$$

$$q = \Delta h - w_s = 1422 \text{ [J/mol]}$$

Solution with T and P

The solution path is depicted in the TP plane in Figure E5.4C. In this case, we need a low pressure to obtain ideal gas behavior. If we write the change in enthalpy in terms of the independent variables T and P, we get

$$dh = \left(\frac{\partial h}{\partial T} \right)_P dT + \left(\frac{\partial h}{\partial P} \right)_T dP = c_P dT + \left[-T \left(\frac{\partial v}{\partial T} \right)_P + v \right] dP \qquad \text{(E5.4G)}$$

where Equation (5.46) was used. The constant-pressure part, step 2, is equivalent to Δh_2 above, so the result is identical. For steps 1 and 3, we must evaluate the second term on the right-hand side of Equation (E5.4G) and integrate. We cannot write the Redlich–Kwong equation explicitly in v to solve. However, we can differentiate Equation (E5.4C) to get

$$dP = \left[-\frac{RT}{(v - b)^2} + \frac{a(2v + b)}{T^{1/2} v^2 (v + b)^2} \right] dv \qquad \text{(E5.4H)}$$

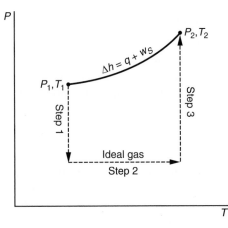

Figure E5.4C Solution path in the TP plane for the compressor in Example 5.4.

At constant T, substituting Equation (E5.4H) into (E5.4G) gives

$$dh = \left[-T\left(\frac{\partial v}{\partial T}\right)_P + v\right]\left[-\frac{RT}{(v-b)^2} + \frac{a(2v+b)}{T^{1/2}v^2(v+b)^2}\right]dv \qquad \text{(E5.4I)}$$

The partial derivative of volume with respect to temperature cannot be obtained directly. However, we can apply the cyclic relation as follows:

$$-1 = \left(\frac{\partial v}{\partial T}\right)_P \left(\frac{\partial P}{\partial v}\right)_T \left(\frac{\partial T}{\partial P}\right)_v \qquad \text{(E5.4J)}$$

We can rewrite Equation (E5.4J) in terms of a form that allows us to use the P explicit Redlich–Kwong equation. Performing this manipulation and taking derivatives gives

$$\left(\frac{\partial v}{\partial T}\right)_P = -\frac{\left(\frac{\partial P}{\partial T}\right)_v}{\left(\frac{\partial P}{\partial v}\right)_T} = \frac{-\dfrac{R}{v-b} - \dfrac{a}{2T^{3/2}v(v+b)}}{\left[-\dfrac{RT}{(v-b)^2} + \dfrac{a(2v+b)}{T^{1/2}v^2(v+b)^2}\right]} \qquad \text{(E5.4K)}$$

Substitution of Equation (E5.4K) into (E5.4I) and simplification gives

$$dh = \left[-\frac{bRT}{(v-b)^2} + \frac{a}{T^{1/2}}\left(\frac{1}{2v(v+b)} + \frac{v(2v+b)}{[v(v+b)]^2}\right)\right]dv \qquad \text{(E5.4L)}$$

If we integrate Equation (E5.4L), we get a result identical to Equation (E5.4F), so the rest of the problem is equivalent to the part above where we used T and v as independent variables. We come up with the same result applying the thermodynamic web to each path; the form $h = h(T,v)$ yields an equivalent result to $h = h(T,P)$. However, the first choice of independent variables made the math easier. This result is not surprising in light of the discussion after Equation (5.37); that is, T and v are the convenient independent variables when we have a pressure-explicit equation of state.

Solution Using Δu
A third alternative for calculating the enthalpy change is to apply the definition for h:

$$\Delta h = \Delta u + \Delta(Pv)$$

We have already calculated the volumes of states 1 and 2, so it is straightforward to obtain $\Delta(Pv)$. We can calculate Δu from Equation (5.42) in conjunction with the path shown in Figure (E5.4B). The result we obtain is equivalent to those presented above. ◀

▶ 5.4 DEPARTURE FUNCTIONS

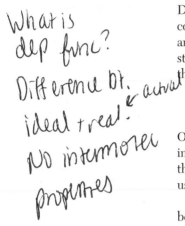

What is
dep func?

Difference of
ideal + real. *actual*

No intermolec
properties

Departure functions often provide us a convenient path for calculating the nonideal contribution to property changes for real gases (or liquids). The departure function of any thermodynamic property is the difference in that property between the real, physical state in which it exists and that of a hypothetical ideal gas at the same T and P. For example, the enthalpy departure is given by

$$\Delta h_{T,P}^{\text{dep}} = h_{T,P} - h_{T,P}^{\text{ideal gas}} \tag{5.49}$$

On a molecular level, we can consider this departure function to represent the change in enthalpy if we could "turn off" the intermolecular interactions in the real fluid. In this section, we will specifically explore how to calculate changes in enthalpy and entropy using departure functions; however, this methodology can be expanded to any property.[6]

We can use the enthalpy departure function to calculate the enthalpy difference between a species in an initial state 1 at temperature T_1 and pressure P_1 and final state 2 at T_2 and P_2. Figure 5.3 shows a calculation path constructed using departure functions. The PT diagram to the left side of the figure represents real, physical space, while the TP diagram on the right-hand side is the hypothetical, ideal gas state in which all the intermolecular interactions are turned off. We first "turn off" the intermolecular interactions the species exhibits at T_1 and P_1. We thus transform from a real fluid into a hypothetical ideal gas. The ideal gas then undergoes an isobaric temperature change to T_2 followed by an isothermal change in pressure to P_2. Finally, we "turn on" the intermolecular interactions, returning to the real, physical state 2. Adding together the four steps in Figure 5.3, we get:

$$h_2 - h_1 = -\Delta h_{T_1,P_1}^{\text{dep}} + \left[\Delta h_{T_1 \rightarrow T_2}^{\text{ideal gas}} + \Delta h_{P_1 \rightarrow P_2}^{\text{ideal gas}} \right] + \Delta h_{T_2,P_2}^{\text{dep}} \tag{5.50}$$

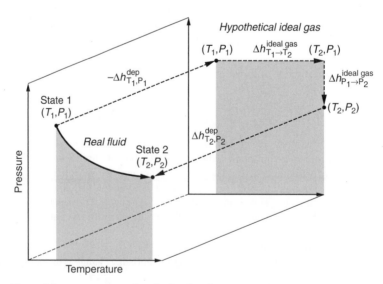

Figure 5.3 Computational paths for the change in enthalpy from state 1 to state 2 using departure functions. The PT diagram on the left is for the real fluid while that on the right represents the hypothetical ideal gas in which all the intermolecular interactions are "turned off."

[6] For example, in Problem 5.23, you will develop the departure function for internal energy.

Using ideal gas heat capacity data for the temperature dependence and recognizing that the enthalpy of an ideal gas does not depend on pressure, we can simplify Equation (5.50) to

$$
h_2 - h_1 = -\Delta h_{T_1,P_1}^{\text{dep}} + \left[\int_{T_1}^{T_2} c_P \mathrm{d}T + 0 \right] + \Delta h_{T_2,P_2}^{\text{dep}}
\tag{5.51}
$$

We now need to come up with an expression for the enthalpy departure function so that we can solve Equation (5.51). Since enthalpy departure at a given state is related to the intermolecular forces involved, we will need to use the PvT relation developed in Chapter 4 and then apply the relationships of the thermodynamic web to come up with an expression for the enthalpy departure function. In the development that follows, we will use the generalized compressibility charts and tables discussed in Section 4.4 to develop values for the generalized enthalpy departure function based on corresponding states. If we wanted to apply departure functions using other property data, we might have to appropriately modify our approach.

First, we add and subtract $h_{T,P=0}^{\text{ideal gas}}$ to Equation (5.49):

$$
\Delta h_{T,P}^{\text{dep}} = h_{T,P} - h_{T,P}^{\text{ideal gas}} = \left(h_{T,P} - h_{T,P=0}^{\text{ideal gas}} \right) - \left(h_{T,P}^{\text{ideal gas}} - h_{T,P=0}^{\text{ideal gas}} \right) = h_{T,P} - h_{T,P=0}^{\text{ideal gas}}
\tag{5.52}
$$

We have simplified Equation (5.52), since the enthalpy of an ideal gas is independent of pressure, that is,

$$
h_{T,P}^{\text{ideal gas}} - h_{T,P=0}^{\text{ideal gas}} = 0
\tag{5.53}
$$

At constant T, Equation (5.46) becomes

$$
\mathrm{d}h_T = \left[-T \left(\frac{\partial v}{\partial T} \right)_P + v \right] \mathrm{d}P
\tag{5.54}
$$

We wish to put PvT data in terms of the compressibility factor so that we can use the generalized compressibility charts. Thus, we want to put Equation (5.54) in terms of z. By the definition of the compressibility factor, the molar volume is written as

$$
v = \frac{zRT}{P}
\tag{5.55}
$$

Applying the product rule to Equation (5.55), the partial derivative with respect to temperature becomes

$$
\left(\frac{\partial v}{\partial T} \right)_P = \left[\frac{RT}{P} \left(\frac{\partial z}{\partial T} \right)_P + \frac{zR}{P} \right]
\tag{5.56}
$$

Substituting Equations (5.55) and (5.56) into (5.54) gives

$$
\mathrm{d}h_T = \left[-\frac{RT^2}{P} \left(\frac{\partial z}{\partial T} \right)_P - \frac{zRT}{P} + \frac{zRT}{P} \right] \mathrm{d}P = \left[-\frac{RT^2}{P} \left(\frac{\partial z}{\partial T} \right)_P \right] \mathrm{d}P
\tag{5.57}
$$

Since we are applying the corresponding states relation, we write Equation (5.57) in reduced coordinates:

$$
\frac{\mathrm{d}h_{T_r}}{RT_c} = \left[-\frac{T_r^2}{P_r} \left(\frac{\partial z}{\partial T_r} \right)_P \right] \mathrm{d}P_r
\tag{5.58}
$$

Integrating between 0 and P, and plugging into Equation (5.52) gives:

$$\frac{\Delta h_{T_r,P_r}^{\text{dep}}}{RT_c} = \frac{h_{T_r,P_r} - h_{T_r,P_r}^{\text{ideal gas}}}{RT_c} = \frac{h_{T_r,P_r} - h_{T_r,P_r=0}^{\text{ideal gas}}}{RT_c} = T_r^2 \int_0^P \left[-\frac{1}{P_r} \left(\frac{\partial z}{\partial T_r} \right)_P \right] dP_r \quad (5.59)$$

If we have a relation for z from 0 to P_r, we can integrate Equation (5.59) to give the enthalpy departure. For example, we can use the generalized compressibility results of the form given in Section 4.4 to get simple fluid and correction enthalpy departure terms. The generalized enthalpy departure can then be determined according to:

$$\frac{h_{T_r,P_r} - h_{T_r,P_r}^{\text{ideal gas}}}{RT_c} = \left[\frac{h_{T_r,P_r} - h_{T_r,P_r}^{\text{ideal gas}}}{RT_c} \right]^{(0)} + \omega \left[\frac{h_{T_r,P_r} - h_{T_r,P_r}^{\text{ideal gas}}}{RT_c} \right]^{(1)} \quad (5.60)$$

While the data presented in Section 4.4 can be numerically integrated, they were generated by Lee and Kesler according to their equation of state. This equation can be analytically integrated to find values for enthalpy departure that can be used in Equation (5.60). The results are shown in Appendix E. Plots of the simple fluid and correction terms that result are presented in Figures 5.4 and 5.5, respectively. Tables of their values are presented in Appendix C (Tables C.3 and C.4).

Like the enthalpy departure function, the entropy departure function can be used to find the entropy change of a real fluid. It is defined as the difference in that property between the real, physical state and that of a hypothetical ideal gas at the same T and P:

$$\Delta s_{T,P}^{\text{dep}} = s_{T,P} - s_{T,P}^{\text{ideal gas}} \quad (5.61)$$

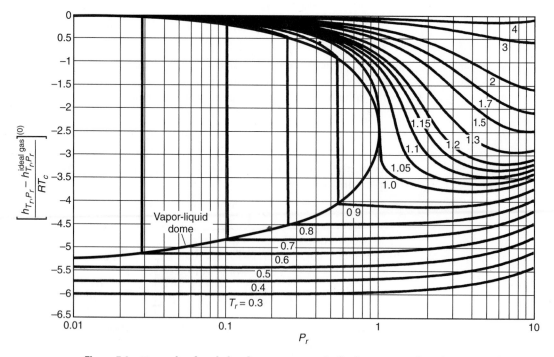

Figure 5.4 Generalized enthalpy departure—simple fluid term. Based on the Lee–Kesler equation of state.

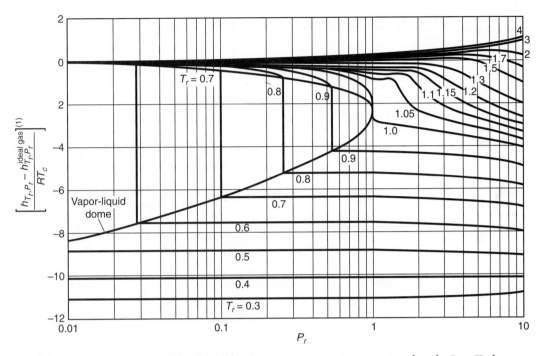

Figure 5.5 Generalized enthalpy departure—correction term. Based on the Lee–Kesler equation of state.

The change in entropy from state 1 to state 2 can be written in analogy to Figure (5.3):

$$s_2 - s_1 = -\Delta s_{T_1,P_1}^{\text{dep}} + \left[\Delta s_{T_1 \to T_2}^{\text{ideal gas}} + \Delta s_{P_1 \to P_2}^{\text{ideal gas}}\right] + \Delta s_{T_2,P_2}^{\text{dep}} \qquad (5.62)$$

Substituting Equation (3.62) for the two ideal gas terms gives

$$s_2 - s_1 = \Delta s_{T_2,P_2}^{\text{dep}} + \left[\int_{T_1}^{T_2} \frac{c_P}{T} dT - R \ln \frac{P_2}{P_1}\right] - \Delta s_{T_1,P_1}^{\text{dep}} \qquad (5.63)$$

As opposed to enthalpy, the entropy change with pressure of an ideal gas is nonzero. To calculate the entropy departure, we add and subtract $s_{T,P=0}^{\text{ideal gas}}$ to Equation (5.61):

$$s_{T,P} - s_{T,P}^{\text{ideal gas}} = \left(s_{T,P} - s_{T,P=0}^{\text{ideal gas}}\right) - \left(s_{T,P}^{\text{ideal gas}} - s_{T,P=0}^{\text{ideal gas}}\right) \qquad (5.64)$$

Using the thermodynamic web, we can determine each difference on the right-hand side of Equation (5.64). At constant temperature, the entropy can be written in terms of the independent property P. According to Equation (5.36),

$$ds_T = -\left(\frac{\partial v}{\partial T}\right)_P dP \qquad (5.65)$$

In the case of an ideal gas, the ideal gas law can be differentiated to give

$$ds_T^{\text{ideal gas}} = -R\frac{dP}{P} \qquad (5.66)$$

Integrating Equation (5.66) gives

$$s_{T,P}^{\text{ideal gas}} - s_{T,P=0}^{\text{ideal gas}} = -\int_0^P R\frac{dP}{P} \tag{5.67}$$

For the real fluid, we can substitute Equation (5.56) into (5.65):

$$ds_T = -\left(\frac{\partial v}{\partial T}\right)_P dP = -\left[\frac{zR}{P} + \frac{RT}{P}\left(\frac{\partial z}{\partial T}\right)_P\right]dP \tag{5.68}$$

Integrating from zero pressure to the pressure of the state of interest gives

$$s_{T,P} - s_{T,P=0}^{\text{ideal gas}} = \int_0^P -\left[\frac{zR}{P} + \frac{RT}{P}\left(\frac{\partial z}{\partial T}\right)_P\right]dP \tag{5.69}$$

Finally, we get the entropy departure by subtracting Equation (5.67) from Equation (5.69):

$$s_{T,P} - s_{T,P}^{\text{ideal gas}} = R\int_0^P -\left[\frac{z-1}{P} + \frac{T}{P}\left(\frac{\partial z}{\partial T}\right)_P\right]dP \tag{5.70}$$

To use the generalized compressibility, we rewrite Equation (5.70) in reduced coordinates:

$$\frac{\Delta s_{T_r,P_r}^{\text{dep}}}{R} = \frac{s_{T_r,P_r} - s_{T_r,P_r}^{\text{ideal gas}}}{R} = \int_0^P -\left[\frac{z-1}{P_r} + \frac{T_r}{P_r}\left(\frac{\partial z}{\partial T_r}\right)_P\right]dP_r \tag{5.71}$$

Again, Equation (5.71) can be integrated with the appropriate data or equation of state for z. Using the Lee–Kesler equation of state gives results that include a simple fluid term and a correction term. The form of entropy departure using this equation of state is given in Appendix E. The entropy departure can then be calculated by

$$\frac{s_{T_r,P_r} - s_{T_r,P_r}^{\text{ideal gas}}}{R} = \left[\frac{s_{T_r,P_r} - s_{T_r,P_r}^{\text{ideal gas}}}{R}\right]^{(0)} + \omega\left[\frac{s_{T_r,P_r} - s_{T_r,P_r}^{\text{ideal gas}}}{R}\right]^{(1)} \tag{5.72}$$

Plots of the simple fluid and correction terms for entropy departure are given in Figures 5.6 and 5.7, respectively. Tables of values generated in the same way are presented in Appendix C (Tables C.5 and C.6).

▶ **EXAMPLE 5.5**
Alternative Solution to Example 5.4 Using Departure Functions

Repeat Example 5.4 using the Lee–Kesler generalized correlation data for enthalpy departure to account for nonideal behavior.

SOLUTION   The schematic of the process is drawn in Figure E5.4A. To find the heat, we need to calculate the enthalpy difference between the outlet (state 2) and the inlet (state 1). From Equation (5.51), we get

$$q + w_s = h_2 - h_1 = -\Delta h_{T_1,P_1}^{\text{dep}} + \left[\int_{T_1}^{T_2} c_p dT + 0\right] + \Delta h_{T_2,P_2}^{\text{dep}} \tag{E5.5A}$$

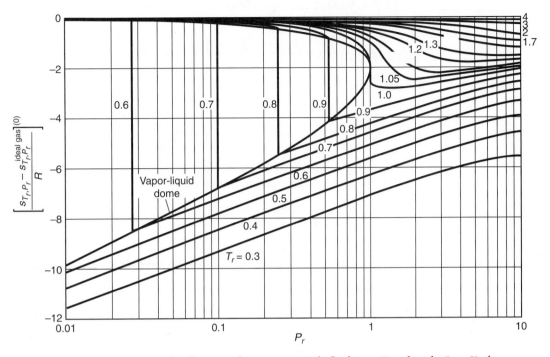

Figure 5.6 Generalized entropy departure—simple fluid term. Based on the Lee–Kesler equation of state.

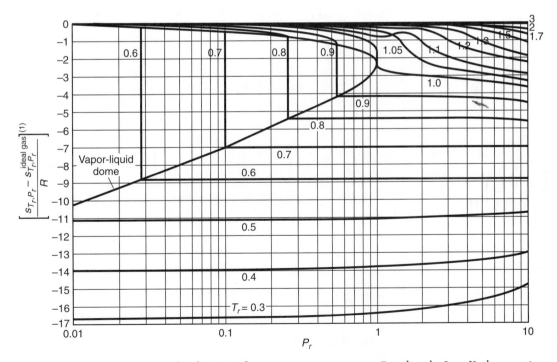

Figure 5.7 Generalized entropy departure—correction term. Based on the Lee–Kesler equation of state.

To find values of enthalpy departure, we can use the Lee–Kesler tables. They have the form

$$\frac{\Delta h_{T_r,P_r}^{dep}}{RT_c} = \frac{h_{T_r,P_r} - h_{T_r,P_r}^{ideal\ gas}}{RT_c} = \left[\frac{h_{T_r,P_r} - h_{T_r,P_r}^{ideal\ gas}}{RT_c}\right]^{(0)} + \omega\left[\frac{h_{T_r,P_r} - h_{T_r,P_r}^{ideal\ gas}}{RT_c}\right]^{(1)}$$

Looking up the critical properties and acentric factor for n-butane from Appendix A.1 gives

$$T_c = 425.2\,[K]$$

$$P_c = 37.9\,[bar]$$

$$\omega = 0.199$$

Thus, the reduced coordinates for state 1 and state 2 are

$$T_{1,r} = \frac{T_1}{T_c} = \frac{353.15\,[K]}{425.3\,[K]} = 0.83 \quad \text{and} \quad P_{1,r} = \frac{P_1}{P_c} = \frac{7.47\,[bar]}{38.0\,[bar]} = 0.20$$

$$T_{2,r} = \frac{T_2}{T_c} = \frac{393.15\,[K]}{425.2\,[K]} = 0.925 \quad \text{and} \quad P_{2,r} = \frac{P_2}{P_c} = \frac{18.9\,[bar]}{37.9\,[bar]} = 0.50$$

We can find the values for the enthalpy departure terms in Tables C.3 and C.4 in Appendix C. For state 1, by interpolation:

$$\left[\frac{h_{T_1,r,P_1,r} - h_{T_1,r,P_1,r}^{ideal\ gas}}{RT_c}\right]^{(0)} = -0.413 \quad \left[\frac{h_{T_1,r,P_1,r} - h_{T_1,r,P_1,r}^{ideal\ gas}}{RT_c}\right]^{(1)} = -0.622$$

so

$$\frac{h_{T_1,r,P_1,r} - h_{T_1,r,P_1,r}^{ideal\ gas}}{RT_c} = \left[\frac{h_{T_1,r,P_1,r} - h_{T_1,r,P_1,r}^{ideal\ gas}}{RT_c}\right]^{(0)} + \omega\left[\frac{h_{T_1,r,P_1,r} - h_{T_1,r,P_1,r}^{ideal\ gas}}{RT_c}\right]^{(1)} = -0.536 \quad (E5.5B)$$

and for state 2:

$$\left[\frac{h_{T_2,r,P_2,r} - h_{T_2,r,P_2,r}^{ideal\ gas}}{RT_c}\right]^{(0)} = -0.771 \quad \text{and} \quad \left[\frac{h_{T_2,r,P_2,r} - h_{T_2,r,P_2,r}^{ideal\ gas}}{RT_c}\right]^{(1)} = -0.994$$

so

$$\frac{h_{T_2,r,P_2,r} - h_{T_2,r,P_2,r}^{ideal\ gas}}{RT_c} = \left[\frac{h_{T_2,r,P_2,r} - h_{T_2,r,P_2,r}^{ideal\ gas}}{RT_c}\right]^{(0)} + \omega\left[\frac{h_{T_2,r,P_2,r} - h_{T_2,r,P_2,r}^{ideal\ gas}}{RT_c}\right]^{(1)} = -0.969 \quad (E5.5C)$$

The value for the integral of the heat capacity is the same as in Example 5.4:

$$\Delta h_{T_1 \to T_2}^{ideal\ gas} = \int_{T_1}^{T_2} c_P dT = R\int_{353}^{393} [1.935 + 36.915 \times 10^{-3}T - 11.402 \times 10^{-6}T^2]dT = 4{,}696\,[J/mol]$$

$$(E5.5D)$$

Plugging in the values from Equations (E5.5B), (E5.5C), and (E5.5D) into Equation (E5.5A) gives

$$h_2 - h_1 = 0.536\,RT_c + 4{,}696 - 0.969\,RT_c = 3{,}167\,[J/mol]$$

The value calculated using the Redlich–Kwong equation of state differs from this value by 11.0%. What value do you think is more accurate? Solving for heat, we get

$$\boxed{q = \Delta h - w_s = 1067\,[J/mol]}$$

$$\mu J = \left(\frac{\partial P}{\partial T}\right)_H$$

Develop an expression for the enthalpy departure function for a gas that obeys the van der Waals equation of state. Write it in terms of reduced coordinates.

SOLUTION Since the van der Waals equation is explicit in pressure, it is convenient to choose T and v as the independent variables.[7] Consequently, we use infinite volume as the limit of an ideal gas. In analogy to Equation (5.52), we write:

$$\Delta h_{T,v}^{dep} = h_{T,v} - h_{T,P}^{\text{ideal gas}} = \left(h_{T,v} - h_{T,v=\infty}^{\text{ideal gas}} \right) - \left(h_{T,P}^{\text{ideal gas}} - h_{T,v=\infty}^{\text{ideal gas}} \right) = h_{T,v} - h_{T,v=\infty}^{\text{ideal gas}}$$

Thus, we want to find the difference between the enthalpy at the volume of the state of interest and an infinite volume at constant temperature. We write the change in the dependent variable h as

$$dh = \left(\frac{\partial h}{\partial T} \right)_v dT + \left(\frac{\partial h}{\partial v} \right)_T dv$$

At constant temperature,

$$dh_T = \left(\frac{T\partial s + v\partial P}{\partial v} \right)_T dv = \left[T\left(\frac{\partial s}{\partial v} \right)_T + v\left(\frac{\partial P}{\partial v} \right)_T \right] dv = \left[T\left(\frac{\partial P}{\partial T} \right)_v + v\left(\frac{\partial P}{\partial v} \right)_T \right] dv$$

where we have used the fundamental property relation for h, Equation (5.7), and a Maxwell relation, Equation (5.18). The derivatives can be found by differentiating the van der Waals equation $\left(P = \frac{RT}{v-b} - \frac{a}{v^2} \right)$ to give

$$dh_T = \left[T\frac{R}{v-b} + v\left(-\frac{RT}{(v-b)^2} + \frac{2a}{v^3} \right) \right] dv = \left[-\frac{RTb}{(v-b)^2} + \frac{2a}{v^2} \right] dv \qquad \text{(E5.6)}$$

Integrating Equation (E5.6) from the ideal gas state of infinite volume to volume, v gives

$$\Delta h^{dep} = \int_{v=\infty}^{v} \left[-\frac{RTb}{(v-b)^2} + \frac{2a}{v^2} \right] dv = \frac{RTb}{(v-b)} - \frac{2a}{v}$$

We can write the enthalpy departure in terms of reduced coordinates if we substitute in Equations (4.39) and (4.40) for the van der Waals constants a and b:

$$\frac{\Delta h^{dep}}{RT_c} = \frac{T_r}{3v_r - 1} - \frac{9}{4v_r}$$

◀

▶ 5.5 JOULE-THOMSON EXPANSION AND LIQUEFACTION

The pressure dependence of the thermodynamic property enthalpy leads to interesting phenomena in the unrestrained, free expansion of real gases. Figure 5.8 shows a schematic of a gas flowing through a porous plug. It enters the system in state 1 at P_1 and T_1 and it exits at a significantly lower pressure, P_2. We wish to study the effect of this so-called Joule–Thomson expansion on the temperature of the gas at the exit, T_2.

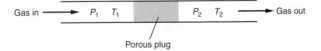

Figure 5.8 Schematic of Joule–Thomson expansion through a porous plug.

[7] See the discussion after Equation (5.37); additionally, inspection of Figure 5.2 shows that derivatives of entropy with volume reduce to P explicit equations of state.

Since the gas spends so little time in the plug, there is no opportunity for heat transfer; thus, we consider this process adiabatic. Additionally, the shaft work is zero. If kinetic energy effects are negligible, the first law for this steady-state, adiabatic throttling process reduces to

$$h_2 - h_1 = \Delta h = 0 \tag{5.73}$$

We call a process that occurs at constant enthalpy, such as this one, **isenthalpic**.

We can determine the change in temperature that results as the pressure decreases in the isenthalpic throttling process if we know the derivative, $(\partial T/\partial P)_h$. We call this relation the **Joule–Thomson coefficient**, μ_{JT}.

$$\mu_{JT} \equiv \left(\frac{\partial T}{\partial P} \right)_h \tag{5.74}$$

Figure 5.9 plots characteristic lines of constant enthalpy (isenthalps) on a TP diagram. In the shaded region, the slopes of the curves are positive; therefore, $\mu_{JT} > 0$, as defined by Equation (5.74). In this region, the temperature will decrease as the pressure decreases during the throttling process. Since the decrease in temperature as the pressure drops corresponds to a decrease in molecular kinetic energy, the molecular potential energy must be increasing or else energy conservation would be violated. We can say the molecules are more stable when they are closer together at the higher pressure and, consequently, that attractive forces are dominant in this region. Conversely, in the nonshaded region, the slopes of the isenthalps are negative; therefore $\mu_{JT} < 0$. The temperature will increase as pressure decreases, indicating that repulsive forces dominate the behavior in this region. These two regions are separated by the **inversion line**, where the slope of T vs. P is zero and where attractive and repulsive interactions exactly balance. For a given pressure, the temperature at which these interactions balance is known as the **Boyle temperature**.

We can use the thermodynamic web to develop an expression for μ_{JT} in terms of PvT property relations and heat capacities. We begin with Equation (5.46):

$$dh = c_p dT + \left[-T \left(\frac{\partial v}{\partial T} \right)_P + v \right] dP \tag{5.46}$$

To liq a gas,
$\mu_{JT} > 0$

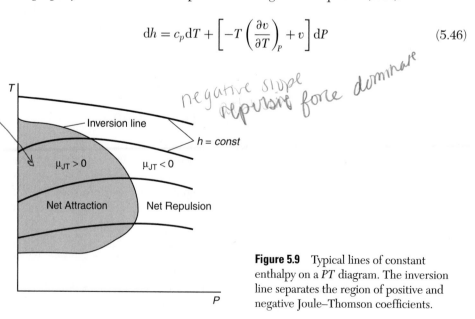

negative slope
repulsive force dominate

Figure 5.9 Typical lines of constant enthalpy on a PT diagram. The inversion line separates the region of positive and negative Joule–Thomson coefficients.

We must use the real heat capacity given by Equation (5.48). During the Joule–Thomson expansion, dh is zero; thus, we can rewrite Equation (5.46) as

use EOS

$$\mu_{JT} = \left(\frac{\partial T}{\partial P}\right)_h = \frac{\left[T\left(\frac{\partial v}{\partial T}\right)_P - v\right]}{c_P} = \frac{\left[T\left(\frac{\partial v}{\partial T}\right)_P - v\right]}{c_P^{\text{ideal gas}} - \int_{P_{\text{ideal gas}}}^{P_{\text{real}}} \left[T\left(\frac{\partial^2 v}{\partial T^2}\right)_P\right] dP} \tag{5.75}$$

where we have substituted in Equation (5.48) for the real heat capacity, since intermolecular interactions are important. If we have an equation of state or other appropriate PvT data and the heat capacity of a fluid, we can evaluate Equation (5.75) for μ_{JT}. Alternatively, we could measure μ_{JT} experimentally using the porous plug in Figure 5.8 to find an unknown c_P.

► **EXAMPLE 5.7**
Joule–Thomson
Coefficient from
the Virial Equation

Develop an expression for the Joule–Thomson coefficient using the pressure-based expansion of the virial equation truncated at the second virial coefficient. Use the corresponding state relationships presented in Chapter 4 for the temperature dependence of B to develop a generalized correlation for μ_{JT}.

SOLUTION To use Equation (5.75), we need to write the virial equation in a form that is explicit in volume. This can be done by rearranging Equation (4.59):

$$v = \frac{RT}{P}(1 + B'P) = \frac{RT}{P} + B'RT \tag{E5.7A}$$

We then take the first and second derivatives:

$$\left(\frac{\partial v}{\partial T}\right)_P = \frac{R}{P} + RB' + RT\left(\frac{dB'}{dT}\right) \tag{E5.7B}$$

and

$$\left(\frac{\partial^2 v}{\partial T^2}\right)_P = R\left(\frac{dB'}{dT}\right) + RT\left(\frac{d^2 B'}{dT^2}\right) \tag{E5.7C}$$

Using Equations (E5.7A), (E5.7B), and (E5.7C) in Equation (5.75) gives:

$$\mu_{JT} = \left(\frac{\partial T}{\partial P}\right)_h = \frac{\left[T\left(\frac{\partial v}{\partial T}\right)_P - v\right]}{c_P} = \frac{\left[T\left(\frac{\partial v}{\partial T}\right)_P - v\right]}{c_P^{\text{ideal gas}} - \int_{P_{\text{ideal gas}}}^{P_{\text{real}}} \left[T\left(\frac{\partial^2 v}{\partial T^2}\right)_P\right] dP}$$

$$= \frac{RT^2\left(\frac{dB'}{dT}\right)}{c_P^{\text{ideal gas}} - \int_{P_{\text{ideal gas}}}^{P_{\text{real}}} \left\{T\left[R\left(\frac{dB'}{dT}\right) + RT\left(\frac{d^2 B'}{dT^2}\right)\right]\right\} dP} \tag{E5.7D}$$

To evaluate this expression for μ_{JT}, we can use the corresponding state relations given in Chapter 4. Using Equations (4.58), (4.62), and (4.63), we get

$$B' = \frac{B}{RT} = \frac{B_r T_c}{P_c T} = \frac{B_r}{P_c T_r} = \frac{B^{(0)} + \omega B^{(1)}}{P_c T_r} \tag{E5.7E}$$

Substituting the generalized relations of Equations (4.64) and (4.65) into Equation (E5.7E) gives

$$B' = \frac{1}{P_c}\left[\left(\frac{0.083}{T_r} - \frac{0.422}{T_r^{2.6}}\right) + \omega\left(\frac{0.139}{T_r} - \frac{0.172}{T_r^{5.2}}\right)\right] \tag{E5.7F}$$

We then take the first and second derivatives:

$$\left(\frac{dB'}{dT}\right) = \frac{1}{P_c}\left[\left(-\frac{0.083}{TT_r} + \frac{1.097}{TT_r^{2.6}}\right) + \omega\left(-\frac{0.139}{TT_r} + \frac{0.894}{TT_r^{5.2}}\right)\right] \tag{E5.7G}$$

and

$$\left(\frac{d^2B'}{dT^2}\right) = \frac{1}{P_c}\left[\left(\frac{0.166}{T^2T_r} - \frac{3.950}{T^2T_r^{2.6}}\right) + \omega\left(\frac{0.278}{T^2T_r} - \frac{5.545}{T^2T_r^{5.2}}\right)\right] \tag{E5.7H}$$

Substituting Equations (E5.7G) and (E5.7H) into (E5.7D) and simplifying gives

$$\mu_{JT} = \cfrac{-\cfrac{RT}{P_c}\left[\left(-\cfrac{0.083}{T_r} + \cfrac{1.097}{T_r^{2.6}}\right) + \omega\left(-\cfrac{0.139}{T_r} + \cfrac{0.994}{T_r^{5.2}}\right)\right]}{c_P^{\text{ideal gas}} - \displaystyle\int_0^P \left[\cfrac{R}{P_c}\left[\left(\cfrac{0.083}{T_r} - \cfrac{2.853}{T_r^{2.6}}\right) + \omega\left(\cfrac{0.139}{T_r} - \cfrac{4.651}{T_r^{5.2}}\right)\right]\right]dP} \tag{E5.7I}$$

where we have set P_{ideal} to zero. Finally, integrating and substituting in $T = T_r T_c$, we get:

$$\mu_{JT} = \cfrac{-\cfrac{T_c}{P_c}\left[\left(-0.083 + \cfrac{1.097}{T_r^{1.6}}\right) + \omega\left(-0.139 + \cfrac{0.994}{T_r^{4.2}}\right)\right]}{\cfrac{c_P^{\text{ideal gas}}}{R} - P_r\left[\left(\cfrac{0.083}{T_r} - \cfrac{2.853}{T_r^{2.6}}\right) + \omega\left(\cfrac{0.139}{T_r} - \cfrac{4.651}{T_r^{5.2}}\right)\right]} \tag{E5.7J}$$

Equation (E5.7J) presents a generalized relation for μ_{JT}. If the critical properties and acentric factor of a species are known, we can use Equation (E5.7J) to calculate the Joule–Thomson coefficient at a specified state at temperature T and pressure P. ◀

Joule–Thomson expansion can be used to liquefy gases if it is performed in the region where $\mu_{JT} > 0$ to the left of the inversion line in Figure 5.9. Liquefaction is an important process industrially: there is a significant market for liquefied gases. For example, liquid nitrogen, helium, and hydrogen are often used to remove energy from cryogenic systems. Additionally, separation of nitrogen and oxygen from air can be accomplished by liquefaction. However, the temperatures at which these gases condense are quite low. He condenses at 4.4 K while N_2 condenses at 77 K. The liquefaction of these gases can require a significant amount of refrigeration.

A schematic of such a liquefaction process is shown in Figure 5.10a. The gas is first compressed from state 1 to 2 to increase its pressure. However, during compression, the temperature of the gas also rises. It is then cooled from state 2 to state 3 to lower its temperature. These two processes are intended to bring it to the shaded region in Figure 5.9 and to put it in a state where a throttling process will bring it into the two-phase region. It now goes through an isenthalpic Joule–Thomson expansion, from state 3 to state 4, where the temperature drops low enough to lead to condensation. The vapor and liquid streams at states 5 and 6, respectively, are then separated. An improvement to the liquefaction process is shown in Figure 5.10b. In this process, an additional heat exchanger is employed to recover the refrigeration from the noncondensed gas. This gas is then recycled. The process depicted in Figure 5.10b is known as the Linde process.

*Combination
of process equip*

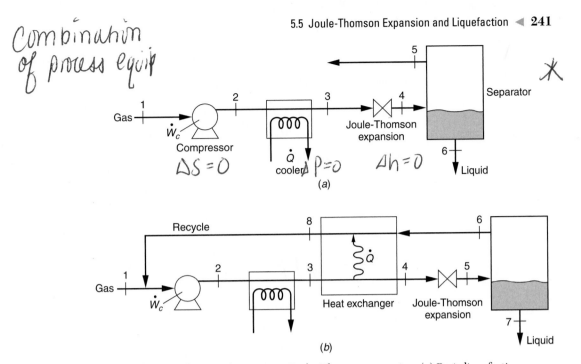

Figure 5.10 Liquefaction of gases using Joule–Thomson expansion. (*a*) Basic liquefaction process using Joule–Thomson expansion and (*b*) Linde process.

▶ **EXAMPLE 5.8**
Liquefaction of N₂ by
Joule–Thomson
Throttling

Consider the liquefaction of N_2 by Joule–Thomson throttling. If the inlet to the expansion valve is at $T_1 = -122°C$ and $P_1 = 100$ bar and the outlet is at 1 bar, determine the percentage of N_2 that is liquefied.

SOLUTION We can solve this problem using the generalized charts for enthalpy departure. Looking up properties for N_2 from Appendix A.1, we get

$$T_c = 126.2\,[\text{K}]$$

$$P_c = 33.8\ \text{bar}$$

$$\omega = 0.039$$

Since the acentric factor, ω, is small, we will use only the simple fluid term in the generalized correlations. The reduced coordinates for state 1 are

$$T_{1,r} = \frac{T_1}{T_c} = \frac{151\,[\text{K}]}{126.2\,[\text{K}]} = 1.20 \quad \text{and} \quad P_{1,r} = \frac{P_1}{P_c} = \frac{100\,[\text{bar}]}{33.8\,[\text{bar}]} = 3.0$$

We can then find the values for the enthalpy departure term in Table C.1 in Appendix C:

$$\left[\frac{h_{T_{1,r},P_{1,r}} - h_{T_{1,r},P_{1,r}}^{\text{ideal gas}}}{RT_c} \right] = -2.81 \tag{E5.8A}$$

For state 2, we have

$$P_{2,r} = \frac{P_2}{P_c} = \frac{1\,[\text{bar}]}{33.8\,[\text{bar}]} = 0.03$$

Looking at Figure 5.4, we see that the two-phase region is at

$$T_{2,r} = 0.61$$

so
$$T_2 = T_{2,r}T_c = 0.61 \times 126.2 \, [\text{K}] = 77 \, [\text{K}]$$

Applying Equation (5.50) for this isenthalpic process:

$$\frac{h_2 - h_1}{RT_c} = 0 = \frac{-\Delta h_{T_1,P_1}^{\text{dep}}}{RT_c} + \frac{\Delta h_{T_1 \to T_2}^{\text{ideal gas}}}{RT_c} + \frac{\Delta h_{T_2,P_2}^{\text{dep}}}{RT_c} \qquad (E5.8B)$$

To find the ideal gas enthalpy change, we need to look up the heat capacity from Appendix A.2:

$$\frac{c_P}{R} = 3.28 + 0.593 \times 10^{-3}T$$

Therefore,

$$\frac{\Delta h_{T_1 \to T_2}^{\text{ideal gas}}}{RT_c} = \frac{1}{T_c} \int_{T_1}^{T_2} \frac{c_P}{R} dT = \frac{1}{126.2} \int_{151}^{77} [3.28 + 0.593 \times 10^{-3}T] dT = -1.28 \qquad (E5.8C)$$

Rearranging Equation (E5.8B) and plugging in values from Equations (E5.8A) and (E5.8C), we get

$$\frac{\Delta h_{T_2,P_2}^{\text{dep}}}{RT_c} = \frac{\Delta h_{T_1,P_1}^{\text{dep}}}{RT_c} - \frac{\Delta h_{T_1 \to T_2}^{\text{ideal gas}}}{RT_c} = -1.53$$

It is useful to view this process on an hP generalized enthalpy chart, as shown in Figure E5.8. The "lever rule" is illustrated.

The quality can be found by:

$$\Delta h^{\text{dep}} = (1 - x)\Delta h_l^{\text{dep}} + x\Delta h_v^{\text{dep}}$$

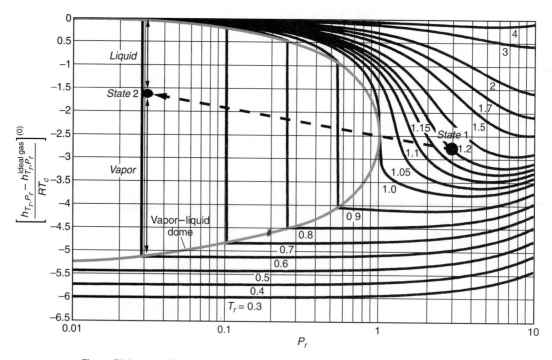

Figure E5.8 Liquefaction process of N_2 from state 1 to state 2.

Solving for x, we get

$$x = \frac{\Delta h^{\text{dep}} - \Delta h_l^{\text{dep}}}{\Delta h_v^{\text{dep}} - \Delta h_l^{\text{dep}}} = \frac{-1.5 + 5.1}{-0.1 + 5.1} = 0.72$$

Since the quality represents the fraction of vapor, we conclude that roughly 28% the inlet stream is liquefied. ◄

[handwritten note: ✕ Read! this]

► **5.6 SUMMARY**

[handwritten note: Remember Second laws energy balance]

In this chapter, we developed a **thermodynamic web** to relate measured, fundamental, and derived thermodynamic properties. The web allows us to use available property data to solve first- and second-law problems for nonideal behavior. Typically, we want relationships between fundamental and derived thermodynamic properties, such as u, s, and h, and things we can measure, such as measured properties P, v, T, or quantities in which measured data are typically reported, such as c_v, c_P, β, κ, and equations of state. In solving these problems, it is often necessary to construct **hypothetical paths** to calculate the change in a desired property between two states. Similarly, the approach for developing solutions to the phase equilibria and chemical reaction equilibria problems in the second half of the text will rely on an ability to exploit property relations and form paths that allow us to use appropriate measured data.

For a system with constant composition, the two properties that we choose to constrain the state of the system become the **independent variables**. We can write the differential change of any other property, the **dependent variable**, in terms of these two properties, as illustrated by Equation (5.4). From a combined form of the first and second laws, we developed the fundamental property relations. We then used the rigor of mathematics to allow us to form this intricate *web* of thermodynamic relationships. Included in the web are the Maxwell relations, the chain rule, derivative inversion, the cyclic relation, and Equations (5.22) through (5.24). A set of useful relationships relating partial derivatives with T, P, s, and v is summarized in Figure 5.2. We use these relationships to solve first- and second-law problems similar to those in Chapters 2 and 3, but for real fluids.

Departure functions often provide us a convenient path to calculate the nonideal contribution to property changes for real gases (or liquids). The departure function of any thermodynamic property is the difference in that property between the real, physical state in which the species exists and that of a hypothetical ideal gas at the same T and P. On a molecular level, we can consider the departure function to represent the change in the value of a property if we could "turn off" the intermolecular interactions in the real fluid. Plots of the simple fluid and correction terms for the enthalpy departure function are presented in Figures 5.4 and 5.5, respectively. Tables of their values are presented in Appendix C (Tables C.3 and C.4). Analogous data for entropy departure are presented in Figures 5.6 and 5.7 and in Appendix C (Tables C.5 and C.6). These values are obtained from the Lee–Kesler equation of state.

Joule–Thomson expansion results from the unrestrained, free expansion of real gases. Such a process occurs at constant enthalpy and is termed **isenthalpic**. We can determine the change in temperature that results as the pressure decreases in the isenthalpic throttling process if we know the **Joule–Thomson coefficient**, $\mu_{\text{JT}} = (\partial T / \partial P)_h$. Joule–Thomson expansion is the basis for **liquefaction** processes, such as those shown in Figure 5.10.

► **5.7 PROBLEMS**

5.1 Write equations analogous to Equation (5.5a) for the exact differential of internal energy, du, in terms of each of the following sets of independent variables:

(a) $u = u(T, P)$

(b) $u = u(T, s)$

(c) $u = u(h, s)$

5.2 Using the thermodynamic web, show that for an ideal gas

$$u = u(T \text{ only})$$

5.3 For an ideal gas, show that

$$c_P = c_v + R$$

5.4 Show that an ideal gas follows the cyclic relationship in P, v, and T.

5.5 Evaluate the derivative:

$$\left(\frac{\partial h}{\partial v}\right)_{T,P}$$

for a pure species that follows the Peng–Robinson equation of state. The subscript T,P indicates that both temperature and pressure are held constant.

5.6 Using the van der Waals equation, find an expression for the derivative

$$\left(\frac{\partial h}{\partial T}\right)_{s}$$

in terms of a, b, c_P, R, v, and T.

5.7 Consider the following equation of state:

$$\frac{Pv}{RT} = 1 + B'P + C'P^2$$

where B' and C' are constant parameters with no temperature dependence. In terms of B', C', R, P, T, and c_P, find the following expressions:

$$\left(\frac{\partial h}{\partial P}\right)_{T}, \left(\frac{\partial h}{\partial P}\right)_{s}, \left(\frac{\partial h}{\partial T}\right)_{P}, \left(\frac{\partial h}{\partial T}\right)_{s}$$

5.8 Consider the following property relation:

$$\left(\frac{\partial u}{\partial P}\right)_{s}$$

(a) Come up with a physical process on a system which is described by the relation above. Sketch the process and describe it as completely as needed so that this relation holds.

(b) Based on the process you chose in part (a), do you think the relation has a positive value, has a negative value, or is 0. **Justify your answer**.

5.9 Verify that

(a) $\left(\dfrac{\partial P}{\partial T}\right)_{v} = \dfrac{\beta}{\kappa}$

(b) $c_P - c_v = \dfrac{v T \beta^2}{\kappa}$

5.10 Use the result of Problem 5.9(b) to calculate the difference, $c_P - c_v$, for liquid acetone at 20°C and 1 bar. Data can be found in Table 4.4. Repeat for benzene and copper. How do the values you obtain compare to the value for c_P?

5.11 Verify that

$$\left(\frac{\partial s}{\partial v}\right)_{T} = \frac{\beta}{\kappa}$$

and

$$\left(\frac{\partial s}{\partial P}\right)_{T} = -\beta v$$

5.12 Before the proliferation of personal computers, it was often convenient to summarize thermodynamic property data in the form of graphic diagrams. The Mollier diagram presents h (y-axis) vs. s (x-axis). Obtain an expression for the slope of an isochor (constant-volume line) on a Mollier diagram for (a) an ideal gas (in terms of T, v, c_v, c_P, and R); (b) a van der Waals gas (in terms of T, v, c_v, c_P, a, b and R).

5.13 Develop a general relationship for the change in temperature with respect to pressure at constant entropy:

$$\left(\frac{\partial T}{\partial P} \right)_s$$

(a) Evaluate the expression for an ideal gas.

(b) From the result in part (a), show that for an ideal gas with constant c_P, an isentropic expansion from state 1 and state 2 yields Equation (2.92).

(c) Evaluate the expression for a gas that obeys the van der Waals equation of state.

5.14 Derive Equation (5.48).

5.15 Your company has just developed a new refrigeration process. This process uses a secret gas, called Gas A. You are told that you need to come up with thermodynamic property data for this gas. The following data have already been obtained for the superheated vapor:

	$P = 10$ bar		$P = 12$ bar	
$T\,[°C]$	$v\,[\text{m}^3/\text{kg}]$	$s\,[\text{kJ}/(\text{kg K})]$	$v\,[\text{m}^3/\text{kg}]$	$s\,[\text{kJ}/(\text{kg K})]$
80	0.16270	5.4960	0.13387	?
100	0.17389		0.14347	

As *accurately* as you can, come up with a value for s in the table above. Clearly indicate your approach and state any assumptions that you make. Do *not* assume ideal gas behavior.

5.16 Propane at 350°C and 600 cm^3/mol is expanded in an isentropic turbine. The exhaust pressure is atmospheric. What is the exhaust temperature? PvT behavior has been fit to the van der Waals equation with

$$a = 92 \times 10^5 \, [(\text{atm cm}^6)/\text{mol}^2]$$

$$b = 91 \, [\text{cm}^3/\text{mol}]$$

(a) Solve this using T and v as the independent variables, that is,

$$s = s(T, v)$$

(b) Solve this using T and P as the independent variables.

5.17 You need to design a heater to preheat a gas flowing into a chemical reactor. The inlet temperature is 27°C and the inlet pressure is 50 bar. You desire to heat the gas to 227°C and 50 bar. You are provided with an equation of state for the gas:

$$z = 1 + \frac{aP}{\sqrt{T}} \qquad \text{with } a = -0.070 \, [\, K^{1/2}/\text{bar}]$$

and with *ideal gas* heat capacity data:

$$\frac{c_P}{R} = 3.58 + 3.02 \times 10^{-3} T \qquad \text{where } T \text{ is in } [K]$$

(a) Under these conditions, do attractive forces or repulsive forces dominate the behavior of this gas. **Explain**.

(b) As accurately as you can, calculate, in [J/mol], the amount of heat required.

5.18 Consider the piston–cylinder assembly shown below; 250 moles of gas expand isothermally after the removal of a 10,000-kg block.

(a) What is the internal energy change for the expansion process?

(b) What is the entropy change of the universe for this process?

Assume that the PvT behavior can be described by the van der Waals equation with $a = 0.5\,[\mathrm{Jm^3/mol^2}]$ $b = 4 \times 10^{-5}\,[\mathrm{m^3/mol}]$ and that the ideal gas heat capacity has a constant value of $c_P^{\text{ideal gas}} = 35\,\mathrm{J/(mol\ K)}$

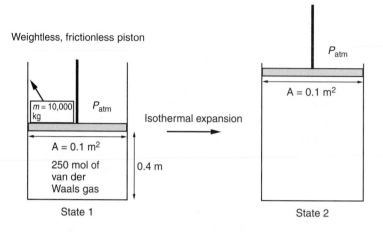

Weightless, frictionless piston

Isothermal expansion

State 1

State 2

5.19 Consider the piston–cylinder assembly shown below. It is well insulated and initially contains two 10,000-kg blocks at rest on the 0.05-m² piston. The initial temperature is 500 K. The ambient pressure is 10 bar. Two moles of gas A are contained in the cylinder. This gas is compressed in a process where another 10,000-kg block is added. The following data are available for gas A:

(i) *Ideal gas* heat capacity of gas A at constant pressure:

$$c_P = 20 + 0.05T$$

where c_p is in $[\mathrm{J/(mol\ K)}]$ and T is in $[\mathrm{K}]$.

(ii) Gas A is can be described by the following equation of state:

$$P = \frac{RT}{v - b} - \frac{aP}{T}$$

with constants

$$a = 25\,\mathrm{K} \quad \text{and} \quad b = 3.2 \times 10^{-5}\,\mathrm{m^3/mol}$$

Determine the temperature of gas A after this process. *Note*: This compression process is *not* isentropic. What is the entropy change of the universe for this process?

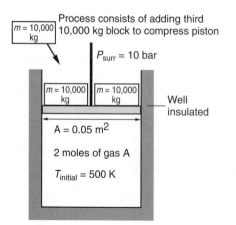

Process consists of adding third 10,000 kg block to compress piston

5.20 One mole of CO is initially contained on one-half of a well-insulated, rigid tank. Its temperature is 500 K. The other half of the tank is initially at vacuum. A diaphragm separates the two compartments. Each compartment has a volume of 1 L. Suddenly, the diaphragm ruptures. Use the van der Waals equation for any nonideal behavior. Answer the following questions:

(a) What is c_v at the initial state?

(b) Do you expect the temperature to increase, decrease, or remain constant. Justify your answer with molecular arguments. Be specific about the nature of the forces involved.

(c) What is the temperature of the final state?

(d) What is the entropy change of the universe for this process?

5.21 A well-insulated, rigid vessel is divided into two compartments by a partition. The volume of each compartment is 0.1 m³. One compartment initially contains 400 moles of gas A at 300 K, and the other compartment is initially evacuated. The partition is then removed and the gas is allowed to equilibrate. Gas A is not ideal under these circumstances but can be described well by the following equation of state:

$$P = \frac{RT}{v - b} - \frac{a}{Tv^2}$$

with constants

$$a = 42\,[(\text{J K m}^3)/\text{mol}^2] \quad \text{and} \quad b = 3.2 \times 10^{-5}\,[\text{m}^3/\text{mol}].$$

You may take the ideal gas heat capacity of gas A to be

$$c_v = (3/2)R$$

Calculate the final temperature.

5.22 Consider filling a gas cylinder with ethane from a high-pressure supply line. Before filling, the cylinder is empty (vacuum). The valve is then opened, exposing the tank to a 3-MPa line at 500 K until the pressure of the cylinder reaches 3 MPa. The valve is then closed. The volume of the cylinder is 50 L. For ethane, use the truncated virial equation of state, in pressure:

$$z = \frac{Pv}{RT} = 1 + B'P$$

with
$$B' = -2.8 \times 10^{-8}\,[\text{m}^3/\text{J}]$$

(a) What is the temperature *immediately* after the valve is closed?

(b) If the cylinder then sits in storage at 293 K for a long time, what is the entropy change of the universe (from the original *unfilled*, state)?

5.23 In analogy to Equation (5.58), develop an expression for the internal energy departure function in dimensionless coordinates:

$$\frac{\Delta u_{T_r,v_r}^{\text{dep}}}{RT_c} = \frac{u_{T_r,v_r} - u_{T_r,v_r}^{\text{ideal gas}}}{RT_c} = ?$$

5.24 Come up with an expression for the enthalpy and entropy departure functions for a gas that follows the Redlich–Kwong equation of state.

5.25 Calculate the enthalpy and entropy departure for water at 400°C and 30 MPa using generalized correlations. Compare these values to those in the steam tables. The ideal heat capacity of steam is useful in this calculation.

5.26 Calculate the enthalpy and entropy change of C_2H_6 from a state at 300 K and 30 bar to a state at 400 K and 50 bar using departure functions.

5.27 Repeat Problem 5.16 using the entropy departure function.

5.28 Methane flowing at 2 mol/min is adiabatically compressed from 300 K and 1 bar to 10 bar. What is the minimum work required?

5.29 Develop a relationship for the Joule–Thomson coefficient in terms of only the thermal expansion coefficient, the heat capacity at constant pressure, and measured thermodynamic properties.

5.30 What is μ_{JT} for an ideal gas?

5.31 Determine expressions for the thermal expansion coefficient, the isothermal compressibility, and the Joule–Thomson coefficient for a gas that obeys the van der Waals equation of state, in terms of T, v, c_v, a, b, and R.

5.32 Determine μ_{JT} for steam at 1 MPa and 300°C using data from the steam tables.

5.33 Use the van der Waals equation of state to plot the inversion line for N_2 on a PT diagram, as schematically shown in Figure 5.9.

5.34 Ethylene is liquefied by a Joule–Thomson expansion. It enters the throttling process at 50 bar and 0°C and leaves at 10 bar. What is the fraction of the inlet stream that is liquefied?

5.35 The speed of sound, V_{sound} [m/s], is formally equal to the partial derivative of pressure with respect to density at constant entropy:

$$V_{sound}^2 = \left(\frac{\partial P}{\partial \rho}\right)_s$$

Show that

$$\left(\frac{\partial P}{\partial \rho}\right)_s = -\left(\frac{\partial P}{\partial v}\right)_s \left(\frac{v^2}{MW}\right)$$

where MW is the molecular weight.

5.36 Based on the definition in Problem 5.35, use the thermodynamic web to come up with an expression for $[V_{sound}]$ in air. What is the value of $[V_{sound}]$ in air at 20 °C? You may consider air to be an ideal gas with $c_P = (7/2)R$. Based on this result, how far away is a bolt of lightning if you hear the thunder four seconds after you see the lightning.

5.37 Based on the definition in Problem 5.35, use the thermodynamic web to come up with an expression and a value for $[V_{sound}]$ in water at 20°C. Use the steam tables for thermodynamic property data of liquid water.

5.38 We are interested in the thermodynamic properties of a strip of rubber as it is stretched (see below). Consider n moles of *pure* ethylene propylene rubber (EPR) that has an unstretched length z_0. If it is stretched by applying a force F, it will obtain an equilibrium length z, given by

$$F = kT(z - z_0)$$

where k is a positive constant.

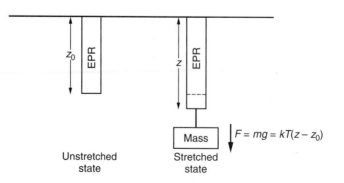

The heat capacity of unstretched EPR is given by

$$c_z = \left(\frac{\partial u}{\partial T}\right)_z = a + bT \qquad \text{where } a \text{ and } b \text{ are constants}$$

(a) Come up with fundamental property relations for dU and dA for this system, where A is the Helmholtz energy $(A = U - TS)$. Recall from mechanics that the work required for a reversible elastic extension is given by

$$\delta W_{rev} = Fdz$$

(b) Develop an expression that relates the change in entropy to the changes in temperature and length, that is, the independent variables z and T (and constants a, b, k, n and z_0). In other words, for $S = S(T, z)$, find dS. *Hint*: You will need to derive an expression for $(\partial S/\partial z)_T$. This can be done with the appropriate Maxwell relationship.

(c) Develop an expression that relates the change in internal energy to the changes in temperature and length. For $U = U(T, z)$, find dU.

(d) Consider the relative energetic and entropic contributions to the isothermal extension of EPR. The energetic force (the component of the force that, on isothermal extension of the rubber, increases the internal energy) is given by

$$F_U = \left(\frac{\partial U}{\partial z}\right)_T$$

while the entropic force is given by

$$F_S = -T\left(\frac{\partial S}{\partial z}\right)_T$$

Come up with expressions for F_U and F_S for EPR.

(e) If the change from the unstretched state to the stretched state above occurred adiabatically, would the temperature of the EPR go up, stay the same, or go down? Explain.

5.39 The process in Example 5.2 indicates that we need to put work *into* the system during an expansion process. Determine whether this result is possible (in a thermodynamic sense); if it is, explain this result physically.

5.40 Gas A expands through an *adiabatic* turbine. The inlet stream flows in at 100 bar and 600 K while the outlet is at 20 bar and 445 K. Calculate the work produced by the turbine. The following data are available for gas A. The *ideal* gas heat capacity for this process is

$$c_P = 30.0 + 0.02T$$

where c_P is in $[J/(mol\ K)]$ and T is in $[K]$. PvT data has been fit to the following equation

$$P(v - b) = RT + \frac{aP^2}{T}$$

where $\qquad a = 0.001\ [(m^3\ K)/(bar\ mol)] \qquad$ and $\qquad b = 8 \times 10^{-5}\ [m^3/mol]$

CHAPTER 6

Phase Equilibria I: Problem Formulation

Learning Objectives

To demonstrate mastery of the material in Chapter 6, you should be able to:

▶ Explain why it is convenient to use the thermodynamic property Gibbs energy to determine pure species phase equilibrium. Discuss the balance between energetic and entropic effects at equilibrium.

▶ Apply the fundamental property relation for Gibbs energy and other tools of the thermodynamic web to predict how the pressure of a pure species in phase equilibrium changes with temperature and how other properties change in relation to one another. Write the Clapeyron equation and use it to relate T and P for a pure species in phase equilibrium. Derive the Clausius–Clapeyron equation for vapor–liquid mixtures, and state the assumptions used. Relate the Clausius–Clapeyron equation to the Antoine equation.

▶ Apply thermodynamics to mixtures. Write the differential for any extensive property, dK, in terms of $m + 2$ independent variables, where m is the number of species in the mixture. Define and find values for pure species properties, total solution properties, partial molar properties, and property changes of mixing.

▶ Define a partial molar property and describe its role in working with mixtures. Calculate the value of a partial molar property for a species in a mixture from analytical and graphical methods. Apply the Gibbs–Duhem equation to relate partial molar properties of different species.

▶ Relate volume, enthalpy, and entropy changes of mixing to the relative intermolecular interactions of like (*i-i*) and unlike (*i-j*) interactions. Define the enthalpy of solution. Calculate the enthalpy of mixing from the enthalpy of solution or vice versa.

▶ Identify the role of the chemical potential—that is, the partial molar Gibbs energy—as the chemical criteria for equilibrium.

▶ 6.1 INTRODUCTION

So far, we have used thermodynamics to form relationships between the states of a system that undergoes certain processes. We can apply the first and second law to both reversible and irreversible processes to get information about (1) how much power is needed or obtained, (2) how much heat has been absorbed or dissipated, or (3) the value of an unknown property (e.g., T) of the final (or initial) state. In the remaining chapters, we examine another type of problem that we can also use thermodynamics to address—the composition a mixture obtains when it reaches equilibrium between coexisting phases or in the presence of chemical reactions.

Chemical engineers routinely deal with processes through which species **chemically react** to form a desired product(s). This product must then be **separated** from the other by-products as well as any reactants that remain. Typical separation schemes involve contact or formation of different phases through which one species of a mixture preferentially segregates. Separations technology is also a major concern in cleaning contaminated environments. Therefore, it is desirable to be able to estimate the degree to which species will react and the degree to which a given species will transfer into a different phase as a function of process conditions.

These problems lead to the second major branch of thermodynamics, which we will now formulate. It deals *only* with equilibrium systems. It should be pointed out that this branch still uses the same observations of nature (conservation of energy and directionality) that we have already studied. In these problems, however, we wish to calculate how species distribute among phases when more than one phase is present (**phase equilibria**) or which type of species distributions systems obtain as they approach equilibrium when molecules in the system chemically react (**chemical reaction equilibria**). We will consider phase equilibria first. These calculations are restricted to equilibrium systems; therefore, they give information on the direction of the driving force for a given system (i.e., the system will spontaneously move toward its equilibrium state) but no information on the rate at which it will reach equilibrium.

A generic representation of the phase equilibria problem is illustrated in Figure 6.1. In this picture, α and β can represent any phase: solid, liquid, or vapor. We may be interested in any of the following: vapor–liquid, liquid–liquid, liquid–solid, gas–solid, or solid–solid equilibrium. Can you think of an example of each type? We consider a closed system, since strictly speaking only closed systems can be in thermodynamic equilibrium. In an open system, mass flows into and out of the boundary. For mass to flow, some type of driving force, such as a pressure gradient, is necessary. However, we cannot simultaneously have a pressure gradient and mechanical equilibrium (equal pressure). Thus, we will develop our formalism for closed systems. Equilibrium analysis still plays an important part in open systems, since it tells us the driving force for the transfer of species from one phase to another.

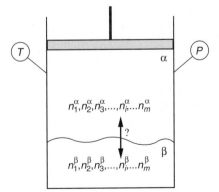

Figure 6.1   Generic phase equilibria problem.

As we learned in Chapter 1, for a system to be in mechanical equilibrium there cannot be a pressure gradient. Similarly, in thermal equilibrium, there are no temperature gradients in the system. Therefore, we can write

$$T^\alpha = T^\beta \qquad \text{Thermal equilibrium} \qquad (6.1)$$

and

$$P^\alpha = P^\beta \qquad \text{Mechanical equilibrium} \qquad (6.2)$$

These two criteria for equilibrium are obvious and therefore straightforward to formulate; They deal with measurable properties. To see that these conditions represent criteria for equilibrium, you may ask, for example, "What would happen if $T^\alpha > T^\beta$?" Energy flows from hot to cold and will, therefore, flow from phase α to β until the temperatures equilibrate. A similar argument can be made for pressure and mechanical equilibrium.[1] We will take thermal and mechanical equilibrium for granted in the following discussion; hence, in formulating the phase equilibria problem, we need only measure the temperature and pressure of one phase, and these values must apply to the entire system. This concept is illustrated schematically in Figure 6.1, where temperature and pressure measurements that are made only to phase α apply to the entire system. A piston–cylinder assembly is used to remind us that the system must be able to change in volume to accommodate thermal and mechanical equilibrium.

The driving force for species transfer is not so apparent. This chapter will focus on the following questions:

- What are the criteria for chemical equilibria of species i?

$$?^\alpha_i = ?^\beta_i \qquad \text{for chemical equilibrium?}$$

- How do I use these criteria to solve phase equilibria problems (with T and P known)?

Before we begin, note that neither mole fraction nor concentration, both measurable properties, represents the driving force for species transfer between phases (as temperature difference represents the driving force for energy transfer between phases). For example, consider an air–water system in phase equilibrium between the vapor and liquid phases. It would be absurd to think that oxygen will transfer from the air into the water until the mole fraction was 0.21 in the water, or conversely that water would transfer into the air until the vapor was almost all water! Unfortunately, the thermodynamic property that drives a system toward chemical equilibrium, unlike thermal or mechanical equilibrium, is not a measurable property.

There are two distinct issues imbedded in the to the problem described above. (1) We must learn why different phases coexist, be it for pure species or for mixtures. (2) We must also develop a formalism to account for mixtures and changing composition; up to this point, we have only dealt with constant-composition systems. Rather than tackle both these tasks simultaneously, our approach is to isolate each problem, solve it, and then integrate them together to solve the entire phase equilibria problem. Figure 6.2 illustrates our solution strategy. In Section 6.2, we will learn why two phases coexist at equilibrium. We will do this for the simplest case possible—a pure species. In Section 6.3, we will learn how to carefully (and formally) describe the thermodynamics of mixtures. Once we have learned these two concepts, they will be integrated together in Section 6.4 to formulate the solution to the problem posed in Figure 6.1.

---

[1] These criteria can be shown formally through the maximization of entropy.

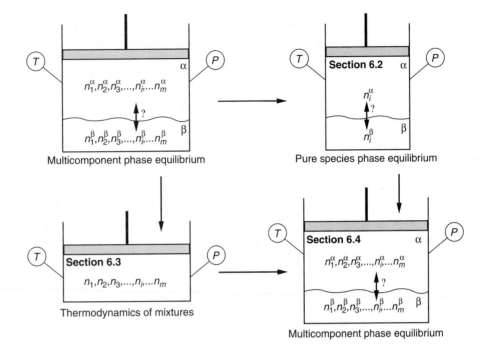

Figure 6.2 Reduction of the multicomponent phase equilibria problem.

▶ 6.2 PURE SPECIES PHASE EQUILIBRIUM

Gibbs Energy as a Criterion for Chemical Equilibrium

We will begin the discussion of equilibrium systems by considering phase equilibrium of a pure species (as shown in the upper right of Figure 6.2). Some elements of this topic were discussed in Chapter 1. In this section, we will develop a criterion that tells us when two phases coexist at equilibrium. The second law of thermodynamics can give us some insight into answering this question. We saw in Chapter 3 that entropy relates to the directionality of a process. Equilibrium represents the state at which the system has no tendency to change (i.e., no further directionality). The equilibrium state occurs when the entropy of the *universe* is a maximum. Thus a pure species at temperature T and pressure P will exist in equilibrium between two phases if the entropy of the *universe* is greater with both phases present than that if the system were in either single phase exclusively. However, this approach is inconvenient since in dealing with the entropy of the entire *universe*, we need to calculate changes in entropy of the surroundings as well as the system. We would prefer a way to tell if we have phase equilibrium by considering *only* properties of the system. In this section, a combination of the first law and second law is used to develop a new property, Gibbs energy, G. The Gibbs energy is useful because by just looking at G of each phase in the *system*, we can determine when two phases coexist at equilibrium. We will now explore how G provides this information.

Consider a closed system composed of pure species i in mechanical and thermal equilibrium and, therefore, at constant T and P. In most phase equilibria problems, the only work is Pv work. If we assume there is *only Pv work*, the first law in differential form can be written as:

$$dU_i = \delta Q + \delta W = \delta Q - P dV_i \tag{6.3}$$

We introduce the subscript "i" to denote that the analysis is performed for pure species i. The rationale for this nomenclature will be discussed further when mixtures are addressed in Section 6.3. For mechanical equilibrium, we have constant pressure, that is, $dP = 0$. Therefore, we can subtract $V_i dP$ from the right-hand side of Equation 6.3:

$$dU_i = \delta Q - P dV_i - V_i dP = \delta Q - d(PV_i) \tag{6.4}$$

Bringing the $d(PV_i)$ term to the left-hand side and applying the definition of enthalpy, Equation (6.4) reduces to

$$dH_i = \delta Q \tag{6.5}$$

The second law can be written

$$dS_i \geq \frac{\delta Q}{T} \tag{6.6}$$

where the inequality holds for irreversible (directional) processes while the equality holds for reversible processes. Solving for δQ, combining Equation (6.6) with (6.5), and rearranging gives

$$0 \geq dH_i - T dS_i \tag{6.7}$$

Thermal equilibrium requires that we be at constant temperature ($dT = 0$); thus, the term $S dT$ may be subtracted from the right-hand side of Equation (6.7):

$$0 \geq dH_i - T dS_i - S_i dT = d(H_i - TS_i) \tag{6.8}$$

We recognize the group of variables on the right-hand side of Equation (6.8) from Chapter 5 as the derived thermodynamic property Gibbs energy, G_i:

$$G_i \equiv H_i - TS_i \tag{5.3}$$

Hence, combination of the first and second laws attributes the following behavior to our closed system:

$$\boxed{0 \geq (dG_i)_{T,P}} \tag{6.9}$$

The subscripts "T" and "P" remind us that this expression is valid only at constant temperature and pressure, which are the criteria for thermal and mechanical equilibrium, respectively. Equation (6.9) says that for a *spontaneous* process, the **Gibbs energy** of a **system** at constant pressure and temperature always gets smaller (or stays the same); it never increases.[2] The system wants to minimize its Gibbs energy. Equilibrium is the

---

[2] Alternatively, the development of Gibbs energy can also include the more general case of non-Pv work, W^*. The analog to Equation (6.9) would then read:

$$\delta W^* \geq (dG_i)_{T,P} \tag{6.9*}$$

Integration of Equation (6.9*) shows that a process that raises the Gibbs energy is possible, but it will *not* occur spontaneously; rather, it will only occur with a commensurate input of work, W^*, of at least the same amount that the Gibbs energy rises. This relationship can tell us, for example, the minimum amount of work required to separate two gases that have spontaneously mixed. Conversely, Equation (6.9*) provides us the maximum useful work we can get from a spontaneous process at constant T and P. Section 9.6 treats a case where non-Pv work is important.

state at which the system no longer changes properties; therefore, equilibrium occurs at minimum Gibbs energy. If we have two phases α and β, we can write the total Gibbs energy of pure species i as

$$G_i = n_i^\alpha g_i^\alpha + n_i^\beta g_i^\beta \tag{6.10}$$

where n_i^α and n_i^β refer to the number of moles of i in phases α and β, respectively. Differentiating this expression and applying the inequality represented in Equation (6.9), we get:

$$dG = d(n_i^\alpha g_i^\alpha + n_i^\beta g_i^\beta) = n_i^\alpha \overset{0}{dg_i^\alpha} + g_i^\alpha dn_i^\alpha + n_i^\beta \overset{0}{dg_i^\beta} + g_i^\beta dn_i^\beta \leq 0 \tag{6.11}$$

Two independent properties constrain the state of each phase of the pure substance i; thus, at a given T and P, g_i^α and g_i^β are constant. Consequently, the first and third terms in Equation (6.11) go to zero.

Since we have a closed system, a species leaving one phase must be added to the other phase, so

$$dn_i^\alpha = -dn_i^\beta \tag{6.12}$$

Substituting Equation (6.12) into Equation (6.11) gives

$$(g_i^\beta - g_i^\alpha)dn_i^\beta \leq 0 \tag{6.13}$$

From Equation (6.13), we can infer how the species in a system respond to approach equilibrium. Consider a system that initially has species in both phases α and β. If g_i^β is larger than g_i^α, dn_i^β must be less than zero to satisfy the inequality. Physically, species i will transfer from phase β to phase α, lowering the Gibbs energy of the system. Species will transfer until only phase α is present, and we will not have phase equilibrium. Conversely if g_i^α is larger than g_i^β, only phase β will be present at equilibrium. However, if the Gibbs energies of both phases are equal, Equation (6.13) becomes an equality and the system has no impetus to change. This condition represents equilibrium. Thus, the **criterion for chemical equilibrium** is when the Gibbs energy is at the minimum:

$$g_i^\alpha = g_i^\beta \tag{6.14}$$

Note the similarity between Equation (6.14) and Equations (6.1) and (6.2). *The derived variable, Gibbs energy, although not directly measurable, provides the same information with regard to chemical equilibrium that temperature does for thermal equilibrium and pressure for mechanical equilibrium!*

Roles of Energy and Entropy in Phase Equilibria

We will look at the simple system shown in Figure 6.3 to explore the implications of the criterion discussed above. Specifically, we will examine the interplay between energetic effects and entropic effects in determining the Gibbs energy. Figure 6.3 shows a system in which an ideal gas and a perfect crystal coexist at temperature T and pressure P. The solid consists of pure a, while the ideal gas is comprised of both a and b. Species b is noncondensable and is not incorporated into the solid lattice. We will explore how the Gibbs energy of this system depends on the fraction of a that is vaporized. As we have just seen, the system obtains equilibrium in the state where the Gibbs energy is minimized. However, Gibbs energy can be lowered by either lowering h or increasing s. The right side

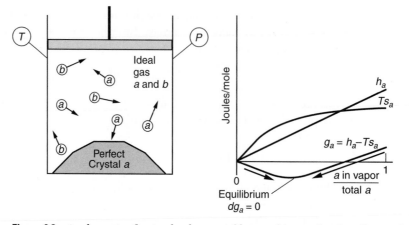

Figure 6.3 A schematic of a simple phase equilibria problem and a plot of some thermodynamic properties at a temperature where both solid and vapor phases coexist at equilibrium.

of Figure 6.3 shows plots of various properties of a in this system vs. fraction of species a in the vapor. In these plots, enthalpy is related to the energetics of the system. Recall from Section 2.6 that enthalpy is the appropriate property to quantify the energetics of a closed system at constant pressure, as it combines internal energy with Pv work. The entropy is multiplied by T to give it *units of energy*, so we can plot it on the same graph. The Gibbs energy is found by subtracting Ts from h.

First, consider the enthalpy of the system. Since we have an ideal gas, there is no intermolecular potential energy in the gas phase, just molecular kinetic energy, which is solely a function of temperature. On the other hand, the solid is more stable than the gas and has lower enthalpy due to the attractive forces between bonded a atoms. The enthalpy in the entire system can be related to the fraction of atoms in the solid. Consider the vaporization process. As a bond in the solid is broken to increase the fraction of a in the vapor, the system increases in molecular potential energy by an amount proportionate to the bond enthalpy. Each atom that leaves the solid increases the enthalpy in the system by the same amount. This proportionate increase manifests itself in a linear relation between the enthalpy and the fraction of a in the vapor phase, as illustrated in Figure 6.3. Now consider the entropy of the system. The entropy will not increase linearly, as enthalpy did. In the case that all a is in the solid and all the vapor is b, we have as ordered a system as we possible can. The first atom of solid that evaporates will cause a large increase in disorder, since the gas will be able to take on many more different configurations. As more and more atoms vaporize, they enter a gas that has more and more a; therefore, for each additional atom, the additional increase in entropy is less and less. Intuitively, we can see that there is a lot more *randomness* introduced by the first atom leaving a pure solid and going into the vapor than when there are already many a atoms in the gas phase.[3] This relationship was described in Chapter 3 and can be formally developed using statistical mechanics. It is also plotted in Figure 6.3.

In this system, there are two tendencies that oppose each other. The perfect crystal with no fraction of a in vapor is the system of minimum energy (or enthalpy), but the

---

[3] Alternatively, we can view the increase in entropy when species a goes into the gas without reference to species b. Since the gas is ideal, species a does not know that b is there. However, the partial pressure of a increases as more and more a goes into vapor phase. In Example 6.8, we will quantitatively show that the increase in entropy goes as $y_a \ln y_a$ and is, indeed, not linear.

gas with all *a* vaporized maximizes entropy. To see the state that the system will obtain at equilibrium, we need to determine the extent to which each effect dominates. To do so, we need only consider the Gibbs energy ($g_a = h_a - Ts_a$) for the system. The optimal compromise between minimum enthalpy and maximum entropy occurs with part solid and part vapor. At low vapor fractions, the increase in entropy caused by introducing *a* into the gas phase more than offsets the increase in enthalpy caused by breaking a bond in the solid. Therefore, the Gibbs energy is lowered, so this process will occur spontaneously, and the system will tend to sublimate. Conversely, at high vapor fractions, the increase in stability caused by forming a solid more than adequately compensates for the decrease in entropy associated with losing an atom from the vapor; thus, the gas will tend to crystallize. These two effects are indicated by arrows in Figure 6.3. The system will exist with two phases in equilibrium in between these extremes with $g^s = g^{v4}$, as illustrated.

Now consider a much lower temperature. Since entropy is multiplied by T, the effect of randomness (entropy) cannot compete with the minimization of enthalpy; in this case, the equilibrium state (minimum g) occurs when the system exists as a pure solid with no *a* in vapor. Hence all the vapor crystallizes. Conversely, at very high temperatures, the effect of entropy dominates and only vapor exists. This behavior is consistent with our experience. Solids exist at low temperature and as the temperature is increased, they sublime (or melt).

The second law states that entropy goes to a maximum; yet if our system were at maximum entropy, the solid phase would not exist. How do you resolve this paradox? [*Hint*: How is the temperature being kept constant?]

▶ **EXAMPLE 6.1**
Role of Gibbs Energy
in Biological Systems

We have seen that the Gibbs energy determines whether a process can occur spontaneously. This concept can be applied to understand aspects of biological systems. Use the Gibbs energy to show why the proteins that control complex living organisms are not stable at high ambient temperatures.

SOLUTION The structure of proteins can be considered in different levels of organization. Proteins are long-chain polymers in which a sequence of amino acids are linked by peptide bonds—a bond between the carbon atom in the carboxyl group from one amino acid and the nitrogen atom in the amino group from the neighboring amino acid. This chain forms the primary structure of the protein. The polymer molecule then folds back upon itself, forming intramolecular coulombic, hydrogen, and van der Waals bonds between different, non-neighboring amino acids in the chain. The first folding level forms the secondary structure of the protein. Common secondary structures include α-helices and β-pleated sheets. The third and fourth folding levels form the tertiary and quaternary structure of the protein. The resulting well-defined structure has the precise chemical properties that enable the protein to perform its specific function.

We consider a native protein (n) with its structure intact and a denatured protein (d) that no longer has its well-defined structure and cannot perform its specified function. The native protein remains intact only within a limited temperature range above room temperature. We can understand this result in terms of the Gibbs energy of the protein. The intramolecular bonds that define the structure of a protein make it energetically favorable relative to its denatured analog in which some of the bonds of the higher levels of organization are broken; thus,

$$h_d > h_n \qquad\qquad\qquad \text{(E6.1A)}$$

---

[4] Intuitively, we can see that this equilibrium problem can be cast as a "pure" species problem since the gas phase is ideal and *a* does not know *b* is there. After we treat mixtures, in general, we will formally see that this relation holds in this special case. It is presented here to help form a conceptual framework for *g*.

On the other hand, the denatured protein is no longer limited by its specific, constrained three-dimensional structure and may undertake many more possible configurations. Hence, its entropy is much higher, that is

$$s_d > s_n \tag{E6.1B}$$

To see whether a protein will spontaneously denature, we consider the Gibbs energy difference between its native and denatured forms. Applying the definition of Gibbs energy:

$$\Delta g = g_d - g_n = (h_d - h_n) - T(s_d - s_n) \tag{E6.1C}$$

The sign of the Gibbs energy change in Equation (E6.1C) depends on the temperature of the system. At lower temperature, the first term in Equation (E6.1C) dominates and inspection of Equation (E6.1A) shows $\Delta g > 0$. Thus, the protein will not spontaneously denature. As the temperature becomes higher, the second term becomes more important and $\Delta g < 0$ since the entropy of the denatured protein is higher. Thus, at higher temperatures, proteins spontaneously denature. The trade-off between the energetically favored hydrogen bonds and electrostatic and van der Waals interactions and the entropically favored randomness of the denatured state determines the temperature at which a protein is no longer stable. In biological systems, this balance commonly occurs between 50° and 70°C.

The above analysis is simplified since we have not considered the interactions of the protein with the solution in which it sits; yet it is essentially valid. However, in many cases the solvent–protein interactions form an important component in understanding the behavior of these systems; we will have to wait until we learn about the thermodynamics of mixtures to address this more complicated case. ◀

The Relationship Between Saturation Pressure and Temperature: The Clapeyron Equation

In Section 1.5, we learned that P and T are not independent for a pure species that exists in two phases at equilibrium. We now wish to come up with an expression relating the pressure at which two phases can coexist to the temperature of the system. This expression will allow us to calculate, for example, how the saturation pressure changes with temperature. Before we begin, we will qualitatively look at the issues of this problem in the context discussed above. We begin with the criterion for equilibrium between two phases:

$$g_i^\alpha = g_i^\beta \tag{6.14}$$

for pure species

Again, α and β can represent the vapor, liquid, or solid phases. Since two intensive variables completely specify the state of the system, the value of g for each phase is constrained at a given T and P. Thus, we can plot surfaces of Gibbs energy for each phase, as Figure 6.4 illustrates. The intersection of the two surfaces, the so-called coexistence line, represents the conditions where Equation (6.14) is satisfied and the two phases are in equilibrium. To the left of the coexistence line, phase β has lower Gibbs energy and will represent the phase at which the Gibbs energy is at a minimum. Conversely, to the right, only phase α will exist. At the given set of conditions for P and T on the coexistence line, however, both phases exhibit identical Gibbs energies; thus, phases α and β coexist in phase equilibrium. This is the same line you see on the PT projection of the PvT surface in Figure 1.6.

Which phase is more random? Which phase has stronger intermolecular attraction?

We calculate the relationship between P and T on the coexistence line as follows: Consider a system with two phases in equilibrium at a given P and T. For a small change in the equilibrium temperature dT, the change in equilibrium pressure, dP, can be

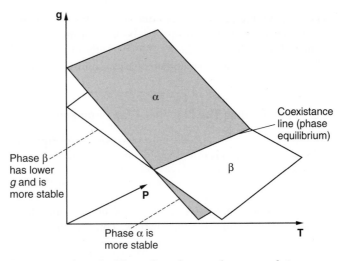

Figure 6.4 Plots of Gibbs surfaces for two phases, α and β.

calculated, since Equation (6.14) must be valid along the entire coexistence curve. Hence, the differential changes in Gibbs energy of each phase must be equal:

$$\mathrm{d}g_i^\alpha = \mathrm{d}g_i^\beta \tag{6.15}$$

As we saw in Chapter 5, the thermodynamic web provides a useful vehicle for relating derived thermodynamic properties to measured properties. Applying the fundamental property relation for g [Equation (5.9)] to each phase, we get:

$$v_i^\alpha \mathrm{d}P - s_i^\alpha \mathrm{d}T = v_i^\beta \mathrm{d}P - s_i^\beta \mathrm{d}T \tag{6.16}$$

The phases have been omitted on T and P. Why? Rearrangement yields:

$$\frac{\mathrm{d}P}{\mathrm{d}T} = \frac{s_i^\alpha - s_i^\beta}{v_i^\alpha - v_i^\beta} \tag{6.17}$$

Now we can also apply Equation (6.14):

$$g_i^\alpha = g_i^\beta \tag{6.14}$$

or by the definition of Gibbs energy, Equation (5.3):

energy directionality

$$h_i^\alpha - Ts_i^\alpha = h_i^\beta - Ts_i^\beta \tag{6.18}$$

Solving for the difference in entropy:

$$s_i^\alpha - s_i^\beta = \frac{h_i^\alpha - h_i^\beta}{T} \tag{6.19}$$

Substitution of Equation (6.19) into Equation (6.17) yields the Clapeyron equation:

$$\frac{\mathrm{d}P}{\mathrm{d}T} = \frac{h_i^\alpha - h_i^\beta}{\left(v_i^\alpha - v_i^\beta\right)T} \tag{6.20}$$

$g = g$

The Clapeyron equation relates the slope of the coexistence curve to the enthalpy and volume changes of phase transition, both experimentally accessible properties! Can you think of how to measure these values? In other words, it tells us the pressure increase, dP, which is necessary to maintain phase equilibria of a substance when the temperature has increased by dT.

Pure Component Vapor–Liquid Equilibrium: The Clausius–Clapeyron Equation

Next we consider the specific case of vapor–liquid equilibrium. In this case, the molar volume of the liquid is often negligible compared to the volume of the vapor:

$$v_i^l \ll v_i^v \quad \text{or} \quad v^l \approx 0 \qquad \text{(Assumption I)}$$

This assumption implies we are down the liquid–vapor dome shown in Figure 1.6, away from the critical point. If additionally we consider the vapor to obey the ideal gas model,

$$v_i^v = \frac{RT}{P} \qquad \text{(Assumption II)}$$

the coexistence equation for vapor–liquid equilibrium becomes:

$$\frac{dP_i^{\text{sat}}}{dT} = \frac{P_i^{\text{sat}} \Delta h_{\text{vap},i}}{RT^2} \qquad (6.21)$$

where $\Delta h_{\text{vap},i} = h_i^v - h_i^l$ and P_i^{sat} represent the enthalpy of vaporization and the saturation pressure, respectively, of species i at temperature T. Separating variables,

$$\frac{dP_i^{\text{sat}}}{P_i^{\text{sat}}} = \frac{\Delta h_{\text{vap},i} dT}{RT^2} \qquad (6.22)$$

Equation (6.22) is called the Clausius–Clapeyron equation. It can be rewritten in the form

$$d \ln P_i^{\text{sat}} = -\frac{\Delta h_{\text{vap},i}}{R} d\left(\frac{1}{T}\right) \qquad (6.23)$$

If we assume the enthalpy of vaporization is **independent of temperature**, that is,

$$\Delta h_{\text{vap},i} \neq \Delta h_{\text{vap},i}(T) \qquad \text{(Assumption III)}$$

we can either definitely integrate Equation (6.23) between state 1 and state 2 to get:

$$\ln \frac{P_2^{\text{sat}}}{P_1^{\text{sat}}} = -\frac{\Delta h_{\text{vap},i}}{R} \left[\frac{1}{T_2} - \frac{1}{T_1}\right] \qquad (6.24)$$

or write the indefinite integral of Equation (6.23):

$$\ln P_i^{\text{sat}} = \text{const} - \frac{\Delta h_{\text{vap},i}}{RT} \qquad (6.25)$$

In fact, Assumption III is not valid over large temperature ranges. The enthalpy of vaporization decreases as temperature increases, so Equations (6.24) and (6.25) can be used only over a limited temperature range. However, in surprisingly many cases, the error introduced by Assumption III is approximately offset by the errors of Assumptions I and II, leading to linear behavior of $\ln P^{\text{sat}}$ vs. T^{-1} over a larger range than would be originally surmised.

Saturation pressure correlations are commonly reported in terms of the Antoine equation, which is commonly used to correlate vapor pressures:

$$\ln P_i^{\text{sat}} = A - \frac{B}{C + T} \tag{6.26}$$

Here A, B, and C are empirical parameters that are available for many fluids. Values for Antoine constants can be found in Appendix A.1. The Antoine equation, an empirical equation, is strikingly similar to Equation (6.25). The Antoine equation brings back a similar theme to that discussed with equations of state (Chapter 4). When an empirical equation's form reflects the basic physics that it is trying to describe, it tends to work better. Why do you think the Antoine equation works better than the Clausius–Clapeyron equation in correlating saturation pressures? There are several more complex correlations reported in the literature for P^{sat} as a function of T. We will not cover these forms.

▶ **EXAMPLE 6.2**
Estimation of the
Enthalpy of Vaporization
from Measured Data

Trimethyl gallium, $Ga(CH_3)_3$, can be used as a feed gas to grow films of GaAs. Estimate the enthalpy of vaporization of $Ga(CH_3)_3$ from the data of saturation pressure vs. temperature given in Table E6.2.[5]

SOLUTION Examination of Equation (6.25) suggests that if we plot $\ln P^{\text{sat}}$ vs. T^{-1}, the slope will give $-\left(\Delta h_{\text{vap},Ga(CH_3)_3}/R\right)$. The data in Table E6.2 are plotted in such a manner in Figure E6.2. A least-squares linear regression is also shown in Figure E6.2. The high correlation coefficient implies $\Delta h_{\text{vap},Ga(CH_3)_3}$ is constant in this temperature range.

Taking the slope of the line, we get:

$$-\frac{\Delta h_{\text{vap},Ga(CH_3)_3}}{R} = -4222.1 \left[K\right]$$

Solving for the enthalpy of vaporization gives

$$\Delta h_{\text{vap},Ga(CH_3)_3} = 35.1 \left[kJ/mol\right]$$

For comparison, a value measured by static bomb combustion calorimetry has been reported as 33.1 kJ/mol, a difference of 6.0%. ◄

▶ **EXAMPLE 6.3**
Verification of the
"Equal Area" Rule
Presented in Section 4.3

In the discussion of cubic equations of state (Section 4.3), it was stated that the vapor–liquid dome could be constructed from a subcritical isotherm generated by the equation of state. The saturation pressure was identified as the line that divided the isotherm into equal areas, as shown on the plot below. Verify that this statement is consistent with the criteria for equilibrium developed above.

TABLE E6.2 Saturation Pressure Data for $Ga(CH_3)_3$

T [K]	P^{sat} [kPa]
250	2.04
260	3.3
270	7.15
280	12.37
290	20.45
300	32.48
310	49.75

[5] (Via NIST) J. F. Sackman, and L. H. Long, *Trans. Faraday Soc.*, **54**, 1797 (1958).

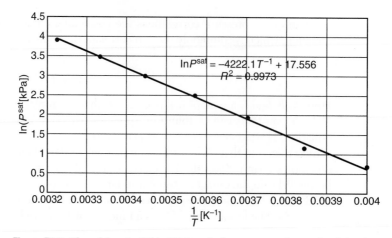

Figure E6.2 Plot of data in Table E6.2 and a least-squares linear fit of the data.

SOLUTION The vapor–liquid equilibrium criterion developed in this section states that the saturated vapor and the saturated liquid have equal Gibbs energies, that is,

$$g_i^v = g_i^l$$

or

$$g_i^v - g_i^l = 0 \tag{E6.3}$$

Another way to write Equation (E6.3) is that the integral, from saturated liquid to saturated vapor, of the differential Gibbs energy must be zero:

$$\int_{g_i^l}^{g_i^v} dg = 0$$

We can apply the fundamental property relation for dg, that is, Equation (5.9):

$$\int_{\text{sat. liq.}}^{\text{sat. vap.}} [v_i dP - s_i \overset{0}{dT}] = 0$$

The second term goes to zero, since we are evaluating the integral along an isotherm; hence, $dT = 0$. Using the product rule, we get

$$\int_{\text{sat. liq.}}^{\text{sat. vap.}} v_i dP = P^{\text{sat}} v_i \Big|_{v_i^l}^{v_i^v} - \int_{v_i^l}^{v_i^v} P dv_i = 0$$

or

$$P^{\text{sat}} \left(v_i^v - v_i^l \right) - \int_{v_i^l}^{v_i^v} P dv_i = 0$$

The resulting expression can be interpreted in regard to the plot in Figure E6.3. The integrated area from v_i^l to v_i^v on the plot of P vs. v is equal to the product $P^{\text{sat}} \left(v_i^v - v_i^l \right)$; therefore, the saturation pressure is the line that divides the isotherm into equal areas, resulting in a net area under the saturation line matching the net area above it. ◀

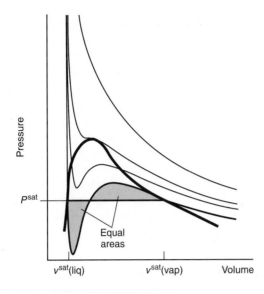

Figure E6.3 Subcritical isotherm given by a cubic equation of state divided into equal areas.

► ## 6.3 THERMODYNAMICS OF MIXTURES

Introduction

In this section, we explore how to formally treat thermodynamic properties of species in mixtures. In terms of possible combinations of intermolecular interactions, mixtures are inherently more complex than pure species. For a pure species i, all the inter-molecular interactions are identical. The resulting thermodynamic properties—such as V_i, U_i, S_i, H_i, and G_i—are a manifestation of those interactions. Since a mixture contains more than one species, its properties are determined only in part by an average of each of the pure species (i-i) interactions. We must now also take into account how each of the species interacts with the other species in the mixture, that is, the unlike (i-j) interactions. Thus the properties of a mixture depend on the nature and amount of each of the species in the mixture. The values of the mixture's properties will be affected not only by how those species behave by themselves but also by how they interact with each other.

Consider, for example, the experiment depicted in Figure 6.5. If we mix 50 ml of ethanol with 20 ml of water at 25°C and measure the resulting volume of solution, as careful as we might be, we get 67 ml! (Try this yourself.) Where has the other 3 ml gone? The solution has "shrunk" because the ethanol and water can pack together more tightly than can each species by itself. This is due to the nature of the hydrogen bonding involved

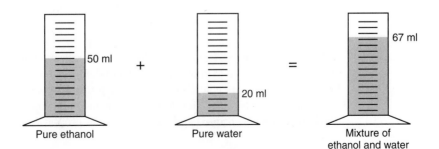

Figure 6.5 Mixing of ethanol and water.

in the structure of the liquid. We still have the same total mass, since mass is a conserved quantity; however, the mixture volume is different from the sum of the pure species volume. The difference in the mixture is based on the nature of the unlike ethanol–water interactions and the fact that they are different from the water–water or ethanol–ethanol pure species interactions. This example shows that in the treatment of multicomponent mixtures, it is important to realize that species in solution can behave quite differently than they do by themselves, depending on the chemical nature of their neighbors in solution. This behavior will affect all the thermodynamic properties of the solution.

When a species becomes part of a mixture, it loses its identity; yet it still contributes to the properties of the mixture, since the total solution properties of the mixture depend on the amount present of each species and its resultant interactions. We wish to develop a way to account for how much of a solution property (V, H, U, S, G...) can be assigned to each species. We do this through a new formalism: *the partial molar property*.

Partial Molar Properties

The state postulate tells us that if we specify two intensive properties for any **pure** species, we constrain the state of a single-phase system. For extensive properties, we must additionally specify the total number of moles.[6] In Chapter 5, we learned how to mathematically describe any intensive thermodynamic property in terms of partial derivatives of two independent, intensive properties. Since we are now concerned with thermal and mechanical equilibrium, it makes sense to choose T and P as the independent, intensive properties. We wish to extend the formulation to **mixtures** with changing composition. In addition to specifying two independent properties, we must also consider the number of moles of each species in the mixture.

We now wish to specify the *extensive* thermodynamic property of the *entire* mixture, K, where we use the symbol K to represent any possible extensive thermodynamic property, that is, $K = V$, H, U, S, G, and so on. In essence, by using K, we avoid repetitive derivations by treating the problem in general. If we were to divide K by the total number of moles in the system, we would get the intensive property $k = v$, h, u, s, g, and so on. We call K (or k) the *total solution property*.

Mathematically, we can write the extensive total solution property K in terms of T, P, and the number of moles of m different species:

$$K = K(T, P, n_1, n_2, \ldots, n_i, \ldots, n_m) \tag{6.27}$$

for example,

$$\left[\begin{aligned} V &= V(T, P, n_1, n_2, \ldots, n_i, \ldots, n_m) \\ H &= H(T, P, n_1, n_2, \ldots, n_i, \ldots, n_m) \\ &\qquad\qquad\qquad \vdots \end{aligned} \right]$$

Note that Equation (6.27) is an extension and generalization of Equation (1.18) to systems of m species. We need to know $m + 2$ independent quantities to completely specify K. In Equation (6.27) we specifically choose the system temperature, pressure, and number of moles of each of the m species. Once these measured properties are specified, the state of the system is constrained and all the extensive properties, K, take specific values.

---

[6] Or an analogous quantity that specifies the size of the system.

The differential of K can then be written as the sum of partial derivatives of each of these independent variables, as follows:

$$dK = \left(\frac{\partial K}{\partial T}\right)_{P,n_i} dT + \left(\frac{\partial K}{\partial P}\right)_{T,n_i} dP + \sum_{i=1}^{m} \left(\frac{\partial K}{\partial n_i}\right)_{T,P,n_{j\neq i}} dn_i \tag{6.28}$$

for example,

$$\left[\begin{array}{l} dV = \left(\dfrac{\partial V}{\partial T}\right)_{P,n_i} dT + \left(\dfrac{\partial V}{\partial P}\right)_{T,n_i} dP + \displaystyle\sum_{i=1}^{m} \left(\dfrac{\partial V}{\partial n_i}\right)_{T,P,n_{j\neq i}} dn_i \\[3ex] dH = \left(\dfrac{\partial H}{\partial T}\right)_{P,n_i} dT + \left(\dfrac{\partial H}{\partial P}\right)_{T,n_i} dP + \displaystyle\sum_{i=1}^{m} \left(\dfrac{\partial H}{\partial n_i}\right)_{T,P,n_{j\neq i}} dn_i \\[2ex] \vdots \end{array}\right]$$

We use the notation $n_{j\neq i}$ to specify that we are holding the number of moles of all $(m-1)$ species except species i constant when we take the partial derivative with respect to n_i. It is convenient to define a new thermodynamic function, the **partial molar property**, $\overline{K}_i$:

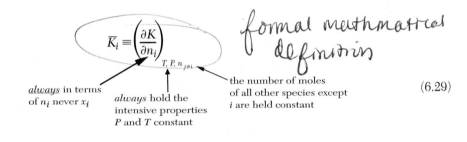

$$\overline{K}_i \equiv \left(\frac{\partial K}{\partial n_i}\right)_{T,P,n_{j\neq i}}$$

formal mathematical definition

always in terms of n_i never x_i

always hold the intensive properties P and T constant

the number of moles of all other species except i are held constant

$$(6.29)$$

for example,

$$\left| \begin{array}{l} \overline{V}_i = \left(\dfrac{\partial V}{\partial n_i}\right)_{T,P,n_{j\neq i}} \\[3ex] \overline{H}_i = \left(\dfrac{\partial H}{\partial n_i}\right)_{T,P,n_{j\neq i}} \\[2ex] \vdots \end{array} \right|$$

A partial molar property is *always* defined at constant temperature and pressure, two of the criteria for phase equilibrium. Partial molar properties are also defined with respect to number of moles. The number of moles of all other j species in the mixture are held constant; it is only the number of moles of species i that is changed. A common mistake in working with partial molar properties is to erroneously replace number of moles with mole fractions. We must realize, however, that Equation (6.29) does not simply convert to mole fraction, that is,

$$\overline{K}_i \neq \frac{1}{n_T} \left(\frac{\partial K}{\partial x_i}\right)_{T,P,x_{j\neq i}} \tag{6.30}$$

In changing the number of moles of species i, we change the mole fractions of all the other species in the mixture as well, since the sum of the mole fractions must equal 1.

Placing Equation (6.29) into Equation (6.28), the total differential of the variable K becomes:

$$dK = \left(\frac{\partial K}{\partial T}\right)_{P,n_i} dT + \left(\frac{\partial K}{\partial P}\right)_{T,n_i} dP + \sum_{i=1}^{m} \overline{K}_i dn_i \tag{6.31}$$

for example,

$$\left[\begin{array}{l} dV = \left(\dfrac{\partial V}{\partial T}\right)_{P,n_i} dT + \left(\dfrac{\partial V}{\partial P}\right)_{T,n_i} dP + \sum_{i=1}^{m} \overline{V}_i dn_i \\[2ex] dH = \left(\dfrac{\partial H}{\partial T}\right)_{P,n_i} dT + \left(\dfrac{\partial H}{\partial P}\right)_{T,n_i} dP + \sum_{i=1}^{m} \overline{H}_i dn_i \\[2ex] \qquad\qquad\qquad\qquad \vdots \end{array}\right]$$

At constant temperature and pressure, Equation (6.31) reduces to

$$dK = \sum \overline{K}_i dn_i \tag{6.32}$$

If, in addition to keeping T and P constant, we also keep the composition of the mixture constant (i.e., the mole fraction of all m species), then the partial molar properties are constant. In this case, we can integrate Equation (6.32) to get:

$$K = \sum \overline{K}_i n_i + C \tag{6.33}$$

where C is a constant of integration To determine C, we can use either a physical argument or a more mathematical argument involving Euler's theorem on homogeneous functions. The physical argument follows; the mathematical argument is presented in Example 6.4.

The intensive property, k, depends only on temperature, pressure, and composition of the m species present. Thus, k depends only on the relative amounts of each species in the system. On the other hand, the *extensive* property K is linearly dependent on the total amount of species present. For example, let's consider volume, a property that can exist in intensive form as $k = v$ or in extensive form as $K = V$. If the number of all species in the system doubles at constant T and P (i.e., $n_1 \longrightarrow 2n_1$, $n_2 \longrightarrow 2n_2, \ldots, n_m \longrightarrow 2n_m$), the molar volume, v, remains the same. The extensive property, V, however, will double. Similarly, if the number of all the m chemical species are cut in half, v remains the same but V will be cut in half. The number of species can be divided in half an arbitrary number of times and v will remain unchanged. However, in the limit of an infinitesimally small number of species, the extensive property, V, will go to 0. In general, the extensive property K must go to zero as $\sum n_i$ goes to zero; thus, the constant of integration in Equation (6.33) becomes 0, and we have

$$K = \sum n_i \overline{K}_i \tag{6.34}$$

for example,

$$\left[\begin{array}{l} V = \sum n_i \overline{V}_i \\[1.5ex] H = \sum n_i \overline{H}_i \\[1.5ex] \quad\ \vdots \end{array}\right]$$

or, dividing by the total number of moles,

$$k = \frac{K}{n_{\text{total}}} = \sum x_i \overline{K}_i \tag{6.35}$$

for example,

$$\left.\begin{array}{c} v = \sum x_i \overline{V}_i \\ h = \sum x_i \overline{H}_i \\ \vdots \end{array}\right|$$

where k is the corresponding intensive property to K and x_i is the mole fraction of species i. Both Equations (6.34) and (6.35) are consistent with the physical argument above.

Equation (6.34) indicates that the extensive total solution property K is equal to the sum of the partial molar properties of its constituent species, each adjusted in proportion to the quantity of that species present. Similarly, Equation (6.35) shows that the intensive solution property k is simply the weighted average of the partial molar properties of each of the species present. The partial molar property, $\overline{K}_i$, can then be thought of as *species i's contribution to the total solution property, K.* We can logically extend this thought to interpret a partial molar property as though it represents the intensive property value of an individual species as it exists in solution. In contrast, the pure species property, k_i, indicates how an individual species acts when it is by itself. The difference $\overline{K}_i - k_I$ compares how the species behaves in the mixture to how it behaves by itself. If this number is zero, the species behaves identically in the mixture to how it behaves as a pure species. In contrast, if this number is large, the species interactions in the mixture are quite different from when it is by itself.

Figure 6.6 presents a schematic representation of a process to measure the partial molar volume of water (w) for the mixture that is shown on the right side of Figure 6.5. To determine the partial molar volume of water for the well-defined composition of n_w (moles of water) and n_e (moles of ethanol) at a temperature T and pressure P, we measure the volume change, ΔV, for an incremental addition of Δn_w moles of water while holding the number of moles ethanol, the temperature, and the pressure constant. The partial molar volume is then given by the change in volume divided by the change in number of moles of water. For the composition in Figure 6.5, we get $\overline{V}_w = 16.9 \, [\text{ml/mol}]$, which is less than the pure species molar volume of water, $v_w = 18 \, [\text{ml/mol}]$. Thus, the contribution of water to the volume of this mixture is less than its contribution to the volume when it exists as a pure species. This result indicates that the inclusion of water–ethanol interactions leads to a closer packing than exclusively water–water interactions.

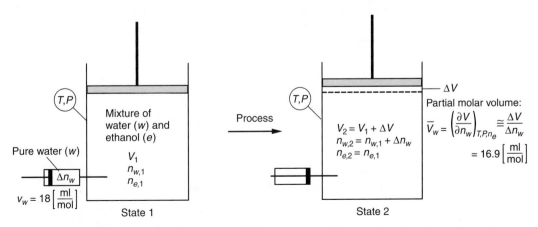

Figure 6.6 Schematic of experimental determination of the partial molar volume of water in a specified mixture of ethanol (e) and water (w).

As we change the composition of the mixture, we change the relative amounts of water–ethanol and water–water interactions, so the value of the partial molar volume will change. While Figure 6.6 specifically illustrates the partial molar volume, similar depictions can be drawn for any partial molar property.

▶ **EXAMPLE 6.4**
Integration of
Equation 6.32

Mathematically verify that integration of Equation (6.32) leads to Equations (6.35) or (6.34) (**a**) starting with Equation (6.31); (**b**) based on applying Euler's theorem to Equation (6.27).

SOLUTION (**a**) We begin with Equation (6.31):

$$dK = \left(\frac{\partial K}{\partial T}\right)_{P,n_i} dT + \left(\frac{\partial K}{\partial P}\right)_{T,n_i} dP + \sum_{i=1}^{m} \overline{K}_i dn_i \qquad (E6.4A)$$

Recognizing that $K = n_T k$ and $n_i = n_T x_i$, we can rewrite Equation (E6.4A) as

$$n_T dk + k dn_T = n_T \left(\frac{\partial k}{\partial T}\right)_{P,n_i} dT + n_T \left(\frac{\partial k}{\partial P}\right)_{T,n_i} dP + \sum_{i=1}^{m} \overline{K}_i x_i dn_T + \sum_{i=1}^{m} \overline{K}_i n_T dx_i \qquad (E6.4B)$$

Collecting terms in Equation (E6.4B) of n_T and dn_T, we get

$$\left[dk - \left(\frac{\partial k}{\partial T}\right)_{P,n_i} dT - \left(\frac{\partial k}{\partial P}\right)_{T,n_i} dP - \sum_{i=1}^{m} \overline{K}_i dx_i\right] n_T + \left[k - \sum_{i=1}^{m} \overline{K}_i x_i\right] dn_T = 0 \qquad (E6.4C)$$

We can now make an argument similar to the physical argument given above. The total size of the system should not affect how the system is affected by changes in composition; thus n_T and dn_T are independent of each other. For Equation (E6.4C) to hold, in general, each of the terms in the bracket must be zero. Hence,

$$k = \sum_{i=1}^{m} \overline{K}_i x_i \qquad (E6.4D)$$

and

$$dk = \left(\frac{\partial k}{\partial T}\right)_{P,n_i} dT + \left(\frac{\partial k}{\partial P}\right)_{T,n_i} dP + \sum_{i=1}^{m} \overline{K} dx_i \qquad (E6.4E)$$

Equation (E6.4D) is identical to Equation (6.35).

(**b**) We start by multiplying the number of moles in a system by an arbitrary amount α. At a given T and P, the extensive property K should also be increased by that amount:

$$\alpha K = K(T, P, \alpha n_1, \alpha n_2, \ldots, \alpha n_i, \ldots, \alpha n_m) \qquad (E6.4F)$$

Euler's theorem would describe Equation (E6.4F) by saying that the extensive thermodynamic quantity K is a **first-order**, homogeneous function of n_i. Differentiating Equation (E6.4F) by α, we get

$$\left[\frac{\partial(\alpha K)}{\partial \alpha}\right]_{T,P} = K = n_1 \left[\frac{\partial K}{\partial(\alpha n_1)}\right]_{T,P,n_i} + n_2 \left[\frac{\partial K}{\partial(\alpha n_2)}\right]_{T,P,n_i} + \cdots$$

$$+ n_i \left[\frac{\partial K}{\partial(\alpha n_i)}\right]_{T,P,n_{j \neq i}} + \cdots + n_m \left[\frac{\partial K}{\partial(\alpha n_m)}\right]_{T,P,n_{j \neq i}} \qquad (E6.4G)$$

where we have applied the chain rule to get the expression on the right-hand side of Equation (E6.4G). Equation (E6.4G) should be valid for any α, so at $\alpha = 1$, we get

$$K = \sum n_i \left[\frac{\partial K}{\partial n_i}\right]_{T,P,n_{j \neq i}} = \sum n_i \overline{K}_i$$

which is identical to Equation (6.34).   ◀

The Gibbs–Duhem Equation

The Gibbs–Duhem equation provides a very useful relationship between the partial molar properties of different species in a mixture. It results from mathematical manipulation of property relations. The approach is similar to that used in Chapter 5 to develop relationships between properties. The reason the Gibbs–Duhem equation is so useful is that it provides constraints between the partial molar properties of different species in a mixture. For example, in a binary mixture, if we know the values for a partial molar property of one of the species, we can apply the Gibbs–Duhem equation to simply calculate the partial molar property values for the other species. The formulation of the Gibbs–Duhem equation follows.

We begin with Equation (6.34), the definition of a partial molar property:

$$K = \sum n_i \overline{K}_i \tag{6.34}$$

We differentiate Equation (6.34) at constant T and P:

$$dK_{PT} = \sum \left[n_i d\overline{K}_i + \overline{K}_i dn_i \right] \tag{6.36}$$

where the subscript "PT" indicates that these properties are held constant. But from Equation (6.32), we know

$$dK_{PT} = \sum \overline{K}_i dn_i \tag{6.32}$$

For both Equations (6.36) and (6.32) to be true, in general:

$$0 = \sum n_i d\overline{K}_i \qquad \text{Const } T \text{ and } P \tag{6.37}$$

for example,

$$\begin{vmatrix} 0 = \sum n_i d\overline{V}_i \\ 0 = \sum n_i d\overline{H}_i \\ \vdots \end{vmatrix}$$

Equation (6.37) is the Gibbs–Duhem equation. Its straightforward derivation should not overshadow its tremendous utility.

To see the usefulness of the Gibbs–Duhem equation, let's examine the scenario where we wish to find the partial molar volume of species b in a binary solution when we know the partial molar volume of species a, $\overline{V}_a$, as a function of composition. If we apply Equation (6.37) to the property volume, we get

$$0 = \sum n_i d\overline{V}_i \qquad \text{Const } T \text{ and } P \tag{6.38}$$

For a binary system, we can differentiate Equation (6.38) with respect to x_a to get

$$0 = n_a \frac{d\overline{V}_a}{dx_a} + n_b \frac{d\overline{V}_b}{dx_a} \tag{6.39}$$

If we divide Equation (6.39) by n_T, rearrange, and integrate:

$$\overline{V}_b = - \int \frac{x_a}{1 - x_a} \left(\frac{d\overline{V}_a}{dx_a} \right) dx_a \tag{6.40}$$

Thus, if we have an expression for (or plot of) the partial molar volume of species a vs. mole fraction, we can apply Equation (6.40) to get the corresponding expression for species b. The expressions for partial molar properties are not independent but rather constrained by the Gibbs–Duhem equation. The partial molar properties are governed by how a species behaves in the mixture. From a molecular point of view, we expect the partial molar properties of a and b to be related since it is the same interaction between a and b that determines the difference between how each of these species behaves in the mixture as compared to by themselves, as pure species.

Summary of the Different Types of Thermodynamic Properties

There are many different types of properties of which to keep track in mixtures. In this section, we review our nomenclature and see how we keep track of the different types of properties. We consider total solution properties, pure species properties, and partial molar properties.

Total Solution Properties

The total solution properties are the properties of the **entire** mixture. They are written as

$$\text{extensive} \quad K: V, G, U, H, S, \ldots$$

$$\text{intensive} \quad k = K/n_{\text{total}} = v, g, u, h, s, \ldots$$

What is V for the experiment in Figure 6.5?

Pure Species Properties

The pure species properties are the properties of any one of the species in the mixture as it exists as a **pure** species **at the temperature, pressure, and phase of the mixture**. We denote a pure species property with a subscript "i". In general, for species i, we have

$$\text{extensive}: \quad K_i: V_i, G_i, U_i, H_i, S_i, \ldots$$

$$\text{intensive}: \quad k_i = \frac{K_i}{n_i} = v_i, g_i, u_i, h_i, s_i, \ldots$$

What is V_{water} for the experiment in Figure 6.5?

Partial Molar Properties

Partial molar properties can be viewed as the specific contribution of species i to the total solution property, as discussed in the previous section. They are written:

$$\overline{K}_i: \overline{V}_i, \overline{G}_i, \overline{U}_i, \overline{H}_i, \overline{S}_i, \ldots$$

Examination of Equations (6.34) and (6.35) shows that these properties must be intensive and, in general,

$$\overline{K}_i \neq k_i$$

We can envision two limiting cases. First, consider when the mole fraction of species i goes to 1. In this case, any given molecule of i will interact only with other molecules of i. Hence, the solution properties will match those of the pure species, and

$$\overline{K}_i = k_i \qquad \lim x_i \longrightarrow 1 \tag{6.41}$$

mostly x_i → pure molar prop

Second, we can imagine the case where species i becomes more and more dilute. In that case, a molecule of species i will not have any like species with which it interacts; rather, it will interact only with unlike species. We call this case that of *infinite dilution* and write the partial molar property as:

$$\overline{K}_i = \overline{K}_i^\infty \qquad \lim x_i \longrightarrow 0 \tag{6.42}$$

Pmp @ infinite dilution

► **EXAMPLE 6.5**
Types of Thermodynamic Properties in Figure 6.5

Label the volumes depicted in Figure 6.5 according to the different types of thermodynamic properties depicted in this section.

SOLUTION The experiment depicted in Figure 6.5 is now labeled in Figure E6.5 according to the different types of properties defined above. Properties identified include the molar volume (v_e, v_w) and the extensive volume of each pure species (V_e, V_w) as well as the partial molar volumes $(\overline{V}_e, \overline{V}_w)$ and total solution volume of the mixture (V). ◄

Property Changes of Mixing

A property change of mixing, ΔK_{mix}, describes how much a given property changes as a result of the mixing process. It is defined as the difference between the total solution property in the mixture and the sum of the pure species properties of its constituents, each in proportion to how much is present in the mixture. Mathematically, the property change of mixing is given by:

$$\Delta K_{\text{mix}} = K - \sum n_i k_i \tag{6.43}$$

for example,

$$\left[\begin{array}{l} \Delta V_{\text{mix}} = V - \sum n_i v_i \\ \Delta H_{\text{mix}} = H - \sum n_i h_i \\ \vdots \end{array} \right]$$

where the pure species properties, k_i, are defined at the temperature and pressure of the mixture. What is ΔV_{mix} for the experiment in Figure 6.5? Substituting Equation (6.34) into Equation (6.43) gives

$$\Delta K_{\text{mix}} = \sum n_i \overline{K}_i - \sum n_i k_i = \sum n_i (\overline{K}_i - k_i) \tag{6.44}$$

$$\underset{n_i}{\quad} \qquad \underset{n_i}{\quad}$$

$\Delta K_{\text{mix}} = \overline{K}_i - k_i$

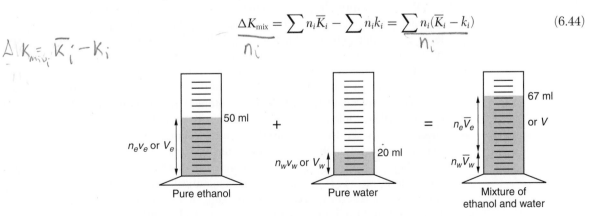

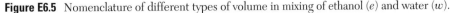
Figure E6.5 Nomenclature of different types of volume in mixing of ethanol (e) and water (w).

for example,

$$
\left[
\begin{aligned}
\Delta V_{\text{mix}} &= \sum n_i(\overline{V}_i - v_i) \\
\Delta H_{\text{mix}} &= \sum n_i(\overline{H}_i - h_i) \\
&\;\;\vdots
\end{aligned}
\right]
$$

Equation (6.44) shows that the property change of mixing is given by the proportionate sum of the difference between the partial molar property and the pure species property for each of the species in the mixture. This result is not surprising, since we interpret a partial molar property of a given species as the contribution of that species to the mixture, while the pure species property is indicative of how that species behaves by itself. In the special case where all the partial molar properties equal the pure species properties—that is $\overline{K}_i = k_i$—the property change of mixing is zero.

Analogously for intensive properties, we get

$$\Delta k_{\text{mix}} = k - \sum x_i k_i \tag{6.45}$$

for example,

$$
\left[
\begin{aligned}
\Delta v_{\text{mix}} &= v - \sum x_i v_i \\
\Delta h_{\text{mix}} &= h - \sum x_i h_i \\
&\;\;\vdots
\end{aligned}
\right]
$$

and,

$$\Delta k_{\text{mix}} = \sum x_i(\overline{K}_i - k_i) \tag{6.46}$$

for example,

$$
\left[
\begin{aligned}
\Delta v_{\text{mix}} &= \sum x_i(\overline{V}_i - v_i) \\
\Delta h_{\text{mix}} &= \sum x_i(\overline{H}_i - h_i) \\
&\;\;\vdots
\end{aligned}
\right]
$$

We can interpret the property change of mixing by considering a hypothetical process in which pure species undergo an isothermal, isobaric mixing process. On a molecular level, the property change of mixing reflects how the interactions between unlike species in the mixture compare to the like interactions of the pure species they replaced. For example, a volume change of mixing results from differences in how closely species in a mixture can pack together in comparison to how they pack as pure species. A negative Δv_{mix} will result when the unlike interactions "pull" the species in the mixture closer together, while a positive Δv_{mix} will result when the unlike interactions "push" the species apart when they are mixed. Negative volume changes of mixing tend to imply that the unlike interactions are more attractive (or less repulsive) than the like interactions they replace, while positive volume changes result when the unlike interactions are not as attractive. If the species behave identically in the mixture as they did as pure species, Δv_{mix} is zero. We see a negative Δv_{mix} in the ethanol–water mixture depicted in Figures 6.6 and E6.5 since the two species can pack together more closely in the mixture as compared to when they are by themselves.

Similarly, we can understand the enthalpy change of mixing in terms of this hypothetical mixing process at constant T and P. Again, we use the idea that at constant pressure, enthalpy is the appropriate thermodynamic property to characterize energetic interactions.[7] Thus, Δh_{mix} quantifies the difference between the energetic interactions of the species in the mixture to those of the pure species. In general, the enthalpy of mixing is negative when the species in the mixture are more stable than their pure species counterparts. Conversely, positive enthalpies of mixing result when the species are less stable in the solution than as pure species. When the energetic interactions of the mixture are identical to those of the pure species, Δh_{mix} is zero.

For example, consider a liquid mixture of water (H_2O) and sulfuric acid (H_2SO_4). Sulfuric acid will dissociate in water, forming positively charged hydrogen ions and single and doubly negative charged sulfate ions. Figure 6.7 schematically compares the energetic interactions of water as a pure species compared to water in the mixture. The top left of the figure shows pure water hydrogen-bonding to other molecules of water at the mixture T and P. The characteristic energy of this interaction is given by its pure species liquid enthalpy, h_{H_2O}, which is depicted on the left of the energy-level diagram at the bottom of the figure. At the top right of the figure, water molecules are shown orienting around a negatively charged bisulfate ion in the mixture. The characteristic enthalpy of water in the mixture is given by its partial molar enthalpy, $\overline{H}_{H_2O}$. The electrostatic interactions that form are more stable, energetically, than the hydrogen bonds depicted for the pure species at the left of the figure. Thus, the energy-level diagram at the bottom of the figure shows $\overline{H}_{H_2O}$ at a lower (more stable) value than as a pure species. Similar arguments can be made about the decrease in enthalpy of sulfuric acid as it dissociates in the mixture.[8] Inspection of Equation (6.46) shows that since $\overline{H}_{H_2O} < h_{H_2O}$ and $\overline{H}_{H_2SO_4} < h_{H_2SO_4}$, then $\Delta h_{\text{mix}} < 0$. In fact, the energetics of this binary mixture have been well studied, and the mixing process is highly exothermic. For example, at 21°C, the enthalpy of mixing of sulfuric acid and water can be fit to the following equation:[9]

$$\Delta h_{\text{mix}} = -74.40\, x_{H_2O_4}\, x_{H_2O}\left(1 - 0.561 x_{H_2SO_4}\right)\ \left[\text{kJ/mol}\right] \tag{6.47}$$

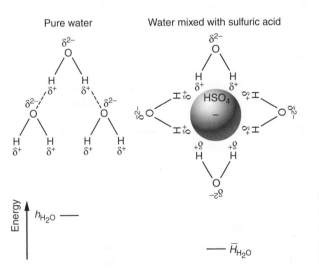

Figure 6.7 Energetic interactions of molecules of water as a pure species and interacting with HSO_4^- in a sulfuric acid – water mixture.

---

[7] Section 2.6 relates enthalpy to energy changes in a closed system at constant P.

[8] The enthalpy of mixing also includes the ionization energy.

[9] Equation fit to data reported by W. D. Ross, *Chem. Eng. Progr.*, **43**, 314 (1952).

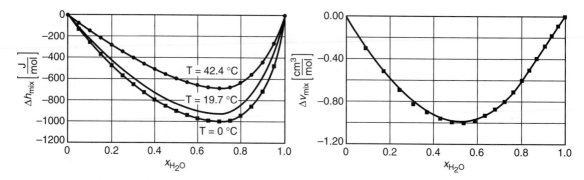

Figure 6.8 Enthalpy and volume changes of mixing of a binary mixture of water and methanol at 0°, 19.7°, and 42°C.

Figure 6.8 shows the enthalpy and volume changes of mixing for water and methanol at 0°, 19.7°, and 42.4°C. The enthalpy of mixing is negative, indicating that the species in the mixture are energetically more stable than they are by themselves, as pure species. Likewise, the volume change of mixing associated with these species is negative, indicating that the species in the mixture pack together more tightly.[10] Again, we can understand the mixing behavior in this system in terms of intermolecular interactions. The strongest energetic interaction is hydrogen bonding; however, these species also manifest van der Waals forces. Both water and methanol can form hydrogen bonds as pure species as well as in the mixture. Apparently, this hydrogen-bonded network can pack more efficiently with both species present in a mixture as compared to when they are separate, as pure species. Since the species are closer, the van der Waals interactions are stronger and the energy is lower. The order of magnitude of this effect is 10^2 [J/mol], far less than the mixing effects of a sulfuric acid–water system, which arises from the presence of point charges of the ions in solution. The enthalpy of mixing decreases as the temperature increases. However, the functionality of Δh_{mix} with mole fraction is similar at all three temperatures. By contrast, the volume change of mixing is practically identical for all three temperatures.

Figure 6.9 shows the enthalpy of mixing for cyclohexane (C_6H_{12}) and toluene at 18°C. In this case, the enthalpy of mixing is positive. Its order of magnitude is 10^2 [J/mol], which is comparable to a methanol–water mixture. The dominant energetic interaction in this nonpolar system is dispersion. The magnitude of unlike dispersion interactions is often well approximated as the geometric mean of the corresponding like interactions. With such a relation, the mixture will always be less stable than the weighted average of the pure species,[11] thus, nonpolar mixtures usually exhibit positive enthalpies of mixing.

Figure 6.10 shows the enthalpy of mixing for chloroform and methanol at 25°C. This figure shows unusual behavior in that the enthalpy of mixing changes sign with changing chloroform mole fraction. At low chloroform mole fractions, the enthalpy of mixing is negative, while at high mole fractions it is positive. This behavior can be understood in terms of the superposition of two competing effects. Over all composition ranges, there is behavior similar to the one described for the system of cyclohexane–toluene, where the average of the like interactions is more favorable than that of the unlike interactions. This effect contributes a positive value to Δh_{mix}. However, the specific and directional

---

[10] In the case of volume changes of mixing, all three curves overlap.

[11] This point will be discussed more in Chapter 7. See Example 7.9.

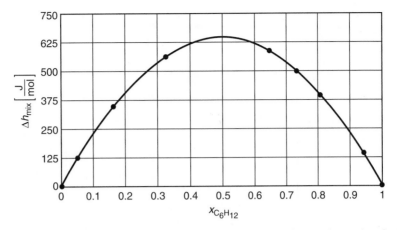

Figure 6.9 Enthalpy of mixing of a binary mixture of cyclohexane (C_6H_{12}) and toluene at 18°C.

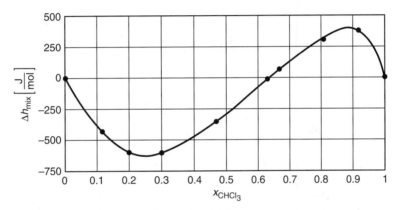

Figure 6.10 Enthalpy of mixing of a binary mixture of chloroform ($CHCl_3$) and methanol at 25°C.

nature of the hydrogen bonds involved also play a role in the energetics of this system. Methanol has both an electronegative O with two sets of lone pairs of electrons as well as one H atom to contribute to H-bonding, while CCl_3H contributes only an H atom. Thus, at a CCl_3H mole fraction of roughly one-third, the number of hydrogen and number of oxygen lone pairs match, and the system can form the most possible H-bonds and, therefore, is most stable. This composition represents the minimum in the enthalpy of mixing in Figure 6.10.

The energetic interactions characteristic of the mixing process are often reported as the enthalpy of solution, $\Delta \tilde{h}_s$, instead of the enthalpy of mixing. The enthalpy of solution corresponds to the enthalpy change when 1 mole of pure solute is mixed in n moles of pure solvent. It is defined per mole of solute as opposed to enthalpy of mixing, which is written per total moles of solution. The value of $\Delta \tilde{h}_s$ can be determined calorimetrically. Since $\Delta \tilde{h}_s$ is convenient to measure in the laboratory, reported values are often in this form. Table 6.1 reports $\Delta \tilde{h}_s$ for several solutes in water at 25°C. Example 6.7 illustrates how to calculate values for the enthalpy of mixing when enthalpy of solution data are available. The enthalpy of solution is also commonly used when the solute exists in the solid or gas phase as a pure species. In this case, the enthalpy difference associated to

Table 6.1 Enthalpy of Solution of Different Species in Water at 25°C

$$\Delta \tilde{h}_s \left[\frac{J}{\text{mol solute}} \right]$$

n [mol H_2O]	HNO_3	H_2SO_4	HCl	HF	H_3PO_4	$C_2H_4O_2$	NH_3	NaOH	C_2H_5OH	$ZnCl_2$	NaCl
1	−13,113	−31,087	−26,225		2,510	628	−29,539		−812		
2	−20,083	−44,936	−48,819	−45,886	−837	669	−32,049		−1,799		
3	−24,301	−52,007	−56,852	−46,798	−4,184	586	−32,761	−28,886	−2,787		
4	−26,978	−57,070	−61,204	−47,187	−6,276		−33,263	−34,430	−3,757	−28,870	
5	−28,727	−61,045	−64,049	−47,384	−7,531	377	−33,598	−37,761	−4,694	−32,677	
10	−31,840	−70,040	−69,488	−47,719	−9,874	−209	−34,267	−42,501	−7,364	−40,083	1,941
20	−32,669	−74,517	−71,777	−47,840	−10,962	−628	−34,434	−42,865	−8,866	−46,191	2,671
50	−32,744	−76,358	−73,729	−47,949	−11,966	−1,130	−34,518	−42,526	−9,975	−55,020	3,732
100	−32,748	−76,986	−73,848	−48,057	−12,468	−1,276	−34,560	−42,334	−10,268	−61,086	4,109
∞	−33,338	−99,203	−75,189	−60,501		−1,435	−34,644	−42,869		−71,463	3,891

Source: F. D. Rossini et al., *Selected Values of Physical and Thermodynamic Properties of Hydrocarbons and Related Compounds* (Pittsburgh: Carnegi Press, 1953).

the phase change is also manifest in $\Delta \tilde{h}_s$. For example, solids held together with ionic bonds—for example, salts such as NaCl—demonstrate positive enthalpies of solution. In this case, the ions in the pure species solid are close together. When they dissolve into the mixture, the energy increases as the ions get father away from one another. Can you think of a practical application for a system with a large positive enthalpy of solution? On the other hand, gases tend to have negative enthalpies of solution, as the species are closer to their neighbors in the mixture and exhibit stronger attraction.

The first six species listed in the Table 6.1 are acids. Their enthalpies of solution differ appreciably, corresponding to different degrees of dissociation. The stronger the acid, the greater the extent to which it disassociates into ionic species and the larger the electrostatic interactions that result. For example, acetic acid, $C_2H_4O_2$, is a weak acid and has a positive enthalpy of solution at higher solute concentrations. On the other hand, $\Delta \tilde{h}_s$ for sulfuric acid is always large and negative due to the energetic interactions discussed earlier. The other species reported in Table 6.1 give a flavor of the energetic mixing effects of other types of solutes. Gaseous NH_3 and solid NaOH form basic solutions; an alcohol (C_2H_5OH) and salts ($ZnCl_2$ and NaCl) are also reported.

While the enthalpy of mixing is characteristic of the how the energetic interactions change upon mixing, the entropy of mixing, Δs_{mix}, characterizes the increase in disorder induced by the mixing process. We can interpret Δs_{mix} by again considering a hypothetical process in which pure species undergo an isothermal, isobaric mixing process. Entropy is proportional to the number of possible *molecular configurations* that a state can exhibit. Since there are always many more ways to configure the species in a mixture as compared to a pure species, Δs_{mix} is always positive. In Example 6.8, we calculate the entropy of mixing for a binary mixture using the ideal gas model. The result from that example can be generalized for an *ideal gas* mixture of m species to give:

Always +ve for
16.   Always (+)

$$\Delta s_{mix} = -R \sum_{i=1}^{m} y_i \ln y_i \qquad (6.48)$$

In contrast, the enthalpy and volume changes of mixing of an *ideal gas* are identically zero, since an ideal gas does not exhibit any intermolecular interactions and the molecules size is negligible. Often it is assumed that liquid solutions mix completely randomly, and so the entropy of mixing follows the ideal gas relation given by Equation (6.48). Such a liquid is said to form a **regular** solution. Species that interact exclusively through van der Waals forces can form regular solutions. However, mixtures whose species significantly differ in size deviate from regular solution behavior. Chemical effects such as association and solvation can also lead to deviation from regular solution behavior. For example, the association reaction between chloroform and acetone leads to mixtures that are structured, and the entropy change of mixing is significantly less than what would be predicted for a regular solution.

▶ **EXAMPLE 6.6**
Calculation of Δh_{mix} from Experimental Data

An experiment is performed to measure the enthalpy of mixing of chloroform, $CHCl_3$, and acetone, C_3H_6O. In this experiment, pure species inlet streams of different compositions are mixed together in an insulated mixer at steady-state. This mixing process is exothermic, and the heat that is removed in order to keep the system at a constant temperature of 14°C is measured. The measured data are presented in Table E6.6A. Based on these data, calculate the enthalpy of mixing vs. mole fraction and plot the result.

SOLUTION A schematic of the system is shown in Figure E6.6A. Chloroform is labeled species 1 and acetone species 2. The weight fractions, w_i, of the inlet streams and enthalpies of all the

TABLE E6.6A   Heat Evolved vs. Weight Percentage Chloroform

Weight % $CHCl_3$	Heat evolved $[J/g], (-\hat{q})$
10	4.77
20	9.83
30	14.31
40	19.38
50	23.27
60	25.53
70	25.07
80	21.55
90	13.56

Source: E. W. Washburn (ed.), *International Critical Tables* (Vol. V) (New York: McGraw-Hill, 1929).

streams are labeled. The mole fraction can be calculated from the weight fraction as follows:

$$x_1 = \frac{\dfrac{w_1}{MW_1}}{\dfrac{w_1}{MW_1} + \dfrac{w_2}{MW_2}} \tag{E6.6A}$$

where MW_i is the molecular weight of species i. A first-law balance gives

$$0 = (\dot{n}_1 h_1 + \dot{n}_2 h_2) - (\dot{n}_1 + \dot{n}_2)h + \dot{Q} \tag{E6.6B}$$

We need to write Equation (E6.6B) in terms of what we are looking for, Δh_{mix}, and what we have measured, $(-\hat{q})$:

$$\Delta h_{mix} = h - (x_1 h_1 + x_2 h_2) = \frac{\dot{Q}}{(\dot{n}_1 + \dot{n}_2)} = q = -(-\hat{q})\overline{MW} \tag{E6.6C}$$

where the average molecular weight is given by $\overline{MW} = x_1 MW_1 + x_2 MW_2$. Table E6.6B gives enthalpy of mixing vs. mole fraction, as calculated from Equations (E6.6A) and (E6.6C), respectively, using data from Table E6.6A. These data are plotted in Figure E6.6B. A large negative enthalpy of mixing results from mixing chloroform and acetone. This result is consistent with the energetic interactions in the system. The dominant unlike interaction results from hydrogen bonding between the H of chloroform and O of acetone (see Section 4.2). Neither of the pure species forms hydrogen bonds; thus, the unlike interactions are more stable than the like interactions they replace, and a large negative enthalpy of mixing results.　◀

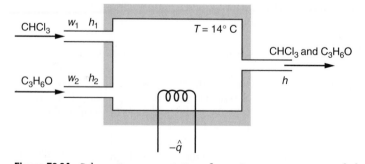

Figure E6.6A   Schematic representation of experiment to measure enthalpy of mixing.

TABLE E6.6B Enthalpy of Mixing vs. Mole Fraction Chloroform, Using Data from Table E6.6A

x_1	$\Delta h_{mix}\ [\text{J/mol}]$
0	0
0.050	−291.9
0.107	−636.6
0.170	−984.1
0.242	−1,420.7
0.323	−1,826.4
0.418	−2,156.1
0.527	−2,291.4
0.657	−2,146.2
0.811	−1,483.5
1.000	0

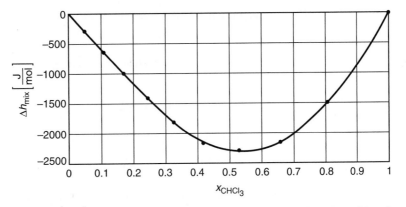

Figure E6.6B Enthalpy of mixing of a binary mixture of chloroform ($CHCl_3$) and acetone at 14°C.

▶ **EXAMPLE 6.7**
Calculation of Δh_{mix} from $\Delta \tilde{h}_s$

Table 6.1 presents data for the enthalpy of solution, $\Delta \tilde{h}_s$, for nitric acid in water at 18°C. Find the corresponding values for the enthalpy of mixing vs. mole fraction of solute.

SOLUTION   Let the solute, HNO_3, be species 1 and the solvent, H_2O, be species 2. The enthalpy of solution can be written by dividing the enthalpy of mixing by the mole fraction solute:

$$\Delta \tilde{h}_s = \frac{\Delta h_{mix}}{x_1} \tag{E6.7A}$$

The mole fraction of solute is defined as the number of moles of solute, 1, divided by the total number of moles in the system, $1 + n$:

$$x_1 = \frac{1}{1+n} \tag{E6.7B}$$

Substituting Equation (E6.7B) in (E6.7A) and rearranging gives:

$$\Delta h_{mix} = \frac{\Delta \tilde{h}_s}{(1+n)} \tag{E6.7C}$$

Table E6.7 presents values for Δh_{mix} as calculated from Equation (E6.7C) vs. mole fraction as calculated from Equation (E6.7B) for the data presented in Table 6.1. The order of table entries has been reversed so that the values for mole fraction appear in ascending order.   ◀

Calculate the entropy change of mixing for a binary ideal gas mixture.

SOLUTION This example is a generalization of Example 3.9, so you should review that example before proceeding. To calculate the entropy change of mixing for an ideal gas, comprised of species a and b, we apply Equation (6.46):

$$\Delta s_{mix} = y_a(\overline{S}_a - s_a) + y_b(\overline{S}_b - s_b) \tag{E6.8A}$$

where the partial molar entropy and the pure species entropy are evaluated at the same total pressure, P. Similar to the path illustrated in Figure E3.9B, we develop a two-step hypothetical process to calculate the change of entropy. In the first step (Step I), each of the *pure* species a and *pure* species b are isothermally expanded to the size of the container of the mixture. During this process, the pressure drops from P to p_a and p_b, respectively. The change in entropy for this process was discussed in Chapter 3, and will be developed below. The next step (Step II) is superimposing both these expanded systems. Since ideal gas a does not know b is there and *vise-versa*, the properties of each individual species do not change. Therefore, the entropy change is determined by the sum of the entropy change of each species in Step I.

For the isothermal decrease in pressure represented by Step I, we can apply Equation 3.62 to give

$$(\overline{S}_a - s_a)^I = -R\ln\frac{p_a}{P} = -R\ln\frac{y_a P}{P} = -R\ln y_a \tag{E6.8B}$$

Similarly for b:

$$(\overline{S}_b - s_b)^I = -R\ln y_b \tag{E6.8C}$$

Substituting Equations E6.8B and E6.8C in E6.8A gives:

$$\Delta s_{mix} = -R\left[y_a\ln y_a + y_b\ln y_b\right]$$

In general, when there are m species, we can sum together similar contributions from each species to get:

$$\Delta s_{mix} = -R\sum_{i=1}^{m} y_i\ln y_i \tag{E6.8D}$$

Equation E6.8D is presented in the text as Equation 6.48. ◀

Determination of Partial Molar Properties

We have introduced a new type of property, the partial molar property. This property tells us about the contribution of a given species to the mixture. Our next question is: How do

TABLE E6.7 Enthalpy of Mixing vs. Mole Fraction HNO_3, Using Data from Table 6.1

x_1	$\Delta h_{mix}\,[\text{J}/\text{mol}]$
0.048	−1,556
0.091	−2,895
0.167	−4,788
0.200	−5,396
0.250	−6,075
0.333	−6,694
0.500	−6,556

we obtain values for these partial molar properties? There are several ways in which to accomplish this task. In this section, we consider two examples of how we might calculate a partial molar property: by analytical means when we have an equation that describes the total solution property or by graphical means from plots of total solution data.

Energy-related properties such as enthalpy or internal energy must be defined with respect to a reference value. In this case, it is often convenient to consider the partial molar property change of mixing, which can be written as

Use definition

$$\overline{\Delta K}_{\text{mix},i} = \left(\frac{\partial \Delta K_{\text{mix}}}{\partial n_i}\right)_{T,P,n_{j\neq i}} = \left(\frac{\partial K}{\partial n_i}\right)_{T,P,n_{j\neq i}} - \left\lfloor\frac{\partial \sum (n_i k_i)}{\partial n_i}\right\rfloor_{T,P,n_{j\neq i}} \tag{6.49}$$

where Equation (6.43) was used. Since the pure species properties are constant at a given temperature and pressure, every term for the derivative in the sum becomes zero except the one associated with n_i. Therefore, Equation (6.49) becomes

$$\overline{\Delta K}_{\text{mix},i} = \overline{K}_i - k_i \tag{6.50}$$

for example,

$$\left[\begin{array}{l} \overline{\Delta H}_{\text{mix},i} = \overline{H}_i - h_i \\ \overline{\Delta G}_{\text{mix},i} = \overline{G}_i - g_i \\ \vdots \end{array}\right]$$

Inspection of Equation (6.50) shows that $\overline{\Delta K}_{\text{mix},i}$ gives the relative value of how species i behaves in a mixture to how it behaves by itself as a pure species. In this form, the pure species property forms a reference state to which the particular partial molar property is referred.

Analytical Determination of Partial Molar Properties

Often an analytical expression for the total solution property, k, is known as a function of composition. In that case, the partial molar property, $\overline{K}_i$, can be found by differentiation of the extensive expression for K with respect to n_i, holding T, P, and the number of moles of the other j species constant, as prescribed by Equation (6.29). We illustrate this method by showing how to calculate partial molar volumes for a binary mixture of species 1 and 2 with the virial equation of state. The virial equation can be written in terms of the total solution molar volume using Equations (4.59) and (4.84):

$$v = \frac{RT}{P}\left[1 + \frac{B_{\text{mix}}P}{RT}\right] = \frac{RT}{P} + y_1^2 B_{11} + 2y_1 y_2 B_{12} + y_2^2 B_{22} \tag{6.51}$$

If we know the virial coefficients, B_{11} and B_{22}, and the cross-virial coefficient, B_{12}, we can solve for the partial molar volumes of each species in the mixture. We first write Equation (6.51) in terms of extensive volume and number of moles:

$$V = (n_1 + n_2)v = (n_1 + n_2)\frac{RT}{P} + \frac{n_1^2 B_{11} + 2n_1 n_2 B_{12} + n_2^2 B_{22}}{(n_1 + n_2)} \tag{6.52}$$

where we have used $y_1 = n_1/(n_1 + n_2)$ and $y_2 = n_2/(n_1 + n_2)$. Differentiation gives:

$$\overline{V}_1 = \left(\frac{\partial V}{\partial n_1}\right)_{T,P,n_2} = \frac{RT}{P} + \frac{2n_1 B_{11} + 2n_2 B_{12}}{(n_1 + n_2)} - \frac{n_1^2 B_{11} + 2n_1 n_2 B_{12} + n_2^2 B_{22}}{(n_1 + n_2)^2} \tag{6.53}$$

We can simplify Equation (6.53) to get

$$\overline{V}_1 = \frac{RT}{P} + (y_1^2 + 2y_1y_2)B_{11} + 2y_2^2B_{12} - y_2^2B_{22} \tag{6.54}$$

Similarly, the partial molar volume of species 2 can be written

$$\overline{V}_2 = \frac{RT}{P} - y_1^2B_{11} + 2y_1^2B_{12} + (y_2^2 + 2y_1y_2)B_{22} \tag{6.55}$$

To obtain a value for the pure species molar volume, we set $y_2 = 0$ in Equation (6.51) to give

$$v_1 = \frac{RT}{P} + B_{11} \tag{6.56}$$

This expression can also be obtained through Equation (6.54), since $\overline{V}_1 = v_1$ in $\lim x_1 \longrightarrow 1$. The volume change of mixing is given by

$$\Delta v_{\text{mix}} = v - (y_1v_1 + y_2v_2) \tag{6.57}$$

Substituting in for v, v_1, and v_2:

$$\Delta v_{\text{mix}} = \left[\frac{RT}{P} + y_1^2B_{11} + 2y_1y_2B_{12} + y_2^2B_{22}\right] - \left[y_1\left(\frac{RT}{P} + B_{11}\right) + y_2\left(\frac{RT}{P} + B_{22}\right)\right] \tag{6.58}$$

or simplifying:

$$\Delta v_{\text{mix}} = 2y_1y_2\left[B_{12} - \frac{B_{11} + B_{22}}{2}\right] \tag{6.59}$$

Inspection of the term in brackets on the right-hand side of Equation (6.59) shows that the magnitude and sign of Δv_{mix} are determined by comparing the strength of the unlike interactions, given by B_{12}, to that of the like interactions, given by the average of B_{11} and B_{22}. If the unlike term is stronger (i.e., has a larger-magnitude negative number), the volume change of mixing is negative, whereas a positive change results when the like interactions are stronger.

▶ **EXAMPLE 6.9**
Calculation of Mixture Properties Using the Virial Equation

Consider a binary mixture of 10 mole% chloroform (1) in acetone (2) at 333 K and 10 bar. The second virial coefficients for this system are reported to be $B_{11} = -910, B_{22} = -1330$, and $B_{12} = -2005$ cm^3/mol. Determine $v_1, \overline{V}_1$, and Δv_{mix}.

SOLUTION Applying Equations (6.56) and (6.54) for the pure species and partial molar volumes, respectively, we get

$$v_1 = \frac{RT}{P} + B_{11} = 1860\left[\text{cm}^3/\text{mol}\right] \tag{E6.9A}$$

and

$$\overline{V}_1 = \frac{RT}{P} + (x_1^2 + 2x_1x_2)B_{11} + x_2^2B_{12} - x_2^2B_{22} = 991\left[\text{cm}^3/\text{mol}\right] \tag{E6.9B}$$

where we represent the liquid mole fractions by x_i instead of y_i. Using Equation (6.59) gives

$$\Delta v_{\text{mix}} = x_1x_2(2B_{12} - B_{11} - B_{22}) = -372\left[\text{cm}^3/\text{mol}\right] \tag{E6.9C}$$

Comparing Equations (E6.9A) and (E6.9B), we see that the molar volume of pure chloroform, v_1, is about double its contribution to the molar volume of the mixture with 10% chloroform, as indicated by $\overline{V}_1$. This result can be explained in terms of hydrogen bonding between H in chloroform and carbonyl in acetone. This association reaction "pulls" the species in the mixture closer to each other (see Section 4.2 and Example 6.6). The result of this interaction between chloroform and acetone is a reduced solution volume; therefore, Δv_{mix} is negative. It should be noted that the interaction between chloroform and acetone is unusually strong in comparison to the pure species interactions and is not typical of most mixtures. ◀

▶ **EXAMPLE 6.10**
Calculation of
$\overline{H}_{H_2SO_4}$ and $\overline{H}_{H_2O}$

Develop expressions for the partial molar enthalpies of sulfuric acid and water in a binary mixture at 21°C. The pure species enthalpies are 1.596 [kJ/mol] and 1.591 [kJ/mol], respectively, and the enthalpy of mixing is given by Equation (6.47). Calculate their values for an equimolar mixture of sulfuric acid and water. Plot $\overline{H}_{H_2SO_4}$ and $\overline{H}_{H_2O}$ vs. $x_{H_2SO_4}$.

SOLUTION We can write the total solution enthalpy as

$$h = h_{H_2SO_4}x_{H_2SO_4} + h_{H_2O}x_{H_2O} + \Delta h_{mix}$$

or, numerically,

$$h = 1.596\,x_{H_2SO_4} + 1.591\,x_{H_2O} - 74.40x_{H_2SO_4}\,x_{H_2O}(1 - 0.561\,x_{H_2SO_4})\,\left[\text{kJ/mol}\right] \qquad \text{(E6.10A)}$$

To apply the definition of a partial molar property, Equation (6.29), we need to write Equation (E6.10A) in terms of the extensive enthalpy and the number of moles of sulfuric acid and water:

$$H = n_T h = 1.596 n_{H_2SO_4} + 1.591 n_{H_2O} - 74.40\frac{n_{H_2SO_4}\,n_{H_2O}}{n_{H_2SO_4} + n_{H_2O}} + 41.74\frac{n_{H_2SO_4}^2\,n_{H_2O}}{(n_{H_2SO_4} + n_{H_2O})^2} \qquad \text{(E6.10B)}$$

Differentiating Equation (E6.10B) with respect to $n_{H_2SO_4}$ gives the partial molar enthalpy of sulfuric acid:

$$\overline{H}_{H_2SO_4} = \left(\frac{\partial H}{\partial n_{H_2SO_4}}\right)_{T,P,n_{H_2O}} = 1.596 - 74.40\frac{n_{H_2O}^2}{(n_{H_2SO_4} + n_{H_2O})^2} + 83.48\frac{n_{H_2SO_4}\,n_{H_2O}^2}{(n_{H_2SO_4} + n_{H_2O})^3}$$

and, simplifying,

$$\overline{H}_{H_2SO_4} = 1.596 - 74.40\,x_{H_2O}^2 + 83.48\,x_{H_2SO_4}\,x_{H_2O}^2\,\left[\text{kJ/mol}\right] \qquad \text{(E6.10C)}$$

Similarly, for water we get

$$\overline{H}_{H_2O} = 1.591 - 74.40x_{H_2SO_4}^2 + 41.74x_{H_2SO_4}^2(1 - 2x_{H_2O})\,\left[\text{kJ/mol}\right] \qquad \text{(E6.10D)}$$

A plot of the partial molar enthalpies calculated from these expressions is presented in Figure E6.10. For an equimolar mixture ($x_{H_2SO_4} = x_{H_2O} = 0.5$), Equations (E6.10C) and (E6.10D) give:

$$\overline{H}_{H_2SO_4} = -6.6\,\left[\text{kJ/mol}\right] \qquad \text{and} \qquad \overline{H}_{H_2O} = -17.0\,\left[\text{kJ/mol}\right] \qquad ◀$$

▶ **EXAMPLE 6.11**
Use of the Gibbs–Duhem
Equation to Relate Partial
Molar Properties

Verify that the expressions developed in Example 6.10 for the partial molar enthalpies of sulfuric acid and water in a binary mixture at 21°C satisfy the Gibbs–Duhem equation.

SOLUTION The Gibbs–Duhem equation can be written for the partial molar enthalpies in this system as

$$0 = n_{H_2SO_4}d\overline{H}_{H_2SO_4} + n_{H_2O}d\overline{H}_{H_2O} \qquad \text{(E6.11A)}$$

Differentiating Equation (E6.11A) with respect to mole fraction of sulfuric acid and dividing by the total number of moles gives

$$0 = x_{H_2SO_4}\frac{d\overline{H}_{H_2SO_4}}{dx_{H_2SO_4}} + x_{H_2O}\frac{d\overline{H}_{H_2O}}{dx_{H_2SO_4}} \qquad \text{(E6.11B)}$$

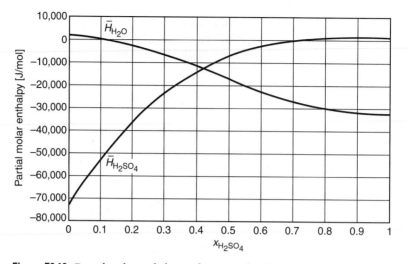

Figure E6.10 Partial molar enthalpies of water and sulfuric acid at 21°C.

We now take the derivatives of the two expressions in Equation (E6.11B) by using Equations (E6.10C) and (E6.10D). First, the derivative of the partial molar enthalpy of sulfuric acid gives

$$\frac{d\overline{H}_{H_2SO_4}}{dx_{H_2SO_4}} = -148.80\,x_{H_2O}\frac{dx_{H_2O}}{dx_{H_2SO_4}} + 166.96\,x_{H_2O}x_{H_2SO_4}\frac{dx_{H_2O}}{dx_{H_2SO_4}} + 83.48x_{H_2O}^2 \qquad \text{(E6.11C)}$$

However, the change in number of moles of water with respect to the number of moles of sulfuric acid is given by

$$\frac{dx_{H_2O}}{dx_{H_2SO_4}} = -1$$

Thus, Equation (E6.11C) becomes

$$\frac{d\overline{H}_{H_2SO_4}}{dx_{H_2SO_4}} = 148.80\,x_{H_2O} - 166.96\,x_{H_2O}x_{H_2SO_4} + 83.48\,x_{H_2O}^2 = -18.16x_{H_2O} + 250.44x_{H_2O}^2$$

$$\text{(E6.11D)}$$

where we have used $x_{H_2SO_4} = (1 - x_{H_2O})$. Finally, multiplying Equation (E6.11D) by mole fraction of sulfuric acid gives

$$x_{H_2SO_4}\frac{d\overline{H}_{H_2SO_4}}{dx_{H_2SO_4}} = -18.16\,x_{H_2O}\,x_{H_2SO_4} + 250.44x_{H_2O}^2\,x_{H_2SO_4} \qquad \text{(E6.11E)}$$

Likewise, for the derivative of the partial molar enthalpy of water, we get

$$\frac{d\overline{H}_{H_2O}}{dx_{H_2SO_4}} = -148.80\,x_{H_2SO_4} + 83.48\big[x_{H_2SO_4}(1 - 2x_{H_2O}) + x_{H_2SO_4}^2\big] = x_{H_2SO4}(18.16 - 250.44x_{H_2O})$$

and

$$x_{H_2O}\frac{d\overline{H}_{H_2O}}{dx_{H_2SO_4}} = 18.16\,x_{H_2O}\,x_{H_2SO_4} - 250.44x_{H_2O}^2\,x_{H_2SO_4} \qquad \text{(E6.11F)}$$

Inspection of Equations (E6.11E) and (E6.11F) shows that Equation (E6.11B) is satisfied. ◀

Graphical Determination of Partial Molar Properties

Say we want to calculate the partial molar volume (or any other partial molar property) for a binary mixture when we have a graph of the molar volume (or whatever molar property) vs. mole fraction of one component, as shown in Figure 6.11. The values in this figure are taken from the chloroform (1)–acetone (2) binary mixture discussed in Example 6.9. Recall that the unlike interaction, in this case, is unusually large.[12] Applying Equation (6.35) to this case:

$$v = x_1 \overline{V}_1 + x_2 \overline{V}_2 \tag{6.60}$$

Substituting $x_1 = 1 - x_2$ yields

$$v = (1 - x_2)\overline{V}_1 + x_2 \overline{V}_2 \tag{6.61}$$

Differentiating with respect to x_2, multiplying by x_2, and applying the Gibbs–Duhem equation gives

$$x_2 \frac{dv}{dx_2} = -x_2 \overline{V}_1 + x_2 \overline{V}_2 \tag{6.62}$$

or

$$x_2 \frac{dv}{dx_2} = -\overline{V}_1 + (x_1 \overline{V}_1 + x_2 \overline{V}_2) \tag{6.63}$$

Using the definition for v in Equation (6.60) and rearranging, we get:

$$v = \overline{V}_1 + x_2 \frac{dv}{dx_2} \tag{6.64}$$

intercept   slope

If v is plotted vs. x_2, Equation (6.64) represents a straight line with slope dv/dx_2 and intercept $\overline{V}_1$. Therefore, the partial molar volume for any composition, x_2, can be found by drawing a line tangent to the curve and taking the value of the intercept, as shown for

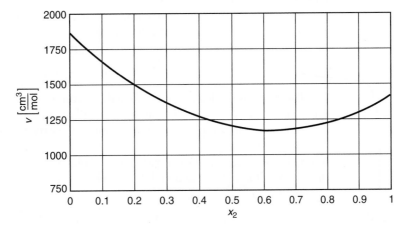

Figure 6.11 Experimental data of the molar volume for a binary system of components 1 and 2.

[12] In fact, this system is chosen for illustration because its pronounced interactions depict the graphical approach well.

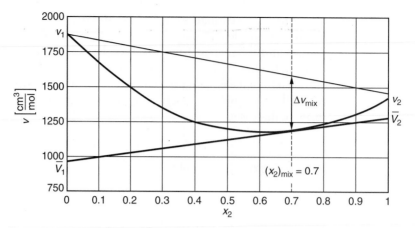

Figure 6.12 Determination of partial molar volumes from graphical data for a binary mixture.

$x_2 = 0.7$ in Figure 6.12. Analogously, $\overline{V}_2$ can be found by taking the value of the line at $x_2 = 1$. This method is descriptively referred to as the **tangent-intercept method**.[13] We can generalize Equation (6.64) to get

$$\overline{K}_1 = k - x_2 \frac{dk}{dx_2} \tag{6.65}$$

only w/ 2 components

for example,

$$\begin{vmatrix} \overline{V}_1 = v - x_2 \dfrac{dv}{dx_2} \\[2mm] \overline{H}_1 = h - x_2 \dfrac{dh}{dx_2} \\[1mm] \vdots \end{vmatrix}$$

Equation (6.65) holds only for binary mixtures. For the generalization to mixtures with more than two components, see Problem 6.33.

It is interesting to consider the limiting cases of the partial molar volume of species 1 in terms of composition of 1. In the limit as x_1 goes to 1, we have all 1 and no 2 in the mixture. In this case, the partial molar volume just equals the pure species molar volume:

$$\overline{V}_1 = v_1 \quad \lim x_1 \longrightarrow 1 \tag{6.66}$$

This result is the logical consequence of our interpretation of a partial molar property. If only species 1 is present, it must contribute entirely to the solution properties. Therefore, it must be equal to the pure solution property. In the other extreme, consider one molecule of species 1 placed in a solution of 2. In this case, 1 interacts only with molecules of 2 and may have a partial molar volume very different from the first case if the nature of the 1-2 interaction differs from the 1-1 interaction. We call this the limit of infinite dilution, $\overline{V}_1^{\infty}$.

$$\overline{V}_1 = \overline{V}_1^{\infty} \quad \lim x_1 \longrightarrow 0 \tag{6.67}$$

Both limits are shown in Figure 6.13. *Note:* $\overline{V}_1^{\infty} \neq v_1$.

---

[13] The tangent-intercept method can also be applied to the partial molar property change of mixing; if Δv_{mix} is plotted vs. x_2, the intercept gives $(\overline{\Delta V}_{mix})_1 = \overline{V}_1 - v_1$.

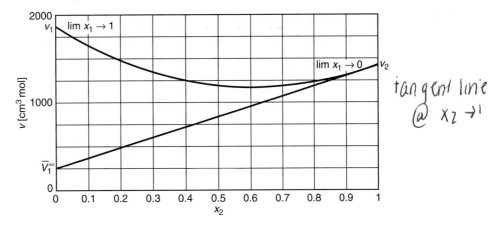

Figure 6.13 Limiting cases of partial molar volumes. (Note that the y-axis scale is different from Figures 6.8 and 6.9.)

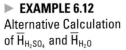

► EXAMPLE 6.12
Alternative Calculation of $\overline{H}_{H_2SO_4}$ and $\overline{H}_{H_2O}$

Develop an expression for the partial molar enthalpy of sulfuric acid in water at 21°C using Equation (6.65).

SOLUTION If we apply Equation (6.65) to this system, we get:

$$\overline{H}_{H_2SO_4} = h - x_{H_2O}\frac{dh}{dx_{H_2O}} \qquad \text{(E6.12A)}$$

Thus, we begin with the expression for the molar enthalpy given by Equation (E6.10A).

$$h = 1.596x_{H_2SO_4} + 1.591x_{H_2O} - 74.40\,x_{H_2SO_4}\,x_{H_2O}(1 - 0.561\,x_{H_2SO_4})\,[kJ/mol] \qquad \text{(E6.12B)}$$

Differentiating Equation (E6.12B) gives

$$\frac{dh}{dx_{H_2O}} = -1.596 + 1.591 - 74.40\big[(x_{H_2SO_4} - x_{H_2O})(1 - 0.561\,x_{H_2SO_4}) + 0.561x_{H_2SO_4}\,x_{H_2O}\big] \qquad \text{(E6.12C)}$$

Substituting Equations (E6.12B) and (E6.12C) into (E6.12A) gives

$$\overline{H}_{H_2SO_4} = 1.596(x_{H_2SO_4} + x_{H_2O}) - 74.40\,x_{H_2O}^2(1 - 0.561\,x_{H_2SO_4}) + 41.74\,x_{H_2SO_4}\,x_{H_2O}^2$$

$$\overline{H}_{H_2SO_4} = 1.596 - 74.40\,x_{H_2O}^2 + 83.48\,x_{H_2SO_4}\,x_{H_2O}^2\,[kJ/mol] \qquad \text{(E6.12D)}$$

Expressions (E6.12D) and (E6.10C) are identical. ◄

► EXAMPLE 6.13
Graphical Determination of $\overline{H}_{H_2SO_4}$ and $\overline{H}_{H_2O}$

Graphically determine values for the partial molar enthalpies of sulfuric acid and water in an equimolar mixture at 21°C by plotting Equation (E6.10A).

SOLUTION A plot of Equation (E6.10A) is shown in Figure E6.13. Also illustrated in the figure is the tangent line to the plot at a mole fraction of 0.5. The values of the intercepts of the tangent lines give

$$\overline{H}_{H_2SO_4} = -6700\,[J/mol] \quad \text{and} \quad \overline{H}_{H_2O} = -17,\!100\,[J/mol]$$

These values agree with the numbers obtained analytically in Example 6.10. ◄

Relations Among Partial Molar Quantities

Partial molar properties can be related to each other, further extending the thermo-dynamic web. For example, consider the total solution enthalpy:

$$H = U + PV \tag{6.68}$$

Differentiating Equation (6.68) with respect to n_i at constant T, P, and n_j, we get

$$\left(\frac{\partial H}{\partial n_i}\right)_{T,P,n_{j\neq i}} = \left(\frac{\partial U}{\partial n_i}\right)_{T,P,n_{j\neq i}} + \left(\frac{\partial PV}{\partial n_i}\right)_{T,P,n_{j\neq i}} \tag{6.69}$$

Since the pressure is constant,

$$\left(\frac{\partial H}{\partial n_i}\right)_{T,P,n_{j\neq i}} = \left(\frac{\partial U}{\partial n_i}\right)_{T,P,n_{j\neq i}} + P\left(\frac{\partial V}{\partial n_i}\right)_{T,P,n_{j\neq i}} \tag{6.70}$$

Applying the definition of a partial molar quantity, we get

$$\overline{H}_i = \overline{U}_i + P\overline{V}_i \tag{6.71}$$

Note the similarity between Equation (6.71), which applies to the partial molar enthalpy, and Equation (6.68), which applies to the total solution enthalpy. It is straightforward to show that the similar derived property relations described in Chapter 5 likewise hold:

How to combine
what we're learning
w/ Stuff from before

$$\overline{G}_i = \overline{H}_i - T\overline{S}_i \tag{6.72}$$

to find G

$$\overline{A}_i = \overline{U}_i - T\overline{S}_i \tag{6.73}$$

We can also find relations for the partial molar Gibbs energy analogous to the Maxwell relations discussed in Chapter 5. We begin by applying Equation (6.31) to Gibbs energy:

$$dG = \left(\frac{\partial G}{\partial T}\right)_{P,n_i} dT + \left(\frac{\partial G}{\partial P}\right)_{T,n_i} dP + \sum_{i=1}^{m} \overline{G}_i dn_i \tag{6.74}$$

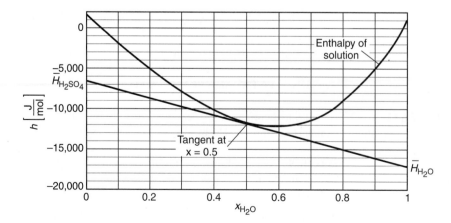

Figure E6.13 Graphical determination of values for the partial molar enthalpies of sulfuric acid and water.

When there is no change in composition, $dn_i = 0$, the above relationship must reduce to Equation (5.9). Therefore, we can write Equation (6.74) as

$$dG = -SdT + VdP + \sum_{i=1}^{m} \overline{G}_i dn_i \tag{6.75}$$

By equating the second derivatives of the first and third terms (as we did for the Maxwell relations), we get

$$\left(\frac{\partial \overline{G}_i}{\partial T} \right)_{P,n_i} = -\overline{S}_i \tag{6.76}$$

A convenient form of the temperature dependence of the partial molar Gibbs energy is given by taking the partial derivative of $(\overline{G}_i/T)$ with respect to T at constant P:

$$\left(\frac{\partial \left(\dfrac{\overline{G}_i}{T} \right)}{\partial T} \right)_{P,n_i} = \frac{1}{T} \left(\frac{\partial \overline{G}_i}{\partial T} \right)_{P,n_i} - \frac{\overline{G}_i}{T^2} = \frac{-T\overline{S}_i - \overline{G}_i}{T^2} = -\frac{\overline{H}_i}{T^2} \tag{6.77}$$

where first the chain rule was applied, then Equation (6.76) was used. Cross-differentiation of the second and third terms of Equation (6.75) yields

$$\left(\frac{\partial \overline{G}_i}{\partial P} \right)_{T,n_i} = \overline{V}_i \tag{6.78}$$

The role of the partial molar Gibbs energy in phase equilibria will be discussed in the next section. For now, we can see that the two equations above are useful in determining the pressure and temperature dependence of this quantity.

► 6.4 MULTICOMPONENT PHASE EQUILIBRIA

The Chemical Potential—The Criteria for Chemical Equilibrium

We are now ready to combine the phase equilibria criteria developed in Section 6.2 with our description for mixtures in Section 6.3 to synthesize the complete phase equilibria problem (see Figure 6.2). We begin with our criterion for chemical equilibria, Equation (6.9). Writing the differential change in the total Gibbs energy as the sum of the differential change in each phase gives

$$dG = 0 = dG^\alpha + dG^\beta \tag{6.79}$$

Substituting the fundamental property relation given by Equation (6.75) to each phase, we get

$$0 = \left[-SdT + VdP + \sum_{i=1}^{m} \overline{G}_i dn_i \right]^\alpha + \left[-SdT + VdP + \sum_{i=1}^{m} \overline{G}_i dn_i \right]^\beta \tag{6.80}$$

We can now apply our criteria for thermal equilibrium and mechanical equilibrium. For thermal equilibrium, we have:

$$T^\alpha = T^\beta \tag{6.81}$$

Thus, if we have a differential change in the temperature of phase α, it must be matched by a corresponding differential change in the temperature of phase β:

$$dT^{\alpha} = dT^{\beta} \tag{6.82}$$

Similarly, the criteria for mechanical equilibrium can be applied to differential changes in pressure to get

$$dP^{\alpha} = dP^{\beta} \tag{6.83}$$

Applying Equations (6.82) and (6.83) to Equation (6.80) gives

$$0 = \left[\sum_{i=1}^{m} \overline{G}_i dn_i \right]^{\alpha} + \left[\sum_{i=1}^{m} \overline{G}_i dn_i \right]^{\beta} \tag{6.84}$$

The partial molar Gibbs energy is such an important quantity in chemical equilibria that it is given a special name: the **chemical potential**, μ_i:

$$\mu_i \equiv \left(\frac{\partial G}{\partial n_i} \right)_{T,P,n_{j \neq i}} \tag{6.85}$$

We will see shortly why. It is important to remember that chemical potential and partial molar Gibbs energy are synonymous. In terms of chemical potential, Equation (6.84) becomes

$$0 = \sum_{m} \mu_i^{\alpha} dn_i^{\alpha} + \sum_{m} \mu_i^{\beta} dn_i^{\beta} \tag{6.86}$$

However, since we have a closed system, any species leaving phase α will enter phase β. Hence,

$$dn_i^{\alpha} = -dn_i^{\beta} \tag{6.87}$$

Applying Equation (6.87) to Equation (6.86), we get

$$0 = \sum_{m} \left(\mu_i^{\alpha} - \mu_i^{\beta} \right) dn_i^{\alpha} \tag{6.88}$$

Now, for Equation (6.88) to be true in general,

$$\boxed{\mu_i^{\alpha} = \mu_i^{\beta}} \tag{6.89}$$

Equation (6.89) applies to all m species in the system; that is, there are m different equations here. By comparing Equation (6.89) to Equation (6.14), it is evident that the criterion for chemical equilibrium in mixtures is obtained by replacing the pure species quantity, the molar Gibbs energy, g_i, with the contribution of species i to the Gibbs energy of the mixture, that is, its partial molar Gibbs energy, $\overline{G}_i = \mu_i$. This realization could be intuited by our interpretation of a partial molar quantity in Section 6.3.

The chemical potential is an abstract concept; it cannot directly be measured. However, the relation between chemical potential and mass transport is identical to the relation between temperature and energy transport or pressure and momentum transport. This concept is illustrated in Figure 6.14. In Figure 6.14a, we see two systems of different temperature. If they are placed in contact, there will be energy transfer via

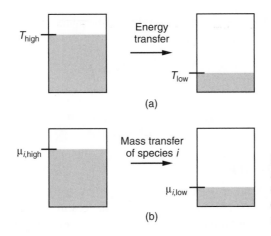

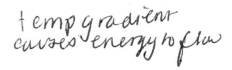

temp gradient causes energy to flow

(a)

(b)

Figure 6.14 Conceptual illustration of the analogy between (*a*) temperature as the driving force for energy transfer and (*b*) chemical potential as the driving force for mass transfer.

heat from high to low temperature until the temperatures become equal and we have reached thermal equilibrium. We could very well call temperature "thermal potential," since it provides the driving force toward thermal equilibrium. However, we already know this property from physical experience as temperature, so we stick with that name. Figure 6.14*b* shows the analogous relation between chemical potential and diffusion. Here we see two systems of different chemical potential for species *i*. In this case, species *i* will transport from high to low chemical potential until the chemical potentials become equal and we have reached chemical equilibrium. If we know μ_i for each phase, we know which way species *i* will tend to transfer. The biggest difficulty in understanding chemical potential is that it is an abstract concept, whereas *T* and *P* are measured properties with which we have direct experience. However, we can apply the concept of chemical potential to learn about driving forces for species transfer in much the same manner as we apply temperature to energy transport.

▶ **EXAMPLE 6.14**
Determination of the
Gibbs Phase Rule for
Nonreacting Systems

Consider a system at temperature *T* and pressure *P* with *m* species present in π phases. How many measurable properties need to be determined (e.g., *T*, *P*, and x_i) to constrain the state of the entire system?

SOLUTION From Equations (6.82), (6.83), and (6.89), we can construct the following set of equations for *m* species and π phases:

$$T^\alpha = T^\beta = \ldots = T^\pi$$
$$P^\alpha = P^\beta = \ldots = P^\pi$$
$$\mu_1^\alpha = \mu_1^\beta = \ldots = \mu_1^\pi$$
$$\mu_2^\alpha = \mu_2^\beta = \ldots = \mu_2^\pi \qquad \text{(E6.14A)}$$
$$\vdots$$
$$\mu_m^\alpha = \mu_m^\beta = \ldots = \mu_m^\pi$$

Each row in the set of equations above has $(\pi - 1)$ equal signs. Thus, there are a total of $(\pi - 1)(m + 2)$ equalities in the set of equations above. The chemical potential in a given phase depends on the temperature, pressure, and mole fraction of each of the species present. Since the sum of mole fractions equals 1, we have to know $(m - 1)$ mole fractions for each phase along with the temperature and pressure to constrain the state of that phase. Thus we must specify $(m + 1)\pi$ variables to determine the state of the system. The number of variables we can independently pick (the so-called degrees of freedom, $\Im$) is obtained by subtracting the total $(m + 1)\pi$ variables we need to specify by the $(\pi - 1)(m + 2)$ equalities in Equations (E6.14A). Thus, we can independently specify:

$$\Im = (m + 1)\pi - (\pi - 1)(m + 2) = m - \pi + 2 \tag{E6.14B}$$

quantities. Equation (E6.14B) is identical to Equation (1.19). This example does not consider chemical reaction. If we have reactions, we place additional constraints due to the reaction stoichiometry. This case is addressed in Chapter 9.   ◀

Temperature and Pressure Dependence of μ_i

not important

As we saw in Section 6.2 for the case of a pure species, we are often interested in how a system in equilibrium responds to changes in measured variables. We can use the thermodynamic web to relate the change in chemical potential and the criteria for chemical equilibrium between phases with changes in pressure, temperature, and mole fraction. It is convenient to begin by dividing Equation (6.89) by T:

$$\frac{\mu_i^\alpha}{T} = \frac{\mu_i^\beta}{T} \tag{6.90}$$

If we choose temperature, pressure, and mole fraction of species i as independent variables, the change in chemical potential divided by temperature can be related by the following partial differentials:

$$\left[\frac{\partial(\mu_i^\alpha/T)}{\partial T}\right]_{P,x_m^\alpha} dT + \left[\frac{\partial(\mu_i^\alpha/T)}{\partial P}\right]_{T,x_m^\alpha} dP + \left[\frac{\partial(\mu_i^\alpha/T)}{\partial x_i}\right]_{T,P} dx_i^\alpha$$

$$= \left[\frac{\partial(\mu_i^\beta/T)}{\partial T}\right]_{P,x_m^\beta} dT + \left[\frac{\partial(\mu_i^\beta/T)}{\partial P}\right]_{T,x_m^\beta} dP + \left[\frac{\partial(\mu_i^\beta/T)}{\partial x_i}\right]_{T,P} dx_i^\beta \tag{6.91}$$

Factoring out $(1/T)$ in the second and third terms on each side of Equation (6.91) gives

$$\left[\frac{\partial(\mu_i^\alpha/T)}{\partial T}\right]_{P,x_m^\alpha} dT + \frac{1}{T}\left[\frac{\partial\mu_i^\alpha}{\partial P}\right]_{T,x_m^\alpha} dP + \frac{1}{T}\left[\frac{\partial\mu_i^\alpha}{\partial x_i}\right]_{T,P} dx_i^\alpha$$

$$= \left[\frac{\partial(\mu_i^\beta/T)}{\partial T}\right]_{P,x_m^\beta} dT + \frac{1}{T}\left[\frac{\partial\mu_i^\beta}{\partial P}\right]_{T,x_m^\beta} dP + \frac{1}{T}\left[\frac{\partial\mu_i^\beta}{\partial x_i}\right]_{T,P} dx_i^\beta \tag{6.92}$$

Applying Equations (6.77) and (6.78) gives

$$-\frac{\overline{H}_i^\alpha}{T^2}dT + \frac{\overline{V}_i^\alpha}{T}dP + \frac{1}{T}\left[\frac{\partial\mu_i^\alpha}{\partial x_i^\alpha}\right]_{T,P} dx_i^\alpha = -\frac{\overline{H}_i^\beta}{T^2}dT + \frac{\overline{V}_i^\beta}{T}dP + \frac{1}{T}\left[\frac{\partial\mu_i^\beta}{\partial x_i^\beta}\right]_{T,P} dx_i^\beta \tag{6.93}$$

Equation (6.73) is valid for any two phases α and β, in general. We now consider the specific case of vapor–liquid equilibrium. Denoting the vapor-phase mole fraction, y_i, and the liquid-phase mole fraction, x_i, Equation (6.73) becomes

$$-\frac{\overline{H}_i^v}{T^2}dT + \frac{\overline{V}_i^v}{T}dP + \frac{1}{T}\left[\frac{\partial\mu_i^v}{\partial y_i}\right]_{T,P} dy_i = -\frac{\overline{H}_i^l}{T^2}dT + \frac{\overline{V}_i^l}{T}dP + \frac{1}{T}\left[\frac{\partial\mu_i^l}{\partial x_i}\right]_{T,P} dx_i \tag{6.94}$$

For the case of an **ideal gas**, we can simplify Equation (6.94) further. The partial molar enthalpy and the partial molar volume equal the pure species molar enthalpy and volume, respectively, since the interactions of species i in the mixture are the same as its interactions as a pure species:

$$\overline{H}_i^v = h_i^v \tag{6.95}$$

$$\overline{V}_i^v = v_i^v = \frac{RT}{P} \tag{6.96}$$

Similarly, the change in chemical potential with respect to mole fraction is given by the change in pure species Gibbs energy with respect to mole fraction:

$$\left(\frac{\partial g_i}{\partial y_i}\right)_{T,P} = \left(\frac{\partial h_i}{\partial y_i}\right)_{T,P} - T\left(\frac{\partial s_i}{\partial y_i}\right)_{T,P} \tag{6.97}$$

For an ideal gas, the enthalpy is independent of mole fraction. To find the dependence of entropy, we apply the analogy of Equation (E6.8B). So at constant pressure:

$$ds_i|_{T,P} = -R\,d\ln(y_i) = -\frac{R}{y_i}dy_i \tag{6.98}$$

Using Equation (6.98) in (6.97) gives

$$\left[\frac{\partial \mu_i^v}{\partial y_i}\right]_{T,P} = \left[\frac{\partial g_i^v}{\partial y_i}\right]_{T,P} = \frac{RT}{y_i} \tag{6.99}$$

Substituting Equations (6.95), (6.96), and (6.99) into Equation (6.94) gives

$$-\frac{h_i^v}{T^2}dT + R\frac{dP}{P} + R\frac{dy_i}{y_i} = -\frac{\overline{H}_i^l}{T^2}dT + \frac{\overline{V}_i^l}{T}dP + \frac{1}{T}\left[\frac{\partial \mu_i^l}{\partial x_i}\right]_{T,P}dx_i \tag{6.100}$$

Equation (6.100) shows how the temperature or pressure of a phase transition for a liquid–ideal gas mixture is related to changes in composition. We will discuss phenomena of boiling point elevation and freezing point depression in this context in Chapter 8.

▶ 6.5 SUMMARY

In this chapter, we formulated the criteria for equilibrium between two phases. We labeled the phases generally as α and β, which can represent the vapor, liquid, or solid phases. We reduced this problem into two parts. First, we addressed pure species phase equilibrium. We determined that the derived property **Gibbs energy** is a minimum at equilibrium, so only when two phases have equal values of Gibbs energy can they coexist. Second, we addressed the thermodynamics of mixtures. We discovered that the **partial molar property**, $\overline{K}_i$, is representative of the contribution of species i to the mixture. In analogy to the case for pure species, the criterion for chemical equilibrium between two phases for species i in a mixture is that the partial molar Gibbs energy, $\overline{G}_i$, is equal in the two phases. Since it sets the criteria for chemical equilibrium, we often call the partial molar Gibbs energy the **chemical potential**, μ_i. In summary, the criteria for equilibrium between phases α and β can be written:

$$T^\alpha = T^\beta \qquad \text{Thermal equilibrium} \tag{6.1}$$

$$P^\alpha = P^\beta \qquad \text{Mechanical equilibrium} \tag{6.2}$$

and

$$\mu_i^\alpha = \mu_i^\beta \qquad \text{Chemical equilibrium} \tag{6.89}$$

Gibbs energy establishes the criteria for pure species phase equilibria by accounting for the balance between the system's tendency to minimize energy and its tendency to maximize entropy. At low temperature, energetic effects become more important, while at higher temperature entropic effects dominate. We can apply the thermodynamic web to relate how the pressure of a system in phase equilibrium varies with temperature. This analysis leads to the **Clapeyron equation**. Application of the Clapeyron equation to vapor–liquid equilibrium, together with the assumptions of a negligibly small liquid volume and an ideal gas, leads to the **Clausius–Clapeyron equation**. The integrated form of the Clausius–Clapeyron equation is functionally similar to the **Antoine equation**, a common empirical equation used to correlate pure species saturation pressures with temperature.

Thermodynamic properties of a **mixture** are affected both by the like (i-i) interactions and by the unlike (i-j) interactions, that is, how each of the species in the mixture interacts with all of the other species it encounters. The **total solution property** K, ($K = V, H, U, S, G, \ldots$), represents a given property of the entire mixture. It can be written as the sum of the partial molar properties of its constituent species, each adjusted in proportion to how much is present:

$$K = \sum_{n_i} n_i \overline{K}_i \tag{6.34}$$

where the partial molar property is defined by

$$\overline{K}_i = \left(\frac{\partial K}{\partial n_i} \right)_{T,P,n_{j \neq i}} \tag{6.29}$$

Additionally, the **pure species property**, k_i, is defined as the value of that property of species i as it exists as a pure species *at the same T and P of the mixture*. Values of a partial molar property for a species in a mixture can be calculated from an analytical expression by applying Equation (6.29) and by graphical methods, as illustrated in Figure 6.12. In the case of **infinite dilution**, species i becomes so dilute that a molecule of species i will not have any like species with which it interacts; rather, it will interact only with unlike species. Additionally, partial molar properties of different species in a mixture can be related to one another by the **Gibbs–Duhem equation**:

$$0 = \sum n_i d\overline{K}_i \quad \text{Const } T \text{ and } P \tag{6.37}$$

A **property change of mixing**, ΔK_{mix}, describes how much a given property changes as a result of a process in which pure species are mixed together. It is written as

$$\Delta K_{\text{mix}} = K - \sum n_i k_i$$

$$= \sum n_i (\overline{K}_i - k_i) \tag{6.43, 6.44}$$

Volume and enthalpy changes of mixing have values of zero when the like and unlike interactions are identical. When the unlike interactions are more attractive, these quantities are negative, while they are positive when unlike interactions are less favorable. Conversely, the entropy of mixing is positive in all cases, since there are many more ways to configure a mixture as compared to the pure species. The energetic interactions characteristic of the mixing process are often reported as the **enthalpy of solution** $\Delta \tilde{h}_s$. The enthalpy of solution corresponds to the enthalpy change when 1 mole of solute is mixed in n moles of solvent.

▶ 6.6 PROBLEMS

6.1

(a) Use the Clausius–Clapeyron equation and data for water at 100°C to develop an expression for the vapor pressure of water as a function of temperature.

(b) Plot the expression you came up with on a PT diagram for temperatures from 0.01°C to 100°C.

(c) Include data from the steam tables on your plot in part (b) and comment on the adequacy of the Clausius–Clapeyron equation.

(d) Repeat parts (b) and (c) for 100°C to 200°C.

(e) Repeat parts (a)–(c), but correct for the temperature dependence of Δh_{vap}, using heat capacity data from Appendix A.2.

6.2 What pressure is needed to isothermally compress ice initially at -5°C and 1 bar so that it changes phase?

6.3 One mole of a pure species exists in liquid–vapor equilibrium in a rigid container of volume $V = 1$ L, a temperature of 300 K, and a pressure of 1 bar. The enthalpy of vaporization and the second virial coefficient in the pressure expansion are:

$$\Delta h_{vap} = 16,628 \; [\text{J/mol}] \quad \text{and} \quad B' = -1 \times 10^{-7} \; [\text{m}^3/\text{J}]$$

Assume the enthalpy of vaporization does not change with temperature. You may *neglect the molar volume of the liquid* relative to that of the gas.

(a) How many moles of vapor are there?

(b) This container is heated until the pressure reaches 21 bar and is allowed to reach equilibrium. Both vapor and liquid phases are still present. Find the final temperature of this system.

(c) How many moles of vapor are there now?

6.4 Tired of studying thermo, you come up with the idea of becoming rich by manufacturing diamond from graphite. To do this process at 25°C requires increasing the pressure until graphite and diamond are in equilibrium. The following data are available at 25°C:

$$\Delta g(25°\text{C}, \; 1 \; \text{atm}) = g_{diamond} - g_{graphite} = 2866 \; [\text{J/mol}]$$

$$\rho_{diamond} = 3.51 \; [\text{g/cm}^3]$$

$$\rho_{graphite} = 2.26 \; [\text{g/cm}^3]$$

Estimate the pressure at which these two forms of carbon are in equilibrium at 25°C.

6.5 You wish to know the melting temperature of aluminum at 100 bar. You find that at atmospheric pressure, Al melts at 933.45 K and the enthalpy of fusion is:

$$\Delta h_{fus} = -10,711 \; [\text{J/mol}]$$

Heat capacity data are given by

$$c_P^l = 31.748 \; [\text{J/(mol K)}], \quad c_P^s = 20.068 + 0.0138T \; [\text{J/(mol K)}]$$

Take the density of solid aluminum to be 2700 [kg/m³] and liquid to be 2300 [kg/m³]. At what temperature does Al melt at 100 bar?

6.6 The vapor pressure of silver (between 1234 K and 2485 K) is given by the following expression:

$$\ln P = -\frac{14,260}{T} - 0.458 \ln T + 12.23$$

with P in torr and T in K. Estimate the enthalpy of vaporization at 1500 K. State the assumptions that you make.

6.7 A pure fluid shows the following s vs. T behavior. Draw schematically how the chemical potential would change with temperature.

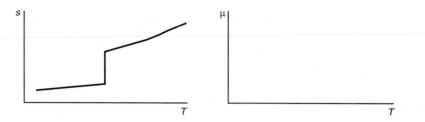

6.8 At room temperature, iron exists in the ferrite phase (α-Fe). At 912°C, it goes through a phase transformation to the austenite phase (γ-Fe). Which phase of iron has stronger bonds? Explain.

6.9 At a temperature of 60.6°C, benzene exerts a saturation pressure of 400 torr. At 80.1°C, its saturation pressure is 760 torr. Using only these data, estimate the enthalpy of vaporization of benzene. Compare it to the reported value of $\Delta h_{vap} = 35\,[\text{kJ/mol}]$.

6.10 Consider the crystallization of species a. The molar Gibbs energy of pure species a vs. temperature at a pressure of 1 bar is shown below. Take the molar volume of species a in the liquid phase to be 20% larger than its molar volume as a solid.

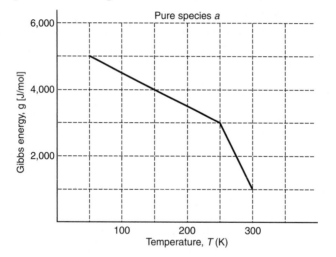

Answer the following questions:

(a) Identify the location of the freezing point on the diagram above. Identify which section of the plot corresponds to liquid and which part corresponds to solid. What is the temperature at which the liquid crystallizes? What is the Gibbs energy of the liquid at this point?

(b) Come up with a value for the entropy of the solid phase and the liquid phase.

(c) Consider this process occurring at a much higher pressure. Sketch how the plot above will change. Will the freezing point be higher or lower? Try to be as accurate as possible with the features of your sketch. Write down all your assumptions.

6.11 A pure substance shows the following v vs. P behavior at constant temperature. Draw schematically how the molar Gibbs energy would change with pressure. Explain your reasoning and describe the important features on your plot.

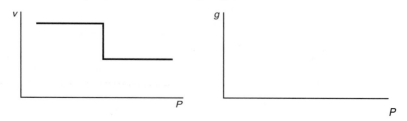

6.12 Pure ethanol boils at a temperature of 63.5°C at a pressure of 400 torr. It also boils at 78.4°C and 760 torr. Estimate the saturation pressure for ethanol at 100°C.

6.13 An alternative criteria for chemical equilibrium between two phases of pure species i can be written:

$$\left(\frac{g_i}{T}\right)^{\alpha} = \left(\frac{g_i}{T}\right)^{\beta}$$

Apply the thermodynamic web to show that the partial derivative of this function with respect to temperature at constant pressure is given by

$$\left[\frac{\partial \left(\frac{g_i}{T}\right)}{\partial T}\right]_P = -\frac{h_i}{T^2}$$

6.14 At 922 K, the enthalpy of liquid Mg is 26.780 [kJ/mol] and the entropy is 73.888 [J/(mol K)]. Determine the Gibbs energy of liquid Mg at 1300 K. The heat capacity of the liquid is constant over this temperature range and has a value of 32.635 [J /(mol K)].

6.15 Solid sulfur undergoes a phase transition between the monoclinic (m) and orthorhombic (o) phases at a temperature of 368.3 K and pressure of 1 bar. Calculate the difference in Gibbs energy between monoclinic sulfur and orthorhombic sulfur at 298 K and a pressure of 1 bar. Which phase is more stable at 298 K? Take the entropy in each phase to be given by the following expressions:

$$\text{Monoclinic phase:} \qquad s_m = 13.8 + 0.066T \; [\text{J}/(\text{mol K})]$$

$$\text{Orthorhombic phase:} \qquad s_o = 11.0 + 0.071T \; [\text{J}/(\text{mol K})]$$

6.16 At 900 K, solid Sr has values of enthalpy and entropy of 20.285 [kJ/(mol)] and 91.222 [J/(mol K)], respectively. At 1500 K, liquid Sr has values of enthalpy and entropy of 49.179 [kJ/mol] and 116.64 [J/(mol K)], respectively. The heat capacity for the solid and liquid phases is given by

$$(c_P)^s_{\text{Sr}} = 37.656 \; [\text{J}/(\text{mol K})] \quad \text{and} \quad (c_P)^l_{\text{Sr}} = 35.146 \; [\text{J}/(\text{mol K})]$$

respectively. Using only these data, determine the temperature of the phase transition between solid and liquid. What is the enthalpy of fusion? The result of Problem 6.13 could be useful.

6.17 At 1100 K, solid SiO_2 has values of enthalpy and entropy of −856.84 [kJ/mol] and 124.51 [J/(mol K)], respectively. At 2500 K, liquid SiO_2 has values of enthalpy and entropy of −738.44 [kJ/mol] and 191.94 [J/(mol K)], respectively. The heat capacities for the solid and liquid phases are given by:

$$(c_P)^s_{\text{SiO}_2} = 53.466 + 0.02706T - 1.27 \times 10^{-5}\,T^2 + 2.19 \times 10^{-9}\,T^3 \; [\text{J}/(\text{mol K})]$$

$$(c_P)^l_{\text{SiO}_2} = 85.772 \; [\text{J}/(\text{mol K})]$$

Using only these data, determine the temperature of the phase transition between solid and liquid. What is the enthalpy of fusion?

6.18 Determine the second virial coefficient, B, for CS_2 at 100°C from the following data. The saturation pressure of carbon disulfide (CS_2) has been fit to the following equation:

$$\ln P^{\text{sat}}_{\text{CS}_2} = 62.7839 - \frac{4.7063 \times 10^3}{T} - 6.7794 \ln T + 8.0194 \times 10^{-3}\,T$$

where T is in $[K]$ and $\ln P_{CS_2}^{sat}$ is in $[Pa]$. The enthalpy of vaporization for CS_2 at $100°C$ has been reported as

$$\Delta h_{vap, CS_2} = 24.050 \, [KJ/mol]$$

Compare to the reported value of

$$B_{CS_2} = -492 \, [cm^3/mol]$$

6.19 Consider an ideal gas mixture at 83.14 kPa and 500 K. It contains 2 moles of species A and 3 moles of species B. Calculate the following: $\overline{V}_A$, $\overline{V}_B$, v_A, v_B, V_A, V_B, V, v, ΔV_{mix}, Δv_{mix}.

6.20 For a given binary system at constant T and P, the molar volume (in cm^3/mol) is given by

$$v = 100y_a + 80y_b + 2.5y_ay_b$$

(a) What is the pure species molar volume for species a, v_a?

(b) Come up with an expression for the partial molar volume, $\overline{V}_a$, in terms of y_b. What is the partial molar volume at infinite dilution, $\overline{V}_a^\infty$?

(c) Is the volume change of mixing, Δv_{mix}, greater than, equal to, or less than 0? Explain.

6.21 Consider a mixture of species 1, 2, and 3. The following equation of state is available for the vapor phase:

$$Pv = RT + P^2[A(y_1 - y_2) + B]$$

where

$$\frac{A}{RT} = -9.0 \times 10^{-5} \left[\frac{1}{atm^2}\right], \frac{B}{RT} = 3.0 \times 10^{-5} \left[\frac{1}{atm^2}\right]$$

and y_1, y_2, and y_3 are the mole fractions of species 1, 2 and 3, respectively. Consider a vapor mixture with 1 mole of species 1, 2 moles of species 2, and 2 moles of species 3 at a pressure of 50 atm and a temperature of 500 K. Calculate the following quantities: v, V, v_1, v_2, v_3, $\overline{V}_1$.

6.22 The molar enthalpy of a ternary mixture of species a, b, and c can be described by the following expression:

$$h = -5000x_a - 3000x_b - 2200x_c - 500x_ax_bx_c \, [J/mol]$$

(a) Come up with an expression for $\overline{H}_a$.

(b) Calculate $\overline{H}_a$ for a solution with 1 mole a, 1 mole b, and 1 mole c.

(c) Calculate $\overline{H}_a$ for a solution with 1 mole a but with no b or c present.

(d) Calculate $\overline{H}_b$ for a solution with 1 mole b but with no a or c present.

6.23 Plot the partial molar volumes of CO_2 and C_3H_8 in a binary mixture at $100°C$ and 20 bar as a function of mole fraction CO_2 using the van der Waals equation of state.

6.24 The Gibbs energy of a binary mixture of species a and species b at 300 K and 10 bar is given by the following expression:

$$g = -40x_a - 60x_b + RT(x_a \ln x_a + x_b \ln x_b) + 5x_ax_b \, [kJ/mol]$$

(a) For a system containing 1 mole of species a and 4 moles of species b, find the following: g_a, $\overline{G}_a$, $\overline{G}_a^\infty$, ΔG_{mix}.

(b) If the pure species are mixed together adiabatically, do you think the temperature of the system will increase, stay the same or decrease. Explain, stating any assumptions that you make.

6.25 Consider a binary mixture of species 1 and species 2. A plot of the partial molar volumes in $[cm^3/mol]$ of species 1 and 2, $\overline{V}_1$ and $\overline{V}_2$, vs. mole fraction of *species 1* is shown below. For a mixture of 1 mole of species 1 and 4 moles of species 2, determine the following quantities for this mixture: $\overline{V}_1$, $\overline{V}_2$, v_1, v_2, V_1, V_2, V, v, ΔV_{mix}.

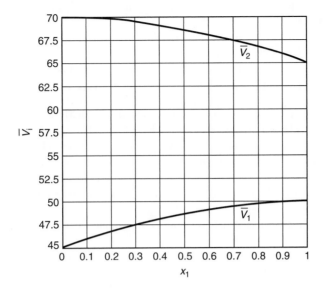

6.26 Enthalpies of mixing for binary mixtures of cadmium (Cd) and tin (Sn) have been fit to the following equation at 500°C:

$$\Delta h_{\text{mix}} = 13,000 X_{\text{Cd}} X_{\text{Sn}} \; [\text{J/mol}]$$

where, X_{Cd} and X_{Sn} are the cadmium and tin mole fractions, respectively. Consider a mixture of 3 moles Cd and 2 moles Sn.

(a) Show that

$$(\overline{\Delta H}_{\text{mix}})_{\text{Cd}} = \overline{H}_{\text{Cd}} - h_{\text{Cd}}$$

(b) *Based on the equations above*, calculate values for $\overline{H}_{\text{Cd}} - h_{\text{Cd}}$ and $\overline{H}_{\text{Sn}} - h_{\text{Sn}}$ at 500°C.

(c) Show that the results are consistent with the Gibbs–Duhem equation.

(d) Data from which the above equation was derived are presented below, along with the model fit. *Graphically* determine values for $\overline{H}_{\text{Cd}} - h_{\text{Cd}}$ and $\overline{H}_{\text{Sn}} - h_{\text{Sn}}$ at 500°C. Compare your answer to part (b). Show your work.

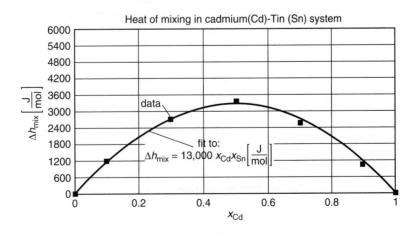

6.27 For a given binary system, the partial molar volume of species 1 is constant. What can you say about species 2? Explain.

6.28 Enthalpies of solution, $\Delta \tilde{h}_s$, are reported in Table 6.1 for 1 mole of HCl diluted in n moles of H_2O at 25°C:

(a) Consider a mixture of 8 moles H_2O and 2 moles HCl. As best you can from these data, estimate $\overline{H}_{H_2O} - h_{H_2O}$ and $\overline{H}_{HCl} - h_{HCl}$.

(b) For a mixture of 80 moles H_2O and 2 moles HCl, estimate $\overline{H}_{H_2O} - h_{H_2O}$.

6.29 Aqueous HCl can be manufactured by gas-phase reaction of H_2 and Cl_2 to form HCl(g), followed by absorption of HCl(g) with water. Consider a steady-state process at 25°C where HCl(g) and pure water are fed to form aqueous acid with 30% HCl by weight. What is the amount of cooling that must be provided per mole of product produced?

6.30 How does the enthalpy of mixing data for H_2SO_4 given by Equation (6.47) compare to the enthalpy of solution data from Table 6.1.? Is the agreement reasonable? What are the reasons they may be different?

6.31 Calculate the enthalpy of mixing for HCl from the enthalpy of solution data reported in Table 6.1.

6.32 What is the heat requirement to dilute an inlet aqueous stream of 50% NaOH, by weight, to a final concentration of 10%?

6.33 Develop an equation for a ternary mixture analogous to Equation (6.65). Then generalize to a mixture with m components.

6.34 The partial molar volume of benzene (1) in cyclohexane (2) at 30°C is given by the following expression:

$$\overline{V}_1 = 92.6 - 5.28x_1 + 2.64x_1^2 \, [\text{cm}^3/\text{mol}]$$

Find an expression for the partial molar volume of cyclohexane. The density of cyclohexane at 30°C is 0.768 $[\text{g/cm}^3]$.

6.35 Using data from Table 6.1, find the partial molar enthalpy of water in a mixture of ethanol (1) and water (2) at 25°C with $x_1 = 0.33$. Use the same reference state as used in the steam tables.

6.36 Consider the isothermal mixing of 20% solute 1 by weight and 80% water, 2, at 25°C. What is the heat transferred for the following mixtures?

(a) pure H_2SO_4 (1) and H_2O (2)

(b) 18 M H_2SO_4 (1) and H_2O (2). (the density of 18 M H_2SO_4 is reported as 1.84 g/cm³)

(c) solid NaOH (1) and H_2O (2)

(d) NH_3 gas (1) and H_2O (2)

6.37 The following data are available for a binary mixture of ethanol and water at 20°C:

Wt % EtOH	ρ [g/ml]
0	0.99823
10	0.98187
20	0.96864
30	0.95382
40	0.93518
50	0.91384
60	0.89113
70	0.86766
80	0.84344
90	0.81797
100	0.78934

E. W. Washburn (ed.), *International Critical Tables* (Vol. V) (New York: McGraw-Hill, 1929).

(a) Make a plot of the *partial* molar volumes of ethanol and water vs. mole fraction ethanol.

(b) What is Δv_{mix} for an equimolar solution?

6.38 The following data have been reported for the density, ρ, vs. mole fraction ethanol of binary mixtures of ethanol (1) and formamide (2) at 25°C and 1 bar.

x_1	ρ, [g/cm^3]
0	1.1314
0.1000	1.0846
0.1892	1.0457
0.2976	1.0042
0.3907	0.9678
0.5009	0.9335
0.5929	0.9022
0.6986	0.8701
0.8009	0.8401
0.8995	0.8126
1	0.7857

E. W. Washburn (ed.), *International Critical Tables* (Vol. V)
(New York: McGraw-Hill, 1929).

Consider a mixture of 3 moles ethanol and 1 mole formamide at 25°C and 1 bar. As best you can, determine the following quantities: v_1, V_1, v_2, V_2, v, V, Δv_{mix}, ΔV_{mix}, $\overline{V}_1$, $\overline{V}_2$. The molecular weights are $MW_1 = 46$ [g/mol] and $MW_2 = 45$ [g/mol].

6.39 Consider a binary mixture of ideal gases, a and b, at temperature T and pressure P. Come up with an expression for $(\overline{\Delta G}_{\text{mix}})_a$ in terms of T, P, and y_a. What is the value of $(\overline{\Delta G}_{\text{mix}})_a^\infty$?

6.40 Consider a system in which liquid water is in phase equilibrium with wet air at 25°C and 1 bar. What is the partial molar Gibbs energy of the water in the vapor phase? You may assume the liquid is pure water and the vapor behaves as an ideal gas.

CHAPTER 7

Phase Equilibria II: Fugacity

Learning Objectives

To demonstrate mastery of the material in Chapter 7, you should be able to:

▶ Find the fugacity and fugacity coefficient of gaseous species i as a pure species and in a mixture using tables, equations of state, and general correlations. Identify the appropriate reference state. Write the Lewis fugacity rule, state the approximation on which it is based, and identify the conditions when it is likely to be valid.

▶ For liquids and solids, determine the activity coefficients for binary and multicomponent mixtures through activity coefficient models, including the two-suffix Margules equation, the three-suffix Margules equation, the van Laar equation, and the Wilson equation. Identify when the symmetric activity coefficient model is appropriate and when you need to use an asymmetric model.

▶ State the molecular conditions when a liquid or solid forms an ideal solution. Identify Lewis/Randall and Henry's law reference states for ideal solutions, including the molecular interactions on which each reference state is based.

▶ Calculate the pure species fugacity of a liquid or solid at high pressure using the Poynting correction. Identify how you correct a value of Henry's law for different pressures or temperatures.

▶ Define *fugacity, fugacity coefficient, activity, activity coefficient*, and *excess Gibbs energy*. State the criteria for chemical equilibrium in terms of fugacity. State why the excess Gibbs energy is useful for empirical models of activity coefficients.

▶ Apply the Gibbs–Duhem equation to relate activity coefficients of different species in a mixture. Evaluate whether a set of activity coefficient data is thermodynamically consistent. Given values of the pure species fugacity and the Henry's law constant, convert between activity coefficients based the Lewis/Randall rule, γ_i, and Henry's Law, $\gamma_i^{\text{Henry's}}$.

▶ 7.1 INTRODUCTION

We have just seen that the chemical potential provides the criteria for chemical equilibrium of species i between phases α and β in a multicomponent system:

$$\mu_i^{\alpha} = \mu_i^{\beta} \tag{7.1}$$

It is a derived thermodynamic property, unlike the measured thermodynamic properties, temperature and pressure, that provide the criteria for thermal and mechanical equilibrium, respectively. Although the chemical potential is an abstract concept, it is useful since it provides a simple criterion for chemical equilibria of each species i.

Unfortunately, in application, it turns out the chemical potential has some inconvenient mathematical behaviors (which we will see shortly). Consequently, it is convenient to define a new derived thermodynamic property that is mathematically better behaved but provides just as simple a criterion for equilibrium: the fugacity.

In a sense, it is hard to get a handle on fugacity since it is an abstraction on the chemical potential, that is, an (abstraction)^2. Moreover, it appears to be almost arbitrarily defined; that is, there are undoubtedly several other types of "mathematical transformations" that would work. These attributes make us uneasy about fugacity. We feel uncomfortable because we do not fully understand it. My suggestion is to consider fugacity as a mathematical transformation of the chemical potential that we use to get a property that is mathematically better behaved. From there, just concentrate on how to apply the concept of fugacity to solve problems. Although fugacity is a very abstract and mysterious concept, it works!

▶ 7.2 THE FUGACITY

Definition of Fugacity

The definition of *fugacity* can be attributed to the thermodynamics giant G. N. Lewis. Unlike the other concepts we have seen so far in this text, it does not follow a straightforward logical development. In fact, fugacity is undoubtedly but one of many ways to get around the mathematical anomalies of the chemical potential; however, it is the way that is used in practice.

To introduce fugacity, we start with Equation (6.78):

$$\left(\frac{\partial \mu_i}{\partial P}\right)_{T,n_i} = \overline{V}_i \tag{7.2}$$

We will begin by restricting ourselves to an ideal gas. We will remove this restriction and include real systems when we introduce fugacity. Equation (7.2) is valid only at constant temperature. In the development of fugacity that follows, we will always be at **constant temperature**. With this restriction, Equation (7.2) can be rewritten as:

$$d\mu_i = \overline{V}_i dP \qquad \text{At } constant \ T \tag{7.3}$$

where we have replaced the partial derivatives in Equation (7.2) with total derivatives, since we are explicitly keeping the temperature constant. Applying the definition of the partial molar volume and then the ideal gas relation to Equation (7.3) yields

$$d\mu_i = \left(\frac{\partial V}{\partial n_i}\right)_{T,P,n_{j\neq i}} dP = \frac{RT}{P}dP \tag{7.4}$$

Since energies never have absolute values, we need a reference state for the partial molar Gibbs energy. The reference state[1] is indicated by a superscript "o". In choosing a

---

[1] The reference state is often termed the *standard* state.

reference state, we must specify the appropriate number of thermodynamic properties as prescribed by the state postulate; the rest of the properties of the reference state are then constrained. The reference chemical potential, μ_i^o, is the chemical potential at the reference pressure, P^o, and at **the same temperature as the chemical potential of interest**, T. The latter constraint derives from our stipulation of constant temperature in Equation (7.2). Integrating between a reference state and the state of the system, we get:

$$\mu_i - \mu_i^o = RT \ln \left[\frac{P}{P^o} \right] \tag{7.5}$$

As we saw in Section 6.3, the chemical potential describes the contribution of species i to the Gibbs energy of the mixture; thus, it is convenient to multiply the top and bottom in the log term by the mole fraction of species i, y_i:

$$\mu_i - \mu_i^o = RT \ln \left[\frac{p_i}{p_i^o} \right] \tag{7.6}$$

where $p_i = y_i P$ is the partial pressure of the gas. The use of partial pressure is valid since we are considering ideal gases. If we examine Equation (7.6), we see that in going from the reference state to the state of the system at constant temperature, the change in the abstract quantity, μ_i, is proportional to the simple log of the partial pressures of species i, p_i. In fact, every property on the right-hand side of Equation (7.6) is a measured property. Hence we have expressed the difference in the abstract quantity, chemical potential, in terms of measured quantities—T, P, and y_i. Equation (7.6) also exposes the mathematical problems associated with the chemical potential in two very important limits: (1) as the mole fraction of species i goes to zero, that is, infinite dilution, and (2) as the pressure goes to zero, that is, the ideal gas limit. In both these cases, the value of μ_i goes to negative infinity.

So far this analysis is relatively straightforward. G. N. Lewis had the tremendous insight to define a new thermodynamic property, the fugacity, $\hat{f}_i$, in analogy to Equation (7.6). Fugacity is *defined* as

$$\mu_i - \mu_i^o \equiv RT \ln \left[\frac{\hat{f}_i}{\hat{f}_i^o} \right] \tag{7.7}$$

Accordingly, fugacity has units of pressure. Comparison of Equations (7.6) and (7.7) shows that fugacity plays the same role in real gases that partial pressure plays in ideal gases. In this sense fugacity can be thought of as a "corrected pressure." In fact, fugacity can roughly be translated from Latin as "the tendency to escape." However, the concept of fugacity goes beyond gases. This defining equation is valid for an isothermal change from the reference state chemical potential to that of the system for all real species. Lewis did not restrict the fugacity to the gas phase! It applies to liquid or solids as well.

The definition above is not complete. The reference state is arbitrary; we are free to choose the most convenient reference state imaginable; however, both μ_i^o and $\hat{f}_i^o$ depend on the single choice of reference state and may not be chosen independently. Let's consider a limiting condition to complete the definition. As the pressure goes to zero, all gases behave ideally; consequently, we define:

$$\lim_{P \to 0} \left(\frac{\hat{f}_i}{p_i} \right) \equiv 1 \qquad \text{(ideal gas)} \tag{7.8}$$

Equations (7.7) and (7.8) together form the abstract but very useful definition for fugacity.

The group, $\hat{f}_i/p_i$, often pops up in our encounters with fugacity; we call it the fugacity coefficient, $\hat{\varphi}_i$:

$$\hat{\varphi}_i \equiv \frac{\hat{f}_i}{p_{i,\text{sys}}} = \frac{\hat{f}_i}{y_i P_{\text{sys}}} \tag{7.9}$$

The fugacity coefficient represents a dimensionless quantity that compares the fugacity of species i to the partial pressure species i would have in the system as an ideal gas. A fugacity coefficient of one represents the case where attractive and repulsive forces balance and is usually indicative of an ideal gas. If $\hat{\varphi}_i < 1$, the corrected pressure, or "tendency to escape," is less than that for an ideal gas. In this case, attractive forces dominate the system behavior. Conversely, when $\hat{\varphi}_i > 1$, repulsive forces are stronger. *Warning*: We define the fugacity coefficient relative to the system partial pressure, **not** the partial pressure of the reference state. A common mistake is to use the wrong pressure here.

Other Forms of Fugacity

The definitions above were based on μ_i, a partial molar property. Hence they describe the contribution of species i to the solution. The fugacity and fugacity coefficient are given a hat instead of a bar to remind us that while they represent the contribution of species i in solution, they do not represent the mathematical definition of a partial molar property, that is,

$$\hat{f}_i \neq \left(\frac{\partial (nf)}{\partial n_i} \right)_{T,P,n_{j \neq i}} \qquad \text{or} \qquad f \neq \sum_i x_i \hat{f}_i$$

We can also define a total solution fugacity, f, and a pure species fugacity, f_i, in analogy with our discussion in Section 6.3:

Total solution fugacity:

$$g - g^o \equiv RT \ln \left[\frac{f}{f^o} \right] \tag{7.10}$$

$$\lim_{P \to 0} \left(\frac{f}{P} \right) \equiv 1 \tag{7.11}$$

$$\varphi \equiv \frac{f}{P_{\text{sys}}} \tag{7.12}$$

Pure species fugacity:

$$g_i - g_i^o \equiv RT \ln \left[\frac{f_i}{f_i^o} \right] \tag{7.13}$$

$$\lim_{P \to 0} \left(\frac{f_i}{P} \right) \equiv 1 \tag{7.14}$$

$$\varphi_i \equiv \frac{f_i}{P_{\text{sys}}} \tag{7.15}$$

Criteria for Chemical Equilibria in Terms of Fugacity

The concept of fugacity works so well because the criterion for chemical equilibria is just as simple as that using chemical potential. To derive this relationship for fugacity, we begin by equating the chemical potentials of phases α and β:

$$\mu_i^{\alpha} = \mu_i^{\beta} \tag{7.16}$$

Substitution of Equation (7.7) into Equation (7.16) gives

$$\mu_i^{\alpha,o} + RT \ln \left[\frac{\hat{f}_i^{\alpha}}{\hat{f}_i^{\alpha,o}} \right] = \mu_i^{\beta,o} + RT \ln \left[\frac{\hat{f}_i^{\beta}}{\hat{f}_i^{\beta,o}} \right] \tag{7.17}$$

Applying a mathematical relationship to the quotient in the logarithms and rearranging gives:

$$\boxed{\mu_i^{\alpha,o} - \mu_i^{\beta,o} = RT \ln \left[\frac{\hat{f}_i^{\alpha,o}}{\hat{f}_i^{\beta,o}} \right]} + RT \ln \left[\frac{\hat{f}_i^{\beta}}{\hat{f}_i^{\alpha}} \right] \tag{7.18}$$

Definition

The first three terms are just a restatement of Equation (7.7); hence the remaining term must be equal to zero, that is,

$$0 = RT \ln \left[\frac{\hat{f}_i^{\beta}}{\hat{f}_i^{\alpha}} \right] \tag{7.19}$$

or

$$\boxed{\hat{f}_i^{\alpha} = \hat{f}_i^{\beta}} \tag{7.20}$$

Equation (7.20) forms the criterion for chemical equilibrium in terms of fugacity. It is just as simple as that for chemical potential. Fugacity is also mathematically much better behaved.

Thus, in practice, we can replace Equation 7.1 with Equation 7.20 in defining our criteria for equilibrium:

$$T^{\alpha} = T^{\beta} \qquad \text{thermal equilibrium}$$

$$P^{\alpha} = P^{\beta} \qquad \text{mechanical equilibrium}$$

$$\mu_i^{\alpha} \cancel{=} \mu_i^{\beta} \quad \hat{f}_i^{\alpha} = \hat{f}_i^{\beta} \qquad \text{chemical equilibrium}$$

Since our equations for chemical equilibria equate fugacities in different phases, we will now explore how to calculate fugacity for (1) the vapor phase (Section 7.3), (2) the liquid phase (Section 7.4), and finally, (3) the solid phase (Section 7.5). We can then equate the fugacity of any phases that coexist (Chapter 8). For example, for vapor–liquid equilibria, we have:

$$\hat{f}_i^v = \hat{f}_i^l \tag{7.21}$$

Once we have examined how to calculate fugacities for the vapor and liquid phases separately, we can merely equate the two expressions and then calculate the composition of species i in each phase.

▶ 7.3 FUGACITY IN THE VAPOR PHASE

The difference in the expression for fugacity between vapor and condensed phases typically lies in the choice of reference state. Since we have learned extensively about the intermolecular forces of gases that cause deviations from ideality, the vapor phase is a logical starting point. We will begin by considering the fugacity of pure species in the vapor phase, f_i^v, and then address the fugacity of species i in a vapor mixture, $\hat{f}_i^v$.

Fugacity and Fugacity Coefficient of Pure Gases

To use fugacity in practice, the first step is to identify an appropriate reference state. There is an obvious choice of reference state for gases: a low enough pressure that the gas behaves as an ideal gas. With this choice, $f_i^o \to P$ and $\varphi_i^o \to 1$. The pressure for the reference state is usually but not always chosen to be 1 bar. Remember, as a result of our definition for fugacity, the reference state *must* be at the same temperature as the system of interest.

We can write the expression for fugacity of pure species i using the ideal gas reference state. In this case, Equation (7.13) becomes:

$$\ast \quad g_i - g_i^o \equiv RT \ln\left[\frac{f_i^v}{P_{\text{low}}}\right] \tag{7.22}$$

use this eqn w/ steam tables

[handwritten: $PV = nRT$; $\frac{P}{P}$ $\frac{P}{P}$; $P = \frac{RT}{V}$]

where we have chosen the following reference state:

[handwritten: $V\,dP \qquad V\left(RT \ln P\right)$]

$$P^o = P_{\text{low}}$$

$$T^o = T_{\text{sys}} \tag{7.23}$$

where P_{low} represents a pressure low enough to be an ideal gas and T_{sys} indicates the system temperature. Similarly Equation (7.15) becomes

$$\varphi_i^v \equiv \frac{f_i^v}{P_{\text{sys}}} \tag{7.24}$$

The pressures in Equations (7.22) and (7.24) have different values. The chemical potential has been replaced with the molar Gibbs energy, since they are equivalent for a pure species.

In order to obtain the fugacity of a real gas, we must have appropriate thermodynamic property data available. We will explore three possible sources of data for pure gases:

1. Tables
2. Equations of state
3. Generalized correlations

Expression for the Fugacity Coefficient of a Pure Gas Using Tables

Tables of thermodynamic properties typically have h, s, T, and P. From the first three properties, g can be calculated. Values for every quantity in Equation (7.22), except fugacity, are obtained. We can, therefore, use these property values to solve for f_i. Example (7.1) illustrates this methodology.

▶ **EXAMPLE 7.1**
Calculation of Fugacity Using the Steam Tables

Determine the fugacity and the fugacity coefficient for saturated steam at 1 atm.

SOLUTION The steam tables provide the appropriate data to solve this problem. It is straightforward to find g_{H_2O} for saturated steam at 1 atm. For example, from Appendix B:

$$\hat{g}_{H_2O} = \hat{h}_{H_2O} - T\hat{s}_{H_2O}$$

$$\hat{h}_{H_2O} = 2676.0 \text{ kJ/kg} \quad \hat{s}_{H_2O} = 7.3548 \text{ kJ/(kg K)} \quad T = 373.15 \text{ K}$$

so that

$$\boxed{\hat{g}_{H_2O} = -68.44 \text{ kJ/kg}}$$

To find the fugacity, we must now choose a reference state. We want as a low a pressure as possible *at the same temperature* as the system of interest (100°C). We can find this state in the superheated steam tables (Appendix B.4). The lowest pressure available in the steam tables is 10 kPa. So for the reference state, we choose 10 kPa, 100°C:

$$\hat{g}^o_{H_2O} = \hat{h}^o_{H_2O} - T\hat{s}^o_{H_2O}$$

Looking up values:

$$\hat{h}^o_{H_2O} = 2687.5 \text{ kJ/kg} \quad \hat{s}^o_{H_2O} = 8.4479 \text{ kJ/(kg K)} \quad T = 373.15 \text{ K}$$

$$\boxed{\hat{g}^o_{H_2O} = -464.83 \text{ kJ/kg}}$$

Now, rearranging Equation (7.22),

$$f^v_{H_2O} = P^o_{H_2O} \exp\left[\frac{\hat{g}_{H_2O} - \hat{g}^o_{H_2O}}{RT}\right]$$

Converting R to mass units and plugging in values:

$$f^v_{H_2O} = 10 \text{ kPa} \exp\left[\frac{-68.44 \text{ kJ/kg} - (-464.83 \text{ kJ/kg})}{8.314 \text{ kJ/(kmol K)}\frac{1}{18} \text{ kmol/kg } 373.15 \text{ K}}\right]$$

$$\boxed{f^v_{H_2O} = 99.73 \text{ kPa}}$$

In solving for the fugacity coefficient, we must use the system pressure:

$$\varphi^v_{H_2O} = \frac{f^v_{H_2O}}{P_{sys}} = \frac{99.73 \text{ kPa}}{101.35 \text{ kPa}} = .984$$

Comment: Even with its strong dipole moment (and associated forces of attraction), the fugacity coefficient of water deviates from ideality by less than 2% at 1 atm. This result gives us confidence that we can use 1 bar as the (ideal gas) reference state for most species. ◀

$$\Delta g = RT \ln\left(\frac{f_i^v}{P_{low}}\right) \qquad \frac{dg}{dP} = v_i$$

Expression for the Fugacity Coefficient of a Pure Gas Using Equations of State

We can also get thermodynamic property data to solve for the left-hand side of Equation (7.22) through an equation of state. At constant temperature (as mandated by the definition of fugacity), we can write the fundamental property relation of the Gibbs energy of pure species i as:

$$dg_i = v_i dP \quad \text{at const } T \qquad (7.25)$$

Hence Equation (7.22) becomes

$$g_i - g_i^o = \int_{P_{low}}^{P} v_i dP = RT \ln\left[\frac{f_i^v}{P_{low}}\right] \qquad (7.26)$$

where we have integrated Equation (7.22) from the reference state to that of the system. We now use our equation of state to relate v_i to P, at constant T, and solve for f_i. Example 7.2 illustrates this methodology.

▶ **EXAMPLE 7.2**
Calculation of Fugacity Using the van der Waals Equation

Determine an expression for the fugacity of a pure gas from the van der Waals equation of state.

SOLUTION We begin with the van der Waals equation:

$$P = \frac{RT}{v_i - b} - \frac{a}{v_i^2} \qquad (E7.2A)$$

At constant temperature, we can differentiate Equation (E7.2A) to get

$$dP = \left[\frac{-RT}{(v_i - b)^2} + \frac{2a}{v_i^3}\right] dv_i \qquad (E7.2B)$$

Plugging Equation (E7.2B) into Equation (7.26) yields

$$\int_{\frac{RT}{P_{low}}}^{v_i} \left[\frac{-v_i RT}{(v_i - b)^2} + \frac{2a}{v_i^2}\right] dv_i = RT \ln\left[\frac{f_i^v}{P_{low}}\right]$$

Integrating the left-hand side, we get

$$\ln\left[\frac{f_i^v}{P_{low}}\right] = -\ln\left[\frac{(v_i - b)}{\frac{RT}{P_{low}} - b}\right] + b\left[\frac{1}{(v_i - b)} - \frac{1}{\frac{RT}{P_{low}} - b}\right] - \frac{2a}{RT}\left[\frac{1}{v_i} - \frac{1}{\frac{RT}{P_{low}}}\right]$$

However, since $RT/P_{low} \gg b$, we can simplify the denominators in the two terms above:

$$\ln\left[\frac{f_i^v}{P_{low}}\right] = -\ln\left[\frac{(v_i - b)P_{low}}{RT}\right] + b\left[\frac{1}{(v_i - b)} - \frac{P_{low}}{RT}\right] - \frac{2a}{RT}\left[\frac{1}{v_i} - \frac{P_{low}}{RT}\right]$$

Adding $\ln(P_{low})$ to each side and then letting $P_{low} \to 0$, we get

$$\ln[f_i^v] = -\ln\left[\frac{(v_i - b)}{RT}\right] + \frac{b}{(v_i - b)} - \frac{2a}{RTv_i}$$

Finally, subtracting $\ln(P)$ from both sides gives

$$\ln\left[\frac{f_i^v}{P}\right] = \ln[\varphi_i^v] = -\ln\left[\frac{(v_i - b)P}{RT}\right] + \frac{b}{(v_i - b)} - \frac{2a}{RTv_i}$$

ALTERNATIVE APPROACH   We have an easier time, mathematically, if we use the virial form of the van der Waals equation truncated to the second term. As shown in Problem 4.20, the van der Waals equation written in virial form becomes

$$PV = n_T\left[RT + \left(b - \frac{a}{RT}\right)P + \text{terms in } P^2, P^3, \dots\right]$$

Recall that for moderate pressures, we can truncate the above expression at P. Plugging into Equation (7.26):

$$\int_{P_{\text{low}}}^{P}\left[\frac{RT}{P} + \left(b - \frac{a}{RT}\right)\right]dP = RT\ln\left[\frac{f_i^v}{P_{\text{low}}}\right]$$

Integrating and rearranging (note that the terms in reference pressure cancel):

$$\left(b - \frac{a}{RT}\right)P = RT\ln\left[\frac{f_i^v}{P}\right] = RT\ln\varphi_i$$

So for pure i, we get

$$\varphi_i^v = \frac{f_i^v}{P} = \exp\left\{\left(b - \frac{a}{RT}\right)\frac{P}{RT}\right\} \tag{E7.2C}$$

The van der Waals equation is not as accurate as more modern cubic equations of state. However, as inspection of Equation (E7.2C) shows, it provides a simple relation between the fugacity coefficient and molecular parameters. For an ideal gas $a = b = 0$, so the fugacity coefficient is 1, as we expect. If attractive forces are greater than repulsive forces, $b < a/RT$ and $\varphi_i^v < 1$; whereas when repulsive forces dominate, $b > a/RT$ and $\varphi_i^v > 1$. As the temperature increases, attractive forces become less important relative to repulsive forces.   ◀

Expression for the Fugacity Coefficient of a Pure Gas Using Generalized Correlations

As we saw in Chapter 4, we can use generalized correlations to relate the compressibility factor to reduced temperature and reduced pressure. In Chapter 5, we extended the use of general correlations to account for nonideal behavior in enthalpy and entropy through the use of departure functions. In this section, we wish to use general correlations to solve for the fugacity coefficient in terms of reduced temperature and reduced pressure. To do this, we will first develop a relationship between the compressibility factor, z_i, and the fugacity coefficient, φ_i, for the pure species i. Since we have already related z_i to T_r and P_r, it is then straightforward to express the fugacity coefficient in terms of a generalized correlation. As in Chapters 4 and 5, we will present results for a simple fluid term and a correction term based on the Lee–Kesler equation of state.

The definition for fugacity is

$$g_i - g_i^o = \int_{P_{\text{low}}}^{P} v_i dP = RT\ln\left[\frac{f_i^v}{P_{\text{low}}}\right] \tag{7.27}$$

Equation (7.27) can be rewritten by dividing by RT and subtracting $\int_{P_{\text{low}}}^{P} (1/P)dP$ from each side to give

$$\int_{P_{\text{low}}}^{P} \frac{v_i}{RT}dP - \int_{P_{\text{low}}}^{P} \frac{1}{P}dP = \ln\left[\frac{f_i^v}{P_{\text{low}}}\right] - \int_{P_{\text{low}}}^{P} \frac{1}{P}dP \qquad (7.28)$$

Combining the integrals on the left-hand side and integrating and rewriting the logarithmic terms on the right-hand side gives

$$\int_{P_{\text{low}}}^{P} \left[\frac{v_i}{RT} - \frac{1}{P}\right]dP = \ln\left[\frac{f_i^v}{P}\right] = \ln\varphi_i^v \qquad (7.29)$$

Expressing Equation (7.29) in terms of the compressibility factor, we get

$$\boxed{\ln\varphi_i^v = \int_{P_{\text{ideal}}}^{P} [z_i - 1]\frac{dP}{P}} \qquad (7.30)$$

or, in terms of reduced variables,

$$\boxed{\ln\varphi_i^v = \int_{P_{r,\text{ideal}}}^{P_r} [z_i - 1]\frac{dP_r}{P_r}} \qquad (7.31)$$

As we saw in Chapter 4, there are charts available for the compressibility factor as a function of the reduced variables, T_r and P_r. It is straightforward, in principle, to graphically integrate the right-hand side of Equation (7.31) and come up with a chart for the fugacity coefficient in reduced coordinates. Alternatively, we can analytically integrate Equation (7.31) with the appropriate equation of state. Figures 7.1 and 7.2 show values for the simple fluid term $\log\varphi^{(0)}$ and the correction term $\log\varphi^{(1)}$, respectively, based on the Lee–Kesler equation of state. Once we have determined the reduced temperature and pressure for a given system, we can then use the form

$$\log\varphi_i = \log\varphi^{(0)} + \omega\log\varphi^{(1)} \qquad (7.32)$$

to solve for φ_i. These same data are reported in tabular form in Appendix C (Tables C.7 and C.8). The expression that was used to generate these values is described in Appendix E.

▶ **EXAMPLE 7.3**
Fugacity Calculation Using Generalized Correlations

Determine the fugacity and the fugacity coefficient of ethane at a pressure of 50 bar and a temperature of 25°C using generalized correlations.

SOLUTION We begin by finding P_r and T_r:

$$P_r = \frac{P}{P_c} = \frac{50\,\text{bar}}{48.7\,\text{bar}} = 1.03 \qquad \text{and} \qquad T_r = \frac{T}{T_c} = \frac{298.2\,\text{K}}{305.5\,\text{K}} = 0.98 \qquad \omega = 0.099$$

where the critical properties and the acentric factor were obtained from Appendix A.1. From Tables C.7 and C.8:

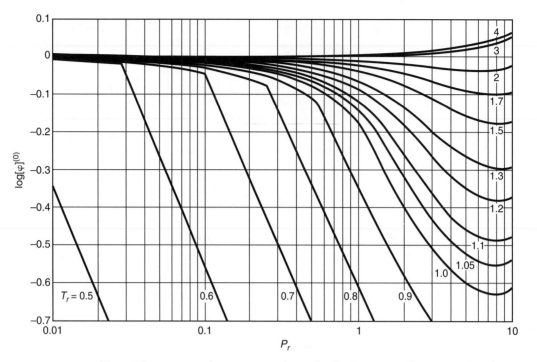

Figure 7.1 Corresponding states correlation for the fugacity coefficient in reduced coordinates—simple fluid term. Based on the Lee–Kesler equation of state.

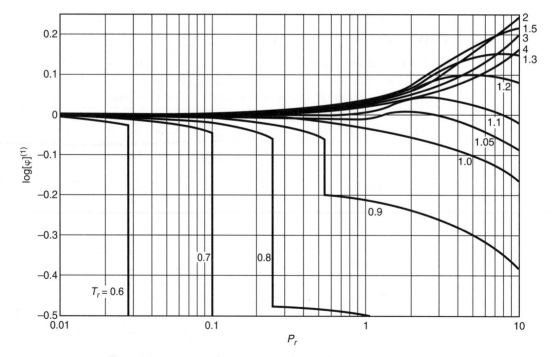

Figure 7.2 Corresponding states correlation for the fugacity coefficient in reduced coordinates—correction term. Based on the Lee–Kesler equation of state.

	$\log \varphi^{(0)}$		$\log \varphi^{(1)}$	
	P_r		P_r	
T_r	1	1.1	1	1.1
0.98	−0.206	−0.240	−0.059	−0.062

By linear interpolation

$$\log \varphi^{(0)} = -0.206 + \frac{0.03}{0.1}(-0.240 + 0.206) = -0.216$$

and

$$\log \varphi^{(1)} = -0.059 + \frac{0.03}{0.1}(-0.062 + 0.059) = -0.060$$

Thus,

$$\log \varphi_{\text{eth}} = \log \varphi^{(0)} + \omega \log \varphi^{(1)} = -0.222$$

So

$$\varphi_{\text{eth}} = 0.60$$

and

$$f_{\text{eth}} = \varphi_{\text{eth}} P = 0.60 \times 50 = 30[\text{bar}]$$

In this case, there are significant deviations from ideality. Since $\varphi_{\text{eth}} < 1$, we surmise that attractive forces dominate. This result is expected since the system is around the critical point of ethane, where intermolecular interactions are important. ◀

Fugacity and Fugacity Coefficient of Species *i* in a Gas Mixture

We now extend our discussion of fugacity in the gas phase to include mixtures. In addition to temperature and pressure, the fugacity of species *i* depends on what *other* species are present in the mixture. In a mixture, the chemical nature of the interactions between species *i* and *all* the other species in the mixture must be taken into account. In other words, the fugacity and fugacity coefficient in the mixture are functions of the composition of the mixture. Equation (7.9) can be rewritten to indicate such dependence as follows:

$$\hat{f}_i^v(T, P, n_1, n_2, \ldots n_m) = y_i \hat{\varphi}_i^v(T, P, n_1, n_2, \ldots n_m)P \qquad (7.33)$$

Recall that we include the hat on the fugacity and fugacity coefficient to indicate species *i* in a mixture.

To illustrate the role of the interactions between the different species in a mixture in our calculations, we compare binary mixtures of (1) ethane–methylcyanide and (2) ethane–propane at identical temperature and pressure. The fugacity and fugacity coefficient of ethane depend not only on the temperature and pressure of the system but also on the nature of the other chemical species with which the ethane is mixed. Both methylcyanide and propane are roughly the same size, with molecular weights of 41 and 44 g/mol, respectively. However, the nature of their attractive forces is very different. Methylcyanide has a strong dipole. Consequently, we expect the potential interactions in the ethane–methylcyanide system to be very different from the interactions of ethane with a nonpolar propane molecule. In the former system, ethane will disrupt the dipole–dipole interactions between methylcyanide molecules. On the other hand, the attractive forces in the ethane–propane system consist entirely of London interactions, so the like interactions are quite similar to the unlike interactions. The fugacity coefficients of ethane calculated using the Redlich–Kwong equation of state for the two mixtures at 50 bar and 550 K are plotted vs. mole fraction ethane in Figure 7.3. We will learn shortly how such a calculation is done. Note that *at the same temperature, pressure, and ethane mole fraction*, the deviation from ideality of the ethane in methylcyanide is greater than when ethane is mixed with propane. Moreover, the differences between the two binary systems become more pronounced at lower ethane mole fraction, where an ethane molecule is more likely to interact with the other component in the mixture. Indeed, the chemical composition of the mixture is a major factor in determining the fugacity coefficient of species i.

Next, we will learn how to calculate the fugacity and fugacity coefficient of a species in a mixture. As with pure species, it is important to define an appropriate reference state for calculating the fugacity in the vapor phase of mixtures. We choose a low enough pressure, P_{low}, so that the mixture behaves as an ideal gas. Again, due to how fugacity is defined, the reference temperature is that of the system of interest, T_{sys}. To completely specify the reference state for a mixture, we must also specify its composition. It turns out to be convenient to specify the composition of the system $n_{i,sys}$. We will see why we chose this composition shortly. In summary, the reference state for a mixture is:

$$P^o = P_{low}$$
$$T^o = T_{sys}$$
$$n_i^o = n_{i,sys} \tag{7.34}$$

By definition, the fugacity for this reference state becomes the partial pressure at the reference state, $\hat{f}_i^o = p_i^o = y_{i,sys} P_{low}$. We will now consider each of the methods discussed above for pure species to see if we can apply them to mixtures.

Expression for the Fugacity Coefficient of a Gas Mixture Using Tables
Tables for the thermodynamic properties of mixtures are uncommon, so we cannot use this approach.

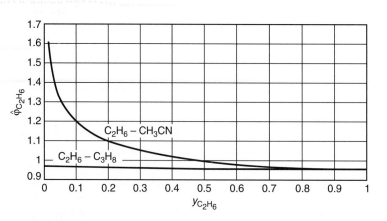

Figure 7.3 Fugacity coefficient of C_2H_6 in CH_3CN and in C_3H_8 as calculated from the Redlich–Kwong equation of state.

**Expression for the Fugacity Coefficient of
a Gas Mixture Using Equations of State**

In order to use an equation of state to calculate the fugacity and fugacity coefficient for species i in a mixture, we need to describe PvT property behavior for the mixture. Unlike with pure species, there are little data available for mixtures. Our approach is therefore to use mixing rules such as those described in Section 4.5. However, recall that the theoretical justification for mixing rules is limited.

We begin by incorporating our reference state in Equation (7.7) to get

$$\mu_i - \mu_i^o = RT \ln \left[\frac{\hat{f}_i^v}{p_i^o} \right] \tag{7.35}$$

where the temperature of the reference state is constrained (by the definition of fugacity) to be the temperature of the mixture and we have chosen the composition of the reference state to be the composition of the mixture. In order to relate Equation (7.35) to an equation of state, we need to apply the thermodynamic web. Using the partial derivative relation given by Equation (6.78), we get

$$\mu_i - \mu_i^o = \int_{P_{\text{low}}}^{P} \overline{V}_i dP = RT \ln \left[\frac{\hat{f}_i^v}{y_i P_{\text{low}}} \right] \tag{7.36}$$

If we have a volume-explicit equation of state, we can determine the partial molar volume directly, as in Example 6.9, and then integrate to determine the fugacity.

Many equations of state are explicit in P but not V (such as the cubic equations discussed in Section 4.3), so it is convenient to try to rewrite the second term in Equation (7.36) in terms of a derivative in P. Recall from the definition of a partial molar property:

$$\overline{V}_i = \left(\frac{\partial V}{\partial n_i} \right)_{T,P,n_{j \neq i}} \tag{7.37}$$

By the cyclic rule, at constant T

$$\left(\frac{\partial V}{\partial n_i} \right)_{T,P} \left(\frac{\partial P}{\partial V} \right)_{T,n_i} \left(\frac{\partial n_i}{\partial P} \right)_{T,V} = -1 \tag{7.38}$$

We can now see why it is so convenient to choose identical compositions for the system and the reference state. With this choice, the integral in Equation (7.36) is carried out with n_i constant. Since both T and n_i are always constant, we can replace the second partial derivative in Equation (7.38) with a total derivative. Consequently, we can rearrange Equation (7.38) to get

$$\left(\frac{\partial V}{\partial n_i} \right)_{T,P} dP = - \left(\frac{\partial P}{\partial n_i} \right)_{T,V} dV \tag{7.39}$$

Substituting Equation (7.39) into Equation (7.36) gives

$$RT \ln \left[\frac{\hat{f}_i^v}{y_i P_{\text{low}}} \right] = - \int_{\left(\frac{n_T RT}{P_{\text{low}}} \right)}^{V} \left(\frac{\partial P}{\partial n_i} \right)_{T,V,n_{j \neq i}} dV \tag{7.40}$$

Equation (7.40) allows us to calculate fugacity when we have an equation of state that is explicit in P.

To take the derivative of a pressure-explicit (e.g., cubic) equation of state in Equation (7.40), we need appropriate mixing rules. For example, recall that the van der Waals equation is

$$P = \frac{RT}{v - b_{mix}} - \frac{a_{mix}}{v^2} \tag{7.41}$$

Since the van der Waals constants a_{mix} and b_{mix} depend on composition, it is necessary to come up with an expression for the composition functionality of these parameters (recall the discussion of mixing rules in Chapter 4). Unfortunately, these cannot be rigorously obtained from molecular theory, so we must "guess" at the appropriate mixing rules. Our best "guess" combines qualitative molecular arguments with mathematical simplicity. Common choices are that the "force" parameter, a, represents the strength of attraction between binary pairs of molecules (two-body interactions), so

$$a_{mix} = \sum\sum y_i y_j a_{ij} \tag{7.42}$$

where
$$a_{ij} = \sqrt{a_{ii}a_{jj}} \tag{7.43}$$

In this interpretation a_{ii} represents the attraction between like i molecules while a_{ij} represents the interaction between one i molecule and one j molecule, the unlike interaction. When more extensive data are available, an empirical fitting parameter, k_{ij}, is often used to obtain better agreement:

$$a_{ij} = \sqrt{a_{ii}a_{jj}}(1 - k_{ij}) \tag{7.44}$$

For the "size" parameter, it is mathematically most convenient to average molecular volumes:

$$b_{mix} = \sum y_i b_i \tag{7.45}$$

What would this look like if we averaged molecular diameters?

In Example 7.4, we develop expressions for the fugacity and fugacity coefficient in a binary mixture using the van der Waals equation of state and the van der Waals mixing rules. The results can be compared to those from Example 7.2, which used the same equation of state to come up with expressions for a pure species.

▶ **EXAMPLE 7.4**
Fugacity in a Mixture Using the van der Waals Equation

Consider a binary gas mixture composed of species a and b. Determine an expression for the fugacity of species a from the van der Waals equation of state. Use mixing rules given by Equations (7.42), (7.43), and (7.45).

SOLUTION We begin by writing Equation (7.40) for species a in a mixture of a and b:

$$RT \ln\left[\frac{\hat{f}_a^v}{y_a P_{low}}\right] = -\int\limits_{\left(\frac{n_T RT}{P_{low}}\right)}^{V} \left(\frac{\partial P}{\partial n_a}\right)_{V, T, n_b} dV \tag{E7.4A}$$

For a binary mixture, the van der Waals equation can be written as

$$P = \frac{RT(n_a + n_b)}{V - (n_a b_a + n_b b_b)} - \frac{n_a^2 a_a + 2n_a n_b \sqrt{a_a a_b} + n_b^2 a_b}{V^2} \tag{E7.4B}$$

where the following expressions are used

$$a_{\text{mix}} = y_a^2 a_a + 2 y_a y_b \sqrt{a_a a_b} + y_b^2 a_b$$

$$b_{\text{mix}} = y_a b_a + y_b b_b$$

and

$$n_T = n_a + n_b$$

Taking the derivative of Equation (E7.4B) gives:

$$\left(\frac{\partial P}{\partial n_a}\right)_{T,V,n_b} = \frac{RT}{V - (n_a b_a + n_b b_b)} + \frac{b_a RT(n_a + n_b)}{[V - (n_a b_a + n_b b_b)]^2} - \frac{2(n_a a_a + n_b \sqrt{a_a a_b})}{V^2} \qquad \text{(E7.4C)}$$

Substituting Equation (E7.4C) into Equation (E7.4A) and integrating we get

$$\ln\left[\frac{\hat{f}_a^v}{y_a P_{\text{low}}}\right] = -\ln\left[\frac{V - n_T b_{\text{mix}}}{\left(\frac{n_T RT}{P_{\text{low}}}\right) - n_T b_{\text{mix}}}\right] + \frac{b_a(n_a + n_b)}{[V - n_T b_{\text{mix}}]} - \frac{b_a(n_a + n_b)}{\left[\left(\frac{n_T RT}{P_{\text{low}}}\right) - n_T b_{\text{mix}}\right]}$$

$$- \frac{2(n_a a_a + n_b \sqrt{a_a a_b})}{RTV} + \frac{2(n_a a_a + n_b \sqrt{a_a a_b})}{RT\left(\frac{n_T RT}{P_{\text{low}}}\right)}$$

However, since $RT/P_{\text{low}} \gg b_{\text{mix}}$, we can simplify the denominators in two terms above. Adding $\ln(P_{\text{low}})$ to each side, simplifying, and then letting $P_{\text{low}} \to 0$, we obtain:

$$\ln\left[\frac{\hat{f}_a^v}{y_a}\right] = -\ln\left[\frac{(V - n_T b_{\text{mix}})}{n_T RT}\right] + \frac{b_a(n_a + n_b)}{[V - n_T b_{\text{mix}}]} - \frac{2(n_a a_a + n_b \sqrt{a_a a_b})}{RTV}$$

Subtracting $\ln(P)$ from both sides leaves

$$\boxed{\ln\left[\frac{\hat{f}_a^v}{y_a P}\right] = \ln[\hat{\varphi}_a^v] = -\ln\frac{P(v - b_{\text{mix}})}{RT} + \frac{b_a}{[v - b_{\text{mix}}]} - \frac{2(y_a a_a + y_b \sqrt{a_a a_b})}{RTv}}$$

ALTERNATIVE APPROACH   We can use the virial form of the van der Waals equation truncated to the second term as in Example (7.2):

$$PV = n_T\left[RT + \left(b_{\text{mix}} - \frac{a_{\text{mix}}}{RT}\right)P + \text{terms in } P^2, P^3, \dots\right] \qquad \text{(E7.4D)}$$

Recall that this is valid for moderate pressures. Differentiating Equation (E7.4D) gives

$$\overline{V}_a = \frac{RT}{P} + \left(b_a - \frac{2(n_a a_a + n_b \sqrt{a_a a_b}) - n_T a_{\text{mix}}}{n_T RT}\right) \qquad \text{(E7.4E)}$$

We must now plug Equation (E7.4E) into Equation (7.36) and integrate:

$$\int_{P_{\text{low}}}^{P}\left[\frac{RT}{P} + \left(b_a - \frac{2(n_a a_a + n_b \sqrt{a_a a_b}) - n_T a_{\text{mix}}}{n_T RT}\right)\right] dP = RT \ln\left[\frac{\hat{f}_a^v}{y_a P_{\text{low}}}\right]$$

so

$$RT \ln\frac{P}{P_{\text{low}}} + \int_{P_{\text{low}}}^{P}\left(b_a - \frac{2(n_a a_a + n_b \sqrt{a_a a_b}) - n_T a_{\text{mix}}}{n_T RT}\right) dP = RT \ln\left[\frac{\hat{f}_a^v}{y_a P_{\text{low}}}\right]$$

Rearranging:

$$\int_{P_{\text{low}}}^{P} \left(b_a - \frac{2(n_a a_a + n_b \sqrt{a_a a_b}) - n_T a_{\text{mix}}}{n_T RT} \right) dP = RT \ln \left[\frac{\hat{f}_a^v}{y_a P} \right] = RT \ln \hat{\varphi}_a^v$$

Integrating and rearranging, we get

$$\hat{\varphi}_a^v = \frac{\hat{f}_a^v}{y_a P} = \exp \left\{ \left(b_a - \frac{a_a}{RT} \right) \frac{P}{RT} \right\} \exp \left\{ \frac{(\sqrt{a_a} - \sqrt{a_b})^2 y_b^2 P}{(RT)^2} \right\} \tag{E7.4F}$$

From Example 7.2, the first exponential term is just equal to the pure species fugacity coefficient, so Equation (E7.4F) becomes

$$\boxed{\hat{f}_a^v = y_a f_a^v \exp \left\{ \frac{(\sqrt{a_a} - \sqrt{a_b})^2 y_b^2 P}{(RT)^2} \right\}} \tag{E7.4G}$$

Again, the van der Waals equation provides us with insight as to the relationship between the fugacity coefficient and molecular parameters. If the intermolecular forces between species a and b are the same, the remaining exponential in Equation (E7.4G) becomes 1, and we get:

$$\hat{f}_a^v = y_a f_a^v$$

or

$$\hat{\varphi}_a^v = \varphi_a^v$$

This approximation is known as the Lewis fugacity rule and will be discussed in more detail shortly. ◀

Expression for the Fugacity Coefficient of a Gas Mixture Using Generalized Correlations

In the case of using generalized correlations, we can apply the same approach as we did for pure species, but we need mixing rules as well. The same concerns discussed in the context of using equations of state apply to the accuracy of the mixing rules. A common approach is to define the mixture as a "pseudo" fluid with corresponding "pseudocritical (pc)" properties. The most common relationship between the pure species critical properties and the pseudocritical properties are given by Kay's rules:

$$T_{\text{pc}} = \sum y_i T_{c,i} \tag{7.46}$$

$$P_{\text{pc}} = \sum y_i P_{c,i} \tag{7.47}$$

$$\omega_{\text{mix}} = \sum y_i \omega_i \tag{7.48}$$

Many other forms have also been explored. Once a pseudocritical temperature, pressure and acentric factor have been defined, the same approach as that described for pure species is used for our pseudocritical fluid. Hence

$$\log \varphi = \log \varphi^{(0)} + \omega_{\text{mix}} \log \varphi^{(1)} \tag{7.49}$$

This approach is limited to calculating the total solution fugacity coefficient, as given by Equation (7.12), rather than the fugacity coefficient of species i in a mixture.

Determine the fugacity and the fugacity coefficient of a mixture of 20% ethane in propane at a pressure of 50 bar and a temperature of 25°C using generalized correlations.

SOLUTION We begin with Equations (7.46), (7.47), and (7.48):

$$T_{pc} = y_{eth}T_{c,eth} + y_{pro}T_{c,pro} = 0.2 \times 305.5 + 0.8 \times 370.0 = 357.1 \, [\text{K}]$$

$$P_{pc} = y_{eth}P_{c,eth} + y_{pro}P_{c,pro} = 0.2 \times 48.7 + 0.8 \times 42.4 = 43.7 \, [\text{bar}]$$

$$\omega_{mix} = y_{eth}\omega_{eth} + y_{pro}\omega_{pro} = 0.2 \times 0.099 + 0.8 \times 0.153 = 0.142$$

where the critical properties and the acentric factor were obtained from Appendix A.1. Hence, for P_r and T_r, we get

$$P_r = \frac{P}{P_{pc}} = \frac{50 \, \text{bar}}{43.7 \, \text{bar}} = 1.14 \quad \text{and} \quad T_r = \frac{T}{T_c} = \frac{298.2 \, \text{K}}{357.1 \, \text{K}} = 0.83$$

From Tables C.7 and C.8,

| | $\log \varphi^{(0)}$ | | $\log \varphi^{(1)}$ | |
| | P_r | | P_r | |
T_r	1.1	1.2	1.1	1.2
0.80	−0.645	−0.684	−0.501	−0.504
0.83	**−0.566**	**−0.600**	**−0.405**	**−0.407**
0.85	−0.513	−0.544	−0.341	−0.343

By double linear interpolation (with the first interpolation performed in the table),

$$\log \varphi^{(0)} = -0.566 + \frac{0.04}{0.1}(-0.600 + 0.566) = -0.579$$

$$\log \varphi^{(1)} = -0.405 + \frac{0.04}{0.1}(-0.407 + 0.405) = -0.406$$

Thus
$$\log \varphi = \log \varphi^{(0)} + \omega_{mix} \log \varphi^{(1)} = -0.637$$

So
$$\varphi = 0.23$$

and
$$f = \varphi P = 0.23 \times 50 = 11.5 \, [\text{bar}]$$

The Lewis Fugacity Rule

As we have seen, the fugacity coefficient (or fugacity) depends not only on T and P but also on the composition (chemical nature) of all the other species in the mixture. In general, we write the fugacity of species i in the mixture as

$$\hat{f}_i^v = y_i\hat{\varphi}_i P \tag{7.50}$$

As shown in Example 7.4, we can sometimes approximate the fugacity coefficient of species i in a mixture by its pure species fugacity coefficient. This approximation is known as the Lewis fugacity rule. It simplifies calculations significantly, since the pure species fugacity coefficient does not depend on the other species in the mixture but rather depends only on T and P (e.g., compare the complexity of Examples 7.2 and 7.4). The Lewis fugacity rule can be written as

$$\hat{\varphi}_i^v = \varphi_i^v \tag{7.51}$$

Applying the Lewis fugacity rule, Equation (7.50) can be simplified to

$$\hat{f}_i^v = y_l \varphi_i^v P \tag{7.52}$$

or
$$\hat{f}_i^v = y_i f_i \tag{7.53}$$

where Equation (7.15) was used

When is the Lewis fugacity rule a good approximation? For some insight, we can look at the expression developed in Example 7.4 for a binary mixture of species a and b using the van der Waals equation of state and van der Waals mixing rules. The fugacity of species a is given by

$$\hat{f}_a^v = y_a f_a^v \exp\left\{\frac{(\sqrt{a_a} - \sqrt{a_b})^2 y_b^2 P}{(RT)^2}\right\} \tag{7.54}$$

The Lewis fugacity rule is a good approximation when the term in the exponential is small. This condition is valid when the following are true:

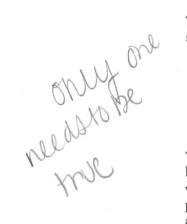 *only one needs to be true*

1. The pressure is low or the temperature is high. This condition corresponds to an ideal gas.
2. Component a is present in large excess (y_b is small).
3. The chemical nature of species a is similar to that of all the other components ($a_a \approx a_b$).

These criteria are true in general; they are not limited to the van der Waals equation or to binary mixtures, the conditions for which Equation (7.54) was derived.[2] For example, if we again compare the two binary mixtures—(1) ethane–methylcyanide and (2) ethane–propane—we would not expect the former mixture to follow the Lewis fugacity rule at all compositions, since the species are chemically dissimilar. On the other hand, ethane and propane are similar, and we may well expect this mixture to be reasonably represented by the Lewis fugacity approximation. Indeed, inspection of Figure 7.3 verifies this supposition.

Property Changes of Mixing for Ideal Gases

The ideal gas provides the reference state for the vapor phase. To set up our discussion of reference states for the liquid phase, we will review the property changes of mixing

---

[2] This analysis illustrates the use of engineering models derived from physical principles. While the van der Waals equation has its quantitative limitations, it provides us with a valuable conceptual guideline in applying the Lewis fugacity rule. While we probably would use a more accurate equation of state to account for the behavior of a species in a mixture, this analysis tells us when we need to concern ourselves with this more difficult calculation and when we may omit it.

for an ideal gas as discussed in Section 6.3. Recall that a property change of mixing is defined as the difference between the total solution property and the sum of the pure species properties apportioned by the amount each species present in the mixture:

$$\Delta k_{\text{mix}} = k - \sum y_i k_i = \sum y_i (\overline{K}_i - k_i) \tag{7.55}$$

We wish to determine $\Delta v_{\text{mix}}, \Delta h_{\text{mix}}, \Delta s_{\text{mix}}, \Delta g_{\text{mix}}$ for an ideal gas. We will represent this case with the superscript "ideal." Under the ideal gas approximation, the species occupy no volume, and we have no intermolecular forces. Therefore,

$$\Delta v_{\text{mix}}^{\text{ideal}} = 0 \qquad \text{Amagat's law} \tag{7.56}$$

$$\Delta h_{\text{mix}}^{\text{ideal}} = 0 \tag{7.57}$$

However, there will be a positive entropy change of mixing, since the ideal gas mixture can undergo more configurations (i.e., is more random) than the separated pure species. The expression for the entropy change of mixing of an ideal gas is given by Equation (6.48):

$$\Delta s_{\text{mix}}^{\text{ideal}} = -R \sum y_i \ln y_i \tag{7.58}$$

See Example 6.8 for the development of this expression for a binary mixture. To find the Gibbs energy of mixing, we apply the definition of this thermodynamic property. At constant temperature,

$$\Delta g_{\text{mix}}^{\text{ideal}} = \Delta h_{\text{mix}}^{\text{ideal}} - T \Delta s_{\text{mix}}^{\text{ideal}} = RT \sum y_i \ln y_i \tag{7.59}$$

Figure 7.4 shows the Gibbs energy of ideal gas species a and b. Curve 1 plots the value of g for a given mixture composition when the two gases exist separately, as pure species, while curve 2 plots the value of g in the mixture. The lowering of g when a and b mix is due to the Gibbs energy of mixing, Δg_{mix}. Note that the Gibbs energy of the mixture is always lower than the weighted average of the pure species. Thus, two ideal gases can lower their Gibbs energy if they are allowed to mix; this indicates that the mixing process occurs spontaneously.

Alternatively, we can calculate the Gibbs energy change of mixing of an ideal gas based on the definition of fugacity, as given by Equation (7.7). Consider applying this equation to species i, where we choose the pure species at the same T and P of the mixture as a reference state. Since we have an ideal gas mixture, the fugacity of the

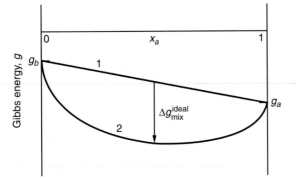

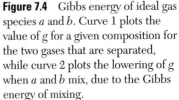

Figure 7.4 Gibbs energy of ideal gas species a and b. Curve 1 plots the value of g for a given composition for the two gases that are separated, while curve 2 plots the lowering of g when a and b mix, due to the Gibbs energy of mixing.

mixture is equal to the partial pressure, whereas the fugacity of pure i is equal to the total pressure. Applying the definition of fugacity, Equation (7.7), we get

$$\mu_i^{\text{ideal}} - g_i = RT \ln \left[\frac{\hat{f}_i}{f_i} \right] = RT \ln \left[\frac{p_i}{P} \right] = RT \ln y_i \tag{7.60}$$

From Equation (7.55), we have

$$\Delta g_{\text{mix}}^{\text{ideal}} = g - \sum y_i g_i = \sum y_i (\overline{G}_i - g_i) = \sum y_i (\mu_i - g_i) \tag{7.61}$$

Plugging Equation (7.60) into Equation (7.61), we get

$$\Delta g_{\text{mix}}^{\text{ideal}} = RT \sum y_i \ln y_i \tag{7.62}$$

Equation (7.62) is identical to Equation (7.59). Again, we see that in thermodynamics, there are many paths to the same destination.

▶ 7.4 FUGACITY IN THE LIQUID PHASE

In this section, we explore how we can calculate fugacity in the liquid phase. In doing so, we will develop an appropriate reference state for the liquid phase, the ideal solution, and then correct for real behavior through the activity coefficient. We will then learn a type of empirical model that correlates experimental data efficiently. The methodology developed in this section can also be applied to solids. Finally, an alternative approach to calculating fugacity in the liquid using PvT equations of state is discussed. This approach is similar to that used for the vapor phase in Section 7.3

Reference States for the Liquid Phase

In the liquid phase, just as in the vapor phase, we need to choose a suitable reference state with a corresponding reference chemical potential and reference fugacity to complete the definition provided by Equation (7.7). We then adjust for the difference between the reference phase and the real system. However, while there is an obvious reference case for gases—the ideal gas—there is no single suitable choice for the liquid phase. There are two common choices for the reference state: (1) the Lewis/Randall rule and (2) Henry's law. The choice of reference state often depends on the system. Both these reference states are limiting cases that result from a natural idealization for condensed phases: the ideal solution.

The Ideal Solution
We wish to define a reference state for the liquid phase to which we can compare the fugacity of a real liquid. Hence, we need something analogous to what an ideal gas provided us for real gases. For liquids, however, we cannot extrapolate to a state where there are no intermolecular interactions as we did at zero pressure for gases. Indeed, it is the very presence of intermolecular forces that makes condensed phases possible; without these forces, only the entropically favored gas phase would exist. Accordingly we will choose an **ideal solution** as our reference state. An ideal solution can be defined in several ways. On a macroscopic level, a solution is ideal when all the mixing rules are the same as for an ideal gas. Analogous to Equations (7.56) through (7.59), an ideal solution

is characterized by the following mixing rules:

$$\Delta v_{\text{mix}}^{\text{ideal}} = 0 \tag{7.63}$$

$$\Delta h_{\text{mix}}^{\text{ideal}} = 0 \tag{7.64}$$

$$\Delta s_{\text{mix}}^{\text{ideal}} = -R \sum x_i \ln x_i \tag{7.65}$$

$$\Delta g_{\text{mix}}^{\text{ideal}} = RT \sum x_i \ln x_i \tag{7.66}$$

In analogy to Equation (7.60) for an ideal gas, the following relation must hold for an ideal solution:

$$\mu_i^{\text{ideal}} - g_i^{\text{ideal}} = RT \ln x_i = RT \ln \left[\frac{\hat{f}_i^{\text{ideal}}}{f_i^{\text{ideal}}} \right] \tag{7.67}$$

Equation (7.67) has interesting implications. For this relation to hold, the fugacity of an ideal solution is linear in mole fraction to the pure species fugacity:

$$\boxed{\hat{f}_i^{\text{ideal}} = x_i f_i^{\text{ideal}}} \tag{7.68}$$

This expression is identical to the Lewis fugacity rule [Equation (7.53)]. Recall that a species obeys the Lewis fugacity rule if all the intermolecular forces are equal.

On a molecular level, we define a solution as ideal when the intermolecular potentials are the *same* between all components of the mixture. Thus, in using the ideal solution as a reference state, we will be comparing the behavior of real mixtures to that where all the species in the mixture interact with equal magnitude. However, there is more than one possible interaction upon which to define f_i^{ideal}. We will now explore two common choices of f_i^{ideal} that we use for reference states: The Lewis/Randall rule and Henry's law.

Consider Figure 7.5, in which the fugacity in the liquid phase of species a in a binary mixture is plotted (solid curve) as a function of mole fraction of a. This curve is linear and, thus, satisfies Equation (7.68) in two places—as x_a approaches 1 and as x_a approaches 0. In the first case ($x_a \to 1$), the solution is almost completely a, as shown in the inset. Thus, species a sees essentially only other species a as it interacts with its neighbors. Therefore, we have an ideal solution because as far as a molecule of a is concerned, all the intermolecular interactions are the same. They are all a-a interactions. The characteristic of this intermolecular interaction is given by the pure species fugacity, f_a. Consequently, one choice of reference state is the pure species fugacity of species a. We call this the **Lewis/Randall rule**:

$$f_a^{\text{ideal}} = f_a^o = f_a \qquad \text{(Lewis / Randall rule)} \qquad \underline{a\text{-}a \text{ interactions}} \tag{7.69}$$

Consider now the case when a is very dilute in b. This case is shown in the inset as x_a goes to zero. As far as a molecule of a is concerned, it essentially sees only species b in solution. Thus, again, we have an ideal solution because all the intermolecular interactions species a has are the same. In this case, however, it is the a-b interaction that is characteristic of the ideal solution. This provides another choice on reference state based on a-b interactions, which we call **Henry's law**:

$$f_a^{\text{ideal}} = f_a^o = \mathcal{H}_a \qquad \text{(Henry's law)} \qquad a\text{-}b \text{ interactions} \tag{7.70}$$

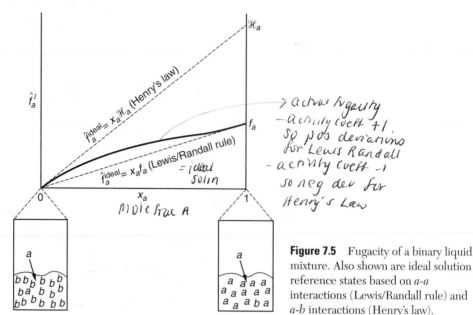

(handwritten annotations)
specie fugacity of component a in mixture

Henry's Law
a = solute

L/R
a = solvent

idea of adding in another variable c

→ actual fugacity
– activity coeff +1
so pos deviations for Lewis Randall
– activity coeff –1
so neg dev for Henry's Law

Figure 7.5 Fugacity of a binary liquid mixture. Also shown are ideal solution reference states based on *a-a* interactions (Lewis/Randall rule) and *a-b* interactions (Henry's law).

The Henry's law limit can be conceptualized as a hypothetical, pure fluid in which the characteristic energy of interaction is that between molecule a and molecule b. If species a is defined by a Lewis/Randall reference state, we call it a **solvent**, whereas when we describe it by Henry's law, it is termed a **solute**. If we had a binary mixture of species a with a different molecule c, the Lewis/Randall reference state for species a would be unchanged. However, the Henry's reference state could differ dramatically. The former is independent of any other species in the mixture, while the latter, by definition, depends on the chemical nature of the other species in the mixture.

In the intermediate concentration ranges in Figure 7.5, we see that the fugacity of species a in the liquid phase is between the two limiting cases given by the Lewis/Randall rule and Henry's law. We expect this behavior because, at intermediate concentrations, molecule a sees some other a molecules (characteristic of the Lewis/Randall rule) and some b molecules (characteristic of Henry's law). Thus, the fugacity of species a will be the appropriate "average" of $a-a$ and $a-b$ interactions: The more a molecules it sees (larger x_a), the closer it will be to the Lewis/Randall reference state; the more b it sees, the closer to Henry's law. To solve problems in phase equilibria, we need to essentially come up with a quantitative formulation to "average" between these limits. The goal is to be able to estimate the fugacity of a liquid at any given composition based on experimentally available data.

▶ **EXAMPLE 7.6**
Relation Between Fugacity and Strength of Intermolecular Interactions

Consider the binary mixture depicted in Figure 7.5. Which interaction is stronger, the like interactions, $a-a$, or the unlike interactions, $a-b$?

SOLUTION At equilibrium, the fugacity of the liquid equals the fugacity of the vapor:

$$f_a^l = f_a^v$$

For this argument, we will consider the specific case of an ideal gas where the fugacity of a in the vapor can be replaced with a partial pressure:

$$f_a^l = p_a$$

However, the conclusion we will reach holds in general. The lower the fugacity of the liquid of species a, the lower its partial pressure in the vapor phase. Having a lower partial pressure in the vapor phase means that species must be held more strongly in the liquid; therefore, the intermolecular interactions are stronger. The like a-a interaction is characterized by the pure species fugacity, f_a, while the a-b interaction is characterized by the Henry's law constant, $\mathcal{H}_a$. In the case of Figure 7.5, f_a, is less than $\mathcal{H}_a$, implying the a-a interaction is stronger.

If we consider the fugacity as a "tendency to escape," the stronger interaction should have the lower fugacity. Since the Henry's constant is larger, the hypothetical ideal species based on the a-b interaction has a greater tendency to escape than the pure a. Thus the *like* interactions are stronger than the *unlike* interaction. ◀

The Activity Coefficient, γ_i

Inspection of Figure 7.5 reveals a natural dimensionless group to express the fugacity of the liquid phase—the activity coefficient, γ_i. The activity coefficient is defined as the ratio of the value of the fugacity in the actual mixture, represented by the solid line in Figure 7.5, to the fugacity that the ideal solution would have at the composition of the mixture, represented by either of the dashed lines in Figure 7.5 (depending on our choice of reference state). Thus, we can write:

$$\gamma_i = \frac{\hat{f}_i^l}{\hat{f}_i^{\text{ideal}}} = \frac{\hat{f}_i^l}{x_i f_i^o} \qquad (7.71)$$

The activity coefficient tells us how "active" the liquid is relative to our choice of reference state. The value of the activity coefficient depends on the specific choice of reference state. For example, consider the binary system represented in Figure 7.5. The activity coefficient for the Lewis/Randall reference state is greater than 1 throughout most of the composition range and equal to 1 in the limit of pure a. Thus,

$$\gamma_a \geq 1 \qquad \text{in Example 7.5} \qquad (7.72)$$

On the other hand, the Henry's reference state represented in Example 7.5 is less than or equal to one:[3]

$$\gamma_a^{\text{Henry's}} \leq 1 \qquad \text{in Example 7.5} \qquad (7.73)$$

In other systems, where the unlike interactions are stronger than the like interactions, the converse of Equations (7.72) and (7.73) is true.

The activity coefficient for the Henry's reference state approaches 1 in the limit as x_a goes to zero, and the activity coefficient for the Lewis/Randall reference state approaches 1 in the limit as x_a goes to 1. We will next consider expressions for the activity coefficients in the other composition limit for each choice of reference state. In the limit as the mole fraction of species a goes to 1, the activity coefficient in the Henry's reference state can be written by applying Equation (7.71):

$$(\gamma_a^{\text{Henry's}})^{\text{pure } a} = \frac{\hat{f}_a^l}{\hat{f}_a^{\text{ideal}}} = \frac{\hat{f}_a^l}{x_a f_a^o} = \frac{f_a}{\mathcal{H}_a} \qquad (7.74)$$

[3] In this text, we use explicitly use the superscript "Henry's" to denote the Henry's law reference state. When the activity coefficient does not have a superscript, we implicitly assume the Lewis/Randall reference state, although, in some cases, the expression may apply to both the Lewis/Randall reference state and the Henry's law reference state.

On the other hand, in the limit of infinite dilution, the activity coefficient in the Lewis/Randall reference state becomes

- opposite end of spectrum ⟶ $\gamma_a^\infty = \dfrac{\mathcal{H}_a}{f_a}$
- still use lewis/Rand $\qquad\qquad\qquad\qquad\qquad\qquad\qquad\qquad$ (7.75)

Comparing Equations (7.74) and (7.75), we see that

$$(\gamma_a^{\text{Henry's}})^{\text{pure } a} = \dfrac{1}{\gamma_a^\infty} \qquad\qquad (7.76)$$

Moreover, by applying the definition of the activity coefficient, Equation (7.71), we see that

$$\hat{f}_a^l = x_a\,\gamma_a f_a = x_a\,\gamma_a^{\text{Henry's}}\mathcal{H}_a = x_a\,\gamma_a^{\text{Henry's}}\gamma_a^\infty f_a \qquad (7.77)$$

where Equation (7.75) was used. Inspection of Equation (7.77) shows that we can relate the activity coefficient in the Henry's law reference state to the activity coefficient in the Lewis/Randall reference state by

$$\gamma_a^{\text{Henry's}} = \dfrac{\gamma_a}{\gamma_a^\infty} \qquad\qquad (7.78)$$

Note the similarity of the activity coefficient represented by Equation (7.71) to the fugacity coefficient given by Equation (7.9):

$$\hat{\varphi}_i^v \equiv \dfrac{\hat{f}_i^v}{p_{i,\text{sys}}} = \dfrac{\hat{f}_i^v}{y_i P_{\text{sys}}} \qquad\qquad (7.9)$$

While the fugacity coefficient can be viewed as a dimensionless quantity expressing how the fugacity in the vapor phase compares to how it would hypothetically behave as an ideal gas, the activity coefficient represents a dimensionless quantity of how the fugacity in the liquid compares relative to whatever ideal solution reference state is chosen. Like the fugacity coefficient, the activity coefficient tells us how far the system is deviating from ideal behavior. In the case of gases, that reference state depicts a unique state where the intermolecular interactions are zero. In the case of liquids, on the other hand, we refer to a state where all the interactions are the same. Consequently, we must remember to what particular reference state we are comparing the fugacity, that is, whether we are comparing the activity of the real liquid to the like interactions or the unlike interactions. Since the reference state for the **vapor** phase, the ideal gas, represents the case where there are no intermolecular potential interactions between the molecules in the system, it is realized in the limit as **pressure** goes to zero. On the other hand, the **liquid**-phase reference state, the ideal solution, occurs when all the intermolecular interactions are the same. This condition is reached in the limit of **composition** when the mole fraction goes either to 1 (Lewis/Randall rule) or to zero (Henry's law). In liquids, therefore, we typically choose the pressure of the reference state equal to that of the system.

The activity of species i in the liquid, a_i, is often used in conjunction with the activity coefficient. It is defined as follows:

$$a_i \equiv \dfrac{\hat{f}_i^l}{f_i^o} \qquad\qquad (7.79)$$

The activity compares the fugacity of species i in the liquid to the fugacity of the *pure* species in its reference state. On the other hand, the activity coefficient is defined with

Equation (7.98) indicates that the activity coefficients in a mixture are interrelated. Consider, for example, a binary mixture of components a and b. Differentiating with respect to x_a at constant T and P, Equation (7.98) becomes

$$x_a \left(\frac{\partial \ln \gamma_a}{\partial x_a} \right)_{T,P} + x_b \left(\frac{\partial \ln \gamma_b}{\partial x_a} \right)_{T,P} = 0 \tag{7.99}$$

The form of the Gibbs–Duhem equation given by Equation (7.99) implies that once you know γ_a as a function of x_a, then γ_b (or, really, the slope of γ_b) is constrained. Example 7.7 illustrates this relationship.

Equation (7.99) says that the activity coefficients of different species in the mixture are interrelated. Thus, when we develop models to fit activity coefficients in the next section, we must make sure that the expressions for the activity coefficients of different species in a mixture are consistent with the Gibbs–Duhem equation. On the other hand, this relationship also suggests an intriguing possibility—perhaps it is possible to consolidate the activity coefficient dependence on composition of all the species into one model expression and, therefore, be able to derive all the activity coefficients from a single expression. Indeed, as we will discover next, we will take just this approach through a new thermodynamic property—the excess Gibbs energy.

▷ **EXAMPLE 7.7**
Application of the Gibbs–Duhem Equation to Determine γ_b Graphically

Consider a binary liquid consisting of species a and b. The activity coefficient, based on the Lewis/Randall rule, for species a vs. mole fraction a is plotted in Figure E7.7A. On the same graph, plot the activity coefficient for species b **(a)** using the Lewis/Randall rule as reference for b; **(b)** using Henry's law as reference for b.

SOLUTION In each case, the Gibbs–Duhem equation must hold, that is,

$$x_a \left(\frac{\partial \ln \gamma_a}{\partial x_a} \right)_{T,P} + (1 - x_a) \left(\frac{\partial \ln \gamma_b}{\partial x_a} \right)_{T,P} = 0 \tag{E7.7}$$

At any value of mole fraction, x_a, we can take the slope of the curve above to get $(\partial \ln \gamma_a)/\partial x_a$. This equation then constrains the slope of the activity coefficient of species b. If we know the value at one point, then we can "graphically integrate" the above equation.
(a) For the Lewis/Randall reference state, species b behaves ideally as $x_b \to 1$, that is, as $x_a \to 0$; thus, at this mole fraction, $\ln \gamma_b = 0$. This fixes one point on the graph, as labeled in Figure E7.7B.

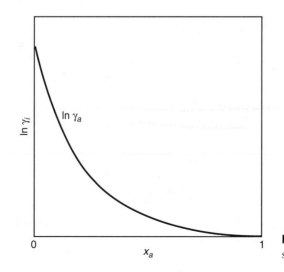

Figure E7.7A Activity coefficient of species a plotted vs. mole fraction a.

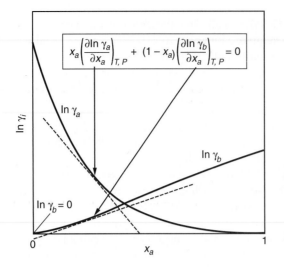

$$x_a\left(\frac{\partial \ln \gamma_a}{\partial x_a}\right)_{T,P} + (1-x_a)\left(\frac{\partial \ln \gamma_b}{\partial x_a}\right)_{T,P} = 0$$

ln γ_a

ln γ_b

ln $\gamma_b = 0$

$\ln \gamma_i$

0   x_a 1

Figure E7.7B Activity coefficient of species b plotted vs. mole fraction a, as determined by the Gibbs–Duhem equation.

From this point, we can plot the curve for the activity coefficient of species b, since the slope at any mole fraction is constrained by the Gibbs–Duhem equation. The result is shown in Figure E7.7B. The solution to Equation (E7.7) is illustrated for $x_a = 0.25$.

(b) For the Henry's law reference state, the Gibbs–Duhem equation must also be satisfied; therefore, the slope of the curve for $\ln \gamma_b^{\text{Henry's}}$ remains the same as the one we determined for the Lewis/Randall reference state and plotted in Figure E7.7B. However, now species b behaves ideally as $x_a \to 1(x_b \to 0)$, and thus $\ln \gamma_b^{\text{Henry's}} = 0$ at that mole fraction. This simply shifts the entire curve as shown in Figure E7.7C. ◀

Henry's Law for species b

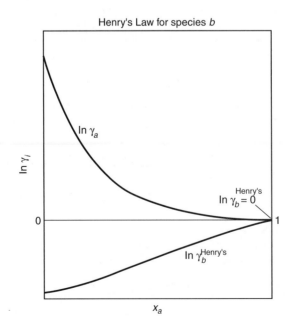

ln γ_a

$\ln \gamma_i$

0

1

ln $\gamma_b = 0$ (Henry's)

$\ln \gamma_b^{\text{Henry's}}$

x_a

Figure E7.7C Activity coefficient of species b plotted vs. mole fraction a, using Henry's law reference state.

Excess Gibbs Energy and Other Excess Properties

Excess Gibbs energy forms the basis from which the activity coefficient of *all* the species in the mixture can be obtained from a single quantitative expression. An excess property, k^E, is defined as the difference between the real value of any thermodynamic property, k, and the hypothetical value it would have as an ideal solution at the same temperature pressure and composition, k^{ideal}:

$$k^E \equiv k(T, P, x_i) - k^{\text{ideal}}(T, P, x_i) \qquad (7.100)$$

for example,

$$\left[\begin{array}{l} v^E = v(T, P, x_i) - v^{\text{ideal}}(T, P, x_i) \\ h^E = h(T, P, x_i) - h^{\text{ideal}}(T, P, x_i) \\ g^E = g(T, P, x_i) - g^{\text{ideal}}(T, P, x_i) \\ \qquad\qquad \vdots \end{array} \right.$$

For k^{ideal}, we must specify which reference state is chosen for the solution, that is, the Lewis/Randall reference state, Henry's law, or the ideal gas. The departure functions discussed in Chapter 5 are a special case of excess properties.

Likewise, we can define a partial molar excess property by applying Equation (6.29) to the extensive excess property as follows:

$$\overline{K}_i^E \equiv \left(\frac{\partial (nk^E)}{\partial n_i} \right)_{T,P,n_{j\neq i}} = \left(\frac{\partial (K - K^{\text{ideal}})}{\partial n_i} \right)_{T,P,n_{j\neq i}} = \overline{K}_i - \overline{K}_i^{\text{ideal}} \qquad (7.101)$$

for example,

$$\left[\begin{array}{l} \overline{V}_i^E = \overline{V}_i - \overline{V}_i^{\text{ideal}} \\ \overline{H}_i^E = \overline{H}_i - \overline{H}_i^{\text{ideal}} \\ \overline{G}_i^E = \overline{G}_i - \overline{G}_i^{\text{ideal}} \\ \qquad \vdots \end{array} \right.$$

Does it make sense to define the pure species excess property, k_i^E?

If we add and subtract $\sum x_i k_i$ to the right side of Equation (7.100), we get

$$k^E = \left(k - \sum x_i k_i \right) - \left(k^{\text{ideal}} - \sum x_i k_i \right) \qquad (7.102)$$

or

$$k^E = \Delta k_{\text{mix}} - \Delta k_{\text{mix}}^{\text{ideal}} \qquad (7.103)$$

Thus, an excess property is also the difference between the real property change of mixing and the ideal property change of mixing.

This suggests two classes of excess properties:

- *Class I.* For class I properties, the ideal property change of mixing is zero:

$$\Delta k_{\text{mix}}^{\text{ideal}} = 0 \qquad (k = u, h, v) \qquad (7.104)$$

For class I properties, the excess property is identical to the property change of mixing:

$$k^E = \Delta k_{\text{mix}} \tag{7.105}$$

for example,

$$\left[\begin{array}{l} v^E = \Delta v_{\text{mix}} \\ h^E = \Delta h_{\text{mix}} \\ \quad \vdots \end{array} \right]$$

- *Class II.* For class II properties, the ideal property change of mixing is not zero:

$$\Delta k_{\text{mix}}^{\text{ideal}} \neq 0 \qquad (k = g, s, a) \tag{7.106}$$

In this case, the excess functions represent a new set of properties, which we can write as follows:

$$k^E = \Delta k_{\text{mix}} - \Delta k_{\text{mix}}^{\text{ideal}} \tag{7.107}$$

for example,

$$\left[\begin{array}{l} s^E = \Delta s_{\text{mix}} - \Delta s_{\text{mix}}^{\text{ideal}} \\ g^E = \Delta g_{\text{mix}} - \Delta g_{\text{mix}}^{\text{ideal}} \\ \quad \vdots \end{array} \right]$$

As illustrated in Figure 7.4, the ideal property change of mixing for Gibbs energy is not zero; hence g^E belongs in class II. If we apply Equation (7.107) to g^E and substitute Equation (7.66), we get

$$\boxed{g^E = \Delta g_{\text{mix}} - RT \sum x_i \ln x_i} \tag{7.108}$$

The partial molar excess Gibbs energy is given by Equation (7.101), which can be extended to read

$$\overline{G}_i^E = \overline{G}_i - \overline{G}_i^{\text{ideal}} = \mu_i - \mu_i^{\text{ideal}} = RT \ln \frac{\hat{f}_i}{\hat{f}_i^{\text{ideal}}} \tag{7.109}$$

If we substitute Equation (7.71) into Equation (7.109), we get the following important result:

$$\boxed{\overline{G}_i^E = RT \ln \gamma_i} \tag{7.110}$$

Equation (7.110) indicates that there is a direct relationship between the partial molar excess Gibbs energy of a species and its activity coefficient in solution. This equation implies that if we have a mathematical expression for the excess Gibbs energy of a mixture as a function of composition, we can get the activity coefficient for any of the m species in the mixture. All we have to do is apply the definition of a partial molar property [Equation (6.29)] and take the partial derivative of the extensive form of the excess Gibbs energy with respect to the number of moles of a given species, holding temperature, pressure, and the number of moles of all the other species constant. Once we have obtained the partial molar Gibbs energy, Equation (7.110) tells us that we now know the activity coefficient for that species. Then we can solve for the fugacity in the liquid via Equation (7.71).

Our approach to come up with values for γ_i to find the fugacity of the liquid phase is to come up with an analytical expression for the excess Gibbs energy, g^E. The activity coefficients for each species i in the mixture can then be found by Equation (7.110). In this way we need only one *model for g^E to obtain activity coefficients for* every *species.*

Since we will be working with excess Gibbs energy to correlate experimental measurements, it is useful to apply thermodynamic property relationships to form expressions for excess Gibbs energy and its derivatives. Applying Equations (6.35) to excess Gibbs energy and substituting in Equation (7.110) gives

$$g^E = \sum x_i \overline{G}_i^E = RT \sum x_i \ln \gamma_i \tag{7.111}$$

Thermodynamic Consistency Tests

The Gibbs–Duhem equation provides a general relation for the partial molar properties of different species in a mixture that must always be true. For example, we just saw how the activity coefficient of different species can be related to one another. In this section, we explore one way to use this interrelation to judge the quality of experimental data. The basic idea is to develop a way to see whether a set of data conform to the constraints posed by the Gibbs–Duhem equation. If the data reasonably match, we say they are thermodynamically consistent. On the other hand, data that do not conform to the Gibbs–Duhem equation are thermodynamically inconsistent and should be considered unreliable. The development that follows is based on the relation between activity coefficients in a binary mixture of species a and b. It serves as an example to this methodology; there are several other ways that have been developed to apply this same type of idea.

According to Equation (7.111), the excess Gibbs energy for a binary mixture can be written as:

$$g^E = RT(x_a \ln \gamma_a + x_b \ln \gamma_b) \tag{7.112}$$

Differentiating Equation (7.112) with respect to x_a at constant T and P gives

$$\frac{dg^E}{dx_a} = RT\left[\ln \gamma_a + x_a \frac{d \ln \gamma_a}{dx_a} - \ln \gamma_b + x_b \frac{d \ln \gamma_b}{dx_a}\right] = RT[\ln \gamma_a - \ln \gamma_b] \tag{7.113}$$

where we have simplified Equation (7.113) using the Gibbs–Duhem equation:

$$x_a \left(\frac{\partial \ln \gamma_a}{\partial x_a}\right)_{T,P} + x_b \left(\frac{\partial \ln \gamma_b}{\partial x_a}\right)_{T,P} = 0$$

Equation (7.113) can be rewritten as

$$dg^E = RT \ln \left(\frac{\gamma_a}{\gamma_b}\right) dx_a \tag{7.114}$$

We next integrate Equation (7.114) over composition from pure b to pure a to get

$$\int_{\text{pure } b}^{\text{pure } a} dg^E = \int_{x_a = 0}^{x_a = 1} RT \ln \left(\frac{\gamma_a}{\gamma_b}\right) dx_a \tag{7.115}$$

Since g^E is a state function, the integral on the left-hand side of Equation (7.115) depends solely on its value at each limit of integration. If we use a Lewis/Randall reference state for both species, the value of g^E is zero for pure a and for pure b. The integral in Equation (7.115) becomes zero, and we can rewrite this equation:

$$\int_0^1 \ln\left(\frac{\gamma_a}{\gamma_b}\right) dx_a = 0 \qquad (7.116)$$

Equation (7.116) suggests a way to evaluate experimental data for thermodynamic consistency. If we plot $\ln(\gamma_a/\gamma_b)$ vs. x_a, the area under the curve should be close to zero. Hence, there should be approximately as much area above the x-axis as below it. This test is often referred to as the *area test*. An example of the application of the area test to experimental data for a binary system of ethanol and water at 60°C is shown in Figure 7.8. In this case, the areas above and below the x-axis are roughly equal and we conclude these data are thermodynamically consistent.

Equation (7.116) was derived assuming constant T and P. However, for real data sets, either T or P must change as composition is varied. In most isothermal data sets, the pressure dependence of the activity coefficient is small over the range considered, so Equation (7.116) can be applied directly to test for thermodynamic consistency. However, in cases where the pressure is held constant, Equation (7.116) often needs to be corrected to include the change in the activity coefficient with temperature. In Problem 7.36, an expression that improves Equation (7.116) by accounting for the variation of activity coefficient with temperature is to be developed.

Models for γ_i using g^E

Once a reference state has been chosen, fugacities in the liquid phase can be expressed via γ_i by rearranging Equation (7.71) as follows:

$$\hat{f}_i^l = x_i \gamma_i f_i^o \qquad (7.117)$$

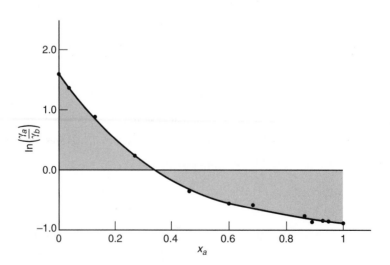

Figure 7.8 Area test for the thermodynamic consistency of a binary mixture of ethanol (a) in water (b) at 60°C.

The Lewis/Randall reference state that is based on the i-i interaction gives

$$\hat{f}_i^l = x_i \gamma_i f_i \tag{7.118}$$

while the Henry's law reference state, based on the i-j interaction, gives

$$\hat{f}_i^l = x_i \gamma_i^{\text{Henry's}} \mathcal{H}_i \tag{7.119}$$

Since accurate equation of state data may be difficult to find for condensed phases, we seek an alternative means for correlating and extending experimental data. This task is most commonly done through another type of model—activity coefficient models using g^E. The nonideality in the liquid phase is most strongly dependent on composition. The goal, then, is to develop an expression for g^E from which the compositional dependence of the activity coefficient can be obtained. Once the compositional dependence of the activity coefficient has been quantified, Equation (7.117) can be used to determine the fugacity of the liquid phase at a given mole fraction i.

The Two-Suffix Margules Equation

One approach to estimating the excess Gibbs energy is to try to fit experimental results to analytical expressions in mole fraction for g^E. This is similar to the equation of state concept. It is more convenient to model the excess Gibbs energy and take the appropriate partial derivatives via Equation (7.110) for the activity coefficient than to model the activity coefficient of each species separately. The goal is to find an expression for g^E (and thus the activity coefficients) over the entire composition range given a limited amount of experimental data. In this section we will look at the simplest nonideal model that we can think of for g^E in depth—the two-suffix Margules equation. After seeing its strengths and limitations, we will extend it to other forms.

Consider a binary solution of species a and b with the Lewis/Randall reference state for both components. The activity coefficient model must satisfy the following two conditions:

1. At mole fractions of $x_a = 1$ and $x_b = 1$, we have an ideal solution; therefore, the excess Gibbs energy is zero:

$$g^E = 0 \begin{cases} x_a = 1; \ x_b = 0 \\ x_b = 1; \ x_a = 0 \end{cases}$$

2. The model for g^E satisfies the Gibbs–Duhem equation. If we attempted to model the activity coefficient of each species separately, we would have much more difficulty with this condition.

What is the simplest nonideal model that satisfies these two conditions?

Let's try

$$\boxed{g^E = A x_a x_b} \tag{7.120}$$

The parameter A is fit to experimental data for a given binary mixture. This parameter may change with temperature or pressure, but it is independent of the composition of the system. Moreover, as we will see shortly, we can apply the thermodynamic web to see how A varies with T and P. Equation (7.120) is called the two-suffix Margules equation; it clearly satisfies condition 1. Example 7.8 shows that it also satisfies condition 2.

In the case of a Lewis/Randall reference state, the deviation from ideality results from the nature of unlike a-b interactions (recall the discussion of Figure 7.5). Consequently, it is not surprising that the compositional dependence of Equation (7.120) is exactly the same as for the mixing term for the van der Waals energy parameter:

$$a_{mix} = x_a^2 a_a + 2\underline{x_a x_b} \sqrt{a_a a_a} + x_b^2 a_b \tag{7.121}$$

Example 7.9 provides further discussion of the molecular origins of the two-suffix Margules equation.

Once we have this expression for g^E, the corresponding activity coefficients for species a and b are given by the appropriate partial molar excess Gibbs energies via Equation (7.110). Applying the definition of a partial molar property to the excess Gibbs energy, we get

$$\overline{G}_a^E = \left(\frac{\partial G^E}{\partial n_a}\right)_{T,P,n_b} = \left(\frac{\partial \left(n_T g^E\right)}{\partial n_a}\right)_{T,P,n_b} = A\left[\frac{\partial \left(\frac{n_a n_b}{n_a + n_b}\right)}{\partial n_a}\right]_{T,P,n_b} \tag{7.122}$$

$$= A\left[\frac{n_b}{n_a + n_b} - \frac{n_a n_b}{(n_a + n_b)^2}\right] = A\frac{n_b^2}{(n_a + n_b)^2}$$

Thus,

$$\boxed{\overline{G}_a^E = Ax_b^2 = RT\ln\gamma_a} \tag{7.123}$$

Similarly,

$$\boxed{\overline{G}_b^E = Ax_a^2 = RT\ln\gamma_b} \tag{7.124}$$

If we have a binary mixture and have determined a value for the two-suffix Margules parameter A, we can find the activity coefficient of both species in the mixture at any composition by Equations (7.123) and (7.124).

▶ **EXAMPLE 7.8**
Thermodynamic Consistency of the Two-Suffix Margules Equation

Show that the two-suffix Margules equation satisfies the Gibbs–Duhem equation.

SOLUTION Applying the Gibbs–Duhem equation, Equation (6.37), to excess Gibbs energy gives

$$x_a d\overline{G}_a^E + x_b d\overline{G}_b^E = 0 \tag{E7.8A}$$

but, from Equation (7.123),

$$\overline{G}_a^E = Ax_b^2$$

so

$$d\overline{G}_a^E = 2Ax_b dx_b \tag{E7.8B}$$

Similarly

$$d\overline{G}_b^E = 2Ax_a dx_a \tag{E7.8C}$$

Plugging Equations (E7.8B) and (E7.8C) into (E7.8A)

$$x_a(2Ax_b dx_b) + x_b(2Ax_a dx_a) = 0$$

Therefore, the two-suffix Margules equation satisfies the Gibbs–Duhem equation since

$$dx_a = -dx_b$$

◄

Provide a molecular explanation for the form of the two-suffix Margules equation:

$$g^E = Ax_ax_b$$

Show how the relative magnitude of the *unlike* interactions to the *like* interactions determines the value of two-suffix Margules parameter, A.

SOLUTION In this example, we assume that all the nonideality is associated with a difference in *energetics* between the species in the mixture and the pure species; that is, the excess entropy is zero. This assumption is valid for species of roughly the same size. Let's consider the difference in energetics between a and b in a mixture vs. a and b as pure species. Figure E7.9A illustrates the possible interactions in the mixture and of pure a and pure b. Pure a and b exhibit only a-a interactions and b-b interactions, respectively. The mixture contains not only these like a-a and b-b interactions but also unlike a-b interactions. In fact, when species a and b are mixed, we can view the process in terms of intermolecular interactions. Some a-a interactions of pure a are replaced by a-b interactions, while some b-b interactions of pure b are also replaced by a-b interactions. To quantify the difference in energy of the mixture relative to the pure species and, therefore, the nonideality of the mixture, we compare the magnitude of the interactions in the mixture with those that existed as pure species.

For two-body interactions, the energetic interactions for the mixture can be written

$$\Gamma_{\text{mix}} = x_a^2\Gamma_{aa} + 2x_ax_b\Gamma_{ab} + x_b^2\Gamma_{bb} \tag{E7.9A}$$

To compare the mixture to the pure species, each pure species has to be scaled in proportion to how much is present in the mixture:

$$\Gamma_{\text{pure},a} = x_a\Gamma_{aa} \quad \text{and} \quad \Gamma_{\text{pure},b} = x_b\Gamma_{bb} \tag{E7.9B}$$

Now the change in energetics upon mixing, at constant pressure, is the difference between these two expressions:

$$\Delta h_{\text{mix}} = N_A\left\{\Gamma_{\text{mix}} - \left[\Gamma_{\text{pure},a} + \Gamma_{\text{pure},b}\right]\right\} \tag{E7.9C}$$

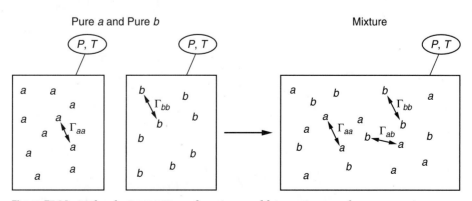

Figure E7.9A Molecular interactions of species a and b in a mixture and as pure species.

We have multiplied Equation (E7.9C) by Avogadro's number for dimensional consistency. Substituting (E7.9A) and (E7.9B) into Equation (E7.9C) gives

$$\frac{\Delta h_{\text{mix}}}{N_A} = x_a^2 \Gamma_{aa} + 2x_a x_b \Gamma_{ab} + x_b^2 \Gamma_{bb} - [x_a \Gamma_{aa} + x_b \Gamma_{bb}]$$

or, rearranging,

$$\frac{\Delta h_{\text{mix}}}{N_A} = (x_a^2 - x_a)\Gamma_{aa} + 2x_a x_b \Gamma_{ab} + (x_b^2 - x_b)\Gamma_{bb}$$

We now factor out x_a from the first term on the right-hand side and x_b from the third:

$$\frac{\Delta h_{\text{mix}}}{N_A} = x_a(x_a - 1)\Gamma_{aa} + 2x_a x_b \Gamma_{ab} + x_b(x_b - 1)\Gamma_{bb}$$

Since the mole fractions sum to 1, we can use $(x_a - 1) = -x_b$ and $(x_b - 1) = -x_a$ to give

$$\frac{\Delta h_{\text{mix}}}{N_A} = [2\Gamma_{ab} - (\Gamma_{aa} + \Gamma_{bb})]x_a x_b$$

For an ideal (liquid) solution,

$$\Delta h_{\text{mix}} = 0$$

So if all "excess" from ideality can be attributed to energetic effects—that is, $g^E = h^E = \Delta h_{\text{mix}}$— then

$$A = N_A [2\Gamma_{ab} - (\Gamma_{aa} + \Gamma_{bb})] \tag{E7.9D}$$

Equation (E7.9D) represents the arithmetic difference between the interaction in the mixture and those of the pure species, as shown in Figure E7.9B.

The magnitude of the two-suffix parameter A can be viewed as the relative importance between the unlike *a-b* interactions and the average of the like *a-a* and *b-b* interactions. This quantity is independent of how much of each species is present in the mixture. If the unlike interactions are stronger, Γ_{ab} is more negative, and $A < 0$. Conversely, weaker unlike interactions lead to $A > 0$. *Thus, when we discuss how the unlike interactions compare to the like interactions, we are essentially comparing two a-b interactions to the sum of an a-a and an b-b interaction, as depicted in Figure E7.9B.*

Certain two-body unlike intermolecular interactions are well approximated by a geometric mean of the like interactions. Consider, for example, spherically symmetric nonpolar species, *a* and *b*. As we saw in Chapter 4, London interactions describe the attractive forces in this system and can be described by

$$\Gamma_{ij} \approx -\frac{3}{2} \frac{\alpha_i \alpha_j}{r^6} \left(\frac{I_i I_j}{I_i + I_j} \right) \tag{4.13}$$

If we assume that the ionization potentials for species *a* and *b* are roughly equal, we get

$$\Gamma_{aa} \approx -\alpha_a^2 \quad \Gamma_{bb} \approx -\alpha_b^2 \quad \text{and} \quad \Gamma_{ab} \approx -\alpha_a \alpha_b \tag{E7.9E}$$

Hence,

$$\Gamma_{ab} = \sqrt{\Gamma_{aa} \Gamma_{bb}}$$

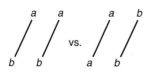

Figure E7.9B Comparison of the difference in molecular interactions that determine the two-suffix Margules parameter, A.

Plugging Equations (E7.9E) back into (E7.9D):

$$A \approx [-2\alpha_a\alpha_b + \alpha_a^2 + \alpha_b^2]$$

or
$$A \approx (\alpha_a - \alpha_b)^2 \qquad \text{(E7.9F)}$$

Equation (E7.9F) shows that if the polarizabilities are equal, we have an ideal solution; in *all* other cases $A > 0$ (for spherically symmetric nonpolar molecules)! Thus, we are much more likely to find in nature a case where like interactions dominate (i.e., have lower energy) and the activity coefficients (based on the Lewis/Randall reference state) are greater than 1; for example, see Equations (7.123) and (7.124). ◀

In Chapter 6 we established that the criterion for spontaneity of a process is the minimization of Gibbs energy. It is interesting to look at the effect of the Margules parameter, A, on the total Gibbs energy of the system. Using Equations (7.108) and (6.45), we get

$$g = \sum x_i g_i + RT \sum x_i \ln x_i + g^E \qquad (7.125)$$

For the case of a binary system that is described by the two-suffix Margules equation, we can write the following expression for molar Gibbs energy:

$$g = \boxed{x_a g_a + x_b g_b} + RT(x_a \ln x_a + x_b \ln x_b) + \boxed{A x_a x_b} \qquad (7.126)$$
$$\;\underset{1}{}\qquad\quad \underset{2}{}\qquad \underset{3}{}$$

Examination of Equation (7.126) [or Equations (7.123) and (7.124)] already shows one limitation in our model for g^E. It is completely symmetric; that is, if we interchanged a and b in this equation, it would be no different. Accordingly, it could not be used to model systems in which the activity coefficients are not symmetric, such as that depicted in Example 7.7. Such asymmetric activity coefficient models will be presented in the next section.

Figure 7.9 plots successive terms on the right-hand side of Equation (7.126) as a function of mole fraction of species a. Species a and b have pure species Gibbs energies g_a and g_b, respectively. The first term just connects the pure species Gibbs energies by a straight line. The second term, the ideal solution Gibbs energy of mixing, shows that the increase in entropy of a mixture leads to a lowering of the Gibbs energy and, therefore, represents a spontaneous process. This case was illustrated in Figure 7.4. The third term

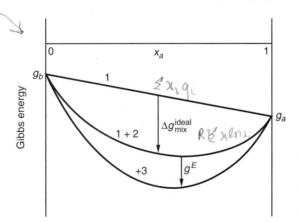

Figure 7.9 Various terms in Equation (7.126) plotted vs. mole fraction of species a. In this case, the Margules parameter, A, is less than zero.

represents the effect of nonideality on the total Gibbs energy of the solution. Consider first the case where $A < 0$. Equation (7.123) shows that this corresponds to the case where $\gamma_a < 1$ for the Lewis/Randall reference state; that is, the case where the unlike interaction is stronger than the like interaction. As discussed in Example 7.9, when we compare unlike interactions to like interactions, we are comparing two a-b interactions to the sum of an a-a and a b-b interaction. As Figure 7.9 illustrates, this increases the tendency of the solution to mix spontaneously by lowering g even further.

We now consider the case where $A > 0$, or the like (a-a and b-b) interaction is stronger than the unlike (a-b) interaction. There are now two opposing tendencies: Mixing is entropically favored, while separating is energetically favored. In other words, if species a and b mix, they increase their randomness; however, if they remain separate, they lower their energy. The magnitude of A tells us which effect will dominate. If A is small, then energetic effects are minor compared to entropic effects and species a and b will completely mix. In this case, the curve for g would be slightly above the $1 + 2$ ideal curve in Figure 7.9. Alternatively, let's look at the case where the like interactions are significantly stronger than the unlike interactions and the magnitude of A is large. In this case, the excess Gibbs energy will drive the $1 + 2$ curve for g up so far that it will actually flip and go through a maximum (with two corresponding minima), as illustrated in Figure 7.10. Figure 7.10a presents the analogous plot to Figure 7.9, while Figure 7.10b magnifies the resultant portion of the g curve, which exhibits the maximum. We have seen systems spontaneously proceed to states in which their Gibbs energy is lowered. There is a composition range in which the mixture can lower its Gibbs energy by splitting into two phases. In Figure 7.10b, a tangent line is drawn below the maximum in Gibbs energy that touches the Gibbs energy curve at two points around the minima. A mixture of composition between these two tangent points can reduce its Gibbs energy by splitting into two phases, an a-rich phase, α, and a b-rich phase, β (see Figure 7.10b). Under these conditions, a and b are only partially miscible (like oil and water), since two liquid phases lead to a lower Gibbs energy. In this case, we can have liquid–liquid equilibrium (LLE) between two coexisting liquid phases. This so-called partial miscibility of liquid phases is often exploited in separations processes. On the other hand, if the total mole

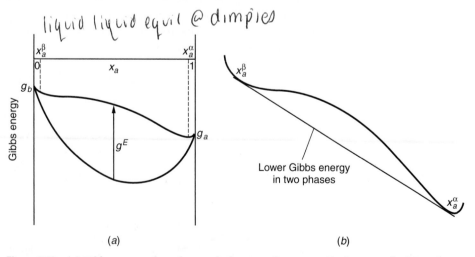

liquid liquid equil @ dimples

(a) (b)

Figure 7.10 (a) Gibbs energy plotted vs. mole fraction of species a. In this case, the Margules parameter, A, is much greater than zero. *Note*: At compositions between x^α and x^β, the system can minimize its free energy by splitting into two phases. (b) Expanded drawing of the Gibbs energy curve.

fraction of species a is less than the first tangent point or greater than the second tangent point, only one phase will be present. While we used the two-suffix Margules model to explore the case of partial miscibility, it is a general phenomenon that occurs when the attractive interactions of the like species dominate over the entropic effects of mixing. Liquid–liquid equilibria will be explored in greater detail in Section 8.2.

▶ **EXAMPLE 7.10**
Calculation of Margules Parameter from Activity Coefficients at Infinite Dilution

Consider a binary mixture of cyclohexane (a) and dodecane (b) at 39.33°C. Activity coefficients at infinite dilution have been reported to be:[4]

$$\gamma_a^\infty = 0.88$$

and

$$\gamma_b^\infty = 0.86$$

Use these data to estimate the value of the two-suffix Margules parameter A.

SOLUTION From Equation (7.123), we have

$$\ln \gamma_a = \frac{A}{RT} x_b^2 \tag{E7.10A}$$

At infinite dilution of a, the mole fraction of b goes to 1 and Equation (E7.10A) becomes

$$\ln \gamma_a^\infty = \frac{A}{RT} = -0.13$$

Solving for the Margules parameter gives

$$A = -332 \, [\text{J/mol}] \tag{E7.10B}$$

We can apply the same methodology to find the value of the activity coefficient of b at infinite dilution. In that case,

$$\ln \gamma_b^\infty = \frac{A}{RT} = -0.15$$

and

$$A = -392 \, [\text{J/mol}] \tag{E7.10C}$$

These values vary by 20%; however, their magnitudes are small, so on an absolute scale they are reasonably close. In solving problems, the best value is given by an average of the Equations (E7.10B) and (E7.10C) to give

$$A = -362 \, [\text{J/mol}]$$

◀

▶ **EXAMPLE 7.11**
Margules Equation for a Henry's Law Reference State

Consider a binary system of dissolved gas a in solvent b. Henry's law is used for the reference state for a and the Lewis/Randall reference state for b. If the activity coefficient for b is given by the two-suffix Margules expression

$$RT \ln\gamma_b = Ax_a^2$$

determine the expression for the activity coefficient of species a in terms of the Margules parameter A.

SOLUTION We can begin with the expression for the activity coefficient of a using the Lewis/Randall reference state:

$$RT \ln \gamma_a = Ax_b^2 \tag{E7.11A}$$

[4]J. Gmehling, U. Onken, and W. Arlt, *Vapor–Liquid Equilibrium Data Collection* (multiple volumes) (Frankfurt: DECHEMA, 1977–1980).

Additionally, by definition:

$$RT \ln \gamma_a = RT \ln \left(\frac{\hat{f}_a}{x_a f_a} \right) \tag{E7.11B}$$

In the limit of infinite dilution—that is, as $x_a \to 0$ and $x_b \to 1$—the fugacity of a must equal the Henry's law constant. If we apply this definition to the expressions in Equations (E7.11B) and (E7.11A), we get

$$\ln \left(\gamma_a^\infty \right) = \ln \left(\frac{\mathcal{H}_a}{f_a} \right) = \frac{A}{RT} \tag{E7.11C}$$

On the other hand, the activity coefficient with the Henry's law reference state is defined as

$$RT \ln \left(\gamma_a^{\text{Henry's}} \right) = RT \ln \left(\frac{\hat{f}_a}{x_a \mathcal{H}_a} \right) = RT \ln \left(\frac{\hat{f}_a}{x_a f_a} \right) + RT \ln \left(\frac{f_a}{\mathcal{H}_a} \right) \tag{E7.11D}$$

where the final equality in Equation (E7.11D) is obtained by adding and subtracting $RT \ln \left(f_a / \mathcal{H}_a \right)$. Substituting Equations (E7.11A) through (E7.11C) into (E7.11D) gives

$$RT \ln \left(\gamma_a^{\text{Henry's}} \right) = A x_b^2 - A = A(x_b^2 - 1) \qquad\qquad ◀$$

Asymmetric Models for g^E

In this section, we discuss other common models used for g^E to fit experimental data and quantify the compositional dependence of the activity coefficients. Similar to our approach to equations of state in Chapter 4, we will restrict the discussion of models for g^E to some common examples. The purpose is to gain some experience with different forms of these equations. It should be recognized, however, that this treatment is not comprehensive and many other models exist. However, many of the other models are derived from those presented here.

The two-suffix Margules equation only has one parameter, A. If we interchange species a and b, the equation for g^E remains unchanged. Thus, it predicts that species a behaves exactly the same in a given proportion of b as species b does in that same amount of a. This representation of the behavior of activity coefficients in a binary system is termed symmetric. The system shown on the left of Figure 7.11 exhibits symmetric behavior. The plot of γ_b vs. x_a is the mirror image of γ_a vs. x_a. Species whose intermolecular interactions are similar in type and in magnitude show symmetric behavior and can be described well by the two-suffix Margules equation. For example, the two-suffix Margules equation could be used to describe a mixture with two species that exhibit only London attractive interactions of roughly the same strength and that are approximately the same size.

The activity coefficients on the right of Figure 7.11 show asymmetric behavior. If we wish to account for such asymmetric activity coefficients, we need to include more than one parameter in our model for g^E. One obvious way to break the symmetry is by adding a term with $(x_a - x_b)$ to the model as follows:

$$g^E = x_a x_b [A + B(x_a - x_b)] \tag{7.127}$$

Inspection of Equation (7.127) shows that if we exchange species a and b, the sign of the parameter B will change. We can find the corresponding activity coefficients through differentiation of Equation (7.127) as prescribed by Equation (7.110). After some math, we get

$$RT \ln \gamma_a = (A + 3B)x_b^2 - 4B x_b^3 \tag{7.128}$$

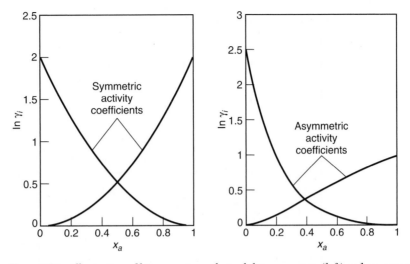

Figure 7.11 Illustration of binary systems that exhibit symmetric (*left*) and asymmetric (*right*) activity coefficients.

and
$$RT \ln \gamma_b = (A - 3B)x_a^2 + 4Bx_a^3 \qquad (7.129)$$

Common expressions used to model g^E for binary systems, and their corresponding activity coefficients, are presented in Table 7.1. These models are ordered in terms of increasing complexity. The model parameters are found empirically through fitting experimental data. Excluding the two-suffix Margules equation, any of these equations can be used to model asymmetric activity coefficients. In fact, it turns out that none of these models exclusively work better than the others in predicting experimental vapor–liquid equilibrium data. While one model may work better in one system, another may work better in a second system, and a third in a third system, and so on. For example, best fits of 3563 pairs of vapor–liquid equilibrium data have been reported for the DECHEMA data collection.[5] Although it demonstrated the greatest success, the Wilson equation showed the best fit for only 30% of the data. Every asymmetric model in Table 7.1 provided best fits for at least 467 of these systems.

The software package accompanying this text allows you to compare the performance of these different models on different data sets. Table 7.2 compares best-fit performance for 16 binary pairs reported in the DECHEMA data collection.[6] Two samples from each of the eight different classes of binary systems are shown. These systems were chosen at random, and the results should not be taken to indicate the performance of a particular model over a larger sample of binary data. Both the results of the software in the text and the ones reported by DECHEMA are shown. The models were tested by independently comparing the predicted values of system pressure using the models with measured values. The average deviation for a given model is reported as "avgDevP," while the maximum deviation is "maxDevP". The model with the best performance, as indicated by the lowest value of "avgDevP," is shown in bold for both the text software and the DECHEMA software. It can be seen that all models perform reasonably well and various models work better for different systems. More detailed discussion of fitting these model parameters using objective functions is presented in Section 8.1.

[5] S. M. Walas, *Phase Equilibria in Chemical Engineering* (Boston: Butterworth, 1985).

[6] J. Gmehling et al., *Vapor–Liquid Equilibrium Data*.

Margules Equations

TABLE 7.1 Common Binary Activity Coefficient Models

Model	g^E	$RT\ln\gamma_a$	$RT\ln\gamma_b$
Two-suffix Margules	Ax_ax_b	Ax_b^2	Ax_a^2
Three-suffix Margules	$x_ax_b[A + B(x_a - x_b)]$	$(A + 3B)x_b^2 - 4Bx_b^3$	$(A - 3B)x_a^2 + 4Bx_a^3$
or	$x_ax_b[A_{ba}x_a + A_{ab}x_b]$	$x_b^2[A_{ab} + 2(A_{ba} - A_{ab})x_a]$	$x_a^2[A_{ba} + 2(A_{ab} - A_{ba})x_b]$
Van Laar	$x_ax_b\left(\dfrac{AB}{Ax_a + Bx_b}\right)$	$A\left(\dfrac{Bx_b}{Ax_a + Bx_b}\right)^2$	$B\left(\dfrac{Ax_a}{Ax_a + Bx_b}\right)^2$
Wilson	$-RT\left[\begin{array}{l}x_a\ln(x_a + \Lambda_{ab}x_b) + \\ x_b\ln(x_b + \Lambda_{ba}x_a)\end{array}\right]$	$-RT\left[\begin{array}{l}\ln(x_a + \Lambda_{ab}x_b) + \\ x_b\left(\dfrac{\Lambda_{ba}}{x_b + \Lambda_{ba}x_a} - \dfrac{\Lambda_{ab}}{x_a + \Lambda_{ab}x_b}\right)\end{array}\right]$	$-RT\left[\begin{array}{l}\ln(x_b + \Lambda_{ba}x_a) + \\ x_a\left(\dfrac{\Lambda_{ab}}{x_a + \Lambda_{ab}x_b} - \dfrac{\Lambda_{ba}}{x_b + \Lambda_{ba}x_a}\right)\end{array}\right]$
NRTL*	$RTx_ax_b\left[\dfrac{\tau_{ba}\mathbf{G}_{ba}}{x_a + x_b\mathbf{G}_{ba}} + \dfrac{\tau_{ab}\mathbf{G}_{ab}}{x_b + x_a\mathbf{G}_{ab}}\right]$	$RTx_b^2\left[\dfrac{\tau_{ba}\mathbf{G}_{ba}^2}{(x_a + x_b\mathbf{G}_{ba})^2} + \dfrac{\tau_{ab}\mathbf{G}_{ab}}{(x_b + x_a\mathbf{G}_{ab})^2}\right]$	$RTx_a^2\left[\dfrac{\tau_{ba}\mathbf{G}_{ba}}{(x_a + x_b\mathbf{G}_{ba})^2} + \dfrac{\tau_{ab}\mathbf{G}_{ab}^2}{(x_b + x_a\mathbf{G}_{ab})^2}\right]$

*Nonrandom two-liquid.

TABLE 7.2 Best-Fit Results for Parameters of the Models Given in Table 7.1 to Experimental Vapor–Liquid Equilibrium[*]

Binary System	T	3-Suffix Margules Text Software avgDevP (Torr)	Text Software maxDevP (Torr)	DECHEMA avgDevP (Torr)	DECHEMA maxDevP (Torr)	van Laar Text Software avgDevP (Torr)	Text Software maxDevP (Torr)	DECHEMA avgDevP (Torr)	DECHEMA maxDevP (Torr)	Wilson Text Software avgDevP (Torr)	Text Software maxDevP (Torr)	DECHEMA avgDevP (Torr)	DECHEMA maxDevP (Torr)	NRTL Text Software avgDevP (Torr)	Text Software maxDevP (Torr)	DECHEMA avgDevP (Torr)	DECHEMA maxDevP (Torr)
Aqueous–Organic																	
Ethanol and water	(75°C)	**3.93**	**11.00**	5.91	11.55	4.61	9.25	7.09	9.78	5.34	8.94	8.08	13.31	4.15	9.83	6.95	9.85
Water and Acetic Acid**	(80°C)	4.79	10.30	3.79	8.11	**4.38**	**11.86**	3.18	9.12	4.70	11.97	2.94	8.36	4.47	11.99	**1.74**	**2.70**
Aliphatic Hydrocarbons (C1–C6)																	
Methylcyclopentane and benzene	(60°C)	2.47	10.17	2.98	10.15	**2.46**	**10.20**	2.97	10.19	2.48	10.22	2.99	10.19	3.00	10.75	3.01	10.00
Hexane and benzene	(25°C)	0.13	0.39	0.48	1.12	0.07	0.31	0.44	0.98	0.07	0.30	**0.43**	**0.92**	**0.06**	**0.31**	0.43	0.99
Aliphatic Hydrocarbons (C7–C18)																	
Chloroform and heptane	(25°C)	0.37	0.81	**0.41**	0.82	0.32	0.72	0.43	0.71	**0.30**	**0.73**	0.44	0.72	0.31	0.72	0.42	0.71
Pyridine and nonane	(97°C)	1.26	**5.14**	5.33	8.14	1.42	3.88	5.67	7.48	1.64	5.10	5.65	8.98	1.63	5.61	5.77	8.98
Alcohols																	
Ethanol and toluene	(60°C)	10.34	46.54	9.95	46.72	**10.32**	**46.42**	9.93	46.64	12.94	51.03	12.48	51.15	10.32	46.45	10.27	48.03
Ethanol and 3-methylbutanol	(70°C)	4.77	9.85	8.13	13.28	4.77	9.85	8.13	13.28	**4.77**	**9.84**	**8.12**	**13.26**	4.77	9.85	8.30	13.67
Carboxylic Acids, Esters																	
Chlorobenzene and propionic acid**	(40°C)	**0.89**	**1.80**	**0.21**	**0.41**	1.01	2.11	0.27	0.58	1.01	2.11	0.28	0.62	0.92	1.91	0.26	0.54
Methyl acetate and ethyl acetate	(40°C)	**1.03**	**2.45**	**0.44**	**0.91**	1.12	2.58	0.53	1.11	1.14	2.62	0.59	1.17	1.09	2.55	0.51	1.09
Aldehydes, Ethers																	
Pentane and acetone	(25°C)	5.86	16.57	4.46	13.43	5.78	16.30	4.43	13.66	**4.00**	**8.14**	1.69	7.10	4.08	8.57	**1.53**	**5.52**
Heptane and butyl ether	(90°C)	20.95	32.21	20.04	31.68	20.88	31.50	20.45	31.18	21.80	32.81	20.87	32.43	**20.42**	**30.89**	20.55	32.45
Nitrogen, Sulfur Compounds																	
Diethylamine and triethylamine	(40°C)	3.04	10.84	2.47	7.64	2.40	5.23	2.66	5.03	2.88	10.70	**2.45**	**7.62**	**2.09**	**4.12**	2.45	7.68
Aniline and n-methylaniline	(145°C)	0.18	0.44	0.23	0.45	**0.18**	**0.40**	0.22	0.42	0.18	0.41	0.22	0.41	0.19	0.43	**0.20**	**0.39**
Aromatic Hydrocarbons																	
Hexafluorobenzene and p-xylene	(40°C)	**0.55**	**1.28**	0.66	1.27	0.63	1.51	0.76	1.51	0.65	1.53	0.78	1.53	0.60	1.39	0.74	1.48
Toluene and chlorobenzene	(70°C)	**0.07**	**0.17**	0.61	1.00	0.07	0.17	0.51	1.07	0.43	0.65	0.59	1.00	0.07	0.17	0.51	1.07

[*] Boldfaced values indicate the pair that has the lowest average deviation and can be considered the best fit. For each binary system, best fits are indicated for both the text software and the results reported with the DECHEMA data.

**The values reported by DECHEMA correct for gas-phase association; this correction is not implemented in the text software.

The three-suffix Margules equation is equivalent in function to Equation (7.125):

$$g^E = x_a x_b [A_{ba} x_a + A_{ab} x_b] \tag{7.130}$$

This equation is algebraically the simplest of the asymmetric models and also gives reasonable predictions of many binary systems; it can even model systems whose activity coefficients show a maximum or a minimum. The van Laar equation is another two-parameter model:

$$g^E = x_a x_b \left(\frac{AB}{A x_a + B x_b} \right) \tag{7.131}$$

While it was developed based on the van der Waals equation of state, its parameters are empirically fit in practice. The two parameters A and B need to have the same sign for the model to work over the entire composition range, and it is unable to represent extrema.

The Wilson equation is given by

$$g^E = -RT[x_a \ln (x_a + \Lambda_{ab} x_b) + x_b \ln(x_b + \Lambda_{ba} x_a)] \tag{7.132}$$

The Wilson equation works well for mixtures of polar and nonpolar species, such as alcohols and alkanes, and is recommended in these cases. It also works well for hydrocarbon mixtures and is readily extended to multicomponent mixtures. However, the Wilson parameters Λ_{ab} and Λ_{ba} must be restricted to positive numbers so that Equation (7.132) is valid in the case of infinite dilution. Additionally, the Wilson equation is unable to describe systems exhibiting partial miscibility, such as the behavior exhibited in Figure 7.10. The Wilson equation is derived from a molecular basis. The Wilson parameters, Λ_{ab} and Λ_{ba}, can be related to molecular parameters as follows:

$$\Lambda_{ab} = \frac{v_b}{v_a} \exp \left(-\frac{\lambda_{ab}}{RT} \right) \tag{7.133}$$

and

$$\Lambda_{ba} = \frac{v_a}{v_b} \exp \left(-\frac{\lambda_{ba}}{RT} \right) \tag{7.134}$$

The terms λ_{ab} and λ_{ba} represent energetic parameters that describe how the a-a or b-b interaction, respectively, varies from the a-b interaction. The size of the molecules is also taken into account through the ratio of their molar volumes. The parameters λ_{ab} and λ_{ba} are relatively insensitive to temperature; therefore, Equations (7.133) and (7.134) can be used to find the Wilson equation parameters at one temperature when they are known at another temperature.

The nonrandom two-liquid (NRTL) model is given by

$$g^E = RTx_a x_b \left[\frac{\tau_{ba} \mathbf{G}_{ba}}{x_a + x_b \mathbf{G}_{ba}} + \frac{\tau_{ab} \mathbf{G}_{ab}}{x_b + x_a \mathbf{G}_{ab}} \right] \tag{7.135}$$

where $\mathbf{G}_{ab} = \exp(-\alpha \tau_{ab})$ and $\mathbf{G}_{ba} = \exp(-\alpha \tau_{ba})$. There are three parameters in the NRTL equation, τ_{ab}, τ_{ba}, and α. Optimally fitting these parameters to experimental data is more complicated than in the models previously discussed. However, the NRTL equation provides an advantage for systems with large deviations from ideality, including partial miscibility and mixtures of organic species with water. Like the Wilson equation, the NRTL equation is derived from a molecular basis. The first two parameters in the

NRTL equation, τ_{ab} and τ_{ba}, represent energetic parameters that describe how the b-b or a-a interaction, respectively, varies from the a-b interaction. The final term, α, represents an entropic parameter related to the *nonrandomness* of the mixture, that is, short-range order and molecular orientation.

More sophisticated models for g^E have been developed from molecular principles. For example, the universal quasi-chemical theory (UNIQUAC) is an extension of the Wilson equation. It divides the excess Gibbs energy into two parts, one due to entropy (the combinatorial part) and one due to energy (the residual part):

$$g^E = g^E_{combinatorial} + g^E_{residual} \tag{7.136}$$

The first term is determined by the sizes and shapes of the molecules and requires only pure component data. The second term is based on intermolecular forces, similar to the approach discussed in Example 7.9. UNIQUAC is mathematically more complicated than the models discussed so far, but it has only two adjustable parameters to fit to experimental data.

Models of g^E for Multicomponent Systems

In chemical systems of interest, we usually have more than two components. In this section we will briefly explore the extension of the activity coefficient models above to multicomponent systems. We begin with an extension of the two-suffix Margules equation to a ternary system. The excess Gibbs energy is written as follows:

$$g^E = A_{ab}x_a x_b + A_{ac}x_a x_c + A_{bc}x_b x_c \tag{7.137}$$

Equation (7.137) quantifies the unlike interaction for the three binary pairs—a-b, a-c, and b-c—independently of the other two. Thus, the parameters that represent these interactions can be found from the corresponding data for the binary pairs in solution. For example, A_{ab} can be found from data for a binary solution of a and b only. Again, the activity coefficients are related to the partial molar excess Gibbs energies by Equation (7.110). Thus,

$$\overline{G}^E_a = \left[\frac{\partial(n_T g^E)}{\partial n_a} \right]_{T,P,n_b,n_c}$$

$$= A_{ab} \left[\frac{\partial\left(\dfrac{n_a n_b}{n_a + n_b + n_c}\right)}{\partial n_a} \right]_{T,P,n_b,n_c} + A_{ac} \left[\frac{\partial\left(\dfrac{n_a n_c}{n_a + n_b + n_c}\right)}{\partial n_a} \right]_{T,P,n_b,n_c}$$

$$+ A_{bc} \left[\frac{\partial\left(\dfrac{n_b n_c}{n_a + n_b + n_c}\right)}{\partial n_a} \right]_{T,P,n_b,n_c} \tag{7.138}$$

Equation (7.138) simplifies to

$$\overline{G}^E_a = A_{ab}\left(x_b^2 + x_b x_c\right) + A_{ac}\left(x_c^2 + x_b x_c\right) - A_{bc}(x_b x_c) \tag{7.139}$$

which can be rewritten as

$$\overline{G}^E_a = RT \ln \gamma_a = A_{ab}x_b^2 + A_{ac}x_c^2 + (A_{ab} + A_{ac} - A_{bc})x_b x_c \tag{7.140}$$

Similarly,
$$\overline{G}_b^E = RT \ln \gamma_b = A_{ab}x_a^2 + A_{bc}x_c^2 + (A_{ab} + A_{bc} - A_{ac})x_a x_c \qquad (7.141)$$

and
$$\overline{G}_c^E = RT \ln \gamma_c = A_{ac}x_a^2 + A_{bc}x_b^2 + (A_{ac} + A_{bc} - A_{ab})x_a x_b \qquad (7.142)$$

If we extend this approach to m components, we write the excess Gibbs energy as

$$g^E = \sum_i \sum_j \frac{A_{ij}}{2} x_i x_j \qquad (7.143)$$

with $A_{ii} = 0$ and $A_{ij} = A_{ji}$.

The Wilson equation is commonly used for multicomponent mixtures. It is relatively simple, represents many systems well, and also depends only on binary pair data. For a system of m components, the excess Gibbs energy is written as

$$g^E = -RT \sum_{i=1}^{m} x_i \ln \left(\sum_{j=1}^{m} x_j \Lambda_{ij} \right) \qquad (7.144)$$

where the Wilson parameter $\Lambda_{ii} = 1$. The activity coefficient of species k in the mixture found from Equation (7.144) is

$$\ln \gamma_k = 1 - \ln \left(\sum_{j=1}^{m} x_j \Lambda_{kj} \right) - \sum_{i=1}^{m} \frac{x_i \Lambda_{ik}}{\sum\limits_{j=1}^{m} x_j \Lambda_{ij}} \qquad (7.145)$$

Other activity coefficient models analogous to those models for binary systems presented in Table 7.1 are also available. In some cases, however, data are needed for ternary a-b-c interactions.

Temperature and Pressure Dependence of g^E

Since fugacity is defined at the same temperature as the reference state, it is often useful to determine the temperature dependence of the activity coefficients through the excess Gibbs energy. The fundamental property relationship for multicomponent systems can be written for excess functions. For excess Gibbs energy, we get

$$dG^E = V^E dP - S^E dT + \sum \overline{G}_i^E dn_i \qquad (7.146)$$

Differentiation of Equation (7.146) leads to

$$\left(\frac{\partial g^E}{\partial P} \right)_{T,n_i} = v^E = \Delta v_{mix} \qquad (7.147)$$

and
$$\left(\frac{\partial g^E/T}{\partial T} \right)_{P,n_i} = \frac{-h^E}{T^2} = \frac{-\Delta h_{mix}}{T^2} \qquad (7.148)$$

Equations (7.147) and (7.148) determine how an expression for Gibbs energy will change with P and T, respectively, in terms of experimentally accessible properties. Thus, a model for g^E fit at one set of experimental conditions can be extended to other values of P and T.

Volume and enthalpy changes of mixing were discussed in Section 6.3. Can you think of how you might measure the quantities on the right-hand side of Equations (7.147) and (7.148), the volume change of mixing and the enthalpy of mixing, respectively, in the laboratory?

When such data are unavailable, it is helpful to consider two limiting cases for the temperature dependence of g^E. We can write the excess Gibbs energy as

$$g^E = h^E - Ts^E \tag{7.149}$$

In the first case, we consider all the nonideality to be associated with energetic effects. We have been doing this implicitly throughout this chapter. In this case, the forces of attraction between the species of the mixture differ, but their size and shape are essentially the same. Hence s^E is negligible and Equation (7.149) becomes

$$g^E = h^E = \Delta h_{\text{mix}} \tag{7.150}$$

Substitution of Equation (7.150) into Equation (7.148) and differentiation gives

$$\left(\frac{\partial g^E/T}{\partial T}\right)_{P,n_i} = \frac{-g^E}{T^2} + \frac{1}{T}\left(\frac{\partial g^E}{\partial T}\right)_{P,n_i} = \frac{-g^E}{T^2} \tag{7.151}$$

Thus, we must have

$$\left(\frac{\partial g^E}{\partial T}\right)_{P,n_i} = 0 \tag{7.152}$$

Integrating Equation (7.152) gives

$$g^E = \text{const} = RT \sum x_i \ln \gamma_i \tag{7.153}$$

We see that the sum of the activity coefficients is inversely proportional to temperature. In this case, we say we have a **regular** solution.

The other extreme consists of the case where the chemical nature of the species is essentially the same but the sizes and shapes differ. Thus, all the nonideality is associated with entropic effects. This might apply, for example, to polymers in solution. In this case, h^E is negligible and Equation (7.149) becomes

$$\frac{g^E}{T} = -s^E \tag{7.154}$$

So,

$$\left(\frac{\partial g^E/T}{\partial T}\right)_{P,n_i} = 0 \tag{7.155}$$

Thus, the excess Gibbs energy is written as

$$\frac{g^E}{T} = \text{const} = R \sum x_i \ln \gamma_i \tag{7.156}$$

We see that the sum of the activity coefficients is independent of temperature. Consequently, systems with this behavior are termed **athermal**. In reality, solutions can deviate due to both energetic and entropic effects, in which case the temperature dependence is between the two cases mentioned above.

In a similar manner, we can also find the temperature and pressure dependencies of the activity coefficient. From Equations (7.109) and (7.71), we have

$$\mu_i - \mu_i^{\text{ideal}} = RT \ln\gamma_i \tag{7.157}$$

Differentiating Equation (7.157) with respect to P at constant T and x gives:

$$\left(\frac{\partial \ln \gamma_i}{\partial P}\right)_{T,x} = \frac{1}{RT}\left[\left(\frac{\partial \mu_i}{\partial P}\right)_{T,x} - \left(\frac{\partial g_i}{\partial P}\right)_T\right] = \frac{\overline{V}_i - v_i}{RT} \tag{7.158}$$

Similarly, differentiating Equation (7.157) with respect to T gives

$$\left(\frac{\partial \ln \gamma_i}{\partial T}\right)_{P,x} = \frac{1}{RT}\left[\left(\frac{\partial \mu_i}{\partial T}\right)_{P,x} - \left(\frac{\partial g_i}{\partial T}\right)_P\right] - \frac{1}{RT^2}(\overline{G}_i - g_i) = \frac{\overline{S}_i - s_i}{RT} - \frac{\overline{G}_i - g_i}{RT^2} \tag{7.159}$$

Applying the definition of Gibbs energy to Equation (7.159), we get

$$\left(\frac{\partial \ln \gamma_i}{\partial T}\right)_{P,x} = -\frac{\overline{H}_i - h_i}{RT^2} \tag{7.160}$$

We can use the methods described in Section 6.3 to get values for the right-hand sides of Equations (7.158) and (7.160).

▶ **EXAMPLE 7.12**
Calculation of the Change in the Margules Parameter with Temperature

Enthalpy of mixing data[7] for benzene (1)–cyclohexane (2) at 18°C has been fit to the following expression:

$$\Delta h_{\text{mix}} = 3250 x_1 x_2 [\text{J/mol}] \tag{E7.12A}$$

The two-suffix Margules parameter A at 10°C is $1401[\text{J/mol}]$. Estimate A at 60°C. Compare to data from the literature.

SOLUTION   From Equation (7.148):

$$\left(\frac{\partial g^E/T}{\partial T}\right)_{P,n_i} = \frac{-h^E}{T^2} = \frac{-\Delta h_{\text{mix}}}{T^2} \tag{7.148}$$

Using the two-suffix Margules equation, $g^E = Ax_1x_2$, and Equation (E7.12A) in Equation (7.148) gives

$$d\left(\frac{A}{T}\right) = \frac{-3250}{T^2} dT \tag{E7.12B}$$

Assuming the enthalpy of mixing is constant with temperature, we can integrate Equation (E7.12B) to get

$$A_{333} = 333\left[3250\left(\frac{1}{333} - \frac{1}{283}\right) + \frac{A_{283}}{283}\right] = 1072\left[\frac{\text{J}}{\text{mol}}\right]$$

The value of A obtained directly from experimental data is (see Problem 8.26):

$$A = 1143[\text{J/mol}]$$

These two values differ by only about 5%.   ◀

[7]E. W. Washburn (ed.), International Critical Tables (Vol. V) (New York: McGraw-Hill, 1929).

Equation of State Approach to the Liquid Phase

In Section 7.3, we explored the use of equations of state to calculate the fugacity in the vapor phase by means of the fugacity coefficient. This approach can be used in the liquid phase, where

$$\hat{f}_i^l = x_i \hat{\varphi}_i^l P \tag{7.161}$$

and

$$\mu_i^l - g_i^0 = - \int\limits_{\left(\frac{n_T RT}{P_{\text{low}}}\right)}^{V} \left(\frac{\partial P}{\partial n_i}\right)_{T,V,n_{j \neq i}} dV = RT\ln\left[\frac{\hat{f}_i^l}{y_i P_{\text{low}}}\right] \tag{7.162}$$

However, its application is much more limited in liquids than in the vapor phase. Equation (7.162) relies on the accuracy of an equation of state to describe liquid behavior for mixtures. Such behavior is very hard to quantify due to uncertainties in mixing rules for liquids, which interact much more than gases. Moreover, in Equation (7.162), the equation of state must be valid over the entire range of integration, from a low-density, ideal gas state to a high-density liquid. Therefore, the activity coefficient approach discussed above is usually preferable for condensed phases.

▶ 7.5 FUGACITY IN THE SOLID PHASE → Skip

Fugacity of the solid phase can be treated similarly to the liquid phase—by how "active" the solid is relative to an ideal solid reference state.

Pure Solids

If the solid is comprised only of a pure species, the solid is, by definition, ideal (all the intermolecular forces are the same). In this case, the activity coefficient of the solid is 1:

$$\Gamma_i^{\text{pure solid}} = 1 \tag{7.163}$$

Note, we use a capital letter, Γ_i, to identify the solid phase. The fugacity of the solid is just equal to the pure species fugacity:

$$\hat{f}_i^s = f_i^s \tag{7.164}$$

Solid Solutions

Solid solutions are finding many applications due to their unique material properties. For example, high-temperature superconductors are being realized through oxide ceramics such as $Ba_x Y_{1-x} Cu_y O_{4-y}$. These *solid solutions* can be treated just like liquid solutions, that is, by defining a reference state and seeking a model for g^E. In this case,

$$\hat{f}_i^s = X_i \Gamma_i f_i^s \tag{7.165}$$

To solve for the fugacity of a species in a solid solution, we can use the same approach as with liquid solutions (Section 7.4).

Interstitials and Vacancies in Crystals

A perfect crystal is made up of a lattice with an atom on every lattice point. In nature, we do not find perfect crystals. When an atom is missing from a lattice point, a *vacancy* is said to exist. If an extra atom is found between lattice points, we have an *interstitial*. The lowest energy state is that of a perfect crystal. Vacancies and interstitials, however, create greater entropy. Again, we can apply the concept of Gibbs energy to quantitatively predict the concentrations of interstitials and vacancies. In this case, we will use the concepts developed in Chapter 9 to determine concentrations of interstitials and vacancies. This material will be covered in Section 9.8.

▶ 7.6 SUMMARY

In this chapter, we introduce **fugacity** as an alternative to chemical potential to write the criteria for chemical equilibrium between species. Fugacity is more amenable to engineering calculations. It is defined as

$$\mu_i - \mu_i^o \equiv RT\ln\left[\frac{\hat{f}_i}{\hat{f}_i^o}\right] \tag{7.7}$$

In applying this definition, we must have a well-defined **reference state** (°). The criteria for chemical equilibrium between phases α and β can be reformulated in terms of fugacity as

$$\hat{f}_i^\alpha = \hat{f}_i^\beta \tag{7.20}$$

In order to solve Equation (7.20) for the composition of species i in each phase at equilibrium, we developed expressions for the fugacity in the vapor phase and in the condensed phases. The difference in the expression for fugacity between vapor and condensed phases typically lies in the choice of reference state.

The fugacity in the **vapor phase** is commonly calculated using an **ideal gas reference state**. Thus, the reference state for pure species i is at a pressure low enough that it behaves as an ideal gas and at the temperature of the system, as restricted by the definition in Equation (7.7). For species i in a mixture, we also specify that the reference state is at the composition of the mixture. We can formulate the fugacity in terms of the **fugacity coefficient**—a dimensionless quantity that compares the fugacity of species i to the partial pressure species i would have in the system as an ideal gas:

$$\hat{\varphi}_i \equiv \frac{\hat{f}_i}{P_{i,\text{sys}}} = \frac{\hat{f}_i}{y_i P_{\text{sys}}} \tag{7.9}$$

A fugacity coefficient of 1 represents the case where attractive and repulsive forces balance and is usually indicative of an ideal gas. If $\hat{\varphi}_i < 1$, attractive forces dominate the system behavior, while $\hat{\varphi}_i > 1$ indicates that the repulsive forces are stronger. The fugacity and fugacity coefficient for pure species and for mixtures can be solved with available data from thermodynamic property tables, equations of state, or generalized correlations. In the case of mixtures, there are three levels of rigor from which to calculate the fugacity coefficient.

1. We can solve the full problem with compositional-dependent fugacity coefficients. This approach is only as accurate as the mixing rules that we use:

$$\hat{f}_i^v = y_i \hat{\varphi}_i^v P \qquad \text{full rigor} \tag{7.33}$$

2. As a first approximation, we can use the **Lewis fugacity rule** and base the fugacity coefficient on the pure species value, as discussed in Section 7.3. The advantage of this approach is that mixing rules are not needed and it is mathematically much easier:

$$\hat{f}_i^v = y_i \varphi_i^v P \qquad \text{Lewis fugacity rule} \qquad \text{first approximation} \tag{7.52}$$

3. As a second approximation, we can assume **ideal gas** behavior, in which case there are no intermolecular forces present, and the fugacity of species i is equal to its partial pressure:

$$\hat{f}_i^v = y_i P \qquad \text{ideal gas} \qquad \text{second approximation}$$

For the liquid or solid phases, we choose an **ideal solution** as our reference state. On a molecular level, we define a solution as ideal when the intermolecular interactions are the *same* between all components of the mixture. This idealization results in a linear relation in mole fraction between the fugacity in the mixture and the pure species fugacity. There are two common choices for the reference state. The **Lewis/Randall rule** reference state is based on the characteristic pure species interactions of species i, that is, the i-i interaction:

$$f_i^{\text{ideal}} = f_i^o = f_i \tag{7.69}$$

The **Henry's law** reference state can be conceptualized as a hypothetical, pure fluid in which the characteristic energy of interaction is that between molecule i and the other molecules j, that is, the i-j interactions:

$$f_i^{\text{ideal}} = f_i^o = \mathcal{H}_i \tag{7.70}$$

To quantify the Lewis/Randall reference state, we must find the value for the fugacity of pure species i, f_i^l at the T and P of the mixture. Equation (7.84), which uses the **Poynting correction**, shows how the fugacity of the pure species changes with pressure. Similarly, we can apply the thermodynamic web to determine how the Henry's law constant changes with temperature and pressure, as given by Equations (7.92) and (7.93), respectively.

The **activity coefficient** is used to quantify how much the fugacity of a given species deviates from the value it would have in an ideal solution. It is defined as follows:

$$\gamma_i = \frac{\hat{f}_i^l}{\hat{f}_i^{\text{ideal}}} = \frac{\hat{f}_i^l}{x_i f_i^o} \tag{7.71}$$

Since the activity coefficient derives from the relative amount of unlike vs. like interactions, its value changes with changing composition. The value also depends on the specific choice of reference state. Similarly, the **activity** of species i, a_i compares the fugacity of species i in the liquid to the fugacity of the pure species at its reference state. The Gibbs–Duhem equation allows us to relate the activity coefficients of different species in a mixture and, thereby, to test for **thermodynamic consistency**.

The **excess Gibbs energy**, g^E, forms the basis from which the activity coefficient of *all* the species in the mixture can be obtained from a single quantitative composition-dependent expression. Excess Gibbs energy is defined as the difference between the real value of the Gibbs energy and the hypothetical value it would have as an ideal solution at the same temperature pressure and composition as the real mixture. Our approach to come up with values for γ_i to find the fugacity of the liquid or solid phases is to come up with an analytical expression for the excess Gibbs energy, g^E. We then empirically fit the parameters in this expression to experimental data. The activity coefficients for each species i in the mixture can then be found by calculating the partial molar excess Gibbs energy, as shown by Equation (7.110). In this way we need only *one* model for g^E to obtain activity coefficients for *every* species.

The **two-suffix Margules equation** forms the simplest nontrivial example of these **activity coefficient models**. On a molecular level, the two-suffix Margules parameter, A, is proportional to the difference between the potential energy of the unlike interactions and the potential energy of the like interactions. It is limited to systems where the activity coefficients are **symmetric**. Other activity coefficient models can be applied to **asymmetric** systems, including the **three-suffix Margules equation**, the **van Laar equation**, the **Wilson equation**, and the **NRTL equation**. Expressions for these models and their associated activity coefficients for binary mixtures are summarized in Table 7.1. The ThermoSolver software that comes with the text allows activity coefficient model parameters to be obtained from experimental data. Expressions that can be used to apply the two-suffix Margules and the Wilson equations to obtain activity coefficients for multicomponent mixtures are given in Section 7.4. The **equation of state approach** can also be applied to the liquid and solid phases. This approach is similar to that used for vapor-phase fugacity.

► **7.7 PROBLEMS**

7.1 Calculate the fugacity and the fugacity coefficient of steam at (a) 2 MPa and 500°C; (b) 50 MPa and 500°C.

7.2 Consider the Berthelot equation of state:

$$P = \frac{RT}{v - b} - \frac{a}{Tv^2}$$

(a) Develop an expression for the fugacity and fugacity coefficient of a pure species.

(b) Use the results of Problem 4.22 to write the result of part (a) in terms of reduced pressure, reduced temperature, and reduced volume.

7.3 Develop an expression for the fugacity and fugacity coefficient of a pure species based on the Redlich–Kwong equation of state.

7.4 Develop an expression for the fugacity and fugacity coefficient of a pure species based on the Peng–Robinson equation of state.

7.5 Calculate the fugacity of water at 647 K and 114 atm using (a) the steam tables; (b) the van der Waals equation; (c) generalized correlations. Which value do you believe the most?

7.6 Calculate the fugacity and fugacity coefficient of the following pure substances at 500°C and 150 bar: (a) CH_4; (b) C_2H_6; (c) NH_3; (d) $(CH_3)_2CO$; (e) C_6H_{12}; (f) CO. Provide an explanation of the relative magnitude of these numbers based on molecular concepts.

7.7 You have a pure gas at 30 bar and 300 K. The compressibility factor (z) under these conditions is 0.9. As best you can, calculate the fugacity and the fugacity coefficient.

7.8 Experimental data taken from 0 to 50 bar give the fugacity of a pure gas to be:

$$f = P\exp(-CP)$$

where P is the pressure in bar and C is a constant that depends only on temperature. For the region of 0°C to 100°C, C is given by:

$$C = -0.065 + \frac{30}{T}$$

where T is in Kelvin.

(a) Find an equation of state for this gas that is valid from 0°C to 100°C.

(b) What is the molar volume (in m^3/mole) at 80°C and 30 bar?

7.9 Consider a system containing pure hydrogen sulfide at 300 K and 20 bar. The following equation of state characterizes the PvT behavior of H_2S well under these conditions:

$$Pv = RT\left[1 + \frac{PT_c}{P_cT}\left(0.083 - \frac{0.422T_c^{1.6}}{T^{1.6}}\right)\right]$$

Using this equation of state, find the fugacity and fugacity coefficient.

7.10 The following data are available for ethylene at 24.95°C. From these data, estimate the fugacity and the fugacity coefficient of ethylene at 50.5 bar and 24.95°C.

P [bar]	v [m³/mol]
1.0	2.45×10^{-2}
5.1	4.78×10^{-3}
10.1	2.32×10^{-3}
15.2	1.50×10^{-3}
20.2	1.08×10^{-3}
25.3	8.34×10^{-4}

(continued)

(Continued)

P [bar]	v [m³/mol]
30.3	6.66×10^{-4}
35.4	5.44×10^{-4}
40.4	4.52×10^{-4}
45.5	3.78×10^{-4}
50.5	3.17×10^{-4}

7.11 Below is a plot of the natural log of the fugacity coefficient, $\ln(\varphi_i)$, of pure NH_3 as a function of pressure at a temperature of 100°C. From this plot, as best you can, determine the molar volume of this species at 500 bar and 100°C.

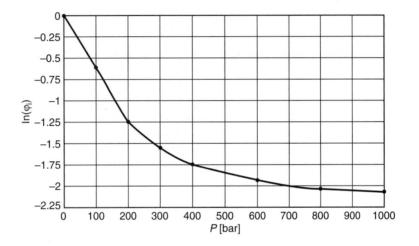

7.12 Consider a mixture of species 1, 2, and 3. The following equation of state is available for the vapor phase:

$$Pv = RT + P^2[A(y_1 - y_2) + B]$$

where

$$\frac{A}{RT} = -9.0 \times 10^{-5}[1/\text{atm}^2], \qquad \frac{B}{RT} = 3.0 \times 10^{-5}[1/\text{atm}^2]$$

and y_1, y_2, and y_3 are the mole fractions of species 1, 2, and 3, respectively.

A vapor mixture of 1 mole of species 1, 2 moles of species 2, and 2 moles of species 3 is cooled to 300 K at constant pressure of 50 atm, where some of it condenses into a liquid phase.

(a) Calculate an expression for the pure species fugacity coefficient for species 1 in the vapor, φ_1^v, and the fugacity coefficient, $\hat{\varphi}_1^v$, of species 1 in the mixture.

(b) If the fugacity of species 1 in the liquid, $\hat{f}_1^l$, is 15 atm, calculate the mole fraction of vapor species 1 in equilibrium with the liquid.

7.13 A mixture of 2 moles propane (1), 3 moles butane (2), and 5 moles pentane (3) is contained at 30 bar and 200°C. The van der Waals constants for these species are:

Species	a[Jm³mol⁻²]	b[m³/mol]
Propane	0.94	9.06×10^{-5}
Butane	1.45	1.22×10^{-4}
Pentane	1.91	1.45×10^{-4}

Determine the fugacity and fugacity coefficient of propane using the following approximations:

(a) the Lewis fugacity rule

(b) the virial form of the van der Waals equation truncated to the second term

7.14 Consider a ternary system of methane (a), ethane (b), and propane (c) at 25°C and 15 bar. Assume this system can be represented by the virial equation truncated at the second term:

$$z = 1 + \frac{B_{mix}}{v}$$

At 25°C, the second virial coefficients [cm^3/mol] are given by:

B_{aa}	−42
B_{bb}	−185
B_{cc}	−399
B_{ab}	−93
B_{ac}	−139
B_{bc}	−274

(a) Develop an expression for the fugacity coefficient of methane in the mixture.

(b) Estimate the fugacity and the fugacity coefficient of methane for a mixture with 20% (mole) methane, 30% ethane, and 50% propane.

(c) Repeat parts (a) and (b) using the Lewis fugacity rule.

7.15 Consider a binary mixture of species a and b that obeys the Redlich–Kwong equation of state with van der Waals mixing rules. Show that the fugacity coefficient of species a in a binary mixture of a and b is given by

$$\ln(\hat{\varphi}_a) = \ln\left[\frac{RT}{Pv}\right] - \ln\left[1 - \frac{b}{v}\right] + b_a\left[\frac{1}{v-b} - \frac{a}{bRT^{1.5}(v+b)}\right]$$
$$+ \frac{1}{bRT^{1.5}}\left[\frac{b_a a}{b} - 2\sqrt{a_a a}\right]\ln\left[1 + \frac{b}{v}\right]$$

7.16 Consider a mixture of CH$_4$ and H$_2$S at 444 K and 70 bar. Use the results of Example 7.4 to plot the fugacity coefficient of methane as a function of methane mole fraction using the van der Waals equation of state and mixing rules. Compare the result with that obtained from the text software, ThermoSolver, using the Peng–Robinson equation of state.

7.17 Consider a mixture of CH$_4$ and H$_2$S at 444 K and 70 bar. Use the results of Problem 7.15 to plot the fugacity coefficient of methane as a function of methane mole fraction using the Redlich–Kwong equation of state with van der Waals mixing rules. Compare the result with that obtained from the text software, ThermoSolver, using the Peng–Robinson equation of state.

7.18 Consider a mixture of CH$_4$ and H$_2$S at 444 K and 70 bar. Calculate the fugacity coefficient of methane in an equimolar mixture using each of the following:

(a) the results of Example 7.4

(b) the results of Problem 7.15

(c) the generalized correlations with Kay's mixing rules

(d) ThermoSolver using the Peng–Robinson equation of state

How do the results from Parts (a)–(d) compare?

7.19 You wish to represent a binary mixture of species a and b at 127°C and 80 bar by the virial equation. At 127°C, the second virial coefficients are given by

$$B_{aa} = -16[cm^3/mol] \quad and \quad B_{bb} = -101[cm^3/mol]$$

You have also found that at infinite dilution—that is, as $y_a \to 0$—the value of the fugacity coefficient is reported as $\hat{\varphi}_a^\infty = 1.08$. As best as you can, estimate B_{ab}.

7.20 A binary mixture of species 1 and 2 can be described by the following equation of state:

$$P = \frac{RT}{v} - \frac{a}{\sqrt{v^3 T}}$$

with the mixing rule

$$a_{\text{mix}} = y_1 a_1 + a_2 y_2$$

The pure species coefficients are given by

$$a_1 = 800\left[(Jm^{1.5}K^{0.5})/\text{mol}^{1.5}\right] \quad \text{and} \quad a_2 = 500\left[(Jm^{1.5}K^{0.5})/\text{mol}^{1.5}\right]$$

Consider a vapor mixture of 1 mole of species 1 and 2 moles of species 2 that occupies a volume of 6 L at 500 K.

(a) Determine the fugacity coefficient of the mixture, $\hat{\varphi}_1$.

(b) By how much does your answer change if you approximate the solution using the Lewis fugacity rule instead?

7.21 Verify that

$$\left(\frac{\partial g_i}{\partial P}\right)_T = v_i = RT\left(\frac{\partial(\ln f_i)}{\partial P}\right)_T$$

7.22 You wish to determine the fugacity of water at 300°C and 300 bar. Using *only* data for saturated steam and superheated vapor from the steam tables, determine, as accurately as you can, the fugacity of water at 300°C and 300 bar. State any assumptions that you make.

7.23 What is the fugacity of pure liquid n-butane at 260 K at the following pressures: (a) 1 bar; (b) 200 bar. You may assume the density of liquid butane, 0.579 g/cm³, is independent of pressure.

7.24 Determine the fugacity of pure liquid acetone at 100 bar and 382 K. The molar volume of the liquid is 73.4 $[\text{cm}^3/\text{mol}]$. You may assume v_i does not change with pressure.

7.25 Consider a binary mixture of a and b at $T = 300$ K and $P = 20$ kPa. A graph of the fugacity of species a as a function of mole fraction is shown below. Use Henry's law as the reference state for species a and the Lewis/Randall rule for species b. Show all your work.

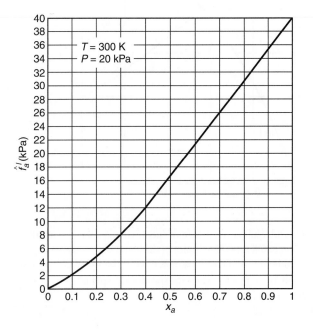

(a) What is the Henry's law constant, $\mathcal{H}_a$, for species a?

(b) What is the activity coefficient for species a at $x_a = 0.4$? At $x_a = 0.8$? (remember the Henry's law reference state!). Show your work.

(c) Is the activity coefficient for species b at $x_a = 0.4$ greater than or less than 1? Explain.

(d) Is the a-b interaction stronger than the pure species interactions? Explain.

(e) Consider the vapor phase to be ideal. What is the vapor-phase mole fraction of a in equilibrium with 40% liquid a?

7.26 Henry's law is often a convenient reference state for the dilute components in the liquid phase in equilibria calculations.

(a) Consider a solution of acetone (1) in water (2) vs. a solution of methane (1) in water (2) Which solution has the larger Henry's law constant, $\mathcal{H}_1$. Explain.

(b) Consider a binary mixture of (a) and (b), in which the Henry's law constant for species a, $\mathcal{H}_a$, at a certain temperature, is equal to its pure species fugacity, f_a. Plot the activity coefficient, γ_a, as a function of mole fraction x_a.

7.27 Consider a binary liquid mixture of species a and b. The activity coefficients for this mixture are adequately described by the two-suffix Margules equation.

(a) If $\Delta h_{mix} = 0$, what can you say about the two-suffix Margules parameter, A?

(b) If $\Delta v_{mix} = 0$, what can you say about the two-suffix Margules parameter, A?

7.28 Below is a plot of the natural log of the activity coefficients ($\ln \gamma_i$) a binary liquid mixture of species a and b vs. mole fraction of species a (x_a) at 300 K.

(a) What is the reference state for each species?

(b) Show that the Gibbs–Duhem equation is satisfied at a mole fraction $x_a = 0.6$.

(c) Come up with an appropriate model for g^E for this system and find the values of the model parameters.

(d) Is it possible for species a and b to separate into two liquid phases? Explain.

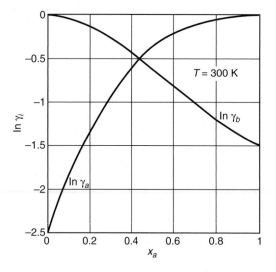

7.29 Consider a binary mixture of species a and species b at 300 K and 1 bar. The vapor pressure of pure a at 300 K is 80 kPa. A plot of the activity coefficient of species a vs. mole fraction of species a is shown below. Based on this plot, answer the following questions:

(a) Specify the reference state for species a. Explain.

(b) What is the value of f_a?

(c) What is the value of $\mathcal{H}_a$?

(d) As best as you can, come up with the Margules parameter A in the two-suffix Margules equation

$$g^E = A x_a x_b$$

(e) Consider a liquid mixture of 2 moles of a and 3 moles of b at 300 K and 1 bar. At equilibrium, what is the mole fraction of a in the vapor phase, y_a?

(f) For the mixture above, determine γ_b based on a Lewis/Randall reference state.

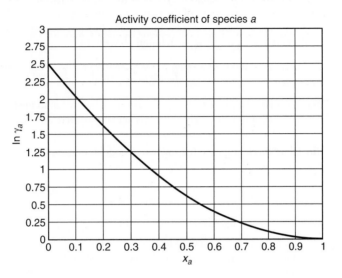

Activity coefficient of species a

7.30 The following plot shows values of excess Gibbs energy over the ideal gas constant, g^E/R in [K], for mixtures of benzene–cyclohexane at 343 K. Answer the following questions:

(a) Estimate the activity coefficient of cyclohexane in benzene (1)–cyclohexane (2) at 343 K for (i) a mole fraction of cyclohexane of $x_2 = 0.25$; (ii) cyclohexane in infinite dilution.

(b) Estimate the Henry's law constant for cyclohexane in benzene.

(c) Are the like interactions stronger or weaker than the unlike interactions? Explain.

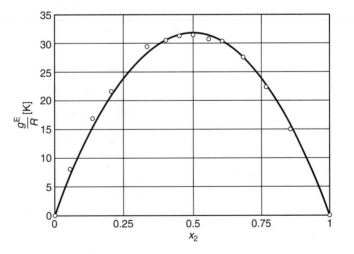

7.31 Derive the expressions for the activity coefficients for a binary mixture from the following models for g^E:

(a) the three-suffix Margules equation

(b) the van Laar equation

(c) the Wilson equation

(d) NRTL

7.32 Consider an equimolar binary mixture of species a and b. The activity coefficients at infinite dilution are given by: $\gamma_a^\infty = 2.0$ and $\gamma_b^\infty = 1.5$. Calculate the activity coefficient of species a and b using the three-suffix Margules equation, the van Laar equation, and the Wilson equation.

7.33 Glycerol (a) and benzyl ethyl amine (b) form two partially miscible liquid phases. At 220°C and 1 atm, the compositions of the two phases are given by $x_a^\alpha = 0.9$ and $x_a^\beta = 0.2$.

From these data, estimate the parameters in the three-suffix Margules equation. At equilibrium, the fugacities in each liquid phase must be equal, that is,

$$\hat{f}_a^{l,\alpha} = \hat{f}_a^{l,\beta} \quad \text{and} \quad \hat{f}_b^{l,\alpha} = \hat{f}_b^{l,\beta}$$

7.34 Calculate the fugacity of liquid water in a binary liquid mixture with 40 mole % water and 60 mole % ethanol at 70°C. The following activity coefficient data, at infinite dilution, are available: $\gamma_{H_2O}^\infty = 2.62$ and $\gamma_{ethanol}^\infty = 7.24$.

7.35 Cline Black proposed the following model for excess Gibbs energy:

$$g^E = \left[\frac{1}{Ax_a} + \frac{1}{Bx_b} \right]^{-1} + Cx_a x_b (x_a - x_b)^2$$

Develop the corresponding expressions for $\ln\gamma_a$ and $\ln\gamma_b$.

7.36 For isobaric data, you want to account for the temperature change of the activity coefficients in the test for thermodynamic consistency. Develop the following equation to use in place of Equation (7.116):

$$\int_0^1 \ln\left(\frac{\gamma_a}{\gamma_b}\right) dx_a = \int_{T_{x_1=0}}^{T_{x_1=1}} \frac{\Delta h_{\text{mix}}}{RT^2} dT = -\int_{(1/T)_{x_1=0}}^{(1/T)_{x_1=1}} \frac{\Delta h_{\text{mix}}}{R} d\left(\frac{1}{T}\right)$$

7.37 The activity coefficients at infinite dilution of a mixture of hexane (a) and toluene (b) at 30°C are $\gamma_a^\infty = 1.27$ and $\gamma_b^\infty = 1.34$. Estimate the fugacity of hexane in mixtures with the following compositions at 1 bar: (a) 20% liquid hexane; (b) 50% liquid hexane; (c) 90% liquid hexane.

7.38 Wilson parameters for mixtures of ethanol (1), 1-propanol (2), and water (3) at 60°C are reported as follows:

$$\Lambda_{12} = 1.216 \quad \Lambda_{21} = 0.617$$

$$\Lambda_{13} = 0.203 \quad \Lambda_{31} = 0.838$$

$$\Lambda_{23} = 0.048 \quad \Lambda_{32} = 0.612$$

Calculate the fugacity of ethanol in a liquid mixture containing 30% ethanol, 20% 1-propanol, and 50% water at 60°C and 1 bar.

7.39 Estimate the fugacity of ethanol in a liquid mixture containing 30% ethanol, 20% 1-propanol, and 50% water at 8°C and 1 bar using the Wilson equation. Wilson parameters at 60°C are given in Problem 7.38.

7.40 Enthalpy of mixing data for binary mixtures of water (1) and acetone (2) have been fit to the following equation:[8]

$$\Delta h_{\text{mix}} = x_1 x_2 \left[-447.8 + 3802(x_2 - x_1) - 1200(x_2 - x_1)^2 + 1554(x_2 - x_1)^3 \right]$$

[8] J. J. Christenson, R. W. Hanks, and R. M. Izatt, *Handbook of Heats of Mixing* (New York: Wiley, 1982).

where Δh_{mix} has units of J/mol. At 60°C, the activity coefficient of water in an equimolar mixture of water and acetone is 1.65. Estimate the activity coefficient of water in an equimolar mixture of water and acetone at 100°C. State any assumptions that you make.

7.41 In Problem 6.26 heats of mixing for binary mixtures of solid-phase cadmium (Cd) and tin (Sn) were reported as:

$$\Delta h_{mix} = 13,000 X_{Cd} X_{Sn} [\text{J/mol}]$$

where, X_{Cd} and X_{Sn} are the cadmium and tin mole fractions, respectively. If we assume cadmium and tin form a regular solution, calculate the activity coefficient of cadmium in a mixture of 2 moles Cd and 3 moles Sn.

7.42 In the ThermoSolver software that comes with the text, go to the **Models for g^E— Parameter Fitting** menu. Find best-fit model parameters for the ethanol (a) and water (b) system at 74.79°C for the following activity coefficient models:

(a) two-suffix Margules

(b) three-suffix Margules

(c) Van Laar

(d) Wilson

(e) NRTL

Use the **Plot Data** ... button to plot ln γ vs. x_a for all the models. Which model do you think best represents the data? Why? The information in the **Statistics** ... button may be useful.

7.43 Repeat Problem 7.42 for pentane (a) and acetone (b) at 25°C.

7.44 Repeat Problem 7.42 for chloroform (a) and heptane (b) at 25°C.

CHAPTER 8

Phase Equilibria III: Phase Diagrams

Now that we have established a criteria for chemical equilibrium:

$$\hat{f}_i^{\alpha} = \hat{f}_i^{\beta} \tag{8.1}$$

and have examined the practical issues in calculating the fugacities of vapor and condensed phases, we are poised to look at some phase equilibria problems.

Learning Objectives

To demonstrate mastery of the material in Chapter 8, you should be able to:

▶ Construct phase diagrams for binary systems in vapor–liquid equilibria (VLE), liquid–liquid equilibria (LLE), vapor–liquid–liquid equilibria (VLLE), solid–liquid equilibria (SLE), solid–solid equilibria (SSE), and solid–solid–liquid equilibria (SSLE), correcting for nonideal behavior in the vapor, liquid, or solid phases using fugacity coefficients and activity coefficients.

▶ Given a phase diagram for a binary mixture, identify what phase or phases are present at a specified state; in the two-phase or three-phase regions, identify the composition of each phase and their relative amounts using the lever rule.

▶ Perform bubble-point and dew-point VLE calculations on your own and with ThermoSolver for binary and multicomponent mixtures when the temperature is known and the pressure is unknown or when the pressure is known and the temperature is unknown. Determine the exit compositions or flow rates for an isothermal flash.

▶ Treat the solubility of gases in liquids using Henry's law for both ideal and nonideal behavior. Correct reported Henry's law coefficients for pressure or temperature. Perform LLE, VLLE, SLE, and SSE phase equilibria calculations. Determine whether a liquid mixture is inherently instable and will split into two liquid phases.

▶ Identify when a binary mixture exhibits an azeotrope. Distinguish between maximum and minimum boiling azeotropes and explain this behavior in terms of intermolecular interactions. Use azeotropic data to determine activity coefficient model parameters.

▶ Define the following terms and explain their context in terms of VLE, LLE, or SSLE: *bubble point, dew point, positive deviation and negative deviation from Raoult's law, binodal curve, spinodal curve, upper* and *lower consulate temperature, eutectic point, peritectic point, congruent melting*, and *incongruent melting.*

▶ For VLE, LLE, and SLE, relate phase diagrams schematically to the Gibbs energy of each phase in the mixture using the minimization of Gibbs energy to determine the equilibrium state of the mixture.

▶ Calculate the following colligative properties of a dilute solution: boiling-point elevation, freezing-point depression, and osmotic pressure.

▶ Fit parameters in binary activity coefficient models on your own and with ThermoSolver from experimental data with objective functions or by linear regression, when appropriate.

▶ 8.1 VAPOR–LIQUID EQUILIBRIUM (VLE)

The most common phase equilibria problems chemical engineers encounter involve vapor–liquid equilibrium (VLE). We can write the general expression for VLE by applying the defining relations in Chapter 7. At equilibrium, the fugacity of species i in the vapor and liquid are equal:

$$\hat{f}_i^v = \hat{f}_i^l \qquad (8.2)$$

If we choose to quantify the vapor-phase nonideality using the fugacity coefficient [Equation (7.9)] and the liquid-phase nonideality using the activity coefficient [Equation (7.71)],[1] we get:

$$\cancel{\times} \quad y_i \hat{\varphi}_i^v P = x_i \gamma_i^l f_i^o \qquad (8.3)$$

Here y_i represents the mole fraction in the vapor phase, while x_i represents the liquid mole fraction. Once we have chosen the appropriate liquid reference state (Lewis/Randall rule or Henry's law), we can solve this problem if we have composition dependencies of the activity coefficient and the fugacity coefficient. Equation (8.3) is actually a set of coupled equations, one for each species i. While Equation (8.3) is completely rigorous (and thus always correct), we have seen that it is not always trivial to calculate these terms.

Hence, we will begin our exploration of VLE by considering the limiting case of Equation (8.3), which is valid only under certain circumstances. When we make approximations to this equation, we must keep in mind that the analysis that follows is valid only for the specific cases where the approximations apply.

Raoult's Law (Ideal Gas and Ideal Solution)

Consider the case when we are at low pressure and all the intermolecular forces are approximately the same. We can treat the vapor as an ideal gas and the liquid as an ideal solution. If we pick the Lewis/Randall reference state ($f_i^o = f_i$), our criteria for equilibrium can be simplified to

$$y_i P = x_i f_i \qquad (8.4)$$

---

[1] Alternatively, we can also use a fugacity coefficient for the liquid phase. This approach, which requires an accurate equation of state for the liquid, is less common than using the activity coefficient.

Applying Equation (7.86) for the pure species fugacity, we get

$$y_i P = x_i P_i^{sat} \tag{8.5}$$

You will recognize Equation (8.5) as Raoult's law, which you have undoubtedly seen before. It directly results from the criteria for equilibrium [Equation (8.1)] under the special circumstances described above (ideal gas, ideal solution, Lewis/Randall reference state). This equation is convenient, since the saturation pressure of species i depends only on the temperature of the system. The relation between P_i^{sat} and T is commonly fit to the Antoine equation. Appendix A.1 provides Antoine equation parameters for several species. Equation (8.6) is often rewritten

$$y_i = K_i x_i \tag{8.6}$$

When Raoult's law applies, we have

$$K_i = \frac{P_i^{sat}(T \text{ only})}{P} \tag{8.7}$$

K_i is termed the K-value[2] of species i and, in this case, depends only on the temperature and the pressure of the system. Equation (8.6) is frequently used in hydrocarbon systems, where extensive data for K-values are available. In fact, this approach can be extended beyond the limiting assumptions of Raoult's law. If we use the Lewis fugacity rule to correct for the fugacity of real gases at higher pressure and apply the Poynting correction to the Lewis/Randall reference state, K remains independent of composition since both these corrections are based on pure species properties. Thus, K can be expressed for any species solely as a function of T and P. In some cases K-values are even extended to real gases, where the ideal gas and ideal solution assumptions discussed above no longer apply. In such a case, the nonideality of the system is buried in K. Now K depends on composition and must be found either empirically or through fugacity and activity coefficients, as described by Equation (8.3).

In order to further explore the implications of Raoult's law, let's consider a binary solution containing species a and species b. We can write the criteria for equilibrium given by Equation (8.5) for each species:

$$y_a P = x_a P_a^{sat} \tag{8.8}$$

and

$$y_b P = x_b P_b^{sat} \tag{8.9}$$

We then obtain an expression for the total system pressure by adding together Equations (8.8) and (8.9):

$$y_a P + y_b P = P = x_a P_a^{sat} + (1 - x_a) P_b^{sat} \tag{8.10}$$

Plugging Equation (8.10) into (8.8) gives

$$y_a = \frac{x_a P_a^{sat}}{x_a P_a^{sat} + (1 - x_a) P_b^{sat}} \tag{8.11}$$

Equation to
↳ use when
tempt more
than one known

---

[2] The K-value is the analogous quantity to phase equilibrium that the equilibrium constant is to chemical reaction equilibrium (Chapter 9).

We can use Equations (8.10) and (8.11) to construct a **phase diagram** for a binary mixture either at constant temperature or at constant pressure. For example, Figure 8.1 shows a phase diagram for a binary mixture of a and b that follows Raoult's law. The liquid- and gas-phase mole fractions are plotted vs. total pressure, while the temperature of the system is held constant. Since data are readily available for the saturation pressure of many species as a function of temperature, P can be calculated via Equation (8.10), given the liquid mole fraction and temperature. Equation (8.11) then provides a way to directly calculate the vapor-phase mole fraction. The linear relationship between pressure and liquid-phase composition, indicated by Equation (8.10), manifests in a straight Px_a line. A binary-phase diagram can also be constructed at constant pressure. However, in this case, calculations are somewhat more involved. Since the temperature changes as the composition changes, we must apply the Antoine equation or a similar relation to describe the variation in P_i^{sat}.

Phase diagrams are useful for identifying the thermodynamic state of a binary mixture. They tell us what phase or phases are present and, in the two-phase region, the composition of the liquid and vapor phases as well as their relative amounts. For any given pressure and overall composition, the phase diagram depicted in Figure 8.1 allows us to identify whether we have only a liquid phase, only a vapor phase, or a combination of two phases—a liquid phase in equilibrium with vapor. By convention, we typically label the *lighter* component, the one that boils more easily, species a. At high pressure, we have a subcooled liquid, as indicated on the top of the diagram. Conversely, at low pressure, the mixture exists as a superheated vapor, as shown at the bottom of the diagram. In between these two regions, we observe a two-phase region where the mixture is in vapor–liquid equilibrium. In this case, the composition and amount of each phase can also be determined from the phase diagram. For example, Figure 8.1 depicts a system with overall composition z_a and pressure P_{sys}. At this pressure and overall composition, we can draw a tie line, as indicated in the figure. The tie line gets its name because it "ties" together the composition of the liquid and vapor phases. It is horizontal because the pressures of the liquid and the vapor are the same. The liquid mole fraction in equilibrium, x_a^{eq}, is obtained from the intersection of the tie line with the liquid line on the left. Similarly, the vapor mole fraction, y_a^{eq}, is obtained from the intersection with the curve on the right. The more volatile species a has a higher concentration in the vapor phase. Additionally, as shown in Example 8.1, the lever rule can be applied to determine

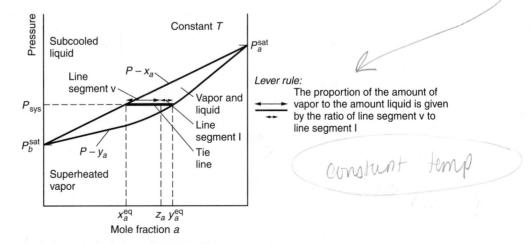

Figure 8.1 Phase diagram for an ideal binary mixture.

lever rule
exp

the percentage liquid vs. percentage vapor. The lever rule says the ratio of the amount of vapor in the system to the amount of liquid is given by the ratio of the line segments from the feed composition to the opposite curve, as shown in the figure. Can you apply a mass balance to develop this relationship in analogy to what we did in Example 1.1? The liquid mole fraction vs. pressure line is often termed the **bubble-point** curve. It gets this name because if we start at high pressure and decrease the system pressure at constant temperature, this curve marks the pressure at which the first bubble of vapor forms. That bubble's composition can be found where the the tie line intersects the Py_a curve. Similarly, the vapor mole fraction pressure curve is termed the **dew-point** curve, since this marks when the first drop of liquid forms when a superheated vapor mixture is isothermally compressed.

Equations (8.10) and (8.11) can be generalized to a system with m components. In this case, the sum of the partial pressures gives

$$P = x_a P_a^{\text{sat}} + x_b P_b^{\text{sat}} \ldots + x_i P_i^{\text{sat}} \ldots + x_m P_m^{\text{sat}} = \sum_{i=1}^{m} x_i P_i^{\text{sat}} \tag{8.12}$$

and the mole fraction of species i is given by

$y_i P = x_i P_i^{\text{sat}}$

$$y_i = \frac{x_i P_i^{\text{sat}}}{\sum_{i=1}^{m} x_i P_i^{\text{sat}}} = P \tag{8.13}$$

Equations (8.12) and (8.13) apply to multicomponent mixtures where the vapor phase is assumed to be an ideal gas and the liquid phase, an ideal solution.

Four common types of vapor–liquid equilibria calculations are illustrated by a grid in Figure 8.2. In a bubble-point calculation, the liquid-phase mole fractions of the system are specified and the vapor mole fractions are solved for. The solution represents the composition of the first *bubble* of vapor that forms when energy is supplied to a saturated liquid. Conversely, in a dew-point calculation, the liquid mole fractions are determined given the vapor mole fractions. This case corresponds to the composition of the first drop of dew that forms from a saturated vapor. Bubble- and dew-point calculations are represented by the two columns in Figure 8.2. In addition to knowing the composition,

$y_i = \dfrac{x_i P_i^{\text{sat}}}{P}$

$y_i = \dfrac{x_i P_i^{\text{sat}}}{\sum x_i P_i^{\text{sat}}}$

$P = \sum x_i P_i^{\text{sat}} (T)$

$y_i = \dfrac{x_i P_i^{\text{sat}}}{P}$ T unknown

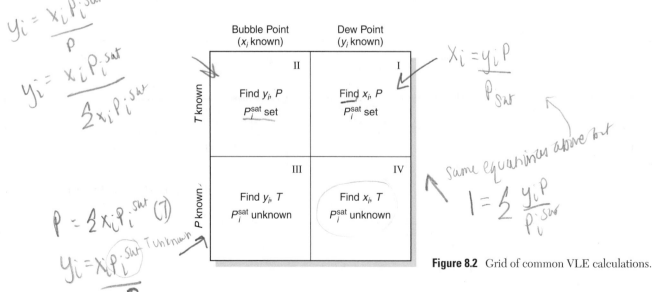

$X_i = \dfrac{y_i P}{P_{\text{sat}}}$

same equations above but

$1 = \sum \dfrac{y_i P}{P_i^{\text{sat}}}$

Figure 8.2 Grid of common VLE calculations.

the value of either the temperature or the pressure needs to be specified to constrain the state of the system. The former case is represented by the first row in Figure 8.2, while the latter case is represented by the second row. Hence, the grid in Figure 8.2 represents four typical combinations of independent and dependant variables found in VLE problems. They are defined by the quadrants I, II, III, and IV for reference in Examples 8.2, 8.3, and 8.5. When faced with such a calculation, it is helpful to recognize the independent and dependant variables appropriately. For binary systems that follow Raoult's law, it is possible to solve for the vapor and liquid mole fractions when T and P are known (see Problem 8.7). However, in mixtures containing three or more components, specifying only T and P underconstrains the problem (i.e., there are multiple possible solutions).

▶ **EXAMPLE 8.1**
Verification of the Lever Rule

Apply the appropriate mass balance equations to verify that the lever rule gives the relative amount of species in each phase along a tie line as depicted in Figure 8.1.

SOLUTION A mass balance of species a gives

$$z_a n = y_a n^v + x_a n^l \tag{E8.1A}$$

while a total mass balance gives

$$n = n^v + n^l \tag{E8.1B}$$

Multiplying Equation (E8.1B) by z_a and equating the result to Equation (E8.1A) gives

$$y_a n^v + x_a n^l = z_a n^v + z_a n^l \tag{E8.1C}$$

We can rearrange Equation (E8.1C) to give the lever rule:

$$\frac{n^v}{n^l} = \frac{(z_a - x_a)}{(y_a - z_a)} = \frac{\text{line segment } \mathbf{v}}{\text{line segment } \mathbf{l}} \tag{E8.1D}$$

Alternatively, we can solve Equation (E8.1B) for n^l and replace in Equation (E8.1A) to find that

$$\frac{n^v}{n} = \frac{(z_a - x_a)}{(y_a - x_a)} = \frac{\text{line segment } \mathbf{v}}{\text{total length of the tie line}} \tag{E8.1E}$$

The mass balances that lead to Equations (E8.1D) and (E8.1E) are general and not limited to the vapor and liquid phases; thus, the lever rule can be applied to find the relative amounts of any two phases in equilibrium. The fraction of material present in one phase can be computed by taking the length of the tie line from the overall composition to the composition of the other phase and then dividing by the total length of the line. ◀

▶ **EXAMPLE 8.2**
Bubble-Point Calculation with P Known

Consider a system with liquid containing 30% n-pentane (1), 30% cyclohexane (2), 20% n-hexane (3), and 20% n-heptane (4) at 1 bar. Determine the temperature at which this liquid develops the first bubble of vapor. What is the vapor composition?

SOLUTION This problem corresponds to quadrant III in the grid of Figure 8.2. Since the components in this system are chemically similar, we will assume an ideal solution. Additionally, at 1 bar, we may assume an ideal gas; thus, we can write the criterion for phase equilibrium in terms of Raoult's law for each of the components, that is:

$$y_i P = x_i P_i^{\text{sat}} \tag{E8.2A}$$

Since the sum of the partial pressures equals the system pressure, we get

$$P = \sum x_i P_i^{\text{sat}} = x_1 P_1^{\text{sat}} + x_2 P_2^{\text{sat}} + x_3 P_3^{\text{sat}} + x_4 P_4^{\text{sat}} \qquad \text{(E8.2B)}$$

The saturation pressure can be found according to Antoine's equation:

$$\ln P_i^{\text{sat}} \, [\text{bar}] = A_i - \frac{B_i}{T\,[\text{K}] + C_i} \qquad \text{(E8.2C)}$$

where the coefficients for parameters A_i, B_i, and C_i are reported in Table E8.2A. Substitution of Equation (E8.2C) into (E8.2B) results in one equation with one unknown—T. This equation can be solved by trial and error, by Excel using "solver," or by other graphical or numerical techniques to give

$$T = 333\,[\text{K}]$$

Saturation pressures and vapor-phase mole fractions for each of the species at this temperature can be calculated according to Equations (E8.2C) and (E8.2A), respectively. Their values are presented in Table E8.2B.

We see that proportionately much more of the *lighter* n-pentane goes into the vapor while very little of the *heavier* n-heptane does. This result forms the basis for separation by distillation.

◀

▶ **EXAMPLE 8.3**
Dew-Point Calculation
with *P* Known

Consider a system with vapor containing 30% n-pentane (1), 30% cyclohexane (2), 20% n-hexane (3), and 20% n-heptane (4) at 1 bar. Determine the temperature at which this vapor develops the first drop of liquid. What is the liquid composition?

SOLUTION   This problem is the dew-point analog to Example 8.2. It corresponds to quadrant IV in the grid of Figure 8.2. We can write Raoult's law for each species as

$$x_i = \frac{y_i P}{P_i^{\text{sat}}} \qquad \text{(E8.3A)}$$

Since the sum of the liquid mole fractions equals 1:

$$1 = \sum \frac{y_i P}{P_i^{\text{sat}}} = \frac{y_1 P}{P_1^{\text{sat}}} + \frac{y_2 P}{P_2^{\text{sat}}} + \frac{y_3 P}{P_3^{\text{sat}}} + \frac{y_4 P}{P_4^{\text{sat}}} \qquad \text{(E8.3B)}$$

TABLE E8.2A   Antoine Coefficients

Species	$n\text{-}C_5H_{12}$	C_6H_{12}	$n\text{-}C_6H_{14}$	$n\text{-}C_7H_{16}$
A_i	9.2131	9.1325	9.2164	9.2535
B_i	2477.07	2766.63	2697.55	2911.32
C_i	−39.94	−50.50	−48.78	−56.51

TABLE E8.2B   Saturation Pressures and Mole Fractions at *T* = 333 [K]

Species	$n\text{-}C_5H_{12}$	C_6H_{12}	$n\text{-}C_6H_{14}$	$n\text{-}C_7H_{16}$
P^{sat} at 333 [K]	2.13 bar	0.514 bar	0.757 bar	0.218 bar
y_i	0.639	0.154	0.151	0.056

Again, the saturation pressures can be written in terms of Antoine's equation:

$$\ln P_i^{\text{sat}} \, [\text{bar}] = A_i - \frac{B_i}{T \, [\text{K}] + C_i} \tag{E8.3C}$$

where the coefficients for parameters A, B, and C are reported in Table E8.2A. Substitution of Equation (E8.3C) into (E8.3B) gives one equation with one unknown—T. The temperature is solved to be

$$T = 349 \, [\text{K}]$$

Saturation pressures and liquid-phase mole fractions for each of the species at this temperature can be calculated according to Equations (E8.3C) and (E8.3A), respectively. Their values are presented in Table E8.3.

In this case, proportionately much more of the *heavier* n-heptane is condensed while little of the *lighter* n-pentane is, again helping us separate by distillation. ◀

▶ **EXAMPLE 8.4**
Isothermal Flash
VLE Calculation

A compressed liquid feed stream containing an equimolar mixture of n-pentane and n-hexane flows into a flash unit as shown in Figure E8.4 at flow rate F. At steady state, 33.3% of the feed stream is vaporized and leaves the drum as a vapor stream with flow rate V. The rest leaves as liquid with flow rate L. If the flash temperature is 20°C, what is the pressure required? What are the composition of the liquid and vapor exit streams?

SOLUTION A mass balance on component a gives

$$x_{a,\text{feed}} F = y_a V + x_a L \tag{E8.4A}$$

If we assume ideal gas and ideal solution, the equilibrium relation in the flash drum can be written according to Equation (8.5):

$$y_a P = x_a P_a^{\text{sat}} \tag{E8.4B}$$

TABLE E8.3 Saturation Pressures and Mole Fractions at $T = 349 \, [\text{K}]$

Species	$n\text{-}C_5H_{12}$	C_6H_{12}	$n\text{-}C_6H_{14}$	$n\text{-}C_7H_{16}$
P^{sat} at 349 [K]	3.296 bar	0.870 bar	1.256 bar	0.494 bar
x_i	0.091	0.345	0.159	0.405

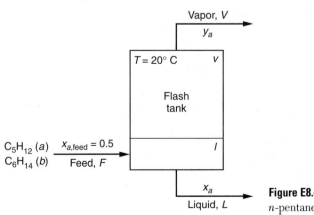

Figure E8.4 Flash vaporization of an n-pentane and n-hexane feed stream.

Substituting Equation (E8.4B) into (E8.4A) gives

$$x_{a,\text{feed}}F = \frac{x_a P_a^{\text{sat}}}{P}V + x_a L = x_a\left(\frac{P_a^{\text{sat}}}{P}V + L\right) \tag{E8.4C}$$

Solving for x_a:

$$x_a = \frac{x_{a,\text{feed}}}{\dfrac{P_a^{\text{sat}}}{P}\left(\dfrac{V}{F}\right) + \left(\dfrac{L}{F}\right)} \tag{E8.4D}$$

Similarly, for the mole fraction of b in the liquid, we get

$$x_b = \frac{x_{b,\text{feed}}}{\dfrac{P_b^{\text{sat}}}{P}\left(\dfrac{V}{F}\right) + \left(\dfrac{L}{F}\right)} \tag{E8.4E}$$

Since the sum of mole fractions must equal 1, we can add Equations (E8.4D) and (E8.4E)

$$1 = \frac{x_{a,\text{feed}}}{\dfrac{P_a^{\text{sat}}}{P}\left(\dfrac{V}{F}\right) + \left(\dfrac{L}{F}\right)} + \frac{x_{b,\text{feed}}}{\dfrac{P_b^{\text{sat}}}{P}\left(\dfrac{V}{F}\right) + \left(\dfrac{L}{F}\right)} \tag{E8.4F}$$

Using the Antoine equation, at 20°C, we get $P_a^{\text{sat}} = 0.56$ [bar] and $P_b^{\text{sat}} = 0.16$ [bar]. Plugging in values to Equation E8.4F gives:

$$1 = \frac{0.5}{\dfrac{0.56}{P}\left(\dfrac{1}{3}\right) + \left(\dfrac{2}{3}\right)} + \frac{0.5}{\dfrac{0.16}{P}\left(\dfrac{1}{3}\right) + \left(\dfrac{2}{3}\right)} \tag{E8.4G}$$

Solving Equation (E8.4G) for pressure gives

$$P = 0.32 \text{ [bar]}$$

At this low pressure, the ideal gas assumption justified. Substituting in Equations (E8.4D) and (E8.4B) gives

$$x_a = 0.40$$

$$y_a = 0.70$$

This example is different in nature from those represented by the grid in Figure 8.2 and illustrated in Examples 8.1 and 8.2. It couples a species mass balance to the VLE phase equilibrium problem. Such calculations are representative of the type encountered in design and analysis of separations processes such as distillation. ◄

Nonideal Liquids

Real systems seldom follow Raoult's law, since it is unlikely that the a-a interaction is identical to the a-b interaction. In this section we consider the behavior of some real binary systems for the case where we are still at low enough pressure for the vapor phase to be an ideal gas. We will explore the cases when the like (a-a and b-b) interaction is stronger than the unlike (a-b) interaction ($\gamma_i > 1$) as well as when the like interaction is weaker ($\gamma_i < 1$).

If the liquid phase consists of chemically dissimilar species and we use a Lewis/Randall reference state, combination of Equations (8.3) and Equation (7.86) yields

$$y_i P = x_i \gamma_i P_i^{\text{sat}} \tag{8.14}$$

Equation (8.14) still assumes that the vapor phase can be represented as an ideal gas and the fugacity of the pure liquid is given by P_i^{sat}. For a binary system we now have

$$y_a P = x_a \gamma_a P_a^{sat} \tag{8.15}$$

and

$$y_b P = x_b \gamma_b P_b^{sat} \tag{8.16}$$

Adding together Equations (8.15) and (8.16):

$$y_a P + y_b P = P = x_a \gamma_a P_a^{sat} + (1 - x_a) \gamma_b P_b^{sat} \tag{8.17}$$

The liquid mole fraction is no longer linear with respect to pressure as it was in Equation (8.10). Plugging Equation (8.17) into (8.15) yields

$$y_a = \frac{x_a \gamma_a P_a^{sat}}{x_a \gamma_a P_a^{sat} + (1 - x_a) \gamma_b P_b^{sat}} \tag{8.18}$$

First consider the case where $\gamma_a > 1$. As we have already seen, this represents the case where like interactions are stronger than unlike interactions. From the Gibbs–Duhem equation, we also know that in this case $\gamma_b > 1$. Comparing Equations (8.10) and (8.17), we see that for a given mole fraction of species a, the system will exhibit a greater pressure than its ideal counterpart. Therefore, we label this case a *positive deviation* from Raoult's law. From a molecular standpoint, this result makes sense. At a given temperature, the molecules in the liquid have a fixed kinetic energy. However, the unlike interactions in the liquid are not as strong as the like interactions. Thus, the two species when mixed are not held as vigorously in the liquid phase. More molecules, therefore, escape to the vapor than in the case of an ideal solution, and the pressure exerted is higher. The dew-point and bubble-point curves will be different from Figure 8.1. An example of a phase diagram binary system exhibiting positive deviations from ideality, methanol and water at 323 K, is shown in Figure 8.3. Such phase diagrams can be constructed using Equations (8.17) and (8.18) in much the same way as we described for Figure 8.1.

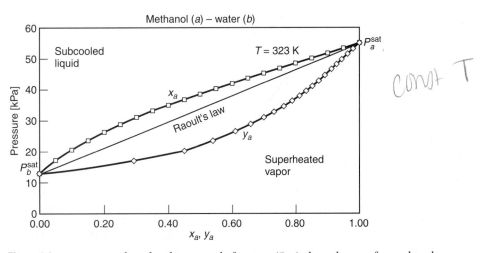

Figure 8.3 Pressure vs. liquid and vapor mole fractions (Pxy) phase diagram for methanol (a)–water (b) binary mixtures at a constant temperature of 323 K. This system shows positive deviations from Raoult's law. The straight line demonstrates what the liquid mole fraction would be in the case of Raoult's law.

In this case, an activity coefficient model is also needed to quantify the liquid-phase nonideality. Can you come up with a model and parameters to describe this data set? The straight line in Figure 8.3 demonstrates what the liquid mole fraction would be in the case of Raoult's law. In contrast, the dew-point curve in the real system occurs at higher pressure. Conversely, we say that systems which have $\gamma_a < 1$ exhibit *negative deviations* from Raoult's law. Correspondingly, the liquid mole fraction is at pressures lower than predicted by Raoult's law. Negative deviations occur when the unlike intermolecular interactions are more attractive than the like interactions of the pure species. Hence, the species in the liquid mixture are pulled toward one another with greater vigor, leading to less molecules in the vapor and a smaller system pressure.

Two other common types of binary-phase diagrams are shown in Figures 8.4 and 8.5. In Figure 8.4, temperature is plotted vs. liquid and vapor mole fraction to construct a *Txy* phase diagram for the water–methanol system. This type of phase diagram is similar to the *Pxy* diagram discussed above; however, instead of holding *T* constant, pressure is held constant at 1 atm. Could you construct such a plot? What data would you need? For any given temperature and composition, we can again identify whether we have only a liquid phase, only a vapor phase, or a combination of two phases—liquid in equilibrium with vapor. The liquid phase is now on the bottom of the phase diagram—that is, at low *T*—while the vapor phase is on the top. For the two-phase region, the composition of each phase can also be determined, and a tie line connects vapor and liquid compositions for a given temperature. For example, the tie line shown in Figure 8.4 indicates that at 358 K and 1 atm, methanol with liquid mole fraction $x_a = 0.15$ is in equilibrium with a vapor of composition $y_a = 0.52$. Again, the lever rule can be applied to discern the relative amounts of liquid and vapor.

In Figure 8.5*a*, the vapor mole fraction vs. liquid mole fraction is shown for the water–methanol system at 1 atm. This so-called *xy* diagram is convenient for illustrating the process of fractional distillation, whereby methanol can be purified by a series of vaporization and condensation stages. The 45° line where $x_a = y_a$ is also plotted in this figure. In essence, a distillation column is a series of flashes of the type described in

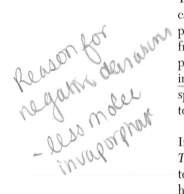

Reason for negative deviation
- less moles in vaporphase

const P

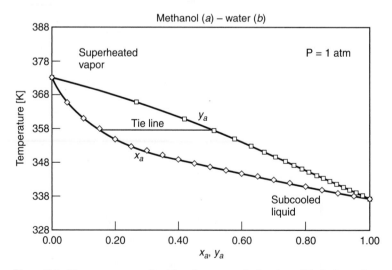

Figure 8.4 Temperature vs. liquid and vapor mole fractions (*Txy*) phase diagram for methanol (*a*)–water (*b*) binary mixtures at a constant pressure of 1 atm. This system shows positive deviations from Raoult's law.

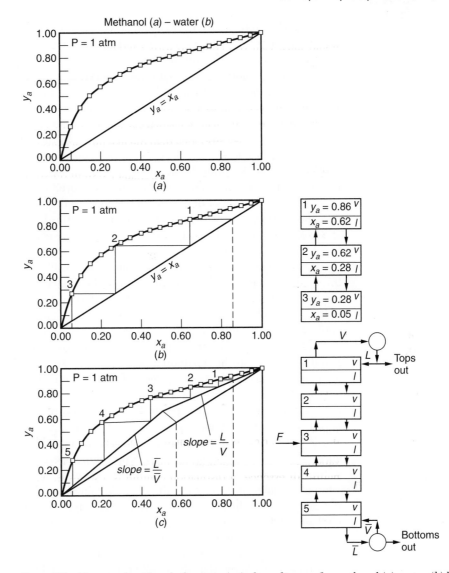

Figure 8.5 Vapor vs. liquid mole fractions (xy) phase diagram for methanol (a)–water (b) binary mixtures at a constant pressure of 1 atm. (a) xy diagram with 45° line indicated (b) stages in distillation at total reflux (c) stages in a five tray distillation column with operating lines shown. Stages in distillation are illustrated on the right of (b) and (c).

Example 8.4. Distillation is closer, in practice, to constant pressure than constant temperature. Figures 8.5b and 8.5c illustrate two cases where a graphical solution is applied to the xy diagram of Figure 8.5a to estimate the separation achieved in distillation.[3] We now give an overview of such a process. Don't worry so much about the details here; you will learn more about these solutions in your Unit Operations class. It is presented here to give you a big picture of how VLE can be applied and the utility of an xy diagram. Figure 8.5b shows the limiting case of total reflux, whereby all the overhead vapor is

---

[3] This approach was first presented by McCabe and Thiele.

condensed and returned to the distillation column. Likewise, none of the liquid bottoms are drawn from the distillation column. Total reflux represents the minimum number of equilibrium stages that are needed to achieve a given separation (unfortunately, in this limit you get nothing out, so it is used merely as a theoretical boundary). The three flash stages to the right of Figure 8.5*b* correspond to the equilibrium "steps" drawn in the figure. The bottom flash tray (labeled "3") takes a liquid containing 5% methanol in water and vaporizes it to 28% methanol. The temperature of this unit is 366 K, as can be inferred from Figure 8.4. This stage is represented by the vertical line labeled "3" in the *xy* diagram on the left. This methanol–water vapor mixture rises up the column and is condensed at 345 K in tray 2. The condensation process is represented by the horizontal line in Figure 8.5*b*. How did we find the temperature? The vaporization process is then repeated, causing a vapor of 62% methanol to leave this tray. Finally, another condensation and vaporization cycle leaves the vapor in the third tray containing 86% methanol at 345 K. Thus, in the limiting case of total reflux, we can purify methanol from 5% in tray 1 to 86% in tray 3.

A more realistic distillation of methanol–water is illustrated in Figure 8.5*c* The five trays are schematically shown to the right of the *xy* diagram. In this case, a fraction of the methanol-rich vapor is removed from the top as labeled "tops out," and a fraction of the water-rich liquid is removed from the bottom as labeled "bottoms out." The feed enters on tray 3 at a flow rate F. Vapor leaves tray 1 at a flow rate V, while the fraction of the liquid that returns from this condensed vapor is at flow rate L. The remaining liquid is collected as the separations product in the stream labeled "tops out." Similarly, liquid leaves tray 5 at flow rate $\overline{L}$, while a fraction of it is returned as vapor with flow rate $\overline{V}$. The rest is collected in the stream "bottoms out." In solving this problem, we must account for both the liquid–vapor phase equilibrium relationships of each tray and the mass balances of vapor and liquid flowing in the tray, as we did in the flash of Example 8.4. It turns out we can represent the constraints posed by the mass balance with two *operating lines* above the 45° line. The slope of the operating lines are given by the ratio of the liquid flow to the vapor flow in each section of the distillation column. We then "step off" equilibrium stages to the operating lines, much as we did to the 45° line in the case of total reflux (Figure 8.5*b*). The five trays in column 8.5*c* take a feed of 58% methanol and separate it into a light stream containing 86% methanol and a heavy stream containing 5% methanol.

We can relate the phase behavior of these binary mixtures to the Gibbs energy of each phase. We know that the equilibrium state is defined where the Gibbs energy is minimized. Since $g = h - Ts$, Gibbs energy can be lowered by either lowering h or increasing s. For systems consisting of liquid and vapors, the resulting phase behavior is determined from the trade-off between the energetically favored liquid phase and the entropically favored vapor. We will illustrate the relationship between Gibbs energy minimization and phase diagrams in terms of the *Txy* diagram shown in Figure 8.4. However, similar arguments can be made for a *Pxy* phase diagram.

The top of Figure 8.6 shows plots of the Gibbs energy of the vapor and liquid phases, g_{vapor} and g_{liquid}, respectively, vs. mole fraction at three temperatures, T_1, T_2, and T_3. On the bottom of the figure, the corresponding *Txy* phase diagram, as reproduced from Figure 8.4, is shown with the three temperatures corresponding to the Gibbs energy plots above identified. First consider the lowest temperature, T_1, below the boiling point of either species. The Gibbs energies of both the liquid and vapor show minima with respect to composition, resulting from the Gibbs energy of mixing. The Gibbs energy of the liquid is less than that of the vapor across the entire composition range. Thus, at equilibrium the system will be in the energetically favored liquid phase. As the

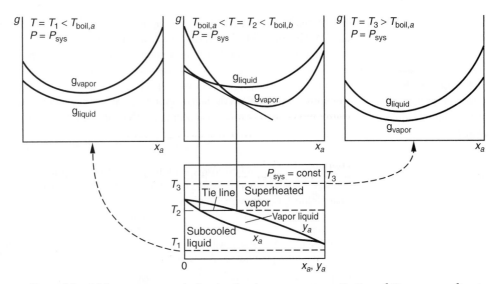

Figure 8.6 Gibbs energy vs. mole fraction (*top*) at temperatures T_1, T_2, and T_3 corresponding to three different isotherms on the phase diagram (*bottom*).

temperature is raised, the contribution of entropy becomes more and more important. The highest temperature shown, T_3, is above the boiling point of both components. At this temperature, the Gibbs energy of the vapor is always lower than that of the liquid, as shown on the top right. Hence, the binary exists as a vapor at all values of x_a. An intermediate temperature, T_2, between the boiling points of species a and b is shown in the Gibbs energy plot in the middle of the figure. At low values of x_a, energetic effects dominate and the Gibbs energy of the liquid is lower than the Gibbs energy of the vapor. Conversely, at high values of x_a, the Gibbs energy of the vapor is lower. The resulting phase behavior shows three regions. When the value of x_a is small, the equilibrium state is as a single liquid phase. Similarly, at large values of x_a, the equilibrium state is as a single vapor phase. Between the minima in Gibbs energy, we can draw a tangent line, as shown in Figure 8.6. The Gibbs energy at mole fractions in between the minima can be lowered if the system separates into two phases in much the same way as was explained in our discussion of two liquid phases in Figure 7.10. In this case, we have vapor–liquid equilibrium, with the composition in each phase determined by the mole fraction at which the tangent line abuts the two curves. We see, once again, that the trade-off between energy and entropy is quantitatively described by Gibbs energy and that phase separation results from minimizing this property.

While Figures 8.3, 8.4, and 8.5 represent Pxy, Txy, and xy diagrams for a system for which the unlike interactions are weaker than the like interactions, such diagrams can also be constructed when the unlike interactions are stronger. Equations (8.17) and (8.18) can be generalized to a system with m components as follows:

$$P = x_a \gamma_a P_a^{\text{sat}} + x_b \gamma_b P_b^{\text{sat}} \ldots + x_i \gamma_i P_i^{\text{sat}} \ldots + x_m \gamma_m P_m^{\text{sat}} = \sum_{i=1}^{m} x_i \gamma_i P_i^{\text{sat}} \qquad (8.19)$$

and

$$y_i = \frac{x_i \gamma_i P_i^{\text{sat}}}{\sum\limits_{i=1}^{m} x_i \gamma_i P_i^{\text{sat}}} \qquad (8.20)$$

Equations (8.19) and (8.20) assume an ideal gas and a fugacity of the pure liquid as given by P_i^{sat}. Example 8.6 illustrates a case where this assumption is no longer valid. In that case, a different VLE relation is developed from Equation (8.3).

▶ **EXAMPLE 8.5**
Dew-Point Calculation of a Nonideal Liquid with T Known

A binary vapor mixture contains 48% ethanol (a) in water (b) at 70°C. Determine the pressure at which this vapor develops the first drop of liquid. What is the liquid composition? The excess Gibbs energy can be described by the three-suffix Margules equation with parameters

$$A = 3590 \, [\text{J/mol}] \quad \text{and} \quad B = -1180 \, [\text{J/mol}]$$

SOLUTION This problem corresponds to quadrant I in the grid of Figure 8.2. Since both ethanol and water exhibit vapor pressures below 1 bar at a temperature of 70°C, we will assume the vapor phase is ideal. Equation (8.19) gives

$$P = x_1 \gamma_1 P_1^{sat} + (1 - x_1) \gamma_2 P_2^{sat} \tag{E8.5A}$$

Applying the expressions in Table 7.1, we can write the activity coefficients in terms of the three-suffix Margules parameters, A and B:

$$\ln \gamma_1 = \frac{(A + 3B)}{RT} x_2^2 - \frac{4B}{RT} x_2^3 \tag{E8.5B}$$

Similarly,

$$\ln \gamma_2 = \frac{(A - 3B)}{RT} x_1^2 + \frac{4B}{RT} x_1^3 \tag{E8.5C}$$

Substituting the expressions given by Equations (E8.5B) and (E8.5C) into Equation (E8.5A), we get

$$P = x_1 \exp\left[\frac{(A + 3B)}{RT} x_2^2 - \frac{4B}{RT} x_2^3\right] P_1^{sat} + x_2 \exp\left[\frac{(A - 3B)}{RT} x_1^2 + \frac{4B}{RT} x_1^3\right] P_2^{sat} \tag{E8.5D}$$

Equation (8.18) can be solved for y_1 using Equation (E8.5D) for pressure:

$$y_1 = \frac{x_1 \exp\left[\dfrac{(A + 3B)}{RT} x_2^2 - \dfrac{4B}{RT} x_2^3\right] P_1^{sat}}{x_1 \exp\left[\dfrac{(A + 3B)}{RT} x_2^2 - \dfrac{4B}{RT} x_2^3\right] P_1^{sat} + x_2 \exp\left[\dfrac{(A - 3B)}{RT} x_1^2 + \dfrac{4B}{RT} x_1^3\right] P_2^{sat}} \tag{E8.5E}$$

We can find the saturation pressures from the appendices. From the Antoine equation, $P_1^{sat} = 0.72$ [bar], and from the steam tables, $P_2^{sat} = 0.31$ [bar]. Since the liquid mole fractions must sum to 1, Equation (E8.5E) can be written in terms of one unknown, x_1. Solving gives

$$x_1 = 0.12$$

Plugging this value into Equation (E8.5D) gives

$$P = 0.55 \, [\text{bar}]$$

These values compare to experimental values of $x_1 = 0.13$ and $P = 0.57$ [bar], respectively. ◀

▶ **EXAMPLE 8.6**
Dew-Point Calculation of a Nonideal Liquid and Nonideal Gas with T Known

At high pressures, both the vapor and liquid phases may be nonideal. Consider a binary mixture of a and b with vapor-phase mole fraction and T known. Develop a set of equations and a solution algorithm to determine the composition in the liquid phase and the system pressure. Use the van der Waals equation to quantify deviations from ideality in the vapor and the three-suffix Margules equation to model the nonideal liquid. Assume that critical properties, liquid volumes, and Antoine coefficients for each species are readily available and that the three-suffix Margules parameters have been determined.

SOLUTION This problem corresponds to quadrant I in the grid of Figure 8.2. We illustrate the solution method using the van der Waals equation of state, since we learned how to solve for the fugacity coefficients in Chapter 7. There are other, more accurate equations of state to use; however, the basis of the solution method remains the same. Additionally, it is straightforward to extend the solution to mixtures with more than two components. For example, the text software uses the Peng–Robinson equation for dew-point calculations of multicomponent mixtures.

Solving Equation (8.3) for the liquid phase mole fraction gives

$$x_i = \frac{y_i \hat{\varphi}_i P}{\gamma_i \varphi_i^{\text{sat}} P_i^{\text{sat}} \exp\left[\dfrac{v_i^l}{RT}\left(P - P_i^{\text{sat}}\right)\right]} \tag{E8.6A}$$

Since the sum of the liquid mole fractions equals 1;

$$1 = \frac{y_a \hat{\varphi}_a P}{\gamma_a \varphi_a^{\text{sat}} P_a^{\text{sat}} \exp\left[\dfrac{v_a^l}{RT}\left(P - P_a^{\text{sat}}\right)\right]} + \frac{y_b \hat{\varphi}_b P}{\gamma_b \varphi_b^{\text{sat}} P_b^{\text{sat}} \exp\left[\dfrac{v_b^l}{RT}\left(P - P_b^{\text{sat}}\right)\right]} \tag{E8.6B}$$

Equation (E8.6B) can be solved for pressure to give

$$P = \left[\frac{y_a \hat{\varphi}_a}{\gamma_a \varphi_a^{\text{sat}} P_a^{\text{sat}} \exp\left[\dfrac{v_a^l}{RT}\left(P - P_a^{\text{sat}}\right)\right]} + \frac{y_b \hat{\varphi}_b}{\gamma_b \varphi_b^{\text{sat}} P_b^{\text{sat}} \exp\left[\dfrac{v_b^l}{RT}\left(P - P_b^{\text{sat}}\right)\right]}\right]^{-1} \tag{E8.6C}$$

At the system T, the saturation pressure of each species can be found from the Antoine coefficients:

$$\ln P^{\text{sat}} = A - \frac{B}{T + C} \tag{E8.6D}$$

The pure species saturation pressures can be found using the results from Example 7.2:

$$\ln\left[\varphi_a^{\text{sat}}\right] = -\ln\left[\frac{(v_a^{\text{sat}} - b)P_a^{\text{sat}}}{RT}\right] + \frac{b}{(v_a^{\text{sat}} - b)} - \frac{2a}{RTv_a^{\text{sat}}} \tag{E8.6E}$$

and $$\ln\left[\varphi_b^{\text{sat}}\right] = -\ln\left[\frac{(v_b^{\text{sat}} - b)P_b^{\text{sat}}}{RT}\right] + \frac{b}{(v_b^{\text{sat}} - b)} - \frac{2a}{RTv_b^{\text{sat}}} \tag{E8.6F}$$

where we can solve for the pure species volumes, v_a^{sat} and v_b^{sat}, using the van der Waals equation:

$$P = \frac{RT}{v - b} - \frac{a}{v^2} \tag{E8.6G}$$

With the critical properties known, the van der Waals parameters, a and b, can be found as follows:

$$a = \frac{27}{64}\frac{(RT_c)^2}{P_c}\left[\frac{\text{Jm}^3}{\text{mol}^2}\right] \tag{E8.6H}$$

$$b = \frac{(RT_c)}{8P_c}\left[\frac{\text{m}^3}{\text{mol}}\right] \tag{E8.6I}$$

We solved for the fugacity coefficients in the mixture in Example 7.4:

$$\ln\left[\hat{\varphi}_a^v\right] = -\ln\frac{P(v - b_{\text{mix}})}{RT} + \frac{b_a}{[v - b_{\text{mix}}]} - \frac{2(y_a a_a + y_b \sqrt{a_a a_b})}{RTv} \tag{E8.6J}$$

and $$\ln\left[\hat{\varphi}_b^v\right] = -\ln\frac{P(v - b_{\text{mix}})}{RT} + \frac{b_b}{[v - b_{\text{mix}}]} - \frac{2(y_b a_b + y_a \sqrt{a_a a_b})}{RTv} \tag{E8.6K}$$

Since we know y_a and y_b, we can find a_{mix} and b_{mix} by

$$a_{mix} = y_a^2 a_a + 2 y_a y_b \sqrt{a_a a_b} + y_b^2 a_b \qquad \text{(E8.6L)}$$

and

$$b_{mix} = y_a b_a + y_b b_b \qquad \text{(E8.6M)}$$

We can solve for the molar volume of the mixture in Equations (E8.6J) and (E8.6K) using the van der Waals equation, (E8.6G).

Using the three-suffix Margules equation for activity coefficients gives

$$\gamma_a = \exp\left[\frac{(A + 3B)}{RT} x_b^2 - \frac{4B}{RT} x_b^3 \right] \qquad \text{(E8.6N)}$$

and

$$\gamma_b = \exp\left[\frac{(A - 3B)}{RT} x_a^2 + \frac{4B}{RT} x_a^3 \right] \qquad \text{(E8.6O)}$$

There are three unknowns we want to solve for: P, x_a, and x_b. These quantities are given by Equations (E8.6C) and (E8.6A), respectively. However, as Equations (E8.6J) and (E8.6K) show, the fugacity coefficients depend on P, which is unknown. Similarly, the activity coefficients presented in Equations (E8.6N) and (E8.6O) depend on the unknowns x_a and x_b. Therefore, an iterative scheme must be used. Figure (E8.6A) shows a flow chart of one possible computational algorithm. We initially set the fugacity coefficients and activity coefficients to 1. Thus, our initial guess treats the vapor as an ideal gas and the liquid as an ideal solution. We first iteratively solve for the fugacity coefficient. The solution is represented by the inner loop in the flow diagram. We successively solve for pressure [Equation (E8.6C)] and fugacity coefficients [Equations (E8.6J) and (E8.6K)] until the pressure difference between calculations falls below a specified convergence criteria. We

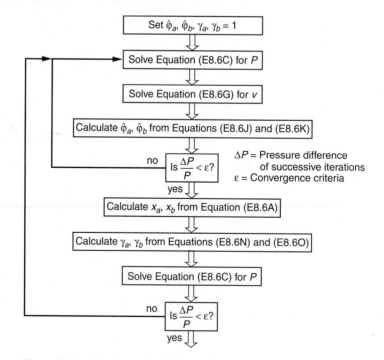

Figure E8.6A Flow diagram of solution algorithm.

then iterate in a similar manner for the activity coefficient in the outer loop. Alternatively, we could have developed an algorithm whereby we iterated for the activity coefficient in the inner loop and the fugacity coefficient in the outer loop. Convergence occurs when values of P, x_a, and x_b give values of $\hat{\varphi}_a$, $\hat{\varphi}_b$, γ_a, and γ_b that satisfy Equations (E8.6C) and (E8.6A).

How would the solution algorithm change if you knew P and not T as in grid IV of Figure 8.2?

◀

Azeotropes

When deviations from Raoult's law are large enough, the Px and Py curves can exhibit extremes. Amazingly, if the Px curve exhibits a maximum, then the Py curve will also exhibit a maximum. Moreover, they will go through the maximum at exactly the same composition! Analogous behavior is observed for minima. We use the term **azeotrope**[4] to describe the point in the phase diagram where the Px and Py curves go through a maximum or a minimum. At the azeotrope, the mole fraction of each species in the liquid phase equals that in the vapor phase:

$$x_i = y_i \qquad \text{at the azeotrope} \tag{8.21}$$

Example 8.8 provides justification of the statements above by showing that thermodynamic property relations dictate that vapor and liquid mole fractions are always equal for maxima or minima in pressure at constant temperature. Likewise, their mole fractions must be equal for extremes in temperature at constant pressure.

For example, binary mixtures of chloroform (a) and n-hexane (b) show large positive deviations from Raoult's law. A phase diagram for this binary system at 318 K is shown in Figure 8.7a. The Px curve goes through a maximum at $x_a = 0.75$. The Py curve exhibits a maximum at exactly the same composition and, consequently, touches the other curve. Hence, the liquid and vapor mole fractions are equal at this point, that is, $x_a = y_a$. We can compare this system with the methanol–water system shown in Figure 8.3. Both these systems show positive deviations from the straight-line behavior of an ideal solution. Apparently, an azeotrope occurs when the deviations are so large that the system pressure is "pushed" above the saturation pressure of the lighter component (P_a^{sat}). The Px curve must then go through a maximum to return to the saturation pressure at pure b. From this argument, we can surmise that azeotropes occur when unlike interactions are very different from like interactions. Additionally, we can induce that when the saturation pressures of the two components are closer in value, an azeotrope is more likely to occur. In other words, the flatter the straight line representing the ideal solution, the more likely a maximum will result. Consequently, azeotropes are less common in binary mixtures with large differences in saturation pressure.

Another interesting ramification of binary systems that exhibit azeotropes is that in some cases T and P no longer uniquely specify the vapor and liquid compositions of the system in phase equilibrium. For example, there are two different states that the chloroform–n-hexane system can take at 59 kPa and 318 K, one on the left side of the azeotrope with $x_a = 0.65$ and $y_a = 0.68$ and another to the right of the azeotrope with $x_a = 0.90$ and $y_a = 0.88$. In contrast, the methanol–water system depicted in Figure 8.3 exhibits a unique solution of liquid and vapor mole fractions for any T and P. Problems 8.7 through 8.9 provide cases where we can solve for vapor and liquid mole fractions given T and P.

---

[4] From Greek, meaning "boiling without changing."

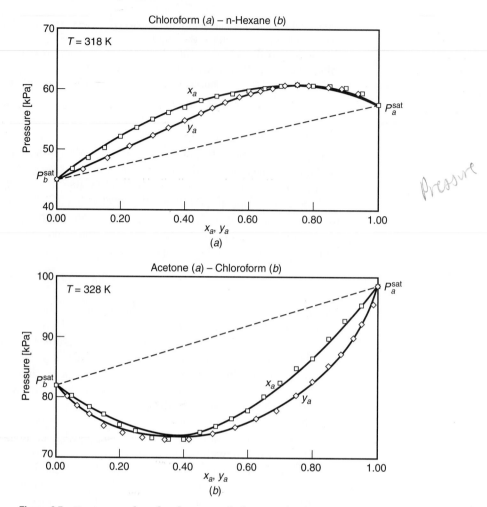

Figure 8.7 Pressure vs. liquid and vapor mole fractions (Pxy) at equilibrium for (a) chloroform (a)–n-hexane (b) at a constant temperature of 318 K and (b) acetone (a)–chloroform (b) at 328 K. Both systems exhibit azeotropes.

If the unlike interaction is stronger than the like interaction, exactly the opposite behavior can occur. In this case, we have *negative deviations* from Raoult's law. In extreme cases, the Px and Py curves exhibit minima at exactly the same composition. The minima occur if the total pressure falls below the saturation pressure of the heavier component. Such behavior is shown in Figure 8.7b for acetone (a)–chloroform (b) at 328 K. We see an azeotrope at a pressure of 73 kPa. At the azeotrope, the mole fraction of liquid and the mole fraction of vapor are both equal to 0.39. Azeotropes showing positive deviations from Raoult's law—that is, maxima in P—are more common than those exhibiting negative deviations.

We can also see azeotropic behavior on Txy phase diagrams at constant pressure. A mixture in which the unlike interactions are weaker than the like interactions will boil more easily than its pure species components. Thus, a system that exhibits a maximum in pressure (positive deviations from Raoult's law) will exhibit a minimum in temperature. As in the case with pressure, a minimum in the Ty curve occurs concurrently with a minimum in the Tx curve and they occur at exactly the same composition. These are termed *minimum boiling azeotropes.* A phase diagram for chloroform (*a*)–*n*-hexane (*b*) at a constant pressure of 1 atm is shown in Figure 8.8*a*. Again, at the azeotrope, the liquid and vapor have the same composition. In this case, an azeotrope becomes more likely as the boiling temperatures of the pure components approach each other. The boiling points of chloroform and *n*-hexane differ by only 7 K. Analogously a binary mixture of acetone (*a*)–chloroform (*b*), which has stronger unlike interactions, exhibits a *maximum boiling azeotrope* at 1 atm, as shown in Figure 8.8*b*. Vapor vs. liquid mole fraction (*xy*) diagrams of the two systems presented in Figure 8.8 are shown in Figure 8.9. Consider the *xy* plot of chloroform and *n*-hexane shown in Figure 8.9*a*. As we perform a fractional distillation process analogous to that described for methanol–water (Figures 8.5*b,c*), we become limited by the azeotrope. A liquid at the azeotrope, $x_a = 0.76$, vaporizes to the same composition. Hence, an azeotrope causes a pinch point in the

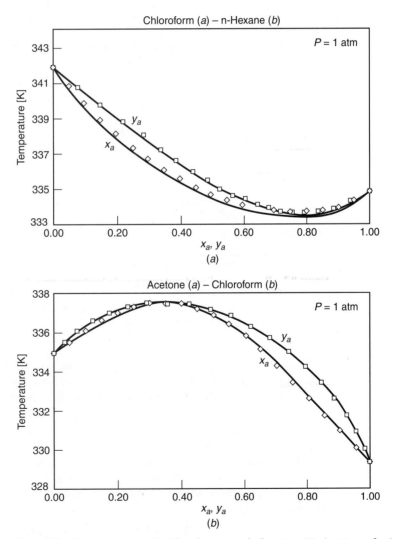

Figure 8.8 Temperature vs. liquid and vapor mole fractions (*Txy*) at 1 atm for (*a*) chloroform (*a*)–*n*-hexane (*b*), which exhibits a minimum boiling azeotrope, and (*b*) acetone (*a*)–chloroform (*b*), which exhibits a maximum boiling azeotrope.

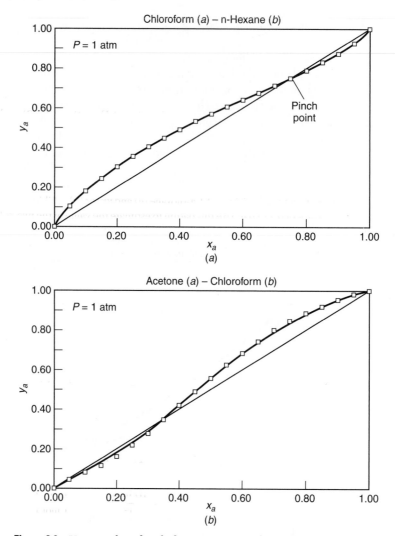

Azeotrope when $x_a = y_a$

Figure 8.9 Vapor vs. liquid mole fractions at 1 atm for (*a*) chloroform (*a*)–*n*-hexane (*b*) and (*b*) acetone (*a*)–chloroform (*b*). A pinch point in the distillation process is illustrated.

Why is this? →

distillation process. Since the vapor and liquid have equal compositions, no further separation is possible. Similarly, the acetone–chloroform system shown in Figure 8.9*b* exhibits a pinch point that disrupts purification by distillation. Thus, azeotropes pose a technical challenge to separation processes. It is possible to shift the azeotrope by large changes in *P* or *T* and, therefore, get around the pinch point or to break the azeotrope by the addition of another component.

While azeotropes are undesirable from a processing point of view, we can take advantage of the phenomena in obtaining parameters for models of g^E. Since the vapor and liquid compositions are equal *at the azeotrope*, our condition for equilibrium for species *a*, Equation (8.15), becomes

$$P = \gamma_a P_a^{sat} \qquad \text{at the azeotrope} \qquad (8.22)$$

Since both the saturation pressure and the azeotrope pressure are experimentally available, Equation (8.22) provides a simple measure of the activity coefficient at the azeotropic composition. Moreover, if we write the corresponding equation for species b, we get

$$P = \gamma_b P_b^{\text{sat}} \qquad \text{at the azeotrope} \qquad (8.23)$$

Example 8.7 illustrates the use of Equations (8.22) and (8.23) to fit the two-suffix Margules equation. It is sometimes convenient to set Equations (8.22) and (8.23) equal to get

$$\frac{\gamma_a}{\gamma_b} = \frac{P_b^{\text{sat}}}{P_a^{\text{sat}}} \qquad (8.24)$$

▶ **EXAMPLE 8.7**
Use of Azeotropic
Data to Calculate
Activity Coefficient
Model Parameters

At 50°C, a binary mixture of 1,4-dioxane (a) and water (b) exhibits an azeotrope at $x_a = 0.554$ and a pressure of 0.223 bar. Use this datum to estimate the value of the two-suffix Margules parameter A.

SOLUTION We need to determine the saturation pressures of each species. From the saturation pressure calculator in ThermoSolver, $P_a^{\text{sat}} = 0.156$ [bar] and from the steam tables, $P_b^{\text{sat}} = 0.124$ [bar]. From Equations (8.21) and (8.22):

$$\gamma_a = \frac{P}{P_a^{\text{sat}}} = \frac{0.223}{0.156} = 1.43$$

so, rewriting Equation (7.123), we get

$$A = \frac{RT\ln \gamma_a}{x_b^2} = 4826 \left[\frac{\text{J}}{\text{mol}} \right]$$

We can analogously use the azeotrope datum to solve for A in terms of species b. In this case,

$$\gamma_b = \frac{P}{P_b^{\text{sat}}} = \frac{0.223}{0.124} = 1.81$$

so

$$A = \frac{RT\ln \gamma_b}{x_a^2} = 5174 \left[\frac{\text{J}}{\text{mol}} \right]$$

Our best estimate, to two significant figures, would be obtained from averaging these two values to get:

$$A = 5.0 \text{ [kJ/mol]}$$

Since the value for A based on species a is close to that from species b, the system is reasonably symmetric and we can use the two-suffix Margules equation. Alternatively, we could use an asymmetric activity coefficient model; the van Laar equation is commonly used for azeotropes. ◀

▶ **EXAMPLE 8.8**
Thermodynamic
Property Restraints
at an Azeotrope

Show that the composition of the vapor and liquid phases must be equal at an azeotrope.

SOLUTION Consider a binary mixture of a and b in vapor–liquid equilibrium. Applying Equation (6.94), we get

$$-\frac{\overline{H}_a^v}{T^2}dT + \frac{\overline{V}_a^v}{T}dP + \frac{1}{T}\left[\frac{\partial \mu_a^v}{\partial y_a}\right]_{T,P} dy_a = -\frac{\overline{H}_a^l}{T^2}dT + \frac{\overline{V}_a^l}{T}dP + \frac{1}{T}\left[\frac{\partial \mu_a^l}{\partial x_a}\right]_{T,P} dx_a \qquad (\text{E8.8A})$$

$$-\frac{\overline{H}_b^v}{T^2}dT + \frac{\overline{V}_b^v}{T}dP + \frac{1}{T}\left[\frac{\partial \mu_b^v}{\partial y_a}\right]_{T,P} dy_a = -\frac{\overline{H}_b^l}{T^2}dT + \frac{\overline{V}_b^l}{T}dP + \frac{1}{T}\left[\frac{\partial \mu_b^l}{\partial x_a}\right]_{T,P} dx_a \qquad (\text{E8.8B})$$

Upon rearranging Equation (E8.8A), we get

$$\left(\frac{\partial \mu_a^l}{\partial x_a}\right)_{T,P} dx_a - \left(\frac{\partial \mu_a^v}{\partial y_a}\right)_{T,P} dy_a = \frac{1}{T}\left(\overline{H}_a^l - \overline{H}_a^v\right) dT - \left(\overline{V}_a^l - \overline{V}_a^v\right) dP \qquad \text{(E8.8C)}$$

Similarly for component b:

$$\left(\frac{\partial \mu_b^l}{\partial x_a}\right)_{T,P} dx_a - \left(\frac{\partial \mu_b^v}{\partial y_a}\right)_{T,P} dy_a = \frac{1}{T}\left(\overline{H}_b^l - \overline{H}_b^v\right) dT - \left(\overline{V}_b^l - \overline{V}_b^v\right) dP \qquad \text{(E8.8D)}$$

The Gibbs–Duhem equation places the following constraint on the chemical potentials of a and b:

$$x_a \frac{\partial \mu_a^l}{\partial x_a} + (1 - x_a)\frac{\partial \mu_b^l}{\partial x_a} = 0 \qquad \text{(E8.8E)}$$

and for the vapor phase, the analogous relation

$$y_a \frac{\partial \mu_a^v}{\partial y_a} + (1 - y_a)\frac{\partial \mu_b^v}{\partial y_a} = 0 \qquad \text{(E8.8F)}$$

Now, if we multiply Equation (E8.8C) by y_a and Equation (E8.8D) by $1 - y_a$ and add them together, Equation (E8.8F) allows for the elimination of μ_a^v and μ_b^v. Finally, applying Equation (E8.8E), after all the algebraic dust settles:

$$\left(\frac{y_a - x_a}{1 - x_a}\right)\frac{\partial \mu_a^l}{\partial x_a} dx_a = \left\{\frac{y_a}{T}\left(\overline{H}_a^l - \overline{H}_a^v\right) + \frac{(1 - y_a)}{T}\left(\overline{H}_b^l - \overline{H}_b^v\right)\right\} dT$$

$$- \left\{y_a\left(\overline{V}_a^l - \overline{V}_a^v\right) + (1 - y_a)\left(\overline{V}_b^l - \overline{V}_b^v\right)\right\} dP \qquad \text{(E8.8G)}$$

For Equation (E8.8G) to be true, in general, when

$$\left(\frac{\partial T}{\partial x_a}\right)_P = 0$$

or

$$\left(\frac{\partial P}{\partial x_a}\right)_T = 0$$

then

$$x_a = y_a!!$$

Hence, when a system goes through an extreme in temperature with respect to liquid-phase mole fraction at constant pressure, or an extreme in pressure with respect to liquid-phase mole fraction at constant temperature, the mole fractions in each phase must be equal. This condition defines an azeotrope. ◀

Fitting Activity Coefficient Models with VLE Data

In Section 7.4, we learned how to quantify liquid-phase nonideality by using models for g^E. Table 7.1 summarizes some commonly used models. These models allow us to come up with expressions for the activity coefficient as a function of composition. The effectiveness of such models rests on our ability to accurately assign values to the model parameters. For example, to use the two-suffix Margules equation effectively, we must have a good representation of A. While it is possible to estimate the model parameters based on limited data (see Examples 7.10 and 8.7), a set of VLE data over the entire composition range provides a more precise estimate. To obtain the best choice for the value of a model parameter, we would like to make use of all the experimental data we have available. In this way, we minimize the inherent error associated with experimentation. We can achieve this objective in several ways. We first look at how to obtain model parameters using **objective functions**. This treatment is general and can be applied to any model equation. Next, we look at how we can rewrite specific model equations to obtain model parameters through averages and linear regression. This approach depends on our ability to successfully manipulate the specific form of the model equation we are using; it cannot be applied to more complicated model equations.

An *objective function* is written in terms of the difference between the calculated value of a given property and the experimental value of the same property. This difference is termed the *residual*. We can change the calculated value by adjusting the model parameters we are trying to determine. To find the best choice of modal parameters, we minimize the sum of the square of the residual over all i measured points. Several choices of variables upon which to base the objective function can be used. For example, we can create an *objective function* based on pressure, OF_P, as

$$OF_P = \sum \left(P_{\mathrm{exp}} - P_{\mathrm{calc}}\right)_i^2 \tag{8.25}$$

where P_{exp} are the experimental values of pressure and P_{calc} are the values calculated using the activity coefficient model. In this way, we compare the value of pressure calculated using the model parameters with the measured value for pressure at every experimental point. By squaring the quantity in Equation (8.25), all numbers become positive, so that large errors in one direction cannot cancel large errors in the other direction. The square also increases the relative importance of calculated values that are farther away from the measured pressure. We now find the value for the model parameters at which the *objective function* 8.25 is a minimum. By minimizing the *objective function*, the parameter that gives the best overall fit is determined. Other common *objective functions* are based on minimizing the excess Gibbs energy:

$$OF_{g^E} = \sum \left(g_{\mathrm{exp}}^E - g_{\mathrm{calc}}^E\right)_i^2 \tag{8.26}$$

or the individual activity coefficients. For example, for a binary mixture of species a and b, we get

$$OF_\gamma = \sum \left[\left(\frac{\gamma_a - \gamma_a^{\mathrm{calc}}}{\gamma_a}\right)^2 + \left(\frac{\gamma_b - \gamma_b^{\mathrm{calc}}}{\gamma_b}\right)^2\right]_i \tag{8.27}$$

The performance of the different model parameters reported in Table 7.2 were evaluated with OF_γ.

In Examples 8.9 and 8.10, we explore different ways to fit model parameters of the two-suffix and three-suffix Margules equations, respectively, using experimental VLE data. We will use an entire data set to find the best value for the two-suffix Margules parameter A or the three-suffix Margules parameters A and B. When a model equation can be written in a linear form, a least-squares linear regression can be employed to determine the model parameters. Example 8.11 uses this method on the same data as used in Examples 8.9 and 8.10. This latter method is restricted to simpler activity coefficient models that can be written in linear form.

▶ **EXAMPLE 8.9**
Calculation of Best Fit of *A* to Data Using Objective Functions

Liquid–vapor equilibrium data have been collected for a binary system of benzene (1)–cyclohexane (2) at 10°C. Mole fraction of liquid and vapor vs. total pressure are reported in Table E8.9A.[5] From these data, determine the value of the two-suffix Margules parameter *A*.

SOLUTION Since the pressures are low, we assume ideal gas and no Poynting correction. Thus, we can relate the pressure to the mole fraction of benzene using Equation (8.19):

$$P = x_1\gamma_1 P_1^{\text{sat}} + (1 - x_1)\gamma_2 P_2^{\text{sat}} \tag{E8.9A}$$

We can use the two-suffix Margules equation to write the activity coefficients in terms of the Margules parameter, *A*. Applying Equations (7.123) and (7.124) gives

$$\ln \gamma_1 = \frac{A}{RT}(1 - x_1)^2 \tag{7.120}$$

Similarly,

$$\ln \gamma_2 = \frac{A}{RT}x_1^2 \tag{7.121}$$

Substituting Equations (7.120) and (7.121) into Equation (E8.9A) gives

$$P_{\text{calc}} = x_1\exp\left[\frac{A}{RT}(1 - x_1)^2\right]P_1^{\text{sat}} + (1 - x_1)\exp\left[\frac{A}{RT}x_1^2\right]P_2^{\text{sat}} \tag{E8.9B}$$

We can compare the value calculated by Equation (E8.9B) with the value for pressure at every experimental point. We then take the square of this difference and sum over all the data to create the *objective function*, OF_P:

$$OF_P = \sum \left(P_{\text{exp}} - P_{\text{calc}}\right)_i^2 \tag{E8.9C}$$

where P_{exp} are the experimental values of pressure reported in Table E8.9A. We find the value for the parameter *A* in Equation (E8.9B) at which the objective function (E8.9C) is a minimum. Table E8.9B summarizes the results that minimize the deviation in pressure for the data in Table E8.9A. The minimum was found for

$$A = 1401 \left[\text{J/mol}\right]$$

TABLE E8.9A Measured *Px* Data for Benzene (1)–Cyclohexane (2) at 10°C

x_1	y_1	P [Pa]
0	0	6344
0.0610	0.0953	6590
0.2149	0.2710	6980
0.3187	0.3600	7140
0.4320	0.4453	7171
0.5246	0.5106	7216
0.6117	0.5735	7140
0.7265	0.6626	6974
0.8040	0.7312	6845
0.8830	0.8200	6617
0.8999	0.8382	6557
1	1	6073

[5] J. Gmehling, U. Onken, and W. Arlt, *Vapor–Liquid Equilibrium Data Collection* (multiple volumes) (Frankfurt: DECHEMA, 1977–1980).

TABLE E8.9B Minimization of OF_P for the Data in Table E8.9A

x_1	P [Pa]	$\gamma_{1,\text{calc}}$	$\gamma_{2,\text{calc}}$	P_{calc} [Pa]	P^2_{err}
0	6344	1.81	1	6344	0
0.061	6590	1.69	1.00	6595	25
0.2149	6980	1.44	1.03	7001	456
0.3187	7140	1.32	1.06	7141	1
0.432	7171	1.21	1.12	7203	1086
0.5246	7216	1.14	1.18	7195	430
0.6117	7140	1.09	1.25	7139	0
0.7265	6974	1.05	1.37	6986	144
0.804	6845	1.02	1.47	6821	589
0.883	6617	1.01	1.59	6586	984
0.8999	6557	1.01	1.62	6525	1026
1	6073	1	1.81	6073	0
				Sum	4740

TABLE E8.9C Comparison of Experimental Vapor-Phase Composition with Those Predicted by the Two-Suffix Margules Equation

y_1	$y_{1,\text{calc}}$	% Difference
0	0	
0.0953	0.0948	0.49%
0.2710	0.2688	0.80%
0.3600	0.3571	0.80%
0.4453	0.4411	0.93%
0.5106	0.5064	0.82%
0.5735	0.5691	0.77%
0.6626	0.6602	0.36%
0.7312	0.7324	−0.16%
0.8200	0.8209	−0.11%
0.8382	0.8426	−0.52%
1	1	

Table E8.9C compares the results to measured vapor mole fractions. The calculated values are always within 1% of the measured values.

The other objective functions described above give values as follows:

$$OF_{g^E} = \sum \left(g^E_{\text{exp}} - g^E_{\text{calc}} \right)^2_i \qquad \text{gives} \qquad A = 1399 \, [\text{J/mol}]$$

and
$$OF_{\gamma} = \sum \left[\left(\frac{\gamma_1 - \gamma_1^{\text{calc}}}{\gamma_1} \right)^2 + \left(\frac{\gamma_2 - \gamma_2^{\text{calc}}}{\gamma_2} \right)^2 \right]_i \qquad \text{gives} \qquad A = 1424 \, [\text{J/mol}]$$

All three choices give values for A that are relatively close. ◀

▶ **EXAMPLE 8.10**
Calculation of Best
Fit of A and B to
Data Using Objective
Functions

Calculate the three-suffix Margules parameters, A and B, for the system of Example 8.9.

SOLUTION As in Example 8.9, we start with the expression for pressure given by

$$P = x_1 \gamma_1 P_1^{sat} + (1 - x_1) \gamma_2 P_2^{sat} \tag{E8.10A}$$

However, we now use the three-suffix Margules equation to write the activity coefficients in terms of the Margules parameters, A and B. Applying the expressions in Table 7.1:

$$\ln \gamma_1 = \frac{(A + 3B)}{RT} x_2^2 - \frac{4B}{RT} x_2^3 \tag{E8.10B}$$

Similarly,

$$\ln \gamma_2 = \frac{(A - 3B)}{RT} x_1^2 + \frac{4B}{RT} x_1^3 \tag{E8.10C}$$

Using the values given by Equations (E8.10B) and (E8.10C) in Equation (E8.10A), we get

$$P_{calc} = x_1 \exp\left[\frac{(A + 3B)}{RT} x_2^2 - \frac{4B}{RT} x_2^3 \right] P_1^{sat} + (1 - x_1) \exp\left[\frac{(A - 3B)}{RT} x_1^2 + \frac{4B}{RT} x_1^3 \right] P_2^{sat}$$

If we minimize the objective function in pressure, we get

$$OF_P = \sum \left(P_{exp} - P_{calc} \right)^2 = 2509 \tag{E8.10D}$$

when

$$A = 1397 \, [\text{J/mol}] \quad \text{and} \quad B = 69 \, [\text{J/mol}]$$

Since the value for A is so much greater than that for B, the system is adequately described by the simpler two-suffix Margules equation. ◀

▶ **EXAMPLE 8.11**
Reestimate the Two-and
Three-Suffix Margules
Parameters by the
Method of Linear
Regression

Determine the model parameters for the two-suffix and the three-suffix Margules equations using the liquid–vapor equilibrium data for a binary system of benzene (1)–cyclohexane (2) at 10°C. Mole fractions of liquid vs. total pressure are reported in Table E8.9A.

SOLUTION The excess Gibbs energy can be written:

$$g^E = RT[x_1 \ln \gamma_1 + x_2 \ln \gamma_2] = RT \left[x_1 \ln \left(\frac{y_1 P}{x_1 P_1^{sat}} \right) + x_2 \ln \left(\frac{y_2 P}{x_2 P_2^{sat}} \right) \right] \tag{E8.11A}$$

where we have rearranged Equation (8.14) to give

$$\gamma_i = \frac{y_i P}{x_i P_i^{sat}}$$

and then substituted for γ_i. Thus, we can use the experimental data from Table E8.9A to obtain values for g^E at every experimental point. We can rewrite the two-suffix and three-suffix Margules equations by dividing g^E by $(x_1 x_2)$. The two-suffix Margules equation becomes

$$\frac{g^E}{x_1 x_2} = A \tag{E8.11B}$$

Thus, the average of $g^E/x_1 x_2$ over all the data will give the best prediction for the value of A. The three-suffix Margules equation becomes

$$\frac{g^E}{x_1 x_2} = A + B(x_1 - x_2) \tag{E8.11C}$$

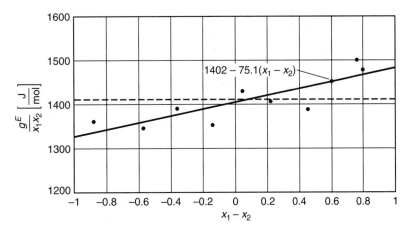

Figure E8.11 Fit of $g^E/x_1 x_2$ vs. $x_1 - x_2$ for benzene (1)–cyclohexane (2) at 10°C to straight line.

A plot of $g^E/x_1 x_2$ vs. $x_1 - x_2$ should give a straight line with slope B and intercept A. The data from Table E8.9A are plotted in this form in Figure E8.11. The average of the data gives the value for the two-suffix Margules parameter:

$$A = 1409 \, [\text{J/mol}]$$

This value is close to those obtained by the *objective functions* in Example 8.9. Linear regression gives the best-fit line to be

$$\frac{g^E}{x_1 x_2} = 1402 - 75.1(x_1 - x_2) \, [\text{J/mol}] \qquad \text{(E8.11D)}$$

Hence, we get

$$A = 1402 \, [\text{J/mol}] \qquad \text{and} \qquad B = 75.1 \, [\text{J/mol}]$$

These values are close to those obtained in Example 8.10.

Solubility of Gases in Liquids

Another important class of VLE problems addresses the solubility of gases in liquids. For example, gases dissolved in seawater are crucial to the biology of marine species. Fish require dissolved oxygen and give off carbon dioxide, while algae performing photosynthesis consume CO_2 and emit O_2. Suppose we wish to calculate the amount of oxygen dissolved in water for a system at 1 bar and 25°C. Pure O_2 has a critical temperature of 154.6 K and exists as a supercritical fluid at the temperature of the system; thus, pure O_2 cannot be a liquid at 25°C. Consequently, it is problematic to calculate its pure species fugacity in the liquid phase and apply a Lewis/Randall reference state. Alternatively, we can use Henry's law to express the liquid fugacity for O_2. In general, it is common to use a Henry's law reference state for a species when the system temperature is well above that species' critical temperature.[6] In this system, the dissolved gas, O_2, is referred to as the solute, while H_2O is termed the solvent.

---

[6] If the system temperature is not too far above that of the critical temperature, an alternative approach is given by extrapolating for a hypothetical saturation pressure using the Clausius–Clapeyron equation and then using the Lewis/Randall reference state.

TABLE 8.1   Henry's Law Constants for Various Gases in Water at 25°C

Gas	$\mathcal{H}_i$ [bar]
Ar	35,987.9
Br$_2$	74,686.8
H$_2$	70,381.1
N$_2$	87,365.0
O$_2$	44,253.9
H$_2$S	54,991.8
CO	58,487.0
CO$_2$	1,651.9
CH$_4$	41,675.8
C$_2$H$_2$	1,342.2
C$_2$H$_4$	11,522.0
C$_2$H$_6$	30,525.9

Source: Modified from E. W. Washburn (ed.), *International Critical Tables* (Vol. III) (New York: McGraw-Hill, 1928).

TABLE 8.2   Henry's Constants in [bar] of H$_2$, N$_2$, O$_2$, CO, and CO$_2$ in Four Different Liquids at 25°C

	H$_2$	N$_2$	O$_2$	CO	CO$_2$
C$_6$H$_6$	3,657.4	2,386.6	1,554.9	1,620.6	114.1
CS$_2$	10,865.9	6,961.9		4,907.3	469.0
CH$_3$OH	6,425.0	4,293.2	3,179.4	3,106.7	158.3
C$_3$H$_6$O	4,373.4	2,288.0	1,478.9	1,502.1	53.1

Source: Modified from E. W. Washburn (ed.), *International Critical Tables* (Vol. III) (New York: McGraw-Hill, 1928).

Values of the Henry's constant for different gases in water at 25°C are given in Table 8.1. Henry's constant for gases in other solvents at 25°C are given in Table 8.2. The Henry's constant is indicative of the unlike *i-j* interaction. Hence, its value depends not only of the identity of the solute but also on the solvent. For example, inspection of Table 8.2 shows that the magnitude of Henry's constant for N$_2$ in C$_6$H$_6$ is roughly three times less than that of CS$_2$. We expect this result, since the polarizability of C$_6$H$_6$ is greater and, therefore, the London interactions are stronger. The stronger unlike attractive interaction leads to a lower "tendency to escape" and, therefore, a lower Henry's law constant.

The values for Henry's law constants are often reported at 25°C and 1 bar. If we are interested in a system at significantly different temperature or pressure, we must correct the value of $\mathcal{H}_i$. Temperature and pressure dependencies of Henry's constant were developed in Section 7.4 as

$$\left(\frac{\partial \ln \mathcal{H}_i}{\partial P}\right)_T = \frac{\overline{V}_i^\infty}{RT} \tag{7.91}$$

and

$$\left(\frac{\partial \ln \mathcal{H}_i}{\partial T}\right)_P = \frac{h_i^v - \overline{H}_i^\infty}{RT^2} \tag{7.92}$$

Equations (7.91) and (7.92) can be used to correct literature values of $\mathcal{H}_i$ for pressure and temperature. Henry's constants usually increase with temperature. We can rearrange Equation (7.92) to get

$$\left(\frac{\partial \ln \mathcal{H}_i}{\partial (1/T)}\right)_P = \frac{\overline{H}_i^\infty - h_i^v}{R} \tag{8.28}$$

Equation (8.28) suggests that if $\overline{H}_i^\infty - h_i^v$ is constant, a plot of $\ln \mathcal{H}_i$ vs. $1/T$ should result in a straight line. Figure 8.10 plots values for Henry's constants of N_2 and O_2 in H_2O as a function of temperature. Figure 8.10a shows that the Henry's constants increase with temperature and then plateau. Figure 8.10b shows a plot of the natural logarithm of Henry's constant vs. $1/T$. In this case, the plot is not linear, suggesting that $\overline{H}_i^\infty - h_i^v$ changes with temperature. This result can be attributed to the effect of hydrogen bonds in water. Problem 8.23 illustrates a case that is linear: O_2 in benzene.

To solve for equilibrium composition of the solute using Henry's law, we again set the fugacity of the vapor equal to the fugacity of the liquid:

$$\hat{f}_i^v = \hat{f}_i^l \tag{8.2}$$

Applying the Henry's law reference state for species i gives:

$$y_i \hat{\varphi}_i^v P = x_i \gamma_i^{\text{Henry's}} \mathcal{H}_i \tag{8.29}$$

Equation (8.29) is true in general and can always be applied. However, we can often apply simplifying approximations.

We will start with a binary system of dissolved gas (solute) a in liquid (solvent) b. We first assume that the gas mixture is well represented by the **ideal gas law**. In the limit of sparing solubility of gas a in the liquid, the liquid consists of almost all b. Thus, solute species a acts ideally in the Henry's law limit; that is, the behavior is dominated by a-b interactions. On the other hand, solvent species b is almost all pure and is ideal in the Lewis/Randall limit (all b-b interactions). With these assumptions, we apply Equation (8.29) to species a to give

$$y_a P = x_a \mathcal{H}_a \tag{8.30}$$

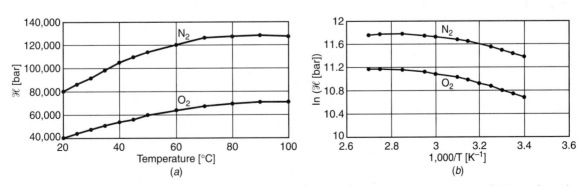

Figure 8.10 (a) Henry's law constants for N_2 and O_2 in H_2O vs. temperature. (b) Natural logarithm of Henry's constant vs. $1/T$.

For species b, we have

$$y_b P = x_b P_b^{\text{sat}} \tag{8.31}$$

where we have approximated the pure species fugacity of species b with its saturation pressure. Adding together Equations (8.30) and (8.31) gives

$$P = x_a \mathcal{H}_a + x_b P_b^{\text{sat}} \tag{8.32}$$

and substituting the value for pressure given by Equation (8.32) into Equations (8.30) and (8.31) gives:

$$y_a = \frac{x_a \mathcal{H}_a}{x_a \mathcal{H}_a + x_b P_b^{\text{sat}}} \tag{8.33}$$

and

$$y_b = \frac{x_b P_b^{\text{sat}}}{x_a \mathcal{H}_a + x_b P_b^{\text{sat}}} \tag{8.34}$$

respectively. Equation (8.30) works, in most systems, up to roughly $x_a = 0.03$ and, in some cases, well beyond. Note the similarities between Equations (8.32) through (8.34) and Equations (8.10) and (8.11).

If there is enough of the lighter component in the liquid, both a-a and a-b interactions become important in describing the fugacity of the liquid phase, and we must account for the nonideality in the liquid. In this case, we can use Equation (8.29) to get the following expressions:

$$P = x_a \gamma_a^{\text{Henry's}} \mathcal{H}_a + x_b \gamma_b P_b^{\text{sat}} \tag{8.35}$$

$$y_a = \frac{x_a \gamma_a^{\text{Henry's}} \mathcal{H}_a}{x_a \gamma_a^{\text{Henry's}} \mathcal{H}_a + x_b \gamma_b P_b^{\text{sat}}} \tag{8.36}$$

and

$$y_b = \frac{x_b \gamma_b P_b^{\text{sat}}}{x_a \gamma_a^{\text{Henry's}} \mathcal{H}_a + x_b \gamma_b P_b^{\text{sat}}} \tag{8.37}$$

Similarly, for *high pressures and ideal liquid*, the vapor is no longer an ideal gas. In this case, our Henry's law expressions become:

$$y_a \hat{\varphi}_a P = x_a \mathcal{H}_a \tag{8.38}$$

and

$$y_b \hat{\varphi}_b P = x_b f_b \tag{8.39}$$

To get the Henry's law constant in Equation (8.38) at high pressure, we must correct $\mathcal{H}$ for P. Integrating Equation (7.91) gives

$$\mathcal{H}_a^{\text{at } P} = \mathcal{H}_a^{\text{at 1 bar}} \exp\left[\int_{1\,\text{bar}}^{P} \frac{\overline{V}_a^{\infty}}{RT} dP \right] \tag{8.40}$$

Similarly, we apply the Poynting correction to get a value for the pure species fugacity of species b:

$$f_b^l = \varphi_b^{\text{sat}} P_b^{\text{sat}} \exp\left[\int_{P_b^{\text{sat}}}^{P} \frac{v_b^l}{RT} dP \right] \tag{8.41}$$

Moreover, since the vapor is mostly solute, we can apply the Lewis fugacity rule to get:

$$\hat{\varphi}_b = \varphi_b \tag{8.42}$$

Substituting Equation (8.40) into (8.38) gives

$$y_a \hat{\varphi}_a P = x_a \mathcal{H}_a^{\text{at 1 bar}} \exp\left[\int_{1\,\text{bar}}^{P} \frac{\overline{V}_a^{\infty}}{RT} dP\right] \tag{8.43}$$

Substituting Equations (8.41) and (8.42) into (8.39) gives

$$y_b \varphi_b P = x_b \varphi_b^{\text{sat}} P_b^{\text{sat}} \exp\left[\int_{P_b^{\text{sat}}}^{P} \frac{v_b^l}{RT} dP\right] \tag{8.44}$$

For the case of *nonideal liquids at high pressure*, we must include activity coefficients in Equations (8.43) and (8.44) to give

$$y_a \varphi_a P = x_a \gamma_a^{\text{Henry's}} \mathcal{H}_a^{\text{at 1 bar}} \exp\left[\int_{1\,\text{bar}}^{P} \frac{\overline{V}_a^{\infty}}{RT} dP\right] \tag{8.45}$$

and

$$y_b \hat{\varphi}_b P = x_b \gamma_b \varphi_b^{\text{sat}} P_b^{\text{sat}} \exp\left[\int_{P_b^{\text{sat}}}^{P} \frac{v_b^l}{RT} dP\right] \tag{8.46}$$

For multicomponent mixtures, ideal mixing gives the following mixing rule for the Henry's constant of species a:

$$\ln \mathcal{H}_a = \sum_j x_j \mathcal{H}_{a,j} \tag{8.47}$$

where $\mathcal{H}_{a,j}$ is the Henry's constant of solute species a in solvent j.

▶ **EXAMPLE 8.12**
Calculation of Dissolved O_2 in H_2O

Calculate the solubility of O_2 from air in the atmosphere in equilibrium with liquid H_2O at 25°C. Report the answer in mole fraction and in molarity.

SOLUTION The partial pressure of oxygen in air is approximately

$$p_{O_2} = y_{O_2} P = 0.21\,[\text{bar}]$$

Using the value for the Henry's law constant in water given in Table 8.1 in Equation (8.30) gives

$$x_{O_2} = \frac{y_{O_2} P}{\mathcal{H}_{O_2}} = \frac{0.21\,[\text{bar}]}{44,253.9\,[\text{bar}]} = 4.75 \times 10^{-6}$$

The concentration of oxygen, $[O_2]$ in units of molality is defined as the number of moles of solute (O_2) per liter of solution, that is,

$$[O_2] = \frac{n_{O_2}}{V}$$

We can relate the number of moles of oxygen to its mole fraction as follows:

$$x_{O_2} = \frac{n_{O_2}}{n_{O_2} + n_{H_2O}} = \frac{n_{O_2}}{n_{H_2O}}$$

where the denominator was simplified since $n_{O_2} << n_{H_2O}$. Since the amount of dissolved gas is so small, we can replace the solution volume with the pure species volume. Solving for the molarity, we get

$$[O_2] = x_{O_2} \frac{n_{H_2O}}{V} = x_{O_2} \frac{n_{H_2O}}{V_{H_2O}} = \frac{x_{O_2}}{v_{H_2O}}$$

$$= \left(\frac{4.75 \times 10^{-6}}{0.001} \left[\frac{kg}{m^3} \right] \right) \left(\frac{1}{0.018} \left[\frac{mol}{kg} \right] \right) \left(0.001 \left[\frac{m^3}{L} \right] \right) = 2.63 \times 10^{-4} \ [M] \quad ◀$$

▶ **EXAMPLE 8.13**
Henry's Law Problem at High Pressure

Determine the solubility of N_2 in H_2O at 300 bar and 25°C. Take

$$\overline{V}_{N_2}^{\infty} = 3.3 \times 10^{-5} \ [m^3/mol]$$

SOLUTION We will assume the solubility is small enough that the liquid solution remains ideal. Solving Equation (8.38) for the mole fraction of N_2 gives

$$x_{N_2} = \frac{y_{N_2} \hat{\varphi}_{N_2} P}{\mathcal{H}_{N_2}} \tag{E8.13A}$$

At 25°C, the vapor pressure of water is small, so we assume

$$y_{N_2} \approx 1$$

To find the Henry's law constant, we must correct the value in Table 8.1 for pressure. Applying Equation (8.40) and assuming that the partial molar volume at infinite dilution does not change with pressure, we get

$$\mathcal{H}_{N_2}^{at \ P} = \mathcal{H}_{N_2}^{at \ 1 \ bar} \exp \left[\int_{1 \ bar}^{P} \frac{\overline{V}_{N_2}^{\infty}}{RT} dP \right] = \mathcal{H}_{N_2}^{at \ 1 \ bar} \exp \left[\frac{\overline{V}_{N_2}^{\infty}}{RT} (P - 1) \right] \tag{E8.13B}$$

Plugging in numbers to Equation (E8.13B) gives

$$\mathcal{H}_{N_2}^{300 \ bar} = (87,365 \ [bar]) \exp \left[\frac{(3.3 \times 10^{-5} \ [m^3/mol]) (299 \times 10^5 \ [J/m^3])}{(8.314 \ [J/(mol \ K)]) (298 \ [K])} \right] = 130,106 \ [bar] \tag{E8.13C}$$

The Henry's constant has increased by approximately 50% at the higher pressure. Finally, to solve for the fugacity coefficient, we assume the Lewis fugacity rule, apply the generalized correlations, and use the values given in Appendix C. The reduced pressure and temperature are as follows:

$$P_r = \frac{P}{P_c} = \frac{300}{33.8} = 8.88 \quad and \quad T_r = \frac{T}{T_c} = \frac{298}{126.2} = 2.36$$

Looking up the acentric factor from Appendix A.1 gives

$$\omega = 0.039$$

So that

$$\log \varphi_{N_2} = \log \varphi^{(0)} + \omega \log \varphi^{(1)} = 0.013 + 0.039(0.210) = 0.021$$

or

$$\varphi_{N_2} = 1.05 \tag{E8.13D}$$

Finally, plugging in values from Equations (E8.13C) and (E8.13D) gives

$$x_{N_2} = \frac{y_{N_2} \varphi_{N_2} P}{\mathcal{H}_{N_2}} = \frac{1.05 \times 300}{130,106} = 0.00242 \quad ◀$$

► 8.2 LIQUID (α)–LIQUID (β) EQUILIBRIUM: LLE

In Chapter 7, we saw that when like (a-a and b-b) interactions are significantly stronger than unlike (a-b) interactions, liquids can split into two different partially miscible phases, which we labeled α and β. They form separate phases to lower the total Gibbs energy of the system. In this case, each species i tends to equilibrate between the two phases, leading to liquid–liquid equilibrium (LLE). We can equate the fugacity of component i for each phase to solve for the equilibrium compositions.

To illustrate how to calculate the equilibrium compositions in LLE, let's consider a binary mixture of species a and b. The fugacities of species a in each liquid phase are equal:

$$\hat{f}_a^{\alpha} = \hat{f}_a^{\beta} \tag{8.48}$$

Applying Equation (7.71) for a Lewis/Randall reference state gives

$$x_a^{\alpha}\gamma_a^{\alpha}\cancel{f_a} = x_a^{\beta}\gamma_a^{\beta}\cancel{f_a} \tag{8.49}$$

and for species b:

$$x_b^{\alpha}\gamma_b^{\alpha} = x_b^{\beta}\gamma_b^{\beta} \tag{8.50}$$

where we have canceled the pure species fugacity in the liquid. To solve Equations (8.49) and (8.50), we need an activity coefficient model for g^E. We will illustrate the approach using the two-suffix Margules equation; however, the same methodology can be applied to any model from Table 7.1. If we substitute Equation (7.123) for the activity coefficient of species a into Equation (8.49), we get

$$x_a^{\alpha}\exp\left[\frac{A}{RT}\left(x_b^{\alpha}\right)^2\right] = x_a^{\beta}\exp\left[\frac{A}{RT}\left(x_b^{\beta}\right)^2\right] \tag{8.51}$$

Similarly for species b, we have

$$x_b^{\alpha}\exp\left[\frac{A}{RT}\left(x_a^{\alpha}\right)^2\right] = x_b^{\beta}\exp\left[\frac{A}{RT}\left(x_a^{\beta}\right)^2\right] \tag{8.52}$$

Additionally, the mole fractions in each phase must sum to 1, so that

$$x_a^{\alpha} + x_b^{\alpha} = 1 \tag{8.53}$$

and

$$x_a^{\beta} + x_b^{\beta} = 1 \tag{8.54}$$

Equations (8.51) through (8.54) form a set of four coupled equations that can be solved for the four unknowns: x_a^{α}, x_a^{β}, x_b^{α}, and x_b^{β}. A similar approach could be used for other models of g^E such as those described in Section 7.4, where the appropriate expressions for activity coefficients would be substituted into Equations (8.49) and (8.50).

A plot of T vs. mole fraction of species a is shown in Figure 8.11 for the case where the two-suffix Margules parameter, A, is independent of temperature. This phase diagram shows regions where only one liquid phase is present and a two-phase region. We typically make such plots at constant pressure, since A is weakly dependent on pressure.[7]

---

[7] Can you justify this statement using the thermodynamic web and physical arguments?

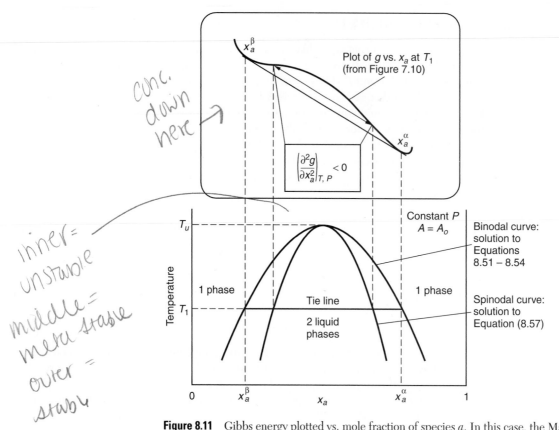

Figure 8.11 Gibbs energy plotted vs. mole fraction of species a. In this case, the Margules parameter, A, is much greater than zero. *Note*: At compositions between x_a^α and x_a^β, the system can minimize its free energy by splitting into two phases.

The curve dividing the two regions, the **binodal curve**, represents the compositions of coexisting liquid phases at any temperature. This curve can be obtained by solving the set of Equations (8.51) through (8.54). For example, at temperature T_1, any liquid with an overall composition in the region labeled "2 phases" will split into an a-rich phase, x_a^α, and a b-rich phase, x_a^β. Again, the composition of these phases can be determined from where the binodal curve intersects the horizontal tie line, and their relative amounts are given by the lever rule. The inset illustrates that the compositions of phase α and phase β occur near the two minima in Gibbs energy, where the tangent line can be drawn, as depicted in Figure 7.10b.

For a given value of A in Figure 8.11, the lower the temperature, the larger the range of partial miscibility. Partial miscibility results when energetic effects dominate entropic effects. In the Gibbs energy, the entropy term is multiplied by temperature, that is, $g = h - Ts$; all other effects being equal, at lower temperatures energetic effects become more significant relative to entropic effects, and the attractive effects of the like interactions become dominant over a larger composition range. We are more likely to have two phases as the mole fractions approach 0.5, another indication that the two-suffix Margules equation is symmetric. The value of the temperature above which the liquid mixture no longer separates into two phases at any composition is termed the **upper consulate temperature**, T_u. It is also shown on this plot.

We now wish to examine when a single phase will spontaneously split into different liquid phases. The criterion for instability of a single liquid phase is given when the curve for the total solution Gibbs energy is *concave down*. Mathematically, this is expressed by

$$\left(\frac{\partial^2 g}{\partial x_a^2}\right)_{T,P} < 0 \qquad (8.55)$$

[handwritten: unstable = likely to split into 2 phases]

Again, we will consider the case when the nonideality is described by the two-suffix Margules equation; however, in general, any model for g^E can be used. From Equation (7.126), we have:

$$g = x_a g_a + x_b g_b + RT(x_a \ln x_a + x_b \ln x_b) + A x_a x_b \qquad (7.126)$$

[handwritten: g^E must be ⊕ for minima to occur]

Differentiating twice and substituting in Equation (8.55) gives

$$\left(\frac{\partial^2 g}{\partial x_a^2}\right)_{T,P} = RT\left(\frac{1}{x_a} + \frac{1}{x_b}\right) - 2A < 0 \qquad (8.56)$$

[handwritten: $A(x)(1-x)$; $A(x - x^2)$; $1 - 2x$; -2]

This simplifies to:

$$\boxed{\frac{RT}{x_a x_b} < 2A} \qquad (8.57)$$

Equation (8.57) tells us how large the two-suffix Margules parameter A has to be for a binary mixture to be unstable and spontaneously separate into two components. The set of solutions to this equation is denoted the **spinodal** curve in Figure 8.11. The compositions denoted by the spinodal curve are different from the set of equilibrium compositions indicated by the binodal curve. At compositions between these two curves, the liquid is metastable. While it is not at its lowest state of Gibbs energy, it will not necessarily spontaneously separate into two liquid phases.

Since the binodal and spinodal curves intersect at the upper consulate temperature, we can solve Equation (8.57) for T_u. The largest value of the product, $x_a x_b$, occurs at $x_a = 0.5$. Using these values for the mole fraction, we get

$$T_u = \frac{A}{2R} \qquad (8.58)$$

Characteristic phase diagrams also are strongly dependent on the temperature dependence of the Margules parameter, A. Recall that parameter A compares the unlike *a-b* interaction to that of the like interactions and, therefore, is intimately related to the chemical nature of the species in the liquid mixture. The liquid–liquid solubility diagram shown in Figure 8.11 is typical of many real systems; however, other systems show very different behavior. Insight into how these systems behave can be obtained by considering the temperature dependence of the intermolecular interactions. If A decreases with temperature, the solubility diagram is qualitatively similar to Figure 8.11. Consider if the case in which A *increases* with temperature, as shown in Figure 8.12*a*. Inspection of Equation (8.57) shows that the mole fraction range in which the homogeneous liquid is unstable may now increase with temperature. This behavior is shown in Figure 8.12*b*. In this case, a temperature exists below which phase separation is impossible at any composition. This temperature is called the **lower consulate** temperature.

For even more complicated systems, solubility behavior as shown in Figure 8.13 has been observed. In these cases, we observe both an upper and a lower consulate temperature. In the phase diagram to the left, two phases exist in the middle temperature range

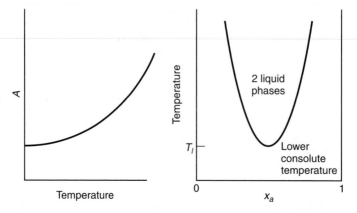

Figure 8.12 Phase stability diagram for a binary mixture described by the two-suffix Margules equation. In this case, the Margules parameter, A, increases with temperature.

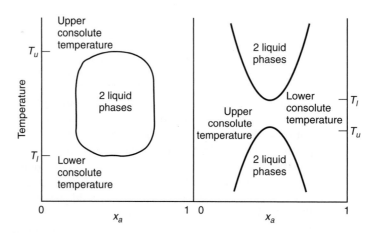

Figure 8.13 Other phase stability diagrams observed for binary mixtures.

in between the lower and upper consulate temperatures. This type of phase diagram is exhibited by mixtures of tetrahydrofuran and water or glycerol and benzyl ethyl amine. Binary mixtures of sulfur and benzene show the uncommon behavior on the right, where the upper consulate temperature is below the lower consulate temperature. Can you think of the temperature dependence of A that would give each of these characteristics? It should be noted that any of the binodal curves on these solubility diagrams can be interrupted from other phase transitions, that is, the formation of a vapor phase as temperature increases or the formation of a solid phase as temperature decreases. An example of such cases will be considered in Sections 8.3 and 8.4.

▶ **EXAMPLE 8.14**
Equilibrium Compositions Between Two Liquid Phases

Calculate the equilibrium composition of the two liquid phases in a binary mixture of methyl diethylamine (a) and water (b) at 1 bar and 20°C. The following three-suffix Margules parameters have been obtained for this binary system:

$$A = 6349 \, [\text{J/mol}] \quad \text{and} \quad B = -384 \, [\text{J/mol}]$$

SOLUTION To solve for the equilibrium mole fractions of a and b in each phase, we set the fugacities of liquid phase α equal to liquid phase β. Since the reference state fugacities are equal, we get

*each compares
one species*

$$x_a^\alpha \gamma_a^\alpha = x_a^\beta \gamma_a^\beta \qquad \text{(E8.14A)}$$

and

$$x_b^\alpha \gamma_b^\alpha = x_b^\beta \gamma_b^\beta \qquad \text{(E8.14B)}$$

We can substitute in the activity coefficient expressions for the three-suffix Margules equation, Equations (7.128) and (7.129), and use $x_b^\alpha = 1 - x_a^\alpha$ and $x_b^\beta = 1 - x_a^\beta$ to get

$$x_a^\alpha \exp\left[\frac{(A+3B)}{RT} \left(1 - x_a^\alpha\right)^2 - \frac{4B}{RT} \left(1 - x_a^\alpha\right)^3 \right] = x_a^\beta \exp\left[\frac{(A+3B)}{RT} \left(1 - x_a^\beta\right)^2 - \frac{4B}{RT} \left(1 - x_a^\beta\right)^3 \right]$$

$$\text{(E8.14C)}$$

and

$$\left(1 - x_a^\alpha\right) \exp\left[\frac{(A-3B)}{RT} \left(x_a^\alpha\right)^2 + \frac{4B}{RT} \left(x_a^\alpha\right)^3 \right] = \left(1 - x_a^\beta\right) \exp\left[\frac{(A-3B)}{RT} \left(x_a^\beta\right)^2 + \frac{4B}{RT} \left(x_a^\beta\right)^3 \right]$$

$$\text{(E8.14D)}$$

Since we know the parameters A and B, Equations (E8.14A) and (E8.14B) represent two equations with two unknowns, x_a^α and x_a^β. However, it helps to have an idea where the solution to these nonlinear equations lies so that we do not obtain the trivial answer $x_a^\alpha = x_a^\beta$. Figure E8.14A plots the quantities $x_a \gamma_a$ and $(1 - x_a)\gamma_b$ vs. x_a. The solution occurs at the two compositions of x_a where Equations (E8.14A) and (E8.14B) are simultaneously satisfied and is indicated in the figure. If you examine the figure, you will see that the solution illustrated is unique and that no other values of x_a will work. For example, if we increase the value of x_a^β, the value of x_a^α that matches $x_a \gamma_a$ occurs at larger and larger values. However, the value of x_a^α that matches $(1 - x_a)\gamma_b$ occurs at smaller and smaller values; hence, they will never meet. A similar argument shows that no solution occurs at smaller values of x_a^β either.

It may not be so straightforward to see where the solution is on a plot like Figure E8.14A. However, we can be more clever in elucidating the solution. Figure E8.14B shows such a way. In this figure, we plot $x_a \gamma_a$ vs. $(1 - x_a)\gamma_b$ as values of x_a increase from 0 to 1. For a similar reasons as we put forward in the discussion of Figure E8.14A, the solution at which equations (E8.14A) and (E8.14B) are satisfied lies at the two compositions where such a plot intersects, as shown in Figure 8.14B. We then determine the values of x_a at this point.

We can obtain the solution either directly from the graphical methods or from use of an equation solver with a reasonable first guess provided by the approximate solution suggested by

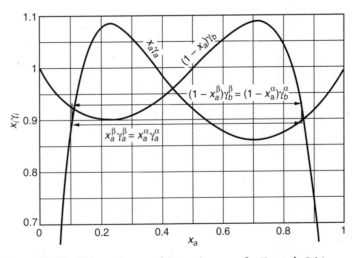

Figure E8.14A Values of $x_a \gamma_a$ and $(1 - x_a)\gamma_b$ vs. x_a for Example 8.14.

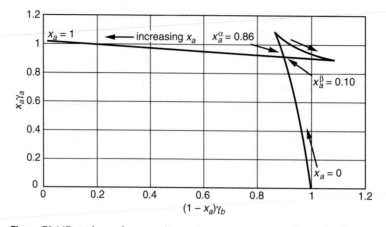

Figure E8.14B Values of $x_a \gamma_a$ vs. $(1 - x_a)\gamma_b$ as we increase values of x_a for Example 8.14. The solution is indicated where the curve crosses itself.

the graphs. In either case, we find the solution to be

$$x_a^\alpha = 0.855 \qquad \text{and} \qquad x_a^\beta = 0.101 \qquad\qquad ◀$$

▶ **EXAMPLE 8.15**
Solution to Example
8.14 by Minimization
of Gibbs Energy

Find a graphical solution to Example 8.14 by determining where the Gibbs energy of the system is at a minimum.

SOLUTION We can solve for the Gibbs energy by applying Equation (7.125) to a binary system:

$$g = (x_a g_a + x_b g_b) + RT[x_a \ln x_a + x_b \ln x_b] + g^E \qquad\qquad \text{(E8.15A)}$$

Applying the three-suffix Margules expression for g^E in Equation (E8.15A) and rearranging gives

$$g - (x_a g_a + x_b g_b) = RT[x_a \ln x_a + x_b \ln x_b] + x_a x_b[A + B(x_a - x_b)] \qquad\qquad \text{(E8.15B)}$$

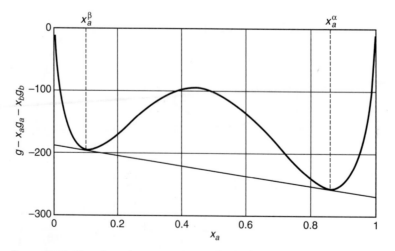

Figure E8.15 Plot of $g - (x_a g_a + x_b g_b)$ vs x_a for the binary system described in Example 8.14. The equilibrium composition is shown.

Figure E8.15 shows a plot of $g-(x_a g_a + x_b g_b)$ vs. x_a. The equilibrium compositions in the two-phase region can be found by drawing a line tangent to the minima. The compositions are found to be

$$x_a^\alpha = 0.85 \qquad \text{and} \qquad x_a^\beta = 0.10$$

Any composition of x_a between these values will lower its Gibbs energy by splitting into two phases. Inspection of Figure E8.15 shows that for $x_a < 0.10$ or $x_a > 0.85$, only one phase exists.

In Example 8.14 we solved for the equilibrium composition of the two liquid phases by setting the fugacities of each species equal. In this example, we came up with the same answer by using another approach—minimization of the Gibbs energy for the entire system. In general, we can solve equilibrium problems in thermodynamics either way. For example, we will see that both approaches are commonly taken in calculating chemical reaction equilibrium with multiple reactions (Section 9.7). We generally take the approach that is the most friendly computationally. Which approach would you favor, that of Example 8.14 or Example 8.15? ◄

► **EXAMPLE 8.16**
Instability in
Example 8.14

Apply the criterion for inherent instability of a single liquid phase to determine the composition range at which the system in Example 8.14 will spontaneously split into two phases.

SOLUTION The instability condition is given by Equation (8.55):

$$\left(\frac{\partial^2 g}{\partial x_a^2}\right)_{T,P} < 0 \tag{8.55}$$

Again, we will consider the case when the nonideality is described by the three-suffix Margules equation with parameters given in Example 8.14. We can write the Gibbs energy according to Equation (E8.15B):

$$g = x_a g_a + x_b g_b + RT(x_a \ln x_a + x_b \ln x_b) + x_a x_b [A + B(x_a - x_b)] \tag{E8.16A}$$

Differentiating Equation (E8.16A) twice gives

$$\left(\frac{\partial^2 g}{\partial x_a^2}\right)_{T,P} = RT\left(\frac{1}{x_a} + \frac{1}{x_b}\right) - 2A + 6B(x_b - x_a) < 0 \tag{E8.16B}$$

Solving Equation (E8.16B) shows the liquid is unstable between

$$0.225 < x_a < 0.706$$

This composition range is tighter than the solution to Example 8.15 and can be seen as the inflection points in Figure E8.15. ◄

► 8.3 VAPOR–LIQUID(α)–LIQUID(β) EQUILIBRIUM: VLLE

In this section, we consider the case when three phases are in equilibrium: a vapor phase and two liquid phases, α and β. A generic diagram of a system with m components in vapor–liquid–liquid (VLLE) equilibrium is shown in Figure 8.14. How does such behavior come about? Let's return to the binary mixture of a and b. Consider the case where we have both an azeotrope in VLE and liquid–liquid equilibrium (LLE). This scenario corresponds to a minimum-boiling azeotrope where the like interactions are stronger than the unlike interactions. Figure 8.15a shows the phase diagram for the case where the azeotrope and the LLE dome are clearly separated. As the system pressure decreases, however, the components can become volatile before the upper consulate temperature is reached.

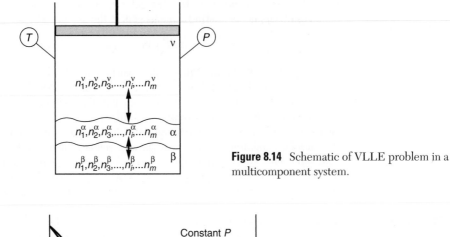

Figure 8.14 Schematic of VLLE problem in a multicomponent system.

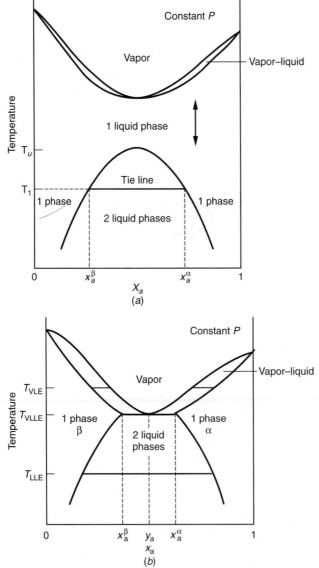

Figure 8.15 (*a*) Phase diagram exhibiting both partial miscibility and an azeotrope. (*b*) Phase diagram exhibiting VLLE.

Figure 8.15*b* shows such a case where the VLE and LLE curves intersect. Three temperatures are delineated. At the lowest temperature, T_{LLE}, three different types of phase behavior are possible. Only liquid-phase β is present at low x_a, while only liquid-phase α is present at high x_a. Two liquid phases in equilibrium manifest at intermediate composition. Their compositions are indicated by the tie line in Figure 8.15*b*. At temperature T_{VLE}, only liquid–phase β is present at low x_a, while only liquid-phase α is present at high x_a. However, as the concentration of x_a increases, a liquid β–vapor phase appears followed by only vapor and then a liquid α–vapor phase. The compositions in the two-phase regions are given by the appropriate tie lines. At the intermediate temperature T_{VLLE}, again only liquid-phase β is present at low x_a, while only liquid-phase α is present at high x_a. However, at mole fractions x_a in between these two single-phase regions, both α and β liquid phases can coexist along with the vapor. The liquid β phase is given by the composition to the left, the liquid α phase by the composition to the right, and the vapor by the point in the center, as illustrated on the *x*-axis on the diagram. Thus, in the intermediate composition region at temperature T_{VLLE}, this binary system exhibits VLLE. More complicated VLLE phase diagrams, where the vapor composition does not fall in between the two liquids, have also been observed. Can you identify the phases and compositions present in Figure 8.15*b* for a given *T* and composition of a system? Can you figure out what the *Pxy* phase diagram would look like?

To calculate the equilibrium compositions in VLLE, we set the fugacities of each species equal:

$$\hat{f}_a^v = \hat{f}_a^\alpha = \hat{f}_a^\beta \quad \text{✷} \tag{8.59}$$

and

$$\hat{f}_b^v = \hat{f}_b^\alpha = \hat{f}_b^\beta \tag{8.60}$$

To solve Equations (8.59) and (8.60), we need a model for g^E. We will illustrate the approach using the two-suffix Margules equation; however, the same methodology can be applied to any model from Table 7.1. If we are at low pressure and the vapor phase is ideal, Equation (8.59) can be written

$$y_a P = x_a^\alpha \exp\left[\frac{A}{RT}\left(x_b^\alpha\right)^2\right] P_a^{\text{sat}} = x_a^\beta \exp\left[\frac{A}{RT}\left(x_b^\beta\right)^2\right] P_a^{\text{sat}} \quad \text{✷} \tag{8.61}$$

Similarly, for species *b*, we have

$$y_b P = x_b^\alpha \exp\left[\frac{A}{RT}\left(x_a^\alpha\right)^2\right] P_b^{\text{sat}} = x_b^\beta \exp\left[\frac{A}{RT}\left(x_a^\beta\right)^2\right] P_b^{\text{sat}} \tag{8.62}$$

Additionally, the mole fractions in each phase must sum to 1, so that

$$y_a + y_b = 1 \tag{8.63}$$

$$x_a^\alpha + x_b^\alpha = 1 \tag{8.64}$$

and

$$x_a^\beta + x_b^\beta = 1 \tag{8.65}$$

The eight unknowns—$y_a, y_b, x_a^\alpha, x_b^\alpha, x_a^\beta, x_b^\beta, P$, and T—are related by seven equations in Equations (8.61) through (8.65). Thus, specifying any one of these variables constrains the other seven and fixes the state of the system. Such a conclusion is also given by the

Gibbs phase rule [see Equation (1.19) and Example 6.14]. In the general m component case specified in Figure 8.14, $m - 1$ variables must be specified.

▶ **EXAMPLE 8.17**
Calculation of
Composition in VLLE

A binary mixture exhibits vapor–liquid–liquid equilibrium at 300 K. The excess Gibbs energy is described by the two-suffix Margules equation with A = 6235 [J/mol]. Determine the composition of the three phases and the total pressure. The saturation pressures are given by $P_a^{\text{sat}} = 100$ [kPa] and $P_b^{\text{sat}} = 50$ [kPa].

SOLUTION We begin by solving for the composition of the two liquid phases at 300 K. We can use the approach of either Example 8.14 or Example 8.15. Using the former, we equate the fugacities of a and b in the liquid phases. In analogy to Equations (E8.14C) and (E8.14D), equating the liquid-phase fugacities and applying the two-suffix Margules equation gives

$$x_a^\alpha \exp\left[\frac{A}{RT}\left(1 - x_a^\alpha\right)^2\right] = x_a^\beta \exp\left[\frac{A}{RT}\left(1 - x_a^\beta\right)^2\right] \tag{E8.17A}$$

and

$$\left(1 - x_a^\alpha\right) \exp\left[\frac{A}{RT}\left(x_a^\alpha\right)^2\right] = \left(1 - x_a^\beta\right) \exp\left[\frac{A}{RT}\left(x_a^\beta\right)^2\right] \tag{E8.17B}$$

Since we know the parameter A, Equations (E8.17A) and (E8.17B) represent two equations with two unknowns, x_a^α and x_a^β. Again, it helps to have an idea where the solution to these nonlinear equations lies, so that we do not obtain the trivial answer $x_a^\alpha = x_a^\beta$. Figure E8.17 plots the quantities $x_a\gamma_a$ and $(1 - x_a)\gamma_b$ vs. x_a. The solution occurs at the two compositions of x_a where Equations (E8.17A) and (E8.17B) are simultaneously satisfied. In this case, both solutions $x_a\gamma_a$ and $(1-x_a)\gamma_b$ occur at the same value, as is indicated in the figure.

In Figure E8.17, we can see that the plot $x_a\gamma_a$ is a mirror image of $(1-x_a)\gamma_b$. This result is not surprising, since the activity coefficient model is symmetric. We can use this feature to formulate an alternative solution to Equations (E8.16A) and (E8.16B). Since the activity coefficients are symmetric, inspection of Figure E8.17 shows that we must have $x_a^\alpha = 1 - x_a^\beta$. Thus Equation (E8.17A) becomes

$$x_a^\alpha \exp\left[\frac{A}{RT}\left(1 - x_a^\alpha\right)^2\right] = \left(1 - x_a^\alpha\right) \exp\left[\frac{A}{RT}\left(x_a^\alpha\right)^2\right] \tag{E8.17C}$$

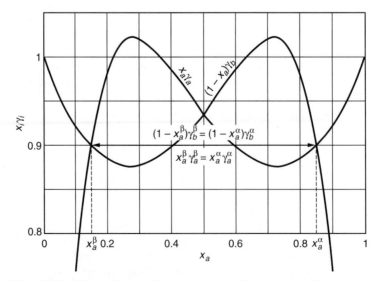

Figure E8.17 Values of $x_a\gamma_b$ and $(1 - x_a)\gamma_b$ vs. x_a for Example 8.17.

Rewriting Equation (E8.17C) gives

$$\ln\left(\frac{x_a^\alpha}{1 - x_a^\alpha}\right) = \frac{A}{RT}\left(2x_a^\alpha - 1\right) \qquad\qquad \text{(E8.17D)}$$

Equations (E8.17C) and (E8.17D) are valid only for the symmetric two-suffix Margules equation. Solving either Equation (E8.17D) or Equations (E8.17A) and (E8.17B):

$$x_a^\alpha = 0.855 \qquad \text{and} \qquad x_a^\beta = 0.145$$

To find the pressure, we add Equations (8.61) and (8.62):

$$P = x_a^\alpha \exp\left[\frac{A}{RT}\left(x_b^\alpha\right)^2\right] P_a^{\text{sat}} + x_b^\alpha \exp\left[\frac{A}{RT}\left(x_a^\alpha\right)^2\right] P_b^{\text{sat}} = 135\,[\text{kPa}]$$

Finally, solving Equation 8.61, we get

$$y_a = \frac{x_a^\alpha \exp\left[\dfrac{A}{RT}\left(x_b^\alpha\right)^2\right] P_a^{\text{sat}}}{P} = 0.667 \qquad\qquad ◀$$

▶ 8.4 SOLID–LIQUID AND SOLID–SOLID EQUILIBRIUM: SLE AND SSE

We next treat the case of solid–liquid equilibria (SLE), solid–solid equilibria (SSE), and solid–solid–liquid equilibria (SSLE). Solids that are in equilibrium with liquids can take two forms: (1) pure solids that are immiscible with other species and (2) solid solutions that, like liquid solutions, contain more than one species. Crystalline solids are formed within a well-defined geometrical lattice structure. While partial miscibility in liquid systems is due solely to the relative strength of like intermolecular interactions compared to unlike intermolecular interactions, the ability of solids to mix depends primarily on how well one atom fits to the lattice structure of the other species. Thus, complete solid miscibility occurs only when species are nearly the same size, have the same crystal structure, and have similar electronegativities and valences. We treat pure solids first and then address solid solutions.

Pure Solids

We consider first a pure solid in equilibrium with a liquid mixture. Figure 8.16 shows two typical phase diagrams for binary systems of a and b. In both cases, the equilibrium between the pure, immiscible solid phases and the completely miscible liquid phase is shown. Figure 8.16a shows a phase diagram in which only solids of pure a or pure b are stable. The phase diagram shows three two-phase regions, solid a–solid b, solid b–liquid, and solid a–liquid. In each of the two-phase regions, the equilibrium compositions at a given temperature are determined in a similar manner to the phase diagrams discussed earlier. Similarly, the amount present in each phase can be determined by the lever rule. At higher temperatures, as in Figure 8.16a, the binary exists as a single liquid phase. We notice that the freezing point of the pure solid decreases as we add a little bit of the other species into the mixture. We will see why shortly. Thus, when a and b are mixed, a single-phase liquid can exist at a lower temperature than the freezing point of either pure solid. The lowest possible temperature in which we have only liquid is called the **eutectic point**. The eutectic point, marked in Figure 8.16a, is the point at which the equilibrium line of liquid for the solid b–liquid binary intersects with that of the solid a–liquid binary. At the eutectic temperature, we have SSLE where three phases can exist in equilibrium

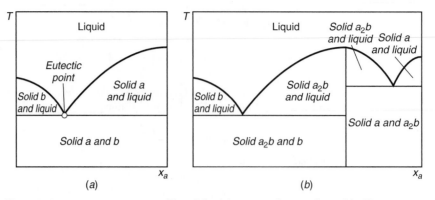

Figure 8.16 SLE in pure, immiscible solids. (*a*) Binary solution of *a* and *b*; (*b*) a compound a_2b forms.

(solid *a*, solid *b*, and liquid) much like the behavior of VLLE in Section 8.3. At this point the binary system is completely constrained, so all the properties are fixed.

The phase diagram in Figure 8.16*b* consists of the case where a solid compound of stoichiometry a_2b forms. Its features are similar to those in Figure 8.16*a*, but now there are five two-phase regions: solid *a*–solid a_2b, solid a_2b–solid *b*, solid *b*–liquid, solid a_2b–liquid, and solid *a*–liquid. Additionally, there are two compositions at which three phases can exist in SSLE: solid *a*–solid a_2b–liquid and solid a_2b–solid *b*–liquid. The former three-phase region occurs at a lower temperature than the latter. While the phase diagram in Figure 8.16*b* may look daunting at first, it can in fact be considered as two simple eutectic diagrams like those shown in Figure 8.16*a* linked together.

In fact, many binary mixtures form more than two compounds; therefore, further linkages of the type shown in Figure 8.16*b* are noticed in their phase diagrams. Figure 8.17 shows the solid–liquid phase diagram of a binary mixture of copper and yttrium. In this system, four compound phases (γ, δ, ϵ, and ζ) exist in addition to pure Cu and two pure Y phases (α and β). Can you identify the stoichiometry of the compounds? The three phases—δ, ϵ, and ζ—possess definite melting points; such compounds are said to have **congruent** melting points. However, the γ phase is not stable all the way up to a well-defined melting point but rather dissociates into a liquid and δ phase solid above around 931°C. Such a compound is said to have an **incongruent** melting point, and the state at which it dissociates is called the **peritectic** point. More complex phase behavior is also observed in phase diagrams containing pure solids. For example, the liquid phase may be only partially miscible.

We now look at how to construct these phase diagrams from thermodynamic property data. Our criteria for species *i* in solid–liquid equilibrium is

$$\hat{f}_i^s = \hat{f}_i^l \qquad (8.66)$$

However, if the solid phase contains only pure species *i*, we can replace the fugacity of solid species *i* in the mixture with its pure species fugacity. Choosing the Lewis/Randall reference state for the liquid, we get:

$$f_i^s = x_i \gamma_i f_i^l \qquad (8.67)$$

Equation (8.67) can be rewritten as

$$x_i \gamma_i = \frac{f_i^s}{f_i^l} \qquad (8.68)$$

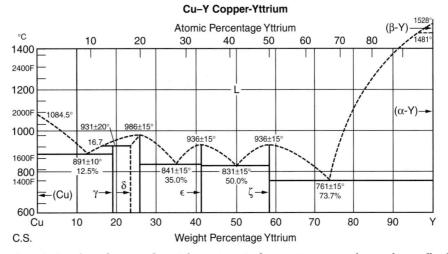

Figure 8.17 Phase diagram of Cu–Y binary system. [From T. Lyman et al., *Metals Handbook, Metalography, Structures, and Phase Diagrams*, 8th ed. (Vol. 8) (Metals Park, OH: American Society for Metals, 1973).] Courtesy of ASM International.

However, we can relate the right-hand side of Equation (8.68) to the definition of fugacity for a pure species:

$$g_i^s - g_i^l = RT \ln \frac{f_i^s}{f_i^l} \tag{8.69}$$

The term on the left-hand side of Equation (8.69) is equal to the Gibbs energy of fusion, Δg_{fus}. Plugging in Equation (8.67) and rearranging:

$$\ln[x_i \gamma_i] = \frac{\Delta g_{\text{fus}}}{RT} = \frac{\Delta h_{\text{fus}}}{RT} - \frac{\Delta s_{\text{fus}}}{R} \tag{8.70}$$

We typically know the enthalpy (heat) and entropy of fusion at a specified temperature—the normal melting point, T_m. Therefore, we need to construct a thermodynamic pathway to find the enthalpy and entropy of fusion at any T. Figure 8.18 illustrates a path for the calculation of Δh_{fus}. Adding together the three steps, we get

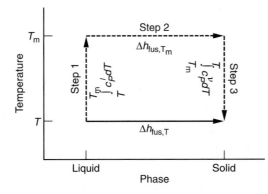

Figure 8.18 Calculation path of Δh_{fus} at temperature T from data available at T_m. Δs_{fus} is calculated by a similar path.

$$\Delta h_{\text{fus},T} = \int_T^{T_m} c_P^l dT + \Delta h_{\text{fus},T_m} + \int_{T_m}^T c_P^s dT = \Delta h_{\text{fus},T_m} + \int_{T_m}^T \Delta c_P^{sl} dT \tag{8.71}$$

where we used the following definition:

$$\Delta c_P^{sl} = c_P^s - c_P^l \tag{8.72}$$

The entropy of fusion can be found by the same procedure:

$$\Delta s_{\text{fus},T} = \int_T^{T_m} \frac{c_P^l}{T} dT + \Delta s_{\text{fus},T_m} + \int_{T_m}^T \frac{c_P^s}{T} dT = \Delta s_{\text{fus},T_m} + \int_{T_m}^T \frac{\Delta c_P^{sl}}{T} dT \tag{8.73}$$

Since $\Delta g_{\text{fus}} = 0$ at the melting temperature, we can rewrite Equation (8.73) as

$$\Delta s_{\text{fus},T} = \frac{\Delta h_{\text{fus},T_m}}{T_m} + \int_{T_m}^T \frac{\Delta c_P^{sl}}{T} dT \tag{8.74}$$

Finally, substituting Equations (8.71) and (8.74) into Equation (8.70) gives

$$\ln[x_i \gamma_i] = \frac{\Delta h_{\text{fus},T_m}}{R} \left[\frac{1}{T} - \frac{1}{T_m} \right] - \frac{1}{R} \int_{T_m}^T \frac{\Delta c_P^{sl}}{T} dT + \frac{1}{RT} \int_{T_m}^T \Delta c_P^{sl} dT \tag{8.75}$$

If Δc_P^{sl} is a constant, Equation (8.75) becomes

$$\ln[x_i \gamma_i] = \frac{\Delta h_{\text{fus},T_m}}{R} \left[\frac{1}{T} - \frac{1}{T_m} \right] + \frac{\Delta c_P^{sl}}{R} \left[1 - \frac{T_m}{T} - \ln\left(\frac{T}{T_m} \right) \right] \tag{8.76}$$

Solid Solutions

It is also possible for solids to form solutions where they mix together in a manner similar to liquids. Consider a solid originally of pure a to which species b is added. A solid solution forms if the crystal structure stays the same upon addition of b. There are two ways in which solid solutions form. In a **substitutional** solid solution, species b occupies the lattice sites where species a once sat. As long as the crystal can accommodate b without altering its basic structure, a solid solution will occur. On the other hand, an **interstitial** solid solution forms when species b sits in interstitial spaces in between the lattice sites where a sits. These spaces are not part of the crystal structure. In this case, the b will be only sparingly soluble in a.

Two examples of phase diagrams of binary solid solutions with liquids are shown in Figure 8.19. Figure 8.19a shows the phase behavior of a solid solution where a and b are miscible in all proportions. It is similar to the type of behavior shown in Figure 8.4 for VLE. Solids that form completely miscible solutions can also show azeotropic behavior analogous to the behavior shown in Figure 8.8. Molybdenum and tungsten exhibit the type of phase behavior depicted in Figure 8.19a. The solid solution can mix in all proportions since the species are similar in size (with nearest-neighbor distances of 2.72 Å and 2.73 Å), are chemically similar (both in Group VIb of the periodic table), and both form body center cubic crystal lattices. In general, a binary pair must have these characteristics to be completely miscible; however, binary pairs with these characteristics

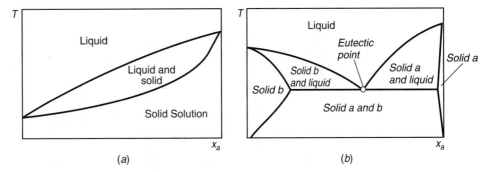

Figure 8.19 SLE in solid solutions. (*a*) The solid solution is miscible in all proportions; (*b*) a partially miscible solid solution.

are relatively rare. Therefore, most solid solutions are only partially miscible. Figure 8.19*b* show the phase diagram for a binary mixture where each solid is partially miscible in the other. This behavior is the analog to Figure 8.15*b* for solid–liquid solutions. In addition to the phases found for pure solids in Figure 8.16*a*, single-phase regions are found for partially miscible solid *a* and partially miscible solid *b*. However, only a limited amount of *b* will dissolve in *a* and only a limited amount of *a* in *b*. Aluminum and silicon exhibit behavior similar to that shown in Figure 8.19*b*.

For the case of a solid solution, we must find the composition of the solid, X_i, and the activity coefficient in the solid, Γ_i. In this case, Equation (8.66) becomes

$$X_i \Gamma_i f_i^s = x_i \gamma_i f_i^l \tag{8.77}$$

Rearranging Equation (8.77) gives

$$\frac{x_i \gamma_i}{X_i \Gamma_i} = \frac{f_i^s}{f_i^l} \tag{8.78}$$

We can rewrite the right-hand side of Equation (8.78) using the same development as above for pure species:

$$\ln\left[\frac{x_i \gamma_i}{X_i \Gamma_i}\right] = \frac{\Delta h_{\text{fus},T_m}}{R}\left[\frac{1}{T} - \frac{1}{T_m}\right] - \frac{1}{R}\int_{T_m}^{T}\frac{\Delta c_P^{sl}}{T}dT + \frac{1}{RT}\int_{T_m}^{T}\Delta c_P^{sl}dT \tag{8.79}$$

If Δc_P is a constant, Equation (8.79) becomes

$$\ln\left[\frac{x_i \gamma_i}{X_i \Gamma_i}\right] = \frac{\Delta h_{\text{fus},T_m}}{R}\left[\frac{1}{T} - \frac{1}{T_m}\right] + \frac{\Delta c_P^{sl}}{R}\left[1 - \frac{T_m}{T} - \ln\left(\frac{T}{T_m}\right)\right] \tag{8.80}$$

We can relate the phase behavior of these binary mixtures to the Gibbs energy of each phase. Figure 8.20 shows plots of Gibbs energy of the solid phase, g_{solid}, and the liquid phase, g_{solid}, for a completely miscible solid solution (Figure 8.20*a*) and a partially miscible solid solution (Figure 8.20*b*). The corresponding phase behavior is shown below the plot of Gibbs energy. The Gibbs energy curves in Figure 8.20*a* correspond to temperature T_1 on the phase diagram, between the melting points of the two pure solids. At low values of x_a, energetic effects dominate and the Gibbs energy of the solid is lower than the Gibbs energy of the liquid. Conversely, at high values of x_a, the Gibbs energy of the liquid is

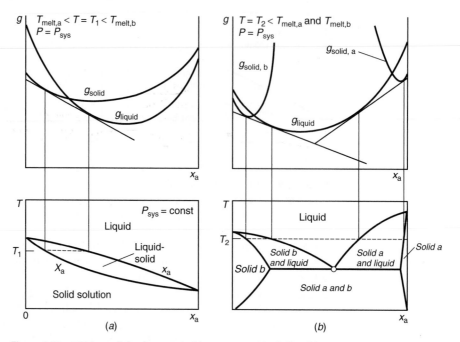

Figure 8.20 SLE in solid solutions. Gibbs energies of solid and liquid phases are plotted on top; phase diagrams are on the bottom. (*a*) The solid solution is miscible in all proportions; (*b*) a partially miscible solid solution.

lower. The resulting phase behavior shows three regions. In low proportion of species a, the solid has a lower Gibbs energy and is favored at equilibrium. At high proportion of a the liquid is favored. Between the minima in Gibbs energy, we can draw a tangent line, as shown in Figure 8.20*a*. The Gibbs energy at mole fractions in between the minima can be lowered if the system separates into two phases, in much the same way as was explained in our discussion of two liquid phases in Figure 7.10 or vapor–liquid equilibrium in Figure 8.6. In this case, we have solid–liquid equilibrium, with the composition in each phase determined by the mole fraction at which the tangent line abuts the two curves.

Figure 8.20*b* shows the case where the solids are only partially miscible. In this case, we see that the Gibbs energy of each solid phase rises dramatically when its lattice can no longer accommodate the other species. The Gibbs energy diagram is drawn for temperature T_2 in the phase diagram below, which corresponds to a temperature below the melting temperature of either of the pure species. In this case, two tangent lines can be drawn between the three corresponding minima in Gibbs energy. At compositions in between those at which either tangent line abuts the two curves, Gibbs energy is minimized by the formation of two phases. The corresponding phases are shown on the phase diagram below.

▶ 8.5 COLLIGATIVE PROPERTIES

In this section, we examine some effects on the properties of a pure liquid when a small amount of solute is added. When the mixture forms an ideal solution, the change in these properties depends only on the amount of solute present, not on the chemical nature of the solute. Such properties, termed **colligative properties**, include boiling-point elevation, freezing-point depression, and osmotic pressure.

We know that if we specify the system pressure, the temperature at which *pure* species a boils is fixed. For example, the normal boiling point of a substance, the temperature at which a pure species boils at a pressure of 1 atm, is well defined. The state of such a system is shown schematically as system I on the left-hand side of Figure 8.21. Consider now the presence of an essentially nonvolatile solute, species b in a liquid phase of mostly solvent a, as shown in system II on the right-hand side of Figure 8.21. System II is at the same pressure as system I. We observe that it always requires a greater temperature for a to boil in system II than in system I. This phenomenon is termed **boiling-point elevation**. Similarly, if we add a small amount of solute b to a liquid solution, solid a will freeze at a lower temperature than the freezing point of the pure liquid. The addition of solute leads to freezing-point depression.

We first qualitatively examine these phenomena in terms of ideal solution behavior; then we will apply the principles of phase equilibrium to quantify these phenomena for both nonideal and ideal solutions. To understand why the boiling point elevates, we can apply the principles of phase equilibrium that we have just developed. We compare system I with pure a to system II, which is at the same pressure but also contains dilute b in the liquid. In system I, the system pressure is, by definition, the saturation pressure, that is,

$$\left(P_a^{\text{sat}}\right)^{\text{I}} = P \tag{8.81}$$

Next, we examine the equilibrium criteria for species a in system II. If we are at low pressure and the solute is dilute enough, Raoult's law applies:

$$y_a P = x_a \left(P_a^{\text{sat}}\right)^{\text{II}} \tag{8.82}$$

However, since the solute is not volatile, the mole fraction of a in the vapor is approximately 1. Thus Equation (8.82) can be written as

$$\left(P_a^{\text{sat}}\right)^{\text{II}} = \frac{P}{x_a} \tag{8.83}$$

↑ P^{sat} = ↑ T_b

Comparing Equations (8.81) and (8.83), we conclude that $\left(P_a^{\text{sat}}\right)^{\text{II}} > \left(P_a^{\text{sat}}\right)^{\text{I}}$ since $x_a < 1$. Since the saturation pressure of a in system II is higher than in system I, the boiling temperature must also be elevated.

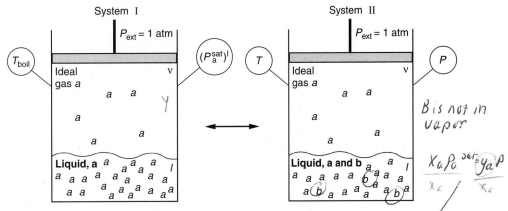

Figure 8.21 Illustration of boiling points for pure species a (*left*) and for species a in the presence of nonvolatile solute b (*right*).

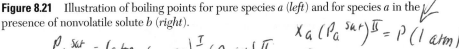

$P_a^{sat} = 1 atm$
$T = T_{boil}$ $\left(P_a^{sat}\right)^I < \left(P_a^{sat}\right)^{II}$

$X_a \left(P_a^{sat}\right)^{II} = P (1 atm)$
@ $T > T_{boil}$

Alternatively, we can examine the colligative properties of species a in terms of its chemical potential. The state at which the liquid and the vapor or the liquid and the solid are in equilibrium is given when the chemical potential in each phase is equal. The chemical potential of a pure species is identical to its molar Gibbs energy. We first assume b is dilute and we have an ideal solution. We will shortly relax this assumption. We have seen that the Gibbs energy quantifies the trade-off between the energetic stability of one phase and the increased randomness of the other. We can compare these two effects for a pure liquid vs. a dilute ideal solution to which a small amount of solute has been added. Since all their intermolecular interactions are the same, the energetic interactions of the dilute solution are identical to the pure liquid and energetics will not affect the trade-off between the liquid and the vapor at the boiling point or the liquid and the solid at the freezing point. On the other hand, the entropy of the dilute liquid is greater than that of the pure liquid since the presence of solute gives it many more possible configurations. Entropy becomes more and more important as T becomes greater. Since the entropy difference between the vapor and the dilute liquid has been reduced relative to that of the vapor and pure liquid, it takes a higher temperature for the more entropic vapor phase to balance the energetic dilute liquid solution, and the boiling point of a is elevated. Similarly, it takes a lower temperature to balance the increased entropy of the dilute liquid in favor of the energetic stability of a solid, and the freezing point is lowered.

We can look at the effect of going from pure species a to a in a mixture quantitatively by examining the effect of adding species b on the chemical potential of a in the liquid. Applying the definition of fugacity, the difference in the chemical potential of a in the mixture and that of pure liquid a can be written as:

$$\mu_a - g_a = RT \ln \frac{\hat{f}_a^{\text{ideal}}}{f_a} = RT \ln x_a \tag{8.84}$$

where the definition of an ideal solution, Equation (7.68), is used. Since the logarithm of the mole fraction is always a negative number, Equation (8.84) mandates that the chemical potential of a in the dilute solution is always less than its pure species molar Gibbs energy.

Figure 8.22 shows a plot of the chemical potential of species a in solid, liquid, and vapor phases. The chemical potentials of the solid and vapor phases are shown as a pure species a, while both pure a and a in a dilute solution are shown for the liquid. We see that the chemical potential of a in a dilute solution is lower than in the pure liquid, as prescribed by Equation (8.84). Phase equilibrium occurs when the lines marking the chemical potential in two phases intersect, as denoted in the figure. The freezing (melting) and boiling points of pure species a are labeled T_m and T_b, respectively. We notice that the shift in chemical potential of a in solution leads to a lower freezing point and a higher boiling point. Therefore, the lowering of the chemical potential of liquid a in solution relative to its pure species value can be seen to give rise to the colligative properties—freezing-point depression and boiling-point elevation.

We now quantify the rise in the boiling point of pure species i when a solute is added to the liquid, in general. If the vapor phase contains only pure species i, we can write the fugacity of the vapor in terms of the pure species fugacity. Choosing the Lewis/Randall reference state for the liquid, we get

$$f_i^v = x_i \gamma_i f_i^l \tag{8.85}$$

Equation (8.85) is identical to Equation (8.67), except that the pure species vapor replaces the pure species solid. Thus, in analogy to the development of Equation (8.76) for solids, we can see that

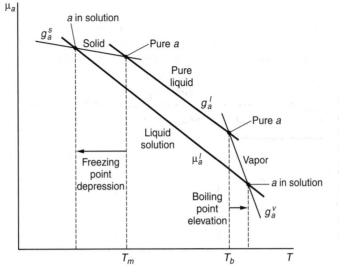

Figure 8.22 Plot of the chemical potential of species a as a pure species and in a liquid solution as a function of temperature. The difference in where the Gibbs energy of the liquid intersects the solid and vapor leads to freezing-point depression and boiling-point elevation, respectively.

$$\ln[x_i\gamma_i] = \frac{\Delta h_{\text{vap},T_b}}{R}\left[\frac{1}{T} - \frac{1}{T_b}\right] + \frac{\Delta c_P^{vl}}{R}\left[1 - \frac{T_b}{T} - \ln\left(\frac{T}{T_b}\right)\right] \tag{8.86}$$

where $\Delta c_P^{vl} = c_P^v - c_P^l$ is assumed constant.

Let's now apply the general result of Equation (8.86) to the system shown in Figure 8.21. If we assume that solute b is dilute enough that the liquid can be treated as an ideal solution and that the boiling point elevation is small, Equation (8.86) becomes

$$\ln(x_a) = \ln(1 - x_b) = \frac{\Delta h_{\text{vap}}}{R}\left[\frac{1}{T} - \frac{1}{T_b}\right] \tag{8.87}$$

Since the mole fraction of b is small, we can write $\ln(1 - x_B) \approx -x_b$. After rearrangement, we see that the boiling point elevation is given by

$$T - T_b \approx \frac{RT_b^2}{\Delta h_{\text{vap}}}x_b \tag{8.88}$$

The boiling-point elevation described by Equation (8.88) depends only on the properties of a; it is independent of the chemical nature of b, depending only on the amount of b present. On the other hand, if there is enough b that the liquid is not ideal, we must account for the activity of species a in solution. In such a case, we can approximate Equation (8.86) by

$$T - T_b \approx \frac{RT_b^2}{\Delta h_{\text{vap}}}\gamma_a x_b \tag{8.89}$$

Equation (8.89) can be rewritten as

$$\gamma_a = \frac{\Delta T \Delta h_{\text{vap}}}{RT_b^2 x_b} \tag{8.90}$$

Where $\Delta T = T - T_b$, Equation (8.90) can be used to find activity coefficient model parameters. Since enthalpies of vaporization are readily available, we can use Equation

(8.90) to calculate the activity coefficient of the solute from a measured boiling-point elevation. We can then use this value to calculate parameters in a model for g^E.

A similar analysis can be applied to **freezing-point depression**. If we consider an ideal solution not too far away from the melting point, we can approximate Equation (8.76) by

$$\ln[x_a] = \frac{\Delta h_{fus,T_m}}{R}\left[\frac{1}{T} - \frac{1}{T_m}\right] \tag{8.91}$$

Again, for a small amount of b, we write $\ln(1 - x_b) \approx -x_b$. Thus, we can rearrange Equation (8.91) to give the freezing-point depression of a for a given mole fraction of b:

$$T - T_m = \frac{x_b R T_m^2}{\Delta h_{fus,T_m}} \tag{8.92}$$

Again, if there is enough solute to have a nonideal solution, we can solve Equation (8.76) for the activity coefficient of a in terms of experimentally measurable quantities:

$$\gamma_a = \frac{\Delta T \Delta h_{fus,T_m}}{R T_m^2 x_b} \tag{8.93}$$

where $\Delta T = T - T_m$. We can then use Equation (8.93) to find the activity coefficient for a. This information can let us fit parameters in a model for g^E.

Osmotic pressure is a third common type of colligative property. Osmosis is the transport of a pure solvent into solution through a semipermeable membrane. The membrane allows passage of solvent but restricts flow of the solute. Figure 8.23 shows the equilibrium state from which we calculate the osmotic pressure. One compartment contains pure solvent a, while a dilute liquid solution of solvent a and solute b is housed on the other side. These two compartments are divided by a semipermeable membrane, which species a can transport through but b cannot. The difference in pressure between the dilute liquid solution on the right and pure species a on the left at equilibrium is given by the osmotic pressure Π. If a pressure less than $(P + \Pi)$ exists on the right compartment, solvent a will spontaneously flow into it from the compartment on the left

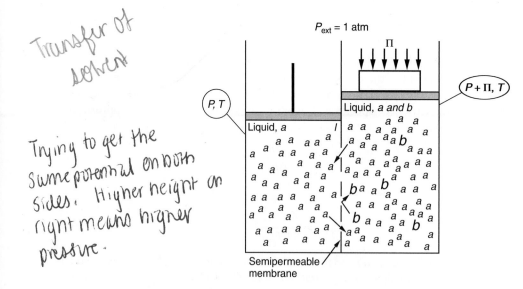

Figure 8.23 Illustration of osmotic pressure at equilibrium. The difference in pressure between the dilute liquid solution on the right and pure species a on the left is given by the osmotic pressure Π.

with pure a. Conversely, if the pressure is greater than $(P + \Pi)$ is applied, solvent a will be forced into the compartment with pure a. This latter case is the basis of separation by reverse osmosis.

To calculate the osmotic pressure, we set the chemical potentials of the pure species on the left equal to the chemical potential of species a on the right:

$$g_a^{T,P} = \mu_a^{T,P+\Pi} \tag{8.94}$$

The superscripts indicate the respective temperature and pressure. Applying the definition of fugacity, we can write the chemical potential at $P + \Pi$ as

$$\mu_a^{T,P+\Pi} = g_a^{T,P+\Pi} + RT \ln \frac{\hat{f}_a}{f_a} = g_a^{T,P+\Pi} + RT \ln x_a \gamma_a \tag{8.95}$$

Substituting Equation (8.95) into (8.94) and rearranging gives

$$g_a^{T,P+\Pi} - g_a^{T,P} = -RT \ln x_a \gamma_a \tag{8.96}$$

We can apply the thermodynamic web to determine the difference in Gibbs energy from P to $P + \Pi$ at constant T:

$$\int_{g_a^{T,P}}^{g_a^{T,P+\Pi}} dg = \int_{P}^{P+\Pi} v_a dP \tag{8.97}$$

Substitution of Equation (8.97) into (8.96) gives

$$\int_{P}^{P+\Pi} v_a dP = -RT \ln x_a \gamma_a \tag{8.98}$$

If we assume the liquid is incompressible, we get

$$\frac{v_a \Pi}{RT} = -\ln x_a \gamma_a \tag{8.99}$$

Equation (8.99) illustrates that the activity coefficient of the solvent can be found through measurement of the osmotic pressure. In the case that species b is dilute enough, the mixture forms an ideal solution. Applying $\ln(1 - x_b) \approx -x_b$, we get

$$\Pi = \frac{x_b RT}{v_a} \tag{8.100}$$

Membranes permeable to water but not large macromolecules are readily available. The osmotic pressure can then be measured to find the average molecular weight, MW_b, of the macromolecules in the solution. In this case we rewrite the mole fraction in Equation (8.100) on a mass concentration, C_i, basis as follows:

$$x_b = \left[\frac{\dfrac{C_b}{MW_b}}{\dfrac{C_a}{MW_a} + \dfrac{C_b}{MW_b}} \right] \approx \left[\frac{\dfrac{C_b}{MW_b}}{\dfrac{C_a}{MW_a}} \right] \tag{8.101}$$

where the approximation made assumes moles of solvent are much greater than moles of solute. Substitution of Equation (8.101) into (8.100) gives

$$MW_b = \frac{RTC_b}{\Pi} \qquad (8.102)$$

Example 8.18 illustrates how we can apply Equation (8.102) to determine the molecular weight of a protein in solution.

▶ **EXAMPLE 8.18**
Determination of the
Boiling-Point Elevation
of Seawater

At what temperature does seawater boil? The concentration of NaCl in seawater is 3.5% by weight.

SOLUTION The mole fraction of salt is given by

$$x_{salt} = \frac{\dfrac{w_{salt}}{MW_{salt}}}{\dfrac{w_{salt}}{MW_{salt}} + \dfrac{w_{water}}{MW_{water}}} = 0.011$$

If we assume that the NaCl completely dissociates into sodium and chloride ions, we get

$$x_b = 2x_{salt} = 0.022$$

To calculate the boiling-point elevation, we can use Equation (8.88):

$$T - T_b \approx \frac{RT_b^2}{\Delta h^{vap}}x_b = \frac{8.314\,[J/(mol\ K)](373.15\,[K])^2}{2257\,[J/g]\,18\,[g/mol]} \times 0.022 = 0.63\,[K]$$

So the temperature that seawater boils is 100.63°C. ◀

▶ **EXAMPLE 8.19**
Determination of the
Molecular Weight of a
Protein Using Osmotic
Pressure

It is desired to measure the osmotic pressure *lactic dehydrogenace* to determine the molecular weight of the protein. The experiment to accomplish this objective is schematically shown in Figure E8.19. When 1.93 g of the protein is dissolved to make 520 cm³ of aqueous solution at 25°C, the height of the dilute solution rises 0.71 cm above the pure solvent. Determine the osmotic pressure and the molecular weight.

SOLUTION The osmotic pressure is given by the barometric formula:

$$\Pi = \rho g h = 10^3\,[kg/m^3]\,9.8\,[m/s^2]0.0071\,[m] = 70\,[Pa]$$

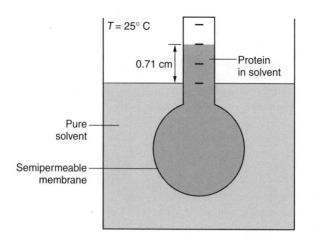

T = 25° C

0.71 cm

Protein
in solvent

Pure
solvent

Semipermeable
membrane

Figure E8.19 Measurement of osmotic pressure of protein in solution.

where the density has been approximated by the liquid density of water. The concentration of protein is given by

$$C_b = \frac{m_b}{V} = 3.71 \left[\text{kg/m}^3\right]$$

From Equation (8.102), we have

$$MW_b = \frac{RTC_b}{\Pi} = \frac{8.314 \left[\text{J/(mol K)}\right] 298 \left[\text{K}\right] 3.71 \left[\text{kg/m}^3\right]}{70 \left[\text{Pa}\right]}$$

$$= 132 \left[\text{kg/mol}\right] = 13,200 \left[\text{g/mol}\right]$$

▶ 8.6 SUMMARY

In Chapter 7, we learned how to calculate the fugacity of a species in the vapor, liquid, and solid phases. In this chapter, we applied these concepts to treat practical phase equilibria problems, including vapor–liquid equilibria (VLE), liquid–liquid equilibria (LLE), vapor–liquid–liquid equilibria (VLLE), solid–liquid equilibria (SLE), solid–solid equilibria (SSE), and solid–solid–liquid equilibria (SSLE). By equating the fugacities of the species in each phase of a mixture, we are able to construct **phase diagrams**. Binary–phase diagrams have been illustrated for P vs. mole fraction at constant T, T vs. mole fraction at constant P, or mole fraction in one phase vs. mole fraction in the other. Phase diagrams can be used to determine what phase or phases are present at a given thermodynamic state and, in the two-phase and three-phase regions, the composition of the phases as well as their relative amounts.

We can write the general expression for **vapor–liquid equilibrium (VLE)** as follows:

$$y_i \hat{\varphi}_i^v P = x_i \gamma_i^l f_i^o \tag{8.3}$$

The simplest type of VLE calculation is given by **Raoult's law**, whereby the vapor is treated as an ideal gas and the liquid as an ideal solution. The pure species fugacity, used as the Lewis/Randall reference state, is given by P_i^{sat}. The saturation pressure is commonly determined at a given T from the Antoine equation. In a **bubble-point** calculation, the composition of the first bubble of vapor that forms when energy is supplied to a saturated liquid is determined. Conversely, in a **dew-point** calculation, the liquid mole fractions are found given the vapor mole fractions. This case corresponds to the composition of the first drop of dew that forms from a saturated vapor.

When intermolecular interactions become important in the liquid phase, we must correct for nonideality using the activity coefficient. In cases when the like interaction is stronger than the unlike interaction—that is, $\gamma_i > 1$—we see a **positive deviation** in pressure from Raoult's law. Conversely, a weaker like interaction—that is, $\gamma_i < 1$—exhibits **negative deviations**. When deviations from Raoult's law are large enough, the Px and Py curves can exhibit extremes. Moreover, they go through the extrema at exactly the same composition. We use the term **azeotrope** to describe the point in the phase diagram where the Px and Py curves or Tx and Ty curves go through an extrema. At the azeotrope, the mole fraction of each species in the liquid phase equals that in the vapor phase. Azeotropes are more likely to occur when the saturation pressures of the two components are close in value. Systems that exhibit a maximum in pressure also exhibit a minimum in temperature and are referred to as **minimum-boiling azeotropes**. Conversely, those that show a minimum in pressure have a maximum in temperature and are referred to a **maximum-boiling azeotropes**.

VLE data can be used to obtain best-fit values of activity coefficient model parameters. A general approach applicable to any model is through the use of **objective functions**. An objective function takes account of the entire measured data set. It is written in terms of the difference between the calculated value of a given property and the experimental value of the same property. The model parameters are determined when the objective function is minimized. Alternatively, equations such as the two-suffix and three-suffix Margules equations can be rewritten to find model parameters through averages and linear regression.

We use **Henry's law** as the reference state for VLE of a dissolved gas in a liquid. The Henry's constant is indicative of the unlike i-j interaction. Hence, its value depends on the identity not only of the solute but also of the solvent. Henry's constants need to be obtained through measured data. If we are interested in a system at significantly different temperature or pressure, we must correct the value of $\mathcal{H}_i$.

When the like interactions are significantly stronger than unlike interactions, liquids can split into two different partially miscible phases. They form separate phases to lower the total Gibbs energy of the system. In this case, each species i tends to equilibrate between the two phases, leading to **liquid–liquid equilibrium (LLE)**. We can equate the fugacity of component i for each phase to solve for the equilibrium compositions:

$$x_i^\alpha \gamma_i^\alpha = x_i^\beta \gamma_i^\beta \tag{8.49}$$

The two-phase region of the resulting phase diagram for such a system is separated from the region where only one phase is present by the **binodal curve**. The temperature above or below which a liquid mixture no longer separates into two phases is termed the **upper consulate** or **lower consulate temperature**, respectively. A liquid is inherently unstable if the Gibbs energy curve is concave down. The **spinodal curve** on a phase diagram distinguishes the region. Similarly, systems with two liquid phases in equilibrium with a vapor phase give rise to **vapor–liquid–liquid equilibrium (VLLE)**.

Equilibrium involving a solid phase, including **solid–liquid equilibrium (SLE)**, **solid–solid equilibrium (SSE)**, and **solid–solid–liquid equilibrium (SSLE)**, can take two forms: (1) **pure solids**, which are immiscible with other species, and (2) **solid solutions**, which, like liquid solutions, contain more than one species. If the solid phase exists as a pure species, the condition for phase equilibrium with a liquid becomes

$$f_i^s = x_i \gamma_i f_i^l \tag{8.67}$$

For the case of a solid solution, we get

$$X_i \Gamma_i f_i^s = x_i \gamma_i f_i^l \tag{8.77}$$

where X_i is the composition of the solid, and Γ_i is the activity coefficient in the solid. Solid miscibility within the rigid crystalline lattice occurs only when species are nearly the same size, have the same crystal structure, and have similar electronegativities and valences.

Liquid mixtures freeze at lower temperatures than the pure components of which they are comprised. The lowest possible temperature in which we only have liquid is called the **eutectic point**. The presence of solid compounds of stoichiometry $a_x b_y$ increases the complexity of their phase diagrams. A compound with a definite melting point is termed **congruent**, while a compound that is not stable all the way up to a well-defined melting point is said to have an **incongruent** melting point. The state at which the latter case dissociates is called the **peritectic point**.

Three **colligative properties**—boiling-point elevation, freezing-point depression, and osmotic pressure—result when a small amount of solute is added to a pure liquid. When the mixture forms an ideal solution, the change in these properties depends only on the amount of solute present, not on the chemical nature of the solute. The lowering of the chemical potential of liquid a in solution relative to its pure species value leads to **boiling-point elevation** and **freezing-point depression**. **Osmotic pressure** is given by the amount of additional pressure that needs to be added to the mixture to prevent a pure solvent from passing through a semipermeable membrane into solution. Expressions for these three colligative properties have been obtained by applying the principles of phase equilibrium.

▶ 8.7 PROBLEMS

8.1 Construct a Txy phase diagram for a binary mixture of cyclohexane and n-hexane at 1 bar. Assume that the liquid forms an ideal solution.

8.2 What is the lowest temperature to which a vapor mixture of 1 mole n-pentane and 2 moles n-hexane at 1 bar can be brought without forming liquid? Assume the liquid forms an ideal solution.

8.3 What is the composition of vapor that is in equilibrium with a liquid mixture with the following composition at 250 K? What is the pressure? Assume ideal behavior.

Species	Mole fraction
Propylene	0.1
Propane	0.25
n-Butane	0.2
Isobutane	0.35
n-Pentane	0.1

8.4 A liquid mixture containing 40% cyclohexane, 20% benzene, 25% toluene, and 15% n-heptane is in equilibrium with its vapor at 1 bar. Determine the temperature and the vapor composition.

8.5 A compressed liquid feed stream containing an equimolar mixture of n-butane and isobutane flows into a flash unit at flow rate F. At steady state, 40% of the feed stream is vaporized and leaves the drum as a vapor stream with flow rate V. The rest leaves as liquid with flow rate L. If the flash pressure is 1 bar, what is the temperature required? What are the composition of the liquid and vapor exit streams?

8.6 A feed stream containing a mixture of 40% n-butane, 30% n-pentane, and 30% n-hexane flows into a flash unit. The flash temperature is 290 K and the flash pressure is 0.6 bar. What is the ratio of the exit vapor flow rate to the feed flow rate? What are the compositions of the exit streams?

8.7 Calculate the liquid and vapor compositions of butane (a) and n-hexane (b) at 0°C and 0.5 bar. You may assume the liquid forms an ideal solution.

8.8 Calculate the liquid and vapor compositions of n-pentane (a) and benzene (b) at 16°C and 0.333 bar. The excess Gibbs energy can be described by the two-suffix Margules equation with $A = 1816 \, [\text{J/mol}]$.

8.9 Calculate the liquid and vapor compositions of a binary mixture of isobutane (a) and hydrogen sulfide (b) at 4.5°C and 8.77 bar. The excess Gibbs energy can be described by the three-suffix Margules equation with these parameters:

$$A = 1918 \, [\text{J/mol}] \quad \text{and} \quad B = -1074 \, [\text{J/mol}]$$

(a) Assume the vapor can be treated as an ideal gas.

(b) Correct for the vapor-phase nonideality using the Lewis fugacity rule.

8.10 At 60°C, ethanol (1) and ethyl acetate (2) exhibit an azeotrope at a pressure of 0.64 bar and $x_1 = 0.4$.

(a) You wish to use the two-suffix Margules equation as a model for g^E. From these data, determine, as *accurately* as you can, the Margules parameter, A.

(b) At 60°C, what is the composition of the vapor in equilibrium with a liquid of composition $x_1 = 0.8$?

8.11 Consider a binary mixture of n-propanol and water in vapor–liquid equilibrium (VLE). Let n-propanol be designated species 1 and water, species 2. A plot of the activity coefficients for this system at 100°C follows. The Lewis/Randall reference state is chosen for both species. The mole fraction of n-propanol in the liquid, x_1, is 0.2, and the temperature is 100°C. The saturation pressure of n-propanol at 100°C is 1.12 bar.

(a) Label the curve that corresponds to the activity coefficient for n-propanol, γ_1, and the curve that corresponds to the activity coefficient for water, γ_2. Explain.

(b) Are like or unlike interactions stronger? Explain.

(c) Find the total pressure of the system.

(d) Find the mole fraction of n-propanol in the vapor phase.

(e) Estimate the value of the Henry's law constant of n-propanol in water, $\mathcal{H}_1$.

(f) Does this system exhibit an azeotrope? Explain.

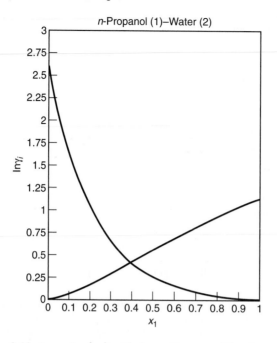

n-Propanol (1)–Water (2)

8.12 An equimolar liquid phase of benzene (1) and m-xylene (2) coexists with its vapor at 260°C. At this temperature, the saturation pressures are $P_1^{sat} = 35.2$ [bar] and $P_2^{sat} = 11.9$ [bar]. Using the van der Waals equation of state, calculate the equilibrium composition and pressure of vapor. You may assume that the liquid acts as an ideal solution.

8.13 Example 8.6 illustrates how you solve a dew-point calculation for a binary mixture of a nonideal liquid and a nonideal gas with T known. This problem corresponds to quadrant I in Figure 8.2. Develop an analogous solution for the bubble point with the liquid-phase mole fractions and T known (quadrant II). As in Example 8.6, use the van der Waals equation for vapor nonideality and the three-suffix Margules equation for liquids. Assume that critical properties, liquid volumes, and Antoine coefficients for each species are readily available and that the three-suffix Margules parameters have been determined.

8.14 From the list of the following species, choose the appropriate binary mixture (group of two species) for each of the following questions.:

Acetone	Chloroform	Ethane
Ethanol	Methane	

Explain your answers.

(a) Consider liquid–vapor equilibrium at 1 bar.

(i) What binary mixture will have the largest positive deviations from ideality?

(ii) What binary mixture will have the largest negative deviations from ideality?

(b) What binary mixture will most closely follow the Lewis fugacity rule?

(c) What binary mixture is most likely to have its second cross-virial coefficient, B_{12}, more negative than that of either of the pure species, B_{11} and B_{22}?

(d) What binary mixture is most likely to have $B_{12} = \sqrt{B_{11}B_{22}}$ as its second cross-virial coefficient?

8.15 Consider the system of ethanol (1)–benzene (2) at 25°C. This mixture exhibits an azeotrope at a mole fraction of $x_1 = 0.28$ and a pressure of 122.3 torr. Determine values for the parameters in the van Laar equation. Estimate the liquid composition and pressure in equilibrium with a vapor of $y_1 = 0.75$ at 25°C.

8.16 The Gibbs energies for the liquid phase and the vapor phase vs. mole fraction a for two systems (system I and system II) follow. These plots are at constant temperature and pressure. What type of behavior does each of these plots correspond to?

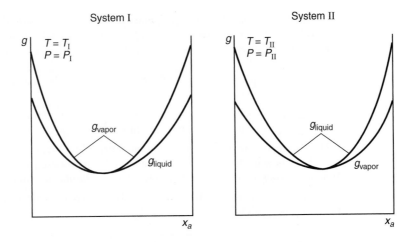

8.17 Consider a mixture of species 1 and 2 in vapor–liquid equilibrium at 25°C and 90 bar. The following equation of state is available for the *vapor* phase:

$$Pv = RT + P^2[Ay_1y_2 + B]$$

where $\quad \dfrac{A}{RT} = -2.0 \times 10^{-4} \left[1/\text{bar}^2\right] \quad$ and $\quad \dfrac{B}{RT} = 8.0 \times 10^{-5} \left[1/\text{bar}^2\right]$

and y_1 and y_2 are the mole fractions of species 1 and 2, respectively. Species 2 is dilute in the *liquid* phase and may be described by *Henry's law* with the following values at 25°C:

$$\mathcal{H}_2 = 7000 \,\text{bar}$$

and $$\ln \gamma_2^{\text{Henry's}} = -7(1 - x_1^2)$$

(a) Consider a vapor mixture with 5 mole of species 1 and 10 moles of species 2. Calculate the following quantities: $v, V, v_2, \overline{V}_2$.

(b) Calculate an expression for the pure species fugacity coefficient, φ_2^v, and the mixture fugacity coefficient, $\hat{\varphi}_2^v$, of species 2 in the vapor.

(c) In the liquid, are like interactions stronger or weaker than unlike interactions? Explain.

(d) Find the mole fraction of species 2 in the liquid in equilibrium with the vapor in part (b).

(e) As best as you can, estimate the saturation pressure of pure species 2 at 25°C. *State any assumptions that you make.*

8.18 A mixture of methanol (a) and ethyl acetate (b) exhibits an azeotrope at 55°C. Their saturation pressures are 68.8 and 46.5 kPa, respectively. The liquid-phase nonideality can be described by the two-suffix Margules equation, with $A = 2900$ J/mol. What is the pressure and composition of the azeotrope? Does this mixture form a maximum-boiling azeotrope or a minimum-boiling azeotrope? Explain.

8.19 The Henry's law constant and an expression for the activity coefficient have been found for the solute for a binary mixture of species 1 (solute)–2 (solvent) at 20°C.

$$\mathcal{H}_1 = 0.5 \,[\text{Pa}]$$

and $$\ln \gamma_1^{\text{Henry's}} = 30.5(1 - x_2^2)$$

(a) Is this expression for the activity coefficient consistent with the Henry's law limit? Are the like interactions stronger than the 1-2 interaction? Explain.

(b) Find an expression for the activity coefficient of 2. Use the Lewis/Randall reference state for species 2.

(c) Consider a system of l-2 in vapor–liquid equilibrium at pressure P. Assume P is low enough for ideal gas behavior. Derive expressions for the vapor mole fraction divided by the liquid mole fraction of each species in terms of P, x_2, P_2^{sat}, and $\mathcal{H}_1$ only. Use the same reference states as in part (b).

(d) Is an azeotrope possible at 20°C? If so, determine the composition and pressure of the azeotrope. Take $P_2^{sat} = 0.02$[bar]

8.20 What is the solubility of oxygen in methanol at 25°C and 1 bar? What is the solubility of oxygen in methanol at 25°C and 100 bar? Take $\overline{V}_{O_2}^{\infty} = 4.5 \times 10^{-5}$ [m³/mol].

8.21 A binary mixture of carbon dioxide and water exists in vapor–liquid equilibrium at 343.15 K and 1 bar. The solubility of CO_2 in the liquid has been measured as $x_{CO_2} = 0.000255$. What is the Henry's law constant for CO_2 at 343.15 K? *State and justify any assumptions that you make.*

8.22 The following data are available for the Henry's law constant of N_2 in H_2O at 19.4°C. From these data, estimate $\overline{V}_{N_2}^{\infty}$.

P_{N_2}[bar]	$\mathcal{H}_{N_2}$ [bar]
1.33	83,391
1.99	83,780
2.66	84,230
3.32	84,627
3.99	85,028
4.65	85,491
5.32	85,959
5.98	86,492
6.64	87,031
7.31	87,639
7.97	88,316
8.64	89,004
9.30	89,703
9.97	90,478
10.63	91,333

Source: E. W. Washburn (ed.), *International Critical Tables* (Vol. III) (New York: McGraw-Hill, 1928).

8.23 The following data are available for the Henry's law constant of O_2 in benzene. From these data, estimate $\overline{H}_{O_2}^{\infty} - h_{O_2}^{v}$:

T(°C)	$\mathcal{H}$ [bar]
14	20.0
18	23.6
22	28.3
26	33.1
30	38.7
35	46.5
40	55.7
45	65.1

Source: Modified from E. W. Washburn (ed.), *International Critical Tables* (Vol. III) (New York: McGraw-Hill, 1928).

8.24 Develop a computer spreadsheet or write a program to verify that the objective function $OF_{g^E} = \sum \left(g^E - g^E_{calc} \right)_i^2$ in Example 8.9 gives the value $A = 1399 \left[J/mol \right]$. What value do you obtain for OF_{g^E}?

8.25 Develop a computer spreadsheet or write a program to verify that the objective function $OF_{\gamma} = \sum \left[\left((\gamma_1 - \gamma_1^{calc})/\gamma_1 \right)^2 + \left((\gamma_2 - \gamma_2^{calc})/\gamma_2 \right)^2 \right]_i$ in Example 8.9 gives the value $A = 1424 \left[J/mol \right]$. What value do you obtain for OF_{γ}?

8.26 Liquid–vapor equilibrium data have been collected for a binary system of benzene (1)–cyclohexane (2) at 60°C. Mole fraction of liquid and vapor vs. total pressure are reported in the table below.

x_1	y_1	P [Pa]
0	0	51,857
0.0672	0.0912	53,431
0.2261	0.267	55,939
0.3201	0.3526	56,741
0.432	0.448	57,527
0.5203	0.5203	57,633
0.6029	0.5895	57,432
0.7095	0.677	56,989
0.7952	0.7563	56,095
0.8752	0.8386	54,934
0.8932	0.86	54,629
1	1	52,190

Source: J. Gmehling, U. Onken, and W. Arlt, *Vapor–Liquid Equilibrium Data Collection* (multiple volumes) (Frankfurt: DECHEMA, 1977–1980).

From these data, determine the value of the two-suffix Margules parameter, A. Compare your result to that obtained in Example 7.12. What value do you think is more accurate?

8.27 Liquid–vapor equilibrium data have been collected for a binary system of methanol (1)–water (2) at 40°C. Mole fraction of liquid vs. total pressure are reported in the table below. Develop a computer spreadsheet or write a program to find the three-suffix Margules parameters, A and B, that best fit the data by doing the following:

x_1	y_1	P [Pa]
0	0	7,295
0.05	0.275	9,562
0.1	0.436	11,695
0.15	0.543	13,682
0.2	0.618	15,536
0.25	0.675	17,256
0.3	0.720	18,883
0.35	0.756	20,390
0.4	0.786	21,817
0.45	0.811	23,150
0.5	0.833	24,404
0.55	0.853	25,604
0.6	0.871	26,751

Continued

x_1	y_1	P [Pa]
0.65	0.888	27,845
0.7	0.903	28,898
0.75	0.918	29,925
0.8	0.933	30,938
0.85	0.949	31,939
0.9	0.964	32,952
0.95	0.981	33,979
1	1	35,032

Source: J. Gmehling, U. Onken, and W. Arlt, *Vapor–Liquid Equilibrium Data Collection* (multiple volumes) (Frankfurt: DECHEMA, 1977–1980).

(a) minimizing the objective function $OF_P = \sum \left(P_{exp} - P_{calc} \right)_i^2$

(b) minimizing the objective function $OF_{g^E} = \sum \left(g^E_{exp} - g^E_{calc} \right)_i^2$

(c) minimizing the objective function $OF_\gamma = \sum \left[\left(\dfrac{\gamma_1 - \gamma_1^{calc}}{\gamma_1} \right)^2 + \left(\dfrac{\gamma_2 - \gamma_2^{calc}}{\gamma_2} \right)^2 \right]_i$

(d) applying linear regression using the method of Example 8.11

Compare the results from parts (a)–(c) to the results using ThermoSolver.

8.28 Test the liquid–vapor equilibrium data for the binary system of methanol (1)–water (2) at 40°C presented in Problem 8.27 for thermodynamic consistency by using the area test.

8.29 The following vapor–liquid equilibrium data have been reported for a binary mixture of acetone (1) in chloroform (2) at 35. 17°C. Test these data for thermodynamic consistency.

x_1	y_1	P [Pa]
0	0	39,086
0.0821	0.05	37,273
0.1953	0.146	35,019
0.2003	0.143	34,926
0.3365	0.317	33,232
0.4182	0.437	33,125
0.4917	0.544	33,726
0.595	0.682	35,593
0.709	0.806	38,100
0.8182	0.897	41,073
0.8768	0.897	42,634
0.938	0.938	42,687
0.972	0.972	44,287
1	1	45,941

Source: J. C. Chu, S. L. Wang, S. L. Levy, and R. Paul, textitVapor–Liquid Equilibrium Data (J. W. Edwards, 1956).

8.30 The following vapor–liquid equilibrium data have been reported for a binary mixture of acetone (1) in water (2) at 1 atm. Test these data for thermodynamic consistency.

x_1	y_1	T [°C]
0	0	100.00
0.015	0.325	89.60
0.036	0.564	79.40
0.074	0.734	68.30
0.175	0.8	63.70
0.259	0.831	61.10
0.377	0.84	60.50
0.505	0.849	59.90
0.671	0.868	59.00
0.804	0.902	58.10
0.899	0.938	57.40

Source: J. Gmehling, U. Onken, and W. Arlt, *Vapor–Liquid Equilibrium Data Collection* (multiple volumes) (Frankfurt: DECHEMA, 1977–1980).

8.31 You wish to fit the benzene (1)–isooctane (2) system to the following model for g^E:

$$g^E = x_1 x_2 (A + B(x_1 - x_2))$$

The system temperature of interest is 200°C. After a literature search, the only vapor–liquid equilibrium data at this temperature that you can find is:

T	P	x_1	y_1
200°C	11.6 bar	0.25	0.37

For the pure components, the Antoine constants

$$\ln P^{\text{sat}} = A - B/(T + C)$$

where P^{sat} sat is in torr (1/760 atm) and T is in K, liquid densities, and second virial coefficients are as follows:

Pure Species Data	Antoine Constants			ρ^l [g/cm³]	B_{ii} [cm³/mol]
	A	B	C		
Benzene (C_6H_6)	15.90	2788.51	−52.36	0.874	−490.0
Isooctane (C_8H_{18})	15.685	2896.3	−52.41	0.688	−833.8

(a) Using only the data given above, as accurately as you can, find the constants A and B (in J/mol). State any assumptions that you make.

(b) Is it ever possible for benzene and isooctane to split into two partially miscible liquid phases? Explain. If so, in what temperature range would you start to look for partial miscibility?

8.32 At a 300 K and 1 bar, a binary mixture of species a and b form two partially miscible liquid phases. The following activity coefficients at infinite dilution are reported: $\gamma_a^\infty = 8$ and $\gamma_b^\infty = 15$. Using the three-suffix Margules equation, determine the composition of two liquid phases in equilibrium.

8.33 Consider a binary liquid mixture of hexane (1) and acetone (2). At 15°C and 300 bar, this mixture forms two partially miscible liquid phases. Phase α has 20 total moles with $x_1^\alpha = 0.2$, while phase β has 10 total moles with $x_1^\beta = 0.8$. The following data are available at 15°C:

Species i	MW [g/mol]	v_i^l [cm^3/mol]	P_i^{sat} [kPa]
Hexane	86	130.5	12.7
Acetone	58	73.4	19.5

(a) Draw a schematic of the system, labeling it with all the information that you have. Make your schematic as *accurate* as possible; for example, consider which phase belongs on top.

(b) Are the like interactions stronger or weaker than the unlike interactions? Explain.

(c) Calculate the value of f_1.

(d) You wish to use the two-suffix Margules equation to describe this system. Based on the data above, come up with a value for the two-suffix Margules parameter, A.

(e) Estimate to what temperature you need to bring the system described above to make it completely miscible, that is, to make it have only one phase present. *State the important assumptions that you make.*

(f) Estimate the value of $\mathcal{H}_1$ at 15°C and 300 bar.

8.34 At 25°C and 1 bar, the following composition has been reported for a liquid–liquid mixture of CHCl$_3$ (a) and H$_2$O (b) : $x_a^\alpha = 0.987$ and $x_a^\beta = 0.0013$. From these data predict the three-suffix Margules parameters, A and B, for this binary mixture.

8.35 The Wilson equation requires positive values for binary parameters Λ_{ab} and Λ_{ba}. Verify that this activity coefficient model is incapable of describing the instability of partially miscible liquids.

8.36 Tetrahydrofuran (a) and water (b) separate into two liquid phases at 1 bar and 50°C. Determine the composition of each liquid phase. The following three-suffix Margules parameters have been obtained for this binary system:

$$A = 7395 \left[J/mol \right] \qquad \text{and} \qquad B = -1380 \left[J/mol \right]$$

8.37 Tetrahydrofuran (a) and water (b) separate into two liquid phases at 1 bar and 50°C. Determine the composition range over which the system is inherently unstable and will spontaneously separate into two phases. The following three-suffix Margules parameters have been obtained for this binary system:

$$A = 7395 \left[J/mol \right] \qquad \text{and} \qquad B = -1380 \left[J/mol \right]$$

8.38 A binary mixture of water (a) and 1-butanol (b) exhibits vapor–liquid–liquid equilibrium at 25°C. The activity coefficients at infinite dilution are given by $\gamma_a^\infty = 7.02$ and $\gamma_b^\infty = 72.37$. Determine the composition of the three phases and the system pressure at which VLLE occurs. At 25°C, the saturation pressure for 1-butanol is 875 Pa.

8.39 Ethanol (1) and n-hexane (2) are in equilibrium at 75°C. There are two liquid phases and one vapor phase present (VLLE). The compositions of the liquid phases are

$$x_1^\alpha = 0.9098 \qquad \text{and} \qquad x_1^\beta = 0.0902$$

The saturation pressures at 75°C are

$$P_1^{sat} = 0.888 \text{ bar} \qquad \text{and} \qquad P_2^{sat} = 1.223 \text{ bar}$$

(a) Is the two-suffix Margules expression ($g^E = Ax_1x_2$) a reasonable model for this system? Explain.

(b) Calculate the constant, A.

(c) What is the total pressure of the system? State and justify any assumptions that you make.

(d) What is the composition of the vapor phase?

8.40 *Para*-xylene (p-xylene) is used as a raw material to make polyester fibers. Billions of pounds are produced every year. It is manufactured by reforming crude oil. p-Xylene must then be separated. In this problem, we consider the final step of the separation process where p-xylene is separated from its isomers, *ortho*-xylene and *meta*-xylene. All the isomers have similar physical properties. The latent heats and the boiling and melting points of the pure xylene isomers are reported below.

	Δh_{vap} [J/mol]	T_b [K]	Δh_{fus} [J/mol]	T_m [K]
Ortho-xylene	36,838	417.3	−13,608	248.1
Meta-xylene	36,413	412.6	−11,577	225.4
Para-xylene	36,094	411.8	−17,125	286.6

(a) From the data given, explain why crystallization of *p*-xylene from a liquid mixture of xylenes works better than distillation to separate *p*-xylene.

(b) Consider a liquid feed containing 1 mol *ortho*-, 2 mol *meta*-, and 1 mol *para*-xylene. You may assume that the liquid forms an ideal solution and neglect Δc_P^{sl} in your calculations. Additionally, assume that each species forms a completely immiscible solid phase.

(i) Estimate the temperature at which the first solid of each isomer will form, at the mole fraction that it exists in the liquid.

(ii) Estimate the temperature where half the *p*-xylene in the liquid has crystallized. Has either of the other isomers formed a solid at this temperature?

(iii) What is the lowest temperature that the system can have in which only *p*-xylene crystallizes? What percentage of feed *p*-xylene has been separated?

(c) Instead, the liquid in part(b) is vaporized at 140°C. The saturation pressures of the isomers at this temperature are as follows:

	P^{sat} [kPa]
Ortho-xylene	89.7
Meta-xylene	103.5
Para-xylene	105.5

(i) What is the composition of vapor in equilibrium with the liquid of part (b).

(ii) Consider successive distillation "stages" where the vapor calculated in part (i) is successively condensed and then evaporated. At total reflux, how many stages would be required to reach a purity of 90% *p*-xylene?

8.41 Your roommate likes really, *really* cold beer. How cold can you set the freezer temperature to have the beer be as cold as possible without its freezing and causing a big mess? Assume beer is 4 mass% ethanol in water. For water, $\Delta h_{fus} = -6.01$ [kJ/mol].

8.42 Bismuth (*a*) and cadmium (*b*) form a eutectic at 144°C and $x_a = 0.45$. They are also completely immiscible in the solid phase. The melting points for bismuth and cadmium are 271°C and 321°C, respectively. The enthalpies of fusion of these elements are −10.46 kJ/mol and −6.1 kJ/mol, respectively. Estimate the parameters of an appropriate model for g^E based on these data.

8.43 In the winter, salt is used as a deicer to improve traction on the roads. Use the phase diagram on the next page to explain how this process works. Note that only part of the phase diagram is illustrated. How much salt would you add? To what temperature is this method effective?

8.44 A binary mixture of solids *a* and *b* is known to form three distinct solid phases: α, β and γ. Gibbs energy is plotted vs. mole fraction *a* for the two systems shown on the next page. Each of these plots is made at constant temperature and pressure. For each system, describe the phases that are present and their composition for the entire range of mole fraction *a*. Explain.

8.45 The excess Gibbs energy for a binary mixture of liquid *a* and liquid *b* is given by

$$g^E = 6000x_a x_b (1 - 0.0005T) \left[J/mol \right]$$

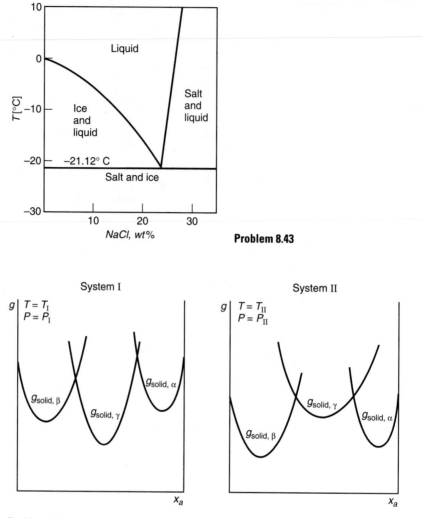

Problem 8.43

Problem 8.44

where T is in $[K]$. The solids of these species are completely immiscible. The enthalpies of fusion and melting temperatures are as follows:

$$\text{species } a: \qquad \Delta h_{\text{fus}} = -12\,[\text{kJ/mol}] \qquad T_m = 1000\,[\text{K}]$$

$$\text{species } b: \qquad \Delta h_{\text{fus}} = -10\,[\text{kJ/mol}] \qquad T_m = 800\,[\text{K}]$$

Determine the temperature and the composition at the eutectic point. You may neglect the change in heat capacity between the solid and liquid phases.

8.46 Antimony and lead form a eutectic at 251°C and 11.2 weight percent antimony. The enthalpy of fusion and melting point of lead are as follows:

$$\Delta h_{\text{fus}} = -5.1\,[\text{kJ/mol}] \qquad T_m = 327.5\,[°\text{C}]$$

As best you can, determine the composition of the lead-rich solid solution at the eutectic. State any assumptions that you make.

8.47 A phase diagram for the solid liquid equilibrium of a binary mixture of silver (Ag) and copper (Cu) is shown below. Answer the following questions. Note that the weight percentage is on the bottom and the mole percentage is on the top.

(a) What is the lowest temperature at which a binary mixture can exist entirely in the liquid phase? What is its composition?

(b) What is the most copper that can be present in a phase of solid silver? At what temperature does this occur?

(c) Consider a liquid mixture of 1 mole Ag and 4 moles Cu at 800°C. At equilibrium, what phases exist and what are their compositions? How many moles are present in each phase?

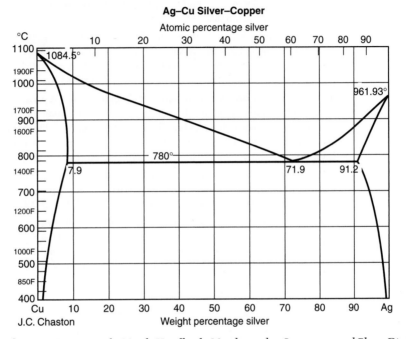

Ag–Cu Silver–Copper

[From T Lyman et al., *Metals Handbook, Metalography, Structures, and Phase Diagrams*, 8th ed. (Vol. 8) (Metals Park, OH: American Society for Metals, 1973).] Courtesy of ASM International.®

8.48 The following solid–liquid equilibrium data are available for a binary mixture of C and metastable γ–Fe.

$T[°C]$	X_c	x_c
1148	0.1000	0.2092
1154	0.0900	0.2072
1200	0.0877	0.1906
1250	0.0718	0.1689
1300	0.0613	0.1450
1350	0.0475	0.1179
1400	0.0333	0.0891
1450	0.0196	0.0570
1495	0.0079	0.0248

Source: T. Lyman et al., *Metal Handbook, Metalography, Structures, and Phase Diagrams*, 8th ed. (Vol. 8) (Metals Park, OH: American Society for Metals, 1973).

From these data estimate the melting point and enthalpy of fusion of γ–Fe.

8.49 When 9 g of urea (CH_4N_2O) are added to 1 kg of acetone at 1 bar, the boiling point of acetone raises 0.24 K. The normal boiling point of acetone is 329.2 K. From this datum, estimate acetone's enthalpy of vaporization.

8.50 Ethylene glycol, $C_2H_6O_2$, is used as an antifreeze to keep the water in the radiator of your car from freezing in the winter. Estimate the fraction of antifreeze, by volume, that you need to keep from freezing at $-10°C$. For water, $\Delta h_{fus} = -6.01$ [kJ/mol].

8.51 What is the minimum pressure required to desalinate seawater by reverse osmosis?

8.52 Find the osmotic pressure of a solution of 0.5 g of sucrose ($C_{12}H_{22}O_{11}$) in 500 g water at 25°C and 1 bar.

8.53 Use ThermoSolver to find the activity coefficient model parameters for the data presented in Problems 8.29 and 8.30.

8.54 Use ThermoSolver to determine the dew-point temperature and composition of a vapor mixture of 0.2 mole fraction n-hexane, 0.25 cyclohexane, 0.25 benzene, and 0.3 toluene at pressures of 1 bar and at 20 bar, using (a) Raoult's law; (b) liquid-phase nonideality but keeping the gas ideal; (c) the best answer that you can get. How do cases (a)–(c) compare at 1 bar? At 20 bar?

8.55 Use ThermoSolver to determine the bubble-point temperature and composition of a liquid mixture of 0.2 mole fraction n-hexane, 0.25 cyclohexane, 0.25 benzene, and 0.3 toluene at 1 bar and at 20 bar, using (a) Raoult's law; (b) liquid-phase nonideality, but keeping the gas ideal; (c) the best answer that you can get. How do cases (a)–(c) compare at 1 bar? At 20 bar?

8.56 Use ThermoSolver to determine the dew-point pressure and composition of a vapor mixture of 0.25 mole fraction methanol, 0.35 acetone, and 0.4 n-hexane at temperatures of 40°C and 200°C using (a) Raoult's law; (b) liquid-phase nonideality, but keeping the gas ideal; (c) the best answer that you can get. How do cases (a)–(c) compare at 40°C? At 200°C?

8.57 Use ThermoSolver to determine the bubble-point pressure and composition of a liquid mixture of 0.25 mole fraction methanol, 0.35 acetone, and 0.4 n-hexane at temperatures of 40°C and at 200°C using (a) Raoult's law; (b) liquid-phase nonideality, but keeping the gas ideal; (c) the best answer that you can get. How do cases (a)–(c) compare at 40°C? At 200°C?

CHAPTER 9

Chemical Reaction Equilibria

Learning Objectives

To demonstrate mastery of the material in Chapter 9, you should be able to:

▶ Determine the equilibrium composition for a system with a single chemical reaction and with multiple chemical reactions given the reaction stoichiometry, temperature, and pressure.

▶ Describe the role of thermodynamics vs. the role of kinetics in assessing chemical reactions.

▶ Write a balanced chemical reaction given a complete or incomplete set of reactants and products. Define the extent of reaction and the stoichiometric coefficient.

▶ Use thermochemical data to determine the equilibrium constant for a chemical reaction at any given temperature.

▶ For a determined reaction stoichiometry and initial reactant composition, write the equilibrium constant in terms of the extent of reaction for gas-phase, liquid-phase, and heterogeneous reactions for ideal or nonideal systems.

▶ Given a set of species in a system, apply the Gibbs phase rule to determine how many independent reactions need to be specified to constrain the system. Write an appropriate set of reactions and solve them using the equilibrium constant formulation. Alternatively, solve for the equilibrium composition using the minimization of Gibbs energy.

▶ Describe the role of non-Pv work in electrochemical systems. Define the roles of the anode, cathode, and electrolyte in an electrochemical cell. Given shorthand notation for an electrochemical cell, identify the oxidation and reduction reactions. Use data for the standard half-cell potential for reduction reactions, E^o, to calculate the standard potential of reaction E^o_{rxn}. Apply the Nernst equation to determine the potential in an electrochemical cell given a reaction and reactant concentrations.

▶ Define the following point defects and identify them as atomic defects or electronic defects: *vacancy, interstitial, substitutional impurity, misplaced atoms, electron, hole, dopant.*

▶ Use appropriate nomenclature to write the symbolic form of a defect in a crystal lattice. Write balanced chemical equations for defects in solids, and apply the principles of chemical reaction equilibrium to write expressions for the equilibrium constant in terms of defect concentration. Describe the formation of electronic defects in intrinsic semiconductors. Describe the process of doping by a set of chemical reactions of the appropriate defects.

Construct a Brouwer diagram to illustrate the effect of gas partial pressure on the concentration of defects in a solid.

▶ 9.1 INTRODUCTION

Analysis of chemical reactions is central to the profession of chemical engineering. There is an infinite possibility of arrangements of chemical bonds between elements to form molecules. The goal of applying equilibrium analysis to chemical reactions is to determine the extent to which products are favored given specified elemental composition, temperature, and pressure. However, chemical reaction equilibria tell us *nothing* about how fast a reaction will proceed; to answer that question we must study the reaction kinetics. We can calculate the equilibrium extent of a given reaction without specific knowledge of the mechanism or kinetics. This analysis tells us the farthest a reaction can possibly go. Hence, if thermodynamics tells us a given reaction will not proceed to a significant degree, we do not need to consider it further. On the other hand, only when thermodynamics tells us the reaction is possible do we need to consider whether we can achieve the reaction in a reasonable time to implement it.

Before we learn how to calculate how far a reaction will proceed at equilibrium using thermodynamics, it is instructive to explore the difference between thermodynamics and kinetics in analyzing chemical reactions. To look at the role each can play, consider the reaction of hydrogen bromide with butadiene shown in Figure 9.1a. These species will first react to form a positively charged transition-state carbonium ion, labeled

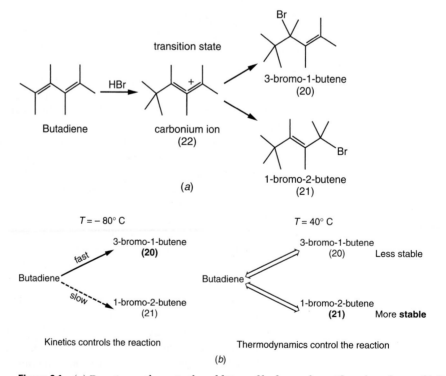

Figure 9.1 (*a*) Reaction pathway in the addition of hydrogen bromide to butadiene. (*b*) Kinetic and thermodynamic control in the addition of hydrogen bromide to butadiene.

"(22)" in the figure, together with a negatively charged bromide ion. The Br⁻ can then *add* into one of two places on the carbonium ion to form either 3-bromo-1-butene or 1-bromo-2-butene, labeled "(20)" and "(21)", respectively. Each of these steps requires that the species overcome a characteristic "activation energy" to proceed along the given reaction pathway and form product. The relative amount of each product depends on the reaction temperature. At −80°C, the reaction products are 80% of (20) and 20% of (21); however, at 40°C, we get 20% of (20) and 80% of (21). How can this difference in product distribution be explained?

There is a strong temperature dependence to reaction rates. Mathematically, reaction rate is usually characterized by an exponential dependence on T as given by the Arrhenius expression. At low temperature (−80°C), we are **kinetically controlled**. The reaction leading to (20) occurs more quickly, so more (20) is formed. Even though (20) is less stable than (21), more (20) forms because this pathway has a lower activation energy than the pathway to (21); that is, it has smaller energetic "hill" to climb to *get* there. Kinetic control is illustrated on the left-hand side of Figure 9.1*b*. In this case, the equilibrium analysis of thermodynamics does not apply. On the other hand, as the temperature is raised, the reacting system has enough energy to sample all the possible states along both reaction pathways. Said another way, the activation energy along the path to (21) does not limit access to this product. In this case, all reactions occur more quickly than the time scale of the process. Thus, all states in the system can be sampled, and the final reaction product distribution represents that which minimizes Gibbs energy. We then get more of the product that is energetically favorable. (As we will discuss in Section 9.2, we still get some of the less stable product due to entropic effects). This regime is **thermodynamically controlled**, as illustrated on the right-hand side of Figure 9.1*b*. This discussion is meant to caution you of the pitfalls of thermodynamic analysis of reacting systems; the calculations we perform in this chapter are valid only in the latter case, when the reaction is thermodynamically controlled.

Figure 9.1*b* illustrates the symbols we will use to distinguish these two cases. For kinetic control, we will use a one-sided arrow (⟶), indicating that the reaction proceeds in one direction, since it is rate limited. For thermodynamic control, we will use a two-sided arrow (⇌), indicating that the reaction samples all available bonding configurations and proceeds to its most favorable (lowest Gibbs energy) state. This chapter deals *only* with the latter. The fundamental question we wish to address is, "What effect do temperature, pressure, and composition have on the **equilibrium conversion** in a chemically reacting system?" This analysis tells us nothing about the **rates** at which a chemical reaction will proceed. It does, however, tell us to what extent a reaction is possible. As in phase equilibria, we will use the Gibbs energy of the system to study chemical reaction equilibria. To illustrate the use of G, we will first consider a specific reaction (Section 9.2). We will then describe the general formalism for a single reaction (Sections 9.3–9.5) and multiple reactions (Sections 9.7–9.8).

▶ 9.2 CHEMICAL REACTION AND GIBBS ENERGY

In this section, we will consider a *specific example* to illustrate how the same principle that we applied to solve phase equilibria problems also applies to chemical equilibria: the minimization of Gibbs energy. As we have seen, Gibbs energy represents a trade-off between reducing the energy of a system and maximizing its entropy.

Consider the following **ideal** gas reaction at low pressure

$$H_2 + Cl_2 \rightleftharpoons 2HCl \qquad\qquad (9.1)$$

First, lets consider the **energetics** of this chemical reaction. We can determine the relative energies of products vs. reactants by looking up the bond dissociation energies (bond strengths), D_{i-j}, of the three different species involved; these values are reported as follows:

$$D_{H-H} = 4.5 \text{ eV}, \qquad D_{Cl-Cl} = 2.5 \text{ eV}, \qquad D_{H-Cl} = 4.5 \text{ eV} \qquad (9.2)$$

Inspection of these bond energies reveals that when two molecules of HCl are formed from Reaction (9.1), the energy of the molecules present will be reduced by 2 eV; therefore HCl is energetically favored (more stable). Does this mean that the reaction will go to completion? As we have seen before, entropy also plays a role. In this case, if we had all product, we would have pure HCl; however, if some reactant remains, we have three species in a mixture. Three species can arrange themselves in many more configurations than one pure species can. Hence, having some H_2 and Cl_2 stay unreacted is more **entropically** favored. How do these two opposing tendencies balance? Again we turn to Gibbs energy.

Before we examine how to calculate the Gibbs energy, it is useful to introduce the concept of **extent of reaction**. This concept is based on the fact that once we specify the initial composition of the system, we are limited by Reaction (9.1) as to the possible composition of the system once the reaction has completed to equilibrium (or to any degree toward equilibrium). For illustration, let's consider a system in which we initially have 1 mole of H_2 (species 1) and 1 mole of Cl_2 (species 2) at 1 bar total pressure. These species can react stoichiometrically to form HCl (species 3). The amount of each species with which we end up is constrained by Reaction (9.1). No matter how much species 1 reacts, it will always consume an equal amount of species 2; therefore,

$$n_2 = n_1 \qquad (9.3)$$

Similarly, the amount of species 3 present is merely the product that results from the amount of species 1 that has reacted. Since we started with 1 mole of species 1, we can mathematically relate how much species 3 is present by knowing how much species 1 is left, as follows:

$$n_3 = 2(1 - n_1) \qquad (9.4)$$

Examination of Equations (9.3) and (9.4) reveals that the composition of all three species can be determined once we know how far the reaction has proceeded. Rather than arbitrarily constraining species 2 and 3 to species 1, as we did in the formulation of Equations (9.3) and (9.4), it is easier if we put all three species' composition in terms of how far the reaction has proceeded, or *the extent of reaction, ξ*. We designate one species, typically the limiting reactant, upon which to base the extent of reaction and then write the other species in relation to the species we have chosen. For example, we can define ξ based on how much H_2 has reacted. The extent of reaction will also tell us how much Cl_2 has reacted. Since we started with 1 mole of each species, the number of moles of each species after the reaction has proceeded by an extent of reaction, $\xi_{(9.1)}$, is given by[1]

$$n_1 = 1 - \xi_{(9.1)} \qquad (9.5)$$

---

[1] The equations for ξ that follow are valid only for the specific reaction stoichiometry given by Reaction (9.1) and are, therefore, denoted with the subscript. The more general treatment follows in Section 9.3.

and
$$n_2 = 1 - \xi \quad (9.1)$$
(9.6)

What are the units for ξ? Similarly, we can relate how much HCl was formed to the extent of reaction:

$$n_3 = 2\xi \quad (9.1)$$
(9.7)

In this case, the possible extent of reaction ranges from $\xi = 0$ mole (no reaction) to $\xi = 1$ mole (complete reaction).

We now wish to calculate the total Gibbs energy of the system whose initial state is described above for all possible extents of reaction; we can then determine at which composition the Gibbs energy is the smallest. This minimum in Gibbs energy will represent the equilibrium conversion. The total Gibbs energy is given by the appropriate proportions of partial molar Gibbs energies, as prescribed by Equation (6.34):

$$G = \sum n_i \overline{G}_i = \sum n_i \mu_i = n_1 \mu_1 + n_2 \mu_2 + n_3 \mu_3 \tag{9.8}$$

where we have replaced the partial molar Gibbs energy with the chemical potential. The chemical potential of each component in an ideal gas is given by Equation (7.6):

$$\mu_i = g_i^o + RT \ln \frac{p_i}{1 \text{ bar}} \tag{9.9}$$

where the reference state is the ideal gas state at the temperature of the reaction and a partial pressure of 1 bar (p_i then has units of bar.). We should also note that since we specified the pressure of the reference state, g_i^o is a function of *only* temperature. Substitution of Equation (9.9) into Equation (9.8) yields

$$G = n_1 g_1^o + n_2 g_2^o + n_3 g_3^o + RT(n_1 + n_2 + n_3) \ln P + RT[n_1 \ln y_1 + n_2 \ln y_2 + n_3 \ln y_3] \tag{9.10}$$

The last term in Equation (9.10) corresponds to a decrease in the Gibbs energy of the system due to mixing $\Delta g_{\text{mix}}^{\text{ideal}}$. This term quantifies the contribution that entropy plays in determining G of this ideal gas. If we now replace each species' composition with the extent of reaction using Equations (9.5) through (9.7), we get

term 1

$$G = \boxed{(1 - \xi)(g_1^o + g_2^o) + \xi g_3^o} + 2RT \ln P + RT[(1 - \xi) \ln y_1 + (1 - \xi) \ln y_2 + \xi \ln y_3] \tag{9.11}$$

Figure 9.2 shows a plot of the Gibbs energy of this chemical reaction, as defined by Equation (9.11), as a function of ξ. The component due to the pure species contribution is labeled "term 1" in the plot and in Equation (9.11). To obtain the Gibbs energy of the system, we must add the Gibbs energy of mixing to term 1, as illustrated in the figure. The product, species 3, has lower Gibbs energy than the reactants, species 1 and 2. However, the equilibrium conversion is not pure species 3, but rather the composition at which the Gibbs energy is a minimum (as labeled). This result comes from the entropy of mixing term in Equation (9.11). Again, we have a trade-off between entropy and energy. If we start with only species 1 and 2, as in the example above, we will follow the arrow to the right until we reach the lowest Gibbs energy. This point will represent the equilibrium conversion. Conversely, if the system starts as all species 3, it can minimize its

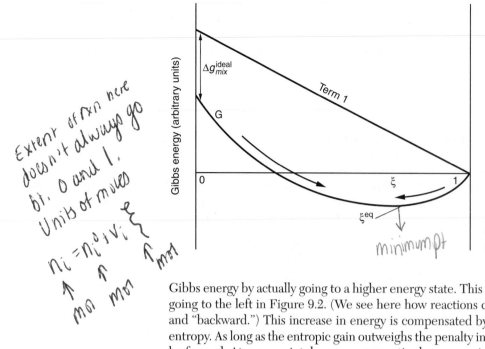

Extent of rxn here doesn't always go bt. 0 and 1. Units of moles

$n_i = n_i^0 + v_i \xi$

mol   mol   mol

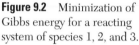

Figure 9.2 Minimization of Gibbs energy for a reacting system of species 1, 2, and 3.

Gibbs energy by actually going to a higher energy state. This is represented by the arrow going to the left in Figure 9.2. (We see here how reactions can proceed both "forward" and "backward.") This increase in energy is compensated by an even larger increase in entropy. As long as the entropic gain outweighs the penalty in energy, species 1 and 2 will be formed. At some point, however, energetics become as important. At that point, the Gibbs energy is a minimum and we are again at the equilibrium conversion. It does not matter where we start! Given identical T, P, and proportion of elements, we will always end up with the same equilibrium composition.

This analysis illustrates the generality of applying equilibrium analysis to chemically reacting systems. Since all the configurations of the system are adequately sampled at equilibrium (as illustrated by the reaction of butadiene shown in Figure 9.1*b*), we need to know neither the starting composition nor the mechanism of reaction to predict what configuration the system will end up with in its equilibrium state. We merely need to specify the amount of each of the elements that are present and the system temperature and pressure.

▶ 9.3 EQUILIBRIUM FOR A SINGLE REACTION

So far, we have examined specific cases to illustrate the effect of kinetics vs. thermodynamics upon reacting systems (butadiene) and how the thermodynamic property Gibbs energy allows us to calculate equilibrium compositions by quantifying the trade-off between energy and entropy (HCl). We now wish to develop a general approach so that we can analyze the chemical reaction equilibria for any system of interest.

The reaction we considered in Reaction (9.1), formation of HCl from H_2 and Cl_2, can be written as follows:

$$A_1 + A_2 = 2A_3 \tag{9.12}$$

where A_i represents compound i of the chemical reaction, that is,

$$A_1 = H_2, \qquad A_2 = Cl_2, \qquad A_3 = HCl$$

and we have replaced the reaction arrows with an equals sign. We can rewrite Equation (9.12) by subtracting $A_1 + A_2$ from each side:

$$0 = 2A_3 - A_1 - A_2 \tag{9.13}$$

To generalize, we can introduce ν_i, the **stoichiometric coefficient**. The stoichiometric coefficient tells us the proportion in which a given species is produced, or reacts, given a particular reaction. Equation (9.13) is then written

$$\nu_1 A_1 + \nu_2 A_2 + \nu_3 A_3 = 0 \tag{9.14}$$

where
$$\nu_1 = -1, \qquad \nu_2 = -1, \qquad \nu_3 = 2$$

By convention, the stoichiometric coefficient is positive for products, negative for reactants, and zero for inerts, that is,

$$\nu_{\text{reactants}} < 0, \qquad \nu_{\text{products}} > 0, \qquad \nu_{\text{inerts}} = 0 \tag{9.15}$$

We now wish to generalize this approach to any possible reaction. This is particularly useful for making complex chemical equilibrium analysis amenable to computer solutions. Using the formalism developed in the example above, we can describe the stoichiometry of any chemical reaction as follows:

$$\sum \nu_i A_i = 0 \tag{9.16}$$

By defining the species, A_i, along with their stoichiometric coefficients, ν_i, in Equation (9.16), a given chemical reaction is completely specified. Moreover, the production or consumption of reacting species is not independent; rather, Equation (9.16) constrains how each of the reacting species changes. To see this point, consider the ratio of the change in the number of moles of species 1 to species 2. It can be defined according to the ratio of the respective stoichiometric coefficients:

$$\frac{(\text{change in moles})_1}{(\text{change in moles})_2} = \frac{dn_1}{dn_2} = \frac{\nu_1}{\nu_2} \tag{9.17}$$

or, rearranging,

$$\frac{dn_1}{\nu_1} = \frac{dn_2}{\nu_2} \tag{9.18}$$

Similarly, between species 1 and 3,

$$\frac{dn_1}{dn_3} = \frac{\nu_1}{\nu_3} \qquad \text{or} \qquad \frac{dn_1}{\nu_1} = \frac{dn_3}{\nu_3} \tag{9.19}$$

The change in all the species that are reacting is determined by the specific reaction stoichiometry defined in Equation (9.16). As we have seen in the HCl example, we can define the *extent of reaction*, ξ,[2] which is formally given by

$$\frac{dn_1}{\nu_1} = \frac{dn_2}{\nu_2} = \frac{dn_3}{\nu_3} \equiv d\xi \tag{9.20}$$

where ξ is zero at the initial state. The extent of reaction has units of moles and is a measure of how far a given reaction has proceeded. The extent of reaction is useful because it provides a single variable to relate how every species in the reaction has changed. It is not restricted to values between 0 and 1; it can even be negative if the

---

[2] ξ is also called the *reaction coordinate*.

reaction proceeds in the reverse direction than we had anticipated when we wrote the reaction stoichiometry.

We can rewrite Equation (9.20) for any species i:

$$dn_i = \nu_i d\xi \tag{9.21}$$

Integrating, with the initial condition specified by the definition of ξ, we get

$$n_i = n_i^o + \nu_i \xi \tag{9.22}$$

where n_i^o is the initial concentration of species i. We can sum all the species that are present in a given phase to get the total number of moles in that phase. For example, for the vapor phase:

$$n^v = \sum_{vapor} n_i = n^o + \nu\xi \tag{9.23}$$

where

$$n^o = \sum_{vapor} n_i^o \quad \text{and} \quad \nu = \sum_{vapor} \nu_i \tag{9.24}$$

Moreover, the mole fraction of each species in the vapor is given by

$$y_i = \frac{n_i}{n^v} = \frac{n_i^o + \nu_i \xi}{n^o + \nu\xi} \tag{9.25}$$

Equations (9.22) and (9.25) allow us to calculate the composition and vapor-phase mole fractions, respectively, of a reacting system based on ξ. As an example of the latter, consider the reaction described in Figure 9.2. In this case, $n^o = 2$ and $\nu = 0$; therefore,

$$y_{H_2} = \frac{1 - \xi}{2}, \quad y_{Cl_2} = \frac{1 - \xi}{2}, \quad \text{and} \quad y_{HCl} = \xi \tag{9.26}$$

Equations analogous to (9.23) and (9.25) can also be written for any liquid or solid phases present. For example, the mole fraction in the liquid phase is given by

$$x_i = \frac{n_i}{n^l} = \frac{n_i^o + \nu_i \xi}{n^o + \nu\xi} \tag{9.27}$$

In applying Equations (9.25) and (9.27) to heterogeneous reactions, it is important to remember to divide only by the total number of moles of species in the same phase.

► **EXAMPLE 9.1**
Extent of Reaction for Fuel Cell Fuel Source

Fuel cells provide an attractive alternative energy source. They require an H_2 feed stream to operate. Consider a fuel cell based on the direct conversion of methanol to form hydrogen:

$$H_2O(g) + CH_3OH(g) \rightleftarrows CO_2(g) + 3H_2(g)$$

The reaction is carried out at 60°C and low pressure, with a feed of twice as much water as methanol. The equilibrium extent of reaction is $\xi = 0.87$. How many moles of H_2 can be produced per mole of CH_3OH in the feed? What is the mole fraction of H_2?

SOLUTION Taking a basis of 1 mole methanol, the initial composition can be written as:

$$n_{CH_3OH}^o = 1 \quad \text{and} \quad n_{H_2O}^o = 2$$

with the number of moles of product equal to zero. Plugging these values into Equations (9.22) and (9.23) gives

$$
\begin{aligned}
n_{CH_3OH} &= n^o_{CH_3OH} + \nu_{CH_3OH}\xi &= 1 - \xi \\
n_{H_2O} &= n^o_{H_2O} + \nu_{H_2O}\xi &= 2 - \xi \\
n_{CO_2} &= n^o_{CO_2} + \nu_{CO_2}\xi &= \xi \\
n_{H_2} &= n^o_{H_2} + \nu_{H_2}\xi &= 3\xi \\
\hline
n^v &= n^o + \nu\xi &= 3 + 2\xi
\end{aligned}
$$

Note that the total number of moles, n^v, can be obtained by adding each of the individual species. Solving for the number of moles of H_2 gives

$$
n_{H_2} = 3\xi = 2.61 \text{ moles}
$$

and a mole fraction of

$$
y_{H_2} = \frac{n_{H_2}}{n^v} = \frac{3\xi}{3 + 2\xi} = 0.55
$$

◀

We can apply the general formalism developed so far to the criteria for chemical equilibrium. At constant temperature and pressure, the condition for equilibrium is the minimization of Gibbs energy. The change in Gibbs energy is given by Equation (6.32):

$$
dG = \sum \mu_i dn_i = \sum \mu_i \nu_i d\xi \tag{9.28}
$$

where we have used Equation (9.21) for dn_i. The system comes to chemical equilibrium at the extent of reaction for which the Gibbs energy is a minimum. Applying this criterion to Equation (9.28) gives

$$
\frac{dG}{d\xi} = 0 = \sum \mu_i \nu_i \tag{9.29}
$$

This relation is the mathematical equivalent of the equilibrium conversion denoted at the minima of Gibbs energy on Figure 9.2.

To solve Equation (9.29), we need an expression for chemical potential. As we did in the case of phase equilibria, we can relate chemical potential to fugacity. To write such an expression we need a well-defined reference state. A particularly common reference state, called the *standard state*, is defined as the pure species at the temperature of the reaction and a pressure of 1 bar (or 1 atm, when appropriate). We will freely interchange 1 atm and 1 bar as the reference state pressure. Since the Gibbs energy has only a weak dependence on P, its value is, for all practical purposes, identical at these two pressures. Using the standard state, the chemical potential is written as

$$
\mu_i = g^o_i + RT \ln\frac{\hat{f}_i}{f^o_i} \tag{9.30}
$$

where the pure species molar Gibbs energy, g^o_i, is a function of *temperature only*. As we will see shortly, using the standard state as a reference state will allow us access to a vast array of tabulated thermochemical data. Substituting Equation (9.30) into Equation (9.29):

$$
0 = \sum \left[g^o_i + RT \ln\frac{\hat{f}_i}{f^o_i} \right] \nu_i \tag{9.31}
$$

and rearranging:

depends only on temp

$$\ln \prod \left(\frac{\hat{f}_i}{f_i^o}\right)^{v_i} = -\frac{\sum v_i g_i^o}{RT} \equiv -\frac{\Delta g_{rxn}^o}{RT} \tag{9.32}$$

where we have used the mathematical identity that the sum of logarithms is equal to the log of the products. We have also defined a new term, $\Delta g_{rxn}^o = \sum v_i g_i^o$, which is the Gibbs energy of reaction. It is based on the pure component Gibbs energies proportioned by the stoichiometric coefficients of the reacting species. For a given reaction stoichiometry, this term is a function *only* of temperature.

Examining Equation (9.32), we see that, in a sense, we have divided our chemical reaction equilibrium problem into two parts—one represented by the left-hand side of the equation and the second by the right-hand side. To solve the left-hand side, we need to employ the extent of reaction formulation developed above. It is here where process variables such as feed composition and reaction pressure affect the value of ξ. The details of this part of the chemical reaction equilibrium calculation will be explored in Section 9.5. The solution of the right-hand side, on the other hand, simply depends on determining the value of Δg_{rxn}^o. As we just saw, Δg_{rxn}^o depends only on the temperature of the system. *Thus, once the reaction stoichiometry has been defined, the reaction temperature is the only variable that needs to be specified to solve the right-hand side of Equation 9.32.*

Since determination of the right-hand side of Equation (9.32) requires only the reaction T, we give it its own name, the equilibrium constant, K, which is defined as follows:

$$\ln K \equiv -\frac{\Delta g_{rxn}^o}{RT} \tag{9.33}$$

where the natural logarithm is used to simplify the left-hand side of Equation (9.32). Note that the equilibrium constant is only "constant" at a given temperature, that is,

$$K = f(T \text{ only}) \qquad \textit{K depends on T only} \tag{9.34}$$

We can now simplify our criteria for chemical reaction equilibrium as follows:

temp, pressure, comp.

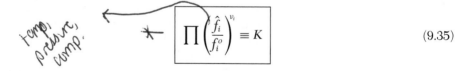

$$\prod \left(\frac{\hat{f}_i}{f_i^o}\right)^{v_i} \equiv K \tag{9.35}$$

In Section 9.4 we will learn how to solve for K at any temperature from available thermochemical data, while we will explore how to relate the left-hand side to the extent of reaction in Section 9.5.

▶ 9.4 CALCULATION OF K FROM THERMOCHEMICAL DATA

To find K, we typically use available thermochemical data (Δg_i, Δh_i, or permutations of these pure species properties), which allow us to calculate the Gibbs energy of reaction. We then solve for K via Equation (9.33). We will first look at how to calculate K from Δg_{rxn}^o at 298 K; then we will examine how to determine Δg_{rxn}^o at any T. Appendix F provides a list of common sources to search for thermodynamic property data.

Calculation of *K* from Gibbs Energy of Formation

The most common thermochemical data available to calculate the equilibrium constant are in the form of the Gibbs energy of formation, Δg_f^o; Appendix A.3 shows some representative values for 25°C and 1 bar. The Gibbs energy of formation is defined analogously to the enthalpy of formation, introduced in Section 2.6. It is equal to the Gibbs energy of reaction when the species of interest is formed from its pure elemental constituents, as found in nature, that is,

$$\text{elements} \overset{\Delta g_f}{\longleftrightarrow} \text{species } i \qquad (9.36)$$

The Gibbs energy of formation of a pure element, as it is found in nature, is identically zero.

With the Gibbs energies of formation available, it is straightforward to calculate the Gibbs energy of reaction. Such a calculation path for the Gibbs energy of reaction at 298 K is illustrated in Figure 9.3. In the dashed (calculation) path, the reactants are first decomposed into their constituent elements, as found in nature. This part of the path is given by Δg_1. The constituent elements are then allowed to react to form products, as given by Δg_2. The stoichiometric coefficients of the reactants are negative, making the signs for Δg_1 consistent with the definition of Gibbs energy of formation above. Equating the two paths yields

$$\Delta g_{rxn,298}^o = \Delta g_1 + \Delta g_2 = \sum_{reactants} v_i\left(\Delta g_{f,298}^o\right)_i + \sum_{products} v_i\left(\Delta g_{f,298}^o\right)_i = \sum v_i\left(\Delta g_{f,298}^o\right)_i$$

$$(9.37)$$

Thus, if Gibbs energies of formation are available for all the species in the chemical reaction of interest, the Gibbs energy of reaction can be determined by scaling each species' Δg_f by its stoichiometric coefficient. In summary,

$$\Delta g_{rxn}^o = \sum v_i g_i^o = \sum v_i(\Delta g_f^o)_i \qquad (9.38)$$

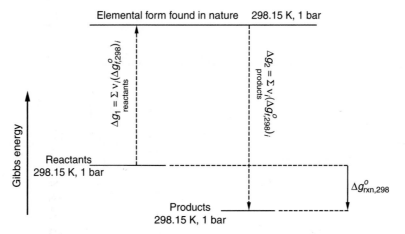

Figure 9.3 Calculation path of Δg^o from the standard Gibbs energies of formation, $\left(\Delta g_f^o\right)_i$.

▶ **EXAMPLE 9.2**
Calculation of K
at 298 K

Calculate the equilibrium constant at 298 K for the reaction of Example 9.1:

$$H_2O(g) + CH_3OH(g) \rightleftarrows CO_2(g) + 3H_2(g)$$

SOLUTION The equilibrium constant can be found from the Gibbs energy of formation. In this case, Equation (9.38) can be written as follows:

Step 1

$$\Delta g^o_{\text{rxn},298} = \sum \nu_i \left(\Delta g^o_{f,298}\right)_i = \left(\Delta g^o_{f,298}\right)_{CO_2} + 3\left(\Delta g^o_{f,298}\right)_{H_2} - \left(\Delta g^o_{f,298}\right)_{H_2O} - \left(\Delta g^o_{f,298}\right)_{CH_3OH}$$

Taking values from Appendix A.3:

$$\Delta g^o_{\text{rxn},298} = (-394.36) + 3(0) - (-228.57) - (-161.96) = -3.83 \left[\text{kJ/mol}\right]$$

While water does not exist as a gas under these conditions, the hypothetical gaseous state was used for water. The practical application could be, for example, as the first step in calculating K for the reaction described in Example 9.1. The next step would be to account for the temperature dependence of K (next section) and find the equilibrium constant at 60°C, where H_2O exists as a gas at low pressure. Plugging the appropriate values (in SI units) into Equation (9.33) gives

$$K_{298} = \exp\left(-\frac{\Delta g^o_{298}}{RT}\right) = \exp\left(-\frac{-3,830}{(8.314)(298.15)}\right) = 4.69$$

◀

The Temperature Dependence of K

Most reactions we wish to analyze are at temperatures other than 25°C. If the Gibbs energies of formation (or other equivalent data) are available at the reaction temperature, the equilibrium constant can be calculated directly from Equation (9.33). However, often the data to calculate the Gibbs energy of reaction are available at a temperature different from the one of interest (typically, 25°C). By determining the temperature dependence of K, we can calculate the equilibrium constant of any reaction at any temperature from one set of Gibbs energy data.

To calculate the temperature dependence of K, once again we take advantage of the thermodynamic web to give us a relationship among the desired properties. We wish to find $d\ln K/dT$. However, by Equation (9.33),

$$\frac{d\ln K}{dT} = -\frac{d(\Delta g^o_{\text{rxn}}/RT)}{dT} \tag{9.39}$$

Applying the product rule to Equation (9.39), we get:

$$\frac{d\ln K}{dT} = \frac{\Delta g^o_{\text{rxn}}}{RT^2} - \frac{1}{RT}\frac{d\Delta g^o_{\text{rxn}}}{dT} = \frac{\Delta g^o_{\text{rxn}}}{RT^2} + \frac{\Delta s^o_{\text{rxn}}}{RT} \tag{9.40}$$

Since we are at constant pressure, the thermodynamic property relation given by Equation (5.14), $\left(\partial \Delta g^o_{\text{rxn}}/\partial T\right)_P = -\Delta s^o_{\text{rxn}}$, was used. By definition,

$$\Delta g^o_{\text{rxn}} = \Delta h^o_{\text{rxn}} - T\Delta s^o_{\text{rxn}} \tag{9.41}$$

so

$$\frac{d\ln K}{dT} = \frac{\Delta h^o_{\text{rxn}}}{RT^2} \tag{9.42}$$

where the enthalpy of reaction, Δh^o_{rxn}, is defined by Equation (2.72) as

$$\Delta h^o_{rxn} = \sum v_i h^o_i = \sum v_i \left(\Delta h^o_f\right)_i \tag{9.43}$$

Examination of Equation (9.42) shows that for exothermic reactions ($\Delta h^o_{rxn} < 0$), the equilibrium constant decreases as temperature increases, since RT^2 is always greater than zero. For endothermic reactions ($\Delta h^o_{rxn} > 0$), the equilibrium constant increases as temperature increases.

$\Delta h^o_{rxn} =$ Constant

For small temperature ranges or approximate calculations, we can assume Δh^o_{rxn} is independent of temperature. If we separate variables in (Equation 9.42), we get

$$d\ln K = \left(\frac{\Delta h^o_{rxn}}{R}\right)\frac{dT}{T^2} \tag{9.44}$$

Integration of Equation (9.44) yields

$$\ln\frac{K_2}{K_1} = -\frac{\Delta h^o_{rxn}}{R}\left(\frac{1}{T_2} - \frac{1}{T_1}\right) \tag{9.45}$$

If we have data for K available at 298 K, Equation (9.45) becomes

$$\ln\frac{K_T}{K_{298}} = -\frac{\Delta h^o_{rxn}}{R}\left(\frac{1}{T} - \frac{1}{298}\right) \tag{9.46}$$

Step 2

▶ **EXAMPLE 9.3**
Calculation of K at
60°C from K_{298}

Calculate the equilibrium constant at 60°C for the reaction of Example 9.1:

$$H_2O(g) + CH_3OH(g) \rightleftarrows CO_2(g) + 3H_2(g)$$

SOLUTION   From Example 9.2,

$$K_{298} = 4.69$$

We can find the enthalpy of reaction from the enthalpies of formation:

$$\Delta h^o_{rxn} = \sum v_i\left(\Delta h^o_f\right)_i = \left(\Delta h^o_f\right)_{CO_2} + 3\left(\Delta h^o_f\right)_{H_2} - \left(\Delta h^o_f\right)_{H_2O} - \left(\Delta h^o_f\right)_{CH_3OH}$$

Using the values from Appendix A.3:

$$\Delta h^o_{rxn,298} = (-393.51) + 3(0) - (-241.82) - (-200.66) = 48.97 \left[kJ/mol\right]$$

Using this value in Equation (9.35):

$$\ln\frac{K_{333}}{4.69} = -\frac{48,970}{8.314}\left(\frac{1}{333} - \frac{1}{298}\right) = 2.08$$

Solving for the equilibrium constant:

$$K_{333} = 37.44$$

The value of the equilibrium constant of this reaction increases by an order of magnitude as the temperature increases from 25° to 60°C. ◀

$\Delta h_{rxn}^o = \Delta h_{rxn}^o(T)$

In general, the enthalpy of reaction is a function of temperature, i.e., $\Delta h_{rxn}^o = \Delta h_{rxn}^o(T)$. This dependence can be quantified using the heat capacity, $c_{P,i}$, of each of the i reacting species. In general, the heat capacity can be expressed in the following form:

$$\frac{(c_p)_i}{R} = A_i + B_i T + C_i T^2 + D_i T^{-2} + E_i T^3 \tag{9.47}$$

The enthalpy of reaction at any temperature T was found in Example 2.14 to be

$$\Delta h_{rxn,T}^o = \Delta h_{rxn,298}^o + \int_{298}^{T} \left(R \sum_i \nu_i \left(A_i + B_i T + C_i T^2 + \frac{D_i}{T^2} + E_i T^3 \right) \right) dT \tag{9.48}$$

where the enthalpy of reaction at 298 K, $\Delta h_{rxn,298}^o$, is available from a table of standard enthalpies of formation (for example, see Appendix A.3). Integrating Equation (9.48) yields

$$\Delta h_{rxn,T}^o = \Delta h_{rxn,298}^o + R \left[\Delta A(T - 298) + \frac{\Delta B}{2}\left(T^2 - 298^2\right) + \frac{\Delta C}{3}\left(T^3 - 298^3\right) \right.$$
$$\left. - \Delta D \left(\frac{1}{T} - \frac{1}{298} \right) + \frac{\Delta E}{4}\left(T^4 - 298^4\right) \right] \tag{9.49}$$

where
$$\Delta A = \sum_i \nu_i A_i, \qquad \Delta B = \sum_i \nu_i B_i, \qquad \Delta C = \sum_i \nu_i C_i,$$
$$\Delta D = \sum_i \nu_i D_i, \qquad \text{and} \qquad \Delta E = \sum_i \nu_i E_i. \tag{9.50}$$

Substituting Equation (9.49) into Equation (9.42) gives

$$\frac{d\ln K}{dT} = \left[\frac{\Delta h_{rxn,298}^o}{R} + \Delta A(T - 298) + \frac{\Delta B}{2}\left(T^2 - 298^2\right) + \frac{\Delta C}{3}\left(T^3 - 298^3\right) \right.$$
$$\left. - \Delta D \left(\frac{1}{T} - \frac{1}{298} \right) + \frac{\Delta E}{4}\left(T^4 - 298^4\right) \right] \Big/ T^2 \tag{9.51}$$

Integrating Equation (9.51):

$$\ln\left(\frac{K_T}{K_{298}}\right) = \left\{ \begin{array}{l} \left[-\dfrac{\Delta h_{rxn,298}^o}{R} + \Delta A(298) + \dfrac{\Delta B}{2}\left(298^2\right) + \dfrac{\Delta C}{3}\left(298^3\right) \right. \\[2ex] \left. - \dfrac{\Delta D}{298} + \dfrac{\Delta E}{4}\left(298^4\right) \right]\left(\dfrac{1}{T} - \dfrac{1}{298} \right) \\[2ex] + \Delta A \ln\left(\dfrac{T}{298}\right) + \dfrac{\Delta B}{2}(T - 298) + \dfrac{\Delta C}{6}\left(T^2 - 298^2\right) \\[2ex] + \dfrac{\Delta D}{2}\left(\dfrac{1}{T^2} - \dfrac{1}{298^2} \right) + \dfrac{\Delta E}{12}\left(T^3 - 298^3\right) \end{array} \right\} \tag{9.52}$$

where the equilibrium constant at 298 K, K_{298} is available from a table of standard Gibbs energies of formation (for example, see Appendix A.3). ThermoSolver can be used to solve Equation (9.52) for reactions using species in the data base.

▶ EXAMPLE 9.4
Calculation of K_T

We wish to produce formaldehyde, CH_2O, by the gas-phase pyrolysis of methanol, CH_4O, according to

$$CH_4O(g) \rightleftharpoons CH_2O(g) + H_2(g)$$

(a) What is the equilibrium constant at room temperature? Would you expect an appreciable yield of product?

(b) Consider the reaction at 600°C and 1 bar. What is the equilibrium constant

(i) assuming $\Delta h^o_{rxn} = $ constant?

(ii) using $\Delta h^o_{rxn} = \Delta h^o_{rxn}(T)$?

SOLUTION First, we must get the appropriate thermochemical data. Data from Appendices A.2 and A.3 are summarized in Table E9.4.

(a) The equilibrium constant at 25°C can be calculated from the Gibbs energy of reaction. From the Table E9.4:

$$\Delta g^o_{rxn,298} = \sum \nu_i \left(\Delta g^o_{f,298}\right)_i = \left(\Delta g^o_{f,298}\right)_{CH_2O} + \left(\Delta g^o_{f,298}\right)_{H_2} - \left(\Delta g^o_{f,298}\right)_{CH_4O}$$

$$= 52.0\,[kJ/mol]$$

so

$$K_{298} = \exp\left(-\frac{\Delta g^o_{rxn,298}}{RT}\right) = \exp\left(-\frac{52,000}{(8.314)(298.15)}\right) = 7.64 \times 10^{-10}$$

This value is 11 orders of magnitude smaller than in Example 9.2. Since the equilibrium constant represents the degree to which products will form, this result tells us that very little formaldehyde will form at 298 K. This example illustrates the wide range of values the equilibrium constant can take. We used a hypothetical gaseous state for methanol, since it exists as a liquid at 298 K and 1 atm but as a gas at 600°C.

(b) (i) To calculate K at 600°C, we need the enthalpy of reaction. This value is available at 298 K from the thermochemical data in Table E9.4. Summing enthalpies of formation:

$$\Delta h^o_{rxn,298} = \sum \nu_i \left(\Delta h^o_{f,298}\right)_i = \left(\Delta h^o_{f,298}\right)_{CH_2O} + \left(\Delta h^o_{f,298}\right)_{H_2} - \left(\Delta h^o_{f,298}\right)_{CH_4O} = 84.7\,[kJ/mol]$$

TABLE E9.4 Summary of Thermochemical Data from Appendices A.2 and A.3

	CH_4O	CH_2O	H_2
$\left(\Delta g^o_{f,298}\right)_i$ [kJ/mol]	−162.0	−110.0	0
$\left(\Delta h^o_{f,298}\right)_i$ [kJ/mol]	−200.7	−116.0	0
ν_i	−1	1	1
A_i	2.211	2.264	3.249
B_i	1.222×10^{-2}	7.022×10^{-3}	0.422×10^{-3}
C_i	-3.450×10^{-6}	-1.877×10^{-6}	—
D_i	—	—	0.083×10^5

Since this reaction is endothermic, K gets larger with T. Applying Equation (9.46):

$$\ln\frac{K_{873}}{7.64 \times 10^{-10}} = -\frac{84,700}{8.314}\left(\frac{1}{873} - \frac{1}{298}\right) = 22.5$$

Solving for the equilibrium constant:

$$K_{873} = 4.63$$

This result indicates that at 600°C we get a noticeable amount of product.

(ii) To account for the change in the enthalpy of reaction with temperature, we must use the heat capacity data. From the Table E9.4, we get

$$\Delta A = \sum_i \nu_i A_i = 3.302$$

$$\Delta B = \sum_i \nu_i B_i = -4.776 \times 10^{-3}$$

$$\Delta C = \sum_i \nu_i C_i = 1.57 \times 10^{-6}$$

$$\Delta D = \sum_i \nu_i D_i = 0.083 \times 10^5$$

Plugging these values into Equation (9.52):

$$\ln\left(\frac{K_{873}}{7.64 \times 10^{-10}}\right) = \left\{ \begin{array}{l} \left[-\dfrac{84,700}{8.314} + (3.302)(298) - \dfrac{4.776 \times 10^{-3}}{2}\left(298^2\right)\right. \\[2mm] \left. + \dfrac{1.57 \times 10^{-6}}{3}\left(298^3\right) - \dfrac{0.083 \times 10^5}{298}\right]\left(\dfrac{1}{873} - \dfrac{1}{298}\right) \\[2mm] + 3.302\ln\left(\dfrac{873}{298}\right) - \dfrac{4.776 \times 10^{-3}}{2}(873 - 298) \\[2mm] + \dfrac{1.57 \times 10^{-6}}{6}\left(873^2 - 298^2\right) + \dfrac{0.083 \times 10^5}{2}\left(\dfrac{1}{873^2} - \dfrac{1}{298^2}\right) \end{array} \right\}$$

Performing the calculation on the right-hand side gives

$$\ln\left(\frac{K_{873}}{7.64 \times 10^{-10}}\right) = 23.2$$

which results in

$$K_{873} = 8.67$$

The assumption of constant heat capacity resulted in an equilibrium constant about half the size of this value. ◀

▶ 9.5 RELATIONSHIP BETWEEN THE EQUILIBRIUM CONSTANT AND THE CONCENTRATIONS OF REACTING SPECIES

We have just learned how to calculate a value for the equilibrium constant at any reaction temperature. Examination of Equation (9.35) shows that the equilibrium constant is also related to the fugacities of the reactants and products. We will now apply what we have learned about fugacities and reference states to relate the equilibrium constant directly to the extent of reaction. As we saw in Example 9.1, with ξ known, it is straightforward to calculate equilibrium concentrations and mole fractions. Since we have already examined how to calculate fugacities for the vapor phase and for condensed phases, we have the formalism by which to accomplish this goal. We will first consider gas-phase reactions.

The Equilibrium Constant for a Gas-Phase Reaction

Recall that our reference state for the gas phase is a pressure low enough that the gas behaves as an ideal gas. For the sake of chemical reaction calculations, we usually choose the standard state pressure to be 1 bar (or 1 atm). For most gases, intermolecular interactions are negligible at 1 bar and the ideal gas assumption is valid. However, if the gas is not ideal at this pressure, we go to a low enough pressure that it is an ideal gas, then extrapolate back to a pressure of 1 bar, assuming the gas is ideal. In this case, the standard state represents that of a hypothetical ideal gas at 1 bar where we have "turned off" the intermolecular interactions of the real gas. In either case, the standard state fugacity becomes

$$f_i^o = 1 \text{ bar} \tag{9.53}$$

For this choice, Equation (9.35) becomes

$$K = \prod \left(\hat{f}_i \, [\text{bar}] \right)^{\nu_i} = \prod (y_i \hat{\varphi}_i \, P[\text{bar}])^{\nu_i} \tag{9.54}$$

The pressure in Equation (9.54) is written in bar since the reference state fugacity has units of bar. From this point on, these units will be implicit.

Recall from Section 7.3 that there are three levels of rigor to solve for the fugacity coefficient:

(a) $\hat{\varphi}_i = \hat{\varphi}_i$

This is the rigorous solution, where the fugacity coefficient depends on concentration.

(b) $\hat{\varphi}_i = \varphi_i$ first approximation (Lewis fugacity rule)

In this case, we approximate the fugacity coefficient of species i in the mixture by the pure species fugacity coefficient, which is therefore independent of concentration.

(c) $\hat{\varphi}_i = 1$ second approximation (ideal gas)

Case (a) requires an iterative solution since equilibrium concentrations must be known to calculate the fugacity coefficient. However, we need the fugacity coefficient to calculate the equilibrium concentrations. Usually we do not approach chemical reaction equilibria with this level of rigor, but rather use case (b) ($\hat{\varphi}_i = \varphi_i$). For case (b), Equation (9.54) can be written

$$K = \prod (y_i)^{\nu_i} \prod (\varphi_i)^{\nu_i} P^{\nu} \tag{9.55}$$

For an *ideal gas* [case (c)], Equation (9.55) can be further simplified to

$$K = \prod (y_i)^{\nu_i} P^{\nu} \quad \text{(ideal gas)} \tag{9.56}$$

▶ **EXAMPLE 9.5**
Effect of Reactor
Conditions on the
Extent of Reaction

Consider the following general gas-phase reaction:

$$aA(g) + bB(g) \rightleftharpoons cC(g) + dD(g) \qquad\qquad \text{(E9.5A)}$$

The constant-pressure reactor also contains an inert species, I. Describe how the following reactor conditions affect yield of reaction products: **(a)** temperature; **(b)** pressure; **(c)** addition of inert; **(d)** additional (nonstoichiometric) reactant in feed.

SOLUTION The total number of moles in the reactor is

$$n_T^v = n_A + n_B + n_C + n_D + n_I$$

Equation (9.55) can be written

$$K = \frac{\left(\dfrac{n_C}{n_T^v}\right)^c \left(\dfrac{n_D}{n_T^v}\right)^d}{\left(\dfrac{n_A}{n_T^v}\right)^a \left(\dfrac{n_B}{n_T^v}\right)^b} \left[\prod (\varphi_i)^{\nu_i}\right] P^\nu$$

and, rearranging,

$$\frac{K(n_T^v)^\nu}{\left[\prod (\varphi_i)^{\nu_i}\right] P^\nu} = \frac{(n_C)^c (n_D)^d}{(n_A)^a (n_B)^b} \qquad\qquad \text{(E9.5B)}$$

We are now ready to consider how changes in reactor conditions can affect the equilibrium conversion of this gas-phase reaction.

(a) A change in *temperature* will most notably affect K (and, to a much lesser extent, φ_i). We can rearrange Equation (E9.5B) as follows:

$$\frac{K}{\left[\prod (\varphi_i)^{\nu_i}\right]} = \left[\frac{P^\nu}{(n_T^v)^\nu}\right] \frac{(n_C)^c (n_D)^d}{(n_A)^a (n_B)^b} \qquad\qquad \text{(E9.5C)}$$

where we have kept the temperature-dependent terms on the left-hand side of Equation (E9.5C). As discussed in Section 9.4, in an exothermic reaction, K decreases as temperature increases; therefore, the right-hand side must also decrease. The equilibrium will shift to the left in Equation (E9.5A), decreasing n_C and n_D while increasing n_A and n_B. Thus, the equilibrium conversion and potential product yield will decrease. Conversely, the equilibrium conversion of an endothermic reaction increases with increasing temperature.

(b) *Pressure* is primarily affected by the term P^ν (and, to a lesser extent, φ_i). K is independent of pressure. Rearranging Equation (E9.5B), we get

$$\frac{1}{P^\nu \left[\prod (\varphi_i)^{\nu_i}\right]} = \left[\frac{1}{K(n_T^v)^\nu}\right] \frac{(n_C)^c (n_D)^d}{(n_A)^a (n_B)^b}$$

Hence, the effect of pressure will primarily depend on the sign of ν. If the number of moles of products is greater than the number of moles of reactants, ν is positive. In this case, using an argument similar to that in part (a), an increase in pressure decreases the equilibrium conversion. Conversely, if there are more reactants than products, conversion increases with increasing pressure. This is a restatement of Le Châtelier's principle, taught in general chemistry. If the number of moles of reactants equals the number of moles of products, then pressure only affects equilibrium conversion weakly through φ_i.

(c) *Addition of inert* will effect n_T^v, which also depends on the sign of ν. Rearranging Equation (E9.5B), we get

$$(n_T^v)^\nu = \left[\frac{P^\nu \left[(\prod \varphi_i)^{\nu_i}\right]}{K}\right] \frac{(n_C)^c (n_D)^d}{(n_A)^a (n_B)^b}$$

exotherm
reactants → prod + heat
↑ T

If there are more moles of products than of reactants ($\nu > 0$), an increase of inert increases the equilibrium conversion. Conversely, if there are more moles of reactants than of products, conversion decreases with added inerts. If the moles of reactants equal the moles of products, there is no effect of adding inerts. This effect can be understood on a molecular scale. If we consider a reaction with more moles of products than of reactants ($\nu > 0$), more individual species must collide (find each other) for the reaction to proceed backward as compared to forward. Addition of an inert will then makes it harder for the backward reaction relative to the forward reaction, since the inerts will make it harder for the greater number of product species to find each other. Thus, the forward reaction will be greater than the reverse reaction relative to the case of no inert, and equilibrium conversion will be greater.

(d) If there is a *reactant in the feed*, the denominator on the right-hand side requires more conversion to satisfy Equation (E9.5B). Thus, a reactant in the feed will increase the conversion of the other (limiting) reactant. ◄

► **EXAMPLE 9.6**
Calculation of Extent of Gas-Phase Reaction with $\Delta h_{rxn}^o = $ Constant

Consider the production of ethylene from unimolecular decomposition of ethane as shown in the following reaction:

$$C_2H_6 \rightleftarrows C_2H_4 + H_2$$

At a temperature of 1000°C and pressure of 1 bar, what is the equilibrium composition of the system? Assume $\Delta h_{rxn}^o = $ constant.

SOLUTION From Appendix A.3, we can construct Table E9.6. The bottom row of Table E9.6 provides the enthalpy and Gibbs energy of reaction by summing the species' values times the stoichiometric coefficient. Only ethane is present in the feed; hence, $n_{C_2H_6}^o = 1$. We can write the number of moles and mole fractions in terms of the extent of reaction according to Equations (9.22) and (9.25), respectively:

$$n_{C_2H_6} = 1 - \xi \qquad\qquad y_{C_2H_6} = \frac{n_{C_2H_6}}{n^v} = \frac{1 - \xi}{1 + \xi}$$

$$n_{C_2H_4} = \xi \qquad\qquad y_{C_2H_4} = \frac{n_{C_2H_4}}{n^v} = \frac{\xi}{1 + \xi}$$

$$n_{H_2} = \xi \qquad\qquad y_{H_2} = \frac{n_{H_2}}{n^v} = \frac{\xi}{1 + \xi}$$

$$\overline{n^v = 1 + \xi}$$

We have simply summed the number of moles of the individual species to obtain the total number of moles in the vapor, n^v. Alternatively we could have used Equation (9.23). Since we are at low pressure, ideal gas behavior is assumed. Thus, the equilibrium constant can be written using Equation (9.56):

$$K = \prod (y_i)^{\nu_i} P^\nu = \frac{y_{C_2H_4} y_{H_2}}{y_{C_2H_6}} P$$

TABLE E9.6 Summary of Thermochemical Data from Appendix A.3

Species	ν_i	Δh_f^o	Δg_f^o
C_2H_6	-1	-84.68 kJ/mol	-32.84 kJ/mol
C_2H_4	1	52.26	68.15
H_2	1	0	0
$\sum \nu_i()_i$	1	136.94	100.99

where $\nu = (-1 + 1 + 1) = 1$. Expressing K in terms of extent of reaction gives

$$K = \frac{\xi^2}{(1 - \xi)(1 + \xi)}P = \frac{\xi^2}{(1 - \xi^2)}P$$

rearranging,
$$\xi = \sqrt{\frac{K}{K + P}} \qquad (E9.6)$$

Equation (E9.6) expresses the extent of reaction in terms of P and T (through K). To numerically evaluate Equation (E9.6), K must be determined from the thermochemical data in Table E9.6. First, we calculate K at 25°C:

$$K_{298} = \exp\left(-\frac{\Delta g^o_{rxn}}{RT}\right) = \exp\left(-\frac{100,990}{[8.314][298.2]}\right) = 2.04 \times 10^{-18}$$

If we assume $\Delta h^o_{rxn} = $ constant, we can correct for T using Equation (9.46):

$$\ln\frac{K_{1273}}{K_{298}} = -\frac{\Delta h^o_{rxn}}{R}\left(\frac{1}{T} - \frac{1}{298}\right) = -\frac{136,940}{8.314}\left(\frac{1}{1273} - \frac{1}{298}\right) = 42.33$$

Solving for the equilibrium constant:

$$K_{1273} = 4.95$$

Using this value in Equation (E9.6) gives

$$\xi = 0.91$$

or, in terms of number of moles,

$$n_{C_2H_6} = 0.09, \qquad n_{C_2H_4} = 0.91, \qquad n_{H_2} = 0.91$$

We see that under these conditions, it is possible to convert 91% of the ethane in the feed to ethylene. ◀

▶ **EXAMPLE 9.7**
Production of Ammonia at 1 Bar

Consider the production of ammonia from the catalytic reaction of a stoichiometric feed of nitrogen and hydrogen. The reaction temperature is 500°C and the reactor pressure is 1 bar.

$$N_2 + 3H_2 \rightleftarrows 2NH_3$$

What is the maximum possible conversion?
(a) Take $\Delta h^o = $ constant.
(b) Take $\Delta h^o = \Delta h^o(T)$.
You may consider the reaction to occur under ideal gas conditions.

SOLUTION From Appendices A.2 and A.3, we can construct the Table E9.7. The bottom row of Table E9.7 provides the sum of the species' values times the stoichiometric coefficient for each quantity listed. Since the feed is stoichiometric, there are three times the number of moles of H_2 than of N_2. Thus, the initial mole fraction can be written as follows:

$$n^o_{H_2} = 3, \quad n^o_{N_2} = 1$$

TABLE E9.7 Summary of Thermochemical Data from Appendices A.2 and A.3

Species	ν_i	Δh_f^o	Δg_f^o	A_i	B_i	D_i
NH_3	2	-46.11 kJ/mol	-16.45 kJ/mol	3.578	3.020×10^{-3}	-0.186×10^5
N_2	-1	0	0	3.280	0.593×10^{-3}	0.040×10^5
H_2	-3	0	0	3.249	0.422×10^{-3}	0.083×10^5
$\sum \nu_i ()_i$	-2	-92.22	-32.90	-5.871	4.180×10^{-3}	-0.661×10^5

We can express the number of moles and mole fractions in terms of the extent of reaction according to Equations (9.22) and (9.25), respectively:

$$n_{N_2} = 1 - 1\xi \qquad\qquad y_{N_2} = \frac{n_{N_2}}{n} = \frac{1 - \xi}{4 - 2\xi}$$

$$n_{H_2} = 3 - 3\xi$$

$$n_{NH_3} = 2\xi \qquad\qquad y_{H_2} = \frac{n_{H_2}}{n} = \frac{3 - 3\xi}{4 - 2\xi}$$

$$\overline{n = 4 - 2\xi}$$

$$y_{NH_3} = \frac{n_{NH_3}}{n} = \frac{2\xi}{4 - 2\xi}$$

Since we are at low pressure, ideal gas behavior is assumed. Thus, the equilibrium constant can be written as

$$K = \Pi(y_i)^{\nu_i} \, \Pi(\varphi_i)^{\nu_i} P^\nu = \frac{y_{NH_3}^2}{y_{N_2} \, y_{H_2}^3} P^{-2}$$

or, in terms of extent of reaction,

$$K = \frac{\left(\dfrac{2\xi}{4 - 2\xi}\right)^2}{\left(\dfrac{1 - \xi}{4 - 2\xi}\right)\left(\dfrac{3 - 3\xi}{4 - 2\xi}\right)^3} P^{-2} = \frac{(2\xi)^2(4 - 2\xi)^2}{(1 - \xi)(3 - 3\xi)^3} P^{-2} \qquad\text{(E9.7)}$$

The equilibrium constant is calculated in the usual manner. First, we use the Gibbs energy of reaction to calculate K at 25°C:

$$K_{298} = \exp\left(-\frac{\Delta g_{rxn}^o}{RT}\right) = \exp\left(-\frac{-32,900}{[8.314][298.2]}\right) = 5.81 \times 10^5$$

(a) For $\Delta h_{rxn}^o = $ constant, we again use Equation (9.46):

$$\ln \frac{K_T}{K_{298}} = -\frac{\Delta h_{rxn}^o}{R}\left(\frac{1}{T} - \frac{1}{298}\right) = \frac{92,220}{8.314}\left(\frac{1}{773} - \frac{1}{298}\right) = -22.88$$

and solving

$$K_T = 6.754 \times 10^{-5}$$

Plugging this into Equation (E9.7) with $P = 1$ bar and solving for the extent of reaction gives

$$\xi = 0.005$$

(b) For $\Delta h_{rxn}^o = \Delta h_{rxn}^o (T)$, we can use Equation (9.52):

$$\ln\left(\frac{K_T}{K_{298}}\right) = \left\{ \begin{array}{l} \left[\dfrac{-\Delta h_{rxn,298}^o}{R} + \Delta A(298) + \dfrac{\Delta B}{2}(298^2) + \dfrac{\Delta C}{3}(298^3)\right. \\[4mm] \left. - \dfrac{\Delta D}{298}\right]\left(\dfrac{1}{T} - \dfrac{1}{298}\right) \\[4mm] + \Delta A \ln\left(\dfrac{T}{298}\right) + \dfrac{\Delta B}{2}(T - 298) + \dfrac{\Delta C}{6}(T^2 - 298^2) \\[4mm] + \dfrac{\Delta D}{2}\left(\dfrac{1}{T^2} - \dfrac{1}{298^2}\right) \end{array} \right\} = -24.37$$

and solving

$$K_T = 1.51 \times 10^{-5}$$

Plugging this into Equation (E9.7) with $P = 1$ bar and solving for the extent of reaction gives

$$\xi = 0.003$$

In summary, under these conditions we do not expect to produce an appreciable amount of ammonia. ◀

▶ **EXAMPLE 9.8**
Strategy to Increase Conversion of Ammonia Production

Consider the production of ammonia from the catalytic reaction of a stoichiometric feed of nitrogen and hydrogen as in Example 9.7. As we saw, when the reaction temperature is 500°C and the reactor pressure is 1 bar, the conversion is very low. We wish to change the reactor conditions to increase conversion. Would you pick T or P? Which way would you change?

SOLUTION   This reaction is exothermic ($\Delta h_{rxn}^o < 0$). As can be deduced from Equation (9.42), lower temperature will lead to higher equilibrium conversions. However, reducing the temperature also reduces the *rate* of reaction, which is also a major industrial concern. In this case, the rates become too slow, so T cannot be reduced to increase conversion. If we next consider P, the primary effect is determined by ν, as shown in Equation (E9.5B). Since ν is negative, an increase in pressure will increase conversion. Example 9.9 examines the equilibrium conversion obtained when the pressure is increased to 300 bar. ◀

▶ **EXAMPLE 9.9**
Production of Ammonia at 300 Bar

Consider the production of ammonia from the catalytic reaction of a stoichiometric feed of nitrogen and hydrogen as in Example 9.7. Consider again the reaction temperature of 500°C. Now the reactor pressure is increased to 300 bar (see Example 9.8 for discussion). What is the maximum possible conversion?

(a) Take the gas to be ideal.

(b) Use the van der Waals equation and the Lewis fugacity rule to account for gas-phase nonideality.

SOLUTION   (a) Considering the gas to be ideal, Equation (E9.7) is still valid. At a pressure of 300 bar:

$$K_T = \frac{\left(\dfrac{2\xi}{4 - 2\xi}\right)^2}{\left(\dfrac{1 - \xi}{4 - 2\xi}\right)\left(\dfrac{3 - 3\xi}{4 - 2\xi}\right)^3}P^{-2} = \frac{(2\xi)^2(4 - 2\xi)^2}{(1 - \xi)(3 - 3\xi)^3}P^{-2}$$

where $K_T = 1.51 \times 10^{-5}$ and $P = 300$ bar

Solving for the extent of reaction gives

$$\xi = 0.37$$

The conversion has increased dramatically at the higher pressure.

(b) At 300 bar, the gas is not ideal. To account for nonideal behavior, we can use Equation (9.55). Recall that this equation uses the Lewis fugacity rule to approximate the fugacity coefficient:

$$K = \prod (y_i)^{\nu_i} \prod (\varphi_i)^{\nu_i} P^{\nu} = \frac{y_{NH_3}^2 \, \varphi_{NH_3}^2}{y_{N_2} \, \varphi_{N_2} \, y_{H_2}^3 \, \varphi_{H_2}^3} P^{-2}$$

Replacing the mole fraction with expressions for ξ determined in Example 9.6, we obtain

$$K = \frac{(2\xi)^2 \varphi_{NH_3}^2 (4 - 2\xi)^2}{(1 - \xi)\varphi_{N_2}(3 - 3\xi)^3 \varphi_{H_2}^3} P^{-2} \qquad (E9.9)$$

A summary of the solution is shown in Table E9.9. The solution algorithm is as follows:

1. Look up P_c, T_c (from Appendix A.1)
2. Calculate van der Waals constants, a and b, from corresponding states

$$a = \frac{27}{64} \frac{(RT_c)^2}{P_c} \qquad b = \frac{(RT_c)}{8P_c}$$

3. Calculate v_i from van der Waals equation of state

$$P = \frac{RT}{v_i - b} - \frac{a}{v_i^2}$$

In practice, we would use a more modern and accurate equation of state; however, the van der Waals equation is used in this example to be consistent with the development in Chapter 7.

4. Calculate φ_i from the result obtained in Example 7.2:

$$\ln\left[\frac{f_i^v}{P}\right] = \ln\left[\varphi_i^v\right] = -\ln\left[\frac{(v_i - b)P}{RT}\right] + \frac{b}{(v_i - b)} - \frac{2a}{RTv_i}$$

Now, plugging values into Equation (E9.9) and solving gives

$$\xi = 0.33$$

A correction of about 10% results from accounting for nonideal behavior; however, the conversion is still significant in comparison to Example 9.7. ◄

TABLE E9.9 Summary of Fugacity Coefficient Calculation

Species	T_c [K]	P_c [atm]	a [Jm^3/mol^2]	b [m^3/mol]	v_i [m^3/mol]	φ_i
NH$_3$	405.5	111.3	0.43	3.75×10^{-5}	2.37×10^{-4}	1.10
N$_2$	126.2	33.5	0.14	3.88×10^{-5}	2.38×10^{-4}	1.11
H$_2$	33.3	12.80	0.02	2.65×10^{-5}	1.92×10^{-4}	0.88

The Equilibrium Constant for a Liquid-Phase (or Solid-Phase) Reaction

The first step in applying Equation (9.35) to the liquid phase is to choose a reference state. Recall that the data from which we calculate K are at 1 atm (or 1 bar), which, therefore, confines the reference pressure. For a Lewis/Randall reference state, Equation (9.29) becomes

$$K = \prod \left(\frac{x_i \gamma_i f_i}{f_i^o} \right)^{\nu_i}$$ (9.57)

Here f_i is at the pressure of the reaction, while f_i^o is at 1 bar. For large differences between these pressures, we must use the Poynting correction described in Chapter 7. *If the pressure dependence of the fugacities is not significant*, the relation between the equilibrium constant and composition becomes

$$K = \prod (x_i \gamma_i)^{\nu_i}$$ (9.58)

where we may have to use a model for g^E to relate the activity coefficient to composition. In the case of an ideal solution, we have

$$K = \prod (x_i)^{\nu_i} \qquad \text{(ideal solution)}$$ (9.59)

▶ **EXAMPLE 9.10**
Extent of an Isomerization Reaction at 298 K

Consider the isomerization reaction of methylcyclopentane ($CH_3C_5H_9$) to cyclohexane (C_6H_{12}) at 298 K. What is the equilibrium conversion? Gibbs energies of formation are as follows:

$$\left(\Delta g_f^o \right)_{CH_3C_5H_9} = 31.72 [kJ/mol] \qquad \text{and} \qquad \left(\Delta g_f^o \right)_{C_6H_{12}} = 26.89 [kJ/mol]$$

SOLUTION The isomerization reaction can be written as

$$CH_3C_5H_9(l) \rightleftarrows C_6H_{12}(l)$$

The Gibbs energy of reaction is given by

$$\Delta g_{rxn}^o = (26.89 - 31.72) = -4.83 \text{ [kJ/mol]}$$

Solving for the equilibrium constant:

$$K = \exp \left(-\frac{\Delta g_{rxn}^o}{RT} \right) = \exp \left(-\frac{-4,830}{(8.314)(298)} \right) = 7.03$$ (E9.10A)

From Equation (9.59), we can write

$$K = \frac{x_{C_6H_{12}}}{x_{CH_3C_5H_9}}$$

or, in terms of extent of reaction,

$$K = \frac{\xi}{1 - \xi}$$ (E9.10B)

Equating Equations (E9.10A) and (E9.10B) and solving gives

$$\xi = 0.875$$

At equilibrium, 87.5% of the liquid exists as cyclohexane.

◀

The Equilibrium Constant for a Heterogeneous Reaction

If we have both vapor and condensed phases present, we simply treat the vapor species and the condensed species as we did before. *We must always remember that the mole fractions in these expressions, however, refer to the mole fraction in a given phase,* **not** *the total mole fraction.*

▶ **EXAMPLE 9.11**
Heterogeneous
Dissociation
of Calcium Carbonate

Calcium carbonate can dissociate according to the following reaction:

$$CaCO_3(s) \rightleftarrows CaO(s) + CO_2(g) \tag{E9.11}$$

Consider a closed system with pure $CaCO_3$ in vacuum at 1000 K. What is the equilibrium pressure of the system? Assume that the two solid phases are completely immiscible. At 1000 K, the following Gibbs energies of formation are reported:

Species	$\left(\Delta g^o_{f,1000}\right)_i$
$CaCO_3$	-951.25
CaO	-531.09
CO_2	-395.81

SOLUTION   Applying the definition of the equilibrium constant given by Equation (9.35):

$$K = \frac{\left(\dfrac{\hat{f}_{CaO}}{f^o_{CaO}}\right)\left(\dfrac{\hat{f}_{CO_2}}{1\text{ bar}}\right)}{\left(\dfrac{\hat{f}_{CaCO_3}}{f^o_{CaCO_3}}\right)}$$

We must treat each of the three pure phases distinctly. If we assume that the pressure will be low enough that we have an ideal gas at equilibrium, we can rewrite the equilibrium constant as

$$K = \frac{\left(\dfrac{f_{CaO}}{f_{CaO}}\right)\left(\dfrac{y_{CO_2}P}{1\text{ bar}}\right)}{\left(\dfrac{f_{CaCO_3}}{f_{CaCO_3}}\right)} = p_{CO_2}$$

At low and moderate pressures, the terms corresponding to the pure species solids go to 1, since the Lewis/Randall reference fugacity equals the pure solid fugacity. At very high pressure, we would have to account for the Poynting correction, since the standard state is defined at 1 bar. Thus, the equilibrium constant equals the CO_2 partial pressure. Solving for K from the thermochemical data provided, we get

$$K = \exp\left(-\frac{\Delta g^o_{rxn}}{RT}\right) = \exp\left[-\frac{(-531.09 - 395.81 + 951.25)(1000)}{(8.314)(1000)}\right] = 0.053$$

Thus, at 1000 K calcium carbonate will dissociate until the pressure reaches

$$p_{CO_2} = K = 0.053 \text{ bar}$$

or until we run out of $CaCO_3$. Note that our ideal gas assumption is valid at this pressure. From this analysis, we can also deduce the following: If we had a system with solid CaO and a CO_2 partial pressure greater than 0.053 bar at 1000 K, Reaction (E9.11) would occur backward and we would consume CO_2 until its partial pressure reached 0.053 bar. ◀

▶ **EXAMPLE 9.12**
Reaction of Acetylene
Vapor to Form Liquid
Benzene

You have just ordered a cylinder of acetylene but are concerned that it might react during shipment to form benzene. Consider the reaction of acetylene to form benzene:

$$3C_2H_2(g) \rightleftarrows C_6H_6(l)$$

For the sake of calculation, take the initial pressure to be 1 bar and the temperature to be 298 K. What is the equilibrium conversion? What is the corresponding final pressure in the system?

SOLUTION First, we will use thermochemical data to determine the equilibrium constant. Looking up the Gibbs energies of formation in Appendix A.3 yields

$$\Delta g^o_{rxn} = \left(\Delta g^o_f\right)_{C_6H_6} - 3\left(\Delta g^o_f\right)_{C_2H_2} = (124.3 - 3 \times 209.2) = -503.3 \, [kJ/mol]$$

Solving for the equilibrium constant:

$$K = \exp\left(-\frac{\Delta g^o_{rxn}}{RT}\right) = \exp\left(-\frac{-503,300}{(8.314)(298)}\right) = 1.7 \times 10^{88}$$

If we assume ideal gas and liquid phases, Equation (9.35) becomes

$$K = \frac{x_{C_6H_6}}{\left(y_{C_2H_2}P\right)^3} \tag{E9.12}$$

As a first approximation, we will assume that no acetylene condenses and that no benzene is volatile; hence,

$$x_{C_6H_6} = 1 \quad \text{and} \quad y_{C_2H_2} = 1$$

Equation (E9.12) becomes

$$K = \frac{1}{P^3}$$

Solving for pressure:

$$P = 3.9 \times 10^{-30} \text{ bar}$$

This pressure corresponds to essentially complete conversion of acetylene. However, this pressure does not represent the final system pressure, since benzene will exert a vapor pressure at 298 K. Benzene's vapor pressure can be determined by the Antoine equation. Looking up the Antoine coefficients in Table A.1 yields

$$\ln(P[\text{bar}]) = A - \frac{B}{T\,[K] + C} = 9.2806 - \frac{2788.51}{298 - 52.36}$$

or

$$P = 0.12 \text{ bar}$$

Thus, at equilibrium, the cylinder is almost completely benzene, with a pressure of 0.12 bar. However, we are still able to ship acetylene, in practice. Even though the thermodynamics favors benzene, this reaction is limited by reaction rates and will not noticeably proceed without a catalyst present. ◀

▶ 9.6 EQUILIBRIUM IN ELECTROCHEMICAL SYSTEMS

So far in our analysis of chemical equilibrium, we have assumed that there is no work other than the Pv work from the changing boundary in a constant-pressure system. In this section, we consider electrochemical systems, where work can be introduced by applying an electric potential to two electrodes. In this case, we must rewrite Equation (6.9) to include non-Pv work, W^*:

$$\delta W^* \geq (dG)_{T,P} \qquad (6.9^*)$$

Equation (6.9*) suggests two general applications of electrochemical systems. Reactions that spontaneously proceed have a negative Gibbs energy change; Equation (6.9*) shows that W^* is negative and that we can generate useful electrical work. This principle forms the basis on which we design batteries and fuel cells. On the other hand, input of a suitable quantity of electrical work can lead to reactions that have a positive change in Gibbs energy and would not spontaneously occur. This aspect is used to advantage in electroplating operations, where a desired metal is grown on a surface, and in manufacturing of metals and other chemicals through electrolysis. Other important electrochemical systems include corrosion processes and electrochemical-based sensors. An electrochemical cell that uses a spontaneous reaction to obtain useful work is termed a **galvanic cell**; conversely, a cell that requires electrical work to induce a reaction that would not occur on its own is called an **electrolytic cell**. We can apply thermodynamics to see how much work we can obtain from a given electrochemical cell or, conversely, the minimum work that is needed to create a desired product.

Electrochemical processes occur within an electrochemical cell. Figure 9.4 shows an electrochemical process to grow copper on a substrate. This process illustrates many common components found in electrochemical systems. The electrochemical cell contains two electrodes in an electrolyte solution. The reactions occurring on each of the two electrodes shown are termed half-cell reactions. The **reduction half-reaction** occurs at the cathode, where electrons are transferred to the reacting species. In Figure 9.1, we have reduction of cupric ions to grow solid copper:

$$Cu^{2+}(l) + 2e^- \longrightarrow Cu(s) \qquad (9.60)$$

In the **oxidation half-reaction** at the **anode**, the reactant loses electrons. In Figure 9.4, the anode is a noble metal, Pt, which does not readily react. Instead, oxidation of

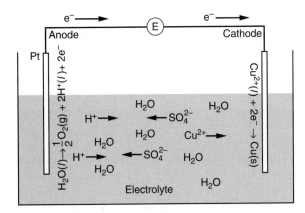

Figure 9.4 Schematic of an electrochemical process.

water occurs as the oxidation half-reaction, as follows:

$$H_2O(l) \longrightarrow \frac{1}{2}O_2(g) + 2H^+(l) + 2e^- \tag{9.61}$$

This reaction is noticeable, as bubbles arise at the anode from the O_2 gas produced. The overall reaction of the electrochemical cell is obtained by adding the two half-reactions. The half-reactions must be balanced so that no net electrons are produced. Additionally, in balancing a half-reaction in an aqueous electrolyte, the appropriate amount of H_2O and either H^+, for acidic solutions, or OH^-, for basic solutions, can be used to account for changes in the stoichiometric amount of O and H between the reactants and the products. The overall reaction depicted in Figure 9.1 is obtained by adding the oxidation and reduction half reactions:

$$H_2O(l) + Cu^{2+}(l) \longrightarrow \frac{1}{2}O_2(g) + 2H^+(l) + Cu(s) \tag{9.62}$$

In this electrolytic cell, we need the input of electrical work to get Reaction (9.62) to proceed.

The overall reaction in an electrochemical cell never contains electrons. In the example above, the electrons produced from oxidizing water flow in the external circuit to the cathode and are consumed in the reduction of cupric ions. To complete the external circuit, charge must be able to flow through the solution. That charge is carried by the transport of ions. A medium with mobile charge carriers is termed an **electrolyte**. Electrolytes are typically liquids, but systems can contain solid electrolytes if they are able to sustain the transport of charged species. The electrolyte used in Figure 9.4 is made from copper sulfate and sulfuric acid. These species dissociate in water to form positively charged cupric and hydrogen ions and negatively charged sulfate ions. Typically, small concentrations of other additives are also used to modify and control the properties of the plated copper. The positive ions flow toward the cathode, while the negative ions flow toward the anode, completing the electric circuit. The cupric ions are depleted from the electrolyte as the copper solid grows, while the hydrogen ion concentration increases due to the anodic reaction. One of the cations, the hydrogen ion, transports through solution much more quickly than either of the other ions. Thus, H^+ initially carries more than its share of the current. Inspection of Figure 9.4 reveals that over time, there will then be a net separation of charge until a potential is set up that retards the flow of H^+ and increases the flow of the anion, SO_4^{2-}. At steady state, a net potential will be established that opposes the potential that we are applying. This added potential due to charge separation is termed a liquid junction.

Figure 9.4 shows the simplest configuration of electrochemical cell where both electrodes are immersed in a common electrolyte. Often electrochemical cells contain different electrolyte compositions at the anode and the cathode. For example, consider an alternative electroplating process for copper in which zinc is oxidized at the anode and goes into solution as Zn^{2+}. A schematic of this process is shown in Figure 9.5. This system is galvanic; that is, the copper growth occurs spontaneously without the input of electrical work. In this case, it is desirable to have the two electrolyte compartments interact through a *salt bridge*. The salt bridge is noted in Figure 9.5. A salt bridge allows a net charge to be transferred from one electrolyte solution to the other but does not allow undesired mixing of the electrolytes. It can be a simple porous disk or a gel saturated with a strong electrolyte such as KCl. Since the mobilities of the K^+ and Cl^- ions are roughly equal, the salt bridge minimizes the liquid junction potential.

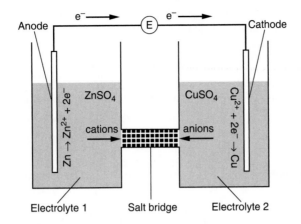

Figure 9.5 Schematic of an electrochemical process with a salt bridge.

To avoid having to sketch an electrochemical cell for every process we are considering, a shorthand notation has been developed for describing electrochemical cells. Always starting at the anode, we pass through the electrolyte to the cathode and indicate the active species in chemical notation. A phase change is indicated by a vertical bar, that is, phase 1|phase 2. When the phases are separated by a salt bridge or other similar device, we use a double bar, that is phase 1||phase 2. Therefore, our shorthand notation for the system depicted in Figure 9.4 is:

$$Pt|O_2(g)|H_2SO_4(l), CuSO_4(l)|Cu(s) \tag{9.63}$$

The copper electroplating process in Figure 9.5 is written:

$$Zn(s)|ZnSO_4(l)||CuSO_4(l)|Cu(s) \tag{9.64}$$

We often designate the concentrations of the active species as well.

We now wish to apply the principles of thermodynamics to determine the electrical work needed or obtained in these systems at equilibrium. The differential electrical work can be related to the electric potential difference between the cathode and the anode, E, and the differential amount of charge transferred, dQ, as follows:

$$\delta W^* = -EdQ \tag{9.65}$$

The sign convention for Equation (9.65) is chosen so that when the cathode has a positive potential with respect to the anode, the system can generate useful work, while a negative potential indicates that work is required for the process to proceed. Electrons that carry the charge in the external circuit result from the oxidation half-reaction; thus, the differential charge transferred can be related to the extent of reaction as follows:

$$\{\text{charge transferred}\} = \left\{\frac{\text{mol e}^- \text{ liberated}}{\text{mole species reacting}}\right\}\left\{\frac{\text{charge}}{\text{mole e}^-}\right\}\{\text{extent of reaction}\} \tag{9.66}$$

$$dQ = zFd\xi$$

where z is the number of moles of electrons liberated per mole of species that reacts and F is Faraday's constant, 96,485 C/(mole e$^-$), which represents the charge of 1 mole of electrons. Substituting Equation (9.66) into (9.65) gives:

$$\delta W^* = -zFE\delta\xi \qquad (9.67)$$

For a *reversible* reaction, we use the equality in Equation (6.9*) to give

$$-zFE\mathrm{d}\xi = \mathrm{d}G = \sum \mu_i v_i \mathrm{d}\xi \qquad (9.68)$$

where Equation (9.28) has been used. Applying Equations (9.29) and (9.30), we get

$$-zFE = \sum \mu_i v_i = \sum \left(g_i^o + RT \ln \frac{\hat{f}_i}{f_i^o} \right) v_i \qquad (9.69)$$

We can apply the fugacity coefficient formulation for gases and the activity coefficient formulation for liquids as we did in Section 9.5. Assuming the activity of the solids in the system is 1, Equation (9.69) becomes:

$$-zFE = \Delta g_{\mathrm{rxn}}^o + RT \ln \left[\underbrace{\prod (y_i\hat{\varphi}_i P)^{v_i}}_{\text{vapors}} \underbrace{\prod (x_i\gamma_i)^{v_i}}_{\text{liquids}} \right] \qquad (9.70)$$

Systems with solid solutions will need a solid term in Equation (9.70) as well. Recall that the fugacity coefficient formulation uses a standard-state fugacity for the vapor of 1 bar, so all units of pressure are in bar. Electrochemical cells typically operate under ideal gas conditions, so we will assume $\hat{\varphi}_i = 1$. For processes at high pressure, the development that follows should be modified to include the fugacity coefficient as well. Additionally, in dilute solutions, the activity of the solvent (γx), is approximately 1.

We need to define the standard state for the liquid in the electrolyte. By convention, we characterize the liquid composition in terms of concentration instead of mole fraction, where the concentration of species i, c_i, has units of molality, m (moles i per 1 kg of solvent). Moreover, we must specify a Henry's law standard state for the ions in solution; they do not exist as pure species, so we cannot use the Lewis/Randall reference state. Our standard state is chosen to be a 1-m ideal solution, in the Henry's law sense. If species i is not ideal at this concentration, we go to a low enough concentration that it obeys Henry's law, then extrapolate back to a hypothetical ideal liquid with 1-m concentration. The l-m standard state for the electrolyte solution is analogous to that of 1 bar for vapors. Thus, all concentrations used should be in units of m. In this case, we can define a molality-based, Henry's law activity coefficient, γ_i^m, to describe the deviations from ideality in the real solution. In other words, if a component in the liquid has a concentration of 1 mole per kg of solvent and its interactions correspond to an ideal solution, it has $c_i\gamma_i^m = 1$. For other concentrations it takes distinct value of c_i, and for real liquids the term γ_i^m differs from 1. Using the molality-based standard state, Equation (9.70) becomes

$$-zFE = \Delta g_{\mathrm{rxn}}^o - RT \ln \left[\underbrace{\prod (p_i)^{v_i}}_{\text{vapors}} \underbrace{\prod \left(c_i\gamma_i^m\right)^{v_i}}_{\text{liquids}} \right] \qquad (9.71)$$

If we divide Equation (9.71) by zF, we get

$$E = E_{rxn}^\circ - \frac{RT}{zF} \ln \left[\prod_{\text{vapors}} (p_i)^{\nu_i} \prod_{\text{liquids}} (c_i \gamma_i^m)^{\nu_i} \right]$$

(9.72)

where the standard potential of reaction is defined as

$$E_{rxn}^\circ = -\frac{\Delta g_{rxn}^o}{zF}$$

(9.73)

Equation (9.72) is known as the Nernst equation. In applying the Nernst equation, we must remember to write concentrations in units of m, just as we have used units of bar for pressure throughout this chapter. This condition results from the standard state that we have chosen. *An important feature of the standard potential is that it does not depend on how many electrons we use in our half-cell reactions.* Inspection of Equation (9.73) shows that if we double the number of electrons, both z and Δg_{rxn}^o double, leaving E_{rxn}° unchanged.

Like the equilibrium constant, E_{rxn}° is obtained from thermochemical data. Since the overall reaction in the electrochemical cell is composed of two half-reactions, we can tabulate electrochemical data in terms of half-reactions. We then simply add together the appropriate reduction and oxidation half-reactions to calculate E_{rxn}° for the entire electrochemical cell. An isolated half-cell reaction cannot occur by itself, so we need to choose a reference half-cell reaction to complete the reference cell. Thermochemical data for electrochemical half-reactions are commonly published as the standard half-cell potential for reduction reactions, E°. Representative values are reported in Table 9.1; however, these are but a few of the extensive sets of values that are available. This potential is measured with reference to a hydrogen–hydrogen ion oxidation reaction, whose potential is defined as zero:

$$H_2(g) \longrightarrow 2H^+(l) + 2e^- \qquad E^\circ = 0.000000 \text{ V}$$

(9.74)

The reactant and product species for both the reduction half-reaction and the hydrogen oxidation half-reaction are specified to be in their standard states. Recall that the standard state of a gas is an ideal gas at 1 bar, a liquid is a $1\,m$ ideal solution in the Henry's law sense, and a solid is the pure solid with an activity of 1. In terms of our shorthand notation, we can measure the standard potential of any reduction half-reaction with a standard hydrogen electrode (S.H.E.):

$$Pt|H_2(g, 1 \text{ bar})|H^+(l, 1 \text{ M})|| \dots$$

(9.75)

It is also possible to obtain the value of oxidation reactions from Table 9.1. The half-cell potential of an oxidation reaction is simply the negative of the reported reduction half-reaction. The half-reaction potential and the hydrogen reduction reaction reference are analogous to our use of Gibbs energy of formation and the elemental form of molecules in the other parts of this chapter.

It is instructive to compare the half-cell reactions from Table 9.1. To construct an electrochemical cell, we use one reduction half-reaction and one oxidation half-reaction. The value of E° for the reduction reaction can be obtained directly from Table 9.1, while the oxidation reaction is the negative of the value listed. Consider the case where all the species in the system are in their standard states. Examination of Equation (9.73)

TABLE 9.1   Standard Half-Cell Potentials at 298 K

Reduction Half-Reaction	E°
$F_2 + 2e^- \longrightarrow 2F^-$	+2.866
$Au^+ + e^- \longrightarrow Au$	+1.692
$PbO_2 + 4H^+ + 2e^- \longrightarrow Pb^{2+} + 2H_2O$	+1.455
$Cl_2 + 2e^- \longrightarrow 2Cl^-$	+1.358
$Pt^{2+} + 2e^- \longrightarrow Pt$	+1.18
$O_2 + 4H^+ + 4e^- \longrightarrow H_2O$	+1.229
$Ag^+ + e^- \longrightarrow Ag$	+0.800
$Cu^+ + e^- \longrightarrow Cu$	+0.521
$O_2 + 2H_2O + 4e^- \longrightarrow 4OH^-$	+0.401
$Cu^{2+} + 2e^- \longrightarrow Cu$	+0.342
$AgCl + e^- \longrightarrow Ag + Cl^-$	+0.222
$Cu^{2+} + e^- \longrightarrow Cu^+$	+0.153
$\mathbf{2H^+ + 2e^- \longrightarrow H_2}$	**0.000**
$Pb^{2+} + 2e^- \longrightarrow Pb$	−0.126
$Fe^{2+} + 2e^- \longrightarrow Fe$	−0.447
$Zn^{2+} + 2e^- \longrightarrow Zn$	−0.762
$2H_2O + 2e^- \longrightarrow H_2 + 2OH^-$	−0.828
$Al^{3+} + 3e^- \longrightarrow Al$	−1.662
$Na^+ + e^- \longrightarrow Na$	−2.71
$Li^+ + e^- \longrightarrow Li$	−3.040

Source: D.R. Lide (ed.), CRC Handbook of Chemistry and Physics, 83rd ed. (Boca Raton, FL: CRC Press, 2002–2003).

reveals that if $E^\circ_{rxn} > 0$, the reaction will spontaneously proceed. Since the half-reactions in Table 9.1 are listed in order of their numerical values, if we choose a given half-reaction as a possible reduction reaction, any oxidation half-reaction that is listed below it will result in a positive value of E°_{rxn} and the reduction-oxidation couple will occur spontaneously. Conversely, any reaction listed above it will require input of electrical work to oxidize. For example, consider again the reduction of cupric ion to form copper; $Cu^{2+} + 2e^- \rightarrow Cu$. It is listed at a standard reduction potential of +0.342 V. Any of the species in reduced form of the half-reactions listed below this half-reaction will spontaneously oxidize under standard-state conditions. For example, let's consider oxidation of zinc metal: $Zn \rightarrow Zn^{2+} + 2e^-$. The oxidation potential is the negative of the reduction potential reported in Table 9.1 and has a value of +0.762 V. Thus, the total cell has a value of +1.104 V, and the oxidation reduction couple will spontaneously occur. This result forms the basis for the electrochemical cell depicted in Figure 9.5. If any species is not in its standard state, the numerical values will change. Similarly, Pb, Fe, Al, . . . will oxidize to couple with copper reduction. Conversely, any reduced form of a species above $Cu^{2+} + 2e^- \rightarrow Cu$ will not spontaneously oxidize. For example, if we have a silver anode, the standard oxidation potential is −0.800 V. Adding this value to the copper reduction half-reaction gives −0.458 V. Thus, at a minimum, a potential of 0.458 V must be applied to get Ag^+. We can see that the higher up a half-reaction is in Table 9.1, the more it will tend to be in its reduced form. Conversely, the lower it is, the more readily it will be oxidized. Thus, Table 9.1 can be quickly scanned to see what oxidation–reduction reactions will spontaneously occur and provide useful work and what reactions will require the input of work.

This section illustrates that the thermodynamic principles we have learned so far can be applied to electrochemical systems. However, to solve Equation (9.72) in general, we need to determine the activity coefficients of the species in solution. The treatment of activity coefficients markedly differs from the nonelectrolyte solutions we have been discussing so far in the text. Charged species in solution have strong ionic interactions that are very different from other interactions in the solution, even in a dilute solution. Moreover, the condition of electroneutrality places another constraint on the relative concentrations of ions in solution. We cannot measure the activity coefficient of the cations, γ_+ or the anions γ_- independently, since they cannot exist by themselves, but rather must exist as an anion–cation pair together in solution. Hence, we refer to the mean activity coefficient of both ions together, $\gamma_\pm$. For example, for NaCl or other "1-1" electrolytes:

$$\gamma_\pm = \sqrt{\gamma_+ \gamma_-}$$

For the general case:

$$X_a Y_b \rightleftharpoons aX^{z_+} + bY^{z_-}$$

we get

$$\gamma_\pm = (\gamma_+^a \gamma_-^b)^{1/(a+b)}$$

where z_+ is the valence of the cation and z_- the anion. Mean activity coefficients for electrolyte solutions are typically obtained from experiment but can be estimated for ions in dilute solution.

We can estimate the nonideality of ions in very dilute solutions by considering the coulombic electrostatic interactions that occur. For a given ion in solution, the presence of an oppositely charged ion is energetically favorable, while the presence of one with like charge is unfavorable; thus, an ion in solution will, on average, have more oppositely charged ions near it than like-charged ions.[3] Quantifying these arguments, Debye and Huckel accounted for the nonideality in the limit of very dilute solutions, coming up with the following expression:[4]

$$\ln \gamma_\pm = -A|z_+ z_-|\sqrt{I} \tag{9.76}$$

The coefficient A groups several terms from the theory and can be considered a solvent parameter that depends on the relative permittivity and temperature. The ionic strength, I, is given by:

$$I = \frac{1}{2}\sum z_i^2 c_i \tag{9.77}$$

where the sum is over all the ions. For water at 25°C, $A = 1.17$, where the 1-m standard-state concentration is applied to I to make A dimensionless. Real systems start to diverge from Equation (9.76) at concentrations above only 0.01 m. The Debye–Huckel theory

---

[3] In Chapter 4, we used this same type of argument to see why there was a net attractive force for species with dipole moments.

[4] See J. O. Bockris and A. K. N. Reddy, *Modern Electrochemistry* (Vol. 1) (New York: Plenum Press, 1970).

has been modified to get better agreement over wider concentration ranges. One such expression adds an adjustable experimental parameter B:

$$\ln \gamma_\pm = -\frac{A|z_+ z_-|\sqrt{I}}{1 + B\sqrt{I}} \tag{9.78}$$

For water at 25°C, $B = 0.33$. Equation (9.78) agrees with experiment up to around 0.1-m concentrations. An alternative approach is to describe the nonideality of the solvent through an activity coefficient model and then determine the activity of the ions through the Gibbs–Duhem equation. More detail is provided elsewhere.[5]

▶ **EXAMPLE 9.13**
Calculation of K
from $E°$

The reverse copper disproportionation reaction has been proposed to etch solid copper:

$$Cu + Cu^{2+}(l) \longrightarrow 2Cu^+(l) \tag{E9.13}$$

Calculate the equilibrium constant of the disproportionation reaction. Will it occur spontaneously?

SOLUTION The standard potential of Reaction (E9.13) is obtained by adding together the two half-cell reactions obtained from Table 9.1, as follows:

$$Cu^{2+}(l) + e^- \longrightarrow Cu^+(l) \qquad E° = 0.153 \text{ V}$$
$$Cu \longrightarrow Cu^+(l) + e^- \qquad E° = -0.521 \text{ V}$$

The sum of the half-cell standard potential gives

$$E°_{rxn} = 0.153 - 0.521 = -0.368 \text{ V}$$

Applying Equation (9.73) gives

$$\Delta g°_{rxn} = -zFE°_{rxn} = \left(-1\left[\frac{\text{mol e}^-}{\text{mol Cu}^{2+}}\right]\right) \times \left(96,485\left[\frac{C}{\text{mol e}^-}\right]\right) \times (-0.37\,[V]) = 35.6\,[\text{kJ/mol}]$$

Using the definition of the equilibrium constant, we get

$$K = \exp\left(-\frac{\Delta g°_{rxn}}{RT}\right) = 5.7 \times 10^{-7}$$

The equilibrium constant is small, and etching will not proceed spontaneously. However, if we apply work through application of an electric potential, we can etch the copper. In fact, this process is used to etch lines in printed circuit board manufacturing. ◀

▶ **EXAMPLE 9.14**
Calculation of Electrode
Potential Needed for
Copper Plating

Consider the plating of copper from the process shown in Figure 9.4. Calculate the minimum electrode potential to achieve copper growth. The aqueous solution has the following composition: 0.07 m CuSO$_4$ and pH = 1. You may assume you have an ideal solution.

SOLUTION $\qquad Cu^{2+}(l) + 2e^- \longrightarrow Cu(s) \qquad E° = 0.34 \text{ V}$

To get the value of the oxidation half-reaction, we take the negative of the value in Table 9.1:

$$H_2O(l) \longrightarrow \frac{1}{2}O_2(g) + 2H^+(l) + 2e^- \qquad E° = -1.23 \text{ V}$$

[5] J. M. Prausnitz, R. N. Lichtenthaler, and E. Gomes de Azeuedo, *Molecular Thermodynamics of Fluid-Phase Equilibria*, 3rd ed. (Upper Saddle River, NJ: Prentice-Hall, 1999).

where we made use of the point that standard cell potentials are independent of the number of electrons on which the reaction is based. Adding together the reduction half-reaction and the oxidation half-reaction gives

$$H_2O(l) + Cu^{2+}(l) \longrightarrow \frac{1}{2}O_2(g) + 2H^+(l) + Cu(s) \qquad \Delta E^\circ_{rxn} = 0.34 - 1.23 = -0.89 \text{ V}$$

Applying Equation (9.72) to this system gives

$$E = E^\circ_{rxn} - \frac{RT}{zF} \ln \left[\prod_{vapors} (p_i)^{\nu_i} \prod_{liquids} (c_i \gamma_i^m)^{\nu_i} \right] = E^\circ_{rxn} - \frac{RT}{zF} \ln \left[\frac{p_{O_2}^{1/2} c_{H^+}^2}{c_{Cu^{2+}}} \right] \qquad \text{(E9.14)}$$

If we assume the O_2 product bubbles up at a partial pressure of 1 bar, Equation (E9.14) becomes

$$E = E^\circ_{rxn} - \frac{2.303 \, RT}{zF} \log c_{H^+}^2 + \frac{RT}{zF} \ln[c_{Cu^{2+}}]$$

Since pH is defined as $pH = -\log(c_{H^+})$, we get

$$E = E^\circ_{rxn} + \frac{2.303 RT}{zF} 2pH + \frac{RT}{zF} \ln[c_{Cu^{2+}}] = -0.90 \text{ V}$$

Thus, we have to apply a potential greater than 0.90 V to get copper to plate out of solution ◀

▶ 9.7 MULTIPLE REACTIONS

Extent of Reaction and Equilibrium Constant for *R* Reactions

In treating chemically reacting systems, we are often faced with cases where there are many possible reaction paths (Figure 9.1 shows one such case). It is straightforward to extend our development of chemical reaction equilibria to more than one reaction. To set up the multiple reaction equilibria problem, we must pick *R* different *independent* chemical reactions to describe the system. A set of reactions is deemed independent if you cannot construct any one of the given reactions by a linear combination of the others. When multiple reactions are considered, each reaction has its own corresponding extent of reaction, ξ_k. Thus, we must keep track of the stoichiometry of each species i for each of the k separate chemical reactions. Equation (9.16) applies to each separate reaction $(1, 2, \ldots k \ldots R)$; thus, we must now sum over all of the i species for each of the k reactions. Mathematically, we accomplish this task by using a double sum, as follows:

$$\sum_{k=1}^{R} \sum_{i=1}^{m} \nu_{ik} A_{ik} = 0 \qquad (9.79)$$

Similarly, Equation (9.21) must be written for each of the k extents of reaction. Again, the mathematics requires a sum over all R reactions:

$$dn_i = \sum_{k=1}^{R} \nu_{ik} d\xi_k \qquad (9.80)$$

If we integrate Equation (9.80), we get:

$$n_i = n_i^o + \sum_{k=1}^{R} \nu_{ik} \xi_k \qquad (9.81)$$

Summing over all i species, in, for example, the vapor phase:

$$n^v = n^o + \sum_{k=1}^{R} v_k \xi_k \tag{9.82}$$

where n^v and n^o are defined analogously to Section 9.3. Finally, dividing Equation (9.81) by Equation (9.82) gives

$$y_i = \frac{n_i}{n^v} = \frac{n_i^o + \sum_{k=1}^{R} v_{ik} \xi_k}{n^o + \sum_{k=1}^{R} v_k \xi_k} \tag{9.83}$$

To find the equilibrium composition in the system, we must determine the equilibrium constants, K_k, for all R reactions. Each equilibrium constant can be found independently, using the appropriate thermochemical data as discussed in Sections 9.4 and 9.5. We can then apply Equation (9.35) to each reaction. The result we obtain is a set of k-coupled nonlinear algebraic equations that we must then solve for the extents of reaction, ξ_k. Once all the extents of reaction are determined, the equilibrium composition can be found by either Equation (9.81) or (9.83). To illustrate how to set up such a problem, an example based on the reaction discussed with Figure 9.1 is illustrative.

▶ **EXAMPLE 9.15**
Specification of Chemical Reactions in Figure 9.1

Consider the addition of 1 mole of HBr to 1 mole of butadiene, as depicted in Figure 9.1. Develop the equations you would need to solve to describe the composition of the system at equilibrium given T and P.

SOLUTION   We need to specify two independent reactions. One possible set we can choose is the reactions described in Figure 9.1:

$$C_4H_6 + HBr \rightleftarrows 1 - BrC_4H_7 \qquad \text{reaction 1} \tag{E9.15A}$$

$$C_4H_6 + HBr \rightleftarrows 3 - BrC_4H_7 \qquad \text{reaction 2} \tag{E9.15B}$$

According to Equations (9.81) through (9.83), we get

$$n_{C_4H_6} = 1 - \xi_1 - \xi_2 \qquad\qquad y_{C_4H_6} = (1 - \xi_1 - \xi_2)/(2 - \xi_1 - \xi_2)$$

$$n_{HBr} = 1 - \xi_1 - \xi_2 \qquad\qquad y_{HBr} = (1 - \xi_1 - \xi_2)/(2 - \xi_1 - \xi_2)$$

$$n_{1-BrC_4H_7} = \xi_1 \qquad\qquad y_{1-BrC_4H_7} = \xi_1/(2 - \xi_1 - \xi_2)$$

$$n_{3-BrC_4H_7} = \xi_2 \qquad\qquad y_{3-BrC_4H_7} = \xi_2/(2 - \xi_1 - \xi_2)$$

$$\overline{\qquad n^v = 2 - \xi_1 - \xi_2 \qquad}$$

The two equilibrium constants are then written:

$$K_1 = \frac{y_{1-BrC_4H_7}}{y_{C_4H_6} y_{HBr} P} = \frac{\xi_1(2 - \xi_1 - \xi_2)}{(1 - \xi_1 - \xi_2)^2 P} \tag{E9.15C}$$

$$K_2 = \frac{y_{3-BrC_4H_7}}{y_{C_4H_6} y_{HBr} P} = \frac{\xi_2(2 - \xi_1 - \xi_2)}{(1 - \xi_1 - \xi_2)^2 P} \tag{E9.15D}$$

Each equilibrium constant can be found by using appropriate thermochemical data, as discussed in Section 9.4. Once values for K_1 and K_2 are obtained, Equations (E9.15C) and (E9.15D) can be solved for the two unknowns, ξ_1 and ξ_2. It is then straightforward to find the number of moles and mole fractions of the species present. ◀

Gibbs Phase Rule for Chemically Reacting Systems

Say we have a system in which the species undergo chemical reaction by rearranging their bonds to minimize the total Gibbs energy and obtain equilibrium. While we have identified the significant species at play and their phases, we do not know what the reaction mechanism is. In fact, there may be many simultaneous reactions that describe these molecular rearrangements. We may be concerned with questions about how to set up the chemical reaction equilibrium problem, such as "What equations should I use to describe the reactions?" and "How do I know if I have included enough reactions?"

It turns out that as far as the thermodynamic (i.e., equilibrium) calculations go, we do not need to pick the actual reactions that the real system undergoes; we are free to choose any set of independent reactions we can think up. As we have seen many times, hypothetical paths can often be convenient for calculating thermodynamic properties. Since the chemical reaction equilibria calculation is based on a thermodynamic property—Gibbs energy—it should not surprise us that it is independent of the specific reaction path used.

The number of chemical reactions we need to specify can be obtained by using the Gibbs phase rule for reacting systems. The phase rule for reacting systems is obtained by counting the total number of variables in the system and making sure we have the same number of independent equations. It is accomplished in much the same way we accounted for variables for nonreacting systems in Example 6.14. We will omit the details of the accounting and merely present the results. The number of independent chemical reactions, R, needed to specify the system is given by:

$$R = m - \Im + 2 - \pi - s \tag{9.84}$$

where

m = the number of chemical species

$\Im$ = the degrees of freedom, that is, the number of intensive properties specified

π = the number of phases in the system

s = stoichiometric constraints

Equation (9.84) tells us the number of independent reactions, R, we need to specify among the m chemical species present. The stoichiometric constraints, s, are dictated by the inlet conditions, since the ratio of elements must stay the same. As long as the reactions we come up with are independent, solution of the multiple reaction equilibria problem will give us the equilibrium composition of the system. Indeed, if we were to chose a different set of equations, as long as they were independent and satisfied the number given by Equation (9.84), we would get the same result for equilibrium composition.

▶ **EXAMPLE 9.16**
Application of Gibbs Phase Rule to Reactions in Figure 9.1

Apply the phase rule to determine the number of independent reactions needed to calculate the equilibrium composition of the butadiene system shown in Figure 9.1, given T, P, and the initial composition of reactants

SOLUTION We need to determine all the quantities on the right-hand side of Equation 9.84. The species present are

$$C_4H_6, HBr, 1 - BrC_4H_7, 3 - BrC_4H_7$$

Thus, $m = 4$. Since this reaction occurs in the gas phase, $\pi = 1$. We have specified 2 degrees of freedom, T and P. The inlet the ratio, $n^o_{C_4H_6}/n^o_{HBr}$, places a stoichiometric constraint. If we know how much product, $1 - BrC_4H_7$ and $3 - BrC_4H_7$ is formed, the number of reactants left is determined. From Equation (9.84),

$$R = m - \Im + 2 - \pi - s = 4 - 2 + 2 - 1 - 1 = 2$$

Hence, we must specify two independent equations.   ◀

▶ **EXAMPLE 9.17**
Alternative Formulation of Example 9.15

Example 9.16 suggests that we need to specify two independent equations to calculate the chemical reaction equilibrium of the butadiene system. Suggest an alternative set to those used in Example 9.15. Show how you would solve the problem with these reactions.

SOLUTION   We now chose the isomerization reaction for reaction 2' as follows:

$$C_4H_6 + HBr \rightleftarrows 1 - BrC_4H_7 \qquad \text{reaction 1} \qquad \text{(E9.17A)}$$

$$1 - BrC_4H_7 \rightleftarrows 3 - BrC_4H_7 \qquad \text{reaction 2'} \qquad \text{(E9.17B)}$$

Since reactions 1 and 2' are also independent, they can be used to solve for the composition of the system shown in Figure 9.1. In this case, we get

$$n_{C_4H_6} = 1 - \xi_1 \qquad\qquad y_{C_4H_6} = (1 - \xi_1)/(2 - \xi_1)$$
$$n_{HBr} = 1 - \xi_1 \qquad\qquad y_{HBr} = (1 - \xi_1)/(2 - \xi_1)$$
$$n_{1-BrC_4H_7} = \xi_1 - \xi_2 \qquad\qquad y_{1-BrC_4H_7} = (\xi_1 - \xi_2)/(2 - \xi_1)$$
$$n_{3-BrC_4H_7} = \xi_2 \qquad\qquad y_{3-BrC_4H_7} = \xi_2/(2 - \xi_1)$$
$$\overline{\qquad n^v = 2 - \xi_1 \qquad}$$

The equilibrium constants can be written using the equilibrium constants:

$$K_1 = \frac{y_{1-BrC_4H_7}}{y_{C_4H_6}y_{HBr}P} = \frac{(\xi_1 - \xi_2)(2 - \xi_1)}{(1 - \xi_1)^2 P} \qquad \text{(E9.17C)}$$

$$K_2 = \frac{y_{3-BrC_4H_7}}{y_{1-BrC_4H_7}} = \frac{\xi_2}{(\xi_1 - \xi_2)} \qquad \text{(E9.17D)}$$

Each equilibrium constant can be found by using appropriate thermochemical data, as discussed in Section 9.4. Once values for K_1 and K_2 are obtained, Equations (E9.17C) and (E9.17D) can be solved for unknowns ξ_1 and ξ_2. Notice that the specific values given by K_2 will differ from that given by Equation E9.15D; hence the value of ξ_2 will also be different. However, the compositions that are calculated will turn out identical to those in Example 9.15. This result illustrates the magic of chemical reaction equilibria; *No matter what set of independent reactions we pick, the equilibrium compositions that are calculated remain the same.*   ◀

▶ **EXAMPLE 9.18**
Cracking of Methane

Steam reforming of methane can be used to make hydrogen gas. CO and CO_2 are observed as by-products. Using a 4:1 H_2O to CH_4 inlet ratio and a pressure of 1 bar, calculate the equilibrium composition obtained in the temperature range of 600–1100 K.

SOLUTION   We first apply the phase rule to determine the number of independent reactions we must specify. For each calculation, we have five species, one phase, and a specific temperature and pressure. We have also specified the inlet feed ratio. Thus, the elements in the system form the stoichiometric constraint; the ratios must be identical to the ratios they have in the feed. The ratio of elemental O:H is constrained to 1:3; similarly, the ratio of C:H is constrained to 1:12. Hence, we have two additional stoichiometric constraints, s. The number of independent reactions is now given by

$$R = m - \Im + 2 - \pi - s = 5 - 2 + 2 - 1 - 2 = 2$$

For a set of two independent reactions, we can choose:

$$CH_4 + H_2O \rightleftarrows CO + 3H_2 \qquad \text{reaction 1} \qquad \text{(E9.18A)}$$

$$CH_4 + 2H_2O \rightleftarrows CO_2 + 4H_2 \qquad \text{reaction 2} \qquad \text{(E9.18B)}$$

Can you do this problem with a different reaction set? According to Equations (9.81) through (9.83), we get

$$n_{CH_4} = 1 - \xi_1 - \xi_2 \qquad\qquad y_{CH_4} = \frac{1 - \xi_1 - \xi_2}{5 + 2\xi_1 + 2\xi_2}$$

$$n_{H_2O} = 4 - \xi_1 - 2\xi_2 \qquad\qquad y_{H_2O} = \frac{4 - \xi_1 - 2\xi_2}{5 + 2\xi_1 + 2\xi_2}$$

$$n_{H_2} = 3\xi_1 + 4\xi_2 \qquad\qquad y_{H_2} = \frac{3\xi_1 + 4\xi_2}{5 + 2\xi_1 + 2\xi_2}$$

$$n_{CO} = \xi_1$$

$$n_{CO_2} = \xi_2 \qquad\qquad y_{CO} = \frac{\xi_1}{5 + 2\xi_1 + 2\xi_2}$$

$$\overline{n^v = 5 + 2\xi_1 + 2\xi_2}$$

$$y_{CO_2} = \frac{\xi_2}{5 + 2\xi_1 + 2\xi_2}$$

At 1 bar, the equilibrium constants can be written assuming ideal gas conditions:

$$K_1 = \frac{y_{CO}\, y_{H_2}^3}{y_{CH_4}\, y_{H_2O}} P^2 = \frac{(\xi_1)(3\xi_1 + 4\xi_2)^3}{(5 + 2\xi_1 + 2\xi_2)^2(1 - \xi_1 - \xi_2)(4 - \xi_1 - 2\xi_2)} P^2 \qquad \text{(E9.18C)}$$

$$K_2 = \frac{y_{CO_2}\, y_{H_2}^4}{y_{CH_4}\, y_{H_2O}^2} P^2 = \frac{(\xi_2)(3\xi_1 + 4\xi_2)^4}{(5 + 2\xi_1 + 2\xi_2)^2(1 - \xi_1 - \xi_2)(4 - \xi_1 - 2\xi_2)^2} P^2 \qquad \text{(E9.18D)}$$

To solve for the equilibrium constants at the elevated temperatures, we need appropriate thermochemical data. Gibbs energy and enthalpy of reaction, as well as heat capacity data available in Appendices A.2 and A.3 are summarized in Table E9.18A. From these data, we can solve for K_1 and K_2 at different temperatures using Equation (9.52). This solution is conveniently done in a spreadsheet. Table E9.18B presents the values obtained at different temperatures. Once the

TABLE E9.18A Summary of Thermochemical Data from Appendices A.2 and A.3

	CH_4	H_2	H_2O	CO	CO_2	Reaction 1	Reaction 2
Δg_f	-50.72	0	-228.57	-137.17	-394.36	142.12	113.50
Δh_f	-74.81	0	-241.82	-110.53	-393.51	206.10	164.94
ν_1	-1	3	-1	1	0		
ν_2	-1	4	-2	0	1		
A_i	1.702	3.249	3.47	3.376	5.457	7.951	9.811
B_i	9.08×10^{-3}	4.22×10^{-4}	1.45×10^{-3}	5.57×10^{-4}	1.05×10^{-3}	-8.708×10^{-3}	-9.243×10^{-3}
C_i	-2.16×10^{-6}	0	0	0	0	2.16×10^{-6}	2.16×10^{-6}
D_i	0	8.30×10^3	1.21×10^4	-3.10×10^3	-1.16×10^5	9.70×10^3	-1.067×10^5

TABLE E9.18B Summary of Solution of Example 9.18

T	K_1	K_2	ξ_1	ξ_2	y_{CH_4}	y_{H_2}	y_{H_2O}	y_{CO}	y_{CO_2}
600	4.91×10^{-7}	1.40×10^{-5}	0.000	0.113	0.170	0.087	0.722	0.000	0.022
650	1.42×10^{-5}	2.24×10^{-4}	0.003	0.191	0.150	0.144	0.671	0.000	0.036
700	2.59×10^{-4}	2.47×10^{-3}	0.011	0.294	0.124	0.215	0.606	0.002	0.052
750	3.24×10^{-3}	2.02×10^{-2}	0.037	0.410	0.094	0.297	0.534	0.006	0.070
800	3.00×10^{-2}	1.29×10^{-1}	0.099	0.515	0.062	0.378	0.461	0.016	0.083
850	2.15×10^{-1}	6.70×10^{-1}	0.207	0.579	0.033	0.447	0.401	0.032	0.088
900	1.25	2.93	0.332	0.584	0.012	0.488	0.366	0.049	0.085
950	6.07	1.11×10^1	0.424	0.551	0.004	0.500	0.356	0.061	0.079
1000	2.52×10^1	3.70×10^1	0.485	0.509	0.001	0.499	0.358	0.069	0.073
1050	9.20×10^1	1.11×10^2	0.540	0.458	0.000	0.493	0.364	0.077	0.065
1100	2.99×10^2	3.02×10^2	0.541	0.458	0.000	0.494	0.363	0.077	0.065
1150	8.78×10^2	7.57×10^2	0.541	0.458	0.000	0.494	0.363	0.077	0.065

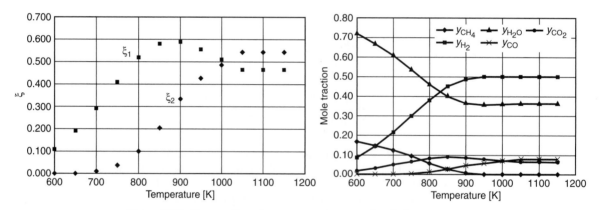

Figure E9.18 Extent of reaction vs. temperature and mole fractions vs. temperature at equilibrium for the steam reforming of methane.

equilibrium constants are found, Equations (E9.18C) and (E9.18D) can be solved simultaneously for ξ_1 and ξ_2. The mole fractions are then straightforward to calculate using the equations above. Table E9.18B presents the results for the extent of reaction and the mole fractions of each species at different temperatures. The extents of reaction vs. temperature and mole fraction vs. temperature are plotted in Figure E9.18. From this analysis, what temperature would you pursue? What other questions would you ask? ◀

▶ **EXAMPLE 9.19**
Growth of Silicon
Thin Films from
Chlorosilane

Single crystal, or *epitaxial*, thin films of silicon are needed to manufacture integrated circuits. These films are grown by chemically reacting a chlorosilane feed gas in the presence of H_2 at elevated temperature. Consider the growth of epitaxial Si at 1300 K and 1 bar. Compare how much Si is produced at equilibrium from $SiCl_4$ vs. $SiCl_3H$ as feed gases, at H_2 dilutions from 1:1 to 150:1 H_2:chlorosilane ratio. Assume the following species are present at equilibrium: Si, $SiCl_2$, $SiCl_4$, $SiCl_3H$, $SiCl_2H_2$, $SiClH_3$, SiH_4, H_2, HCl.

The following data are available:

Gibbs Energies of Formation in [kJ/mol] at 1300 K

Species	$SiCl_2$	$SiCl_4$	$SiCl_3H$	$SiCl_2H_2$	$SiClH_3$	SiH_4	HCl
$\left(\Delta g^o_{f,1300}\right)_i$	-216.012	-492.536	-356.537	-199.368	-28.482	151.897	-102.644

SOLUTION   We first apply the phase rule to see how many independent reactions we need. Since silicon can be present in either the gas or solid phase, its stoichiometry in the vapor is not constrained to the feed ratio. On the other hand, the Cl:H ratio must remain constant. Thus, we have $s = 1$ and can write:

$$R = m - \Im + 2 - \pi - s = 9 - 2 + 2 - 2 - 1 = 6$$

Six independent reactions are then constructed as follows (other choices are also possible):

$$SiCl_4 + H_2 \rightleftarrows SiCl_2 + 2HCl \qquad \text{reaction 1}$$
$$SiCl_4 + H_2 \rightleftarrows SiCl_3H + HCl \qquad \text{reaction 2}$$
$$SiCl_3H + H_2 \rightleftarrows SiCl_2H_2 + HCl \qquad \text{reaction 3}$$
$$SiCl_2H_2 + H_2 \rightleftarrows SiClH_3 + HCl \qquad \text{reaction 4}$$
$$SiClH_3 + H_2 \rightleftarrows SiH_4 + HCl \qquad \text{reaction 5}$$
$$SiCl_4 + 2H_2 \rightleftarrows Si(s) + 4HCl \qquad \text{reaction 6}$$

We then write each species' concentration in terms of the extents of these six reactions:

$$n_{SiCl_4} = n^o_{SiCl_4} - \xi_1 - \xi_2 - \xi_6$$
$$n_{SiCl_2} = \xi_1$$
$$n_{SiCl_3H} = n^o_{SiCl_3H} + \xi_2 - \xi_3$$
$$n_{SiCl_2H_2} = \xi_3 - \xi_4$$
$$n_{SiClH_3} = \xi_4 - \xi_5$$
$$n_{SiH_4} = \xi_5$$
$$n_{H_2} = n^o_{H_2} - \xi_1 - \xi_2 - \xi_3 - \xi_4 - \xi_5 - 2\xi_6$$
$$n_{HCl} = 2\xi_1 + \xi_2 + \xi_3 + \xi_4 + \xi_5 + 4\xi_6$$
$$n^v = n^o_{SiCl_4} + n^o_{SiCl_3H} + n^o_{H_2} + \xi_1 + \xi_6$$
$$n_{Si} = \xi_6$$

The six equilibrium constants are then written assuming ideal gas behavior:

$$K_1 = \frac{y_{SiCl_2} y^2_{HCl}}{y_{SiCl_4} y_{H_2}} P$$

$$= \frac{(\xi_1)(2\xi_1 + \xi_2 + \xi_3 + \xi_4 + \xi_5 + 4\xi_6)^2}{\left(n^o_{SiCl_4} - \xi_1 - \xi_2 - \xi_6\right)\left(n^o_{H_2} - \xi_1 - \xi_2 - \xi_3 - \xi_4 - \xi_5 - 2\xi_6\right)\left(n^o_{SiCl_4} + n^o_{SiCl_3H} + n^o_{H_2} + \xi_1 + \xi_6\right)} P$$

$$K_2 = \frac{y_{SiCl_3H} y_{HCl}}{y_{SiCl_4} y_{H_2}} = \frac{\left(n^o_{SiCl_3H} + \xi_2 - \xi_3\right)(2\xi_1 + \xi_2 + \xi_3 + \xi_4 + \xi_5 + 4\xi_6)}{\left(n^o_{SiCl_4} - \xi_1 - \xi_2 - \xi_6\right)\left(n^o_{H_2} - \xi_1 - \xi_2 - \xi_3 - \xi_4 - \xi_5 - 2\xi_6\right)}$$

$$K_3 = \frac{y_{SiCl_2H_2} y_{HCl}}{y_{SiCl_3H} y_{H_2}} = \frac{(\xi_3 - \xi_4)(2\xi_1 + \xi_2 + \xi_3 + \xi_4 + \xi_5 + 4\xi_6)}{(n^o_{SiCl_3H} + \xi_2 - \xi_3)(n^o_{H_2} - \xi_1 - \xi_2 - \xi_3 - \xi_4 - \xi_5 - 2\xi_6)}$$

$$K_4 = \frac{y_{SiClH_3} \, y_{HCl}}{y_{SiCl_2H_2} \, y_{H_2}} = \frac{(\xi_4 - \xi_5)(2\xi_1 + \xi_2 + \xi_3 + \xi_4 + \xi_5 + 4\xi_6)}{(\xi_3 - \xi_4)(n^o_{H_2} - \xi_1 - \xi_2 - \xi_3 - \xi_4 - \xi_5 - 2\xi_6)}$$

$$K_5 = \frac{y_{SiH_4} \, y_{HCl}}{y_{SiClH_3} \, y_{H_2}} = \frac{(\xi_5)(2\xi_1 + \xi_2 + \xi_3 + \xi_4 + \xi_5 + 4\xi_6)}{(\xi_4 - \xi_5)(n^o_{H_2} - \xi_1 - \xi_2 - \xi_3 - \xi_4 - \xi_5 - 2\xi_6)}$$

$$K_6 = \frac{\left(\dfrac{\hat{f}_{Si}}{f^o_{Si}}\right) y^4_{HCl}}{y_{SiCl_4} \, y^2_{H_2}}$$

$$= \frac{(1)(2\xi_1 + \xi_2 + \xi_3 + \xi_4 + \xi_5 + 4\xi_6)^4}{(n^o_{SiCl_4} - \xi_1 - \xi_2 - \xi_6)(n^o_{H_2} - \xi_1 - \xi_2 - \xi_3 - \xi_4 - \xi_5 - 2\xi_6)^2(n^o_{SiCl_4} + n^o_{SiCl_3H} + n^o_{H_2} + \xi_1 + \xi_6)} P$$

From the values of Gibbs energy, we can calculate the values of the six equilibrium constants. We must then solve for the six unknown extents of reaction to match the equilibrium constants. An example of one solution set taken from the solution spreadsheet for a feed ratio of H_2 : $SiCl_4$ of 1:1 is given below:

T [K]	1300
P [bar]	1
$n^o_{H_2}$	1
$n^o_{SiCl_4}$	1
$n^o_{SiCl_3H}$	0

	$\left(\Delta g^o_{f,1300}\right)_i$	n_i	y_i
H_2	0	0.8214	0.406
$SiCl_2$	−216.012	0.0763	3.77×10^{-2}
$SiCl_4$	−492.536	0.7770	3.84×10^{-1}
$SiCl_3H$	−356.537	0.1912	9.44×10^{-2}
$SiCl_2H_2$	−199.368	0.0066	3.28×10^{-3}
$SiClH_3$	−28.482	0.0001	3.20×10^{-5}
SiH_4	151.897	0.0000	1.30×10^{-7}
HCl	−102.644	0.1525	7.53×10^{-2}
Total		2.0251	
Si		−0.0512	

	K_i	ξ_i	$K_{i,\,calc}$	$K_i - K_{i,\,calc}$
Reaction 1	1.37×10^{-3}	0.0763	1.37×10^{-3}	1.71×10^{-8}
Reaction 2	4.57×10^{-2}	0.1979	4.57×10^{-2}	1.02×10^{-8}
Reaction 3	6.44×10^{-3}	0.0067	6.44×10^{-3}	6.74×10^{-10}
Reaction 4	1.81×10^{-3}	0.0001	1.81×10^{-3}	1.19×10^{-10}
Reaction 5	7.52×10^{-4}	0.0000	7.52×10^{-4}	-2.15×10^{-11}
Reaction 6	5.09×10^{-4}	−0.0512	5.09×10^{-4}	2.84×10^{-10}

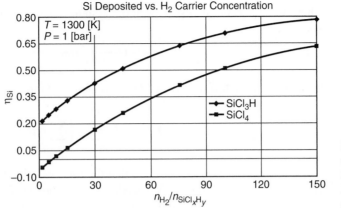

Figure E9.19 Deposition efficiency of $SiCl_4$ and $SiCl_3H$ at 1300 K and 1 bar at different dilutions in H_2.

Solution is obtained by calculating the equilibrium constant K_i in two ways: (1) from Gibbs energy data (labeled K_i) and (2) from the extents of reaction (labeled $K_{i,\text{calc}}$). The extents are changed until the two values matched within a convergence criteria as determined by $K_i - K_{i,\text{calc}}$. In this case the extent of reaction 6 is negative, indicating that solid Si would actually be removed, or etched, from the substrate. Hence, thermodynamics tells us there is no way to deposit Si under these conditions. We define the equilibrium deposition efficiency, η, as

$$\eta = \frac{\text{amount of Si deposited as solid}}{\text{amount of Si in feed gas}}$$

In the case above, we actually get an efficiency of -0.0512, since Si was etched. The complete case study is plotted in Figure E9.19. ◄

Solution of Multiple Reaction Equilibria by Minimization of Gibbs Energy

In this section, we develop an alternative set of coupled nonlinear algebraic equations to solve chemical reaction equilibria problems involving multiple reactions. In Chapter 8, we showed how to solve an LLE problem first by equating fugacities (Example 8.14) and then by minimizing Gibbs energy (Example 8.15). We can essentially do the same for multiple reaction equilibria. We have already seen how to solve this set of equations by explicitly writing a set of independent reactions. We then used the **equilibrium constant formulation** analogous to a single reaction to solve a set of R equations for each extent of reaction. We will now learn another approach to arrive at the same answer. We solve for the equilibrium composition through **minimization of the total Gibbs energy** of the system. This approach is useful in developing computer algorithms to solve reactions for complex systems.

We first introduce the formula coefficient matrix, β_{ij}. This matrix relates the j elements, b_j, in our system to the species i. We can relate these two quantities as follows:

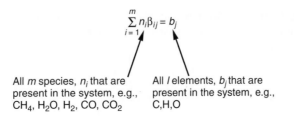

$$\sum_{i=1}^{m} n_i \beta_{ij} = b_j$$

All *m* species, n_i that are present in the system, e.g., CH_4, H_2O, H_2, CO, CO_2

All *l* elements, b_j that are present in the system, e.g., C, H, O

(9.85)

where we sum over all m species in the system. We can write the Gibbs energy of the system as

$$G = \sum_{i=1}^{m} \mu_i n_i \tag{9.86}$$

We next define a new function G' by introduction of the Lagrangian multipliers, λ_j:

$$G' = \sum_{i=1}^{m} \mu_i n_i + \sum_{j=1}^{l} \lambda_j \left(\sum_{i=1}^{m} n_i \beta_{ij} - b_j \right) \tag{9.87}$$

Inspection of Equation (9.85) shows that the term we added in Equation (9.87) is zero. Hence, we can recast our problem as the minimization of G'. To find the composition at which this function is a minimum, we set its derivative with respect to n_i to zero:

$$\left(\frac{\partial G'}{\partial n_i} \right)_{T,P,n_{j \neq i}} = 0 = \mu_i + \sum_{j=1}^{l} \lambda_j \beta_{ij} \tag{9.88}$$

We now insert the phase-appropriate expression for the chemical potential, as we did before. For example, for an ideal gas, we get

$$g_i^o + RT \ln y_i P + \sum_{j=1}^{l} \lambda_j \beta_{ij} = 0 \tag{9.89}$$

We can substitute the Gibbs energy of formation for the standard-state Gibbs energy of species i in Equation (9.89). Since we will write one equation for every species in the system, the pure species Gibbs energies part of the Gibbs energy of formation end up canceling one another. Thus, we get

$$\Delta g_i^f + RT \ln y_i P + \sum_{j=1}^{l} \lambda_j \beta_{ij} = 0 \tag{9.90}$$

Equations (9.85) and (9.90) represent a set of $m + l$ equations that can be solved for the unknowns y_i and λ_j. Example 9.20 shows how we solve the same problem we encountered in Example 9.18 using the minimization of Gibbs energy. Expressions similar to Equation (9.90) can be developed for real gases, liquids, and solids.

▶ **EXAMPLE 9.20**
Cracking of Methane—Revisited

Reformulate the solution of Example 9.18 using the minimization of Gibbs energy. Compare the results at 800 K. The Gibbs energies of formation at 800 K are given by

Species	CH_4	H_2O	CO	CO_2
$\left(\Delta g_{f,800}^o \right)_i$ in [kJ/mol]	−2.057	−192.713	−182.494	−203.595

SOLUTION We first formulate the coefficient matrix and the vector b for a 4:1 H_2O: CH_4 inlet ratio. For the formula coefficient matrix, we write the species in the rows and the elements in the columns:

$$
\begin{matrix}
 & \begin{matrix} \text{C} & \text{H} & \text{O} \end{matrix} \\
\begin{matrix} CH_4 \\ H_2O \\ H_2 \\ CO \\ CO_2 \end{matrix} & \begin{bmatrix} 1 & 4 & 0 \\ 0 & 2 & 1 \\ 0 & 2 & 0 \\ 1 & 0 & 1 \\ 1 & 0 & 2 \end{bmatrix}
\end{matrix}
\quad \text{and} \quad
\begin{matrix}
b_{\text{C}} = 1 \\
b_{\text{H}} = 12 \\
b_{\text{O}} = 4
\end{matrix}
$$

or in matrix form:

$$\beta = \begin{bmatrix} 1 & 4 & 0 \\ 0 & 2 & 1 \\ 0 & 2 & 0 \\ 1 & 0 & 1 \\ 1 & 0 & 2 \end{bmatrix} \qquad \text{and} \qquad b = \begin{bmatrix} 1 \\ 12 \\ 4 \end{bmatrix}$$

Equation (9.85) gives:

$$\begin{bmatrix} n_{CH_4} & n_{H_2O} & n_{H_2} & n_{CO} & n_{CO_2} \end{bmatrix} \begin{bmatrix} 1 & 4 & 0 \\ 0 & 2 & 1 \\ 0 & 2 & 0 \\ 1 & 0 & 1 \\ 1 & 0 & 2 \end{bmatrix} = \begin{bmatrix} 1 & 12 & 4 \end{bmatrix}$$

which can be written as three coupled equations:

$$n_{CH_4} + n_{CO} + n_{CO_2} = 1 \tag{E9.20A}$$

$$4n_{CH_4} + 2n_{H_2O} + 2n_{H_2} = 12 \tag{E9.20B}$$

$$n_{H_2O} + n_{CO} + 2n_{CO_2} = 4 \tag{E9.20C}$$

Equation (9.90) can be written for each of the $N = 5$ species as

$$\Delta g^o_{f,CH_4} + RT \ln \frac{n_{CH_4}}{n_T} + \lambda_C + 4\lambda_H = 0 \tag{E9.20D}$$

$$\Delta g^o_{f,H_2O} + RT \ln \frac{n_{H_2O}}{n_T} + 2\lambda_H + \lambda_O = 0 \tag{E9.20E}$$

$$\Delta g^o_{f,H_2} + RT \ln \frac{n_{H_2}}{n_T} + 2\lambda_H = 0 \tag{E9.20F}$$

$$\Delta g^o_{f,CO} + RT \ln \frac{n_{CO}}{n_T} + \lambda_C + \lambda_O = 0 \tag{E9.20G}$$

$$\Delta g^o_{f,CO_2} + RT \ln \frac{n_{CO_2}}{n_T} + \lambda_C + 2\lambda_O = 0 \tag{E9.20H}$$

where

$$n_T = n_{CH_4} + n_{H_2O} + n_{H_2} + n_{CO} + n_{CO_2}$$

Equations (E9.20A) and (E9.20H) form a set of eight coupled equations with eight unknowns. This set of equations can be solved to give:[6]

λ_C/RT	λ_O/RT	λ_H/RT	n_{CH_4}	n_{H_2O}	n_{H_2}	n_{CO}	n_{CO_2}
0.73	0.44	30.58	0.30	2.66	2.74	0.13	0.57

We have used λ_j/RT in the numerical solution to have values of a reasonable order of magnitude so that the numerical nonlinear solution algorithm is better behaved. ◀

---

[6] This set of equations was solved by a modified Newton–Raphson root finder using the text software ThermoSolver. For more detail, see the ThermoSolver documentation.

▶ 9.8 REACTION EQUILIBRIA OF POINT DEFECTS IN CRYSTALLINE SOLIDS

The structure of a crystalline material is defined by its lattice. The crystal lattice consists of a well-defined geometry of repeating units with an atom at every lattice point. Any disruption of the perfect order of an ideal crystal is termed a **defect**. Even though defect concentrations at equilibrium are quite small (usually less than 1 part per million), it turns out that many important properties of solid materials are controlled by the nature and concentration of its defects. For example, the conductivity, diffusivity, and luminescence of solid materials can all be dramatically altered by such defects. In this section, we examine **point defects**, defects which occur at single atomic site.[7] We apply the principles from this chapter to describe the defect concentrations in a crystal at equilibrium as we change the state of the system. An ability to control the concentration of defects allows us to control the crystal properties related to those defects. Only a brief overview is presented here. Several books can be consulted for a more extensive treatment of this subject.[8]

Atomic Defects

There are two major types of point defects: atomic defects and electronic defects. An atomic defect occurs when atoms are misplaced from their regular position in the crystal lattice. Figure 9.6 shows a two-dimensional representation of a crystal lattice with some common types of atomic point defects. A **vacancy** occurs when an atom is absent from a lattice site that is normally occupied. An **interstitial** occurs when an atom sits in a place in the crystal that is not a distinct lattice site, but rather in between lattice sites. Figure 9.6 shows two types of interstitials. A **self-interstitial** contains an atom of the same type that makes up the host crystal, while an **impurity interstitial** consists of a foreign atom. A **substitutional impurity** occurs when a foreign atom occupies a lattice site normally housed by a host atom. In compound solids, such as AB, we can have **misplaced** atoms, where species A sits in a B site or vice versa.

First, we introduce a nomenclature to identify the different types of point defects in a crystal. We will then write balanced chemical equations for processes that occur in the solid. The principles of thermodynamics that we have learned can then be applied to understand what defects are present at equilibrium and to quantify their concentrations. In writing these balance equations, we must adhere to the following physical principles.

1. The equation must be balanced in terms of lattice sites.

2. In a compound material, the ratio of sites must remain fixed; for example, in the crystal AB, we cannot create an additional A site without also creating an additional B site; in AB_2, we must create two B sites for every A; and so on.

3. We treat neutral crystals. Thus, in treating charged species, we must conserve charge; creation of a positively charged species must concomitantly be associated with the production of a negative charge. This condition is referred to as **electroneutrality**.

---

[7] Higher dimensional defects such as dislocations and grain boundaries are thermodynamically unstable, and their behavior must be left for a class that covers kinetics.

[8] W van Gool, *Principles of Defect Chemistry in Crystalline Solids*, (New York: Academic Press, 1966); F. A. Kroeger, *The Chemistry of Imperfect Crystals* (Vol. 2), (New York: North Holland, 1973); R. A. Swalin, *Thermodynamics of Solids*, (New York: Wiley, (1972). For a ChE example, see, T. J. Anderson, "Examples of Chemical Engineering Principles Applied to the Growth of Semiconductors," in S. L. Sandler and B. A. Finlayson (eds) *Chemical Engineering Education in a Changing Environment*. (New York: United Engineering Trustees, 1988), p. 311.

Atomic Point Defects

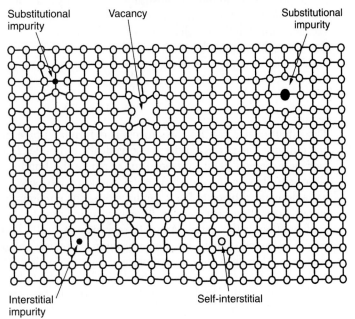

Figure 9.6 Atomic point defects in a monatomic crystal lattice.

We use the following symbols to represent point defects: an i for an interstitial and a V for a vacancy. A zinc self-interstitial, for example, may have the following designation:

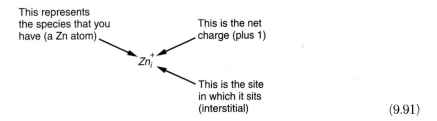

$$(9.91)$$

Equation (9.91) depicts three pieces of information. The center designates the species of interest (zinc). The subscript tells us what crystal site that species sits in (interstitial), while the superscript tells us the effective charge (+1). The effective charge is relative to that of the site in the perfect lattice. For example, consider a crystal of NaCl. The sodium normally gives up a valence electron to chloride to form the ionic bonds in the lattice. However, if a calcium atom exists as a substitutional impurity in a sodium site, it may lose both its valence electrons. Its effective charge, relative to the sodium that would normally sit there, is then +1. The nomenclature for such an impurity would be written as Ca_{Na}^{+} [9]

Let's practice a few; what do the following symbols represent?

$$Al_{Zn}^{+} \qquad V_{Zn} \qquad V_{As} \qquad V_{As}^{+}$$

[9] The widely used nomenclature proposed by Kroeger and Vink uses ', for each unit of negative effective charge and · for positive effective charge. They choose a separate nomenclature to distinguish between effective charge and real charge. In this text, any charges associated with defect equilibria will implicitly be the effective charge.

Al_{Zn}^{+} represents an aluminum substitutional impurity in a zinc site with an effective charge of $+1$. The next three terms represent neutral zinc, arsenic vacancies, and an arsenic vacancy with an effective charge of $+1$, respectively.

Using this nomenclature, we can describe processes involving defects in the crystal lattice. For example, a zinc atom that leaves its lattice site for an interstitial may be described by the following reaction:[10]

$$Zn_{Zn} \rightleftarrows Zn_i + V_{Zn} \tag{9.92}$$

The extent of reaction for this process is determined by a trade-off between enthalpy and entropy. The perfect lattice is energetically favorable. To form a vacancy–interstitial pair via Equation (9.92), we need to overcome the bond energy of a zinc atom in its lattice site. However, when the crystal exists in a perfectly ordered lattice, every atom has a designated place to be; it can have only one configuration. Therefore, its entropy is low. Formation of a vacancy–interstitial pair dramatically increases the number of configurations in the system. Any of the approximately 10^{23} lattice sites may be vacant; we cannot say exactly which one. Similarly, there are many configurations for the interstitial zinc atom. Therefore, the entropy increases by introduction of these defects.[11] The trade-off between these two effects can be quantified by the Gibbs energy. At equilibrium, a crystal exhibits a thermodynamically prescribed defect concentration that minimizes the Gibbs energy of the system. As we increase temperature, the effect of entropy becomes more important relative to enthalpy and the defect concentration increases.

We can quantify the concentration of defects through the equilibrium constant formulation. The equilibrium constant of Equation (9.92) is

$$K = \frac{\left(\dfrac{\hat{f}_{Zn_i}}{f_{Zn_i}^o}\right)\left(\dfrac{\hat{f}_{V_{Zn}}}{f_{V_{Zn}}^o}\right)}{\left(\dfrac{\hat{f}_{Zn}}{f_{Zn}^o}\right)} \tag{9.93}$$

Since the defects are very dilute and are not defined in the Lewis/Randall limit, we choose a Henry's law reference state for them, i.e., $f_{Zn}^o = \mathcal{H}_{Zn}$ and $f_{V_{Zn}}^o = \mathcal{H}_{V_{Zn}}$. This state is the hypothetical pure species characterized by all a-b interactions; its properties are given by those at infinite dilution. Thus, we use the partial molar Gibbs energy at infinite dilution for these terms in the Gibbs energy of reaction. Using Henry's law for the defects, Equation 9.93 becomes

$$K = \frac{\left(\gamma_{Zn_i}^{\text{Henry's}} x_{Zn_i}\right)\left(\gamma_{V_{Zn}}^{\text{Henry's}} x_{V_{Zn}}\right)}{\left(\dfrac{\hat{f}_{Zn}}{f_{Zn}}\right)} \tag{9.94}$$

If the defect concentrations are dilute, the Henry's law activity coefficients go to 1. Additionally, the host crystal has a fugacity nearly equal to its pure species fugacity, so the denominator goes to 1. Equation (9.94) can then be simplified to

$$K = x_{Zn_i} x_{V_{Zn}} \tag{9.95}$$

---

[10] There are also charged versions of this process. For example, the zinc interstitial can leave with a $+1$ charge, leaving behind a vacancy with a -1 charge.

[11] Since the sites in a crystal are well defined, we can use statistical mechanics to come up with a quantitative expression for the increase in entropy; it ends up being analogous to the entropy of mixing given by Example 6.8.

In compounds that deviate significantly from their stoichiometric proportions, high defect concentrations are observed and the activity coefficients from Equation (9.94) must be included. Since the atomic number density of the crystal is well defined, Equation (9.95) is often written in terms of defect concentration:

$$K_c = [Zn_i][V_{Zn}] \tag{9.96}$$

where $[Zn_i]$ and $[V_{Zn}]$ are the zinc interstitial and zinc vacancy number densities in units of $[\#/m^3]$, and K_c is the equilibrium constant for units of concentration given by:

$$K_c = KN^2 \tag{9.97}$$

where N is the number density of the host crystal.

There are many other defect-inducing processes that we may consider. We may have a system that forms a vacancy from diffusion of a zinc atom to the surface, followed by evaporation into the vapor. This process is described by the reaction

$$Zn_{Zn} \rightleftarrows Zn(g) + V_{Zn} \tag{9.98}$$

Assuming an ideal gas, we can write the equilibrium constant corresponding to Equation (9.98) as

$$K_c = p_{Zn}[V_{Zn}] \tag{9.99}$$

Again, the activity of zinc in the crystal has been assumed to be 1.

Electronic Defects

In addition to atomic point defects, **semiconductor** materials can also have electronic point defects. These defects provide mobile charge carriers that move about the crystal lattice. They provide the basis for many useful applications. In fact, the entire microelectronics industry is based on being able to control electronic defects in these materials. For example, consider silicon, the most prevalent semiconductor. Figure 9.7 shows two representations of a pure silicon lattice using Lewis dot structures. Each silicon atom has four valence electrons and is therefore tetrahedrally bonded to four neighboring Si atoms. This repeated structure forms the silicon lattice; its crystal structure is identical to diamond. The diagram in the middle shows a perfect Si lattice at absolute zero. In this case, all the electrons are covalently bonded between a given pair of silicon atoms.

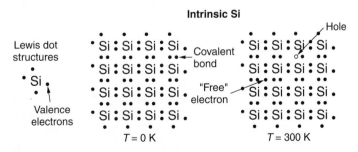

Figure 9.7 Lewis dot structure of pure Si at 0 K and 300 K. At 0 K, Si is insulating. Mobile electrons and holes are created at 300 K. These electronic point defects in the Si lattice make the material semiconducting.

At 0 K, silicon has no mobile charge carriers; it is insulating. The diagram to the right shows pure silicon at 300 K. At finite temperatures, the atoms in the solid vibrate. The energy of vibration varies according to a Boltzmann distribution. A small number of atomic pairs vibrate with enough energy that the electrons shake free of their covalent bonds and are free to move around the crystal lattice. This process creates free electrons, which are mobile negative charge carriers. A free electron is shown in the diagram of pure Si at 300 K. Furthermore, as shown in the figure, the missing negatively charged electron leaves behind a positively charged **hole**. It turns out that holes are also mobile. A neighboring valence electron can tunnel through and exchange places with the hole. By this mechanism, the hole has moved from one bond to the next. Now, in this new position, another neighboring electron can tunnel through, causing the hole to move once again. This process can be repeated over and over, leading to the mobility of holes about the lattice. Of course, when a free electron runs into a hole, it can refill it; we call such a process recombination. Pure silicon, which forms an identical number of holes and electrons, is termed an **intrinsic** semiconductor.

We can again use chemical reactions to describe processes associated with electronic defects. For example, the creation of an electron–hole pair from pure silicon is written as:

$$0 \rightleftarrows h^+ + e^- \tag{9.100}$$

where we use the symbol "e" for a free electron and "h" for a hole. The zero on the left of Reaction (9.100) designates the crystal as it exists in its ideal, perfectly ordered state. Using a similar development as that described in the last section, it can be shown that the equilibrium constant for Reaction (9.100) can be written as:

$$K_c = pn = N^2 \exp\left(-\frac{\Delta g^o_{\text{rxn}}}{RT}\right) \tag{9.101}$$

The concentration of holes is given by "p", and electrons by "n", as is conventionally done in semiconductor physics, that is, $p = [h^+]$ and $n = [e^-]$. Conservation of charge requires:

$$p = n = n_i \tag{9.102}$$

where n_i is termed the intrinsic carrier concentration. Thus Equation (9.101) becomes:

$$K_c = n_i^2 \tag{9.103}$$

Equation (9.101) gives the intrinsic carrier concentration in pure Si as

$$n_i = N \exp\left(-\frac{\Delta g^o_{\text{rxn}}}{2RT}\right) \tag{9.104}$$

Equation (9.104) indicates that for a given semiconductor material, the equilibrium carrier concentration of an intrinsic semiconductor depends only on temperature. Values of n_i for four semiconductors—Si, Ge, GaAs, and InP—at four temperatures are presented in Table 9.2. The values range over 13 orders of magnitude from 5300 (GaAs at 250 K) to 1.9×10^{16} (Ge at 500 K). To put these concentrations in perspective, the number density of Si is on the order of $5 \times 10^{22} [\#/m^3]$. Thus, even the largest value in Table 9.2 represents a total carrier concentration less than one part per million. For any given material, as T gets higher, the entropy becomes more important, shifting Reaction (9.100) to the right. Therefore, the defect concentration rises. As we go down the periodic table, there is less energy to hold in valence electrons, and we also see higher carrier concentration. From the data in Table 9.2, can you determine Δg^o_{rxn} for Si, Ge, GaAs and InP?

The ability of semiconductors to create electrical devices lies in our ability to control the number of positive and negative mobile charge carriers through a process called **doping**. We dope a semiconductor by introducing a specific substitutional impurity into the crystal lattice. This atomic defect then affects the electron and hole concentrations. For example, consider the introduction of a phosphorous atom into the silicon lattice, as shown in Figure 9.8. The dopant P has five valence electrons, of which only four can bond to neighboring silicon atoms. The extra electron is free to move about the lattice and becomes a mobile charge carrier. Thus for every P that we introduce into a Si lattice site, we introduce a negative mobile charge carrier. Once the electron has left, P has an overabundance of protons and is positively charged.

We can again describe this process through a set of chemical equations. First, consider the creation of a silicon vacancy in a manner similar to Equation (9.92):

$$Si_{Si} \rightleftarrows Si_i + V_{Si} \tag{9.105}$$

The phosphorous atom can then be incorporated into the lattice as follows:

$$P(g) + V_{Si} \rightleftarrows P^+_{Si} + e^- \tag{9.106}$$

The creation of electron hole pairs still occurs via Reaction (9.100), so

$$0 \rightleftarrows h^+ + e^- \tag{9.107}$$

TABLE 9.2 Intrinsic Carrier Concentration, n_i [#/m³], in Selected Semiconductors

Species	T [K]			
	250	300	400	500
Si	5.0×10^7	1.0×10^{10}	3.5×10^{12}	1.7×10^{14}
Ge	7.9×10^{11}	2.08×10^{13}	1.4×10^{15}	1.9×10^{16}
GaAs	5.3×10^3	2.8×10^6	8.1×10^9	1.1×10^{12}
InP		1.6×10^7		

Source: From http://jas2.eng.buffalo.edu.

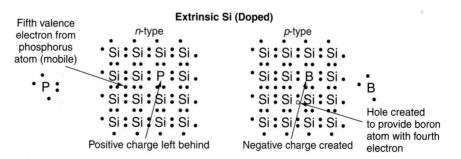

Figure 9.8 Lewis dot structure of doped Si. The phosphorous substitutional impurity forms *n*-Si while boron forms *p*-Si.

Additionally, the charges in the crystal must balance. The condition of electroneutrality can be written by equating the total number of negative species equal to the total number of positive species:

$$n = \left[P_{Si}^{+}\right] + p \tag{9.108}$$

We can see two limiting cases to Equation (9.108). In the limit of large dopant concentrations, $[P_{Si}^{+}] \gg p$, the number of electrons is approximately equal to the number of dopants

$$n \approx \left[P_{Si}^{+}\right] \tag{9.109}$$

Thus we can control the number density of free electrons directly by adjusting the dopant concentration. Moreover, the equilibrium relation of Equations (9.101) and (9.103) still holds:

$$p = \frac{K_c}{n} = \frac{n_i^2}{\left[P_{Si}^{+}\right]} \tag{9.110}$$

For a given semiconductor at temperature T, Equation (9.110) shows that as the number of free electrons increases, the number of holes proportionately decreases, so that their product remains the same. Thus the amount of the phosphorus dopant that is introduced controls the amount of both the electrons and the holes in the semiconductor. We call the carriers of higher concentration the **majority carriers**, while those of lower concentration are the **minority carriers**. Since electrons are the majority carrier when we dope silicon with phosphorous, we call this material an **n-type** semiconductor. When the number densities of minority and majority carriers are controlled by the amount of dopant, we say we have an **extrinsic** semiconductor. In the limit of small dopant concentrations, $[P_{Si}^{+}] \ll p$, there is no effect of the substitutional impurity of the electronic defects in the semiconductor, and it behaves similarly to an intrinsic semiconductor.

We have just seen that the addition of a dopant with one more valence electron than the atom it replaces in the lattice leads to additional free electrons. Similarly, the addition of a dopant with one fewer valence electrons leads to creation of holes. Such a case is shown on the right side of Figure 9.8. In this case, we incorporate boron as follows:

$$B(g) + V_{Si} \rightleftarrows B_{Si}^{-} + h^{+} \tag{9.111}$$

Similarly, electroneutrality requires:

$$p = \left[B_{Si}^{-}\right] + n \tag{9.112}$$

and in the limit $[B_{Si}^{-}] \gg n$, we get

$$p \approx \left[B_{Si}^{-}\right] \tag{9.113}$$

Such a material is called a **p-type** semiconductor, since holes are now the majority charge carriers. The electrons are the minority carriers and can be described by

$$n = \frac{K_3}{p} = \frac{n_i^2}{\left[B_{Si}^{-}\right]} \tag{9.114}$$

By the proper choice of dopant, we can make semiconductor materials with a majority of positive or negative charge carriers, and by controlling the amount of dopant, we can target specific carrier concentrations. This ability forms the basis for engineering devices using these materials.

Effect of Gas Partial Pressure on Defect Concentrations

We can use defect equilibria expressions and equilibrium relations to understand the relation between the properties of solids and the gas environments in which they are processed. In this section, we go through two examples of gas-defect equilibrium involving compound semiconductors, AB. We can consider defects in sublattice A or sublattice B. However, the ratio of site A to site B must remain fixed, as defined by the stoichiometry.

There are two common types of atomic defect pairs that exist in compounds. A vacancy–interstitial defect pair is termed a **Frenkel** defect. If it is formed in sublattice B, it can be written as:

$$B_B \rightleftarrows B_i + V_B \tag{9.115}$$

Frenkel defects can occur in the A sublattice as well. Alternatively, a **Schottky** defect results from the formation of a vacancy–vacancy atomic defect pair, as follows:

$$0 \rightleftarrows V_A + V_B \tag{9.116}$$

▶ **EXAMPLE 9.21**
Growth Mechanism
of ZnO by Oxidation
of Zn[12]

ZnO is a II-VI semiconductor that forms a hexagonal crystal structure with interpenetrating Zn and O sublattices. Consider the growth of ZnO by oxidation of Zn. At 390°C, a ZnO film from zinc that contains 1% Al grows two orders of magnitude slower than when pure Zn is used. On the other hand, with 0.4% Li in the zinc, the growth rate increases two orders of magnitude. Develop a model to describe the defect equilibria in ZnO and, based on the above data, propose a growth mechanism.

SOLUTION We grow ZnO films by placing metallic zinc in a furnace with an oxygen atmosphere. The overall reaction is given by

$$\text{Zn}(s) + \frac{1}{2}\text{O}_2(g) \longrightarrow \text{ZnO}(s)$$

After some ZnO product has been grown, the two reactants are physically separated from each other by the solid product, as schematically shown in Figure E9.21. It is clear that somehow an O-containing species or a Zn-containing species (or both) must diffuse through the ZnO film for the reaction to proceed. We wish to know the growth mechanism. We will see how defect

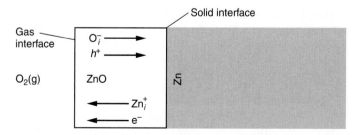

Figure E9.21 Growth of ZnO films by oxidation of Zn.

[12]Adapted from data presented in R. A. Swalin, *Thermodynamics of Solids*. (New York: Wiley, 1972).

equilibria can help us understand how solid zinc becomes oxidized in the presence of oxygen gas. In the analysis that follows, we will assume singly ionized zinc and oxygen interstitials. However, the presence of some doubly ionized or neutral interstitials would not affect the argument.

We begin by writing balanced equations and corresponding equilibrium constants that describe the oxygen and zinc interstitials in the ZnO film. The oxygen is assumed to dissociatively adsorb on the surface and be incorporated into an interstitial, as follows:

$$O_2(g) \rightleftarrows 2O_i \qquad K_1 = \frac{[O_i]^2}{p_{O_2}} \qquad\qquad (E9.21A)$$

The oxygen interstitial may become ionized by grabbing an electron from a neighboring bond, leaving a hole behind:

$$O_i \rightleftarrows O_i^- + h^+ \qquad K_2 = \frac{[O_i^-]p}{[O_i]} \qquad\qquad (E9.21B)$$

At the solid interface, zinc enters the interstitials:

$$Zn(s) \rightleftarrows Zn_i^+ + e^- \qquad K_3 = [Zn_i^+]n \qquad\qquad (E9.21C)$$

We again have the balance between production and consumption of electronic charge carriers in the semiconductor:

$$0 \rightleftarrows e^- + h^+ \qquad K_4 = pn \qquad\qquad (E9.21D)$$

Al and Li act as dopants in the ZnO lattice. The dopants incorporate on the zinc sites. Since Al has an extra valence electron, it forms n-type ZnO as follows:

$$Al + V_{Zn} \rightleftarrows Al_{Zn}^+ + e^- \qquad K_5 = \frac{[Al_{Zn}^+]n}{[Al][V_{Zn}]} \qquad\qquad (E9.21E)$$

On the other hand, Li forms p-type ZnO since it only has one valence electron.

$$Li + V_{Zn} \rightleftarrows Li_{Zn}^- + h^+ \qquad K_6 = \frac{[Li_{Zn}^-]p}{[Li][V_{Zn}]} \qquad\qquad (E9.21F)$$

As stated in the problem, we observe the presence of Al *decreases* the growth rate of ZnO films. Equation (E9.21E) shows that the presence of Al increases the electron concentration; hence, the hole concentration must proportionately decrease [Equation (E9.21.D)]. We can combine Equations (E9.21B) and (E9.21A) to give:

$$[O_i^-] = \frac{\sqrt{K_1 p_{O_2}} K_2}{p} \qquad\qquad (E9.21G)$$

Thus, for constant temperature and oxygen pressure, a decrease in hole concentration translates into increased $[O_i^-]$. We deduce that the growth rate *decreases* when the oxygen interstitial concentration *increases*. Similarly, Reaction (E9.21C) shows that the zinc interstitial concentration decreases as the electron concentration increases. So the growth rate *decreases* when the zinc interstitial concentration *decreases*.

On the other hand, the presence of Li *increases* the growth rate. Equation (E9.21F) shows that the presence of Li increases the hole concentration, decreasing the electron concentration. From Equation (E9.21C), we conclude that $[Zn_i^+]$ increases. We deduce that the growth rate *increases* when the zinc interstitial concentration *increases*. Additionally, Equation (E9.21G) shows that the oxygen interstitial concentration *decreases*. Since the growth rate goes in proportion to the zinc defect concentration, and behaves oppositely to the oxygen interstitial concentration, we conclude that transport of Zn across the ZnO film is the mechanism by which the oxidation reaction proceeds. ◀

▶ **EXAMPLE 9.22**
Construction of a
Brouwer Diagram

Consider a compound semiconductor AB. Species B is volatile and exists as a dimer, B_2, in the vapor phase. Describe how the concentration of defects varies over a wide range of B_2 partial pressure. For the sake of this analysis, consider only atomic defects of B and assume that they either are neutral or singly ionized.

SOLUTION It is useful to be able to look at defect concentrations over a wide range of behavior. In this example, we develop one such way—the Brouwer diagram. We consider the compound semiconductor AB. The anion B (e.g., O, S, P, As) may form a Frankel defect as follows:

$$B_B \rightleftarrows B_i + V_B \qquad K_1 = [B_i][V_B] \qquad\qquad \text{(E9.22A)}$$

The vacant lattice site may be ionized through the following reaction:

$$V_B \rightleftarrows V_B^+ + e^- \qquad K_2 = \frac{[V_B^+]n}{[V_B]} \qquad\qquad \text{(E9.22B)}$$

Similarly, an interstitial can become charged by obtaining an electron. In this process, a hole is created:

$$B_i \rightleftarrows B_i^- + h^+ \qquad K_3 = \frac{[B_i^-]p}{[B_i]} \qquad\qquad \text{(E9.22C)}$$

The gas-phase incorporation of B can be written:

$$B_2(g) \rightleftarrows 2B_i \qquad K_4 = \frac{[B_i]^2}{p_{B_2}} \qquad\qquad \text{(E9.22D)}$$

As always, electrons can be freed from their chemical bonds:

$$0 \rightleftarrows e^- + h^+ \qquad K_5 = pn \qquad\qquad \text{(E9.22E)}$$

Inserting Equation (E9.22C) into Equation (E9.22D) and rearranging:

$$[B_i^-] = \frac{K_3\sqrt{K_4 p_{B_2}}}{p} \qquad\qquad \text{(E9.22F)}$$

Equations (E9.22A), (E9.22B), and (E9.22D) give:

$$[V_B^+] = \frac{K_1 K_2}{n\sqrt{K_4 p_{B_2}}} \qquad\qquad \text{(E9.22G)}$$

and Equation (E9.22E) gives

$$p = \frac{K_5}{n} \qquad\qquad \text{(E9.22H)}$$

Electroneutrality yields

$$n + [B_i^-] = [V_B^+] + p \qquad\qquad \text{(E9.22I)}$$

As Equations (E9.22F) and E(9.22G) show, the atomic defects, $[B_i^-]$ and $[V_B^+]$, depend on the B_2 partial pressure in the system. We consider three regions:

   In *region 1*, we have low p_{B_2}. Inspection of Equation (E9.22G) shows that the B vacancy concentration is large, while Equation (E9.22F) tells us that the interstitial concentration is small. We could deduce the results from physical arguments as well. Thus, for region 1, we take $[V_B^+] >> p$ and $n >> [B_i^-]$, and Equation (E9.22I) reduces to $n \cong [V_B^+]$. With this limit, concentrations of the defects can be solved via Equations (E9.22F) through (E9.22H). Results for the charged defects are shown in the column marked "low p_{B_2}" in Table E9.22. From these results, we can also solve for the neutral atomic defects.

In *region 3*, we have high p_{B_2}. Inspection of Equation (E9.22F) shows that B interstitial concentration is large, while Equation (E9.22G) tells us that the vacancy concentration is small. Thus, for region 3, we take $[B_i^-] >> n$ and $p >> [V_B^+]$, and Equation (E9.22I) reduces to $n \cong [V_B^+]$. Equation (E9.22I) reduces to $p \cong [B_i^-]$. With this limit, concentrations of the defects can be solved via Equations (E9.22F) through (E9.22H). Results are shown in the column marked "high p_{B_2}" in Table E9.22.

In *region 2*, we have intermediate p_{B_2}. We assume that the electronic defect concentration is larger than the atomic defect concentration (although in some systems the opposite may found; see Problem 9.41). Thus, we have $n > [B_i^-]$ and $p > [V_B^+]$, and Equation (E9.22I) reduces to $n \cong p$. Again, solutions are shown in Table E9.22.

Each of the charged defect types in Table E9.22 has a well-defined dependence on B_2 partial pressure. A convenient way to see the behavior of the defect concentrations over many orders of magnitude is to use a log-log plot of the defect concentrations vs. partial pressure. Figure E9.22

TABLE E9.22   Defect Concentrations for Three Regions of p_{B_2}

Region 1: Low p_{B_2}	Region 2: Intermediate p_{B_2}	Region 3: High p_{B_2}
$n \cong [V_B^+]$	$n \cong p$	$p \cong [B_i^-]$
$[V_B^+] = \sqrt{\dfrac{K_1 K_2}{K_4^{1/4}}} p_{B_2}^{-1/4}$	$[V_B^+] = \dfrac{K_1 K_2}{\sqrt{K_4 K_5}} p_{B_2}^{-1/2}$	$[V_B^+] = \dfrac{K_1 K_2 \sqrt{K_3}}{K_5 K_4^{1/4}} p_{B_2}^{-1/4}$
$[B_i^-] = \dfrac{\sqrt{K_1 K_2} K_3 K_4^{1/4}}{K_5} p_{B_2}^{1/4}$	$[B_i^-] = K_3 \sqrt{\dfrac{K_4}{K_5}} p_{B_2}^{1/2}$	$[B_i^-] = \sqrt{K_3} K_4^{1/4} p_{B_2}^{1/4}$
$p = \dfrac{K_5 K_4^{1/4}}{\sqrt{K_1 K_2}} p_{B_2}^{1/4}$	$p = \sqrt{K_5}$	$n = \dfrac{K_5}{\sqrt{K_3} K_4^{1/4}} p_{B_2}^{-1/4}$

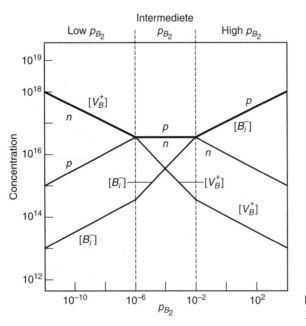

Figure E9.22 Brouwer diagram for Example 9.22.

shows such a Brouwer diagram for the relations established in Table E9.22 for anion Frankel defects. The three regions analyzed above are shown. The limiting behavior of each region is presented; in reality, there would be smooth transitions between the regions. The slope of the lines on this log-log plot are given by the power of the partial pressure. For example, the slopes for $[V_B^+], [B_i^-]$, and p in region 1 are $-1/4$, $1/4$, and $1/4$ respectively. The nature of the semiconductor changes as we increase the partial pressure of B_2. The AB semiconductor goes from n-type (region 1), to intrinsic (region 2), to p-type (region 3). Thus, we can control the doping through the partial pressure of B_2. ◄

► 9.9 SUMMARY

In this chapter, we learned how to calculate the equilibrium composition for a system undergoing a single chemical reaction or a set of reactions at a specified T, P, and initial composition. Our analysis was based on the principle that when a system reaches the equilibrium state, its Gibbs energy is a minimum. For a determined stoichiometry of a chemical reaction, the minimization of Gibbs energy leads to the following expression for the **equilibrium constant**, K:

$$K \equiv \prod \left(\frac{\hat{f}_i}{f_i^o} \right)^{v_i} \tag{9.35}$$

The value of the equilibrium constant can be determined solely from thermochemical data. Additionally, the product of the fugacities on the right-hand side of Equation (9.35) can be related to a single unknown variable, the **extent of reaction**, ξ. Thus, Equation (9.35) can be solved for ξ; it is then straightforward to solve for the composition of species in the system at equilibrium. **Thermodynamics** tells us the possible extent to which a given reaction will proceed; however, the reaction **kinetics** also plays a role. Even when a reaction is thermodynamically favored, if its rate is very slow, it will not noticeably proceed. The analysis we learned in this chapter tells us nothing about reaction kinetics.

The first step in equilibrium analysis of a chemical reaction is to define the reaction stoichiometry and identify the **stoichiometric coefficients**, v_i, of all the species in the system. The stoichiometric coefficient tells us the proportion in which a given species is produced, or reacts, for a particular reaction. It is positive for products, negative for reactants, and zero for inerts. With a known reaction stoichiometry, the equilibrium constant is then determined. The equilibrium constant is related to the Gibbs energy of reaction and can be calculated from available Gibbs energies of formation. For a given reaction stoichiometry, the equilibrium constant depends only on temperature, that is, $K = f$ (T only). If we know the equilibrium constant at one temperature (e.g., 298 K), we can find it at another by applying the thermodynamic web to determine the change of the equilibrium constant with temperature. For limited temperature differences, an approximate solution that assumes the enthalpy of reaction is constant [Equation (9.46)] can be used. A more general expression, Equation (9.52), accounts for the temperature dependence of the enthalpy of reaction.

We can then calculate the equilibrium composition from the product of the fugacities described in Equation (9.35). First, the mole fractions of all the species are written in terms of a single variable, the extent of reaction. The extent of reaction is a measure of how far the reaction has proceeded. It can be related to the relative compositions of reactants and products through the reaction stoichiometry. For example, given an initial composition, the mole fraction of species i in the gas phase is given by:

$$y_i = \frac{n_i}{n^v} = \frac{n_i^o + v_i \xi}{n^o + v \xi} \tag{9.25}$$

The fugacity ratios for each component on the right-hand side of Equation (9.35) can be written by applying the concepts developed in Chapter 7. For the gas phase, we choose a reference state with a pressure low enough that the gas behaves as an ideal gas and a fugacity $f_i^o = 1$ bar. If *only* **gas-phase species** are present and we use the Lewis fugacity approximation, Equation (9.35) becomes

$$K = \prod (y_i)^{v_i} \prod (\varphi_i)^{v_i} P^v \tag{9.55}$$

For an *ideal gas*, Equation (9.55) can be further simplified to

$$K = \prod (y_i)^{\nu_i} P^\nu \quad \text{(ideal gas)} \tag{9.56}$$

Similarly, the liquid-phase (and solid-phase) mole fractions can be written in terms of the extent of reaction. For cases where the pressure dependence of the pure species fugacity is not important, reactions involving *only* **liquid-phase species** reduce to

$$K = \prod (x_i \gamma_i)^{\nu_i} \tag{9.58}$$

In the case of an ideal solution, we have

$$K = \prod (x_i)^{\nu_i} \quad \text{(ideal solution)} \tag{9.59}$$

For **heterogeneous reactions**, where more than one phase is present, we simply treat the fugacities in the vapor as we did for Equation (9.55) or (9.56) and the liquid as we did for Equation (9.58) or (9.59). If a pure solid is present, we use an activity of 1; otherwise, we use a formulation analogous to liquids. When solving problems for heterogeneous systems, we must remember to use only the total number of moles in a given phase when we put the mole fractions into terms with extent of reaction.

Electrochemical systems make use of non-Pv work through application of an electric potential across two electrodes. On one hand, reactions that spontaneously proceed can be used to generate useful electrical work, such as in batteries and fuel cells. On the other hand, input of a suitable quantity of electrical work can lead to reactions that would not spontaneously occur, such as in electroplating or electrolysis. Electrochemical processes occur within an electrochemical cell that contains two electrodes in an electrolyte solution. The oxidation half-reaction occurs at the anode, while the corresponding reduction half-reaction occurs at the cathode. Electrons generated from the oxidation reaction are supplied to the cathode through an external circuit. The equilibrium composition can be related to the applied potential by the Nernst equation, Equation (9.72). Thermochemical data for electrochemical half-reactions are commonly published as the half-cell potential for oxidation reactions, $E°$, with reference to the hydrogen – hydrogen ion reduction reaction. Charged species in solution have strong ionic interactions that are very different from other interactions in the solution, even in a dilute solution. Thus, the treatment of activity coefficients of ions in solution must be approached differently than that of nonelectrolyte solutions. Mean activity coefficients for electrolyte solutions are typically obtained from experiment but can be estimated for ions in dilute solution.

In treating chemically reacting systems, we are often faced with cases where there are many possible reactions. It is straightforward to extend our development of chemical reaction equilibria to more than one reaction. To set up the multiple reaction equilibria problem, we use the **Gibbs phase rule** to determine how many different *independent* chemical reactions are needed to describe the system. Each reaction is then assigned its own corresponding extent of reaction, ξ_k. Thus, we must keep track of the stoichiometry of each species i for each of the k separate chemical reactions. This method is termed the **equilibrium constant formulation**. Alternatively, we can solve for the equilibrium composition through **minimization of the total Gibbs energy** of the system by using the formula coefficient matrix, β_{ij}. This approach is useful in developing computer algorithms to solve reactions for complex systems and is used by ThermoSolver.

We can examine **point defects**, defects that occur at single atomic site, by applying the principles of chemical reaction equilibrium from this chapter. Atomic point defects include **vacancies, interstitials, substitutional impurities**, and misplaced atoms. Electronic point defects include mobile **electrons** and **holes**. From this approach, we can study carrier concentrations in semiconductors and see the effect of gas partial pressure on defect concentrations at equilibrium. The **Brouwer diagram** is a particularly useful tool in seeing the effect of gas partial pressure on defect concentration over many orders of magnitude.

▶ 9.10 PROBLEMS

9.1 At 25°C and 1 atm, the Gibbs energy of reaction to produce liquid hydrogen peroxide (H_2O_2) from liquid water has been measured to be 116.8 kJ/mol. From this value determine the $(\Delta g^o_{f,298})_{H_2O_2}$.

9.2 Consider a system initially charged with 1 mole of pure I_2 that is maintained at 1300 K and 1 bar in which the following dissociation reaction occurs:

$$I_2(g) \rightleftarrows 2I(g)$$

For monatomic iodine

$$\Delta h^o_{f,1300} = 77.5 \,[\text{kJ/mol}]$$

$$\Delta s^o_{f,1300} = 53.4 \,[\text{J/mol K}]$$

Plot ΔH, $T\Delta S$, and ΔG as a function of extent of reaction. What is the equilibrium conversion?

9.3 Calculate the Gibbs energy of formation of NH_3 at 1000 K. Remember that the Gibbs energies of formation of the elements are still zero at this temperature.

9.4 Consider the hydrogenation reaction of 1-butene to butane by the following reaction:

$$C_4H_8 + H_2 \rightleftarrows C_4H_{10}$$

The feed flow ratio is 10 moles H_2 : mole C_4H_8. Consider a reactor temperature of 1000 K and reactor pressure of 5 bar. Calculate the ratio of butane:1-butene at equilibrium. You may assume ideal gas behavior and that Δh^o_{rxn} is constant for the reaction.

9.5 One step in the manufacture of sulfuric acid is the oxidation of sulfur dioxide to sulfur trioxide. Consider performing this oxidation at a pressure of 1 bar with an excess of 100 mole % oxygen, using air as the oxygen source. For the optimal yield of SO_3, it is desirable to maintain the reactor at a constant temperature of 700°C. You may assume Δh^o_{rxn} is independent of temperature.

(a) What is the equilibrium constant at processing conditions?

(b) If equilibrium is attained within the reactor, what is the exit stream composition?

(c) How much heat must be supplied to or removed from the reactor to maintain isothermal operation?

(d) What is the effect of increasing the pressure on the extent of reaction?

(e) What is the effect of increasing pressure on the equilibrium constant?

(f) Would an increase in pressure be justified from a processing standpoint? Explain.

9.6 Calculate the equilibrium constant in Problem 9.5, accounting for the variation of Δh^o_{rxn} with temperature.

9.7 Consider the industrial production of cyclohexane, C_6H_{12}, by the gas-phase hydrogenation of benzene, C_6H_6. Assume that this process is carried out by two reactors in series as shown below. The first reactor is at 340°C and 5 bar, while the second reactor is at 265°C and 5 bar. The feed ratio of hydrogen gas to benzene is 10:1; there is no cyclohexane in the feed. All species are in the gas phase. You may assume ideal gas behavior and that Δh^o_{rxn} does not change with temperature.

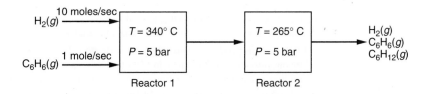

(a) What is the equilibrium composition at the exit of the second reactor?

(b) What is the purpose of the first reactor; that is, why do we use two reactors instead of just one?

(c) Would we get more product if we used a pressure of 1 bar instead of 5 bar. Explain.

(d) Would you recommend diluting the feed with an inert to increase the yield of C_6H_{12}?. Explain.

9.8 You are a process engineer in charge of growing solid silicon from a feed of $SiCl_4$ and H_2 gases. The growth process can be described by the following chemical reaction:

$$SiCl_4(g) + 2H_2(g) \longrightarrow Si(s) + 4HCl(g)$$

The reactor pressure is 100 Pa.

(a) Economics dictates that you need at least a 75% utilization of $SiCl_4$; that is, 75% of the $SiCl_4$ in the feed needs to end up as Si(s). Consider a stoichiometric feed (1 mol $SiCl_4$: 2 mol H_2). What is the minimum possible temperature in the reactor to obtain this objective. State any assumptions that you make.

(b) Your supervisor tells you that the temperature calculated in part (a) is too high. She suggests two possible strategies for decreasing the minimum reactor temperature. Indicate the effect of each of the following process changes on the minimum possible reactor temperature to obtain a utilization of 75%. Explain your reasoning.

(i) Decrease the reactor pressure.

(ii) Dilute the feed stream to a ratio of 1 mol $SiCl_4$: 100 mol H_2.

9.9 Consider the production of 1,1-dichloroethane ($C_2H_4Cl_2$) from ethylene (C_2H_4) and chlorine (Cl_2). This gas-phase reaction is the first step in producing polyvinyl chloride (PVC). The feed ratio of reactants is 2 moles chlorine: 1 mole ethylene. You may assume Δh_{rxn}^o does not change with temperature.

(a) Calculate the maximum temperature at which 90.0% conversion can be obtained at a pressure of 1 bar.

(b) Consider an increase in pressure to 30 bar. What is the conversion obtained at the same temperature as that calculated in part (a)? You may assume that the Lewis fugacity rule applies and use the virial truncated form of the van der Waals equation. The following van der Waals constants are available:

Species	$a\,[\mathrm{J\,m^3\,mol^{-2}}]$	$b\,[\mathrm{m^3/mol}]$
1,1-Dichloroethane	1.71	1.09×10^{-4}
Chlorine	0.61	5.18×10^{-5}
Ethylene	0.46	5.81×10^{-5}

9.10 The following SO_2 partial pressures have been observed from the reaction:[13]

$$CaS(s) + 3CaSO_4(s) \rightleftarrows 4CaO(s) + 4SO_2(g)$$

$T[°C]$	900	960	1000	1040	1080	1120
p_{SO_2} [bar]	5.33×10^{-3}	0.0253	0.0547	0.110	0.206	0.317

From these data, calculate $\Delta h_{rxn,298}^o$ and $\Delta g_{rxn,298}^o$. You may assume that each of the solids forms distinct phases and all are immiscible with one another.

9.11 Fuel cells produce electricity from hydrogen. The life of the fuel cell depends on producing relatively pure hydrogen. Methane (natural gas) is often used as a feed to produce hydrogen. Consider the production of hydrogen (H_2) by the dissociation of methane (CH_4) into solid carbon (C). The process can be described by the following chemical reaction:

$$CH_4(g) \longrightarrow C(s) + 2H_2(g)$$

[13] Ralph A. Wenner, *Thermochemical Calculations* (New York: McGraw-Hill, 1941).

The temperature is 1000 K and the pressure is 1000 Pa. You may assume that Δh^o_{rxn} does not change with temperature.

(a) What is the equilibrium constant at 298 K?

(b) What is the equilibrium constant at 1000 K?

(c) What is the maximum amount of H_2 that can be produced per mole of CH_4 in the feed?

(d) Why is this reaction run at 1000 K instead of 298 K?

(e) Why is this reaction run at 1000 Pa instead of 1 bar?

9.12 A vessel contains a liquid and a vapor phase in equilibrium at a pressure of 0.1 atm. The vapor phase contains 100 moles of species A and 200 moles of species B. The liquid phase contains 500 total moles of species. The saturation pressure of species A is 0.1 atm and of B is 0.5 atm. *Both the vapor and the liquid phases behave ideally*!

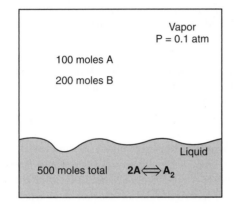

Species A dimerizes in the liquid phase *and A_2 is completely involatile.*

(a) Calculate the equilibrium constant for the dimerization reaction.

(b) Calculate the values for the number of moles of A, B, and A_2 in the liquid phase.

(c) A colleague maintains that the dimerization reaction does *not* occur in the liquid phase; rather, he believes that the liquid phase behaves nonideally. If species A occurs only as a monomer (as A, not A_2) as he thinks, how many moles of species A would exist in the liquid phase [keep the total number of A atoms in the liquid phase the same as for part (b)]?

(d) Given your colleague's model in part (c), what value of γ_B is necessary for his model to fit the data?

(e) Can you explain nonideality in phase equilibria solely by introducing an association reaction to an ideal vapor–liquid system? Can an association reaction be used to explain negative ($\gamma_B < 1$) deviations from ideality?

9.13 Gaseous hydrogen can be produced by the steam cracking of methane in a catalytic reactor at 500°C and 1 bar according to the following reaction:

$$CH_4 + 2H_2O \rightleftarrows CO_2 + 4H_2$$

(a) If 5 moles of steam are fed into the reactor for every mole of methane, at equilibrium, how many moles of hydrogen are produced?

(b) If we build a reactor and run it under these conditions, could we ever get a lower conversion (less hydrogen) than that calculated in part (a)? Can we ever get a higher conversion? Explain.

(c) Would it make sense to increase the pressure in order to increase the equilibrium conversion? Explain.

9.14 You have been tasked with the removal of H_2S and SO_2 from an industrial stack. The following reaction is proposed to dispose of both species at once:

$$2H_2S(g) + SO_2(g) \longrightarrow 3S(s) + 2H_2O(g)$$

Consider this reaction at 500°C. PvT data for H_2S and SO_2 have been fit to the following equation of state:

$$z = \frac{Pv}{RT} = 1 + B'P$$

with values of $B' - 2.2 \times 10^{-9}$ and $-4.4 \times 10^{-9} [Pa^{-1}]$ for pure H_2S and pure SO_2, respectively. For H_2O, use the steam tables for thermodynamic properties. For simplicity, you may use the following equation for the variation of the enthalpy of reaction with temperature:

$$\Delta h^o_{rxn} = \Delta h^o_{rxn,298}[1 + C(T - 298)]$$

with $C = 9 \times 10^{-5} [K^{-1}]$ and T is in K.

(a) Find an expression for the *pure* species fugacity coefficients of H_2S and SO_2 as a function of pressure at 500°C. P should be the only variable in your final expression. Solve for φ explicitly.

(b) Calculate the equilibrium constant at 500°C.

(c) To see if this reaction scheme is plausible, a high-pressure laboratory reactor is set up using an inlet stream of *only* H_2S and SO_2, consisting of 75% H_2S and 25% SO_2. What is the equilibrium conversion, ξ, at 10 MPa? You may approximate the fugacity coefficients in the mixture by their pure species fugacity coefficients.

(d) If a higher conversion is desired than that calculated in part (c), would you increase or decrease P? Explain.

(e) *Quickly and roughly* estimate the approximate pressure needed to obtain $\xi = 0.95$ at 500°C.

(f) Another possible way to obtain higher conversion is to change T. Would you increase or decrease T? Explain. In a real reactor, why would it be better to change P than T?

(g) There are many other species that do not take part in the reaction above. Would the addition of inerts lead to a greater conversion than that calculated in part (d)? Explain.

9.15 Calculate the equilibrium composition of the isomerization reaction between propylene oxide (C_3H_6O) and acetone (C_3H_6O) at 298 K and 1 bar. Under these conditions, they form miscible liquids and can be described by the two-suffix Margules equation with $A = -650$ J/mol.

9.16 Estimate the equilibrium composition at 1 bar of a gas mixture containing the following isomers: 1-butene (1), *cis*-butene (2), and *trans*-butene (3).

(a) at 298 K

(b) at 1000 K

9.17 Determine the equilibrium composition of the following isomers of C_3H_8O: at 500 K and 1 bar: ethyl methyl ether (1), *n*-propyl alcohol (2), and isopropyl alcohol (3). The following Gibbs energies of formation are available at this temperature:

$$\left(\Delta g^o_{f,500}\right)_1 = -47.3, \qquad \left(\Delta g^o_{f,500}\right)_2 = -95.4, \qquad \left(\Delta g^o_{f,500}\right)_3 = -103.2 \left[kJ/mol\right]$$

(a) Use the equilibrium constant approach.

(b) Use the minimization of Gibbs energy.

9.18 You have obtained the following equilibrium data for the reaction:

$$A(g) + 2B(g) \longrightarrow 2C(g)$$

with a stoichiometric feed of A and B. At 200°C and 1 bar, 25% of the species in the reactor were the product C. At 300°C and 1 bar, 53.9% C was produced.

(a) One mole of A and 2 moles of B react at 250°C and 2 bar. Based on the data above, come up with an estimate of the equilibrium concentrations. State any assumptions that you make.

(b) As a process engineer, you wish to maximize the production of C. As completely as you can, discuss the implications of the following strategies:

(i) Increase the temperature.

(ii) Increase the pressure.

(iii) Add an inert to the feed stream.

9.19 Consider the production of hydrogen by the *water gas* shift reaction:

$$H_2O(g) + CO(g) \rightleftarrows H_2(g) + CO_2(g)$$

The feed contains an equal amount of CO and H_2. Water is present at 500% excess of its stoichiometric requirement. Calculate the equilibrium composition at 1000 K and 1 bar.

9.20 Consider the molecular dissociation of diatomic oxygen to monatomic oxygen:

$$O_2 \rightleftarrows 2O$$

What is the minimum temperature required to get 10% O at 1 bar? How can you change the pressure to lower the minimum temperature required?

9.21 Determine the equilibrium composition of NO from air at 1 bar in the temperature range of 1100–3000 K. Plot the mole fraction of NO vs. temperature. You may assume Δh^o_{rxn} is constant for the reaction.

9.22 Determine the equilibrium composition of NO_2 from air at a temperature 3000 [K] and 1 bar. Repeat at 500 bar.

9.23 Consider the formation of NO and NO_2 from air at 3000 K and 500 bar.

(a) Determine the equilibrium conversions and composition from the following set of independent reactions:

$$N_2 + O_2 \rightleftarrows 2NO$$

$$\frac{1}{2}N_2 + O_2 \rightleftarrows NO_2$$

(b) Determine the equilibrium conversions and composition from the following set of independent reactions:

$$N_2 + O_2 \rightleftarrows 2NO$$

$$NO + \frac{1}{2}O_2 \rightleftarrows NO_2$$

(c) Compare the answers obtained in parts (a) and (b).

9.24 Calculate the equilibrium compositions due to the decomposition of 1 mole of nitrogen tetroxide at 25°C and 1 bar in each of the following cases.

$$N_2O_4(g) \rightleftarrows 2NO_2(g)$$

(a) The initial state consists of pure N_2O_4.

(b) The initial state consists of 1 mole of inert in addition to 1 mole of nitrogen tetroxide.

9.25 The Gibbs energy of reaction for the following reaction:

$$Ti(s) + 2Cl_2(g) \rightleftarrows TiCl_4(g)$$

has been reported as:

$$\Delta g^o_{rxn} = -757,000 - 7.5\,T \log T + 145\,T \; [J/mol]$$

with T in [K]. Estimate $\Delta h^o_{rxn,298}$ for this reaction.

9.26 Solve the multiple chemical reaction equilibrium problem in Example 9.18 at 800 K using the following set of independent reactions:

$$CH_4 + H_2O \rightleftarrows CO + 3H_2 \qquad \text{reaction 1}$$

$$CO + H_2O \rightleftarrows CO_2 + H_2 \qquad \text{reaction 2}$$

9.27 Hydrogen cyanide can be manufactured by reaction of acetylene and nitrogen:

$$C_2H_2 + N_2 \rightleftarrows 2HCN$$

Calculate the equilibrium composition at 800 K and 1 bar.

9.28 Consider the equilibrium between copper and its oxide:

$$4Cu(s) + O_2(g) \rightleftarrows 2Cu_2O(s)$$

The Gibbs energy of formation of Cu_2O is given by

$$\Delta g_f^o = -1.70' \times 10^5 - 7.12T \ln T + 124\,T$$

Make a plot of p_{O_2} vs. T, illustrating where Cu is stable and where Cu_2O is stable in the temperature range of 300 K to 1300 K.

9.29 The following reaction is used to produce ethanol:

$$C_2H_4 + H_2O \rightleftarrows C_2H_5OH$$

(a) The process is *not* carried out under conditions of 25°C and 1 bar. Why?
(b) The process is *not* carried out under conditions of 250°C and 1 bar. Why?
(c) The process is carried out under conditions of 250°C and 150 bar. Why?
Support your explanations with equilibrium calculations.

9.30 Consider the reaction of $CrCl_2$ with H_2 to form solid Cr as follows:

$$CrCl_2(s) + H_2(g) \rightleftarrows Cr(s) + 2HCl(g)$$

At $T = 632°C, K = 1.98 \times 10^{-5}$. At $T = 806°C, K = 1.12 \times 10^{-3}$. Answer the following questions:
(a) From these data, estimate the enthalpy of reaction.
(b) In an attempt to increase the extent of reaction, the reaction temperature is raised to 1000°C and 1 bar. At equilibrium, how much Cr is produced for every mole of H_2 in the feed?
(c) Additionally, you wish to increase the extent of reaction by changing the pressure. Would you increase or decrease the pressure? Explain.

9.31 What is the electrode potential of the following electrochemical cell:

$$Zn(s)|ZnSO_4(l, 0.5\ m)||CuSO_4(l, 0.1\ m)|Cu(s)?$$

What is the overall reaction? Is the reaction spontaneous? If the reaction is allowed to proceed until the cell reaches equilibrium, what are the concentrations of $ZnSO_4$ and $CuSO_4$ that result? Take each compartment in the cell to have the same volume.

9.32 Consider the following electrochemical cell:

$$Pt(s)|H_2(g)|(H^+(l, ?\ m)||Pb^{2+}(l, 1\ m)|Pb(s)$$

Answer the following questions:
(a) Draw a schematic of the cell.
(b) Write the oxidation and reduction half-reactions.
(c) Calculate the pH in the electrolyte when the cell potential is $E = 0.244V$; when $E = 0.717V$.
(d) Describe how an electrochemical cell can be used as a pH sensor.

9.33 Corrosion of steel in concrete can be described by the following shorthand notation:

$$Fe(s)|Fe^{2+}(l, 0.1\ m)|O_2(g)|Fe(s)$$

Where oxygen comes from atmospheric air, and the pH is 12. Calculate the electrode potential. Will the corrosion process occur spontaneously?

9.34 Electrolysis of NaCl is used to manufacture NaOH, Cl_2, and H_2. Answer the following questions:

(a) Determine the overall reaction and each half-cell reaction.

(b) Write the process in terms of shorthand notation.

(c) Determine the standard potential of reaction.

9.35 Verify that the standard half-cell potentials reported in Table 9.1 for the reactions between cupric, cuprous, and solid copper—$Cu^{2+} + e^- \rightarrow Cu^+, Cu^{2+} + 2e^- \rightarrow Cu$, and $Cu^+ + e^- \rightarrow Cu$—are self-consistent.

9.36 Hydrogen-based fuel cells show promise as an alternative fuel source. They use a galvanic cell in which oxygen gas is supplied to one compartment and hydrogen gas to another. The reduction of O_2 gas and oxidation of H_2 gas are used to generate power. Consider a fuel cell where the cathodic compartment has an O_2 partial pressure of 0.21 bar and the anodic compartment has an H_2 pressure of 0.4 bar. At room temperature, what is the maximum work produced for every mole of H_2 that reacts? What is the value if the fuel cell operates at 650°C? Assume Δh°_{rxn} = constant.

9.37 Electroplating is used to deposit the copper metal lines in integrated circuit processing. In this case, dissolution of a copper anode is used to provide cupric ions (Cu^{2+}) for reaction; that is, the solid copper–cupric ion reaction is used both at the anode to generate cupric ions and at the cathode to grow the resulting film. Calculate the minimum electrode potential to achieve copper growth for the following cell:

$$Cu(s)|CuSO_4(l,\ 0.1\,m)||CuSO_4(l,\ 0.01\,m)|Cu(s).$$

Assume the following:

(a) It is an ideal solution.

(b) Equation (9.78) represents the activity coefficient of copper.

9.38 The cell potential of the following cell has been measured to be 0.4586 V at 25°C:

$$Pt|H_2(g)|HCl(l, 0.0122\,m)|AgCl(s)|Ag(s)$$

Assume that the fugacity of hydrogen gas is 1 bar. Calculate the activity coefficient, $\gamma_{\pm}$, of HCl in this solution.

9.39 A ZnO semiconductor is exposed to Cl_2 gas. Write the associated chemical reactions for the incorporation of Cl in oxygen sites, with the generation of free electrons. Write the equilibrium constant relations for the reactions you propose.

9.40 Consider a compound semiconductor AB described by the following defect processes:

$$0 \rightleftharpoons V_A^- + V_B^+$$
$$B_2(g) \rightleftharpoons 2B_B + 2V_A^- + 2h^+$$
$$2B_B \rightleftharpoons B_2(g) + 2V_B^+ + 2e^-$$
$$0 \rightleftharpoons h^+ + e^-$$

Construct a Brouwer diagram, including regions of low p_{B_2}, intermediate p_{B_2}, and high p_{B_2}. In the region of intermediate p_{B_2}, assume that the concentration of electronic defects is greater than the vacancy concentration. When is this material intrinsic? When is it n-type? When is it p-type?

9.41 In Example 9.22 we assumed that in the intermediate region, the concentration of electronic defects is greater than the atomic defects. Draw a Brouwer diagram for the case where the concentration of atomic defects is greater than the electronic defects.

9.42 In Example 9.21, we studied the growth of ZnO by oxidation of Zn. The conductivity of these films is proportional to the concentration of free electrons. Develop a relationship between the conductivity of ZnO and the oxygen partial pressure with which it is processed. How does the conductivity change if the oxygen partial pressure is increased by a factor of 10?

9.43 Consider a crystal of Cu_2O in which the majority of the point defects are *copper* vacancies. This crystal is placed in 1 atm of air. At constant temperature, the pressure is reduced to 3 torr. Assuming chemical equilibrium, calculate the ratio of copper vacancies at reduced pressure to copper vacancies at atmospheric pressure. You need not consider oxygen vacancies or any interstitials. Recall that in a compound material the ratio of lattice sites must remain fixed.

9.44 Consider Si uniformly doped with 10^{15} cm^{-3} boron atoms.

(a) What are the carrier concentrations at 27°C?

(b) What are the carrier concentrations at 100°C?

(c) If $(10^{16}/cm^3)$ phosphorous atoms are added, what are the approximate equilibrium electron and hole concentrations at 27°C?

9.45 Copper is a promising interconnect for silicon. However, it is undesirable to have copper impurities in silicon. It is proposed that two types of copper impurities exist: Cu_i^+ and Cu_{Si}^{3-}. It has been reported that the Cu defect concentrations in *intrinsic* Si at 25°C are as follows:

$$Cu_i^+ = 10^6 \ cm^{-3} \quad \text{and} \quad Cu_{Si}^{3-} = 10^2 \ cm^{-3}$$

Your supervisor suggests that the total copper concentration will be reduced if Si is doped with phosphorus. Write defect equilibria equations for the incorporation of each of these species to examine the effect of phosphorus doping on each type of Cu impurity. At what concentration of P will you get the minimum total Cu concentration. By how much will the amount of Cu be reduced over intrinsic Si? Assume that the silicon vacancy concentration does not change with impurity concentration.

9.46 Consider a process whereby a crystal of silicon is placed in a furnace with diborane gas, B_2H_6. Using the concepts of defect equilibria, we wish to understand the effect of diborane pressure in the furnace on doping concentration. Consider the following reactions

$$B_2H_6(g) \rightleftarrows 2B(a) + 3H_2(g) \qquad (1)$$

$$B(a) + V_{Si} \rightleftarrows B_{Si}^- + h^+ \qquad (2)$$

$$0 \rightleftarrows h^+ + e^- \qquad (3)$$

where B(a) is adsorbed boron on the silicon surface.

(a) Write equilibrium constant expressions for the three reactions above.

(b) Based on the species described above, write the condition for charge neutrality (electroneutrality).

(c) Consider the furnace at constant temperature. In this case, the concentration of silicon vacancies is a constant. Consider two regions possible—low diborane partial pressure and high diborane partial pressure. Come up with a qualitative plot of $\log[p]$ and $\log[n]$ vs. $\log p_{B_2H_6}$ and indicate each region.

9.47 Use ThermoSolver to determine the equilibrium composition from the combustion of butane and a stoichiometric amount of air at 2000 K and 50 bar. Consider H_2O, H_2, CO, CO_2, NO, and NO_2 as possible reaction products. Repeat for 2500 K and 50 bar.

APPENDIX A

Physical Property Data

▶ A.1 CRITICAL CONSTANTS, ACENTRIC FACTORS AND, ANTOINE COEFFICIENTS:[1]

The Antoine equation is of the form: $\ln(P^{\text{sat}}[\text{bar}]) = A - \dfrac{B}{T\,[\text{K}] + C}$

TABLE A.1.1 Organic compounds

Formula	Name	$MW_{[\text{g/mol}]}$	T_c [K]	P_c [bar]	ω	A	B	C	$T_{\min}$	$T_{\max}$
CH_2O	Formaldehyde	30.026	408	65.86	0.253	9.8573	2204.13	−30.15	185	271
CH_4	Methane	16.042	190.6	46.00	0.008	8.6041	597.84	−7.16	93	120
CH_4O	Methanol	32.042	512.6	80.96	0.559	11.9673	3626.55	−34.29	257	364
C_2H_4	Acetylene	26.038	308.3	61.40	0.184	9.7279	1637.14	−19.77	194	202
C_2H_3N	Acetonitrile	41.052	548	48.33	0.321	9.6672	2945.47	−49.15	260	390
C_2H_4	Ethylene	28.053	282.4	50.36	0.085	8.9166	1347.01	−18.15	120	182
C_2H_4O	Acetaldehyde	44.053	461	55.73	0.303	9.6279	2465.15	−37.15	210	320
C_2H_4O	Ethylene oxide	44.053	469	71.94	0.200	10.1198	2567.61	−29.01	300	310
$C_2H_4O_2$	Acetic acid	60.052	594.4	57.86	0.454	10.1878	3405.57	−56.34	290	430
C_2H_6	Ethane	30.069	305.4	48.74	0.099	9.0435	1511.42	−17.16	130	199
C_2H_6O	Ethanol	46.068	516.2	63.83	0.635	12.2917	3803.98	−41.68	270	369
C_3H_6	Propylene	42.080	365.0	46.20	0.148	9.0825	1807.53	−26.15	160	240
C_3H_6O	Acetone	58.079	508.1	47.01	0.309	10.0311	2940.46	−35.93	241	350
C_3H_8	Propane	44.096	370.0	42.44	0.152	9.1058	1872.46	−25.16	164	249
C_3H_8O	1-Propanol	60.095	536.7	51.68	0.624	10.9237	3166.38	−80.15	285	400
C_4H_6	1,3-Butadiene	54.090	425	43.27	0.195	9.1525	2142.66	−34.30	215	290
C_4H_8	cis-2-Butene	56.106	435.6	42.05	0.202	9.1969	2210.71	−36.15	200	305
C_4H_8	trans-2-Butene	56.106	428.6	41.04	0.214	9.1975	2212.32	−33.15	200	300
$C_4H_8O_2$	Ethyl acetate	88.105	523.2	38.30	0.363	9.5314	2790.50	−57.15	260	385
C_4H_{10}	n-Butane	58.122	425.2	37.90	0.193	9.0580	2154.90	−34.42	195	290
C_4H_{10}	Isobutane	58.122	408.1	36.48	0.176	8.9179	2032.76	−33.15	187	280
$C_4H_{10}O$	n-Butanol	74.122	562.9	44.18	0.590	10.5958	3137.02	−94.43	288	404
C_5H_{10}	1-Pentene	70.133	464.7	40.53	0.245	9.1444	2405.96	−39.63	220	325
C_5H_{12}	n-Pentane	72.149	469.6	33.74	0.251	9.2131	2477.07	−39.94	220	330
C_6H_6	Benzene	78.112	562.1	48.94	0.212	9.2806	2788.51	−52.36	280	377
C_6H_6O	Phenol	94.111	694.2	61.30	0.440	9.8077	3490.89	−98.59	345	481
C_6H_7N	Aniline	93.127	699	53.09	0.382	10.0546	3857.52	−73.15	340	500
C_6H_{12}	Cyclohexane	84.159	553.4	40.73	0.213	9.1325	2766.63	−50.50	280	380
C_6H_{12}	1-Hexene	84.159	504.0	31.71	0.285	9.1887	2654.81	−47.30	240	360
C_6H_{14}	n-Hexane	86.175	507.4	29.69	0.296	9.2164	2697.55	−48.78	245	370

(Continued)

[1] For a more complete set of compounds, consult ThermoSolver, the text software.

TABLE A.1.1 Continued

Formula	Name	$MW_{[g/mol]}$	T_c [K]	P_c [bar]	ω	A	B	C	T_{min}	T_{max}
C_7H_8	Toluene	92.138	591.7	41.14	0.257	9.3935	3096.52	−53.67	280	410
C_7H_{14}	1-Heptene	98.186	537.2	28.37	0.358	9.2692	2895.51	−53.97	265	400
C_7H_{16}	n-Heptane	100.202	540.2	27.36	0.351	9.2535	2911.32	−56.51	270	400
C_8H_8	Styrene	104.149	647.0	39.92	0.257	9.3991	3328.57	−63.72	305	460
C_8H_{10}	o-Xylene	106.165	630.2	37.29	0.314	9.4954	3395.57	−59.46	305	445
C_8H_{10}	m-Xylene	106.165	617.0	35.46	0.331	9.5188	3366.99	−58.04	300	440
C_8H_{10}	p-Xylene	106.165	616.2	35.16	0.324	9.4761	3346.65	−57.84	300	440
C_8H_{10}	Ethylbenzene	106.165	617.1	36.07	0.301	9.3993	3279.47	−59.95	300	450
C_8H_{16}	1-Octene	112.213	566.6	26.24	0.386	9.3428	3116.52	−60.39	288	420
C_8H_{18}	n-Octane	114.229	568.8	24.82	0.394	9.3224	3120.29	−63.63	292	425
C_9H_{20}	n-Nonane	128.255	594.6	23.10	0.444	9.3469	3291.45	−71.33	312	452
$C_{10}H_8$	Naphthalene	128.171	748.4	40.53	0.302	9.5224	3992.01	−71.29	360	545
$C_{10}H_{22}$	n-Decane	142.282	617.6	21.08	0.490	9.3912	3456.80	−78.67	330	476

TABLE A.1.2 Inorganic Compounds

Formula	Name	$MW_{[g/mol]}$	T_c [K]	P_c [bar]	ω	A	B	C	T_{min}	T_{max}
Ar	Argon	39.948	150.8	48.74	−0.004	8.6128	700.51	−5.84	81	94
BCl_3	Boron trichloride	117.169	451.95	38.71	0.148	9.0985	2242.71	−38.99	182	286
B_2H_6	Diborane	27.670	289.80	40.50	0.138	8.7074	1377.84	−22.18	118	181
Br_2	Bromine	159.808	584	103.35	0.132	9.2239	2582.32	−51.56	259	354
CCl_3F	Trichlorofluoromethane	137.367	471.2	44.08	0.188	9.2314	2401.61	−36.3	240	300
CF_4	Carbon tetrafluoride	88.004	227.6	37.39	0.191	9.4341	1244.55	−13.06	93	148
C_2F_6	Hexafluoroethane	138.012	292.8	30.42	0.255	9.1646	1559.11	−24.51	180	195
$CHCl_3$	Chloroform	119.377	536.4	54.72	0.216	9.3530	2696.79	−46.16	260	370
CO	Carbon monoxide	28.010	132.9	34.96	0.049	7.7484	530.22	−13.15	63	108
CO_2	Carbon dioxide	44.010	304.2	73.76	0.225	15.9696	3103.39	−0.16	154	204
CS_2	Carbon disulfide	76.143	552	79.03	0.115	9.3642	2690.85	−31.62	228	342
Cl_2	Chlorine	70.905	417	77.01	0.073	9.3408	1978.32	−27.01	172	264
F_2	Fluorine	37.997	144.3	52.18	0.048	9.0498	714.10	−6.00	59	91
H_2	Hydrogen	2.016	33.2	12.97	−0.22	7.0131	164.90	3.19	14	25
HBr	Hydrogen bromide	80.912	363.2	85.52	0.063	7.8485	1242.53	−47.86	184	221
HCN	Hydrogen cyanide	27.025	456.8	53.90	0.407	9.8936	2585.80	−37.15	234	330
HCl	Hydrogen chloride	36.461	324.6	83.09	0.12	9.8838	1714.25	−14.45	137	200
H_2O	Water	18.015	647.3	220.48	0.344	11.6834	3816.44	−46.13	284	441
H_2S	Hydrogen sulfide	34.082	373.2	89.37	0.100	9.4838	1768.69	−26.06	190	230
NH_3	Ammonia	17.031	405.6	112.77	0.250	10.3279	2132.50	−32.98	179	261
He	Helium-4	4.003	5.19	2.27	−0.387	5.6312	33.7329	1.79	3.7	4.3
HF	Hydrogen fluoride	20.006	461	64.85	0.372	11.0756	3404.49	15.06	206	313
Kr	Krypton	83.800	209.4	55.02	−0.002	8.6475	958.75	−8.71	113	129
N_2	Nitrogen	28.013	126.2	33.84	0.039	8.3340	588.72	−6.60	54	90
NF_3	Nitrogen trifluoride	71.002	234	45.29	0.132	8.9905	1155.69	−15.37	103	155
N_2O	Nitrous oxide	44.013	309.6	72.45	0.160	9.5069	1506.49	−25.99	144	200
NO	Nitric oxide	30.006	180	64.85	0.607	13.5112	1572.52	−4.88	95	140
NO_2	Nitrogen dioxide	46.006	431.4	101.33	0.86	13.9122	4141.29	3.65	230	320
Ne	Neon	20.180	44.4	27.56	0.00	7.3897	180.47	−2.61	24	29
O_2	Oxygen	31.999	154.6	50.46	0.021	8.7873	734.55	−6.45	63	100
PH_3	Phosphene	33.998	324.45	65.35	0.042	9.2700	1617.91	−11.07	144	186
SF_6	Sulfur hexafluoride	146.056	318.7	37.59	0.286	12.7583	2524.78	−11.16	159	220
SO_2	Sulfur dioxide	64.065	430.8	78.83	0.251	10.1478	2302.35	−35.97	195	280

(Continued)

TABLE A.1.2 Continued

Formula	Name	$MW_{[g/mol]}$	T_c [K]	P_c [bar]	ω	A	B	C	T_{min}	T_{max}
SO_3	Sulfur trioxide	80.064	491.0	82.07	0.41	14.2201	3995.70	−36.66	290	332
$SiCl_3H$	Trichlorosilane	135.452	479.0	41.7	0.203	9.7079	2694.02	−27.00	275	305
$SiCl_4$	Silicon tetrachloride	169.896	507.0	37.49	0.264	9.1817	2634.16	−43.15	238	364
SiF_4	Silicon tetrafluoride	104.079	259.09	37.15	0.456	16.3709	2810.45	−6.88	129	128
SiH_4	Silane	32.117	269.69	48.43	0.089	9.7222	1620.99	5.35	94	162
WF_6	Tungsten hexafluoride	297.830	444.0	43.40	0.231	10.4899	2351.42	−64.70	202	290

Sources: Mostly from R. C. Reid, J. M. Prausnitz, and T. K. Sherwood. *The Properties of Gases and Liquids*, 3rd ed. (New York: McGraw-Hill, 1977). Also from: CRC Handbook of Chemistry and Physics (Boca Raiton, FL CRC Press, (various) years); P. J. Linstrom and W. G. Mallard, Eds., **NIST Chemistry WebBook, NIST Standard Reference Database Number 69,** March 2003, National Institute of Standards and Technology, Gaithersburg MD, 20899 (http://webbook.nist.gov/chemistry/fluid).; C. L. Yaws, *Handbook of Vapor Pressure* (vol. 4) (Houston: Gulf Publishing, 1995).

► A.2 HEAT CAPACITY DATA

$$\frac{c_p}{R} = A + BT + CT^2 + DT^{-2} + ET^3 \text{ with } T \text{ in [K]}$$

TABLE A.2.1 Heat Capacity of Ideal gases: Organic Compounds

Formula	Name	A	$B \times 10^3$	$C \times 10^6$	$D \times 10^{-5}$	$E \times 10^9$	T_{min}	T_{max}	Source
CH_2O	Formaldehyde	2.264	7.022	−1.877			298	1500	1
CH_4	Methane	1.702	9.081	−2.164			298	1500	1
CH_4O	Methanol	2.211	12.216	−3.45			298	1500	1
C_2H_2	Acetylene	6.132	1.952		−1.299		298	1500	1
C_2H_4	Ethylene	1.424	14.394	−4.392			298	1500	1
C_2H_4O	Acetaldehyde	1.693	17.978	−6.158			298	1000	1
C_2H_4O	Ethylene oxide	−0.385	23.463	−9.296			298	1000	1
C_2H_6	Ethane	1.131	19.225	−5.561			298	1500	1
C_2H_6O	Ethanol	3.518	20.001	−6.002			298	1500	1
C_3H_6	Propylene	1.637	22.706	−6.915			298	1500	1
C_3H_8	Propane	1.213	28.785	−8.824			298	1500	1
C_4H_6	l.3-Butadiene	2.734	26.786	−8.882			298	1500	1
C_4H_8	1-Butene	1.967	31.63	−9.873			298	1500	1
C_4H_{10}	n-Butane	1.935	36.915	−11.402			298	1500	1
C_4H_{10}	Isobutane	1.677	37.853	−11.945			298	1500	1
C_5H_{10}	1-Pentene	2.691	39.753	−12.447			298	1500	1
C_5H_{12}	n-Pentane	2.464	45.351	−14.111			298	1500	1
C_6H_6	Benzene	−0.206	39.064	−13.301			298	1500	1
C_6H_{12}	Cyclohexane	−3.876	63.249	−20.928			298	1500	1
C_6H_{12}	1-Hexene	3.220	48.189	−15.157			298	1500	1
C_6H_{14}	n-Hexane	3.025	53.722	−16.791			298	1500	1
C_7H_8	Toluene	0.290	47.052	−15.716			298	1500	1
C_7H_{14}	1-Heptene	3.768	56.588	−17.847			298	1500	1
C_7H_{16}	n-Heptane	3.570	62.127	−19.468			298	1500	1
C_8H_8	Styrene	2.050	50.192	−16.662			298	1500	1
C_8H_{10}	Ethylbenzene	1.124	55.38	−18.476			298	1500	1
C_8H_{16}	1-Octene	4.324	64.96	−20.521			298	1500	1
C_8H_{18}	n-Octane	8.163	70.567	−22.208			298	1500	1

Sources:
1. J. M. Smith, H. C. Van Ness, and M. M. Abbott, *Introduction to Chemical Engineering Thermodynamics*, 5th ed. (New York: McGraw-Hill, 1996).
2. P. J. Linstrom and W. G. Mallard, Eds., **NIST Chemistry WebBook, NIST Standard Reference Database Number 69,** March 2003, National Institute of Standards and Technology, Gaithersburg MD, 20899 (http://webbook.nist.gov/chemistry/fluid).

TABLE A.2.2 Heat Capacity of Ideal Gases: Inorganic Compounds

Formula	Name	A	$B \times 10^3$	$C \times 10^6$	$D \times 10^{-5}$	$E \times 10^9$	T_{min}	T_{max}	Source
	Air	3.355	0.575		−0.016		298	2000	1
BCl_3	Boron trichloride	4.245	16.539	−18.969	−0.176	8.031	298	700	2
		9.882	0.078	−0.018	−3.374	0.001	700	6000	2
B_2H_6	Diborane	−1.494	32.188	−18.314	0.361	3.988	298	1200	2
		19.440	1.351	−0.262	−44.224	0.018	1200	6000	2
Br_2	Bromine	4.493	0.056		−0.154		298	3000	1
CF_4	Carbon tetrafluoride	1.921	25.299	−22.789	−0.261	7.482	298	1000	2
		12.776	0.129	−0.027	−10.032	0.002	1000	6000	2
CO	Carbon monoxide	3.376	0.557		−0.031		298	2500	1
CO_2	Carbon dioxide	5.457	1.045		−1.157		298	2000	1
CS_2	Carbon disulfide	6.311	0.805		−0.906		298	1800	1
C_2F_6	Hexafluoroethane	8.389	27.106	−20.948	−1.751	5.671	298	1400	2
		21.284	0.123	−0.023	−13.447	0.001	1400	6000	2
Cl_2	Chlorine	4.442	0.089		−0.344		298	3000	1
H_2	Hydrogen	3.249	0.422		0.083		298	3000	1
HBr	Hydrogen bromide	3.815	−1.648	2.809	−0.035	−1.084	298	1100	2
		3.956	0.339	−0.057	−3.819	0.004	1100	6000	2
HCN	Hydrogen cyanide	4.736	1.359		−0.725		298	2500	1
HCl	Hydrogen chloride	3.156	0.623		0.151		298	2000	1
HF	Hydrogen fluoride	3.622	−0.390	0.345	−0.030	0.055	298	1000	2
		2.955	0.829	−0.150	−0.282	0.010	1000	6000	2
H_2O	Water	3.470	1.45		0.121		298	2000	1
H_2S	Hydrogen sulfide	3.931	1.49		−0.232		298	2300	1
N_2	Nitrogen	3.280	0.593		0.04		298	2000	1
NH_3	Ammonia	3.5778	3.02		−0.186		298	1800	1
N_2O	Nitrous oxide	5.328	1.24		−0.928		298	2000	1
NO	Nitric oxide	3.387	0.629		0.014		298	2000	1
NO_2	Nitrogen dioxide	4.982	1.195		−0.792		298	2000	1
N_2O_4	Dinitrogen tetroxide	11.660	2.257		−2.787		298	2000	1
O_2	Oxygen	3.639	0.506		−0.227		298	2000	1
PH_3	Phosphene	1.431	10.160	−4.576	0.348	0.685	298	1200	2
SF_6	Sulfur hexafluoride	7.085	30.736	−30.343	−1.935	10.676	298	1000	2
		18.901	0.058	−0.012	−9.959	0.001	1000	6000	2
SO_2	Sulfur dioxide	5.699	0.801		−1.015		298	2000	1
SO_3	Sulfur trioxide	8.06	1.056		−2.028		298	2000	1
$SiCl_4$	Tetrachlorosilane	12.700	0.255	−0.069	−1.744	0.006	298	6000	2
$SiClH_3$	Chlorosilane	2.977	14.807	−9.231	−0.432	2.242	298	1100	2
		11.954	0.572	−0.114	−17.303	0.008	1100	6000	2
$SiCl_2H_2$	Dichlorosilane	6.026	10.145	−6.014	−0.959	1.328	298	1500	2
		12.603	0.195	−0.035	−15.277	0.002	1500	6000	2
$SiCl_3H$	Trichlorosilane	7.732	10.262	−8.671	−0.908	2.818	298	1000	2
		12.552	0.253	−0.052	−7.179	0.004	1000	6000	2
SiF_4	Silicon tetrafluoride	5.170	19.158	−18.179	−0.514	6.207	298	1000	2
		12.903	0.057	−0.012	−6.482	0.001	1000	6000	2
SiH_4	Silane	0.729	16.835	−9.368	0.163	1.953	298	1300	2
		12.010	0.511	−0.097	−24.525	0.006	1300	6000	2
WF_6	Tungsten hexafluoride	18.137	0.730	−0.197	−3.690	0.017	1000	6000	2

Sources:

1. J. M. Smith, H. C. Van Ness, and M. M. Abbott, *Introduction to Chemical Engineering Thermodynamics*, 5th ed. (New York: McGraw-Hill, 1996).

2. P. J. Linstrom and W. G. Mallard, Eds., **NIST Chemistry WebBook, NIST Standard Reference Database Number 69,** March 2003, National Institute of Standards and Technology, Gaithersburg MD, 20899 (http://webbook.nist.gov/chemistry/fluid).

TABLE A.2.3 Heat Capacity of Liquids and Solids

Formula	Name	Phase	A	$B \times 10^3$	$D \times 10^{-5}$	Source
CH_4O	Methanol	L, $\bar{c}_P$	9.815			2
C_2H_6O	Ethanol	L, $\bar{c}_P$	13.592			2
C_3H_6O	Acetone	L	11.184	13.375		2
C_5H_{12}	Pentane	L	18.691	5.254		3
C_6H_6	Benzene	L	16.310	0.000		2
C_6H_{14}	Hexane	L	23.695			2
Al	Aluminum	L	3.819			1
Al	Aluminum	S	2.486	1.490		1
Al_2O_3	Aluminum oxide	S	23.154			4
C	Graphite	S	2.063	0.514	−1.057	1
C	Diamond	S	0.782			4
Cu	Copper	L	3.950			4
Cu	Copper	S	2.723			1
Cu_2O	Cuprous oxide	S, alpha	7.498			1
CuO	Cupric oxide	S	4.666			1
Fe	Iron	S, alpha	2.104	2.979		1
Fe_3O_4	Iron oxide	S	11.012	24.260		1
GaAs	Gallium arsenide	S	5.438	0.730		1
Ni	Nickel	S	1.508	4.308	0.297	1
Si	Silicon	L	3.272			4
Si	Silicon	S	2.879	0.297	−0.498	1
SiO_2	Silicon dioxide	S	5.647	4.127	−1.359	1
$SiCl_3H$	Trichlorosilane	L, $\bar{c}_P$	15.678			2
$SiCl_4$	Tetrachlorosilane	L, $\bar{c}_P$	16.117			2
H_2O	Water	L, $\bar{c}_P$	9.069			2
H_2O	Water (ice)	S, $\bar{c}_P$	4.196			5
H_2SO_4	Sulfuric acid	L	16.731	1.875		3
HNO_3	Nitric acid	L, $\bar{c}_P$	13.315			2
NH_3	Ammonia	L	6.880	9.682		2

Sources:
1. O. Kubaschewski and C. B. Alcock, *Metallurgical Thermochemistry*, 5th ed. (New York: Peramon Press, 1979).
2. Milan Zabransky et al., *Heat Capacity of Liquids* (Washington, DC: American Chemical Society; Woodbury, NY: National Bureau of Standards, 1996).
3. Richard M. Felder and Ronald W. Rousseau, *Elementary Principles of Chemical Processes*, 3rd ed. (New York: Wiley, 2000).
4. M. W. Chase et al., *JANAF Themochemical Tables*, 4th ed. (Washington, DC: American Chemical Society; National Bureau of Standards, 1998).
5. K. Ranjevic, *Handbook of Thermodynamic Tables and Charts* (New York: McGraw-Hill, 1976).

▶ A.3 ENTHALPY AND GIBBS ENERGY OF FORMATION AT 298 K AND 1 BAR

TABLE A.3.1 Organic Compounds

Formula	Name	Phase	$\Delta h^o_{f,298}$ [kJ/mol]	$\Delta g^o_{f,298}$ [kJ/mol]	Source
CH_2O	Formaldehyde	G	−115.97	−109.99	1
CH_4	Methane	G	−74.81	−50.72	1
CH_4O	Methanol	L	−238.73	−166.34	1
CH_4O	Methanol	G	−200.66	−161.96	1
C_2H_2	Acetylene	G	226.88	209.24	1
C_2H_3N	Acetonitrile	L	53.17	98.93	1
C_2H_3N	Acetonitrile	G	87.92	105.67	1
C_2H_4	Ethylene	G	52.26	68.15	1

(*Continued*)

TABLE A.3.1 Contunued

Formula	Name	Phase	$\Delta h^o_{f,298}$ [kJ/mol]	$\Delta g^o_{f,298}$ [kJ/mol]	Source
$C_2H_4Cl_2$	1,1-Dichloroethane	L	−160.86	−76.20	1
$C_2H_4Cl_2$	1,1-Dichloroethane	G	−130.00	−73.14	1
C_2H_4O	Acetaldehyde	G	−166.47	−133.39	1
C_2H_4O	Ethylene oxide	L	−77.46	−11.43	1
C_2H_4O	Ethylene oxide	G	−52.67	−13.10	1
$C_2H_4O_2$	Acetic acid	L	−484.41	−389.62	1
$C_2H_4O_2$	Acetic acid	G	−435.13	−376.94	1
C_2H_6	Ethane	G	−84.68	−32.84	1
C_2H_6O	Ethanol	L	−277.17	−174.25	1
C_2H_6O	Ethanol	G	−234.96	−168.39	1
C_3H_6	Propylene	G	20.43	62.76	1
C_3H_6O	Acetone	L	−248.28	−155.50	1
C_2H_6O	Acetone	G	−217.71	−153.15	1
C_3H_6O	Propylene oxide	L	−120.75	−26.75	1
C_3H_6O	Propylene oxide	G	−92.82	−25.79	1
C_3H_8	Propane	G	−103.85	−23.49	1
C_3H_8O	1-Propanol	L	−304.76	−170.78	1
C_3H_8O	1-Propanol	G	−257.70	−163.08	1
C_4H_6	1,3-Butadiene	L	85.41	149.68	1
C_4H_6	1,3-Butadiene	G	110.24	150.77	1
C_4H_8	1-Butene	G	−0.13	71.34	1
C_4H_8	cis-2-Butene	G	−6.99	65.90	1
C_4H_8	trans-2-Butene	G	−11.18	63.01	1
$C_4H_8O_2$	Ethyl acetate	L	−479.35	−332.93	1
$C_4H_8O_2$	Ethyl acetate	G	−443.21	−327.62	1
C_4H_{10}	n-Butane	L	−147.75	−15.07	1
C_4H_{10}	n-Butane	G	−126.23	−17.17	1
C_4H_{10}	Isobutane	L	−158.55	−21.98	1
C_4H_{10}	Isobutane	G	−134.61	−20.89	1
$C_4H_{10}O$	n-Butanol	L	−326.03	−161.19	1
$C_4H_{10}O$	n-Butanol	G	−274.61	−150.77	1
C_5H_{10}	1-Pentene	L	−46.72	78.25	1
C_5H_{10}	1-Pentene	G	−20.93	79.17	1
C_5H_{12}	n-Pentane	L	−173.33	−9.46	1
C_5H_{12}	n-Pentane	G	−146.54	−8.37	1
C_6H_6	Benzene	L	49.07	124.34	1
C_6H_6	Benzene	G	82.98	129.75	1
C_6H_6O	Phenol	S	−165.13	−50.45	1
C_6H_6O	Phenol	G	−96.42	−32.91	1
C_6H_7N	Aniline	L	31.11	149.18	1
C_6H_7N	Aniline	G	86.92	166.80	1
C_6H_{12}	Cyclohexane	L	−156.34	26.89	1
C_6H_{12}	Cyclohexane	G	−123.22	31.78	1
C_6H_{12}	1-Hexene	L	−72.43	83.44	1
C_6H_{12}	1-Hexene	G	−41.70	87.50	1
C_6H_{14}	n-Hexane	L	−198.96	−4.35	1
C_6H_{14}	n-Hexane	G	−167.30	−0.25	1
C_7H_8	Toluene	L	12.02	113.84	1
C_7H_8	Toluene	G	50.03	122.09	1
C_7H_{14}	1-Heptene	L	−98.01	88.84	1
C_7H_{14}	1-Heptene	G	−62.34	95.88	1
C_7H_{16}	n-Heptane	L	−224.54	1.00	1

(Continued)

TABLE A.3.1 Continued

Formula	Name	Phase	$\Delta h_{f,298}^o$ [kJ/mol]	$\Delta g_{f,298}^o$ [kJ/mol]	Source
C_7H_{16}	*n*-Heptane	G	−187.90	8.00	1
C_8H_{10}	*o*-Xylene	L	−24.45	110.53	1
C_8H_{10}	*o*-Xylene	G	19.01	122.17	1
C_8H_{10}	*m*-Xylene	L	−25.41	107.73	1
C_8H_{10}	*m*-Xylene	G	17.25	118.95	1
C_8H_{10}	*p*-Xylene	L	−24.45	110.03	1
C_8H_{10}	*p*-Xylene	G	17.96	121.21	1
C_8H_{10}	Ethylbenzene	L	−12.48	119.78	1
C_8H_{10}	Ethylbenzene	G	29.81	130.67	1
C_8H_{16}	1-Octene	L	−123.59	94.16	1
C_8H_{16}	1-Octene	G	−82.98	104.29	1
C_8H_{18}	*n*-Octane	L	−250.12	6.49	1
C_8H_{18}	*n*-Octane	G	−208.59	16.41	1
C_9H_{20}	*n*-Nonane	L	−275.66	11.76	1
C_9H_{20}	*n*-Nonane	G	−229.19	24.83	1
$C_{10}H_8$	Naphthalene	S	78.13	201.18	1
$C_{10}H_8$	Naphthalene	G	151.06	223.74	1
$C_{10}H_{22}$	*n*-Decane	L	−301.24	17.25	1
$C_{10}H_{22}$	*n*-Decane	G	−249.83	33.24	1

Sources:
1. Daniel R. Stull, Edgar F. Westrum, and Gerard C. Sinke, *The Chemical Thermodynamics of Organic Compounds*, (New York: Wiley, 1969).

TABLE A.3.2 Inorganic Compounds

Formula	Name	Phase	$\Delta h_{f,298}^o$ [kJ/mol]	$\Delta g_{f,298}^o$ [kJ/mol]	Source
BCl_3	Boron trichloride	G	−402.96	−387.96	2
B_2H_6	Diborane	G	35.61	86.77	2
BN	Boron nitride	S	−254.387	−228.501	2
B_2O_3	Boron oxide	S	−1271.94	−1192.8	2
CCl_3F	Trichlorofluoromethane	G	−284.70	−245.51	1
CF_4	Carbon tetrafluoride	G	−975.52	−889.03	1
C_2F_6	Hexafluoroethane	G	−1343.06	−1257.3	2
$CHCl_3$	Chloroform	L	−132.30	−71.89	1
$CHCl_3$	Chloroform	G	−101.32	−68.58	1
CHN	Hydrogen cyanide	G	130.63	120.20	1
CO	Carbon monoxide	G	−110.53	−137.17	2
CO_2	Carbon dioxide	G	−393.51	−394.36	2
CS_2	Carbon disulfide	G	116.94	66.82	2
CaS	Calcium sulfide	S	−473.2	−468.178	2
$CaSO_4$	Calcium sulfate	S	−1434.11	−1321.68	2
CaO	Calcium oxide	S	−635.09	−603.51	2
$CuCl$	Copper chloride	S	−155.65	−138.66	2
CuO	Copper monoxide	S	−156.06	−128.29	2
Cu_2O	Dicopper oxide	S	−170.71	−147.88	2
CuS	Copper sulfide	S	−53.1	−53.47	2
$CuSO_4$	Copper sulfate	S	−771.36	−662.08	2
Fe_3C	Triiron carbide	S	25.104	20.029	2

(*Continued*)

TABLE A.3.2 Continued

Formula	Name	Phase	$\Delta h^o_{f,298}$ [kJ/mol]	$\Delta g^o_{f,298}$ [kJ/mol]	Source
Fe_2O_3	Hematite	S	−824.25	−742.29	2
Fe_2O_4	Magnetite	S	−1118.38	−1015.23	2
HBr	Hydrogen bromide	G	−36.38	−53.45	2
HCl	Hydrogen chloride	G	−92.312	−95.29	2
HF	Hydrogen fluoride	G	−272.55	−274.65	2
HNO_3	Nitric acid	G	−134.31	−73.96	2
H_2O	Water	L	−285.83	−237.14	2
H_2O	Water	G	−241.82	−228.57	2
H_2S	Hydrogen sulfide	G	−20.5	−33.33	2
H_2SO_4	Sulfuric acid	L	−813.99	−689.89	2
H_2SO_4	Sulfuric acid	G	−735.13	−653.37	2
NH_3	Ammonia	G	−46.11	−16.45	2
N_2O	Nitrous oxide	G	82.05	104.17	2
NO	Nitric oxide	G	90.29	86.6	2
NO_2	Nitrogen dioxide	G	33.1	51.26	2
N_2O_4	Dinitrogen tetraoxide	L	−19.564	97.51	2
N_2O_4	Dinitrogen tetraoxide	G	9.079	97.79	2
NaCl	Sodium chloride	S	−411.12	−384.02	2
NaF	Sodium fluoride	S	−573.48	−545.08	2
NaOH	Sodium hydroxide	S	−425.93	−379.73	2
NaOH	Sodium hydroxide	G	−197.49	−200.19	2
O	Oxygen	G	249.17	231.74	2
PH_3	Phosphene	G	5.57	13.59	2
TaN	Tantalum nitride	S	−252.3	−226.58	2
TiC	Titanium carbide	S	−184.5	−180.84	2
TiN	Titanium nitride	S	−337.86	−309.16	2
SiC	Silicon carbide	S	−73.22	−70.85	2
$SiCl_2$	Dichlorosilylene	G	−168.61	−180.36	2
$SiCl_4$	Silicon tetrachloride	G	−662.75	−622.76	2
SiF_4	Silicon tetrafluoride	G	−1614.94	−1572.7	2
$SiCl_3H$	Trichlorosilane	G	−496.22	−464.9	3
$SiCl_2H_2$	Dichlorosilane	G	−320.49	−294.9	3
SiH_3Cl	Chlorosilane	G	−141.838	−119.29	3
SiH_4	Silane	G	34.31	56.82	2
Si_3N_4	Silicon nitride	S	−744.75	−647.34	2
SiO_2	Silicon dioxide, trigonal	S	−910.86	−856.44	2
SiO_2	Silicon dioxide, hexagonal	S	−906.34	−757.11	2
SiO_2	Silicon dioxide, cristobalite	S	−902.53	−716.46	2
SiO_2	Silicon dioxide	L	−935.34	−551.67	2
SF_6	Sulfur hexafluoride	G	−1220.47	−1116.5	2
SO_2	Sulfur dioxide	G	−296.813	−300.1	2
SO_3	Sulfur trioxide	G	−395.77	−371.02	2
WF_6	Tungsten hexafluoride	G	−1721.72	−1632.29	2
ZnO	Zinc oxide	S	−350.46	−320.48	2
ZnS	Zinc sulfide, wurtzite	S	−191.84	−190.14	2
ZnS	Zinc sulfide, sphalerite	S	−205.18	−200.4	2
$ZnSO_4$	Zinc sulfate	S	−982.8	−871.45	2

Sources:

1. Daniel R. Stull, Edgar F. Westrum, and Gerard C. Sinke, *The Chemical Thermodynamics of Organic Compounds* (New York: Wiley, 1969).
2. Ihsan Barin, *Thermochemical Data of Pure Substances*, 3rd ed. (vol. I and II) (New York: VCH, 1995).
3. M. W. Chase et al., *JANAF Thermochemical Tables*, 3rd ed. (Washington, DC: American Chemical Society; (New York: National Bureau of Standards, 1986).

APPENDIX B

Steam Tables

▶ TABLE B.1: Saturated Water: Temperature Table   [508]
▶ TABLE B.2: Saturated Water: Pressure Table   [510]
▶ TABLE B.3: Saturated Water: Solid-Vapor   [512]
▶ TABLE B.4: Superheated Water Vapor   [513]
▶ TABLE B.5: Subcooled Liquid Water   [519]

Symbols Used in the Steam Tables

T	Temperature	°C
P	Pressure	kPa or MPa
$\hat{v}$	Specific volume	m³/kg
$\hat{u}$	Specific internal energy	kJ/kg
$\hat{h}$	Specific enthalpy	kJ/kg
$\hat{s}$	Specific entropy	kJ/kg K

Subscripts

l	Liquid in equilibrium with vapor
s	Solid in equilibrium with vapor
v	Vapor in equilibrium with liquid or solid
lv	Change by evaporation
sv	Change by sublimation

Source: New York: Wiley J. H. Keenan, F. G. Keys, P. G. Hill, and J. G. Moore, *Steam Tables* (1969), as used by G. J. Van Wylen, R. E. Sonntag, and C. Borgnakke, *Fundamentals of Classical Thermodynamics*, 4th ed., (New York: Wiley, 1994).

TABLE B.1 Saturated Water: Temperature Table

T °C	P kPa, MPa	$\hat{v}_l$ m³/kg	$\hat{v}_v$ m³/kg	$\hat{u}_l$ kJ/kg	$\Delta\hat{u}_{lv}$ kJ/kg	$\hat{u}_v$ kJ/kg	$\hat{h}_l$ kJ/kg	$\Delta\hat{h}_{lv}$ kJ/kg	$\hat{h}_v$ kJ/kg	$\hat{s}_l$ kJ/kg K	$\Delta\hat{s}_{lv}$ kJ/kg K	$\hat{s}_v$ kJ/kg K
0.01	0.6113	0.001000	206.132	0.00	2375.3	2375.3	0.00	2501.3	2501.3	0.0000	9.1562	9.1562
5	0.8721	0.001000	147.118	20.97	2361.3	2382.2	20.98	2489.6	2510.5	0.0761	8.9496	9.0257
10	1.2276	0.001000	106.377	41.99	2347.2	2389.2	41.99	2477.7	2519.7	0.1510	8.7498	8.9007
15	1.7051	0.001001	77.925	62.98	2333.1	2396.0	62.98	2465.9	2528.9	0.2245	8.5569	8.7813
20	2.3385	0.001002	57.790	83.94	2319.0	2402.9	83.94	2454.1	2538.1	0.2966	8.3706	8.6671
25	3.1691	0.001003	43.359	104.86	2304.9	2409.8	104.87	2442.3	2547.2	0.3673	8.1905	8.5579
30	4.2461	0.001004	32.893	125.77	2290.8	2416.6	125.77	2430.5	2556.2	0.4369	8.0164	8.4533
35	5.6280	0.001006	25.216	146.65	2276.7	2423.4	146.66	2418.6	2565.3	0.5052	7.8478	8.3530
40	7.3837	0.001008	19.523	167.53	2262.6	2430.1	167.54	2406.7	2574.3	0.5724	7.6845	8.2569
45	9.5934	0.001010	15.258	188.41	2248.4	2436.8	188.42	2394.8	2583.2	0.6386	7.5261	8.1647
50	12.350	0.001012	12.032	209.30	2234.2	2443.5	209.31	2382.7	2592.1	0.7037	7.3725	8.0762
55	15.758	0.001015	9.568	230.19	2219.9	2450.1	230.20	2370.7	2600.9	0.7679	7.2234	7.9912
60	19.941	0.001017	7.671	251.09	2205.5	2456.6	251.11	2358.5	2609.6	0.8311	7.0784	7.9095
65	25.033	0.001020	6.197	272.00	2191.1	2463.1	272.03	2346.2	2618.2	0.8934	6.9375	7.8309
70	31.188	0.001023	5.042	292.93	2176.6	2469.5	292.96	2333.8	2626.8	0.9548	6.8004	7.7552
75	38.578	0.001026	4.131	313.87	2162.0	2475.9	313.91	2321.4	2635.3	1.0154	6.6670	7.6824
80	47.390	0.001029	3.407	334.84	2147.4	2482.2	334.88	2308.8	2643.7	1.0752	6.5369	7.6121
85	57.834	0.001032	2.828	355.82	2132.6	2488.4	355.88	2296.0	2651.9	1.1342	6.4102	7.5444
90	70.139	0.001036	2.361	376.82	2117.7	2494.5	376.90	2283.2	2660.1	1.1924	6.2866	7.4790
95	84.554	0.001040	1.982	397.86	2102.7	2500.6	397.94	2270.2	2668.1	1.2500	6.1659	7.4158
100	0.10135	0.001044	1.6729	418.91	2087.6	2506.5	419.02	2257.0	2676.0	1.3068	6.0480	7.3548
105	0.12082	0.001047	1.4194	440.00	2072.3	2512.3	440.13	2243.7	2683.8	1.3629	5.9328	7.2958
110	0.14328	0.001052	1.2102	461.12	2057.0	2518.1	461.27	2230.2	2691.5	1.4184	5.8202	7.2386
115	0.16906	0.001056	1.0366	482.28	2041.4	2523.7	482.46	2216.5	2699.0	1.4733	5.7100	7.1832
120	0.19853	0.001060	0.8919	503.48	2025.8	2529.2	503.69	2202.6	2706.3	1.5275	5.6020	7.1295
125	0.2321	0.001065	0.77059	524.72	2009.9	2534.6	524.96	2188.5	2713.5	1.5812	5.4962	7.0774
130	0.2701	0.001070	0.66850	546.00	1993.9	2539.9	546.29	2174.2	2720.5	1.6343	5.3925	7.0269
135	0.3130	0.001075	0.58217	567.34	1977.7	2545.0	567.67	2159.6	2727.3	1.6869	5.2907	6.9777
140	0.3613	0.001080	0.50885	588.72	1961.3	2550.0	589.11	2144.8	2733.9	1.7390	5.1908	6.9298
145	0.4154	0.001085	0.44632	610.16	1944.7	2554.9	610.61	2129.6	2740.3	1.7906	5.0926	6.8832
150	0.4759	0.001090	0.39278	631.66	1927.9	2559.5	632.18	2114.3	2746.4	1.8417	4.9960	6.8378
155	0.5431	0.001096	0.34676	653.23	1910.8	2564.0	653.82	2098.6	2752.4	1.8924	4.9010	6.7934
160	0.6178	0.001102	0.30706	674.85	1893.5	2568.4	675.53	2082.6	2758.1	1.9426	4.8075	6.7501
165	0.7005	0.001108	0.27269	696.55	1876.0	2572.5	697.32	2066.2	2763.5	1.9924	4.7153	6.7078
170	0.7917	0.001114	0.24283	718.31	1858.1	2576.5	719.20	2049.5	2768.7	2.0418	4.6244	6.6663

TABLE B.4 Superheated Water Vapor

$P = 10$ kPa

T °C	$\hat{v}$ m³/kg	$\hat{u}$ kJ/kg	$\hat{h}$ kJ/kg	$\hat{s}$ kJ/kg K
sat	14.674	2437.9	2584.6	8.1501
50	14.869	2443.9	2592.6	8.1749
100	17.196	2515.5	2687.5	8.4479
150	19.513	2587.9	2783.0	8.6881
200	21.825	2661.3	2879.5	8.9037
250	24.136	2736.0	2977.3	9.1002
300	26.445	2812.1	3076.5	9.2812
400	31.063	2968.9	3279.5	9.6076
500	35.679	3132.3	3489.0	9.8977
600	40.295	3302.5	3705.4	10.1608
700	44.911	3479.6	3928.7	10.4028
800	49.526	3663.8	4159.1	10.6281
900	54.141	3855.0	4396.4	10.8395
1000	58.757	4053.0	4640.6	11.0392
1100	63.372	4257.5	4891.2	11.2287
1200	67.987	4467.9	5147.8	11.4090
1300	72.603	4683.7	5409.7	11.5810

$P = 50$ kPa

T °C	$\hat{v}$ m³/kg	$\hat{u}$ kJ/kg	$\hat{h}$ kJ/kg	$\hat{s}$ kJ/kg K
sat	3.240	2483.8	2645.9	7.5939
100	3.418	2511.6	2682.5	7.6947
150	3.889	2585.6	2780.1	7.9400
200	4.356	2659.8	2877.6	8.1579
250	4.821	2735.0	2976.0	8.3555
300	5.284	2811.3	3075.5	8.5372
400	6.209	2968.4	3278.9	8.8641
500	7.134	3131.9	3488.6	9.1545
600	8.058	3302.2	3705.1	9.4177
700	8.981	3479.5	3928.5	9.6599
800	9.904	3663.7	4158.9	9.8852
900	10.828	3854.9	4396.3	10.0967
1000	11.751	4052.9	4640.5	10.2964
1100	12.674	4257.4	4891.1	10.4858
1200	13.597	4467.8	5147.7	10.6662
1300	14.521	4683.6	5409.6	10.8382

$P = 100$ kPa

T °C	$\hat{v}$ m³/kg	$\hat{u}$ kJ/kg	$\hat{h}$ kJ/kg	$\hat{s}$ kJ/kg K
sat	1.6940	2506.1	2675.5	7.3593
100	1.6958	2506.6	2676.2	7.3614
150	1.9364	2582.7	2776.4	7.6133
200	2.1723	2658.0	2875.3	7.8342
250	2.4060	2733.7	2974.3	8.0332
300	2.6388	2810.4	3074.3	8.2157
400	3.1026	2967.8	3278.1	8.5434
500	3.5655	3131.5	3488.1	8.8341
600	4.0278	3301.9	3704.7	9.0975
700	4.4899	3479.2	3928.2	9.3398
800	4.9517	3663.5	4158.7	9.5652
900	5.4135	3854.8	4396.1	9.7767
1000	5.8753	4052.8	4640.3	9.9764
1100	6.3370	4257.3	4890.9	10.1658
1200	6.7986	4467.7	5147.6	10.3462
1300	7.2603	4683.5	5409.5	10.5182

$P = 200$ kPa

T °C	$\hat{v}$ m³/kg	$\hat{u}$ kJ/kg	$\hat{h}$ kJ/kg	$\hat{s}$ kJ/kg K
sat	0.88573	2529.5	2706.6	7.1271
150	0.95964	2576.9	2768.8	7.2795
200	1.08034	2654.4	2870.5	7.5066
250	1.19880	2731.2	2971.0	7.7085
300	1.31616	2808.6	3071.8	7.8926
400	1.54930	2966.7	3276.5	8.2217
500	1.78139	3130.7	3487.0	8.5132
600	2.01297	3301.4	3704.0	8.7769
700	2.24426	3478.8	3927.7	9.0194
800	2.47539	3663.2	4158.3	9.2450
900	2.70643	3854.5	4395.8	9.4565
1000	2.93740	4052.5	4640.0	9.6563
1100	3.16834	4257.0	4890.7	9.8458
1200	3.39927	4467.5	5147.3	10.0262
1300	3.63018	4683.2	5409.3	10.1982

$P = 300$ kPa

T °C	$\hat{v}$ m³/kg	$\hat{u}$ kJ/kg	$\hat{h}$ kJ/kg	$\hat{s}$ kJ/kg K
sat	0.60582	2543.6	2725.3	6.9918
150	0.63388	2570.8	2761.0	7.0778
200	0.71629	2650.7	2865.5	7.3115
250	0.79636	2728.7	2967.6	7.5165
300	0.87529	2806.7	3069.3	7.7022
400	1.03151	2965.5	3275.0	8.0329
500	1.18669	3130.0	3486.0	8.3250
600	1.34136	3300.8	3703.2	8.5892
700	1.49573	3478.4	3927.1	8.8319
800	1.64994	3662.9	4157.8	9.0575
900	1.80406	3854.2	4395.4	9.2691
1000	1.95812	4052.3	4639.7	9.4689
1100	2.11214	4256.8	4890.4	9.6585
1200	2.26614	4467.2	5147.1	9.8389
1300	2.42013	4683.0	5409.0	10.0109

$P = 400$ kPa

T °C	$\hat{v}$ m³/kg	$\hat{u}$ kJ/kg	$\hat{h}$ kJ/kg	$\hat{s}$ kJ/kg K
sat	0.46246	2553.6	2738.5	6.8958
150	0.47084	2564.5	2752.8	6.9299
200	0.53422	2646.8	2860.5	7.1706
250	0.59512	2726.1	2964.2	7.3788
300	0.65484	2804.8	3066.7	7.5661
400	0.77262	2964.4	3273.4	7.8984
500	0.88934	3129.2	3484.9	8.1912
600	1.00555	3300.2	3702.4	8.4557
700	1.12147	3477.9	3926.5	8.6987
800	1.23722	3662.5	4157.4	8.9244
900	1.35288	3853.9	4395.1	9.1361
1000	1.46847	4052.0	4639.4	9.3360
1100	1.58404	4256.5	4890.1	9.5255
1200	1.69958	4467.0	5146.8	9.7059
1300	1.81511	4682.8	5408.8	9.8780

(Continued)

TABLE B.4 Continued

P = 500 kPa

T °C	v̂ m³/kg	û kJ/kg	ĥ kJ/kg	ŝ kJ/kg K
sat	0.37489	2561.2	2748.7	6.8212
200	0.42492	2642.9	2855.4	7.0592
250	0.47436	2723.5	2960.7	7.2708
300	0.52256	2802.9	3064.2	7.4598
350	0.57012	2882.6	3167.6	7.6328
400	0.61728	2963.2	3271.8	7.7937
500	0.71093	3128.4	3483.8	8.0872
600	0.80406	3299.6	3701.7	8.3521
700	0.89691	3477.5	3926.0	8.5952
800	0.98959	3662.2	4157.0	8.8211
900	1.08217	3853.6	4394.7	9.0329
1000	1.17469	4051.8	4639.1	9.2328
1100	1.26718	4256.3	4889.9	9.4224
1200	1.35964	4466.8	5146.6	9.6028
1300	1.45210	4682.5	5408.6	9.7749

P = 600 kPa

T °C	v̂ m³/kg	û kJ/kg	ĥ kJ/kg	ŝ kJ/kg K
sat	0.31567	2567.4	2756.8	6.7600
200	0.35202	2638.9	2850.1	6.9665
250	0.39383	2720.9	2957.2	7.1816
300	0.43437	2801.0	3061.6	7.3723
350	0.47424	2881.1	3165.7	7.5463
400	0.51372	2962.0	3270.2	7.7078
500	0.59199	3127.6	3482.7	8.0020
600	0.66974	3299.1	3700.9	8.2673
700	0.74720	3477.1	3925.4	8.5107
800	0.82450	3661.8	4156.5	8.7367
900	0.90169	3853.3	4394.4	8.9485
1000	0.97883	4051.5	4638.8	9.1484
1100	1.05594	4256.1	4889.6	9.3381
1200	1.13302	4466.5	5146.3	9.5185
1300	1.21009	4682.3	5408.3	9.6906

P = 800 kPa

T °C	v̂ m³/kg	û kJ/kg	ĥ kJ/kg	ŝ kJ/kg K
sat	0.24043	2576.8	2769.1	6.6627
200	0.26080	2630.6	2839.2	6.8158
250	0.29314	2715.5	2950.0	7.0384
300	0.32411	2797.1	3056.4	7.2327
350	0.35439	2878.2	3161.7	7.4088
400	0.38426	2959.7	3267.1	7.5715
500	0.44331	3125.9	3480.6	7.8672
600	0.50184	3297.9	3699.4	8.1332
700	0.56007	3476.2	3924.3	8.3770
800	0.61813	3661.1	4155.7	8.6033
900	0.67610	3852.8	4393.6	8.8153
1000	0.73401	4051.0	4638.2	9.0153
1100	0.79188	4255.6	4889.1	9.2049
1200	0.84974	4466.1	5145.8	9.3854
1300	0.90758	4681.8	5407.9	9.5575

P = 1 MPa

T °C	v̂ m³/kg	û kJ/kg	ĥ kJ/kg	ŝ kJ/kg K
sat	0.19444	2583.6	2778.1	6.5864
200	0.20596	2621.9	2827.9	6.6939
250	0.23268	2709.9	2942.6	6.9246
300	0.25794	2793.2	3051.2	7.1228
350	0.28247	2875.2	3157.7	7.3010
400	0.30659	2957.3	3263.9	7.4650
500	0.35411	3124.3	3478.4	7.7621
600	0.40109	3296.8	3697.9	8.0289
700	0.44779	3475.4	3923.1	8.2731
800	0.49432	3660.5	4154.8	8.4996
900	0.54075	3852.2	4392.9	8.7118
1000	0.58712	4050.5	4637.6	8.9119
1100	0.63345	4255.1	4888.5	9.1016
1200	0.67977	4465.6	5145.4	9.2821
1300	0.72608	4681.3	5407.4	9.4542

P = 1.2 MPa

T °C	v̂ m³/kg	û kJ/kg	ĥ kJ/kg	ŝ kJ/kg K
sat	0.16333	2588.8	2784.8	6.5233
200	0.16930	2612.7	2815.9	6.5898
250	0.19235	2704.2	2935.0	6.8293
300	0.21382	2789.2	3045.8	7.0316
350	0.23452	2872.2	3153.6	7.2120
400	0.25480	2954.9	3260.7	7.3773
500	0.29463	3122.7	3476.3	7.6758
600	0.33393	3295.6	3696.3	7.9434
700	0.37294	3474.5	3922.0	8.1881
800	0.41177	3659.8	4153.9	8.4149
900	0.45051	3851.6	4392.2	8.6272
1000	0.48919	4050.0	4637.0	8.8274
1100	0.52783	4254.6	4888.0	9.0171
1200	0.56646	4465.1	5144.9	9.1977
1300	0.60507	4680.9	5406.9	9.3698

P = 1.4 MPa

T °C	v̂ m³/kg	û kJ/kg	ĥ kJ/kg	ŝ kJ/kg K
sat	0.14084	2592.8	2790.0	6.4692
200	0.14302	2603.1	2803.3	6.4975
250	0.16350	2698.3	2927.2	6.7467
300	0.18228	2785.2	3040.4	6.9533
350	0.20026	2869.1	3149.5	7.1359
400	0.21780	2952.5	3257.4	7.3025
500	0.25215	3121.1	3474.1	7.6026
600	0.28596	3294.4	3694.8	7.8710
700	0.31947	3473.6	3920.9	8.1160
800	0.35281	3659.1	4153.0	8.3431
900	0.38606	3851.0	4391.5	8.5555
1000	0.41924	4049.5	4636.4	8.7558
1100	0.45239	4254.1	4887.5	8.9456
1200	0.48552	4464.6	5144.4	9.1262
1300	0.51864	4680.4	5406.5	9.2983

TABLE B.4 Continued

$P = 1.6$ MPa

T °C	$\hat{v}$ m³/kg	$\hat{u}$ kJ/kg	$\hat{h}$ kJ/kg	$\hat{s}$ kJ/kg K
sat	0.12380	2595.9	2794.0	6.4217
225	0.13287	2644.6	2857.2	6.5518
250	0.14184	2692.3	2919.2	6.6732
300	0.15862	2781.0	3034.8	6.8844
350	0.17456	2866.0	3145.4	7.0693
400	0.19005	2950.1	3254.2	7.2373
500	0.22029	3119.5	3471.9	7.5389
600	0.24998	3293.3	3693.2	7.8080
700	0.27937	3472.7	3919.7	8.0535
800	0.30859	3658.4	4152.1	8.2808
900	0.33772	3850.5	4390.8	8.4934
1000	0.36678	4049.0	4635.8	8.6938
1100	0.39581	4253.7	4887.0	8.8837
1200	0.42482	4464.2	5143.9	9.0642
1300	0.45382	4679.9	5406.0	9.2364

$P = 2.5$ MPa

T °C	$\hat{v}$ m³/kg	$\hat{u}$ kJ/kg	$\hat{h}$ kJ/kg	$\hat{s}$ kJ/kg K
sat	0.07998	2603.1	2803.1	6.2574
225	0.08027	2605.6	2806.3	6.2638
250	0.08700	2662.5	2880.1	6.4084
300	0.09890	2761.6	3008.8	6.6437
350	0.10976	2851.8	3126.2	6.8402
400	0.12010	2939.0	3239.3	7.0147
450	0.13014	3025.4	3350.8	7.1745
500	0.13998	3112.1	3462.0	7.3233
600	0.15930	3288.0	3686.2	7.5960
700	0.17832	3468.8	3914.6	7.8435
800	0.19716	3655.3	4148.2	8.0720
900	0.21590	3847.9	4387.6	8.2853
1000	0.23458	4046.7	4633.1	8.4860
1100	0.25322	4251.5	4884.6	8.6761
1200	0.27185	4462.1	5141.7	8.8569
1300	0.29046	4677.8	5404.0	9.0291

$P = 1.8$ MPa

T °C	$\hat{v}$ m³/kg	$\hat{u}$ kJ/kg	$\hat{h}$ kJ/kg	$\hat{s}$ kJ/kg K
sat	0.11042	2598.4	2797.1	6.3793
225	0.11673	2636.6	2846.7	6.4807
250	0.12497	2686.0	2911.0	6.6066
300	0.14021	2776.8	3029.2	6.8226
350	0.15457	2862.9	3141.2	7.0099
400	0.16847	2947.7	3250.9	7.1793
500	0.19550	3117.8	3469.7	7.4824
600	0.22199	3292.1	3691.7	7.7523
700	0.24818	3471.9	3918.6	7.9983
800	0.27420	3657.7	4151.3	8.2258
900	0.30012	3849.9	4390.1	8.4386
1000	0.32598	4048.4	4635.2	8.6390
1100	0.35180	4253.2	4886.4	8.8290
1200	0.37761	4463.7	5143.4	9.0096
1300	0.40340	4679.4	5405.6	9.1817

$P = 3$ MPa

T °C	$\hat{v}$ m³/kg	$\hat{u}$ kJ/kg	$\hat{h}$ kJ/kg	$\hat{s}$ kJ/kg K
sat	0.06668	2604.1	2804.1	6.1869
250	0.07058	2644.0	2855.8	6.2871
300	0.08114	2750.0	2993.5	6.5389
350	0.09053	2843.7	3115.3	6.7427
400	0.09936	2932.7	3230.8	6.9211
450	0.10787	3020.4	3344.0	7.0833
500	0.11619	3107.9	3456.5	7.2337
600	0.13243	3285.0	3682.3	7.5084
700	0.14838	3466.6	3911.7	7.7571
800	0.16414	3653.6	4146.0	7.9862
900	0.17980	3846.5	4385.9	8.1999
1000	0.19541	4045.4	4631.6	8.4009
1100	0.21098	4250.3	4883.3	8.5911
1200	0.22652	4460.9	5140.5	8.7719
1300	0.24206	4676.6	5402.8	8.9442

$P = 2$ MPa

T °C	$\hat{v}$ m³/kg	$\hat{u}$ kJ/kg	$\hat{h}$ kJ/kg	$\hat{s}$ kJ/kg K
sat	0.09963	2600.3	2799.5	6.3408
225	0.10377	2628.3	2835.8	6.4146
250	0.11144	2679.6	2902.5	6.5452
300	0.12547	2772.6	3023.5	6.7663
350	0.13857	2859.8	3137.0	6.9562
400	0.15120	2945.2	3247.6	7.1270
500	0.17568	3116.2	3467.6	7.4316
600	0.19960	3290.9	3690.1	7.7023
700	0.22323	3471.0	3917.5	7.9487
800	0.24668	3657.0	4150.4	8.1766
900	0.27004	3849.3	4389.4	8.3895
1000	0.29333	4047.9	4634.6	8.5900
1100	0.31659	4252.7	4885.9	8.7800
1200	0.33984	4463.2	5142.9	8.9606
1300	0.36306	4679.0	5405.1	9.1328

$P = 3.5$ MPa

T °C	$\hat{v}$ m³/kg	$\hat{u}$ kJ/kg	$\hat{h}$ kJ/kg	$\hat{s}$ kJ/kg K
sat	0.05707	2603.7	2803.4	6.1252
250	0.05873	2623.7	2829.2	6.1748
300	0.06842	2738.0	2977.5	6.4460
350	0.07678	2835.3	3104.0	6.6578
400	0.08453	2926.4	3222.2	6.8404
450	0.09196	3015.3	3337.2	7.0051
500	0.09918	3103.7	3450.9	7.1571
600	0.11324	3282.1	3678.4	7.4338
700	0.12699	3464.4	3908.8	7.6837
800	0.14056	3651.8	4143.8	7.9135
900	0.15402	3845.0	4384.1	8.1275
1000	0.16743	4044.1	4630.1	8.3288
1100	0.18080	4249.1	4881.9	8.5191
1200	0.19415	4459.8	5139.3	8.7000
1300	0.20749	4675.5	5401.7	8.8723

(Continued)

TABLE B.4 Continued

P = 4 MPa

T °C	$\hat{v}$ m³/kg	$\hat{u}$ kJ/kg	$\hat{h}$ kJ/kg	$\hat{s}$ kJ/kg K
sat	0.04978	2602.3	2801.4	6.0700
275	0.05457	2667.9	2886.2	6.2284
300	0.05884	2725.3	2960.7	6.3614
350	0.06645	2826.6	3092.4	6.5820
400	0.07341	2919.9	3213.5	6.7689
450	0.08003	3010.1	3330.2	6.9362
500	0.08643	3099.5	3445.2	7.0900
600	0.09885	3279.1	3674.4	7.3688
700	0.11095	3462.1	3905.9	7.6198
800	0.12287	3650.1	4141.6	7.8502
900	0.13469	3843.6	4382.3	8.0647
1000	0.14645	4042.9	4628.7	8.2661
1100	0.15817	4248.0	4880.6	8.4566
1200	0.16987	4458.6	5138.1	8.6376
1300	0.18156	4674.3	5400.5	8.8099

P = 4.5 MPa

T °C	$\hat{v}$ m³/kg	$\hat{u}$ kJ/kg	$\hat{h}$ kJ/kg	$\hat{s}$ kJ/kg K
sat	0.04406	2600.0	2798.3	6.0198
275	0.04730	2650.3	2863.1	6.1401
300	0.05135	2712.0	2943.1	6.2827
350	0.05840	2817.8	3080.6	6.5130
400	0.06475	2913.3	3204.7	6.7046
450	0.07074	3004.9	3323.2	6.8745
500	0.07651	3095.2	3439.5	7.0300
600	0.08765	3276.0	3670.5	7.3109
700	0.09847	3459.9	3903.0	7.5631
800	0.10911	3648.4	4139.4	7.7942
900	0.11965	3842.1	4380.6	8.0091
1000	0.13013	4041.6	4627.2	8.2108
1100	0.14056	4246.8	4879.3	8.4014
1200	0.15098	4457.4	5136.9	8.5824
1300	0.16139	4673.1	5399.4	8.7548

P = 5 MPa

T °C	$\hat{v}$ m³/kg	$\hat{u}$ kJ/kg	$\hat{h}$ kJ/kg	$\hat{s}$ kJ/kg K
sat	0.03944	2597.1	2794.3	5.9733
275	0.04141	2631.2	2838.3	6.0543
300	0.04532	2697.9	2924.5	6.2083
350	0.05194	2808.7	3068.4	6.4492
400	0.05781	2906.6	3195.6	6.6458
450	0.06330	2999.6	3316.1	6.8185
500	0.06857	3090.9	3433.8	6.9758
600	0.07869	3273.0	3666.5	7.2588
700	0.08849	3457.7	3900.1	7.5122
800	0.09811	3646.6	4137.2	7.7440
900	0.10762	3840.7	4378.8	7.9593
1000	0.11707	4040.3	4625.7	8.1612
1100	0.12648	4245.6	4878.0	8.3519
1200	0.13587	4456.3	5135.7	8.5330
1300	0.14526	4672.0	5398.2	8.7055

P = 6 MPa

T °C	$\hat{v}$ m³/kg	$\hat{u}$ kJ/kg	$\hat{h}$ kJ/kg	$\hat{s}$ kJ/kg K
sat	0.03244	2589.7	2784.3	5.8891
300	0.03616	2667.2	2884.2	6.0673
350	0.04223	2789.6	3043.0	6.3334
400	0.04739	2892.8	3177.2	6.5407
450	0.05214	2988.9	3301.8	6.7192
500	0.05665	3082.2	3422.1	6.8802
550	0.06101	3174.6	3540.6	7.0287
600	0.06525	3266.9	3658.4	7.1676
700	0.07352	3453.2	3894.3	7.4234
800	0.08160	3643.1	4132.7	7.6566
900	0.08958	3837.8	4375.3	7.8727
1000	0.09749	4037.8	4622.7	8.0751
1100	0.10536	4243.3	4875.4	8.2661
1200	0.11321	4454.0	5133.3	8.4473
1300	0.12106	4669.6	5396.0	8.6199

P = 7 MPa

T °C	$\hat{v}$ m³/kg	$\hat{u}$ kJ/kg	$\hat{h}$ kJ/kg	$\hat{s}$ kJ/kg K
sat	0.02737	2580.5	2772.1	5.8132
300	0.02947	2632.1	2838.4	5.9304
350	0.03524	2769.3	3016.0	6.2282
400	0.03993	2878.6	3158.1	6.4477
450	0.04416	2977.9	3287.0	6.6326
500	0.04814	3073.3	3410.3	6.7974
550	0.05195	3167.2	3530.9	6.9486
600	0.05565	3260.7	3650.3	7.0894
700	0.06283	3448.6	3888.4	7.3476
800	0.06981	3639.6	4128.3	7.5822
900	0.07669	3835.0	4371.8	7.7991
1000	0.08350	4035.3	4619.8	8.0020
1100	0.09027	4240.9	4872.8	8.1933
1200	0.09703	4451.7	5130.9	8.3747
1300	0.10377	4667.3	5393.7	8.5472

P = 8 MPa

T °C	$\hat{v}$ m³/kg	$\hat{u}$ kJ/kg	$\hat{h}$ kJ/kg	$\hat{s}$ kJ/kg K
sat	0.02352	2569.8	2757.9	5.7431
300	0.02426	2590.9	2785.0	5.7905
350	0.02995	2747.7	2987.3	6.1300
400	0.03432	2863.8	3138.3	6.3633
450	0.03817	2966.7	3272.0	6.5550
500	0.04175	3064.3	3398.3	6.7239
550	0.04516	3159.8	3521.0	6.8778
600	0.04845	3254.4	3642.0	7.0205
700	0.05481	3444.0	3882.5	7.2812
800	0.06097	3636.1	4123.8	7.5173
900	0.06702	3832.1	4368.3	7.7350
1000	0.07301	4032.8	4616.9	7.9384
1100	0.07896	4238.6	4870.3	8.1299
1200	0.08489	4449.4	5128.5	8.3115
1300	0.09080	4665.0	5391.5	8.4842

TABLE B.4 Continued

P = 9 MPa

T °C	v̂ m³/kg	û kJ/kg	ĥ kJ/kg	ŝ kJ/kg K
sat	0.02048	2557.8	2742.1	5.6771
325	0.02327	2646.5	2855.9	5.8711
350	0.02580	2724.4	2956.5	6.0361
400	0.02993	2848.4	3117.8	6.2853
450	0.03350	2955.1	3256.6	6.4843
500	0.03677	3055.1	3386.1	6.6575
550	0.03987	3152.2	3511.0	6.8141
600	0.04285	3248.1	3633.7	6.9588
650	0.04574	3343.7	3755.3	7.0943
700	0.04857	3439.4	3876.5	7.2221
800	0.05409	3632.5	4119.4	7.4597
900	0.05950	3829.2	4364.7	7.6782
1000	0.06485	4030.3	4613.9	7.8821
1100	0.07016	4236.3	4867.7	8.0739
1200	0.07544	4447.2	5126.2	8.2556
1300	0.08072	4662.7	5389.2	8.4283

P = 10 MPa

T °C	v̂ m³/kg	û kJ/kg	ĥ kJ/kg	ŝ kJ/kg K
sat	0.01803	2544.4	2724.7	5.6140
325	0.01986	2610.4	2809.0	5.7568
350	0.02242	2699.2	2923.4	5.9442
400	0.02641	2832.4	3096.5	6.2119
450	0.02975	2943.3	3240.8	6.4189
500	0.03279	3045.8	3373.6	6.5965
550	0.03564	3144.5	3500.9	6.7561
600	0.03837	3241.7	3625.3	6.9028
650	0.04101	3338.2	3748.3	7.0397
700	0.04358	3434.7	3870.5	7.1687
800	0.04859	3629.0	4114.9	7.4077
900	0.05349	3826.3	4361.2	7.6272
1000	0.05832	4027.8	4611.0	7.8315
1100	0.06312	4234.0	4865.1	8.0236
1200	0.06789	4444.9	5123.8	8.2054
1300	0.07265	4660.4	5387.0	8.3783

P = 12.5 MPa

T °C	v̂ m³/kg	û kJ/kg	ĥ kJ/kg	ŝ kJ/kg K
sat	0.01350	2505.1	2673.8	5.4623
350	0.01613	2624.6	2826.2	5.7117
400	0.02000	2789.3	3039.3	6.0416
450	0.02299	2912.4	3199.8	6.2718
500	0.02560	3021.7	3341.7	6.4617
550	0.02801	3124.9	3475.1	6.6289
600	0.03029	3225.4	3604.0	6.7810
650	0.03248	3324.4	3730.4	6.9218
700	0.03460	3422.9	3855.4	7.0536
800	0.03869	3620.0	4103.7	7.2965
900	0.04267	3819.1	4352.5	7.5181
1000	0.04658	4021.6	4603.8	7.7237
1100	0.05045	4228.2	4858.8	7.9165
1200	0.05430	4439.3	5118.0	8.0987
1300	0.05813	4654.8	5381.4	8.2717

P = 15 MPa

T °C	v̂ m³/kg	û kJ/kg	ĥ kJ/kg	ŝ kJ/kg K
sat	.010338	2455.4	2610.5	5.3097
350	.011470	2520.4	2692.4	5.4420
400	.015649	2740.7	2975.4	5.8810
450	.018446	2879.5	3156.2	6.1403
500	.020800	2996.5	3308.5	6.3442
550	.022927	3104.7	3448.6	6.5198
600	.024911	3208.6	3582.3	6.6775
650	.026797	3310.4	3712.3	6.8223
700	.028612	3410.9	3840.1	6.9572
800	.032096	3611.0	4092.4	7.2040
900	.035457	3811.9	4343.8	7.4279
1000	.038748	4015.4	4596.6	7.6347
1100	.042001	4222.6	4852.6	7.8282
1200	.045233	4433.8	5112.3	8.0108
1300	.048455	4649.1	5375.9	8.1839

P = 17.5 MPa

T °C	v̂ m³/kg	û kJ/kg	ĥ kJ/kg	ŝ kJ/kg K
sat	.0079204	2390.2	2528.8	5.1418
400	.0124477	2685.0	2902.8	5.7212
450	.0151740	2844.2	3109.7	6.0182
500	.0173585	2970.3	3274.0	6.2382
550	.0192877	3083.8	3421.4	6.4229
600	.0210640	3191.5	3560.1	6.5866
650	.0227372	3296.0	3693.9	6.7356
700	.0243365	3398.8	3824.7	6.8736
800	.0273849	3601.9	4081.1	7.1245
900	.0303071	3804.7	4335.1	7.3507
1000	.0331580	4009.3	4589.5	7.5588
1100	.0359695	4216.9	4846.4	7.7530
1200	.0387605	4428.3	5106.6	7.9359
1300	.0415417	4643.5	5370.5	8.1093

P = 20 MPa

T °C	v̂ m³/kg	û kJ/kg	ĥ kJ/kg	ŝ kJ/kg K
sat	.0058342	2293.1	2409.7	4.9269
400	.0099423	2619.2	2818.1	5.5539
450	.0126953	2806.2	3060.1	5.9016
500	.0147683	2942.8	3238.2	6.1400
550	.0165553	3062.3	3393.5	6.3347
600	.0181781	3174.0	3537.6	6.5048
650	.0196929	3281.5	3675.3	6.6582
700	.0211311	3386.5	3809.1	6.7993
800	.0238532	3592.7	4069.8	7.0544
900	.0264463	3797.4	4326.4	7.2830
1000	.0289666	4003.1	4582.5	7.4925
1100	.0314471	4211.3	4840.2	7.6874
1200	.0339071	4422.8	5101.0	7.8706
1300	.0363574	4638.0	5365.1	8.0441

(Continued)

TABLE B.4 Continued

$P = 25$ MPa

T °C	$\hat{v}$ m³/kg	$\hat{u}$ kJ/kg	$\hat{h}$ kJ/kg	$\hat{s}$ kJ/kg K
375	.001973	1798.6	1847.9	4.0319
400	.006004	2430.1	2580.2	5.1418
450	.009162	2720.7	2949.7	5.6743
500	.011124	2884.3	3162.4	5.9592
550	.012724	3017.5	3335.6	6.1764
600	.014138	3137.9	3491.4	6.3602
650	.015433	3251.6	3637.5	6.5229
700	.016647	3361.4	3777.6	6.6707
800	.018913	3574.3	4047.1	6.9345
900	.021045	3783.0	4309.1	7.1679
1000	.023102	3990.9	4568.5	7.3801
1100	.025119	4200.2	4828.2	7.5765
1200	.027115	4412.0	5089.9	7.7604
1300	.029101	4626.9	5354.4	7.9342

$P = 30$ MPa

T °C	$\hat{v}$ m³/kg	$\hat{u}$ kJ/kg	$\hat{h}$ kJ/kg	$\hat{s}$ kJ/kg K
375	.001789	1737.8	1791.4	3.9303
400	.002790	2067.3	2151.0	4.4728
450	.006735	2619.3	2821.4	5.4423
500	.008679	2820.7	3081.0	5.7904
550	.010168	2970.3	3275.4	6.0342
600	.011446	3100.5	3443.9	6.2330
650	.012596	3221.0	3598.9	6.4057
700	.013661	3335.8	3745.7	6.5606
800	.015623	3555.6	4024.3	6.8332
900	.017448	3768.5	4291.9	7.0717
1000	.019196	3978.8	4554.7	7.2867
1100	.020903	4189.2	4816.3	7.4845
1200	.022589	4401.3	5079.0	7.6691
1300	.024266	4616.0	5344.0	7.8432

$P = 35$ MPa

T °C	$\hat{v}$ m³/kg	$\hat{u}$ kJ/kg	$\hat{h}$ kJ/kg	$\hat{s}$ kJ/kg K
375	.001700	1702.9	1762.4	3.8721
400	.002100	1914.0	1987.5	4.2124
450	.004962	2498.7	2672.4	5.1962
500	.006927	2751.9	2994.3	5.6281
550	.008345	2920.9	3213.0	5.9025
600	.009527	3062.0	3395.5	6.1178
650	.010575	3189.8	3559.9	6.3010
700	.011533	3309.9	3713.5	6.4631
800	.013278	3536.8	4001.5	6.7450
900	.014883	3754.0	4274.9	6.9886
1000	.016410	3966.7	4541.1	7.2063
1100	.017895	4178.3	4804.6	7.4056
1200	.019360	4390.7	5068.4	7.5910
1300	.020815	4605.1	5333.6	7.7652

$P = 40$ MPa

T °C	$\hat{v}$ m³/kg	$\hat{u}$ kJ/kg	$\hat{h}$ kJ/kg	$\hat{s}$ kJ/kg K
375	.0016406	1677.1	1742.7	3.8289
400	.0019017	1854.5	1930.8	4.1134
450	.0036931	2365.1	2512.8	4.9459
500	.0056225	2678.4	2903.3	5.4699
600	.0080943	3022.6	3346.4	6.0113
700	.0099415	3283.6	3681.3	6.3750
800	.0115228	3517.9	3978.8	6.6662
900	.0129626	3739.4	4257.9	6.9150
1000	.0143238	3954.6	4527.6	7.1356
1100	.0156426	4167.4	4793.1	7.3364
1200	.0169403	4380.1	5057.7	7.5224
1300	.0182292	4594.3	5323.5	7.6969

$P = 50$ MPa

T °C	$\hat{v}$ m³/kg	$\hat{u}$ kJ/kg	$\hat{h}$ kJ/kg	$\hat{s}$ kJ/kg K
375	.0015593	1638.6	1716.5	3.7638
400	.0017309	1788.0	1874.6	4.0030
450	.0024862	2159.6	2283.9	4.5883
500	.0038924	2525.5	2720.1	5.1725
600	.0061123	2942.0	3247.6	5.8177
700	.0077274	3230.5	3616.9	6.2189
800	.0090761	3479.8	3933.6	6.5290
900	.0102831	3710.3	4224.4	6.7882
1000	.0114113	3930.5	4501.1	7.0146
1100	.0124966	4145.7	4770.6	7.2183
1200	.0135606	4359.1	5037.2	7.4058
1300	.0146159	4572.8	5303.6	7.5807

$P = 60$ MPa

T °C	$\hat{v}$ m³/kg	$\hat{u}$ kJ/kg	$\hat{h}$ kJ/kg	$\hat{s}$ kJ/kg K
375	.0015027	1609.3	1699.5	3.7140
400	.0016335	1745.3	1843.4	3.9317
450	.0020850	2053.9	2179.0	4.4119
500	.0029557	2390.5	2567.9	4.9320
600	.0048345	2861.1	3151.2	5.6451
700	.0062719	3177.3	3553.6	6.0824
800	.0074588	3441.6	3889.1	6.4110
900	.0085083	3681.0	4191.5	6.6805
1000	.0094800	3906.4	4475.2	6.9126
1100	.0104091	4124.1	4748.6	7.1194
1200	.0113167	4338.2	5017.2	7.3082
1300	.0122155	4551.4	5284.3	7.4837

TABLE B.5 Subcooled Liquid Water

$P = 5$ MPa

T °C	$\hat{v}$ m³/kg	$\hat{u}$ kJ/kg	$\hat{h}$ kJ/kg	$\hat{s}$ kJ/kg K
0	.0009977	0.03	5.02	0.0001
20	.0009995	83.64	88.64	0.2955
40	.0010056	166.93	171.95	0.5705
60	.0010149	250.21	255.28	0.8284
80	.0010268	333.69	338.83	1.0719
100	.0010410	417.50	422.71	1.3030
120	.0010576	501.79	507.07	1.5232
140	.0010768	586.74	592.13	1.7342
160	.0010988	672.61	678.10	1.9374
180	.0011240	759.62	765.24	2.1341
200	.0011530	848.08	853.85	2.3254
220	.0011866	938.43	944.36	2.5128
240	.0012264	1031.34	1037.47	2.6978
260	.0012748	1127.92	1134.30	2.8829

$P = 10$ MPa

T °C	$\hat{v}$ m³/kg	$\hat{u}$ kJ/kg	$\hat{h}$ kJ/kg	$\hat{s}$ kJ/kg K
0	.0009952	0.10	10.05	0.0003
20	.0009972	83.35	93.32	0.2945
40	.0010034	166.33	176.36	0.5685
60	.0010127	249.34	259.47	0.8258
80	.0010245	332.56	342.81	1.0687
100	.0010385	416.09	426.48	1.2992
120	.0010549	500.07	510.61	1.5188
140	.0010737	584.67	595.40	1.7291
160	.0010953	670.11	681.07	1.9316
180	.0011199	756.63	767.83	2.1274
200	.0011480	844.49.	855.97	2.3178
220	.0011805	934.07	945.88	2.5038
240	.0012187	1025.94	1038.13	2.6872
260	.0012645	1121.03	1133.68	2.8698
280	.0013216	1220.90	1234.11	3.0547
300	.0013972	1328.34	1342.31	3.2468

$P = 15$ MPa

T °C	$\hat{v}$ m³/kg	$\hat{u}$ kJ/kg	$\hat{h}$ kJ/kg	$\hat{s}$ kJ/kg K
0	.0009928	0.15	15.04	0.0004
20	.0009950	83.05	97.97	0.2934
40	.0010013	165.73	180.75	0.5665
60	.0010105	248.49	263.65	0.8231
80	.0010222	331.46	346.79	1.0655
100	.0010361	414.72	430.26	1.2954
120	.0010522	498.39	514.17	1.5144
140	.0010707	582.64	598.70	1.7241
160	.0010918	667.69	684.07	1.9259
180	.0011159	753.74	770.48	2.1209
200	.0011433	841.04	858.18	2.3103
220	.0011748	929.89	947.52	2.4952
240	.0012114	1020.82	1038.99	2.6770
260	.0012550	1114.59	1133.41	2.8575
280	.0013084	1212.47	1232.09	3.0392
300	.0013770	1316.58	1337.23	3.2259
320	.0014724	1431.05	1453.13	3.4246
340	.0016311	1567.42	1591.88	3.6545

$P = 20$ MPa

T °C	$\hat{v}$ m³/kg	$\hat{u}$ kJ/kg	$\hat{h}$ kJ/kg	$\hat{s}$ kJ/kg K
0	.0009904	0.20	20.00	0.0004
20	.0009928	82.75	102.61	0.2922
40	.0009992	165.15	185.14	0.5646
60	.0010084	247.66	267.82	0.8205
80	.0010199	330.38	350.78	1.0623
100	.0010337	413.37	434.04	1.2917
120	.0010496	496.75	517.74	1.5101
140	.0010678	580.67	602.03	1.7192
160	.0010885	665.34	687.11	1.9203
180	.0011120	750.94	773.18	2.1146
200	.0011387	837.70	860.47	2.3031
220	.0011693	925.89	949.27	2.4869
240	.0012046	1015.94	1040.04	2.6673
260	.0012462	1108.53	1133.45	2.8459
280	.0012965	1204.69	1230.62	3.0248
300	.0013596	1306.10	1333.29	3.2071
320	.0014437	1415.66	1444.53	3.3978
340	.0015683	1539.64	1571.01	3.6074
360	.0018226	1702.78	1739.23	3.8770

$P = 30$ MPa

T °C	$\hat{v}$ m³/kg	$\hat{u}$ kJ/kg	$\hat{h}$ kJ/kg	$\hat{s}$ kJ/kg K
0	.0009856	0.25	29.82	0.0001
20	.0009886	82.16	111.82	0.2898
40	.0009951	164.01	193.87	0.5606
60	.0010042	246.03	276.16	0.8153
80	.0010156	328.28	358.75	1.0561
100	.0010290	410.76	441.63	1.2844
120	.0010445	493.58	524.91	1.5017
140	.0010621	576.86	608.73	1.7097
160	.0010821	660.81	693.27	1.9095
180	.0011047	745.57	778.71	2.1024
200	.0011302	831.34	865.24	2.2892
220	.0011590	918.32	953.09	2.4710
240	.0011920	1006.84	1042.60	2.6489
260	.0012303	1097.38	1134.29	2.8242
280	.0012755	1190.69	1228.96	2.9985
300	.0013304	1287.89	1327.80	3.1740
320	.0013997	1390.64	1432.63	3.3538
340	.0014919	1501.71	1546.47	3.5425
360	.0016265	1626.57	1675.36	3.7492
380	.0018691	1781.35	1837.43	4.0010

$P = 50$ MPa

T °C	$\hat{v}$ m³/kg	$\hat{u}$ kJ/kg	$\hat{h}$ kJ/kg	$\hat{s}$ kJ/kg K
0	.0009766	0.20	49.03	−0.0014
20	.0009804	80.98	130.00	0.2847
40	.0009872	161.84	211.20	0.5526
60	.0009962	242.96	292.77	0.8051
80	.0010073	324.32	374.68	1.0439
100	.0010201	405.86	456.87	1.2703
120	.0010348	487.63	539.37	1.4857
140	.0010515	569.76	622.33	1.6915
160	.0010703	652.39	705.91	1.8890
180	.0010912	735.68	790.24	2.0793
200	.0011146	819.73	875.46	2.2634
220	.0011408	904.67	961.71	2.4419
240	.0011702	990.69	1049.20	2.6158
260	.0012034	1078.06	1138.23	2.7860
280	.0012415	1167.19	1229.26	2.9536
300	.0012860	1258.66	1322.95	3.1200
320	.0013388	1353.23	1420.17	3.2867
340	.0014032	1451.91	1522.07	3.4556
360	.0014838	1555.97	1630.16	3.6290
380	.0015883	1667.13	1746.54	3.8100

APPENDIX C

Lee–Kesler Generalized Correlation Tables[1]

TABLE C.1   Values for $z^{(0)}$

T_r	P_r 0.01	0.025	0.05	0.075	0.1	0.25	0.5	0.6	0.7	0.8	0.9	1	1.1
0.3	0.0029	0.0072	0.0145	0.0217	0.0290	0.0724	0.1447	0.1737	0.2026	0.2315	0.2604	0.2892	0.3181
0.35	0.0026	0.0065	0.0130	0.0196	0.0261	0.0652	0.1303	0.1564	0.1824	0.2084	0.2344	0.2604	0.2863
0.4	0.0024	0.0060	0.0119	0.0179	0.0239	0.0596	0.1191	0.1429	0.1667	0.1904	0.2142	0.2379	0.2616
0.45	0.0022	0.0055	0.0110	0.0166	0.0221	0.0552	0.1102	0.1322	0.1542	0.1762	0.1981	0.2200	0.2420
0.5	0.0021	0.0052	0.0103	0.0155	0.0207	0.0516	0.1031	0.1236	0.1441	0.1647	0.1851	0.2056	0.2261
0.55	0.9804	0.0049	0.0098	0.0146	0.0195	0.0487	0.0972	0.1166	0.1360	0.1553	0.1746	0.1939	0.2131
0.6	0.9849	0.9614	0.0093	0.0139	0.0186	0.0463	0.0925	0.1109	0.1293	0.1476	0.1660	0.1842	0.2025
0.65	0.9881	0.9697	0.9377	0.0134	0.0178	0.0445	0.0887	0.1063	0.1239	0.1415	0.1590	0.1765	0.1939
0.7	0.9904	0.9757	0.9504	0.9238	0.8958	0.0430	0.0857	0.1027	0.1197	0.1366	0.1535	0.1703	0.1871
0.75	0.9922	0.9802	0.9598	0.9386	0.9165	0.0420	0.0836	0.1001	0.1166	0.1330	0.1493	0.1656	0.1819
0.8	0.9935	0.9837	0.9669	0.9497	0.9319	0.8093	0.0823	0.0985	0.1147	0.1307	0.1467	0.1626	0.1784
0.85	0.9946	0.9864	0.9725	0.9582	0.9436	0.8465	0.0823	0.0983	0.1143	0.1301	0.1458	0.1614	0.1769
0.9	0.9954	0.9885	0.9768	0.9649	0.9528	0.8739	0.7019	0.1006	0.1164	0.1321	0.1476	0.1630	0.1783
0.93	0.9959	0.9896	0.9790	0.9683	0.9573	0.8871	0.7420	0.6635	0.1204	0.1359	0.1512	0.1664	0.1814
0.95	0.9961	0.9902	0.9803	0.9703	0.9600	0.8948	0.7637	0.6967	0.6107	0.1410	0.1557	0.1705	0.1852
0.97	0.9963	0.9908	0.9815	0.9721	0.9625	0.9018	0.7825	0.7240	0.6538	0.5580	0.1648	0.1779	0.1916
0.98	0.9965	0.9911	0.9821	0.9730	0.9637	0.9051	0.7910	0.7360	0.6714	0.5887	0.1748	0.1844	0.1966
0.99	0.9966	0.9914	0.9826	0.9738	0.9648	0.9082	0.7990	0.7471	0.6873	0.6138	0.5070	0.1959	0.2041
1	0.9967	0.9916	0.9832	0.9746	0.9659	0.9112	0.8065	0.7574	0.7017	0.6353	0.5477	0.2918	0.2167
1.01	0.9968	0.9919	0.9837	0.9754	0.9669	0.9141	0.8136	0.7671	0.7149	0.6542	0.5785	0.4648	0.2486
1.02	0.9969	0.9921	0.9842	0.9761	0.9679	0.9168	0.8204	0.7761	0.7271	0.6710	0.6038	0.5146	0.3597
1.03	0.9969	0.9924	0.9846	0.9768	0.9689	0.9194	0.8267	0.7846	0.7383	0.6863	0.6256	0.5501	0.4439
1.05	0.9971	0.9928	0.9855	0.9781	0.9707	0.9243	0.8385	0.8002	0.7586	0.7130	0.6618	0.6026	0.5312
1.1	0.9975	0.9937	0.9874	0.9811	0.9747	0.9350	0.8634	0.8323	0.7996	0.7649	0.7278	0.6880	0.6449
1.15	0.9978	0.9945	0.9891	0.9835	0.9780	0.9438	0.8833	0.8576	0.8309	0.8032	0.7744	0.7443	0.7129
1.2	0.9981	0.9552	0.9904	0.9856	0.9808	0.9511	0.8994	0.8779	0.8557	0.8330	0.8097	0.7858	0.7613
1.3	0.9985	0.9963	0.9926	0.9889	0.9852	0.9626	0.9240	0.9083	0.8924	0.8764	0.8602	0.8438	0.8275
1.4	0.9988	0.9971	0.9942	0.9913	0.9884	0.9710	0.9416	0.9298	0.9180	0.9062	0.8945	0.8827	0.8710
1.5	0.9991	0.9977	0.9954	0.9932	0.9909	0.9772	0.9546	0.9456	0.9367	0.9278	0.9190	0.9103	0.9018
1.6	0.9993	0.9982	0.9964	0.9946	0.9928	0.9820	0.9644	0.9575	0.9507	0.9439	0.9373	0.9308	0.9243
1.7	0.9994	0.9986	0.9971	0.9957	0.9943	0.9858	0.9721	0.9667	0.9614	0.9563	0.9512	0.9463	0.9414
1.8	0.9995	0.9989	0.9977	0.9966	0.9955	0.9888	0.9780	0.9739	0.9698	0.9659	0.9620	0.9583	0.9546
1.9	0.9996	0.9991	0.9982	0.9973	0.9964	0.9912	0.9828	0.9796	0.9765	0.9735	0.9706	0.9678	0.9650
2	0.9997	0.9993	0.9986	0.9979	0.9972	0.9931	0.9866	0.9842	0.9819	0.9796	0.9774	0.9754	0.9734
2.25	0.9999	0.9996	0.9993	0.9989	0.9986	0.9965	0.9935	0.9924	0.9913	0.9904	0.9895	0.9887	0.9879
2.5	0.9999	0.9999	0.9997	0.9996	0.9994	0.9987	0.9977	0.9975	0.9972	0.9971	0.9970	0.9969	0.9969
2.75	1.0000	1.0000	1.0000	1.0000	1.0000	1.0001	1.0005	1.0008	1.0011	1.0014	1.0018	1.0022	1.0027
3	1.0000	1.0001	1.0002	1.0003	1.0004	1.0011	1.0024	1.0030	1.0036	1.0043	1.0050	1.0057	1.0065
3.5	1.0001	1.0002	1.0004	1.0006	1.0008	1.0022	1.0045	1.0055	1.0065	1.0075	1.0086	1.0097	1.0108
4	1.0001	1.0003	1.0005	1.0008	1.0010	1.0027	1.0055	1.0066	1.0078	1.0090	1.0102	1.0115	1.0127
5	1.0001	1.0003	1.0006	1.0009	1.0012	1.0030	1.0060	1.0073	1.0085	1.0098	1.0111	1.0124	1.0137

[1] As calculated by text software.

TABLE C.1 Continued

T_r	P_r 1.2	1.3	1.4	1.5	1.75	2	2.5	3	4	5	7.5	10
0.3	0.3470	0.3758	0.4047	0.4335	0.5055	0.5775	0.7213	0.8648	1.1512	1.4366	2.1463	2.8507
0.35	0.3123	0.3382	0.3642	0.3901	0.4549	0.5195	0.6487	0.7775	1.0344	1.2902	1.9251	2.5539
0.4	0.2853	0.3090	0.3327	0.3563	0.4154	0.4744	0.5921	0.7095	0.9433	1.1758	1.7519	2.3211
0.45	0.2638	0.2857	0.3076	0.3294	0.3840	0.4384	0.5470	0.6551	0.8704	1.0841	1.6128	2.1338
0.5	0.2465	0.2669	0.2873	0.3077	0.3585	0.4092	0.5103	0.6110	0.8110	1.0094	1.4989	1.9801
0.55	0.2323	0.2515	0.2707	0.2899	0.3377	0.3853	0.4803	0.5747	0.7620	0.9475	1.4042	1.8520
0.6	0.2207	0.2390	0.2571	0.2753	0.3206	0.3657	0.4554	0.5446	0.7213	0.8959	1.3247	1.7440
0.65	0.2113	0.2287	0.2461	0.2634	0.3065	0.3495	0.4349	0.5197	0.6872	0.8526	1.2573	1.6519
0.7	0.2038	0.2205	0.2372	0.2538	0.2952	0.3364	0.4181	0.4991	0.6588	0.8161	1.1999	1.5729
0.75	0.1981	0.2142	0.2303	0.2464	0.2863	0.3260	0.4046	0.4823	0.6352	0.7854	1.1508	1.5047
0.8	0.1942	0.2099	0.2255	0.2411	0.2798	0.3182	0.3942	0.4690	0.6160	0.7598	1.1087	1.4456
0.85	0.1924	0.2077	0.2230	0.2382	0.2759	0.3132	0.3868	0.4591	0.6007	0.7388	1.0727	1.3943
0.9	0.1935	0.2085	0.2235	0.2383	0.2751	0.3114	0.3828	0.4527	0.5892	0.7220	1.0421	1.3496
0.93	0.1963	0.2112	0.2259	0.2405	0.2766	0.3122	0.3822	0.4507	0.5841	0.7138	1.0261	1.3257
0.95	0.1998	0.2144	0.2288	0.2432	0.2787	0.3138	0.3827	0.4501	0.5815	0.7092	1.0164	1.3108
0.97	0.2055	0.2195	0.2334	0.2474	0.2821	0.3164	0.3841	0.4504	0.5796	0.7052	1.0073	1.2968
0.98	0.2097	0.2231	0.2366	0.2503	0.2843	0.3182	0.3851	0.4508	0.5789	0.7035	1.0030	1.2901
0.99	0.2154	0.2278	0.2407	0.2538	0.2871	0.3204	0.3864	0.4514	0.5784	0.7018	0.9989	1.2835
1	0.2237	0.2342	0.2459	0.2583	0.2904	0.3229	0.3880	0.4522	0.5780	0.7004	0.9949	1.2772
1.01	0.2370	0.2432	0.2529	0.2640	0.2944	0.3260	0.3899	0.4533	0.5778	0.6991	0.9912	1.2710
1.02	0.2629	0.2568	0.2624	0.2715	0.2993	0.3297	0.3921	0.4547	0.5778	0.6980	0.9875	1.2650
1.03	0.3168	0.2793	0.2760	0.2813	0.3053	0.3340	0.3948	0.4563	0.5780	0.6970	0.9841	1.2592
1.05	0.4437	0.3630	0.3246	0.3131	0.3219	0.3452	0.4014	0.4604	0.5790	0.6956	0.9776	1.2481
1.1	0.5984	0.5492	0.5003	0.4580	0.4026	0.3953	0.4277	0.4770	0.5851	0.6950	0.9639	1.2232
1.15	0.6803	0.6468	0.6129	0.5798	0.5116	0.4760	0.4718	0.5042	0.5972	0.6987	0.9538	1.2021
1.2	0.7363	0.7110	0.6856	0.6605	0.6029	0.5605	0.5295	0.5425	0.6155	0.7069	0.9471	1.1844
1.3	0.8111	0.7947	0.7784	0.7624	0.7243	0.6908	0.6467	0.6344	0.6681	0.7358	0.9427	1.1580
1.4	0.8595	0.8480	0.8367	0.8256	0.7992	0.7753	0.7387	0.7202	0.7299	0.7761	0.9486	1.1419
1.5	0.8933	0.8850	0.8768	0.8689	0.8499	0.8328	0.8052	0.7887	0.7884	0.8200	0.9619	1.1339
1.6	0.9180	0.9119	0.9059	0.9000	0.8863	0.8738	0.8537	0.8410	0.8386	0.8617	0.9795	1.1320
1.7	0.9367	0.9321	0.9277	0.9234	0.9133	0.9043	0.8899	0.8809	0.8798	0.8984	0.9986	1.1343
1.8	0.9511	0.9477	0.9444	0.9413	0.9339	0.9275	0.9176	0.9118	0.9129	0.9297	1.0174	1.1391
1.9	0.9624	0.9599	0.9575	0.9552	0.9500	0.9456	0.9391	0.9359	0.9396	0.9557	1.0348	1.1452
2	0.9715	0.9697	0.9680	0.9664	0.9628	0.9599	0.9561	0.9550	0.9611	0.9772	1.0503	1.1516
2.25	0.9873	0.9867	0.9861	0.9857	0.9849	0.9846	0.9854	0.9880	0.9986	1.0157	1.0805	1.1661
2.5	0.9970	0.9971	0.9973	0.9976	0.9984	0.9996	1.0031	1.0080	1.0215	1.0395	1.1003	1.1763
2.75	1.0033	1.0038	1.0045	1.0051	1.0070	1.0092	1.0143	1.0205	1.0357	1.0543	1.1125	1.1823
3	1.0074	1.0082	1.0091	1.0101	1.0126	1.0153	1.0215	1.0284	1.0446	1.0635	1.1196	1.1848
3.5	1.0120	1.0131	1.0143	1.0156	1.0187	1.0221	1.0292	1.0368	1.0537	1.0723	1.1249	1.1834
4	1.0140	1.0153	1.0166	1.0179	1.0214	1.0249	1.0323	1.0401	1.0567	1.0747	1.1239	1.1773
5	1.0150	1.0163	1.0176	1.0190	1.0224	1.0259	1.0331	1.0405	1.0559	1.0722	1.1153	1.1611

TABLE C.2 Values for $z^{(1)}$

T_r	P_r												
	0.01	0.025	0.05	0.075	0.1	0.25	0.5	0.6	0.7	0.8	0.9	1	1.1
0.3	-0.0008	-0.0020	-0.0040	-0.0061	-0.0081	-0.0202	-0.0403	-0.0484	-0.0564	-0.0645	-0.0725	-0.0806	-0.0886
0.35	-0.0009	-0.0023	-0.0046	-0.0069	-0.0093	-0.0231	-0.0462	-0.0554	-0.0646	-0.0738	-0.0830	-0.0921	-0.1013
0.4	-0.0010	-0.0024	-0.0048	-0.0071	-0.0095	-0.0238	-0.0475	-0.0570	-0.0664	-0.0758	-0.0852	-0.0946	-0.1040
0.45	-0.0009	-0.0023	-0.0047	-0.0070	-0.0094	-0.0234	-0.0467	-0.0560	-0.0652	-0.0745	-0.0837	-0.0929	-0.1021
0.5	-0.0009	-0.0023	-0.0045	-0.0068	-0.0090	-0.0226	-0.0450	-0.0539	-0.0628	-0.0716	-0.0805	-0.0893	-0.0981
0.55	-0.0009	-0.0022	-0.0043	-0.0065	-0.0086	-0.0215	-0.0428	-0.0513	-0.0598	-0.0682	-0.0766	-0.0849	-0.0932
0.6	-0.0314	-0.0022	-0.0041	-0.0062	-0.0082	-0.0204	-0.0406	-0.0487	-0.0566	-0.0646	-0.0725	-0.0803	-0.0882
0.65	-0.0205	-0.0543	-0.0772	-0.0059	-0.0078	-0.0194	-0.0385	-0.0461	-0.0536	-0.0611	-0.0685	-0.0759	-0.0833
0.7	-0.0137	-0.0357	-0.0507	-0.0809	-0.1161	-0.0185	-0.0366	-0.0438	-0.0509	-0.0579	-0.0649	-0.0718	-0.0787
0.75	-0.0093	-0.0240	-0.0339	-0.0531	-0.0744	-0.1650	-0.0350	-0.0417	-0.0484	-0.0550	-0.0616	-0.0681	-0.0745
0.8	-0.0064	-0.0163	-0.0228	-0.0353	-0.0487	-0.0963	-0.0337	-0.0401	-0.0464	-0.0526	-0.0588	-0.0648	-0.0708
0.85	-0.0044	-0.0111	-0.0152	-0.0234	-0.0319	-0.0577	-0.0331	-0.0391	-0.0451	-0.0509	-0.0566	-0.0622	-0.0677
0.9	-0.0029	-0.0075	-0.0099	-0.0151	-0.0205	-0.0420	-0.1777	-0.0396	-0.0451	-0.0503	-0.0554	-0.0604	-0.0653
0.93	-0.0019	-0.0049	-0.0075	-0.0114	-0.0154	-0.0336	-0.1088	-0.1662	-0.0469	-0.0514	-0.0558	-0.0602	-0.0645
0.95	-0.0015	-0.0037	-0.0062	-0.0093	-0.0126	-0.0264	-0.0805	-0.1110	-0.1747	-0.0540	-0.0572	-0.0607	-0.0642
0.97	-0.0012	-0.0031	-0.0050	-0.0075	-0.0101	-0.0233	-0.0594	-0.0770	-0.1023	-0.1647	-0.0616	-0.0623	-0.0643
0.98	-0.0010	-0.0025	-0.0044	-0.0067	-0.0090	-0.0203	-0.0507	-0.0641	-0.0812	-0.1100	-0.0690	-0.0641	-0.0644
0.99	-0.0009	-0.0022	-0.0039	-0.0059	-0.0079	-0.0176	-0.0429	-0.0531	-0.0646	-0.0796	-0.1143	-0.0680	-0.0641
1	-0.0008	-0.0020	-0.0034	-0.0052	-0.0069	-0.0151	-0.0360	-0.0435	-0.0511	-0.0588	-0.0665	-0.0789	-0.0607
1.01	-0.0007	-0.0017	-0.0030	-0.0045	-0.0060	-0.0127	-0.0297	-0.0351	-0.0398	-0.0429	-0.0421	-0.0223	0.0082
1.02	-0.0006	-0.0015	-0.0026	-0.0039	-0.0051	-0.0105	-0.0241	-0.0277	-0.0301	-0.0303	-0.0256	-0.0062	0.0895
1.03	-0.0005	-0.0013	-0.0022	-0.0033	-0.0043	-0.0066	-0.0190	-0.0211	-0.0217	-0.0198	-0.0130	0.0053	0.0588
1.05	-0.0004	-0.0011	-0.0015	-0.0022	-0.0029	-0.0066	-0.0100	-0.0097	-0.0078	-0.0032	0.0056	0.0220	0.0528
1.1	-0.0003	-0.0007	0.0000	0.0001	0.0001	0.0012	0.0066	0.0106	0.0161	0.0236	0.0338	0.0476	0.0659
1.15	0.0000	0.0000	0.0011	0.0017	0.0023	0.0068	0.0177	0.0237	0.0309	0.0396	0.0500	0.0625	0.0772
1.2	0.0002	0.0005	0.0019	0.0029	0.0040	0.0108	0.0254	0.0326	0.0407	0.0499	0.0603	0.0719	0.0848
1.3	0.0004	0.0009	0.0030	0.0045	0.0061	0.0159	0.0345	0.0429	0.0518	0.0612	0.0713	0.0819	0.0931
1.4	0.0006	0.0015	0.0036	0.0054	0.0072	0.0185	0.0390	0.0477	0.0567	0.0661	0.0757	0.0857	0.0959
1.5	0.0007	0.0018	0.0039	0.0058	0.0078	0.0198	0.0409	0.0497	0.0586	0.0677	0.0770	0.0864	0.0959
1.6	0.0008	0.0019	0.0040	0.0060	0.0080	0.0204	0.0415	0.0501	0.0589	0.0677	0.0766	0.0855	0.0945
1.7	0.0008	0.0020	0.0040	0.0061	0.0081	0.0204	0.0413	0.0497	0.0582	0.0667	0.0752	0.0838	0.0923
1.8	0.0008	0.0020	0.0040	0.0060	0.0081	0.0202	0.0406	0.0488	0.0570	0.0652	0.0734	0.0816	0.0897
1.9	0.0008	0.0020	0.0040	0.0059	0.0079	0.0199	0.0397	0.0477	0.0556	0.0635	0.0714	0.0792	0.0870
2	0.0008	0.0019	0.0039	0.0058	0.0078	0.0194	0.0387	0.0464	0.0541	0.0617	0.0692	0.0767	0.0842
2.25	0.0007	0.0018	0.0036	0.0055	0.0073	0.0181	0.0360	0.0431	0.0501	0.0570	0.0640	0.0708	0.0776
2.5	0.0007	0.0017	0.0034	0.0051	0.0068	0.0168	0.0334	0.0399	0.0464	0.0528	0.0591	0.0654	0.0716
2.75	0.0006	0.0016	0.0032	0.0047	0.0063	0.0156	0.0310	0.0370	0.0430	0.0490	0.0548	0.0607	0.0664
3	0.0006	0.0015	0.0029	0.0044	0.0059	0.0146	0.0289	0.0345	0.0401	0.0456	0.0511	0.0565	0.0619
3.5	0.0005	0.0013	0.0026	0.0039	0.0052	0.0128	0.0254	0.0303	0.0352	0.0401	0.0449	0.0497	0.0544
4	0.0005	0.0012	0.0023	0.0034	0.0046	0.0114	0.0226	0.0270	0.0314	0.0357	0.0400	0.0443	0.0485
5	0.0004	0.0009	0.0019	0.0028	0.0038	0.0094	0.0186	0.0222	0.0258	0.0294	0.0329	0.0365	0.0400

TABLE C.2 Continued

T_r	P_r 1.2	1.3	1.4	1.5	1.75	2	2.5	3	4	5	7.5	10
0.3	-0.0966	-0.1047	-0.1127	-0.1207	-0.1408	-0.1608	-0.2008	-0.2407	-0.3203	-0.3996	-0.5965	-0.7915
0.35	-0.1105	-0.1196	-0.1287	-0.1379	-0.1607	-0.1834	-0.2287	-0.2738	-0.3634	-0.4523	-0.6713	-0.8863
0.4	-0.1134	-0.1228	-0.1321	-0.1414	-0.1647	-0.1879	-0.2341	-0.2799	-0.3707	-0.4603	-0.6799	-0.8936
0.45	-0.1113	-0.1204	-0.1296	-0.1387	-0.1614	-0.1840	-0.2289	-0.2734	-0.3612	-0.4475	-0.6577	-0.8606
0.5	-0.1069	-0.1156	-0.1243	-0.1330	-0.1547	-0.1762	-0.2189	-0.2611	-0.3440	-0.4253	-0.6216	-0.8099
0.55	-0.1015	-0.1098	-0.1180	-0.1263	-0.1467	-0.1669	-0.2070	-0.2465	-0.3238	-0.3991	-0.5800	-0.7521
0.6	-0.0960	-0.1037	-0.1115	-0.1192	-0.1383	-0.1572	-0.1945	-0.2312	-0.3026	-0.3718	-0.5369	-0.6929
0.65	-0.0906	-0.0978	-0.1051	-0.1123	-0.1301	-0.1476	-0.1822	-0.2160	-0.2816	-0.3447	-0.4943	-0.6346
0.7	-0.0855	-0.0923	-0.0990	-0.1057	-0.1222	-0.1385	-0.1703	-0.2013	-0.2611	-0.3184	-0.4531	-0.5785
0.75	-0.0808	-0.0871	-0.0934	-0.0996	-0.1149	-0.1298	-0.1590	-0.1872	-0.2414	-0.2929	-0.4134	-0.5250
0.8	-0.0767	-0.0825	-0.0883	-0.0940	-0.1080	-0.1217	-0.1481	-0.1736	-0.2222	-0.2682	-0.3752	-0.4740
0.85	-0.0731	-0.0784	-0.0837	-0.0888	-0.1015	-0.1138	-0.1375	-0.1602	-0.2032	-0.2439	-0.3383	-0.4254
0.9	-0.0701	-0.0748	-0.0795	-0.0840	-0.0951	-0.1059	-0.1265	-0.1463	-0.1839	-0.2195	-0.3022	-0.3788
0.93	-0.0687	-0.0729	-0.0770	-0.0810	-0.0910	-0.1007	-0.1194	-0.1374	-0.1718	-0.2045	-0.2808	-0.3516
0.95	-0.0678	-0.0715	-0.0751	-0.0788	-0.0878	-0.0967	-0.1141	-0.1310	-0.1634	-0.1943	-0.2666	-0.3339
0.97	-0.0669	-0.0698	-0.0728	-0.0759	-0.0840	-0.0921	-0.1082	-0.1240	-0.1545	-0.1837	-0.2524	-0.3163
0.98	-0.0661	-0.0685	-0.0712	-0.0740	-0.0816	-0.0893	-0.1049	-0.1202	-0.1499	-0.1783	-0.2452	-0.3075
0.99	-0.0646	-0.0665	-0.0688	-0.0715	-0.0787	-0.0861	-0.1013	-0.1162	-0.1451	-0.1728	-0.2381	-0.2989
1	-0.0609	-0.0628	-0.0652	-0.0678	-0.0750	-0.0824	-0.0972	-0.1118	-0.1401	-0.1672	-0.2309	-0.2902
1.01	-0.0473	-0.0545	-0.0586	-0.0621	-0.0702	-0.0778	-0.0927	-0.1072	-0.1349	-0.1615	-0.2237	-0.2816
1.02	0.0227	-0.0310	-0.0452	-0.0524	-0.0635	-0.0722	-0.0876	-0.1021	-0.1295	-0.1556	-0.2165	-0.2731
1.03	0.1159	0.0318	-0.0152	-0.0343	-0.0540	-0.0649	-0.0818	-0.0966	-0.1239	-0.1495	-0.2092	-0.2646
1.05	0.1059	0.1357	0.0951	0.0451	-0.0195	-0.0432	-0.0671	-0.0838	-0.1118	-0.1370	-0.1946	-0.2476
1.1	0.0897	0.1185	0.1468	0.1630	0.1300	0.0698	-0.0033	-0.0373	-0.0751	-0.1021	-0.1572	-0.2056
1.15	0.0943	0.1137	0.1345	0.1548	0.1835	0.1667	0.0906	0.0332	-0.0272	-0.0611	-0.1185	-0.1642
1.2	0.0991	0.1146	0.1310	0.1477	0.1840	0.1990	0.1651	0.1095	0.0304	-0.0141	-0.0782	-0.1231
1.3	0.1048	0.1169	0.1294	0.1420	0.1729	0.1991	0.2223	0.2079	0.1435	0.0875	0.0053	-0.0423
1.4	0.1063	0.1169	0.1276	0.1383	0.1648	0.1894	0.2259	0.2397	0.2171	0.1737	0.0870	0.0350
1.5	0.1055	0.1152	0.1248	0.1345	0.1582	0.1806	0.2186	0.2433	0.2525	0.2309	0.1584	0.1058
1.6	0.1035	0.1124	0.1214	0.1303	0.1521	0.1729	0.2098	0.2381	0.2654	0.2631	0.2147	0.1673
1.7	0.1008	0.1092	0.1176	0.1259	0.1463	0.1658	0.2013	0.2305	0.2672	0.2788	0.2556	0.2179
1.8	0.0978	0.1058	0.1137	0.1216	0.1408	0.1593	0.1932	0.2224	0.2639	0.2846	0.2834	0.2576
1.9	0.0947	0.1023	0.1099	0.1173	0.1356	0.1532	0.1858	0.2144	0.2582	0.2848	0.3013	0.2876
2	0.0916	0.0989	0.1061	0.1133	0.1307	0.1476	0.1789	0.2069	0.2517	0.2820	0.3120	0.3096
2.25	0.0843	0.0909	0.0975	0.1039	0.1198	0.1350	0.1638	0.1901	0.2346	0.2691	0.3198	0.3392
2.5	0.0778	0.0839	0.0899	0.0958	0.1104	0.1245	0.1511	0.1757	0.2189	0.2542	0.3146	0.3475
2.75	0.0721	0.0778	0.0833	0.0888	0.1023	0.1154	0.1403	0.1635	0.2049	0.2399	0.3045	0.3454
3	0.0672	0.0724	0.0776	0.0828	0.0954	0.1076	0.1310	0.1529	0.1925	0.2268	0.2930	0.3385
3.5	0.0591	0.0637	0.0683	0.0728	0.0840	0.0949	0.1158	0.1356	0.1719	0.2042	0.2700	0.3194
4	0.0527	0.0569	0.0610	0.0651	0.0751	0.0849	0.1038	0.1219	0.1554	0.1857	0.2493	0.2994
5	0.0434	0.0469	0.0503	0.0537	0.0621	0.0703	0.0863	0.1016	0.1306	0.1573	0.2155	0.2637

TABLE C.3 Values for $\left[\dfrac{h_{T_r,P_r} - h_{T_r,P_r}^{\text{ideal gas}}}{RT_c}\right]^{(0)}$

T_r \ P_r	0.01	0.025	0.05	0.075	0.1	0.25	0.5	0.6	0.7	0.8	0.9	1	1.1
0.3	−6.046	−6.045	−6.043	−6.042	−6.040	−6.031	−6.017	−6.011	−6.005	−5.999	−5.993	−5.987	−5.981
0.35	−5.907	−5.906	−5.904	−5.903	−5.901	−5.892	−5.876	−5.870	−5.864	−5.858	−5.852	−5.845	−5.839
0.4	−5.763	−5.762	−5.761	−5.759	−5.757	−5.748	−5.732	−5.726	−5.719	−5.713	−5.707	−5.700	−5.694
0.45	−5.615	−5.614	−5.612	−5.611	−5.609	−5.600	−5.583	−5.577	−5.571	−5.564	−5.558	−5.551	−5.545
0.5	−5.465	−5.464	−5.462	−5.461	−5.459	−5.450	−5.434	−5.427	−5.421	−5.414	−5.408	−5.401	−5.395
0.55	−0.032	−5.314	−5.312	−5.311	−5.309	−5.300	−5.284	−5.277	−5.271	−5.265	−5.258	−5.252	−5.245
0.6	−0.027	−0.068	−5.162	−5.161	−5.159	−5.150	−5.135	−5.129	−5.122	−5.116	−5.110	−5.104	−5.098
0.65	−0.023	−0.058	−0.118	−5.009	−5.008	−5.000	−4.985	−4.980	−4.974	−4.968	−4.962	−4.956	−4.951
0.7	−0.020	−0.050	−0.101	−0.156	−0.213	−4.846	−4.834	−4.828	−4.823	−4.818	−4.813	−4.808	−4.802
0.75	−0.017	−0.043	−0.088	−0.135	−0.183	−4.685	−4.676	−4.672	−4.668	−4.664	−4.659	−4.655	−4.651
0.8	−0.015	−0.038	−0.078	−0.118	−0.160	−0.451	−4.505	−4.504	−4.501	−4.499	−4.496	−4.494	−4.491
0.85	−0.014	−0.034	−0.069	−0.105	−0.141	−0.387	−4.311	−4.313	−4.315	−4.316	−4.316	−4.316	−4.316
0.9	−0.012	−0.031	−0.062	−0.094	−0.126	−0.339	−0.813	−4.074	−4.085	−4.094	−4.101	−4.108	−4.113
0.93	−0.011	−0.029	−0.058	−0.088	−0.118	−0.315	−0.729	−0.960	−3.898	−3.920	−3.938	−3.953	−3.965
0.95	−0.011	−0.028	−0.056	−0.084	−0.113	−0.300	−0.684	−0.885	−1.150	−3.763	−3.798	−3.825	−3.847
0.97	−0.011	−0.027	−0.054	−0.081	−0.109	−0.287	−0.645	−0.824	−1.044	−1.356	−3.599	−3.658	−3.700
0.98	−0.010	−0.026	−0.053	−0.079	−0.107	−0.281	−0.627	−0.797	−1.002	−1.273	−3.434	−3.544	−3.607
0.99	−0.010	−0.026	−0.052	−0.078	−0.105	−0.275	−0.610	−0.773	−0.964	−1.206	−1.579	−3.376	−3.491
1	−0.010	−0.025	−0.051	−0.076	−0.103	−0.269	−0.594	−0.750	−0.930	−1.151	−1.455	−2.574	−3.326
1.01	−0.010	−0.025	−0.050	−0.075	−0.101	−0.264	−0.579	−0.728	−0.899	−1.102	−1.366	−1.796	−3.014
1.02	−0.010	−0.024	−0.049	−0.074	−0.099	−0.258	−0.565	−0.708	−0.870	−1.060	−1.295	−1.627	−2.318
1.03	−0.009	−0.024	−0.048	−0.072	−0.097	−0.253	−0.551	−0.689	−0.844	−1.022	−1.235	−1.515	−1.953
1.05	−0.009	−0.023	−0.046	−0.070	−0.094	−0.243	−0.525	−0.654	−0.796	−0.955	−1.138	−1.359	−1.642
1.1	−0.008	−0.021	−0.042	−0.064	−0.086	−0.221	−0.471	−0.581	−0.699	−0.827	−0.966	−1.120	−1.292
1.15	−0.008	−0.019	−0.039	−0.059	−0.079	−0.202	−0.426	−0.523	−0.625	−0.732	−0.846	−0.968	−1.098
1.2	−0.007	−0.018	−0.036	−0.054	−0.073	−0.186	−0.388	−0.474	−0.564	−0.657	−0.755	−0.857	−0.964
1.3	−0.006	−0.016	−0.031	−0.047	−0.063	−0.159	−0.328	−0.399	−0.471	−0.545	−0.620	−0.698	−0.778
1.4	−0.005	−0.014	−0.027	−0.041	−0.055	−0.138	−0.282	−0.341	−0.402	−0.463	−0.525	−0.588	−0.652
1.5	−0.005	−0.012	−0.024	−0.036	−0.048	−0.121	−0.246	−0.297	−0.348	−0.400	−0.452	−0.505	−0.558
1.6	−0.004	−0.011	−0.021	−0.032	−0.043	−0.107	−0.216	−0.261	−0.305	−0.350	−0.395	−0.440	−0.485
1.7	−0.004	−0.010	−0.019	−0.029	−0.038	−0.096	−0.192	−0.231	−0.270	−0.309	−0.348	−0.387	−0.427
1.8	−0.003	−0.009	−0.017	−0.026	−0.034	−0.086	−0.172	−0.206	−0.240	−0.275	−0.309	−0.344	−0.378
1.9	−0.003	−0.008	−0.015	−0.023	−0.031	−0.077	−0.154	−0.185	−0.216	−0.246	−0.277	−0.307	−0.338
2	−0.003	−0.007	−0.014	−0.021	−0.028	−0.070	−0.139	−0.167	−0.194	−0.222	−0.249	−0.276	−0.303
2.25	−0.002	−0.006	−0.011	−0.017	−0.022	−0.055	−0.109	−0.130	−0.152	−0.173	−0.194	−0.215	−0.236
2.5	−0.002	−0.004	−0.009	−0.013	−0.018	−0.044	−0.087	−0.104	−0.121	−0.137	−0.154	−0.170	−0.187
2.75	−0.001	−0.004	−0.007	−0.011	−0.014	−0.035	−0.070	−0.083	−0.097	−0.110	−0.123	−0.136	−0.149
3	−0.001	−0.003	−0.006	−0.009	−0.011	−0.028	−0.056	−0.067	−0.078	−0.088	−0.099	−0.109	−0.119
3.5	−0.001	−0.002	−0.004	−0.006	−0.007	−0.018	−0.036	−0.043	−0.049	−0.056	−0.063	−0.069	−0.075
4	0.000	−0.001	−0.002	−0.003	−0.005	−0.011	−0.022	−0.026	−0.030	−0.033	−0.037	−0.041	−0.044
5	0.000	0.000	0.000	−0.001	−0.001	−0.002	−0.003	−0.003	−0.004	−0.004	−0.004	−0.004	−0.004

TABLE C.3 Continued

T_r	P_r 1.2	1.3	1.4	1.5	1.75	2	2.5	3	4	5	7.5	10
0.3	−5.975	−5.969	−5.963	−5.957	−5.942	−5.927	−5.898	−5.868	−5.808	−5.748	−5.598	−5.446
0.35	−5.833	−5.827	−5.821	−5.814	−5.799	−5.783	−5.752	−5.721	−5.658	−5.595	−5.437	−5.278
0.4	−5.687	−5.681	−5.675	−5.668	−5.652	−5.636	−5.604	−5.572	−5.507	−5.442	−5.278	−5.113
0.45	−5.538	−5.532	−5.525	−5.519	−5.502	−5.486	−5.453	−5.420	−5.354	−5.288	−5.120	−4.950
0.5	−5.388	−5.382	−5.375	−5.369	−5.352	−5.336	−5.303	−5.270	−5.203	−5.135	−4.964	−4.791
0.55	−5.239	−5.233	−5.226	−5.220	−5.203	−5.187	−5.154	−5.121	−5.054	−4.986	−4.814	−4.638
0.6	−5.091	−5.085	−5.079	−5.073	−5.057	−5.041	−5.008	−4.976	−4.909	−4.842	−4.669	−4.492
0.65	−4.945	−4.939	−4.933	−4.927	−4.911	−4.896	−4.865	−4.833	−4.769	−4.702	−4.531	−4.353
0.7	−4.797	−4.792	−4.786	−4.781	−4.767	−4.752	−4.723	−4.693	−4.631	−4.566	−4.397	−4.221
0.75	−4.646	−4.641	−4.637	−4.632	−4.620	−4.607	−4.581	−4.554	−4.495	−4.434	−4.269	−4.095
0.8	−4.488	−4.485	−4.481	−4.478	−4.469	−4.459	−4.437	−4.413	−4.361	−4.303	−4.145	−3.974
0.85	−4.316	−4.315	−4.314	−4.312	−4.308	−4.302	−4.287	−4.269	−4.225	−4.173	−4.024	−3.857
0.9	−4.118	−4.121	−4.125	−4.127	−4.131	−4.132	−4.129	−4.119	−4.086	−4.043	−3.905	−3.744
0.93	−3.976	−3.985	−3.993	−4.000	−4.012	−4.020	−4.026	−4.024	−4.001	−3.963	−3.834	−3.678
0.95	−3.865	−3.880	−3.893	−3.904	−3.925	−3.939	−3.955	−3.958	−3.943	−3.910	−3.788	−3.634
0.97	−3.732	−3.758	−3.779	−3.796	−3.830	−3.853	−3.879	−3.890	−3.883	−3.856	−3.741	−3.591
0.98	−3.652	−3.686	−3.714	−3.736	−3.778	−3.806	−3.840	−3.854	−3.853	−3.829	−3.717	−3.569
0.99	−3.558	−3.605	−3.641	−3.670	−3.723	−3.758	−3.799	−3.818	−3.823	−3.801	−3.694	−3.548
1	−3.441	−3.510	−3.560	−3.598	−3.664	−3.706	−3.757	−3.782	−3.792	−3.774	−3.670	−3.526
1.01	−3.283	−3.395	−3.465	−3.516	−3.600	−3.652	−3.713	−3.744	−3.760	−3.746	−3.647	−3.505
1.02	−3.039	−3.246	−3.352	−3.422	−3.530	−3.595	−3.668	−3.705	−3.729	−3.718	−3.624	−3.484
1.03	−2.657	−3.043	−3.213	−3.313	−3.454	−3.534	−3.621	−3.665	−3.696	−3.690	−3.600	−3.462
1.05	−2.034	−2.497	−2.831	−3.030	−3.277	−3.398	−3.521	−3.583	−3.630	−3.632	−3.553	−3.420
1.1	−1.487	−1.709	−1.955	−2.203	−2.686	−2.965	−3.231	−3.353	−3.456	−3.484	−3.435	−3.315
1.15	−1.239	−1.389	−1.550	−1.719	−2.139	−2.479	−2.888	−3.091	−3.268	−3.329	−3.315	−3.211
1.2	−1.076	−1.193	−1.315	−1.443	−1.770	−2.079	−2.537	−2.807	−3.065	−3.166	−3.194	−3.107
1.3	−0.860	−0.943	−1.029	−1.116	−1.338	−1.560	−1.964	−2.274	−2.645	−2.825	−2.947	−2.899
1.4	−0.716	−0.782	−0.848	−0.915	−1.084	−1.253	−1.576	−1.857	−2.255	−2.486	−2.696	−2.692
1.5	−0.611	−0.665	−0.719	−0.774	−0.910	−1.046	−1.309	−1.549	−1.926	−2.175	−2.449	−2.486
1.6	−0.531	−0.576	−0.622	−0.667	−0.781	−0.894	−1.114	−1.318	−1.659	−1.904	−2.213	−2.285
1.7	−0.466	−0.505	−0.544	−0.583	−0.681	−0.777	−0.964	−1.139	−1.441	−1.672	−1.994	−2.091
1.8	−0.413	−0.447	−0.481	−0.515	−0.600	−0.683	−0.844	−0.996	−1.264	−1.476	−1.794	−1.908
1.9	−0.368	−0.398	−0.429	−0.458	−0.533	−0.606	−0.747	−0.880	−1.117	−1.309	−1.614	−1.736
2	−0.330	−0.357	−0.384	−0.411	−0.476	−0.541	−0.665	−0.782	−0.993	−1.167	−1.453	−1.577
2.25	−0.257	−0.277	−0.297	−0.318	−0.367	−0.416	−0.509	−0.597	−0.755	−0.889	−1.120	−1.232
2.5	−0.203	−0.219	−0.234	−0.250	−0.289	−0.326	−0.398	−0.465	−0.585	−0.687	−0.866	−0.954
2.75	−0.162	−0.174	−0.187	−0.199	−0.229	−0.258	−0.314	−0.365	−0.457	−0.534	−0.667	−0.729
3	−0.129	−0.139	−0.149	−0.159	−0.182	−0.205	−0.248	−0.288	−0.357	−0.415	−0.509	−0.545
3.5	−0.081	−0.087	−0.093	−0.099	−0.113	−0.127	−0.152	−0.174	−0.211	−0.239	−0.272	−0.264
4	−0.048	−0.051	−0.054	−0.058	−0.065	−0.072	−0.085	−0.095	−0.109	−0.116	−0.105	−0.061
5	−0.004	−0.004	−0.004	−0.003	−0.002	−0.001	0.003	0.009	0.024	0.045	0.117	0.213

TABLE C.4 Values for $\left[\dfrac{h_{T_r,P_r} - h^{\text{ideal gas}}_{T_r,P_r}}{RT_c}\right]^{(1)}$

T_r \ P_r	0.01	0.025	0.05	0.075	0.1	0.25	0.5	0.6	0.7	0.8	0.9	1	1.1
0.3	−11.101	−11.101	−11.100	−11.099	−11.098	−11.093	−11.084	−11.081	−11.078	−11.074	−11.071	−11.067	−11.064
0.35	−10.652	−10.652	−10.651	−10.651	−10.651	−10.648	−10.645	−10.643	−10.642	−10.640	−10.639	−10.637	−10.636
0.4	−10.120	−10.120	−10.120	−10.120	−10.120	−10.120	−10.120	−10.120	−10.120	−10.120	−10.120	−10.120	−10.120
0.45	−9.513	−9.514	−9.514	−9.514	−9.514	−9.516	−9.519	−9.520	−9.521	−9.522	−9.523	−9.525	−9.526
0.5	−8.867	−8.867	−8.868	−8.868	−8.869	−8.872	−8.877	−8.879	−8.881	−8.883	−8.885	−8.888	−8.889
0.55	−0.059	−8.213	−8.213	−8.213	−8.214	−8.218	−8.224	−8.227	−8.230	−8.232	−8.235	−8.238	−8.241
0.6	−0.045	−0.155	−0.247	−7.569	−7.570	−7.574	−7.582	−7.585	−7.588	−7.592	−7.595	−7.598	−7.601
0.65	−0.034	−0.116	−0.185	−0.292	−6.949	−6.954	−6.962	−6.966	−6.969	−6.973	−6.976	−6.980	−6.984
0.7	−0.027	−0.088	−0.142	−0.220	−0.415	−6.361	−6.370	−6.373	−6.377	−6.381	−6.385	−6.388	−6.392
0.75	−0.021	−0.068	−0.110	−0.170	−0.306	−5.798	−5.806	−5.809	−5.813	−5.816	−5.820	−5.824	−5.828
0.8	−0.017	−0.054	−0.087	−0.134	−0.234	−0.753	−5.268	−5.271	−5.274	−5.277	−5.281	−5.285	−5.288
0.85	−0.014	−0.043	−0.070	−0.106	−0.182	−0.534	−4.753	−4.754	−4.756	−4.758	−4.760	−4.763	−4.767
0.9	−0.012	−0.034	−0.061	−0.093	−0.144	−0.400	−1.133	−4.254	−4.250	−4.248	−4.248	−4.249	−4.251
0.93	−0.011	−0.030	−0.056	−0.085	−0.126	−0.341	−0.855	−1.236	−3.952	−3.941	−3.936	−3.934	−3.934
0.95	−0.010	−0.028	−0.052	−0.078	−0.115	−0.309	−0.736	−0.994	−1.448	−3.737	−3.720	−3.713	−3.711
0.97	−0.010	−0.026	−0.050	−0.075	−0.105	−0.280	−0.643	−0.837	−1.100	−1.616	−3.496	−3.471	−3.465
0.98	−0.009	−0.025	−0.048	−0.072	−0.101	−0.267	−0.603	−0.776	−0.994	−1.324	−3.397	−3.332	−3.322
0.99	−0.009	−0.024	−0.046	−0.069	−0.097	−0.254	−0.568	−0.722	−0.908	−1.154	−1.618	−3.164	−3.150
1	−0.009	−0.023	−0.044	−0.066	−0.093	−0.243	−0.535	−0.675	−0.836	−1.034	−1.307	−2.382	−2.888
1.01	−0.008	−0.022	−0.042	−0.063	−0.089	−0.232	−0.505	−0.632	−0.775	−0.940	−1.138	−1.375	−1.866
1.02	−0.008	−0.021	−0.040	−0.061	−0.085	−0.221	−0.478	−0.594	−0.721	−0.863	−1.020	−1.180	−1.078
1.03	−0.007	−0.020	−0.037	−0.056	−0.082	−0.211	−0.452	−0.559	−0.674	−0.797	−0.928	−1.052	−1.080
1.05	−0.006	−0.019	−0.030	−0.046	−0.075	−0.193	−0.406	−0.498	−0.593	−0.691	−0.789	−0.877	−0.928
1.1	−0.005	−0.015	−0.025	−0.037	−0.061	−0.155	−0.316	−0.381	−0.445	−0.507	−0.566	−0.617	−0.655
1.15	−0.004	−0.012	−0.020	−0.030	−0.050	−0.124	−0.248	−0.296	−0.342	−0.385	−0.425	−0.459	−0.486
1.2	−0.003	−0.010	−0.013	−0.020	−0.040	−0.100	−0.196	−0.232	−0.266	−0.297	−0.325	−0.349	−0.368
1.3	−0.002	−0.007	−0.008	−0.012	−0.026	−0.064	−0.122	−0.142	−0.161	−0.177	−0.191	−0.203	−0.212
1.4	−0.002	−0.004	−0.005	−0.007	−0.016	−0.039	−0.072	−0.083	−0.093	−0.100	−0.107	−0.111	−0.114
1.5	0.000	−0.002	−0.002	−0.003	−0.009	−0.022	−0.038	−0.042	−0.046	−0.048	−0.049	−0.049	−0.048
1.6	0.000	−0.001	0.000	0.000	−0.004	−0.009	−0.013	−0.013	−0.012	−0.011	−0.008	−0.005	−0.001
1.7	0.001	0.000	0.001	0.002	0.000	0.001	0.006	0.009	0.012	0.017	0.021	0.027	0.033
1.8	0.001	0.001	0.003	0.004	0.003	0.008	0.019	0.025	0.031	0.037	0.044	0.051	0.059
1.9	0.001	0.001	0.004	0.005	0.005	0.014	0.030	0.037	0.045	0.053	0.061	0.070	0.079
2	0.001	0.002	0.005	0.008	0.007	0.018	0.039	0.047	0.056	0.065	0.075	0.085	0.094
2.25	0.001	0.003	0.006	0.009	0.010	0.026	0.053	0.064	0.075	0.087	0.098	0.110	0.121
2.5	0.001	0.003	0.007	0.010	0.012	0.031	0.062	0.075	0.087	0.100	0.112	0.125	0.138
2.75	0.001	0.003	0.007	0.011	0.014	0.034	0.068	0.081	0.095	0.108	0.122	0.135	0.149
3	0.002	0.004	0.008	0.012	0.014	0.036	0.072	0.086	0.100	0.114	0.128	0.142	0.156
3.5	0.002	0.004	0.008	0.012	0.016	0.039	0.077	0.092	0.107	0.122	0.137	0.152	0.167
4	0.002	0.004	0.008	0.012	0.016	0.040	0.08	0.096	0.112	0.127	0.143	0.158	0.173
5	0.002	0.004	0.009	0.013	0.017	0.042	0.084	0.101	0.117	0.133	0.15	0.166	0.182

TABLE C.4   Continued

T_r	P_r 1.2	1.3	1.4	1.5	1.75	2	2.5	3	4	5	7.5	10
0.3	−11.061	−11.057	−11.054	−11.051	−11.042	−11.034	−11.017	−11.001	−10.968	−10.936	−10.857	−10.782
0.35	−10.634	−10.633	−10.632	−10.630	−10.627	−10.623	−10.616	−10.610	−10.597	−10.584	−10.555	−10.529
0.4	−10.120	−10.120	−10.120	−10.120	−10.120	−10.121	−10.121	−10.122	−10.124	−10.127	−10.137	−10.150
0.45	−9.527	−9.528	−9.529	−9.531	−9.534	−9.537	−9.544	−9.550	−9.564	−9.579	−9.619	−9.663
0.5	−8.891	−8.893	−8.895	−8.897	−8.903	−8.908	−8.919	−8.931	−8.954	−8.979	−9.043	−9.111
0.55	−8.243	−8.246	−8.249	−8.252	−8.259	−8.266	−8.281	−8.296	−8.327	−8.359	−8.444	−8.531
0.6	−7.605	−7.608	−7.611	−7.615	−7.623	−7.632	−7.650	−7.668	−7.705	−7.744	−7.845	−7.950
0.65	−6.987	−6.991	−6.995	−6.999	−7.008	−7.018	−7.038	−7.059	−7.102	−7.147	−7.263	−7.383
0.7	−6.396	−6.400	−6.404	−6.408	−6.419	−6.430	−6.452	−6.475	−6.523	−6.573	−6.703	−6.837
0.75	−5.832	−5.836	−5.841	−5.845	−5.856	−5.868	−5.892	−5.918	−5.971	−6.027	−6.170	−6.317
0.8	−5.292	−5.297	−5.301	−5.306	−5.317	−5.330	−5.356	−5.384	−5.444	−5.506	−5.664	−5.824
0.85	−4.771	−4.775	−4.779	−4.784	−4.796	−4.810	−4.840	−4.871	−4.939	−5.008	−5.184	−5.358
0.9	−4.255	−4.258	−4.263	−4.268	−4.282	−4.298	−4.333	−4.371	−4.450	−4.530	−4.727	−4.916
0.93	−3.937	−3.940	−3.945	−3.951	−3.968	−3.987	−4.029	−4.073	−4.163	−4.251	−4.463	−4.662
0.95	−3.713	−3.717	−3.723	−3.730	−3.750	−3.773	−3.822	−3.873	−3.972	−4.068	−4.291	−4.498
0.97	−3.467	−3.473	−3.482	−3.492	−3.521	−3.551	−3.611	−3.670	−3.782	−3.886	−4.122	−4.336
0.98	−3.327	−3.337	−3.349	−3.363	−3.399	−3.434	−3.503	−3.568	−3.686	−3.795	−4.039	−4.257
0.99	−3.164	−3.183	−3.203	−3.222	−3.270	−3.313	−3.392	−3.464	−3.591	−3.705	−3.956	−4.178
1	−2.952	−2.997	−3.033	−3.065	−3.131	−3.186	−3.279	−3.358	−3.495	−3.615	−3.874	−4.100
1.01	−2.595	−2.743	−2.824	−2.880	−2.979	−3.051	−3.162	−3.251	−3.399	−3.525	−3.792	−4.023
1.02	−1.723	−2.326	−2.537	−2.650	−2.809	−2.906	−3.041	−3.142	−3.303	−3.435	−3.711	−3.947
1.03	−0.978	−1.630	−2.108	−2.345	−2.612	−2.748	−2.915	−3.031	−3.206	−3.346	−3.631	−3.871
1.05	−0.878	−0.835	−1.113	−1.496	−2.110	−2.381	−2.645	−2.800	−3.010	−3.167	−3.473	−3.722
1.1	−0.673	−0.662	−0.631	−0.617	−0.854	−1.261	−1.853	−2.167	−2.507	−2.720	−3.086	−3.362
1.15	−0.503	−0.509	−0.502	−0.487	−0.474	−0.604	−1.083	−1.497	−1.990	−2.275	−2.713	−3.019
1.2	−0.381	−0.388	−0.388	−0.381	−0.353	−0.361	−0.591	−0.934	−1.489	−1.840	−2.355	−2.692
1.3	−0.218	−0.221	−0.221	−0.218	−0.200	−0.178	−0.182	−0.300	−0.693	−1.066	−1.691	−2.086
1.4	−0.115	−0.115	−0.112	−0.108	−0.092	−0.070	−0.034	−0.044	−0.228	−0.504	−1.117	−1.547
1.5	−0.046	−0.042	−0.038	−0.032	−0.014	0.008	0.052	0.078	0.023	−0.142	−0.654	−1.080
1.6	0.004	0.009	0.016	0.023	0.043	0.065	0.113	0.151	0.163	0.082	−0.299	−0.689
1.7	0.040	0.047	0.055	0.063	0.085	0.109	0.158	0.202	0.248	0.223	−0.036	−0.369
1.8	0.067	0.076	0.084	0.094	0.117	0.143	0.194	0.241	0.306	0.317	0.157	−0.112
1.9	0.088	0.098	0.107	0.117	0.143	0.169	0.221	0.271	0.347	0.381	0.299	0.092
2	0.105	0.115	0.125	0.136	0.163	0.190	0.244	0.295	0.379	0.428	0.406	0.255
2.25	0.133	0.145	0.156	0.168	0.197	0.227	0.284	0.338	0.433	0.505	0.578	0.534
2.5	0.150	0.163	0.176	0.188	0.219	0.250	0.310	0.367	0.469	0.552	0.678	0.704
2.75	0.162	0.175	0.189	0.202	0.234	0.266	0.328	0.387	0.494	0.585	0.744	0.817
3	0.170	0.184	0.198	0.211	0.245	0.278	0.342	0.403	0.514	0.611	0.793	0.899
3.5	0.181	0.196	0.210	0.224	0.260	0.294	0.361	0.425	0.544	0.650	0.864	1.015
4	0.188	0.203	0.218	0.233	0.27	0.306	0.375	0.442	0.567	0.68	0.917	1.097
5	0.198	0.214	0.229	0.245	0.283	0.321	0.395	0.466	0.601	0.726	0.997	1.219

TABLE C.5 Values for $\left[\dfrac{s_{T_r,P_r} - s_{T_r,P_r}^{ideal\ gas}}{R}\right]^{(0)}$

T_r	P_r 0.01	0.025	0.05	0.075	0.1	0.25	0.5	0.6	0.7	0.8	0.9	1	1.1
0.3	-11.613	-10.699	-10.008	-9.605	-9.319	-8.417	-7.747	-7.574	-7.429	-7.304	-7.196	-7.099	-7.013
0.35	-11.185	-10.270	-9.579	-9.176	-8.890	-7.986	-7.314	-7.140	-6.995	-6.869	-6.760	-6.663	-6.576
0.4	-10.802	-9.887	-9.196	-8.792	-8.506	-7.602	-6.929	-6.755	-6.608	-6.483	-6.373	-6.275	-6.188
0.45	-10.453	-9.538	-8.847	-8.443	-8.158	-7.253	-6.579	-6.405	-6.258	-6.132	-6.022	-5.924	-5.837
0.5	-10.137	-9.222	-8.531	-8.127	-7.842	-6.937	-6.263	-6.089	-5.942	-5.816	-5.706	-5.608	-5.520
0.55	-0.038	-8.936	-8.245	-7.841	-7.555	-6.651	-5.978	-5.803	-5.657	-5.531	-5.421	-5.324	-5.236
0.6	-0.029	-0.075	-7.983	-7.580	-7.294	-6.391	-5.719	-5.544	-5.399	-5.273	-5.163	-5.066	-4.979
0.65	-0.023	-0.059	-0.122	-7.338	-7.052	-6.150	-5.479	-5.306	-5.161	-5.036	-4.927	-4.830	-4.743
0.7	-0.018	-0.047	-0.096	-0.149	-0.206	-5.922	-5.254	-5.082	-4.938	-4.814	-4.706	-4.610	-4.524
0.75	-0.015	-0.038	-0.078	-0.120	-0.164	-5.700	-5.036	-4.866	-4.723	-4.600	-4.494	-4.399	-4.314
0.8	-0.013	-0.032	-0.064	-0.098	-0.134	-0.390	-4.817	-4.649	-4.508	-4.388	-4.283	-4.191	-4.108
0.85	-0.011	-0.027	-0.054	-0.082	-0.111	-0.312	-4.581	-4.418	-4.282	-4.166	-4.065	-3.976	-3.897
0.9	-0.009	-0.023	-0.046	-0.069	-0.094	-0.257	-0.648	-4.145	-4.019	-3.912	-3.820	-3.738	-3.665
0.93	-0.008	-0.021	-0.042	-0.063	-0.085	-0.231	-0.556	-0.750	-3.815	-3.723	-3.641	-3.569	-3.503
0.95	-0.008	-0.019	-0.039	-0.059	-0.080	-0.215	-0.508	-0.671	-0.897	-3.556	-3.493	-3.433	-3.378
0.97	-0.007	-0.018	-0.037	-0.056	-0.075	-0.202	-0.467	-0.607	-0.787	-1.056	-3.286	-3.259	-3.224
0.98	-0.007	-0.018	-0.036	-0.054	-0.073	-0.195	-0.449	-0.580	-0.743	-0.971	-3.116	-3.142	-3.129
0.99	-0.007	-0.017	-0.035	-0.053	-0.071	-0.189	-0.432	-0.555	-0.705	-0.903	-1.228	-2.972	-3.011
1	-0.007	-0.017	-0.034	-0.051	-0.069	-0.183	-0.416	-0.532	-0.671	-0.847	-1.104	-2.167	-2.846
1.01	-0.007	-0.016	-0.033	-0.050	-0.067	-0.178	-0.401	-0.510	-0.640	-0.799	-1.015	-1.391	-2.535
1.02	-0.006	-0.016	-0.032	-0.049	-0.065	-0.172	-0.386	-0.491	-0.611	-0.757	-0.945	-1.225	-1.850
1.03	-0.006	-0.015	-0.031	-0.047	-0.063	-0.167	-0.373	-0.472	-0.586	-0.720	-0.887	-1.116	-1.493
1.05	-0.006	-0.015	-0.030	-0.045	-0.060	-0.158	-0.349	-0.439	-0.540	-0.656	-0.794	-0.965	-1.194
1.1	-0.005	-0.013	-0.026	-0.039	-0.053	-0.138	-0.298	-0.371	-0.450	-0.537	-0.633	-0.742	-0.867
1.15	-0.005	-0.011	-0.023	-0.035	-0.047	-0.121	-0.258	-0.319	-0.383	-0.452	-0.527	-0.607	-0.695
1.2	-0.004	-0.010	-0.021	-0.031	-0.042	-0.107	-0.226	-0.277	-0.331	-0.389	-0.449	-0.512	-0.580
1.3	-0.003	-0.008	-0.017	-0.025	-0.033	-0.086	-0.178	-0.217	-0.257	-0.298	-0.341	-0.385	-0.431
1.4	-0.003	-0.007	-0.014	-0.021	-0.027	-0.070	-0.144	-0.174	-0.205	-0.237	-0.270	-0.303	-0.337
1.5	-0.002	-0.006	-0.011	-0.017	-0.023	-0.058	-0.118	-0.143	-0.168	-0.194	-0.220	-0.246	-0.272
1.6	-0.002	-0.005	-0.010	-0.015	-0.019	-0.049	-0.099	-0.120	-0.141	-0.162	-0.183	-0.204	-0.225
1.7	-0.002	-0.004	-0.008	-0.012	-0.017	-0.042	-0.085	-0.102	-0.119	-0.137	-0.154	-0.172	-0.190
1.8	-0.001	-0.004	-0.007	-0.011	-0.014	-0.036	-0.073	-0.088	-0.102	-0.117	-0.132	-0.147	-0.162
1.9	-0.001	-0.003	-0.006	-0.009	-0.013	-0.032	-0.063	-0.076	-0.089	-0.102	-0.115	-0.127	-0.140
2	-0.001	-0.003	-0.006	-0.008	-0.011	-0.028	-0.056	-0.067	-0.078	-0.089	-0.100	-0.111	-0.123
2.25	-0.001	-0.002	-0.004	-0.006	-0.008	-0.021	-0.041	-0.050	-0.058	-0.066	-0.074	-0.083	-0.091
2.5	-0.001	-0.002	-0.003	-0.005	-0.006	-0.016	-0.032	-0.038	-0.045	-0.051	-0.057	-0.064	-0.070
2.75	-0.001	-0.001	-0.003	-0.004	-0.005	-0.013	-0.026	-0.031	-0.036	-0.041	-0.046	-0.050	-0.055
3	0.000	-0.001	-0.002	-0.003	-0.004	-0.010	-0.021	-0.025	-0.029	-0.033	-0.037	-0.041	-0.045
3.5	0.000	-0.001	-0.001	-0.002	-0.003	-0.007	-0.015	-0.017	-0.020	-0.023	-0.026	-0.029	-0.031
4	0.000	-0.001	-0.001	-0.002	-0.002	-0.005	-0.011	-0.013	-0.015	-0.017	-0.019	-0.021	-0.023
5	0.000	0.000	-0.001	-0.001	-0.001	-0.003	-0.007	-0.008	-0.009	-0.010	-0.012	-0.013	-0.014

TABLE C.5 Continued

T_r	P_r 1.2	1.3	1.4	1.5	1.75	2	2.5	3	4	5	7.5	10
0.3	-6.935	-6.864	-6.799	-6.740	-6.608	-6.497	-6.319	-6.182	-5.983	-5.847	-5.657	-5.578
0.35	-6.497	-6.426	-6.360	-6.299	-6.165	-6.052	-5.870	-5.728	-5.521	-5.376	-5.163	-5.060
0.4	-6.109	-6.036	-5.970	-5.909	-5.774	-5.660	-5.475	-5.330	-5.117	-4.967	-4.738	-4.619
0.45	-5.757	-5.685	-5.618	-5.557	-5.421	-5.306	-5.120	-4.974	-4.757	-4.603	-4.364	-4.234
0.5	-5.441	-5.368	-5.302	-5.240	-5.105	-4.989	-4.802	-4.656	-4.438	-4.282	-4.036	-3.899
0.55	-5.157	-5.084	-5.018	-4.956	-4.821	-4.706	-4.519	-4.373	-4.154	-3.998	-3.750	-3.607
0.6	-4.900	-4.828	-4.762	-4.700	-4.566	-4.451	-4.266	-4.120	-3.902	-3.747	-3.498	-3.353
0.65	-4.665	-4.593	-4.527	-4.467	-4.333	-4.220	-4.036	-3.892	-3.677	-3.523	-3.276	-3.131
0.7	-4.446	-4.375	-4.310	-4.250	-4.118	-4.007	-3.826	-3.684	-3.473	-3.322	-3.079	-2.935
0.75	-4.238	-4.168	-4.104	-4.046	-3.916	-3.807	-3.630	-3.491	-3.286	-3.138	-2.902	-2.761
0.8	-4.034	-3.966	-3.904	-3.846	-3.721	-3.615	-3.444	-3.310	-3.112	-2.970	-2.741	-2.605
0.85	-3.825	-3.760	-3.701	-3.646	-3.526	-3.425	-3.262	-3.135	-2.947	-2.812	-2.594	-2.463
0.9	-3.599	-3.539	-3.484	-3.434	-3.324	-3.231	-3.081	-2.963	-2.789	-2.663	-2.458	-2.334
0.93	-3.444	-3.390	-3.341	-3.295	-3.194	-3.108	-2.969	-2.860	-2.696	-2.577	-2.381	-2.262
0.95	-3.326	-3.279	-3.235	-3.193	-3.102	-3.023	-2.893	-2.790	-2.634	-2.520	-2.331	-2.215
0.97	-3.188	-3.151	-3.115	-3.081	-3.002	-2.932	-2.814	-2.719	-2.572	-2.463	-2.283	-2.170
0.98	-3.106	-3.078	-3.049	-3.019	-2.949	-2.884	-2.774	-2.682	-2.541	-2.436	-2.259	-2.148
0.99	-3.010	-2.995	-2.975	-2.953	-2.893	-2.835	-2.732	-2.646	-2.510	-2.408	-2.235	-2.126
1	-2.893	-2.900	-2.893	-2.879	-2.834	-2.784	-2.690	-2.609	-2.479	-2.380	-2.211	-2.105
1.01	-2.736	-2.785	-2.799	-2.798	-2.770	-2.730	-2.647	-2.571	-2.448	-2.352	-2.188	-2.083
1.02	-2.495	-2.639	-2.688	-2.706	-2.702	-2.673	-2.602	-2.533	-2.416	-2.325	-2.165	-2.062
1.03	-2.122	-2.441	-2.552	-2.599	-2.628	-2.614	-2.556	-2.494	-2.385	-2.297	-2.142	-2.042
1.05	-1.523	-1.915	-2.185	-2.328	-2.457	-2.483	-2.461	-2.415	-2.322	-2.242	-2.097	-2.001
1.1	-1.012	-1.180	-1.368	-1.557	-1.908	-2.081	-2.191	-2.202	-2.159	-2.104	-1.987	-1.903
1.15	-0.790	-0.894	-1.007	-1.126	-1.421	-1.649	-1.885	-1.968	-1.992	-1.966	-1.881	-1.810
1.2	-0.651	-0.727	-0.807	-0.890	-1.106	-1.308	-1.587	-1.727	-1.820	-1.827	-1.777	-1.722
1.3	-0.478	-0.527	-0.576	-0.628	-0.759	-0.891	-1.127	-1.299	-1.484	-1.554	-1.580	-1.556
1.4	-0.372	-0.407	-0.442	-0.478	-0.570	-0.663	-0.839	-0.990	-1.194	-1.303	-1.394	-1.402
1.5	-0.299	-0.326	-0.353	-0.381	-0.450	-0.520	-0.654	-0.777	-0.967	-1.088	-1.223	-1.260
1.6	-0.247	-0.268	-0.290	-0.312	-0.367	-0.421	-0.528	-0.628	-0.794	-0.913	-1.071	-1.130
1.7	-0.208	-0.225	-0.243	-0.261	-0.306	-0.350	-0.437	-0.519	-0.662	-0.773	-0.938	-1.013
1.8	-0.177	-0.192	-0.207	-0.222	-0.259	-0.296	-0.369	-0.438	-0.561	-0.661	-0.824	-0.908
1.9	-0.153	-0.166	-0.179	-0.191	-0.223	-0.255	-0.316	-0.375	-0.481	-0.570	-0.726	-0.815
2	-0.134	-0.145	-0.156	-0.167	-0.194	-0.221	-0.274	-0.325	-0.417	-0.497	-0.644	-0.733
2.25	-0.099	-0.107	-0.115	-0.123	-0.143	-0.162	-0.200	-0.237	-0.305	-0.366	-0.486	-0.570
2.5	-0.076	-0.082	-0.088	-0.094	-0.109	-0.124	-0.153	-0.181	-0.233	-0.281	-0.379	-0.453
2.75	-0.060	-0.065	-0.070	-0.075	-0.087	-0.098	-0.121	-0.143	-0.184	-0.222	-0.303	-0.367
3	-0.049	-0.053	-0.057	-0.061	-0.070	-0.080	-0.098	-0.116	-0.150	-0.181	-0.248	-0.303
3.5	-0.034	-0.037	-0.040	-0.042	-0.049	-0.056	-0.068	-0.081	-0.104	-0.126	-0.175	-0.216
4	-0.025	-0.027	-0.029	-0.031	-0.036	-0.041	-0.050	-0.059	-0.077	-0.093	-0.130	-0.162
5	-0.015	-0.017	-0.018	-0.019	-0.022	-0.025	-0.031	-0.036	-0.047	-0.057	-0.080	-0.100

T_r	P_r 0.01	0.025	0.05	0.075	0.1	0.25	0.5	0.6	0.7	0.8	0.9	1	1.1
0.3	-16.790	-16.787	-16.783	-16.778	-16.773	-16.744	-16.695	-16.675	-16.656	-16.637	-16.617	-16.598	-16.578
0.35	-15.408	-15.406	-15.402	-15.399	-15.395	-15.375	-15.341	-15.328	-15.314	-15.301	-15.288	-15.274	-15.261
0.4	-13.989	-13.987	-13.985	-13.983	-13.980	-13.966	-13.942	-13.932	-13.923	-13.914	-13.904	-13.895	-13.885
0.45	-12.562	-12.561	-12.559	-12.557	-12.556	-12.545	-12.528	-12.521	-12.514	-12.508	-12.501	-12.494	-12.488
0.5	-11.201	-11.200	-11.198	-11.197	-11.196	-11.188	-11.176	-11.171	-11.166	-11.161	-11.156	-11.151	-11.146
0.55	-0.115	-9.950	-9.949	-9.948	-9.947	-9.941	-9.932	-9.928	-9.924	-9.921	-9.917	-9.914	-9.910
0.6	-0.078	-0.207	-8.828	-8.827	-8.827	-8.822	-8.814	-8.811	-8.808	-8.806	-8.803	-8.800	-8.798
0.65	-0.055	-0.143	-0.309	-7.833	-7.832	-7.828	-7.822	-7.819	-7.817	-7.815	-7.812	-7.810	-7.808
0.7	-0.040	-0.102	-0.216	-0.343	-0.491	-6.949	-6.943	-6.941	-6.939	-6.937	-6.935	-6.933	-6.932
0.75	-0.029	-0.075	-0.156	-0.244	-0.340	-6.172	-6.165	-6.162	-6.160	-6.158	-6.156	-6.155	-6.153
0.8	-0.022	-0.056	-0.116	-0.179	-0.246	-0.812	-5.471	-5.467	-5.462	-5.462	-5.460	-5.458	-5.456
0.85	-0.017	-0.043	-0.088	-0.135	-0.183	-0.545	-4.846	-4.841	-4.836	-4.832	-4.829	-4.826	-4.824
0.9	-0.013	-0.033	-0.068	-0.103	-0.140	-0.392	-1.137	-4.269	-4.258	-4.250	-4.243	-4.238	-4.235
0.93	-0.011	-0.029	-0.058	-0.089	-0.120	-0.328	-0.834	-1.219	-3.933	-3.914	-3.902	-3.893	-3.888
0.95	-0.010	-0.026	-0.053	-0.081	-0.109	-0.293	-0.706	-0.961	-1.418	-3.697	-3.672	-3.658	-3.650
0.97	-0.010	-0.024	-0.048	-0.073	-0.099	-0.263	-0.609	-0.797	-1.055	-1.570	-3.438	-3.406	-3.394
0.98	-0.009	-0.023	-0.046	-0.070	-0.094	-0.250	-0.569	-0.734	-0.946	-1.270	-3.337	-3.264	-3.248
0.99	-0.009	-0.022	-0.044	-0.067	-0.090	-0.237	-0.533	-0.680	-0.859	-1.098	-1.556	-3.093	-3.073
1	-0.008	-0.021	-0.042	-0.064	-0.086	-0.225	-0.500	-0.632	-0.787	-0.977	-1.242	-2.311	-2.810
1.01	-0.008	-0.020	-0.040	-0.061	-0.082	-0.214	-0.470	-0.590	-0.726	-0.883	-1.074	-1.306	-1.795
1.02	-0.008	-0.019	-0.039	-0.058	-0.078	-0.204	-0.443	-0.552	-0.673	-0.807	-0.958	-1.113	-1.015
1.03	-0.007	-0.018	-0.037	-0.056	-0.075	-0.194	-0.418	-0.518	-0.627	-0.744	-0.868	-0.989	-1.017
1.05	-0.007	-0.017	-0.034	-0.051	-0.069	-0.177	-0.374	-0.460	-0.549	-0.642	-0.735	-0.820	-0.872
1.1	-0.005	-0.014	-0.028	-0.041	-0.055	-0.141	-0.289	-0.350	-0.411	-0.470	-0.527	-0.577	-0.617
1.15	-0.005	-0.011	-0.023	-0.034	-0.045	-0.114	-0.229	-0.275	-0.319	-0.361	-0.401	-0.437	-0.467
1.2	-0.004	-0.009	-0.019	-0.028	-0.037	-0.094	-0.185	-0.220	-0.254	-0.286	-0.316	-0.343	-0.366
1.3	-0.003	-0.007	-0.013	-0.020	-0.026	-0.065	-0.125	-0.148	-0.170	-0.190	-0.209	-0.226	-0.241
1.4	-0.002	-0.005	-0.010	-0.014	-0.019	-0.046	-0.089	-0.104	-0.119	-0.133	-0.146	-0.158	-0.168
1.5	-0.001	-0.004	-0.007	-0.011	-0.014	-0.034	-0.065	-0.076	-0.087	-0.097	-0.106	-0.115	-0.123
1.6	-0.001	-0.003	-0.005	-0.008	-0.011	-0.026	-0.049	-0.057	-0.065	-0.073	-0.080	-0.086	-0.092
1.7	-0.001	-0.002	-0.004	-0.006	-0.008	-0.020	-0.038	-0.044	-0.050	-0.056	-0.061	-0.067	-0.071
1.8	0.001	-0.002	-0.003	-0.005	-0.006	-0.016	-0.030	-0.035	-0.040	-0.044	-0.049	-0.053	-0.057
1.9	0.001	-0.001	-0.003	-0.004	-0.005	-0.013	-0.024	-0.028	-0.032	-0.036	-0.039	-0.043	-0.046
2	0.000	-0.001	-0.002	-0.003	-0.004	-0.010	-0.019	-0.023	-0.026	-0.029	-0.032	-0.035	-0.038
2.25	0.000	-0.001	-0.001	-0.002	-0.003	-0.007	-0.013	-0.015	-0.017	-0.019	-0.021	-0.023	-0.025
2.5	0.000	0.000	-0.001	-0.001	-0.002	-0.005	-0.009	-0.010	-0.012	-0.014	-0.015	-0.017	-0.018
2.75	0.000	0.000	-0.001	-0.001	-0.001	-0.003	-0.007	-0.008	-0.009	-0.010	-0.012	-0.013	-0.014
3	0.000	0.000	-0.001	-0.001	-0.001	-0.003	-0.005	-0.006	-0.007	-0.008	-0.009	-0.010	-0.011
3.5	0.000	0.000	0.000	-0.001	-0.001	-0.002	-0.004	-0.004	-0.005	-0.006	-0.007	-0.007	-0.008
4	0.000	0.000	0.000	0.000	-0.001	-0.001	-0.003	-0.003	-0.004	-0.005	-0.005	-0.006	-0.006
5	0.000	0.000	0.000	0.000	0.000	-0.001	-0.002	-0.002	-0.003	-0.003	-0.004	-0.004	-0.004

TABLE C.6 Continued

T_r	P_r 1.2	1.3	1.4	1.5	1.75	2	2.5	3	4	5	7.5	10
0.3	-16.559	-16.540	-16.521	-16.501	-16.453	-16.405	-16.310	-16.214	-16.025	-15.838	-15.377	-14.927
0.35	-15.248	-15.234	-15.221	-15.208	-15.175	-15.142	-15.077	-15.012	-14.884	-14.757	-14.450	-14.154
0.4	-13.876	-13.867	-13.858	-13.848	-13.825	-13.803	-13.757	-13.713	-13.626	-13.540	-13.337	-13.144
0.45	-12.481	-12.475	-12.468	-12.461	-12.445	-12.429	-12.398	-12.367	-12.308	-12.251	-12.119	-11.998
0.5	-11.142	-11.137	-11.132	-11.128	-11.116	-11.105	-11.083	-11.063	-11.023	-10.986	-10.905	-10.836
0.55	-9.907	-9.903	-9.900	-9.897	-9.889	-9.881	-9.866	-9.853	-9.828	-9.806	-9.763	-9.732
0.6	-8.795	-8.793	-8.790	-8.788	-8.783	-8.777	-8.768	-8.759	-8.746	-8.735	-8.721	-8.720
0.65	-7.807	-7.805	-7.803	-7.801	-7.798	-7.794	-7.789	-7.784	-7.779	-7.778	-7.788	-7.811
0.7	-6.930	-6.929	-6.928	-6.926	-6.924	-6.922	-6.919	-6.919	-6.921	-6.928	-6.959	-7.002
0.75	-6.152	-6.151	-6.150	-6.149	-6.147	-6.146	-6.147	-6.149	-6.159	-6.174	-6.224	-6.285
0.8	-5.455	-5.454	-5.453	-5.452	-5.452	-5.452	-5.455	-5.461	-5.478	-5.501	-5.570	-5.648
0.85	-4.822	-4.821	-4.820	-4.820	-4.820	-4.822	-4.828	-4.839	-4.866	-4.898	-4.988	-5.083
0.9	-4.232	-4.231	-4.230	-4.230	-4.232	-4.236	-4.250	-4.267	-4.307	-4.351	-4.465	-4.578
0.93	-3.885	-3.883	-3.883	-3.883	-3.888	-3.896	-3.917	-3.941	-3.993	-4.046	-4.177	-4.300
0.95	-3.647	-3.646	-3.646	-3.648	-3.657	-3.669	-3.697	-3.728	-3.790	-3.851	-3.994	-4.125
0.97	-3.391	-3.392	-3.396	-3.401	-3.418	-3.437	-3.477	-3.517	-3.592	-3.661	-3.818	-3.957
0.98	-3.247	-3.252	-3.259	-3.268	-3.293	-3.318	-3.366	-3.412	-3.494	-3.569	-3.732	-3.875
0.99	-3.082	-3.096	-3.111	-3.126	-3.162	-3.195	-3.254	-3.306	-3.397	-3.477	-3.648	-3.795
1	-2.868	-2.908	-2.940	-2.967	-3.022	-3.067	-3.140	-3.200	-3.301	-3.387	-3.565	-3.717
1.01	-2.513	-2.657	-2.732	-2.784	-2.871	-2.933	-3.024	-3.094	-3.206	-3.297	-3.484	-3.640
1.02	-1.655	-2.246	-2.450	-2.557	-2.703	-2.790	-2.904	-2.986	-3.110	-3.209	-3.405	-3.565
1.03	-0.927	-1.567	-2.031	-2.259	-2.512	-2.636	-2.781	-2.878	-3.016	-3.121	-3.326	-3.492
1.05	-0.831	-0.800	-1.073	-1.443	-2.030	-2.283	-2.522	-2.655	-2.827	-2.949	-3.174	-3.348
1.1	-0.640	-0.639	-0.620	-0.618	-0.857	-1.241	-1.786	-2.067	-2.360	-2.534	-2.814	-3.013
1.15	-0.489	-0.502	-0.506	-0.502	-0.518	-0.654	-1.100	-1.471	-1.900	-2.138	-2.483	-2.708
1.2	-0.385	-0.399	-0.408	-0.412	-0.415	-0.447	-0.680	-0.991	-1.473	-1.767	-2.178	-2.430
1.3	-0.254	-0.265	-0.275	-0.282	-0.292	-0.300	-0.351	-0.481	-0.835	-1.147	-1.646	-1.944
1.4	-0.178	-0.186	-0.194	-0.200	-0.212	-0.220	-0.240	-0.290	-0.488	-0.730	-1.220	-1.544
1.5	-0.130	-0.136	-0.142	-0.147	-0.158	-0.166	-0.181	-0.206	-0.315	-0.479	-0.900	-1.222
1.6	-0.098	-0.103	-0.108	-0.112	-0.121	-0.129	-0.142	-0.159	-0.224	-0.334	-0.671	-0.969
1.7	-0.076	-0.080	-0.084	-0.087	-0.095	-0.102	-0.114	-0.127	-0.173	-0.248	-0.511	-0.775
1.8	-0.060	-0.064	-0.067	-0.070	-0.077	-0.083	-0.094	-0.105	-0.140	-0.195	-0.401	-0.628
1.9	-0.049	-0.052	-0.054	-0.057	-0.063	-0.069	-0.079	-0.089	-0.117	-0.160	-0.323	-0.518
2	-0.040	-0.043	-0.045	-0.048	-0.053	-0.058	-0.067	-0.077	-0.101	-0.136	-0.269	-0.434
2.25	-0.027	-0.029	-0.031	-0.032	-0.036	-0.040	-0.048	-0.056	-0.075	-0.100	-0.187	-0.302
2.5	-0.020	-0.021	-0.022	-0.024	-0.027	-0.031	-0.037	-0.044	-0.060	-0.080	-0.145	-0.230
2.75	-0.015	-0.016	-0.017	-0.019	-0.021	-0.024	-0.030	-0.037	-0.050	-0.067	-0.120	-0.187
3	-0.012	-0.013	-0.014	-0.015	-0.018	-0.020	-0.026	-0.031	-0.044	-0.058	-0.103	-0.158
3.5	-0.009	-0.010	-0.010	-0.011	-0.013	-0.015	-0.019	-0.024	-0.034	-0.046	-0.081	-0.122
4	-0.007	-0.008	-0.008	-0.009	-0.010	-0.012	-0.016	-0.020	-0.028	-0.038	-0.066	-0.100
5	-0.005	-0.005	-0.006	-0.006	-0.007	-0.009	-0.011	-0.014	-0.020	-0.028	-0.048	-0.073

TABLE C.7 Values for log [$\varphi^{(0)}$]

T_r \ P_r	0.01	0.025	0.05	0.075	0.1	0.25	0.5	0.6	0.7	0.8	0.9	1	1.1
0.3	-3.708	-4.104	-4.402	-4.575	-4.697	-5.076	-5.346	-5.412	-5.467	-5.512	-5.551	-5.584	-5.613
0.35	-2.472	-2.868	-3.166	-3.339	-3.461	-3.842	-4.115	-4.183	-4.239	-4.285	-4.325	-4.359	-4.390
0.4	-1.566	-1.962	-2.261	-2.434	-2.557	-2.939	-3.214	-3.283	-3.340	-3.387	-3.428	-3.464	-3.495
0.45	-0.879	-1.276	-1.574	-1.748	-1.871	-2.254	-2.531	-2.601	-2.658	-2.707	-2.748	-2.784	-2.816
0.5	-0.344	-0.741	-1.040	-1.214	-1.336	-1.721	-1.999	-2.070	-2.128	-2.177	-2.219	-2.256	-2.288
0.55	-0.007	-0.315	-0.614	-0.788	-0.911	-1.296	-1.576	-1.647	-1.705	-1.755	-1.798	-1.835	-1.868
0.6	-0.007	-0.016	-0.269	-0.443	-0.566	-0.952	-1.233	-1.304	-1.363	-1.413	-1.456	-1.494	-1.527
0.65	-0.005	-0.013	-0.026	-0.160	-0.283	-0.670	-0.951	-1.023	-1.082	-1.132	-1.176	-1.214	-1.248
0.7	-0.004	-0.010	-0.021	-0.032	-0.043	-0.435	-0.717	-0.789	-0.848	-0.899	-0.942	-0.981	-1.015
0.75	-0.003	-0.009	-0.017	-0.026	-0.035	-0.237	-0.520	-0.592	-0.652	-0.703	-0.746	-0.785	-0.819
0.8	-0.003	-0.007	-0.014	-0.021	-0.029	-0.076	-0.354	-0.426	-0.486	-0.537	-0.581	-0.619	-0.654
0.85	-0.002	-0.006	-0.012	-0.018	-0.024	-0.062	-0.213	-0.285	-0.345	-0.396	-0.440	-0.479	-0.513
0.9	-0.002	-0.005	-0.010	-0.015	-0.020	-0.052	-0.111	-0.166	-0.225	-0.276	-0.320	-0.359	-0.393
0.93	-0.002	-0.005	-0.009	-0.014	-0.018	-0.047	-0.099	-0.122	-0.163	-0.214	-0.258	-0.296	-0.330
0.95	-0.002	-0.004	-0.008	-0.013	-0.017	-0.044	-0.092	-0.113	-0.136	-0.176	-0.220	-0.258	-0.292
0.97	-0.002	-0.004	-0.008	-0.012	-0.016	-0.041	-0.086	-0.105	-0.126	-0.148	-0.185	-0.223	-0.256
0.98	-0.002	-0.004	-0.008	-0.012	-0.016	-0.040	-0.083	-0.101	-0.121	-0.142	-0.168	-0.206	-0.240
0.99	-0.001	-0.004	-0.007	-0.011	-0.015	-0.038	-0.080	-0.098	-0.117	-0.137	-0.159	-0.191	-0.224
1	-0.001	-0.004	-0.007	-0.011	-0.015	-0.037	-0.077	-0.095	-0.113	-0.132	-0.152	-0.176	-0.209
1.01	-0.001	-0.004	-0.007	-0.011	-0.014	-0.036	-0.075	-0.091	-0.109	-0.127	-0.146	-0.168	-0.195
1.02	-0.001	-0.003	-0.007	-0.010	-0.014	-0.035	-0.073	-0.088	-0.105	-0.122	-0.141	-0.161	-0.184
1.03	-0.001	-0.003	-0.007	-0.010	-0.013	-0.034	-0.070	-0.086	-0.101	-0.118	-0.136	-0.154	-0.175
1.05	-0.001	-0.003	-0.006	-0.009	-0.013	-0.032	-0.066	-0.080	-0.095	-0.110	-0.126	-0.143	-0.161
1.1	-0.001	-0.002	-0.005	-0.008	-0.011	-0.028	-0.057	-0.069	-0.081	-0.093	-0.106	-0.120	-0.133
1.15	-0.001	-0.002	-0.005	-0.007	-0.009	-0.024	-0.049	-0.059	-0.069	-0.080	-0.091	-0.102	-0.113
1.2	-0.001	-0.002	-0.004	-0.006	-0.008	-0.021	-0.042	-0.051	-0.060	-0.069	-0.078	-0.088	-0.097
1.3	-0.001	-0.002	-0.003	-0.005	-0.006	-0.016	-0.033	-0.039	-0.046	-0.052	-0.059	-0.066	-0.073
1.4	-0.001	-0.001	-0.003	-0.004	-0.005	-0.013	-0.025	-0.030	-0.035	-0.040	-0.046	-0.051	-0.056
1.5	0.000	-0.001	-0.002	-0.003	-0.004	-0.010	-0.020	-0.024	-0.028	-0.032	-0.035	-0.039	-0.043
1.6	0.000	-0.001	-0.002	-0.002	-0.003	-0.008	-0.016	-0.019	-0.022	-0.025	-0.028	-0.031	-0.034
1.7	0.000	-0.001	-0.001	-0.002	-0.002	-0.006	-0.012	-0.015	-0.017	-0.020	-0.022	-0.024	-0.027
1.8	0.000	0.000	-0.001	-0.001	-0.002	-0.005	-0.010	-0.012	-0.014	-0.015	-0.017	-0.019	-0.021
1.9	0.000	0.000	-0.001	-0.001	-0.002	-0.004	-0.008	-0.009	-0.011	-0.012	-0.014	-0.015	-0.016
2	0.000	0.000	-0.001	-0.001	-0.001	-0.003	-0.006	-0.007	-0.008	-0.009	-0.011	-0.012	-0.013
2.25	0.000	0.000	0.000	0.000	-0.001	-0.002	-0.003	-0.004	-0.004	-0.005	-0.005	-0.006	-0.006
2.5	0.000	0.000	0.000	0.000	0.000	-0.001	-0.001	-0.001	-0.002	-0.002	-0.002	-0.002	-0.002
2.75	0.000	0.000	0.000	0.000	0.000	0.000	0.000	0.000	0.000	0.000	0.000	0.000	0.001
3	0.000	0.000	0.000	0.000	0.000	0.001	0.001	0.001	0.001	0.002	0.002	0.002	0.002
3.5	0.000	0.000	0.000	0.000	0.000	0.001	0.002	0.002	0.003	0.003	0.003	0.004	0.004
4	0.000	0.000	0.000	0.000	0.000	0.001	0.002	0.003	0.003	0.004	0.004	0.004	0.005
5	0.000	0.000	0.000	0.000	0.001	0.001	0.003	0.003	0.004	0.004	0.005	0.005	0.006

TABLE C.7 Continued

T_r	P_r 1.2	1.3	1.4	1.5	1.75	2	2.5	3	4	5	7.5	10
0.3	-5.638	-5.660	-5.680	-5.697	-5.733	-5.759	-5.793	-5.810	-5.810	-5.782	-5.647	-5.462
0.35	-4.416	-4.440	-4.460	-4.479	-4.518	-4.548	-4.588	-4.611	-4.623	-4.608	-4.505	-4.352
0.4	-3.522	-3.546	-3.568	-3.588	-3.629	-3.661	-3.707	-3.735	-3.757	-3.752	-3.673	-3.545
0.45	-2.845	-2.870	-2.892	-2.913	-2.956	-2.990	-3.039	-3.071	-3.101	-3.104	-3.046	-2.938
0.5	-2.317	-2.343	-2.366	-2.387	-2.432	-2.468	-2.520	-2.555	-2.592	-2.601	-2.559	-2.468
0.55	-1.897	-1.924	-1.947	-1.969	-2.015	-2.052	-2.107	-2.145	-2.187	-2.201	-2.173	-2.095
0.6	-1.557	-1.584	-1.608	-1.630	-1.677	-1.715	-1.773	-1.812	-1.859	-1.878	-1.861	-1.795
0.65	-1.278	-1.305	-1.329	-1.352	-1.400	-1.439	-1.498	-1.539	-1.589	-1.612	-1.604	-1.549
0.7	-1.045	-1.073	-1.097	-1.120	-1.169	-1.208	-1.269	-1.312	-1.365	-1.391	-1.391	-1.344
0.75	-0.850	-0.877	-0.902	-0.925	-0.974	-1.015	-1.076	-1.121	-1.176	-1.204	-1.212	-1.172
0.8	-0.684	-0.712	-0.737	-0.760	-0.810	-0.851	-0.913	-0.958	-1.016	-1.046	-1.060	-1.026
0.85	-0.544	-0.572	-0.597	-0.620	-0.670	-0.711	-0.774	-0.820	-0.879	-0.911	-0.929	-0.901
0.9	-0.424	-0.452	-0.477	-0.500	-0.550	-0.591	-0.654	-0.700	-0.761	-0.794	-0.817	-0.793
0.93	-0.361	-0.389	-0.414	-0.437	-0.486	-0.527	-0.591	-0.637	-0.698	-0.732	-0.756	-0.735
0.95	-0.322	-0.350	-0.375	-0.398	-0.447	-0.488	-0.552	-0.598	-0.659	-0.693	-0.719	-0.699
0.97	-0.287	-0.314	-0.339	-0.362	-0.411	-0.452	-0.515	-0.561	-0.622	-0.657	-0.683	-0.665
0.98	-0.270	-0.297	-0.322	-0.344	-0.393	-0.434	-0.497	-0.543	-0.604	-0.639	-0.666	-0.649
0.99	-0.254	-0.281	-0.305	-0.328	-0.377	-0.417	-0.480	-0.526	-0.587	-0.622	-0.650	-0.633
1	-0.238	-0.265	-0.289	-0.312	-0.360	-0.401	-0.463	-0.509	-0.570	-0.605	-0.634	-0.617
1.01	-0.224	-0.250	-0.274	-0.297	-0.345	-0.385	-0.447	-0.493	-0.554	-0.589	-0.618	-0.602
1.02	-0.210	-0.236	-0.260	-0.282	-0.330	-0.370	-0.432	-0.477	-0.538	-0.573	-0.603	-0.588
1.03	-0.199	-0.223	-0.246	-0.268	-0.315	-0.355	-0.417	-0.462	-0.523	-0.558	-0.588	-0.573
1.05	-0.180	-0.201	-0.222	-0.242	-0.288	-0.327	-0.388	-0.433	-0.493	-0.529	-0.559	-0.546
1.1	-0.148	-0.163	-0.178	-0.193	-0.232	-0.267	-0.324	-0.368	-0.427	-0.462	-0.493	-0.482
1.15	-0.125	-0.136	-0.148	-0.160	-0.191	-0.220	-0.272	-0.312	-0.369	-0.403	-0.435	-0.426
1.2	-0.106	-0.116	-0.126	-0.135	-0.160	-0.184	-0.229	-0.266	-0.319	-0.352	-0.384	-0.377
1.3	-0.080	-0.086	-0.093	-0.100	-0.117	-0.134	-0.167	-0.195	-0.239	-0.269	-0.299	-0.293
1.4	-0.061	-0.066	-0.071	-0.076	-0.089	-0.101	-0.125	-0.146	-0.181	-0.205	-0.231	-0.226
1.5	-0.047	-0.051	-0.055	-0.059	-0.068	-0.077	-0.095	-0.111	-0.138	-0.157	-0.178	-0.173
1.6	-0.037	-0.040	-0.043	-0.046	-0.053	-0.060	-0.073	-0.085	-0.105	-0.120	-0.136	-0.129
1.7	-0.029	-0.031	-0.033	-0.036	-0.041	-0.046	-0.056	-0.065	-0.081	-0.092	-0.102	-0.094
1.8	-0.023	-0.024	-0.026	-0.028	-0.032	-0.036	-0.044	-0.050	-0.061	-0.069	-0.075	-0.066
1.9	-0.018	-0.019	-0.020	-0.022	-0.025	-0.028	-0.033	-0.038	-0.046	-0.052	-0.054	-0.043
2	-0.014	-0.015	-0.016	-0.017	-0.019	-0.021	-0.025	-0.029	-0.034	-0.037	-0.036	-0.024
2.25	-0.007	-0.007	-0.008	-0.008	-0.009	-0.010	-0.011	-0.012	-0.013	-0.013	-0.005	0.010
2.5	-0.002	-0.002	-0.002	-0.002	-0.003	-0.003	-0.003	-0.002	0.000	0.003	0.014	0.031
2.75	0.001	0.001	0.001	0.001	0.001	0.002	0.003	0.004	0.008	0.012	0.026	0.044
3	0.003	0.003	0.003	0.003	0.004	0.005	0.007	0.009	0.013	0.018	0.034	0.053
3.5	0.005	0.005	0.006	0.006	0.007	0.008	0.011	0.013	0.019	0.025	0.042	0.061
4	0.006	0.006	0.007	0.007	0.009	0.010	0.013	0.016	0.022	0.028	0.045	0.064
5	0.006	0.007	0.007	0.008	0.009	0.011	0.014	0.016	0.022	0.029	0.045	0.062

TABLE C.8 Values for log $[\varphi^{(1)}]$*

T_r	P_r 0.01	0.025	0.05	0.075	0.1	0.25	0.5	0.6	0.7	0.8	0.9	1	1.1
0.3	-8.779	-8.779	-8.780	-8.781	-8.782	-8.787	-8.796	-8.799	-8.803	-8.806	-8.810	-8.813	-8.817
0.35	-6.526	-6.527	-6.528	-6.529	-6.530	-6.536	-6.546	-6.550	-6.554	-6.558	-6.562	-6.566	-6.570
0.4	-4.912	-4.913	-4.914	-4.915	-4.916	-4.922	-4.932	-4.936	-4.941	-4.945	-4.949	-4.953	-4.957
0.45	-3.726	-3.726	-3.727	-3.728	-3.729	-3.736	-3.746	-3.750	-3.754	-3.758	-3.762	-3.766	-3.770
0.5	-2.838	-2.838	-2.839	-2.840	-2.841	-2.847	-2.857	-2.861	-2.865	-2.868	-2.872	-2.876	-2.880
0.55	-0.013	-2.163	-2.164	-2.165	-2.166	-2.171	-2.181	-2.184	-2.188	-2.192	-2.196	-2.199	-2.203
0.6	-0.009	-0.023	-1.644	-1.645	-1.646	-1.651	-1.660	-1.664	-1.667	-1.671	-1.674	-1.678	-1.681
0.65	-0.006	-0.015	-0.031	-1.241	-1.241	-1.247	-1.255	-1.258	-1.262	-1.265	-1.268	-1.272	-1.275
0.7	-0.004	-0.010	-0.021	-0.032	-0.044	-0.929	-0.937	-0.940	-0.943	-0.946	-0.949	-0.952	-0.955
0.75	-0.003	-0.007	-0.014	-0.022	-0.030	-0.677	-0.685	-0.688	-0.691	-0.694	-0.697	-0.700	-0.703
0.8	-0.002	-0.005	-0.010	-0.015	-0.020	-0.056	-0.484	-0.487	-0.490	-0.493	-0.496	-0.498	-0.501
0.85	-0.001	-0.003	-0.006	-0.010	-0.013	-0.036	-0.324	-0.327	-0.330	-0.332	-0.335	-0.338	-0.341
0.9	-0.001	-0.002	-0.004	-0.006	-0.009	-0.023	-0.053	-0.199	-0.202	-0.204	-0.207	-0.210	-0.212
0.93	-0.001	-0.002	-0.003	-0.005	-0.007	-0.017	-0.037	-0.048	-0.138	-0.141	-0.143	-0.146	-0.149
0.95	-0.001	-0.001	-0.003	-0.004	-0.005	-0.014	-0.030	-0.037	-0.046	-0.103	-0.106	-0.108	-0.111
0.97	0.000	-0.001	-0.002	-0.003	-0.004	-0.011	-0.023	-0.029	-0.034	-0.042	-0.072	-0.075	-0.077
0.98	0.000	-0.001	-0.002	-0.003	-0.004	-0.010	-0.020	-0.025	-0.030	-0.035	-0.056	-0.059	-0.062
0.99	0.000	-0.001	-0.002	-0.003	-0.003	-0.009	-0.018	-0.021	-0.025	-0.030	-0.034	-0.044	-0.047
1	0.000	-0.001	-0.001	-0.002	-0.003	-0.008	-0.015	-0.018	-0.022	-0.025	-0.028	-0.031	-0.034
1.01	0.000	-0.001	-0.001	-0.002	-0.003	-0.007	-0.013	-0.016	-0.018	-0.021	-0.023	-0.024	-0.023
1.02	0.000	-0.001	-0.001	-0.002	-0.002	-0.006	-0.011	-0.013	-0.015	-0.017	-0.018	-0.019	-0.018
1.03	0.000	0.000	-0.001	-0.001	-0.002	-0.005	-0.009	-0.011	-0.012	-0.013	-0.014	-0.014	-0.013
1.05	0.000	0.000	-0.001	-0.001	-0.001	-0.003	-0.006	-0.006	-0.007	-0.007	-0.007	-0.007	-0.005
1.1	0.000	0.000	0.000	0.000	0.000	0.000	0.001	0.002	0.003	0.004	0.005	0.007	0.009
1.15	0.000	0.000	0.000	0.001	0.001	0.003	0.006	0.008	0.009	0.011	0.014	0.016	0.019
1.2	0.000	0.000	0.001	0.001	0.002	0.004	0.009	0.012	0.014	0.017	0.020	0.023	0.026
1.3	0.000	0.001	0.001	0.002	0.003	0.007	0.014	0.017	0.020	0.023	0.027	0.030	0.034
1.4	0.000	0.001	0.002	0.002	0.003	0.008	0.016	0.020	0.023	0.027	0.030	0.034	0.038
1.5	0.000	0.001	0.002	0.003	0.003	0.008	0.017	0.021	0.024	0.028	0.032	0.036	0.039
1.6	0.000	0.001	0.002	0.003	0.003	0.009	0.018	0.021	0.025	0.029	0.032	0.036	0.040
1.7	0.000	0.001	0.002	0.003	0.004	0.009	0.018	0.021	0.025	0.029	0.032	0.036	0.039
1.8	0.000	0.001	0.002	0.003	0.003	0.009	0.018	0.021	0.025	0.028	0.032	0.035	0.039
1.9	0.000	0.001	0.002	0.003	0.003	0.009	0.017	0.021	0.024	0.028	0.031	0.034	0.038
2	0.000	0.001	0.002	0.003	0.003	0.008	0.017	0.020	0.024	0.027	0.030	0.034	0.037
2.25	0.000	0.001	0.002	0.002	0.003	0.008	0.015	0.019	0.022	0.025	0.028	0.031	0.034
2.5	0.000	0.001	0.001	0.002	0.003	0.007	0.014	0.018	0.020	0.023	0.026	0.029	0.032
2.75	0.000	0.001	0.001	0.002	0.003	0.007	0.013	0.016	0.019	0.022	0.024	0.027	0.030
3	0.000	0.001	0.001	0.002	0.003	0.006	0.011	0.015	0.018	0.020	0.023	0.025	0.028
3.5	0.000	0.001	0.001	0.002	0.002	0.006	0.010	0.013	0.016	0.018	0.020	0.022	0.024
4	0.000	0.001	0.001	0.001	0.002	0.005	0.008	0.012	0.014	0.016	0.018	0.020	0.022
5	0.000	0.000	0.001	0.001	0.002	0.004	0.006	0.010	0.011	0.013	0.015	0.016	0.018

* Many of these values have a different last decimal place from those published by Lee–Kesler.

TABLE C.8 Continued

T_r	P_r 1.2	1.3	1.4	1.5	1.75	2	2.5	3	4	5	7.5	10
0.3	-8.820	-8.824	-8.827	-8.831	-8.840	-8.848	-8.866	-8.883	-8.918	-8.953	-9.039	-9.126
0.35	-6.574	-6.578	-6.582	-6.586	-6.596	-6.606	-6.625	-6.645	-6.685	-6.724	-6.822	-6.919
0.4	-4.961	-4.965	-4.969	-4.973	-4.984	-4.994	-5.014	-5.034	-5.075	-5.115	-5.214	-5.312
0.45	-3.774	-3.778	-3.782	-3.786	-3.796	-3.806	-3.826	-3.846	-3.885	-3.924	-4.020	-4.115
0.5	-2.884	-2.888	-2.892	-2.896	-2.905	-2.915	-2.934	-2.953	-2.990	-3.027	-3.119	-3.208
0.55	-2.207	-2.210	-2.214	-2.218	-2.227	-2.236	-2.254	-2.272	-2.307	-2.342	-2.427	-2.510
0.6	-1.685	-1.688	-1.692	-1.695	-1.704	-1.712	-1.729	-1.746	-1.779	-1.812	-1.891	-1.967
0.65	-1.278	-1.281	-1.285	-1.288	-1.296	-1.304	-1.320	-1.336	-1.367	-1.397	-1.470	-1.540
0.7	-0.959	-0.962	-0.965	-0.968	-0.975	-0.983	-0.998	-1.013	-1.041	-1.069	-1.137	-1.201
0.75	-0.705	-0.708	-0.711	-0.714	-0.721	-0.728	-0.742	-0.756	-0.783	-0.809	-0.870	-0.929
0.8	-0.504	-0.507	-0.510	-0.512	-0.519	-0.526	-0.539	-0.551	-0.576	-0.600	-0.656	-0.709
0.85	-0.343	-0.346	-0.348	-0.351	-0.357	-0.364	-0.376	-0.388	-0.410	-0.432	-0.483	-0.530
0.9	-0.215	-0.217	-0.220	-0.222	-0.228	-0.234	-0.245	-0.256	-0.277	-0.296	-0.342	-0.384
0.93	-0.151	-0.154	-0.156	-0.158	-0.164	-0.170	-0.180	-0.191	-0.210	-0.228	-0.270	-0.310
0.95	-0.114	-0.116	-0.118	-0.121	-0.126	-0.132	-0.142	-0.151	-0.170	-0.187	-0.227	-0.265
0.97	-0.080	-0.082	-0.084	-0.087	-0.092	-0.097	-0.107	-0.116	-0.133	-0.150	-0.188	-0.223
0.98	-0.064	-0.067	-0.069	-0.071	-0.076	-0.081	-0.090	-0.099	-0.116	-0.132	-0.169	-0.203
0.99	-0.050	-0.052	-0.054	-0.056	-0.061	-0.066	-0.075	-0.084	-0.100	-0.115	-0.151	-0.184
1	-0.036	-0.038	-0.040	-0.042	-0.047	-0.052	-0.060	-0.069	-0.084	-0.099	-0.134	-0.166
1.01	-0.024	-0.026	-0.028	-0.030	-0.034	-0.038	-0.047	-0.054	-0.069	-0.084	-0.117	-0.149
1.02	-0.015	-0.015	-0.016	-0.018	-0.022	-0.026	-0.033	-0.041	-0.055	-0.069	-0.102	-0.132
1.03	-0.010	-0.007	-0.007	-0.008	-0.011	-0.014	-0.021	-0.028	-0.042	-0.055	-0.086	-0.116
1.05	-0.002	0.002	0.006	0.008	0.009	0.007	0.001	-0.005	-0.017	-0.029	-0.058	-0.085
1.1	0.012	0.016	0.020	0.025	0.035	0.041	0.044	0.042	0.035	0.026	0.004	-0.019
1.15	0.022	0.026	0.030	0.034	0.046	0.056	0.069	0.074	0.074	0.069	0.054	0.036
1.2	0.029	0.033	0.037	0.041	0.052	0.064	0.082	0.093	0.101	0.102	0.093	0.081
1.3	0.038	0.041	0.045	0.049	0.060	0.071	0.091	0.109	0.131	0.142	0.150	0.148
1.4	0.041	0.045	0.049	0.053	0.063	0.074	0.094	0.112	0.141	0.161	0.183	0.191
1.5	0.043	0.047	0.051	0.055	0.064	0.074	0.094	0.112	0.143	0.167	0.202	0.218
1.6	0.043	0.047	0.051	0.055	0.064	0.074	0.092	0.110	0.142	0.167	0.210	0.234
1.7	0.043	0.047	0.050	0.054	0.063	0.072	0.090	0.107	0.138	0.165	0.213	0.242
1.8	0.042	0.046	0.049	0.053	0.062	0.070	0.087	0.104	0.134	0.161	0.212	0.246
1.9	0.041	0.045	0.048	0.052	0.060	0.068	0.085	0.101	0.130	0.157	0.209	0.246
2	0.040	0.044	0.047	0.050	0.058	0.066	0.082	0.097	0.126	0.152	0.205	0.244
2.25	0.037	0.040	0.043	0.046	0.054	0.061	0.076	0.090	0.116	0.141	0.193	0.234
2.5	0.035	0.037	0.040	0.043	0.050	0.057	0.070	0.083	0.108	0.130	0.181	0.222
2.75	0.032	0.035	0.037	0.040	0.046	0.053	0.065	0.077	0.100	0.121	0.169	0.210
3	0.030	0.032	0.035	0.037	0.043	0.049	0.061	0.072	0.093	0.114	0.159	0.199
3.5	0.026	0.028	0.031	0.033	0.038	0.043	0.053	0.063	0.082	0.101	0.142	0.179
4	0.023	0.025	0.027	0.029	0.034	0.038	0.048	0.057	0.074	0.090	0.128	0.163
5	0.019	0.021	0.022	0.024	0.028	0.032	0.039	0.047	0.061	0.075	0.108	0.138

APPENDIX D

Unit Systems

The most common system of units for scientific work is the Systeme Internationale, or SI. The SI unit system uses seven *primary* dimensions: *m, s, kg, kgmol, K, amp,* and *cd.* These are presented in Table D.1. Each of these primary dimensions is defined in terms of some measured standard. For example, the length of a meter is defined as the length given by 1,650,763.73 times the wavelength of a particular emission in the spectrum of ^{86}Kr. Additionally, the derived *secondary* units are also presented in terms of the primary dimensions of that measuring system, in the column labeled "SI units." There are several other unit systems we encounter as engineers. The units of common thermodynamic variables for three common unit systems are presented in Table D.1: SI, CGS, and English units.

TABLE D.1   Common Variables Used in Thermodynamics and Their Associated Units

Variable	SI Units	Primary SI Dimensions	CGS Units	English Units
Length	meter [m]	M	centimeter [cm]	foot [ft]
Time	second [s]	S	second [s]	second [s]
Mass	kilogram [kg]	kg	gram [g]	pound mass [lb$_m$] or slug [sl]
Moles	kgmole	kgmol	gmole	lb mole
Temperature				
Absolute	Kelvin [K]	K	Kelvin [K]	Rankine [°R]
Relative	Celsius[°C]		Celsius [°C]	Fahrenheit [°F]
Force	newton [N]	kgm/s^2	dyne [dyne]	pound force [lb$_F$]
energy	joule	kgm^2/s^2	erg	foot pound [ft lb$_F$] or British Thermal Unit [BTU]
pressure	pascal [Pa]	kg/ms^2	dyne/cm^2	pound force per square inch [PSI]
power	watt [W]	kgm^2/s^3	erg/s	ft lb$_f$/s BTU/s
concentration		kgmol/m^3	gmol/cm^3	lbmol/ft^3
density		kg/m^3	g/cm^3	lb/ft^3

In most cases, it is easy to convert between SI and CGS unit systems. For example, the form of Newton's second law that force, F, equals mass, m, times acceleration, a, is the same in either system:

$$F = ma \qquad \text{SI or CGS units} \qquad (D.1)$$

Thus the fundamental unit for force is defined in the SI system as $1[\text{N}] = 1[\text{kg m/s}^2]$, as shown in Table D.1. Similarly, in the CGS system, it is defined as $1[\text{dyne}] = 1[\text{g cm/s}^2]$. With electric and magnetic units, however, changing between the two unit systems is not as straight-forward. The physical laws of nature can actually change form. For example Coulomb's law presented in Chapter 4 has a different form depending on what unit system is used. In the SI system of units, we have:

$$F_{12} = \frac{Q_1 Q_2}{4\pi \varepsilon_0 r^2} \qquad \text{SI units} \qquad (D.2)$$

On the other hand, the CGS unit system gives a simpler form:

$$F_{12} = \frac{Q_1 Q_2}{r^2} \qquad \text{CGS units} \qquad (D.3)$$

The forces affecting the behavior of molecules are often caused by electric interactions. In this section, we explore the differences that manifest in the SI and CGS unit systems when we treat electric and magnetic quantities. The origin of the differences lies in how units are defined, and it actually leads to different forms of the fundamental equations, as illustrated above. The premise is that by understanding these differences, you will make fewer mistakes when calculating quantities associated with electric or magnetic properties. We further specify the unit system associated with CGS as *Gaussian* units to distinguish it from other ways electric and magnetic units have been incorporated into the CGS unit system. For electromagnetics, the Gaussian unit system is simpler and better pedagogically than SI units. Thus, we use Gaussian units when we relate the electrical characteristics of matter to thermodynamic properties in Chapter 4 – unlike the rest of the text where SI units are employed.

The two basic forces on charges are electric and magnetic in nature. The electric force given by Coulomb's law can be written in general as:

$$F_{12} = k_E \frac{Q_1 Q_2}{r^2} \qquad (D.4)$$

where k_E is a proportionality constant. Similarly, the magnetic force per length, f, between two wires carrying currents I_1 and I_2 is given by:

$$f_{12} = 2k_M \frac{I_1 I_2}{r} \qquad (D.5)$$

where k_M is a proportionality constant. The two proportionality constants, k_E and k_M determine the system of units, i.e., they relate units of charge and current on the right hand side of Equations D.4 and D.5, respectively, to units of force. Moreover, their ratio is fixed by the laws of physics to be

$$\frac{k_E}{k_M} = c^2 \qquad (D.6)$$

where c is the speed of light. In both SI and CGS (Gaussian) units systems, units of force and distance are well defined. Furthermore, we choose the unit of current to be charge per second, or vice versa; thus, Equations D.4 – D.6 have only one parameter left to specify. The difficulty in converting between SI vs. CGS (Gaussian) unit systems is that each system chooses a different parameter to specify these equations. *In CGS units, k_E has a value of 1 and is unitless.* Inspection of Equation D.4 shows the units of charge are then defined as:

$$\text{unit of charge in cgs system} = \sqrt{\text{dyne cm}} = \text{g}^{1/2}\,\text{cm}^{3/2}/\text{s} \equiv \text{esu}$$

The definition above defines the unit of charge as the electrostatic unit [esu][1]. It is directly related to the primary units in the CGS unit system. Equation D.4 reduces to:

$$F_{12} = \frac{Q_1 Q_2}{r^2} \qquad \text{CGS units} \tag{D.7}$$

Equation D.7 is identical to Equation 4.8 in the text. In this unit system, currents are 1 esu/s. Using Equation D.6, the magnetic force per length f between two wires carrying currents I_1 and I_2 is given by

$$f_{12} = \frac{2 I_1 I_2}{c^2 r} \qquad \text{CGS units} \tag{D.8}$$

On the other hand, in order to specify the one remaining parameter in Equations D.4 – D.6, the SI unit system defines a *new* unit for current, the ampere [A]. The ampere is defined so that the proportionality constant in Equation D.5 becomes:

$$k_M = 10^{-7}[\text{N/A}^2] \equiv \frac{\mu_0}{4\pi} \qquad \text{SI units} \tag{D.9}$$

The definition given in Equation D.9 also includes the value of k_M in terms of the permeability of free space, μ_0. The unit for charge in the SI unit system becomes 1 [As] and is defined as the coulomb [C]. Using Equation D.6, the proportionality constant in Equation D.4 becomes:

$$k_E = c^2 \left(10^{-7}[\text{N/A}^2] \right) \equiv \frac{1}{4\pi\varepsilon_0} \qquad \text{SI units} \tag{D.10}$$

The definition in Equation D.10 includes the value of the permittivity of free space, ε_0[2]. Thus, in the SI system, Equation D.4 become

$$F_{12} = \frac{Q_1 Q_2}{4\pi\varepsilon_0 r^2} \qquad \text{SI units} \tag{D.11}$$

while the magnetic force per length f between two wires carrying currents I_1, and I_2 is given by

$$f_{12} = \frac{2\mu_0 I_1 I_2}{4\pi r} \qquad \text{SI units} \tag{D.12}$$

Example D.1 illustrates calculations for of the Coulombic potential energy in each unit system and shows the two unit systems are, indeed, consistent.

---

[1] This unit is often alternatively called the "statcoulomb."

[2] The terms permeability and permittivity of free space have their origins back when scientists viewed space as containing a material-type substance called "ether."

TABLE D.2 **Conversion between CGS (Gaussian) units and SI units**

Quantity		CGS (Gaussian) Units	SI Units	Converion Factor: SI = CGS multiplied by
Charge	Q	$\sqrt{\text{dyne}}\,\text{cm} = \text{g}^{1/2}\,\text{cm}^{3/2}/\text{s} \equiv$ **esu**	C	3.34×10^{-10}
Current	I	$\sqrt{\text{dyne}}\,\text{cm/s} = \text{g}^{1/2}\,\text{cm}^{3/2}/\text{s}^2 = \text{esu/s}$	$\text{A} = \text{C/s}$	3.34×10^{-10}
Power	$\dot{W}$	$\text{erg/s} = \text{dyne cm/s} = \text{gcm}^2/\text{s}^3$	$\text{J/s} = \text{W}$	1.00×10^{-7}
Electric potential	E	$\sqrt{\text{dyne}}$	$\text{V} = \text{J/C}$	300
Resistance	R	s/cm	$\Omega = \text{V/J}$	8.99×10^{11}
Dipole moment	μ^*	$\text{g}^{1/2}\text{cm}^{5/2}/\text{s} = \text{esu cm}$	C m	3.34×10^{-12}
Polarizability	α	cm^3	$\text{C}^2\text{m}^2/\text{J}$	1.11×10^{-16}
Magnetic Field	B	$\sqrt{\text{dyne}}/\text{cm} \equiv$ **gauss**	Tesla	1.00×10^{-4}

*The commonly used unit for the dipole moment is the Debye [D]. 1 [D] = 10^{-18} [esu cm].

For free space, it is straight-forward to translate between CGS (Gaussian) and SI units, by substituting $\varepsilon_0 \to 1/(4\pi)$ and $\mu_0 \to 4\pi/c^2$. For dielectric and magnetic materials, conversion between the two quantities is not so easy. For details, see a treatise on electromagnetism.[3] An abbreviated set of conversion factors between CGS and SI units that are relevant to the quantities of interest in this text is given in Table D.2. The convenience of Gaussian units is illustrated by the fact that all quantities presented in Table D.2 can be related to the three fundamental units of the CGS system – cm, g, and s.

▶ **EXAMPLE D.1**

Electrostatic calculations in Gaussian and SI units:

Consider a singly ionized negative ion and a singly ionized positive ion that are separated by 1 nm. Calculate the potential energy between the two ions using SI units and CGS (Gaussian) units

SOLUTION In SI units, the charge of a singly ionized ion is approximately 1.60×10^{-19}[C]. The permitivity of free space has the value, $\varepsilon_0 = 8.85 \times 10^{-12}$[C^2/(Jm)]. Thus, Equation D.11 becomes:

$$\Gamma_{12} = \frac{Q_1 Q_2}{4\pi\varepsilon_0 r} = \frac{\left(1.60 \times 10^{-19}\,[\text{C}]\right)\left(1.60 \times 10^{-19}\,[\text{C}]\right)}{(4\pi)\left(8.85 \times 10^{-12}\,[\text{C}^2/(\text{Jm})]\right)\left(10^{-9}[\text{m}]\right)} = -2.30 \times 10^{-19}\,[\text{J}] \qquad \text{(ED.1)}$$

The charge of a singly ionized ion in CGS units can be found from Table D.2:

$$\left(1.60 \times 10^{-19}\,[\text{C}]\right)\left(\frac{[\text{esu}]}{3.34 \times 10^{-10}\,[\text{C}]}\right) = 4.80 \times 10^{-10}\,[\text{esu}]$$

Thus, Equation D.7 becomes:

$$\Gamma_{12} = \frac{Q_1 Q_2}{r} = \frac{\left(-4.80 \times 10^{-10}\,[\text{esu}]\right)\left(4.80 \times 10^{-10}\,[\text{esu}]\right)}{10^{-7}\,[\text{cm}]} = -2.30 \times 10^{-12}\,[\text{erg}] \qquad \text{(ED.2)}$$

Since 1 [erg] = 1 g cm^2/s^2 = 1 $\times$ 10^{-7} [J], Expressions ED.1 and ED.2 are equivalent. ◀

---

[3] John D. Jackson, *Classical Electrodynamics*, 3rd ed., New York: Wiley (1999).

► APPENDIX E

ThermoSolver Software

► E.1 SOFTWARE DESCRIPTION

Program Installation

Requirements: Windows 95 or later
Recommended: Pentium 200 MHz or faster, 16 MB RAM
Features

- Thermodynamic properties of 300+ compounds are provided.
- Saturation pressure calculator is provided for any species in the database.
- Solver for the Peng–Robinson and Lee–Kesler equations of state is provided.
- Fugacity coefficients can be solved for pure species or mixtures.
- Models for Gibbs energy can be fit to isobaric or isothermal vapor–liquid equilibrium data. Sample data sets are provided. The results can be plotted.
- Bubble-point and dew-point calculations are provided.
- Equilibrium constant (K_T) solver is provided.
- General chemical reaction equilibria calculations are provided.
- Equations used in the calculation process can be viewed.

Installation is a one-time process. Simply download the software from http://www.wiley.com/college/koretsky. If the setup process does not start automatically, double-click **Setup.exe**. Once the setup process has started, follow the on-screen instructions. This process needs to be completed only for the first-time installation. Once the software has been installed, you may consult the Documentation program for more detailed documentation, including screenshots and descriptions of the numerical methods used to solve these problems.

Program Usage

Click **Start, Programs, ThermoSolver**, and click the ThermoSolver program icon to begin. The Thermodynamics Menu will appear. From here, eight programs are available to choose from:

Species Database
The thermodynamic properties of more than 300 species are available from here. Choose a species from the drop-down list at the top of the screen; the list is sorted by reduced chemical formula, so "Ethanol" will be found under "C2H6O." *Use the scrollbar at the side of the drop-down list to choose a species quickly.*

The Species Database provides all of the thermodynamic data used by the rest of the software. Thus, if a species is not available elsewhere in ThermoSolver, there are two causes: Either the given species does not have all of the fields filled in that are required for the calculation, or the species is not in the database. Edits can be made to the database. The program will ask if the changes are to be saved when the Species Database is closed.

Once a species is chosen, the thermodynamic properties are displayed. Choose one of the three tabs to view **General Properties, Energy Properties,** or **Heat Capacity Properties**. If a field is blank, it is not provided for the given species.

In the **General Properties** tab, the Antoine constants are shown for the chosen logarithm base and units. Click the **Antoine Eqn** button at the bottom of the window to be shown the general form of the Antoine equation. The critical temperature, critical pressure, and acentric factor are also reported.

In the **Heat Capacity** tab, the heat capacity constants are reported for a generalized heat capacity equation. Click the **Cp/R Eqn** button at the bottom of the view to display the general heat capacity equation. The quantity c_p/R is unitless and the temperature has units of K.

Saturation Pressure Calculator

This program uses the Antoine equation to calculate either a saturation pressure at a given temperature or a saturation temperature at a given pressure. Select a species, enter either a pressure or temperature, and click **Solve** next to the unspecified variable.

Equation of State Solver

The equation of state solver uses two of the measure properties (P, v, T) to solve for the third, using either the Peng–Robinson or Lee–Kesler equation of state. For example, given the pressure and temperature of a species, the molar volume can be found. To use the program, first choose a species, enter two of the values out of (P, v, T), and click **Solve** next to the third value.

▶ **EXAMPLE E.1**
Using Equation of
State Solver to find v

Find the molar volume of helium at 0°C and 1 atm.

SOLUTION   From the Select a Species drop-down list, scroll down and select **He—Helium-4**. Enter "1" for pressure and choose the units **atm**. Enter "273.15 K" for temperature. Choose **L/mol** for the molar volume unit and click **Solve** next to the molar volume field. The result is 22.4201 L/mol. This is expected, since ideal gases at STP should have molar volumes of 22.4 L/mol.

◀

Fugacity Coefficient Solver

The fugacity coefficient solver uses either the Peng–Robinson or the Lee–Kesler equation of state to calculate fugacity coefficients. At the main Fugacity Coefficient Solver window, add one or more species by clicking the **Add** button. The program will prompt for the number of moles and allow the species to be chosen from a drop-down list. Continue to add species until the entire-system is represented. Now choose a temperature and pressure at the bottom of the Fugacity Coefficient Solver window. On the right side of the window, the fugacity coefficients will be displayed for the current system. The mole values can be changed directly in the summary table.

If the Peng–Robinson option is selected, the Peng–Robinson equation of state will be used to calculate both the pure species fugacity coefficients (φ_i) and the fugacity coefficients in the mixture ($\hat{\varphi}_i$). The Lee–Kesler equation of state can calculate only pure fugacity coefficients.

Models for g^E—Parameter Fitting

This software program solves for the parameters of two-suffix Margules, three-suffix Margules, van Laar, Wilson, and NRTL models. First, choose whether P or T will be held constant in the data. At the next window, the experimental data are listed on the left, and the activity coefficient model and parameters are listed on the right.

Any one of the currently saved data sets can be selected in the Experimental Data frame, via a drop-down list. To create a new data set, click **New**, and enter the data in the table. Data can be copied and pasted to and from Microsoft Excel. Make sure to select the appropriate units and enter the constant pressure or temperature at the bottom of the window. Choose **Save** to save an entered data set—saved data sets will be available in the drop-down list of experimental data in all future sessions of the program. The built-in data sets are read-only and cannot be saved.

▷ **EXAMPLE E.2**
Example 8.9 Using ThermoSolver

Verify the activity coefficient parameter for the two-suffix Margules equation found for the binary system benzene (a) and cyclohexane (b) found in Example 8.9.

SOLUTION Choose the **Isothermal** button since constant temperature data will be used. The main binary mixture VLE coefficient solver window will appear next. From the drop-down list in the upper-left corner of the window, select **Benzene (a) and Cyclohexane (b)**. The pressure, x_a, and y_a data will load into the data grid. Click **Solve**. Coefficient A is now 1400.75 J/mol, which is the optimal two-suffix Margules parameter for objective function pressure. To plot the curve that was just fit to the data, click **Plot Data**. Choose **Pressure vs Xa** as the property to plot, and then click **OK**. ◀

Bubble-Point / Dew-Point Calculations

This program performs bubble-point and dew-point calculations, with various fugacity and activity coefficient corrections. After the appropriate bubble-point or dew-point calculation type has been selected, the main window will be presented.

Add a species to the mixture by using the **Add** button. Choose the desired fugacity and activity coefficient corrections at the bottom of the window. If the multicomponent Wilson model is used, the model parameters should entered by choosing **Edit** in the Wilson Model Parameters frame. Once everything is set, choose **Solve Unknowns** to perform the bubble-point or dew-point calculation. Choose **More Information** to see the values of the correction factors at equilibrium.

▷ **EXAMPLE E.3**
Dew-Point Calculation

A system with vapor contains 30% n-pentane (1), 30% cyclohexane (2), 20% n-hexane (3), and 20% n-heptane (4) at 1 bar. Determine the temperature and liquid composition at which the vapor develops the first drop of liquid.

SOLUTION This system corresponds to quadrant IV, since the vapor composition and pressure are known. Choose **Add**, select **C5H12—n-Pentane**, and enter "0.3" for the vapor mole fraction. Next, choose **Add**, select **C6H12—Cyclohexane**, and enter "0.3" for the mole fraction. Add **C6H14—n-Hexane** with 0.2 mole fraction and, finally, **C7H16—n-Heptane** with a mole fraction of 0.2. Enter "1 bar" for the pressure, and click **Solve Unknowns**. The liquid mole fractions and equilibrium temperature will be displayed. The dew point temperature is 75.7°C, and the liquid composition is:

Species	n-Pentane	Cyclohexane	n-Hexane	n-Heptane
x_i	0.10	0.35	0.16	0.40

◀

Equilibrium Constant (K_T) Calculator
Choose **Chemical Reaction Equilibria** from the ThermoSolver main menu, then choose the
Equilibrium Constant Calculation. The equilibrium constant calculator solves for the equilibrium constant K at a given temperature, as described in Section 9.4. The reactants and products
for an equation can be added by choosing the Add button on either the reactant or product side
of the window. *Make sure to specify the stoichiometric coefficient of each species as it is added.*

Once all reactants and products have been added, the correct chemical equation should be
displayed at the bottom of the window, and the equation status should read "Balanced." If not,
select a reactant or product and choose **Edit** to adjust the stoichiometric coefficient. Fractions
may be used for stoichiometric coefficients. Finally, enter a temperature, and the corresponding
equilibrium constant will be displayed.

Reaction Equilibria Calculations
Choose **Chemical Reaction Equilibria** from the ThermoSolver main menu, then choose the
Reaction Equilibria Calculations. The reaction equilibria program solves for the composition
of a reacting system at equilibrium, using the Gibbs energy minimization method discussed in
Section 9.6. The software is limited to gases and solids.

Click the **Add** button to add one or more species to the reaction. In the **Add** dialog box, a
method is needed to calculate the Gibbs energy of formation at a particular temperature. If the
species is selected from the database, the Gibbs energy of formation will be computed automatically. If the species is manually entered, its Gibbs energy of formation must be known for the
temperature at which the reaction takes place. Enter the number of moles initially present, and
click **Add** to add the species to the reaction.

At the main window, the temperature and pressure of the reaction can be adjusted, and the
vapor phase correction can be chosen. Click **Calculate EQ** when everything is set. The results of
the equilibrium calculation will be displayed in a pop-up window.

▶ E.2 CORRESPONDING STATES USING THE LEE–KESLER EQUATION OF STATE[1]

Below the solution algorithm to the Lee–Kesler equation of state is described. Pick a reduced
temperature and pressure

$$P_r = \frac{P}{P_c} \quad \text{and} \quad T_r = \frac{T}{T_c}$$

Solve for v^*

$$z = \frac{Pv}{RT} = \frac{P_r v^*}{T_r} = 1 + \frac{B}{v^*} + \frac{C}{(v^*)^2} + \frac{D}{(v^*)^5} + \frac{c_4}{T_r^3 (v^*)^2} \left(\beta + \frac{\gamma}{(v^*)^2} \right) \exp \left(-\frac{\gamma}{(v^*)^2} \right)$$

where

$$v^* = \frac{P_c v}{RT_c}$$

$$B = b_1 - \frac{b_2}{T_r} - \frac{b_3}{T_r^2} - \frac{b_4}{T_r^3}, \qquad C = c_1 - \frac{c_2}{T_r} + \frac{c_3}{T_r^3}, \qquad D = d_1 + \frac{d_2}{T_r}$$

Simple:

$$z = z^{(0)}$$

Correction:

$$z^{(1)} = \frac{z^{(c)} - z^{(0)}}{0.3978}$$

---

[1] From B. I. Lee and M. G. Kesler, *AIChE Journal*, **21**, 510 (1975).

Departure functions:

$$\frac{h_{T_r,P_r} - h_{T_r,P_r}^{\text{ideal gas}}}{RT_c} = T_r \left\{ z - 1 - \frac{1}{T_r v^*} \left(b_2 + \frac{2b_3}{T_r} + \frac{3b_4}{T_r^2} \right) \right.$$

$$- \frac{1}{2T_r(v^*)^2} \left(c_2 - \frac{3c_3}{T_r^2} \right) + \frac{d_2}{5T_r(v^*)^5} + \frac{3c_4}{2T_r^3 \gamma}$$

$$\left. \times \left[\beta + 1 - \left(\beta + 1 + \frac{\gamma}{(v^*)^2} \right) \exp \left(-\frac{\gamma}{(v^*)^2} \right) \right] \right\}$$

$$\frac{s_{T_r,P_r} - s_{T_r,P_r}^{\text{ideal gas}}}{R} = \ln \frac{z}{P[\text{atm}]} - \frac{1}{v^*} \left(b_1 + \frac{b_3}{T_r^2} + \frac{2b_4}{T_r^3} \right)$$

$$- \frac{1}{2(v^*)^2} \left(c_1 - \frac{2c_3}{T_r^2} \right) - \frac{d_1}{5(v^*)^5} + \frac{c_4}{T_r^3 \gamma}$$

$$\times \left[\beta + 1 - \left(\beta + 1 + \frac{\gamma}{(v^*)^2} \right) \exp \left(-\frac{\gamma}{(v^*)^2} \right) \right]$$

Fugacity coefficient:

$$\ln \varphi = z - 1 - \ln z + \frac{B}{v^*} + \frac{C}{2(v^*)^2} + \frac{D}{5(v^*)^5} + \frac{c_4}{2T_r^3 \gamma}$$

$$\times \left[\beta + 1 - \left(\beta + 1 + \frac{\gamma}{(v^*)^2} \right) \exp \left(-\frac{\gamma}{(v^*)^2} \right) \right]$$

	Simple (0)	Correction (c)
b_1	0.1181193	0.2026579
b_2	0.265728	0.331511
b_3	0.154790	0.027655
b_4	0.030323	0.203488
c_1	0.0236744	0.0313385
c_2	0.0186984	0.0503618
c_3	0	0.06901
c_4	0.042724	0.041577
d_1	1.55488×10^{-5}	4.8736×10^{-5}
d_2	6.23689×10^{-5}	7.40336×10^{-6}
β	0.65392	1.226
γ	0.060167	0.03754

APPENDIX F

References

▶ F.1 SOURCES OF THERMODYNAMIC DATA

General

J. H. Keenan, F. G. Keys, P. G. Hill, and J. G. Moore, *Steam Tables* (New York: Wiley, 1969).

David R. Lide (ed.), *CRC Handbook of Chemistry and Physics*, 83rd ed. (Boca Raton, FL: CRC Press, 2002–2003).

P. J. Linstrom and W. G. Mallard, Eds., **NIST Chemistry WebBook, NIST Standard Reference Database Number 69**, March 2003, National Institute of Standards and Technology, Gaithersburg MD, 20899 (http://webbook.nist.gov/chemistry/fluid).

Taylor Lyman et al., *Metals Handbook, Metalography, Structures, and Phase Diagrams*, 8th ed. (Vol. 8) (Metals Park, OH: American Society for Metals, 1973).

R. H. Perry, D. W. Green, and J. O. Maloney (eds.), *Perry's Chemical Engineers' Handbook*, 7th ed., (New York: McGraw-Hill, 1997).

Robert C. Reid, John M. Prausnitz, and Thomas K. Sherwood, *The Properties of Gases and Liquids*, 3rd ed. (New York: McGraw-Hill, 1977).

Frederick D. Rossini et al., *Selected Values of Physical and Thermodynamic Properties of Hydrocarbons and Related Compounds* (American Petroleum Institute Research Project 44) (Pittsburgh Carnegie Press, 1953).

R. W. Rowley et al., *Physical and Thermodynamic Properties of Pure Chemicals: Evaluated Process Design Data* (Vol. 1–5) (Philadelphia: Taylor and Francis, 1989–2003).

Edward W. Washburn (ed.), *International Critical Tables* (Vol. III and V) (New York: McGraw-Hill, 1928, 1929).

Milan Zabransky et al., *Heat Capacity of Liquids* (Washington, DC: American Chemical Society; (Woodbury, NY: National Bureau of Standards, 1996).

Journal of Physical and Chemical Reference Data (New York, American Chemical Society.

Enthalpy of Mixing

James J. Christenson, Richard W. Hanks, and Reed M. Izatt, *Handbook of Heats of Mixing* (New York: Wiley, 1982).

Frederick D. Rossini et al., *Selected Values of Chemical Thermodynamic Properties* (circular of the National Bureau of Standards 500), Washington, DC: United States Printing Office (1952).

Phase Equilibrium

Ju Chin Chu, Shu Lung Wang, Sherman L. Levy, and Rejendra Paul, *Vapor–Liquid Equilibrium Data* (Ann Arbor: MI J. W. Edwards, 1956).

J. H. Dymond and E. B. Smith, *The Virial Coefficients of Pure Gases and Mixtures* (Oxford: Clarendon Press, 1980).

J. Grehling, U. Onken, and W. Alrt, *Vapor–Liquid Equilibrium Data Collection* (Multiple volumes) (Frankfort: DECHEMA; 1977–1980).

Shuzo Ohe, *Vapor–Liquid Equilibrium Data* (New York: Elsevier, 1989).

Stanley M. Walas, *Phase Equilibria in Chemical Engineering* (Boston: Butterworth, 1985).

Jaime Wisniak, *Phase Diagrams: A literature source book* (New York: Elsevier, 1981).

Jaime Wisniak and Mordechay Herskowitz, *Solubility of Gases and Solids: A Literature Source Book* (New York: Elsevier, 1984).

Jaime Wisniak and Abraham Tamir, *Liquid–Liquid Equilibrium and Extraction: A Literature Source Book* (New York: Elsevier, 1981).

Reaction Thermochemistry

Ihsan Barin, *Thermochemical Data of Pure Substances*, 3rd ed. (Vol. I and II) (New York: VCH, 1995).

M. W. Chase et al., *JANAF Thermochemical Tables*, 3rd ed. (Washington, DC: American Chemical Society; Woodbury, NY: National Bureau of Standards, 1986).

O. Knacke, O. Kubaschewski, and K. Hesselmann (eds.), *Thermochemical Properties of Inorganic Substances*, 2nd ed. (Vol. I and II) (Düsseldorf): 2nd Ed. (Springer-Verlag, 1991).

Daniel R. Stull, Edgar F. Westrum, and Gerard C. Sinke, *The Chemical Thermodynamics of Organic Compounds* (New York: Wiley, 1969).

Richard A. Robie and Bruce S. Hemingway, *Thermodynamic Properties of Minerals and Related Substances at 298.15 K and 1 Bar (10^5 Pascals) Pressure and Higher Temperatures* (US Geological Survey Bulletin 2131) (Washington, DC: United States Government Printing Office, 1995).

▶ F.2 TEXTBOOKS AND MONOGRAPHS

Introductory

Richard P. Feynman, Robert B. Leighton, and Mathew Sands, *The Feynman Lectures on Physics*, (Menlo Park, CA: Addison-Wesley, 1963).

Olaf A. Hougen and Kennith M. Wilson, *Chemical Process Principles, Part One, Material and Energy Balances* and *Part Two, Thermodynamics* (New York: Wiley, 1943, 1947).

Octave Levenspiel, *Understanding Engineering Thermodynamics* (Upper Saddle River, NJ: Prentice-Hall, 1996).

Michael J. Moran and Howard M. Shapiro, *Fundamentals of Engineering Thermodynamics*, 4th ed. (New York: Wiley, 2000).

George C. Pimentel and Richard D. Spratley, *Understanding Chemical Thermodynamics* (San Francisco: Holden-Day, 1969).

Gordon J. Van Wylen, Richard E. Sonntag, and Claus Borgnakke, *Fundamentals of Classical Thermodynamics*, 4th ed. (New York: Wiley, 1994).

Intermediate

J. Richard Elliot and Carl T. Lira, *Introductory Chemical Engineering Thermodynamics* Upper Saddle River, NJ: Prentice-Hall, 1999).

B. G. Kyle, *Chemical and Process Thermodynamics*, 3rd ed. (Upper Saddle River, NJ: Prentice-Hall, 1999).

C. H. P. Lupis, *Chemical Thermodynamics of Materials* (New York: North-Holland, 1983).

Frederick D. Rossini, *Chemical Thermodynamics* (New York: Wiley, 1950).

Stanley I. Sandler, *Chemical and Engineering Thermodynamics*, 3rd ed. (New York: Wiley, 1999).

R. A. Swalin, *Thermodynamics of Solids* (New York: Wiley, 1972).

J. M. Smith, H. C. Van Ness, and M. M. Abbott, *Introduction to Chemical Engineering Thermodynamics*, 5th ed. (New York: McGraw-Hill, 1996).

Advanced

Kennith J. Denbigh, *The Principles of Chemical Equilibrium*, 3rd ed. (New York: Cambridge University Press, 1971).

Joseph O. Hirshfelder, Charles F. Curtiss, and R. Byron Bird, *Molecular Theory of Gases and Liquids* (New York: Wiley, 1954).

Kenneth S. Pitzer, *Thermodynamics*, 3rd ed. (New York: McGraw-Hill, 1995).

John M. Prausnitz, Ruediger N. Lichtenthaler, and Edmundo Gomes de Azevedo, *Molecular Thermodynamics of Fluid-Phase Equilibria*, 3rd ed. (Upper Saddle River, NJ: Prentice-Hall, 1999).

Jefferson W. Tester and Michael Modell, *Thermodynamics and its Applications*, 3rd ed., (Upper Saddle River, NJ: Prentice-Hall, 1997).

Hendrick. C. Van Ness and Michael M. Abbot, *Classical Thermodynamics of Nonelectrolyte Solutions* (New York: McGraw-Hill, 1982).

Defect Equilibria

F. A. Kroeger, *The Chemistry of Imperfect Crystals* (Vol. 2) (New York: North Holland Publishing Company, 1973).

W. Van Gool, *Principles of Defect Chemistry in Crystalline Solids* (New York: Academic Press, 1966).

Index

A

Absolute temperature 10
Absolute zero 10
Acentric factor 181–182, 499–501
Activation energy 435
Activity 326
Activity coefficient 325–327, 330–332
  Henry's law reference 325, 394, 462, 480
  infinite dilution 326
  Lewis/Randall reference 325
  mean 465
  molality based 462
  solid 353, 411
  T and P dependence 352
Activity coefficient models 336–350
  asymmetric 344–349
  from Azeotropic data 385
  best fit of data 386–391, 542
  molecular origins 339–341
  multicomponent 349–350
  NRTL 346–349, 542
      symmetric 341, 345
  two-suffix Margules 337–344, 349, 542
  three-suffix Margules 344–347, 542
  UNIQUAC 349
  van Laar 346–348
  Wilson 346–348, 350, 542–543
Adiabatic 5, 40
Adiabatic demagnetization 151–152
Adiabatic flame temperature 73–75
Anode 459
Antoine equation 185, 261, 329, 366, 499–501
Area test 336
Association 183–185
Athermal solution 351
Azeotrope 381–386
  maximum boiling 383
  minimum boiling 383

B

Beattie-Bridgeman equation of state 195
Benedict-Webb-Rubin equation of state 195

Bernoulli equation 135–137
Binary interaction parameter 200, 202
Binodal curve 398
Boiling-point elevation 413–416
Boltzmann's constant 10
Bond dissociation energy 70–71, 436
Boundary 3
Boyle temperature 238
Brouwer diagram 487–489
Bubble-point 368

C

Carnot cycle 86–92
Cathode 459
CGS units, 167–168, 536–539
Chain rule 217
Charge 167, 168, 461
Chemical forces 182–186
Chemical potential 290–291
  mathematical anomalies 304
  T and P dependence 292–293
Chemical reaction
  association 183–185
  bond dissociation energy 70–71, 436
  enthalpy 70–76, 444–447
  equilibrium constant 442–448
  extent 436, 439
  Gibbs energy 442
  solvation 183–185
  stoichiometry 71, 439
Chemical reaction equilibrium 13–15
  electrochemical systems 459–467
  cases 449–455
  Gibbs phase rule 469
  liquids 456
  heterogeneous systems 457–458
  independent reactions 469–470
  multiple reactions 467–475
  point defects 478–489
Clapeyron equation 258–260
Clausius-Clapeyron equation 260–261
Class of molecules 179, 181–182, 197
Closed system 3
  first law 49–52, 76–80
  second law 119–123

Coefficient of performance 88–89, 144–145
Colligative properties 412–419
  boiling-point elevation 413–416
  freezing-point depression 416
  osmotic pressure 416–418
Compound solid 408–409, 478, 485–489
Compressibility factor 165, 197–200, 520–523
Concentration
  carrier 483
  defect 481
  mass 417
  molal 462
  molar 481
Congruent melting 408
Consulate temperature 398–400
Coulombic interactions 168–169, 461
Coulomb's law 168, 537–539
Critical point 21, 181, 187–188
Critical properties 21, 181, 188, 499–501
Crystal lattice 407, 478, 481
  effective charge 479
Cubic equation of state 187–188, 192–194
Cycles 86
  Carnot 86–92
  Rankine 138–142
  refrigeration 88–89, 143–145
  vapor-compression 138–145
Cyclic relation 217

D

Debye-Huckel theory 465–466
DECHEMA data collection 345, 347
Deep learning 3
Defect 478
Defect equilibrium 478–489
  Brouwer diagram 487–489
Degrees of freedom 15, 469
Departure functions 230–237, 524–531, 544
Dependent property 15, 212–214
Derivative inversion 217
Derived property 212

Dew-point 368
Differential
    partial 213
    total 213
Differential volume element 5
Diffuser 80–81
Dipole 169–173
Dipole moment 170–173
Directionality 104–105
Dispersion 174–175
Distillation 374–376
Doping 483
Driving force 12, 42

E
Efficiency 49, 88, 114–115
    isentropic 137, 141–142
    Rankine 139–142
Electric field 167, 168
Electric potential difference 461
Electrochemical cell 459
Electrochemical systems 459–467
    Nernst equation 463
    shorthand notation 461
Electrolyte 459–460
Electrolytic cell 459
Electroneutrality 465, 478, 484
Electrostatic forces 168–172
Element vector 475–477
Endothermic 71, 445
Energetically favored 258–259, 342,
    376–377, 436, 480
Energy 32
English units 536
Enthalpy 54, 61
    departure function 230–233,
        524–527, 544
    of formation 72–73, 503–506
    of fusion 67
    ideal gas 54
    of mixing 273–275
    of reaction 70–76, 444–447
    of solution 275–277
    of sublimation 67
    of vaporization 67
Entropically favored 258–259, 342,
    376–377, 436, 480
Entropy 103–163
    definition 108
    departure function 232–235,
        528–531, 544
    information 108
    of mixing 277
    molecular 108, 146–152

Equal-area rule 187–188, 261–263
Equation of state 164–210
    Beattie-Bridgeman 195
    Benedict-Webb-Rubin 195
    cubic 187–188, 192–194
    definition 165
    gas phase fugacity 309–310,
        315–318
    Lee-Kesler 195, 197–199, 232–235,
        310–313, 541, 543–544
    liquid fugacity 353
    Peng-Robinson 192–194, 541
    Rackett 197
    Redlich-Kwong 192–194
    Soave- Redlich-Kwong 192
    van der Waals 186–191
    virial 194–195
Equilibrium 12–15
    chemical 13, 252 , 289–291, 306
    liquid-liquid 397–403
    mechanical 13, 252 ,306
    thermal 13, 252, 306
    solid-liquid 407–412
    solid-solid 407–412
    vapor-liquid 365–396
Equilibrium constant 442
    temperature dependence 444–448
    gas phase reaction 449–455
    heterogeneous reaction 457–458
    liquid phase reaction 456
    point defects 480–488
Equilibrium conversion 435
Eutectic point 407
Excess properties 333–335
    class I 334–335
    class II 334–335
    Gibbs energy 334–352
Exothermic 71, 445
Extensive property 4
Extent of reaction 436, 439
    effect of pressure 450
    effect of temperature 450
    effect of inerts 450–451
    multiple reactions 467

F
Faraday's constant 462
First law of thermodynamics 31–102
    closed systems 49–52, 76–80
    open systems 52–58, 80–86
Flash 371–372
Formation
    enthalpy of 72–73, 503–506
    Gibbs energy of 443, 503–506

Freezing-point depression 416
Fugacity 302–363
    criteria for chemical equilibrium
        306–307
    definition 303–305
    equations of state 309–310,
        315–318
    mixtures of gases 313–319
    pure gases 307–313
    pure liquid or solid 327–329
    reference state 303–304, 307, 314,
        322–324
    solid phase 353
    total solution 305
    vapor phase 307–322
Fugacity coefficient
    definition 305
    equations of state 309–310, 315–318
    generalized correlations 310–313,
        318–319, 532–535, 544
    mixtures 313–319
    pure gases 307–313
Fundamental property relations
    214–215
Fundamental grouping 216
Formula coefficient matrix 475

G
Galvanic cell 459
Gas 5
    vs. vapor 22
Gas constant 26
Gaussian units 167–168, 537–539
Generalized Correlations 181,
    197–200, 520–535, 543–544
    compressibility charts 197–200,
        520–523, 543
    enthalpy departure 232–233,
        524–527, 544
    entropy departure 234–235,
        528–531, 544
    fugacity coefficient 310–313,
        318–319, 532–535, 544
Geometric mean 202
Gibbs-Duhem equation 269–270
    activity coefficients 330–332
Gibbs energy 212, 253–258
    excess 334–352
    of formation 443, 503–506
    minimization 402–403, 435–438,
        543
    of mixing 321–322
    of reaction 442
    relation to vle 376–377

Gibbs phase rule 15, 291–292
    reacting systems 469

H
Half-cell reaction 459
Hard sphere potential 177–178
Heat 32, 37, 39–41
Heat capacity 58–66, 501–503
    constant pressure 61
    constant volume 58–60
    ideal gas 61, 214
    liquids 62
    mean 62
    molecular modes 59
    real 223–225
    solids 62
Heat exchanger 82–85
Helmholz energy 212
Henry's law 322–324, 391–396
    T and P dependence 329–330
    values 392
Hole 482
Hydrogen bond 182–183

I
Ideal gas 26–28, 166
    energy balance of 76–80
    enthalpy 54
    entropy change of 126–134
    internal energy 35
    property change of mixing 320–322
Ideal solution 322–324
Incongruent melting 408
Independent chemical reactions
    469–470
Independent property 15, 212–214
Induction 173–174
Intensive property 4
Intermolecular force 166–186
Intermolecular potential functions
    177–180
    hard sphere 177–178
    Lennard-Jones 178–180
    Sutherland 177–178
Internal energy 32–36
    departure function 247
    ideal gas 35
Interpolation 25
Inversion line 238
Ionic strength 465
Ionization potential 175
Irreversible 42–49, 105–107
Isenthalpic 85, 238

Isentropic 110
Isentropic efficiency 137, 141–142
Isobaric 5
Isochoric 5
Isothermal 5
Isothermal compressibility 196, 219

J
Joule-Thomson coefficient 238–239
Joule-Thomson expansion 237–243

K
Kay's rules 202, 318
K value 366
Kinetically controlled 435
Kinetic energy 32–34
    molecular 8–10, 34–36, 166

L
Lagrangian multiplier 476
Latent heat 34, 40, 67–70
Law 2
Le Chatelier's principle 450
Lee-Kesler equation of state 195,
    197–199, 232–235, 310–313, 541,
    543–544
Lennard-Jones potential 178–180
Lever rule 22, 242, 367–369
Lewis fugacity rule 318, 319–320
Lewis/Randall rule 322–324, 327–329
Liquid 5
    saturated 19
    subcooled 19
Liquid junction 460
Liquid-liquid equilibrium (LLE)
    397–403
    binodal curve 398
    Gibbs energy minimization
    402–403
    spinodal curve 399
Liquifaction 240–243
Linde process 240–241

M
Macroscopic 5
Margules equation
    three-suffix 344–347, 542
    two-suffix 337–344, 349, 542
Maxwell-Boltzmann distribution 9, 34
Maxwell relations 216–217
Mean activity coefficients 465

Measured property 7, 211–212
Mechanical energy balance 135–137
Microscopic 5
Minimization of Gibbs energy
    402–403, 435–438, 543
Mixing rules 200–203, 316
    binary interaction parameter 200,
    202
Mixtures 263–289
Molecular 5
    activity coefficient models, origin
    339–341
    kinetic energy 8–10, 34–36, 166
    modes of heat capacity 59
    potential energy 10, 34–36,
    166–186
Molecular configurations 146
Molecular weight 499–501
Multiple chemical reactions 467–477
    equilibrium constant formulation
    467–475
    minimization of Gibbs energy
    475–477

N
Nernst equation 463
Newton's second law 536
Normal boiling point 20
Nozzle 80–81
NRTL equation 346–349, 542

O
Objective function 387
Open system 3
    first law 52–58, 80–86
    second law 123–126
Osmotic pressure 416–418
Oxidation half reaction 459

P
Path 4
    function 5
    hypothetical 4, 41–42, 214
Partial miscibility 342, 397–398, 411
Partial molar property 264–271
    analytical determination of 281–284
    graphical determination of 285–287
    infinite dilution 271, 286
    property change of mixing 281
    relations among 288–289
Peng-Robinson equation of state
    192–194, 541

Peritectic point 408
Permeability of free space 538–539
Permittivity of free space 538–539
Phase 5
  boundary 5
Phase diagram 18, 367
  Lever rule 22, 242, 367–369
  Pxy 367, 373
  Txy 374
  xy 375
Phase Equilibrium 5, 13–14, 251–263
  pure species 253–263
Piston-cylinder assembly 3
Point charges 168–169
Point defect 478
  atomic 478–481
  effect of partial pressure 485–489
  electron 482
  electronic 481–485
  Frenkel 485
  hole 482
  interstitial 353–354, 478
  misplaced atom 478
  nomenclature 479
  Schottky 485
  substitutional impurity 478
  vacancy 352–354, 478
Polarizability 174
Polytropic process 80
Potential energy 32–34
  molecular 10, 34–36, 166–186
Potential function 177
Poynting correction 328
Pressure 4, 11–12
  external 38
  partial 21, 304
  saturation 14, 20–21
  vapor 20–21
Primary SI dimensions 536
Principle of corresponding states
  180–182
Process 5
  adiabatic 5, 40
  irreversible 42–49, 105–107
  isenthalpic 85, 238
  isentropic 110
  isobaric 5
  isochoric 5
  isothermal 5
  polytropic 80
  reversible 42–49, 105–107
Properties 3, 41
  dependent 15, 212–214
  derived 212
  extensive 4

fundamental 212
independent 15, 212–214
intensive 4
measured 7, 211–212
pure species 265, 270
pseudocritical 202
specific 7, 23
total solution 264, 270
Property change of mixing 271–280,
  320–322
  enthalpy 273–275
  entropy 277
  Gibbs energy 321–322
  ideal gases 277, 320–322
  ideal solutions 323
  molecular 272
  partial molar property 281
Pseudocritical acentric factor 202, 318
Pseudocritical pressure 202, 318
Pseudocritical temperature 202, 318
PT diagram 17–19
Pump 81–82
Pure species property 265, 270
Pv diagram 17–19
PvT Surface 17–18

Q
Quality 16

R
Rackett equation 197
Rankine cycle 138–142
Raoult's law 365–372
  negative deviation 374
  positive deviation 373–374
Reaction coordinate 439
Redlich-Kwong equation of state
  192–194
Reduced pressure 181
Reduced temperature 181
Reduced volume 181
Reduction half-reaction 459
Reference state
  energy 33
  gas fugacity 303–304, 307, 314
  liquid fugacity 322–324
  reaction 71–72
  standard state 441, 449, 456, 462
Refrigerant 143
Refrigeration cycle 88–89, 143–145
Regular solution 277, 351
Repulsive forces 177–178
Reservoir (thermal) 76, 88

Reverse osmosis 417
Reversible 42–49, 105–107

S
Salt bridge 460
Second law of thermodynamics
  103–163
  closed systems 119–123
  open systems 123–126
Semiconductor 481
  carrier concentration 482–483
  extrinsic 484
  intrinsic 482
  majority carriers 484
  minority carriers 484
  n-type 484
  p-type 484
Sensible Heat 34, 40, 59
Shorthand electrochemical cell nota-
  tion 461
SI units 5, 536–539
Simple molecules 179, 197
Soave- Redlich-Kwong equation of
  state 192
Solid 5
Solid-liquid equilibrium (SLE)
  407–412
  pure solids 407–410
  solid solutions 410–412
Solid-solid equilibrium (SSE) 407–412
Solid solutions 410–412
  interstitial 410
  substitutional 410
Solubility of gases in liquids 391–396
Solute 275, 324, 391
Solvation 183–185
Solvent 275, 324, 391
Spring assembled piston-cylinder
  65–66
Spinodal curve 399
Standard half-cell potential 463–464
Standard hydrogen electrode 463
Standard potential of reaction 463
Standard state 441
  electrolyte 462
  gases 449
  liquids 456
State 4
State function 5, 41
State postulate 15, 213
Steady-state 12, 53–54, 80–86,
  123–124, 135–137
Steam tables 23–25, 507–519
Stoichiometric coefficient 71, 439

Stoichiometric constraint 469
Supercritical fluid 22
Surroundings 3
Sutherland potential 177–178
System 3
  closed 3
  open 3
  isolated 3

T
Tangent-intercept method 286
Temperature 3, 8–11
  dependence of the equilibrium
    constant 444–448
  eutectic 407
  lower consulate 398–400
  normal boiling point 20
  normal melting point 409
  upper consulate 398–400
  saturation 20
  scale 10
Thermal conductivity 40
Thermal expansion coefficient 196,
  219
Thermochemical data 442–445,
  463–464
Thermodynamically controlled 435
Thermodynamic consistency 335–336
Thermodynamic web 211–249
  accessing reported data 218–220
  calculation of Dh 224–229
  calculation of Ds 220–221
  calculation of Du 221–224
  property relationships 212–220
  roadmap 219
ThermoSolver 193, 199, 209–210,
  345–347, 358, 363, 379, 385, 426,
  432, 447, 477, 490, 498, 540–543

bubble-point / dew-point
  calculations 542
equation of state solver 541
equilibrium constant calculator 543
fugacity coefficient solver 541
installation 540
models for g^E - parameter fitting 542
reaction equilibria calculations 543
saturation pressure calculator 541
species database 540
Throttling device 85–86
Tie line 22, 367
Total solution property 264, 270
Triple line 19
Triple point 17
Turbine 81–82

U
Unit systems 536–539
  conversion between CGS and SI
    539
Universe 3

V
van der Waals equation of state
  186–191
  fugacity in a mixture 316–318
  pure species fugacity 309–310
  virial form 208, 310, 317–318
van der Waals forces 175
  dipole 169–173
  dispersion 174–175
  Induction 173–174
van Laar equation 346–348
Vapor
  saturated 20
  superheated 20

Vapor-compression cycle 138–145
Vapor-liquid equilibrium 365–396
  azeotrope 381–386
  DECHEMA data collection 345,
    347
  distillation 374–376
  fitting activity coefficient models
    386–391
  Gibbs energy minimization
    376–377
  negative deviations 374
  positive deviations 373–374
  Raoult's law 365–372
  solubility of gases in liquids
    391–396
Vapor-liquid-liquid equilibrium
  (VLLE) 403–407
Virial equation of state 194–195
Volume 4, 7–8
  of mixing 272
  molar 4, 7–8
  specific 4, 7–8

W
Water 16
Wilson equation 346–348, 350,
  542–543
Work 32, 37–39
  electrical 459, 461–462
  Flow 53
  non Pv 254, 459
  Pv 38
  shaft 38–39, 53